HANDBOOK OF AGRICULTURAL ECONOMICS
VOLUME 3

HANDBOOKS IN ECONOMICS

18

Series Editors

KENNETH J. ARROW
MICHAEL D. INTRILIGATOR

ELSEVIER

AMSTERDAM • BOSTON • HEIDELBERG • LONDON
NEW YORK • OXFORD • PARIS • SAN DIEGO
SAN FRANCISCO • SINGAPORE • SYDNEY • TOKYO
North-Holland is an imprint of Elsevier

N·H

HANDBOOK OF AGRICULTURAL ECONOMICS

VOLUME 3

AGRICULTURAL DEVELOPMENT: FARMERS, FARM PRODUCTION AND FARM MARKETS

Edited by

ROBERT EVENSON
Yale University, New Haven, CT, USA

and

PRABHU PINGALI
Agricultural and Development Economics Division (ESA), Food and Agriculture Organization (FAO), United Nations

ELSEVIER

AMSTERDAM • BOSTON • HEIDELBERG • LONDON
NEW YORK • OXFORD • PARIS • SAN DIEGO
SAN FRANCISCO • SINGAPORE • SYDNEY • TOKYO
North-Holland is an imprint of Elsevier

N·H

North-Holland is an imprint of Elsevier
Radarweg 29, PO Box 211, 1000 AE Amsterdam, The Netherlands
The Boulevard, Langford Lane, Kidlington, Oxford OX5 1GB, UK

First edition 2007

Library of Congress Cataloging-in-Publication Data
A catalog record for this book is available from the Library of Congress

British Library Cataloguing in Publication Data
A catalogue record for this book is available from the British Library

ISBN-13: 978-0-444-51873-6

ISSN: 0169-7218 (Handbooks in Economics series)
ISSN: 1574-0072 (Handbook of Agricultural Economics series)

For information on all North-Holland publications
visit our website at books.elsevier.com

Printed and bound in Great Britain by MPG Books Ltd, Bodmin, Cornwall

07 08 09 10 11   10 9 8 7 6 5 4 3 2 1

INTRODUCTION TO THE SERIES

The aim of the *Handbooks in Economics* series is to produce Handbooks for various branches of economics, each of which is a definitive source, reference, and teaching supplement for use by professional researchers and advanced graduate students. Each Handbook provides self-contained surveys of the current state of a branch of economics in the form of chapters prepared by leading specialists on various aspects of this branch of economics. These surveys summarize not only received results but also newer developments, from recent journal articles and discussion papers. Some original material is also included, but the main goal is to provide comprehensive and accessible surveys. The Handbooks are intended to provide not only useful reference volumes for professional collections but also possible supplementary readings for advanced courses for graduate students in economics.

KENNETH J. ARROW and MICHAEL D. INTRILIGATOR

CONTENTS OF THE HANDBOOK

VOLUME 1A

PART 1: AGRICULTURAL PRODUCTION

VOLUME 1B

PART 2: MARKETING, DISTRIBUTION AND CONSUMERS

VOLUME 2A

PART 3: AGRICULTURE, NATURAL RESOURCES AND THE ENVIRONMENT

PART 4: AGRICULTURE IN THE MACROECONOMY

VOLUME 2B

PART 5: AGRICULTURAL AND FOOD POLICY

VOLUME 3

PART 1: INTRODUCTION

PART 2: HUMAN RESOURCES AND TECHNOLOGY MASTERY

PART 3: INVENTION AND INNOVATION

PART 4: MARKETS, INSTITUTIONS AND TRANSACTION COSTS

CONTENTS OF VOLUME 3

PART 1

INTRODUCTION

OVERVIEW

ROBERT E. EVENSON
Economic Growth Center, Yale University, New Haven, CT

PRABHU PINGALI
Agricultural and Development Economics Division, Food and Agriculture Organization, Rome, Italy

This overview of Volume 3 of the Handbook of Agricultural Economics is organized as follows:

Section 1 offers insights from the economic growth literature;
Section 2 addresses the Green Revolution and its impact on developing countries;
Section 3 addresses the Gene Revolution and its limited impact on developing countries;
Section 4 addresses returns to research studies;
Section 5 discusses the decline in aid effectiveness for agriculture;
Section 6 offers comments on the 18 chapters covered in Volume 3.

1. Insights from economic growth theory

Economic growth theory can be divided into early growth theory and newer models of "endogenous" growth.

The chief insight from the early growth theory [Solow (1956)] was that an efficient economy leads to a steady state solution where product per worker does not grow without invention and innovation. When exogenous technological change is introduced in these models, product per worker does grow. Invention and innovation are required for economic growth.

Jones (2002) reports a "Malthusian" extension of early growth theory. Under the assumption that cultivable land is in fixed supply, Jones shows that the steady state solution is

$$\text{Growth in product/worker} = \text{Growth in technology} - \beta n,$$

where β is the coefficient on land in the agricultural production function, and n is the rate of population growth. This casts growth as a race between technology and population growth. The designers of the International Agricultural Research Center (IARC) system recognized that technological gains could offset the negative consequences of population growth. The designers of the IARC system also evaluated the National Agricultural Research Systems (NARS) in place in the 1950s and concluded that they were not up to the task of meeting the challenge of high population growth rates.

Demographic transition models showed that the decline in death rates, particularly child mortality rates, associated with improved public health and immunization programs were producing high population growth rates in the 1950s and 1960s. By 2000 many countries, particularly in Sub-Saharan Africa, had tripled their populations since 1950. By 2000, however, most countries, even in Sub-Saharan Africa, were well along in their demographic transitions.

The more recent endogenous growth models treat R&D as a variable endogenously determined by incentive structures, particularly regarding intellectual property rights. Endogenous growth models treat population growth as a positive inducement to invention and innovation. The reasoning is that invention and innovation is proportional to population size and that invention and innovation produce externalities that benefit all members of the population. Data on patents granted certainly do not bear this out. The number of patents granted to inventors in Sub-Saharan Africa and even in South Asia is negligible. Inventions are not proportional to population. This is because R&D is not proportional to population.

Jones (2002) develops a model where invention and innovation is undertaken in high-income countries and where developing countries devote effort to "mastering the World Technology frontier". But as will be noted in the section on the Green Revolution, many developing countries invest nothing in industrial R&D. Almost all countries invest in public sector agricultural research in agricultural experiment stations. But as noted below, several have failed to produce a Green Revolution.

2. The Green Revolution

More than 40 years ago, Theodore W. Schultz wrote an influential book *Transforming Traditional Agriculture* (Yale, 1964) in which he argued that "traditional" agricultural economies were "poor but efficient" and "efficient but poor". Traditional agriculture was defined to be an agriculture where the development of improved technology in the form of improved crop varieties and improved animals was proceeding at a very slow pace. Implicit in this definition is the notion that agricultural technology has a high degree of "location specificity". Crop varieties, for example, require breeding programs in the regions served by the program.[1]

[1] This was first noted in the study of hybrid maize (corn) by Zvi Griliches (1957, 1958). Griliches noted that farmers in Alabama did not have hybrid maize varieties until 20 years after farmers in Iowa had access to hybrid maize. It was not until breeding programs were established in Alabama, selecting varieties for Alabama farm conditions that farmers in Alabama had access to hybrid maize. Farmers in West Africa did not have hybrid maize until 75 years after farmers in Iowa had hybrid maize. Farmers in Central Africa still do not have access to hybrid maize.

The Schultz argument implicitly suggests that agricultural extension programs cannot effectively "transform traditional agriculture", because traditional agriculture is already efficient. Note that this statement regarding efficiency holds the transaction costs associated with institutions constant. Thus, markets may be inefficient with high levels of transaction costs, but given this, farmers are efficient largely because they have had time to experiment with technological improvements under conditions of slow delivery.

We now have an opportunity to reassess the Schultz argument in the context of the Green Revolution. Agricultural extension programs might not be effective in improving the efficiency of farmers in a setting where farmers are already efficient, but agricultural extension programs could be successful in facilitating the transfer of technology produced in a foreign country to the country in question. Many countries have counted on this technology transfer function. In many Sub-Saharan African countries the number of agricultural extension personnel far exceeds the number of agricultural scientists.[2] (See Table 1 below.)

The Schultz position on agricultural extension and agricultural research was that the technology transfer function of agricultural extension was not realized because of the inherent "localness" of agricultural extension programs. Ultimately Schultz indicated that only a "Green Revolution" could "transform" traditional agriculture, and a Green Revolution depends primarily on competently-managed plant breeding programs in National Agricultural Research Systems (NARS) programs supported by International Agricultural Research Centers (IARCs).

Figure 1 lists 87 countries classified according to aggregate Green Revolution Modern Variety (GRMV) adoption rates in 2000. The 12 countries in the first column report negligible GRMV adoption in the year 2000. All other classes are based on area weighted GRMV adoption rates for the 11 crops included in the GRMV study.[3]

Table 1 lists indicators by Green Revolution cluster. The clusters can be roughly categorized as non-performing (Cluster 1), underperforming (Clusters 2, 3 and 4) and performing (Clusters 5, 6, 7 and 8). Economic and social indicators by cluster are reported in Table 1.

The economic indicators show the following:

1. Crop value (in US dollars) per hectare is very low for countries not realizing a Green Revolution and rises to high levels for countries realizing the highest levels of GRMV adoption.
2. Fertilizer application per hectare is negligible for the first four clusters and significant for the highest GRMV clusters.
3. Crop TFP growth is negligible for countries not realizing a Green Revolution and highest for countries with the highest levels of GRMV adoption.[4]

[2] Evenson and Kislev (1975) report relative price ratios of 20 to 1 for the cost of scientists vs the cost of extension workers. This is partly related to the relative prices of extension personnel relative to the price of agricultural scientists.

[3] The 11 crops were rice, wheat, maize, sorghum, millets, barley, groundnuts, lentils, beans, potatoes and cassava [Evenson and Gollin (2003a, 2003b)].

[4] Crop TFP growth is reported in Avila and Evenson (forthcoming).

LT 2%	2–10%	10–20%	20–30%	30–40%	40–50%	50–65%	GT 65%
Afghanistan	Burkina Faso	Bolivia	Colombia	Cuba	Dominican Republic	Algeria	Argentina
Angola	Cambodia	Benin	Costa Rica	Egypt	Iran	Bangladesh	Chile
Burundi	Chad	Botswana	Ecuador	Mexico	Kenya	Brazil	China
Central African Republic	El Salvador	Cameroon	Ghana	Namibia	Morocco	Myanmar	India
Congo (B)	Gabon	Congo (Z)	Laos	Paraguay	Nepal	Tunisia	Indonesia
Gambia	Guatemala	Côte d'Ivoire	Madagascar	Peru	Thailand		Malaysia
Guinea Bissau	Guinea	Ethiopia	Mali	Saudi Arabia	Turkey		Pakistan
Mauritania	Haiti	Liberia	Sierra Leone	South Africa			Philippines
Mongolia	Jamaica	Honduras		Syria			Sri Lanka
Niger	Libya	Mauritius					Vietnam
Somalia	Malawi	Nicaragua					
Yemen	Mozambique	Nigeria					
	Panama	Rwanda					
	Senegal	Sudan					
	Swaziland	Tanzania					
	Togo	Uruguay					
	Uganda	Venezuela					
	Zambia	Zimbabwe					

Figure 1. Green Revolution clusters by GRMV adoption level.

Table 1
Green Revolution cluster indicators

Economic indicators

Clusters by GRMV adoption	Crop value per ha (dollars)	Fertilizer per hectare (kg/ha)	Crop TFP growth (1961–2000)	Scientists per million ha cropland		Extension work per million ha		Industrial competitiveness (UNIDO)	
				1960	2000	1960	2000	1985	1998
LT 2%	78	2	0.09	0.019	0.030	0.230	0.461	0.002	0.002
2–10%	128	22	0.72	0.018	0.093	0.392	0.402	0.020	0.028
10–20%	94	6	1.07	0.013	0.033	0.149	0.220	0.028	0.029
20–30%	112	12	0.87	0.033	0.076	0.245	0.416	0.037	0.051
30–40%	180	40	1.30	0.033	0.179	0.070	0.371	0.050	0.076
40–50%	227	52	0.96	0.023	0.063	0.287	0.827	0.038	0.072
50–60%	300	68	1.36	0.050	0.063	0.070	0.140	0.060	0.080
GT 65%	488	166	1.56	0.079	0.120	0.150	0.442	0.047	0.111

Social indicators

Clusters by GRMV adoption	Countries in class	Population in 2000 (millions)	Average population (millions)		Birth rates		Child mortality rates		Dietary energy sufficiency		GDP per capita	
			1960	2000	1960	2000	1960	2000	1960	2000	1960	2000
LT 2%	12	90	2.5	7.5	47	41	293	160	2029	2192	361	388
2–10%	18	153	3.1	8.5	45	36	236	118	2074	2387	815	1291
10–20%	18	385	7.0	21.4	44	36	214	134	1983	2282	866	1295
20–30%	8	115	9.0	14.3	46	32	238	124	2070	2384	695	1156
30–40%	9	337	14.3	37.4	42	26	156	27	2050	2574	1169	3514
40–50%	2	284	15.5	40.3	46	26	221	61	2084	2506	805	1660
50–60%	5	385	34.9	76.7	46	23	240	50	2038	2391	1096	2153
GT 65%	10	2886	135.1	288.6	39	22	165	43	2100	2719	1049	2305

4. Countries without a Green Revolution did have both agricultural scientists and extension workers. Scientists per million hectares of cropland rise with higher levels of GRMV adoption.
5. Extension workers per million hectares of cropland are roughly 20 times as great as scientists per million hectares of cropland. The number of extension workers increased in every cluster. No correlation between extension workers per million hectares of cropland and GRMV adoption exists.
6. None of the countries without a Green Revolution has industrial competitiveness. A UNIDO index of 0.05 or greater indicates industrial competitiveness. Only countries in 30–40% GRMV clusters and above have industrial competitiveness. Improvement in industrial competitiveness is greatest for the highest GRMV clusters.[5]

The social indicators show the following:

1. 63% of the 4.65 billion people living in developing countries are located in the ten countries in the highest Green Revolution cluster. 84% live in performing clusters. Countries without a Green Revolution make up less than 2% of the population in developing countries.
2. The average population of countries in 1960 and 2000 rises as GRMV adoption levels rise. This suggests a strong bias against small countries.
3. In 1960, birth rates were similar across GRMV clusters. By 2000, birth rates had declined in all GRMV clusters, with highest declines in the highest GRMV clusters.
4. Child mortality rates in 1960 were similar in most GRMV clusters. By 2000, they had declined in all GRMV clusters with highest declines in the highest GRMV clusters. In the top two GRMV clusters, child mortality rates in 2000 were only 24% of their 1960 levels.
5. Dietary Energy Sufficiency (DES) was similar for all GRMV clusters in 1960. By 2000, improvements were achieved in all clusters with highest improvements in highest GRMV clusters. DES improvement is highly correlated with child mortality reduction.
6. GDP per capita (using exchange rate conversion to dollars, Atlas method) was lowest in countries without a Green Revolution in 1960 and did not improve in 2000. GDP per capita for the next three GRMV clusters rose from 1960 to 2000 by 56%. GDP per capita for the highest four GRMV clusters rose by 140% from 1960 to 2000.

NARS programs in specific countries bear the ultimate responsibility for failing to deliver GRMVs to their farmers. But IARC programs are not immune from criticism.

[5] None of the countries without a Green Revolution reported investing in R&D in 1970. The Central African Republic reported industrial R&D in 1990. Of the 18 countries in the 2–10% cluster, 5 reported industrial R&D in 1970, 12 reported industrial R&D in 1990.

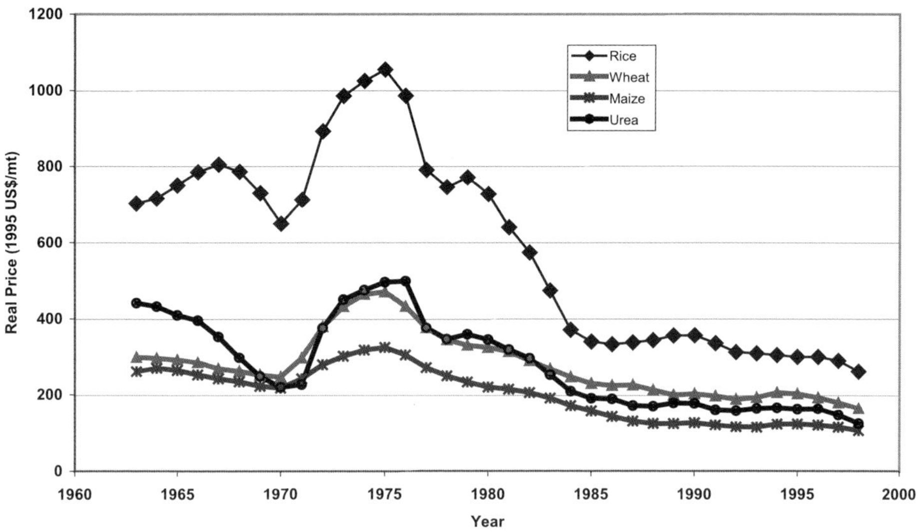

Figure 2. Real world prices of rice, wheat, maize and urea (1961–2000, 5-yr moving average). *Source*: IFPRI.

There are three IARCs located in Africa – ICRAF in Kenya, ILRI in Ethiopia and Kenya, and IITA in Nigeria. ICRAF has had little impact because agroforestry generates little income for farmers. ILRI has also had little impact although it does not deal with crops. IITA has had an impact only after developing breeding programs with CIMMYT for maize and with CIAT for cassava. Similarly, ICRISAT had little impact until sorghum, millet and groundnut breeding programs were developed in Africa.

Why did twelve countries fail to produce a Green Revolution? A closer examination suggests three explanations. The first is the "failed state" explanation. The second is the "small state" explanation. The third is the "civil conflict" explanation. Many of the countries failing to deliver a Green Revolution to their farmers are effectively failed states. But they are also small states with an average population of 2.5 million people in 1960 (Angola and Yemen had 5 million people in 1960). None have universities to train agricultural scientists. Many have been in civil conflict for much of the past 40 years. Given low GDP per capita, limited taxing power and civil conflict, it is not surprising that they did not produce a Green Revolution.

The second GRMV cluster did have a small Green Revolution, but they too are small countries (Mozambique and Uganda being largest with populations around 7 million in 1960). Most of these countries have also been in civil conflict. Few have universities to train agricultural scientists, but they did manage a small Green Revolution.

Figure 2 depicts "real" prices for the 1960 to 2000 period (a 5-year moving average). The prices of rice, wheat and maize in 2000 were approximately 45% of their 1960 level

(35% of their 1950 level). The real prices of the world's major cereal grains have been declining by more than 1% per year for the past 50 years.[6]

In the OECD developed countries, it is estimated that total factor productivity rates (a measure of cost reduction in agriculture) have been roughly 1% per year higher than in the rest of the economy. For developing countries, crop TFP growth rates have been high except for countries in the lowest GRMV clusters. A few of the industrially competitive countries have had industrial TFP growth rates that are higher than agricultural TFP growth rates.

Why then do we have "hunger in a world awash with grain". For this we need only look at crop value per hectare in Table 1. With low crop yields, crop value per hectare is low. The highest GRMV cluster produces more than six times as much crop value per hectare as does the lowest cluster. At 1960 prices, farmers in Sub-Saharan Africa with 1.2 hectares could earn $2 per day per capita. At 2000 prices with 0.8 hectares, farmers in Sub-Saharan Africa can earn only $1 per day per capita. Farmers in a number of countries have been delivered price declines without cost declines, and many have moved from mass poverty to extreme poverty.

3. The Gene Revolution

In 1953 Watson and Crick reported the "double helix" structure of DNA and showed that DNA conveyed inheritance from one generation to another. In 1974 Boyer and Cohen achieved the first "transformation" by inserting alien DNA from a source organism into a host organism and the field of genetic engineering was born.

The first genetically modified (GM) products (ice minus and the flavor-saver tomato) were not commercially successful. Monsanto introduced Bovine Somatotrophin Hormone (BST) in 1993 to dairy farmers. In 1995, several crop GM products were introduced to the market. One class of GM products provided herbicide tolerance enabling farmers to control weeds and practice low tillage methods with conventional herbicides (Roundup, Liberty). A second class of products conveyed insect resistance to plants (from *Bacillus thuriengensis*).[7]

Scientific reviews for food safety show no serious food safety issues for GM crops (or foods). Environmental studies show that environmental issues can be managed using existing management technology. Thus, existing GM products convey cost reduction advantages to farmers in countries where they are approved for sale. Because farmers using GM products increase their supply, world market prices are lower. This means

[6] Note that the real price of urea fertilizer has also been falling. In Asia and Latin America increased fertilizer use over the period of the Green Revolution was realized. In Africa this was not the case [Evenson and Gollin (2003a, 2003b)].

[7] Seven multinational firms now dominate the GM product market. Three are based in the US (Monsanto, Dupont and Dow), three are based in Europe (Bayer, BASF and Syngentia), and one is based in Mexico (Savia). These seven firms now spend $3 billion per year on R&D.

that farmers in countries not approving GM crops for sale suffer a double penalty. They do not realize cost reductions and they face lower prices.

The political economy of GM crops (foods) over recent years has resulted in a significant divergence between North America (the US and Canada) and the European Union (EU, before expansion). North America advises developing countries to take advantage of cost reducing opportunities. The EU countries urge developing countries to follow the "precautionary principle" in science policy.[8]

Are developing countries taking advantage of cost reduction potential from GM crop products? Table 2 reports data for developed and developing countries on both the potential cost reduction gains from GM crops[9] and on cost reduction gains realized as of 2004.[10]

Table 2 shows that potential gains vary greatly from country to country, being highest in the US, Argentina, Paraguay and Costa Rica, and lowest in the European Union. Several African countries have high cost reduction potential (largely because they are cotton producers) but no African countries, except South Africa, have taken advantage of cost reduction potential.

The European Union has little cost reduction potential because European countries do not produce significant quantities of cotton, soybeans, canola or rice. Thus, European Union countries have little at stake in terms of cost reduction potential. But they do have very significant influence on developing countries because they threaten to ban GM crop imports.[11]

Nonetheless several developing countries, Mexico, Argentina, Brazil, Paraguay, Bolivia, Costa Rica, China, and India have realized some cost reduction for GM crops. The potential for cost reduction in cotton producing countries in Africa is large, but no African country has taken advantage of this potential. None of the countries not realizing a Green Revolution has realized a Gene Revolution.[12]

4. Returns to research

Two sets of returns to agricultural research investments have been reported. The first is reported in Evenson (2001), Volume 1A of the Handbook of Agricultural Economics.

[8] The precautionary principle is usually interpreted as requiring a high level of proof that food safety and environmental safety rules are being met. When applied to regulatory policies, this requirement is problematic. When applied to science, it effectively halts scientific progress.

[9] Estimates of cost reduction gains are from Bennett, Morse and Ismael (2003), Falck-Zepeda, Traxler and Nelson (1999), Gianessi et al. (2002), Pray and Huang (2003), Qaim and Zilberman (2003), Qaim and de Janvry (2003), Qaim and Traxler (2004), and Traxler et al. (2003).

[10] Estimates of potential cost reduction gains presume 80% adoption rates for GM crops. Maize, cotton, soybeans, canola (rapeseed) and rice. Actual gains are based on GM adoption in 2004.

[11] Actually most of the countries in Sub-Saharan Africa export little or nothing to the European Union.

[12] It is unlikely that unimproved crop varieties benefit from genetic modifications.

Table 2
Potential and realized (as of 2004) cost reduction gains, selected countries

	Potential cost reduction (%)	Realized cost reduction (2004) (%)
Developed countries		
Canada	5	2
USA	9	6
Japan	1.5	0
European Union – Northern	0.6	0
European Union – Southern	1.5	0.1
Eastern Europe	3	0.1
Former Soviet Union	4	0
Developing countries		
Latin America:		
Mexico	3	0.5
Argentina	9	8
Brazil	7	2
Paraguay	9	2
Bolivia	7	1
Costa Rica	10	2
Other Latin American countries	4	0
Asia:		
China	4	1
Southeast Asia	4	0
Bangladesh	5	0
India	3	0.5
Pakistan	5	0
Africa:		
Egypt	3	0
Kenya	3	0
Central Africa	3	0
Mali	12	0
Benin	11	0
Burkina Faso	11	0
Malawi	4	0
South Africa	5	1
Zimbabwe	11	0

The methods for estimating returns to research range from project evaluation methods for cases where technology adoption rates are available to statistical methods utilizing research stock variables with time and spatial weights. Table 3 summarizes studies of returns to research as measured by Internal Rates of Return (IRRs).[13]

[13] Internal rates of return are the rates for which the present value of benefits equals the present value of costs.

Table 3
Return to agricultural research studies

	Distribution of internal rates of return (percent distribution)							Median IRR
	# of IRRs	0–20	21–40	41–60	61–80	81–100	100+	
Project evaluation methods	121	0.25	0.31	0.14	0.18	0.06	0.07	38
Statistical methods	254	0.14	0.20	0.213	0.12	0.10	0.20	46
Aggregate programs	126	0.16	0.27	0.29	0.10	0.09	0.09	42
Pre-invention science	12	0	0.17	0.38	0.17	0.17	0.17	51
Private sector R&D	11	0.18	0.09	0.45	0.09	0.18	0	45
By region								
OECD countries	146	0.15	0.35	0.21	0.10	0.07	0.11	40
Asian developing	120	0.08	0.18	0.21	0.15	0.11	0.26	61
Latin American developing	80	0.15	0.29	0.29	0.15	0.07	0.06	45
Africa developing	44	0.27	0.27	0.18	0.11	0.11	0.05	35

Source: Evenson (2001).

Table 4
Green Revolution returns to research

Countries	IARCs	NARS
Latin America	39	31
Asia	115	33
West Asia–North Africa	165	22
Sub-Saharan Africa	68	9

Source: Evenson's estimates.

Pre-invention science IRRs are for basic research investments. Private sector R&D programs do not reflect returns to R&D in the private companies but measure returns that spill-in to the agricultural sector.

Table 4 reports IRRs for IARCs and NARS programs for the Green Revolution. They are based on GRMV adoption rates. The low rates for Sub-Saharan Africa reflect the fact that many Sub-Saharan NARS have been spending significant funds for many years, often with few benefits.[14]

5. The decline in aid effectiveness

In 1985 USAID offered aid programs to small farms in Asia, Africa and Latin America totaling $2.5 billion (in 2005 constant dollars). Some of these projects supported

[14] Most studies in Table 3 were statistical and failed to capture the full costs of NARS programs.

research and extension programs, some supported rural credit programs, some supported rural infrastructure programs. The USAID budget for agriculture for 2005 is only $400 million.[15] World Bank lending to agriculture has also declined over this period. Support for Ph.D. level training is no longer provided by USAID.[16] The period of decline in aid to farmers took place during the period when "sustainable development" strategies were in high favor.

The Millenium Challenge Account (MCA) policy strategy of the U.S. State Department argues that aid is ineffective in countries below a certain institutional/governance threshold. The first MCA grant went to Madagascar (see Figure 1). Most countries with lower levels of institutional development than Madagascar are ineligible for MCA grants.

Table 1 shows that there is a sectoral sequence to development. In the 1960s only 25 or so of the developing countries in Figure 1 could be considered to be industrialized. Since 1960, virtually all countries in Figure 1 realized productivity gains in the agricultural sector before they realized productivity gains in the industrial sector. The abandonment of the agricultural sector by USAID and, to a lesser extent by the World Bank, is thus a serious matter.

The decline in aid effectiveness and in aid support is related to the end of the Cold War. Prior to the early 1990s, both the West and the East (the Soviet Union) vied for influence in developing countries. Many developing countries initiated Marxist-style revolutions only to find that the economic model underlying these revolutions, the centrally planned economy, collapsed in both the Soviet Union and China.

6. Comments on chapters

Part 2. Schooling, Nutrition and Extension

Part 2 includes two chapters. Chapter 43 focuses on schooling and health issues. Chapter 44 focuses on agricultural extension.

Chapter 43 of the Volume addresses two questions. The first is whether increased schooling of farmers makes farmers more productive. The second is whether improved nutrition (i.e., improved consumption of calories) of farmers makes farmers and farm workers more productive.

The first question is addressed in two ways. First, a two-sector model of schooling impacts on agriculture and non-agriculture is developed. Then micro- and macro-evidence

[15] Much of the 2005 aid is in support of biodiversity programs, entailing expansions of protected areas. Charles Geisler of Cornell estimates that the expansion of protected areas in Africa from 3.5% of areas in 1985 to 7% in 2000 has driven as many as 15 million African "slash and burn" farmers from their livelihoods. Most have ended up in refugee camps.

[16] USAID actually supported the Ph.D. programs of many of the scientists who produced the Green Revolution.

on schooling and economic growth is reviewed. This review of evidence is undertaken for both boys and girls in developing countries.

The second question is related to work by Fogel (1989) for the 18th and 19th centuries in Europe. Prior to the industrial revolution in the late 1700s Fogel argued that caloric consumption was too low in many countries to allow workers to achieve full work effort. Fogel attributed significant gains in worker productivity to improvement in calorie consumption in 18th and 19th century Europe.

FAO reports estimates of Dietary Energy Sufficiency (DES) (effectively average calories consumed per capita). Table 1 (above) reports DES data for 1960 and 2000 by GRMV adoption class. As can be seen, DES did improve significantly in all GRMV clusters. Did this lead to economic growth comparable to the growth in labor productivity in 18th and 19th century Europe? Chapter 43 discusses this issue and finds evidence supporting the proposition that labor productivity did improve in many developing countries because of nutritional improvements.

The contributions of agricultural extension to farm productivity are addressed in Chapter 44. One of the continuing themes in agricultural development policy is that farmers are inefficient. If so, agricultural extension programs can make them more efficient. It is also widely expected that agricultural extension programs can achieve "technology transfer" by bringing "best practice" technology developed in one country to another country. For countries without a Green Revolution, technology transfer has clearly not taken place.

Chapter 44 discusses organizational, political and bureaucratic factors affecting the performance of public extension systems. This discussion clarifies why traditional agricultural extension services were often judged to be ineffective. It explores how various innovations and adaptations such as training and visit (T&V) systems, farmers field schools and privatized "fee for service" systems have attempted, with various degrees of success, to resolve some of the deficiencies, at times exacerbating other problems.

Chapter 44 also reviews extension impact studies critically. Many studies have found significant extension impacts [see Evenson (2001)], but many studies find these impacts for short periods only and many studies find a high degree of recidivism in extension impacts. That is, farmers may adopt "best practice" technologies for a period and then return to previous practices (as noted in the Green Revolution discussion, extension programs appear not to have been effective in "jump starting" agriculture onto a sustained growth path).

Part 3. Invention and Innovation in Agriculture

Part 3 includes six chapters. Chapter 45 addresses the international agricultural research centers. Chapter 46 addresses the Green Revolution. Chapter 47 addresses livestock productivity. Chapter 48 addresses intellectual property rights. Chapter 49 addresses private sector R&D. Chapter 50 addresses the Gene Revolution.

Chapter 45 reviews International Agricultural Research Centers (IARC) contributions to agricultural development. These contributions take two forms. The first is in the form of IARC-crossed Green Revolution Modern Varieties (GRMVs). The second form is the delivery of advanced breeding lines to National Agricultural Research System (NARS) breeding programs. Both have been important (see Chapter 46).

Chapter 45 reviews the history of development for the IARCs and levels of support. In recent decades, support for the IARCs through the Consultative Group for International Agricultural Research (CGIAR) has plateaued in spite of abundant evidence for high returns to investment in the IARCs. The chapter notes that returns to Natural Resource Management (NRM) research programs are relatively low.

In view of the evidence for wide divergence in GRMV adoption rates reported above, the IARC contributions have to be qualified. IARC programs have often made the mistake of concluding that similar Agro-Ecology Zone (AEZ) classifications on different continents meant that GRMVs developed in Asia or Latin America could be transferred to Africa. This was simply not the case. Asian IARCs (IRRI and ICRISAT), Latin American IARCs (CIMMYT, CIAT and CIP) and Middle East IARCs (ICARDA) eventually set up breeding programs in Sub-Saharan Africa. But they were 20 years late in doing so. The IARCs in Sub-Saharan Africa (ICRAF, ILRI and IITA) did not achieve the same degree of success as did the IARCs located in Asia and Latin America.

Chapter 46 is the chapter describing the Green Revolution in developing countries. The extreme unevenness of the Green Revolution has been noted in the early part of this overview. The Green Revolution was based on the "practical" judgments of plant breeders in making crosses between parent cultivars. One of the early studies of genetic laws were the experiments of Gregor Mendel in 1869. Mendel's papers were not actually rediscovered until 1900. By then, however, many formal plant breeding programs had been established in many countries.

Two major developments important to the Green Revolution had been established by 1920. The first was the development of "hybrid" varieties based on "heterosis" breeding. The second was the development of "wide-crossing" or interspecific hybridization techniques enabling plant breeders to partially broach the "breeding barrier" between species.

Heterosis-based hybridization techniques were actually developed in New Haven, Connecticut where Donald Jones of the Connecticut Agricultural Experiment Station developed the "double-cross" method for hybrid seed production. This was achieved before 1920. Since Connecticut was not a major production region for corn (maize), a decade or so passed before hybrid corn varieties suited to production conditions in Iowa were developed. Farmers in Iowa had hybrid corn varieties at the end of the 1920s. Farmers in Asia did not get hybrid varieties until the 1980s. Farmers in West Africa got hybrid maize in the 1990s, and farmers in Central Africa still do not have hybrid maize varieties.

It is instructive to consider which agencies produced Green Revolution Modern Varieties (GRMVs) as noted in Chapter 46. The following observations may be made:

- NGOs did not produce GRMVs.

- Developed country plant breeding programs did not produce GRMVs for developing country farmers.[17]
- Private sector firms produced only hybrid maize, sorghum and millet GRMVs, and then only after IARCs produced higher yielding open-pollinated varieties. Few, if any, GRMVs were produced by private breeding companies in response to "Breeders Rights" intellectual property rights. Private sector GRMVs were about 5% of GRMVs.[18]
- IARC-crossed GRMVs accounted for 35% of all GRMVs. IARC-crossed GRMVs were generally released in a number of countries.
- NARS-crossed GRMVs accounted for 60% of all GRMVs. Most NARS-crossed GRMVs were released only in the home country. IARC-crossed GRMVs were widely used as parent varieties in NARS breeding programs.[19]

IARC-crossed GRMVs were generally the initiating force in the Green Revolution. In the case of wheat and rice, the first generation modern varieties were new "plant type" GRMVs with more fertilizer responsiveness to take advantage of falling fertilizer prices. Figure 2 (above) shows two features of the Green Revolution in terms of real prices of food grains and of urea fertilizer. The first feature is that the outcome of the Green Revolution in terms of increased supply of rice, wheat and maize is that the prices of these grains declined in real terms to approximately 40% of their 1960 levels. For most developed countries, agricultural Total Factor Productivity (TFP) growth over the 1960 to 2000 period was roughly 1% greater than in the rest of the economy. For developing countries, TFP experience ranged from countries with little or no Green Revolution to countries with major supply increases.

But the second feature of Figure 2 is that the real price of urea fertilizer has also fallen over time. Many critics of the Green Revolution fail to understand that the falling price of urea (the major nitrogen fertilizer) makes breeding efforts to achieve higher fertilizer responsiveness in GRMVs more productive. This is the induced innovation model at work.

Chapter 47 addresses global livestock development. The chapter begins by noting that as incomes rise, particularly from low levels, the demand for meat, milk and eggs rises rapidly. Some observers have treated this demand as creating a "Livestock Revolution" akin to the Green Revolution. But the Green Revolution was a supply-driven revolution, not a demand-driven revolution.

Chapter 47 does report Malmquist TFP indexes that are quite variable with Asian countries realizing highest TFP growth rates, Latin America realizing intermediate TFP

[17] This generalization holds even for the Francophone African countries where most GRMVs were developed in African countries.

[18] The WTO and TRIPS requirements that a *sui generis* system for plant varieties be in place is expected to be a Breeders Rights system.

[19] A study of the complementarity of IARC-crossed breeding lines and NARS breeding success confirmed the hypothesis that IARC breeding materials made NARS breeding programs more productive. The WTO-TRIPS agreement specifies that a "*sui generis*" system for protection of plant varieties be introduced. It is widely expected that this will be a Plant Breeders Rights system.

growth rates, and Sub-Saharan Africa realizing lowest rates. The Malmquist indexes show faster TFP growth in the 1981–2000 period than in the 1961–1980 period.

A second set of TFP growth rates, computed by Avila and Evenson (to be included as Chapter 73 in Volume 4 of this Handbook) found similar patterns. Highest TFP growth rates were realized in Asia in both 1961–1980 and 1981–2000. Latin America had higher TFP growth in the 1981–2000 period than in the 1961–1980 period. Africa and Latin America had comparable TFP growth in the 1961–1980 period, but Africa had a very disappointing 1981–2000 period. Livestock TFP productivity growth exceeded crop TFP growth in Asia in both periods. Crop TFP growth was higher than livestock TFP growth in Latin America in both periods and in Africa in the 1981–2000 period.

This evidence does suggest that Asia clearly did have a supply side livestock revolution and that Latin America, while not matching crop TFP growth did realize livestock TFP growth. Africa had very disappointing livestock TFP growth in the 1981–2000 period.[20]

Chapter 47 notes that livestock production systems range from "backyard" systems to highly commercial industrial systems with high degrees of specialization. As countries move from the backyard systems to the commercial systems, markets became more efficient and "structural change" occurs. Vertical coordination of markets and high levels of contracting characterize commercial systems.

As Chapter 55 notes, the growth of supermarkets and supermarket procurement practices is forcing changes in livestock production. This is an extraordinary case of "marketing technology" driving production technology.[21]

Chapter 48 deals with technological institutions, particularly those associated with incentives for private sector R&D. Chapter 48 first compares public and private R&D associated with the agricultural sector in developing and developed countries. Before 1980 virtually all research in developing countries was public sector R&D. By the 1970s, private sector R&D for agriculture exceeded public sector R&D in developed economies. Developing economies were also increasing private sector R&D, and by 1995 more than 5% of R&D expenditures in developing countries was in the private sectors. In general, an expansion in private sector R&D tends not to be associated with reductions in public sector R&D in developed countries. In developing countries, it appears that if incentive systems for private sector R&D can be developed, the range of R&D options is expanded.

Intellectual Property Rights (IPRs) are key incentives for private sector R&D in all countries. Several IPRs have been developed for plants and animals including plant patents, plant breeders rights, utility patents, trade secrets, trademark and Application

[20] Production costs include the transaction costs associated with inefficient markets. The Green Revolution was not accompanied by increased intensity of cultivation in Sub-Saharan Africa because of inefficient markets [Evenson and Gollin (2003a, 2003b)].

[21] Maize yields in Sub-Saharan Africa were only 20 to 25% of OECD maize yields in 1960. In 2000 they were less than 15% of OECD maize yields.

of Origin rights. Each of these IPRs is discussed in Chapter 48. The chapter also discusses alternatives to IPRs. An important discussion of the complexities associated with the Convention on Biological Diversity (CBD), the WTO-TRIPS agreement and the International Treaty on Plant Genetic Resources for Food and Agriculture is undertaken in this chapter.

The discussion in Chapter 48 anticipates Chapter 50 on biotechnology developments. The field of biotechnology inventions is now dominated by seven multinational firms, three from the US – Monsanto, DuPont and Dow – three from Europe – Bayer, Syngenta and BASF – and one from Mexico and the US (Savia). These seven firms now expend more than $3 billion annually on R&D. This is more than double total expenditures in public sector agricultural research in Sub-Saharan Africa. And, as noted above, much of the public expenditure in Sub-Saharan Africa is unproductive.

The WTO-TRIPS agreement calls for a *sui generis* arrangement for the protection of "plant varieties". It is widely expected that this will be a Breeders Rights system, but a number of court rulings have already established that the term plant varieties refers to conventionally bred varieties, not to biotechnology-based plant varieties.[22] Biotechnology varieties (i.e., genetically engineered using recombinant DNA technologies) are likely to be subject to patent protection, not Breeders Rights protection.

Chapter 49 addresses the magnitude and impact of private sector R&D programs in agriculture in developing countries. There is general agreement that in developed countries, private sector R&D expenditures now exceed public sector agricultural research expenditures. In developing countries, private sector R&D remains only 5% of public sector agricultural research. Much of private sector R&D is on chemicals and biotechnology, although the farm machinery sector also conducts significant R&D.

As a practical matter, most low-income countries have no industrial R&D undertaken by private firms. For example, in the Green Revolution clusters discussed above, none of the countries not realizing a Green Revolution engage in private sector R&D. Very few of the countries in the 2–10% and 10–20% clusters engage in any industrial R&D. It appears to be the case that all failed states (LT 2%) fail because they do not have functioning pubic sector agricultural research programs, and none of these failed states are even remotely industrially competitive. Even the "underperforming" clusters (2–10%, 10–20% and 20–30%) have many countries with little or no private sector R&D. Most of the "performing" clusters (30–40% and above) have private sector R&D and industrial competitiveness.

Chapter 49 introduces the concept of R&D "spillovers" in which one country benefits from the R&D of another country. Spillovers are high between OECD countries because all OECD countries have high levels of industrial R&D. Spillovers are negligible between OECD countries and countries that do not engage in private sector industrial

[22] The U.S. Supreme Court ruled that plant varieties could be protected either by Breeders Rights or Patents on both.

R&D. They are intermediate between OECD countries and advanced developing countries that do engage in industrial R&D.[23]

Chapter 50 addresses the Gene Revolution or more accurately, the "recombinant DNA" or genetic engineering revolution. The Gene Revolution was a case of science-enabled technology. The science in question was achieved by Watson and Crick in 1953 when they discovered the double helix structure of DNA and established that DNA was the chief carrier of genetic information.[24]

Today, the biotechnology industries are engaged in the broad categories of genetically modified (GM) products, medical GM products and agricultural GM products.[25] The Rockefeller Foundation (RF) supported early work on agricultural biotechnology (particularly on rice) and Chapter 50 describes the RF program in some detail. The chapter also describes RF initiatives to develop PIPRA (a public clearinghouse for IPRs) and the evaluation of regulatory systems for agricultural biotechnology.

A strong "political economy" dimension of agricultural biotechnology has emerged in recent years. This reflects a conflict between the European Union[26] and North America. Scientists have evaluated both food safety and environmental safety issues. On food safety, no evidence exists that GM foods are less safe than their non-GM counterparts. Scientists also conclude that while many environmental issues have emerged, these issues can be managed. However, in Europe GM foods have been "politicized". Scientists' judgment have been ignored. North America, by contrast, accepts scientific opinion. Thus, European advice to developing country research programs is that developing countries should follow the "precautionary principle" at least in regulatory policies.[27] The North American position is that developing countries should develop the regulatory framework to take advantage of cost reductions associated with GM crops.

Early GM products were not commercially successful (the ice-minus product and the flavor-saver tomato). The first commercially successful product was Bovine Somatotrophin Hormone (BST), released in 1993. In 1995, several GM crop products were released. These included insect toxicity products (*Bacillus thuriengensis*, Bt) and glyfosate (glufosate) tolerance products (Roundup Ready, Liberty, etc.). GM products have been installed on cotton, soybeans, canola and maize varieties.[28] To date, these prod-

[23] In fact, these spillovers can be so high as to explain the phenomenon of "super growth" in a few countries [Ruttan (2001)].

[24] Just 20 years later, Berg at Stanford produced recombinant DNA in his lab and in 1974, Boyer at UCSF and Cohen at Stanford achieved the first rDNA "transformation" by moving rDNA from a source organism to a host organism. With this achievement, the biotechnology industry was born [Cohen et al. (1973)].

[25] Most medical GM products are broadly accepted by consumers. Agricultural GM products are much more controversial.

[26] Particularly the original members of the European Union.

[27] The precautionary principle has some currency in regulatory policy but none in science policy. European crop science appears to have been damaged by the European position.

[28] China is releasing a Bt rice product in 2005.

ucts do not have quality enhancement features, but they are widely adopted because they reduce costs of production.[29]

It is expected that the second-generation GM products will include quality-enhanced products, and this is likely to diffuse some of the intense political hostility to GM foods.

Interestingly, the IARCs who clearly led the Green Revolution have not led the Gene Revolution. Most IARCs have invested little in the relevant skills and the IARCs have not seen themselves as trainers of modern plant breeders (as they did for conventional breeding skills).[30] Several developing countries, notably, China, India and Brazil have developed strong capabilities in GM product development.

Part 4. Markets and Transactions Costs

Part 4 includes six chapters. Chapter 51 addresses land markets and tenancy relationships. Chapter 52 addresses the evolution of labor contracts and labor relations. Chapter 53 deals with the fertilizer and agricultural chemicals industries. Chapter 54 addresses the farm machinery industry. Chapter 55 addresses product markets and supermarkets. Chapter 56 addresses financial markets.

Chapter 51 addresses land markets and land tenancy. Table 1 in Chapter 51 reports widely varying farm sizes and degrees of tenancy between Asia, Africa and Latin America. Farm size is smallest in Africa and was reduced between 1970 and 1990. Share tenancy is relatively low in Africa. Farm sizes are larger in Asia but have changed less than in Africa. Asia has a high proportion of land in share tenancy. Latin America has much larger farms and intermediate levels of share tenancy.

The Marshallian view of share tenancy is that share tenancy is inefficient. Because tenants get only a share of the product, they will tend to utilize too little labor. Yet share tenancy persists. Chapter 51 suggests that share tenancy has advantages (risk-sharing and increased interlinkage with credit markets) that explain this persistence. Empirical studies and land reforms are reviewed in Chapter 51. It is noted that land reforms have an inherent limitation in that they do not take the next generation into account.

Chapter 52 addresses labor market contracts and organizations. The chapter reviews several features of labor markets (efficiency wages, casual vs permanent workers, interdependency of markets, farm size, separability of labor markets).

The major contribution of the chapter is a view of the evolution of contractual forms in labor markets. The evolutionary view considers family workers as the original contractual form in labor markets. We observe that the family has advantages in many labor markets including the predominance of family farms in agriculture throughout the world. The next evolution stage is to add exchange labor to family labor. Exchange labor enables families to meet peak demand and related conditions.

[29] Cost reduction potential in the European Union countries is low (see Table 2). This is because the production of cotton, soybeans and canola in Europe is low. Europe thus has little stake in cost reduction but an important stake in trade.

[30] Modern breeding skills include marker-aided selection techniques to facilitate conventional breeding.

Unskilled hired workers are next added as labor markets became more sophisticated. Piece rate labor is added next and exchange labor effectively disappears from the mix. In the next stage piece rates with team supervisors are added. Specialized skilled laborers also emerge. In the final stage family workers (mostly engaged in supervision), piece rate teams and specialized skilled workers remain.

Throughout this evolution, transaction costs decline. As transaction costs decline the number of transactions increases. In the final stage, labor markets are efficient and labor market transaction costs are low.

Chapter 53 addresses the fertilizer and farm chemicals markets. Fertilizer use has been a critical part of the Green Revolution, except in Africa (see Table 2). As noted in Figure 2, the real (price deflated) price of urea, the major nitrogen fertilizer, has been declining for a number of years. This price decline is associated with technological improvements in urea production (i.e., improvements in the Haber process) and in decreases in the real price of natural gas, a necessary ingredient in the production of urea.

The induced innovation model of invention and innovation calls for increases in fertilizer use by plants when the real price of fertilizer falls. Both rice and wheat GRMVs achieved higher fertilizer responsiveness by incorporating dwarfing genes in GRMV to forestall "lodging". This did increase fertilizer responsiveness in most varieties.

Chapter 53 documents the changes in fertilizer consumption as these changes took place. In 1961 fertilizer application rates were less than 10 kg per hectare in most crops. They were less than 5 kg per hectare in Sub-Saharan Africa. For all developing countries, fertilizer consumption in 2002 was 135 kg per hectare. For all crops, application rates are highest in Asia, next highest in Latin America, next highest in the Middle East/North Africa region and lowest in Sub-Saharan Africa. The low rates of consumption in Sub-Saharan Africa reflect low rates of GRMV adoption and high transaction costs in markets. High transaction costs in markets are the result of poor infrastructure and poor institutions.[31]

Chapter 54 addresses the market for farm machines (planters, tillage implements and harvesters). It is useful to remind ourselves that different parts of the world have had rather specialized experiences with mechanization.

In 1000 AD, most farm work was undertaken by hand. By 1500 AD animal-drawn implements were being introduced. These were often simple plows and tillage implements drawn by oxen and water buffaloes. Animal breeders at this time began to breed

[31] A tonne of grain can be shipped 9000 km from the US to the port of Mombassa for $50 per tonne. The same tonne of grain can be shipped from Mombassa to Kampala, a distance of 500 km, for an additional $100 per tonne. Fertilizer prices rose in 1974–1975 and 1978 but have trended steadily downward since. Pesticides (insecticides and herbicides) have been increasingly utilized in many countries. In general, herbicide use is low in the low-wage countries because weeds are controlled by hand weeding. Insecticide use has increased as farmers perceive insecticides to be a low-cost control measure. Integrated Pest Management (IPM) techniques have reduced insecticide use on crops. For cotton and maize insect-toxic GM crops (Bt) products have also reduced insecticide use.

more versatile draft animals. Harnesses were first adapted for urban uses. With the development of mowers and reapers in the late 17th and early 18th centuries in the United States and Europe, the work-horse came to dominate farm production.

Thus, plows and tillage equipment drawn by work oxen and later by horses came to dominate agriculture in developed countries. The tractor was not really developed until late in the 19th century when stationary tractors and steam engines came into wide use on farms. The "row crop" tractor was not developed until the early 20th century (along with automobiles and trucks). As these developments took place, agriculture in the industrialized countries was rapidly mechanized. By 1950 mechanized planters, tillage equipment and harvesters were widely adopted in industrialized countries.

This hand, animal, machine sequence was pursued in low-income countries as well but at different rates. In some parts of Africa, animal health problems slowed the introduction of animal power. Parts of Latin America, South Africa and a few parts of Asia were mechanized by 1950. But for most developing countries, mechanization took place after World War II.

In the 1950s and 1960s there was a concern that implicit subsidies, lowering the real cost of purchasing planting, tillage and harvesting equipment, led to "premature" mechanization with consequent unemployment outcomes. It is generally true that many governments subsidized the cost of purchasing machinery. Most of the subsidy was in the form of subsidized credit. Some credit in Latin America was available to farmers at negative real interest rates (the rate of inflation exceeded the nominal interest rate on the loan).

These subsidies to machinery purchases have tended to disappear as inefficient and distorted credit markets have undergone reform. Fewer governments have such credit market distortions today. The concern with premature mechanization has lessened considerably in recent years as a consequence.

Today, farmers in many countries make machine adoption decisions by comparing the cost of hand planting, tillage and harvesting with the cost of machine planting, tillage and harvesting. As farmers make these comparisons, rising real wages tend to be the major forces triggering machine purchases. In some cases, machines can outperform hand processes, as in precision planting and these factors influence machine purchases as well.

Most countries in Latin America and the Caribbean have high levels of mechanization. Most Asian countries have lower levels of mechanization but are mechanizing rapidly. The same can be said for the Middle East–North Africa region. Many parts of Sub-Saharan Africa have low levels of mechanization.

Chapter 55 traces the evolution in agricultural output markets in developing countries since 1950. Traditional food markets, characterized by many small producers selling undifferentiated commodities to rural markets or to urban wholesalers, are giving way to modern agrifood systems that emphasize highly differentiated products supplied by large scale processors, supermarkets, and food service chains. The chapter describes the differential speed and magnitude of the change that is taking place across the developing world, although the authors argue that the trend is generally in the direction of increased consolidation. The transformation in food systems is seen as a direct consequence of the overall economic development, induced by demand side factors, such as income growth and urbanization and technology drivers, such as information, telecommunications, shipping and storage technologies. The transformation of markets is further influenced by organizational changes, such as the shift from spot markets to the use of specialized wholesalers, and institutional changes, such as the adoption of contracts and the imposition of private standards.

The authors emphasize that the changes in domestic food markets came not just from trade liberalization but also from the liberalization of Foreign Direct Investment (FDI) in the food sector. The latter had far more influence on the structure of the domestic food system than had the "enclave-type" investments targeted toward the export market. It is through FDI in processing and retailing that globalization is changing domestic food markets in developing countries.

The transformation of food systems poses enormous challenges for smallholders in developing countries. In particular, there is evidence that small farmers are particularly challenged to meet the volume, cost, quality, and consistency requirements of the increasingly dominant supermarket chains and large-scale agro-processors. Identifying organizational and institutional mechanisms for reducing the transaction costs of small farmers is crucial for the benefits of the change in food systems to be shared more equitably.

Chapter 56 documents the role of rural financial markets in the development process and provides a framework within which the evolution of financial intermediation in rural economies can be understood. The chapter provides evidence from a wide variety of rural settings that financial markets are highly fragmented and imperfect. Borrowers are systematically sorted out across different types of financial contracts according to their characteristics and activities. The diversity in contract forms and intermediary structures can be largely explained by imperfect information, extreme inequality in asset distribution, and the high cost of contract enforcement.

The above market imperfections have resulted in financial arrangements that are associated with close involvement of the lender in the activities of the borrower. Lenders in rural financial markets, usually input suppliers or product traders, tend to invest heavily in monitoring their borrowers, and generally become more involved in choices of technology and product decisions. The ability to move beyond financial arrangements that are primarily short term and tied to product outcomes, is determined by the extent to which public policy can reduce information imperfections, establish property rights, and enforce legal contracts.

While the public good nature of government intervention in rural financial markets is often desirable, the chapter argues that in fact, government intervention has often been heavy-handed and resulted in financial repression. State sponsored directed credit programs have been the norm in developing countries for decades. These programs were generally subject to political capture and often bypassed the majority of small farmers who needed them the most. Moreover, interest rate caps and excessive regulation and state involvement in banking has been a principle culprit of the lack of effective intermediation in developing countries, the authors argue that a movement away from bad policies is as important as any potential policy innovation.

Part 5. Resources

Part 5 includes four chapters. Chapter 57 addresses soil degradation. Chapter 58 addresses irrigation system management. Chapter 59 addresses land use issues. Chapter 60 addresses global warming.

In Chapter 57 Pierre Crosson reports a range of estimates of soil degradation. One of the difficulties in the measurement of soil degradation is that many soils are degraded but we do not know the rate at which they have been degraded. There are estimates of recent rates of soil degradation reported by Oldeman (1994) and his colleagues at the Agricultural University in Wageningen in the Netherlands. Crosson chooses the estimate of *bona fide* soil scientists over those of non-scientists.

Chapter 58 addresses the management of irrigation systems. Irrigation investments have been an important part of the Green Revolution. A study for India concluded that the availability of GRMVs stimulate significant investment in irrigation. Irrigation investment enables multiple cropping, and multiple cropping was also an important factor in the Green Revolution. Chapter 58 documents costs and benefits for irrigation systems.

Irrigation systems can be based on river and canal systems or on tube wells. Much of the Asian investment in irrigation stimulated by GRMVs was tubewell irrigation. It is often said that Sub-Saharan Africa has less potential for irrigation expansion than Asia. This may be a factor in the delayed Green Revolution achieved in Sub-Saharan Africa.

Many irrigation systems designed to irrigate a specific number of hectares often fail to deliver adequate irrigation for those hectares. Many systems are afflicted with poor management and inadequate penalties for water hoarding. Virtually all irrigation water in developing countries is unpriced or underpriced.

Chapter 58 evaluates both benefits and costs of irrigation systems including environmental costs. The chapter covers irrigation systems as well as drainage systems. Management issues are covered in the chapter.

Chapter 59 addresses two concerns. The first is whether the population-driven expansion of demand for forest products is creating intense competition between forest land and agricultural land (both cropland and pasture). Deforestation estimates are reported in the chapter. The second question is whether the world is in a period of species extinction and the implications of expanding protected areas.

The first question is addressed in Figure 1 in Chapter 59 where it is shown that world industrial roundwood production has increased very little over the 1979–2000 period when the population of developing countries approximately doubled. In fact, coniferous species of roundwood production actually declined over the period. This was also a period when major advances in the use of "waste" roundwood products were achieved. Thus, the increased demand for industrial roundwood did not increase with population growth. With the technological advances in industrial product development, forest use has not competed with agricultural land uses.

The notion that we are in a sixth wave of species extinction was created by Wilson (1992). Wilson argued that many species remain undiscovered (he uses 10 million species as the estimated number of species, 1.46 million have been classified). He then applies the "island model" of species loss combined with deforestation estimates from Myers (1991) to conclude that we may be losing 2500 species per year. However, actual rates of species discovery in recent decades are not consistent with the estimate that 10 million species exist, nor are data on species extinctions.

This would matter little to human populations were it not for the proposed remedies in the form of protected areas. Charles Geisler of Cornell notes that in 1985, 3.5% of the area of Sub-Saharan Africa was protected. By 2000, 7% of the area of Sub-Saharan Africa was protected. Geisler and de Sousa (2001) calculates that from 1 million to 15 million people have been displaced as a result of the expansion of protected areas. Most of these displaced people have lost their livelihoods as a result.

Chapter 60 addresses global warming issues. The first point made in Chapter 60 is that temperatures did increase by 0.25 °C between 1960 and 2000. In addition, the level of atmospheric carbon dioxide increased from 317 ppm in 1960 to 367 ppm in 2000. Temperature increases depend on latitude. Warming was higher in high latitudes, lower in mid-latitudes and lowest in low latitudes. Warming has also been higher in winter months than in the rest of the year. Precipitation has also increased differentially in different regions.

Two methods were used to compute temperature effects on agricultural productivity. The first method is based on experimental data where crops were subjected to temperature increases and productivity changes (crop yield changes) were measured in these experiments. This method has the limitation that crop choice is held constant. In actuality when temperature change occurs, farmers may adapt by changing crops.

The second method, termed the cross-sectional method (sometimes referred to as the Ricardian method), allows for farmer adaptation. This method was pioneered by the chapter author, Robert Mendelsohn and William Nordhaus, both at Yale University. The method estimates the relationship between land values per hectare, and long-run temperature and rainfall estimates.

Table 2 in Chapter 60 reports estimated temperature and rainfall impacts of actual temperature and rainfall changes on land values (or crop revenues) per hectare. As expected, the experimental method yields more extreme estimates than the cross-section estimates. Both methods show that regions with already high temperatures are damaged most by temperature increases. African productivity is reduced by 1.7% in the experimental method and 0.26% in the cross section methods. Both estimates show that the poorest countries in the world will lose most from climate change. Temperate zone climates gain from global warming.

However, when CO_2 fertilization effects (note that CO_2 fertilization occurs at the same rate everywhere) are factored in as reported in Table 3 in Chapter 60, none of the major regions of the world actually lose from climate change (Appendix A reports country data showing that using the experimental estimates, some countries lose from climate change). This is an extraordinary result.[32]

Nonetheless, both cross-sectional and experimental estimates show differential gains and losses by latitude. High latitude countries gain most and low latitude countries lose most from climate change. Productivity growth in many countries has "swamped" the climate change effects.

References

Avila, A.F., Evenson, R.E. (forthcoming). "Total factor productivity growth in agriculture: The role of technological capital". In: Evenson, R.E., Pingali, P. (Eds.), Handbook of Agricultural Economics, vol. 4, Elsevier Science, Amsterdam. In press.

Bennett, R., Morse, S., Ismael, Y. (2003). "The benefits of Bt cotton to small-scale producers in developing countries: The case of South Africa", Paper presented at the 7th ICABR International Conference on Public Goods Public Policy for Agricultural Biotechnology, Ravello, Italy, June 29–3, 2003. Available at http://www.economia.uniroma2.it/conferenze/icabr2003/papers/papers.htm; accessed March 2004.

Cohen, S.N., Chang, A.C.Y., Boyer, H., Helling, R.B. (1973). "Construction of biologically functional bacterial plasmids in vitro". Proceedings of the National Academy of Sciences U.S.A. 70, 3240–3244.

Evenson, R.E. (2001). "Economic impacts of agricultural research and extension". In: Gardner, B., Rausser, G. (Eds.), Handbook of Agricultural Economics, vol. 1A. Elsevier Science, Amsterdam. Ch. 11.

Evenson, R.E., Gollin, D. (Eds.) (2003a). Crop Variety Improvement and Its Effect on Productivity: The Impact of International Agricultural Research. CABI Publishing, Wallingford, UK.

Evenson, R.E., Gollin, D. (2003b). "Assessing the impact of the Green Revolution, 1960 to 2000". Science 300, 758–762.

Evenson, R.E., Kislev, Y. (1975). Agricultural Research and Productivity. Yale University Press, New Haven, CT.

Falck-Zepeda, J.B., Traxler, G., Nelson, R.G. (1999). "Rent creation and distribution from the first three years of planting Br cotton", ISAAA Briefs No. 14. International Service for the Acquisition of Agri-Biotech Applications, Ithaca, NY.

Fogel, R.W. (1989). "Second thoughts on the European escape from hunger: Famines, price elasticities, entitlements, chronic malnutrition and mortality rates". National Bureau of Economic Research, Cambridge, MA.

Geisler, C., de Sousa, R. (2001). "From refuge to refugee: The African case". Journal of Public Administration and Development 21, 159–170.

Gianessi, L.P., Silvers, C.S., Sankula, S., Carpenter, J.E. (2002). Plant Biotechnology: Current and Potential Impact for Improving Pest Management in US Agriculture: An Analysis of 40 Case Studies. National Center for Food and Agricultural Policy, Washington, DC.

Griliches, Z. (1957). "Hybrid corn: An exploration in the economics of technical change". Econometrica 25, 501–522.

Griliches, Z. (1958). "Research costs and social returns: Hybrid corn and related innovations". Journal of Political Economy 66, 419–431.

Jones, C.I. (2002). Introduction to Economic Growth, second ed. W.W. Norton and Co., New York.

Myers, N. (1991). "Tropical deforestation: The latest situation". Bioscience 41 (5), 282.

Oldeman, R. (1994). "The global extent of soil degradation". In: Greenland, D.J., Szabolics, I. (Eds.), Soil Resilience and Sustainable Land Use. CAB International, Wallingford, UK, pp. 99–118.

[32] Many observers would argue that the CO_2 fertilization impacts are overstated.

Pray, C.E., Huang, J. (2003). "The impact of Bt cotton in China". In: Kalaitzandonakes, N. (Ed.), The Economic and Environmental Impacts of Agbiotech: A Global Perspective. Kluwer–Plenum Academic Publishers, New York.
Qaim, M., de Janvry, A. (2003). "Genetically modified crops, corporate pricing strategies, and farmers' adoption: The case of Bt cotton in Argentina". American Journal of Agricultural Economics 85 (4), 814–828.
Qaim, M., Traxler, G. (2004). "Roundup Ready soybeans in Argentina: Farm level, environmental, and welfare effect". Agricultural Economics. In press.
Qaim, M., Zilberman, D. (2003). "Yield effects of genetically modified corps in developing countries". Science 299, 900–902.
Ruttan, V. (2001). Technology, Growth, and Development: An Induced Innovation Perspective. Oxford University Press, Oxford, UK.
Solow, R.M. (1956). "A contribution to the theory of economic growth". Quarterly Journal of Economics 70, 65–95.
Traxler, G., Godoy-Avila, S., Falck-Zepeda, J., Espinoza-Arellano, J. (2003). "Transgenic cotton in Mexico: Economic and environmental impacts". In: Kalaitzandonakes, N. (Ed.), The Economic and Environmental Impacts of Agbiotech: A Global Perspective. Kluwer–Plenum Academic Publishers, New York.
Wilson, E.O. (1992). The Diversity of Life. W.W. Norton & Co., New York.

PART 2

HUMAN RESOURCES AND TECHNOLOGY MASTERY

Chapter 43

AGRICULTURE AND HUMAN CAPITAL IN ECONOMIC GROWTH: FARMERS, SCHOOLING AND NUTRITION*

WALLACE E. HUFFMAN and PETER F. ORAZEM

Department of Economics, Iowa State University, Ames, IA

Contents

* The paper was presented at the Bretton Woods Conference on Agricultural Producers and Productivity organized by Yale University, September 29–October 1, 2002, Bretton Woods, NH. We acknowledge helpful comments on earlier drafts from Jonathan Conning, Bob Evenson, Bruce Gardner, Roger Sedjo, Chris Udry and other attendees at the Bretton Woods conference. Financial support from the Iowa Agricultural and Home Economics Experiment Station is acknowledged. Victoria Gunnarsson and Luis Gonzales provided excellent research assistance.

Handbook of Agricultural Economics, Volume 3
Edited by Robert Evenson and Prabhu Pingali

DOI: 10.1016/S1574-0072(06)03043-X

Abstract

This survey reviews the existing literature, identifying the contribution of agriculture, schooling, and nutrition to economic growth and development over time and across countries. Particular attention is paid to the roles of improvements in agricultural technology and of the human capital of farmers and farm people. Macroeconomic and microeconomic evidence related to the interactions between human capital, productivity and health are explored. Most of the world's growth in population, labor productivity and real income per capita have occurred over the past 250 years. We show that for most countries, development is a process of conversion from primarily agrarian economies to urban industrial and service economies. The evidence is that positive technology shocks to agriculture have played a key role in igniting a transition from traditional to modern agriculture and to long-term economic growth in almost all countries. Improvements in agricultural technologies improve labor productivity and create surplus agricultural labor that can provide workers for the growing urban areas. In some cases, improved nutrition helps raise labor productivity and allows individuals to work for longer hours, which makes human capital investments more attractive. The induced improvements in the skill level of a population have major implications for raising living standards, improving health standards, and altering time allocation decisions. In most currently poor and middle income countries, improved schooling has been more important than improved nutrition or caloric intake in explaining recent economic growth. Nevertheless, the poorest countries of the world continue to have a large share of their labor force in agriculture, and growth cannot occur until they experience their own agricultural transformation.

Keywords

farmers, schooling, health, nutrition, human capital, economic growth, agricultural household models, agricultural transformation, two-sector models

JEL classification: O15, O18, O33, O40

1. Introduction

This paper examines the important role played by agriculture and human capital of farmers and farm people in economic growth and development. In particular, we place great emphasis on the importance of positive agricultural technology shocks for igniting what may become long-term economic growth with increased per capita food production, an improved standard of living, migration of labor from the farm to the nonfarm sector and the rise of cities. We provide a critique of the existing literature, identify the contributions of agriculture, schooling, and nutrition to economic growth and development of countries, and provide recommendations about gaps and puzzles that exist. We place the analysis in the context of long-term economic growth, starting from an economy and labor force that is primarily agricultural [Johnson (2000)] and then consider productivity shocks to agriculture as an essential event before modern economic growth with industrialization can occur. We show that agriculture and human capital in schooling and nutrition are important to the growth process. In particular, as economies are transformed from traditional low income societies, the farm sector becomes a major source of labor for the nonfarm sector, but inter-sector and occupational mobility require a skilled labor force. Also, the skills of women, even if they work primarily as unpaid workers or in housework, are an important source of human capital production as they nurture children and families.

This chapter: (i) provides a conceptual framework for visualizing economic growth from agricultural technology shocks and human capital production and investments, (ii) summarizes 2000 years of world economic growth and development experiences, (iii) presents an overview of the econometric evidence that schooling contributes to economic growth, (iv) reviews the evidence on production of health, nutrition, and work, and (v) summarizes schooling outcomes in agriculture. Finally, some conclusions are formulated.

2. A conceptual framework for visualizing economic growth from agriculture and human capital investments

This section first provides a conceptual framework for viewing the contribution of agriculture to economic growth. Next we present a three-period model of optimal human capital production and investment and develop a few implications.

2.1. Two-sector model: Agriculture and non-agriculture

As shown in Jorgenson (1965) and Huffman (1977), a two-sector model of the linkages between the farm and nonfarm sectors can generate useful insights into how technical change in agriculture can foster economic development. It also provides useful insights on labor mobility or migration. In poor countries, trade barriers are frequently ubiquitous, and they do not have sufficient foreign exchange to regularly purchase large quantities of food from abroad. Hence, closed economy models may provide a satisfactorily close approximation to conditions faced by many developing countries.

The following model allows us to illustrate how various economic linkages transmit technology shocks in agriculture across the economy, altering relative farm and nonfarm prices, incomes and populations. Define $X_i^d = D_i(P_1, P_2, Y)$, $i = 1, 2$, as the demand function for farm output ($i = 1$) and for nonfarm output ($i = 2$). P_i, $i = 1, 2$, is the price of farm and nonfarm output, respectively, and I is total income of domestic demanders. In the supply equation, $X_i^s = AS_i(P_i)$, $i = 1, 2$, where A is the coefficient of disembodied technical change.

Because we are concerned with growth, the two-sector model is expressed in time-rate of change form:

$$x_i^d = \varepsilon_{11} p_1 + \varepsilon_{12} p_2 + \eta_i y, \quad \varepsilon_{ii} < 0,\ i = 1, 2 \quad \text{(demand equation)}, \tag{1}$$

$$x_i^s = \varphi_{ii} p_i + a_i, \quad \varphi_{ii} \geqslant 0,\ i = 1, 2 \quad \text{(supply function)}, \tag{2}$$

where $x_i = \frac{\mathrm{d}X_i}{\mathrm{d}t}\frac{1}{X_i}$, $p_i = \frac{\mathrm{d}P_i}{\mathrm{d}t}\frac{1}{P_i}$, $i = 1, 2$, are the percentage rates of change in the output of sector i and of the price of output of sector i, respectively. The response elasticities are own-price demand elasticity, $\varepsilon_{ii} = \frac{\mathrm{d}X_i}{\mathrm{d}P_i}\frac{P_i}{X_i}$; cross-price demand elasticity, $\varepsilon_{ij} = \frac{\mathrm{d}X_i}{\mathrm{d}P_j}\frac{P_j}{X_i}$, $i, j = 1, 2$, $i \neq j$; and the income elasticity, $\eta_i = \frac{\mathrm{d}X_i}{\mathrm{d}Y}\frac{Y}{X_i}$, $i = 1, 2$. In addition, $a_i = \frac{\mathrm{d}A_i}{\mathrm{d}t}\frac{1}{A_i}$, $i = 1, 2$, is the rate of disembodied technical change in sector i.

We assume that the two markets are initially in equilibrium, $X_i^d = X_i^s$, $i = 1, 2$, i.e. the markets clear, and we maintain the neoclassical assumption that markets clear even when shocks to demand and (or) supply occur:

$$\begin{cases} \varphi_{11} p_1 + a_1 = \varepsilon_{11} p_1 + \varepsilon_{12} p_2 + \eta_i y & \text{(farm sector)}, \\ \varphi_{22} p_2 + a_2 = \varepsilon_{21} p_1 + \varepsilon_{22} p_2 + \eta_2 y & \text{(nonfarm sector)}, \end{cases} \tag{3}$$

$$\begin{cases} (\varepsilon_{11} - \varphi_{11}) p_1 + \varepsilon_{12} p_2 = a_1 - \eta_1 y = c_1, \\ \varepsilon_{21} p_1 + (\varepsilon_{22} - \varphi_{22}) p_2 = a_2 - \eta_2 y = c_2. \end{cases} \tag{4}$$

Now Equation (4) is arranged to emphasize that income growth (y) and disembodied technical change (a_1, a_2) are driving changes in the prices of farm and nonfarm output, p_1 and p_2. This set of equations can be solved for the equilibrium rate of change in the prices of the farm (X_1) and nonfarm (X_2) outputs due to $c_i = a_i - \eta_i y \neq 0$:

$$p_1 = \frac{(a_1 - \eta_1 y)(\varepsilon_{22} - \varphi_{22}) - (a_2 - \eta_2 y)\varepsilon_{12}}{(\varepsilon_{11} - \varphi_{11})(\varepsilon_{22} - \varphi_{22}) - \varepsilon_{12}\varepsilon_{21}}, \tag{5}$$

$$p_2 = \frac{(a_2 - \eta_2 y)(\varepsilon_{11} - \varphi_{11}) - (a_1 - \eta_1 y)\varepsilon_{21}}{(\varepsilon_{11} - \varphi_{11})(\varepsilon_{22} - \varphi_{22}) - \varepsilon_{12}\varepsilon_{21}}. \tag{6}$$

Therefore the rate of change in equilibrium market prices of farm and nonfarm output due to income growth and technical change are a function of the two own-price elasticities ($\varepsilon_{11}, \varepsilon_{22}$), two cross-price elasticities ($\varepsilon_{12}, \varepsilon_{21}$), two-income elasticities (η_1, η_2) and two rates of technical change (a_1, a_2).

Now assume both farm and nonfarm outputs are normal goods ($\eta_i > 0, i = 1, 2$), and the income elasticity of farm output is less than for nonfarm output ($\eta_1 < \eta_2$). If the rate of disembodied technical change is the same in the two sectors, ($a_1 = a_2 = a$) and if the cross-price elasticities of demand are zero ($\varepsilon_{12} = \varepsilon_{21} = 0$), then $a - \eta_1 y > 0$ and $a - \eta_2 y < 0$. Consequently,

$$\mathrm{sign}(p_1 - p_2) = \mathrm{sign}\big[(a - \eta_1 y)(\varepsilon_{22} - \varphi_{22}) - (a - \eta_2 y)(\varepsilon_{11} - \varphi_{11})\big] < 0, \qquad (7)$$

so that equal rates of technical change in the two sectors will cause the relative price of farm output to decline.

If the cross-price elasticities of demand are of opposite signs and the other conditions hold, then condition (7) still holds. If the cross-price elasticities are of the same sign but $|(\varepsilon_{12}\varepsilon_{21})| < |(\varepsilon_{11} - \varphi_{11})(\varepsilon_{22} - \varphi_{22})|$, then condition (7) also holds. If we impose homogeneity of degree zero in prices and income so that $\varepsilon_{11} + \varepsilon_{12} + \eta_1 = \varepsilon_{21} + \varepsilon_{22} + \eta_2 = 0$, then the denominators in (5) and (6) will still be positive, and so condition (7) remains satisfied. Thus, if the income elasticity for nonfarm output is larger than for farm output, positive growth of income (or technical change) causes the relative price of farm output to decline under very general conditions. The result in (7) becomes even stronger if TFP growth is faster in the agricultural than in the non-agricultural sector, as has been found by Jorgenson and Stiroh (2000) for the U.S.

Consistent with this simple theoretical argument and evidence of relative farm and nonfarm productivity growth, economic development in the United States since 1900 has generally been accompanied by falling relative prices of farm commodities [Huffman and Evenson (2006, p. 251)]. This pattern has held generally as economic development has occurred in other countries and has profound implications for the proportion of the population engaged in farm production over the long run. Assume that labor is the only variable input in the farm and nonfarm sectors and that farm and nonfarm labor markets are in equilibrium. Then, the real wage or its equivalent is approximately equal across the two sectors. Again assume that the rate of disembodied technical change and population growth is the same in both the farm and nonfarm sectors. With the farm output price falling relative to the nonfarm sector output price, the real cost of food will fall. To maintain equilibrium real wage rates between sectors, labor must move from the farm to the non-farm sector, and this migration is a form of human capital investment. If the natural population growth rate is faster in the farm than the nonfarm sector, the rate of mobility from the farm/agricultural sector must be even faster. These migrants become a potentially important supply of labor for growing nonagricultural sectors such as manufacturing and services. See Floyd (1967) for a detailed framework.

Gollin (2000) reemphasizes that the share of the labor force which is self-employed, working on own account, or unpaid family labor is largest in low income countries. Gollin, Parente and Rogerson (2002) also provide supporting evidence that growth in agriculture is central to economic growth and development in poor countries.

The argument thus far assumes a closed economy. In open economies in which domestic agricultural prices are set exogenously by world markets, technological change in agriculture can raise rural incomes relative to urban incomes. The lack of either rural to urban migration or a reduction in food prices breaks the linkage between advances in agricultural technologies and overall growth [Matsuyama (1992)]. This scenario, however, may not fit actual experiences of most developing countries. To the extent significant subsectors of agriculture are in the nontraded sector (especially milk and fresh meat) technological advances will carry through as in the closed economy model. As we discuss in detail below, the simple closed economy model appears to conform well to the past development experiences of many countries.

2.2. *A multiperiod agricultural household model*

When human capital investment decisions are the central focus (e.g., schooling, informal training, migration, information search, technology adoption, nutrition, health), multiperiod household utility maximizing models provide a useful guide for empirical work. Once household members have obtained their human capital and the focus is on choice of occupation, hours of work, purchased-input use, wage rates, or income, one-period static agricultural household models provide a useful guide to researchers [Singh, Squire and Strauss (1986)]. In particular, behavioral models provide one useful guide to researchers for deciding which variables should be treated as endogenous and which are to be held exogenous or causal.

2.2.1. *A three-period model of optimal production and investment*

Consider a risk-neutral farm household living three periods. In each period, the household consumes human capital services as leisure (L_{1t}) and goods purchased in the market (X_{1t}), and these goods are the source of household utility. The household has an initial endowment of human capital (K_0) coming into the initial period (0), and this stock is translated at a constant rate (α) into human capital services (a flow) available for use in the initial period. In addition to leisure, a household's human capital services are potentially allocated each period to human capital production (L_{2t}), farm production (L_{3t}), and to wage work L_t^W. In addition to human capital services, the household's production of human capital uses inputs purchased in the market (X_{2t}), and a time-invariant individual or household-specific genetic or innate ability factor (A_2). This technology is assumed to exhibit decreasing returns to scale in production. Furthermore, K_0, the initial human capital endowment and A_2 are different; including that A_2 does not change over time. The household production of farm output uses human capital services plus inputs purchased in the market (X_{3t}) and a time-invariant farm-specific factor (A_3), e.g., agro-climatic conditions. The farm production technology is assumed to exhibits decreasing returns to scale in the variable inputs.

In this model, human capital produced in one period increases the stock of human capital and available human capital services in later periods. Thus, for those who are accustomed to thinking of a household having the same fixed time (hours) endowment in each period, this model takes a different approach. The "endowment" is variable over time, and it is in units of human capital services or quality adjusted hours. Moreover, additions to an individual's or household's human capital services does not change the wage per unit of human capital service [Ben-Porath (1967)]. Also, we assume that human capital depreciates at some constant rate, $0 \leqslant \delta < 1$, due to deterioration in health from the toll of diseases, smoking, or excess weight which inevitable lead to the breakdown of chromosomes and organ systems [Fogel (2004); Valdes et al. (2005)]. The household receives cash income from the sale of farm output and from supplying human capital services to the labor market in the form of wage work. It spends this income on the purchased inputs for utility, human capital production, and farm production.

The farm household has a well-behaved three-period utility function:

$$U = U(L_{10}, X_{10}, L_{11}, X_{11}, L_{12}, X_{12}). \tag{8}$$

The household's technology for human capital production in each period is represented as

$$Z_{2t} = F_2(L_{2t}, X_{2t}, A_2), \quad t = 0, 1, 2. \tag{9}$$

Hence, when the variable input prices are fixed to the household within any time period, the assumption of decreasing returns to scale implies that the marginal cost is rising with added human capital produced. If input prices are the same across time periods, the shape of the marginal cost curve will also be exactly the same in each time period. The rising marginal cost of human capital production in each time period reflects, for example, the reality of an upper limit on mental capacity of an individual to learn in each period.

The household's technology for farm output production in each period is represented as

$$Z_{3t} = F_3(L_{3t}, X_{3t}, A_3), \quad t = 0, 1, 2. \tag{10}$$

It has decreasing returns to scale in the region of an optimal solution due to natural limitations placed on the production process by the agro-climatic conditions.

The equations giving the quantity of human capital services available to the household in each period is summarized in Equations (11a)–(11c):

$$L_0 = \alpha K_0 = L_{10} + L_{20} + L_{30} + L_0^W, \quad L_{20}, L_{30}, L_0^W \geqslant 0, \tag{11a}$$

$$\begin{aligned} L_1 &= \alpha K_1 = \alpha\big[(1-\delta)K_0 + \gamma Z_{20}\big] \\ &= L_{11} + L_{21} + L_{31} + L_1^W, \quad L_{21}, L_{31}, L_1^W \geqslant 0, \end{aligned} \tag{11b}$$

$$\begin{aligned} L_2 &= \alpha K_2 = \alpha\big[(1-\delta)^2 K_0 + (1-\delta)\gamma Z_{20} + \gamma Z_{21}\big] \\ &= L_{12} + L_{22} + L_{32} + L_2^W, \quad L_{22}, L_{32}, L_2^W \geqslant 0. \end{aligned} \tag{11c}$$

In each equation and time period, the number of human capital services available is determined by the size of the human capital stock coming into the period (and the constant α). This stock is given by the second term in Equations (11a)–(11c). Furthermore, given that human capital production may occur in period 1 and 2, the third term in Equations (11b) and (11c) give the exact size of the stock of human capital in terms of the initial endowment and human capital that has been produced in periods 0 and 1. In Equations (11b) and (11c), the parameter γ just converts stocks to flows and is constant over time. Human capital services available in each time period are allocated as specified in the right side of Equations (11a)–(11c).[1] The quantity of human capital services allocated to human capital production, farm production, and wage work might be zero each period, and this is reflected in the non-negativity constraints in these equations.

The household discounted cash budget constrain over three periods is

$$\sum_{t=0}^{2} \frac{P_{3t} Z_{3t} + W_t L_t^W}{(1+r)^t} = \sum_{i=1}^{3} \sum_{t=0}^{2} \frac{P_{it} X_{it} + C_t}{(1+r)^t}, \tag{12}$$

where P_{3t} is the expected price of farm output and P_{it} is the expected price of the purchased consumption goods, inputs into human capital production, or inputs into farm production, respectively. The expected wage rate for per unit of human capital service is W_t; $C_t \geqslant 0$ is any fixed cost associated with the household's production or consumption activities, and r is a fixed discount rate and no borrowing or lending constraint exists for the farm household.

Now if Equation (10) is substituted into Equation (12), the farm production and multiperiod budget constraint are combined into one equation:

$$\sum_{t=0}^{2} \frac{P_{3t} F_3(L_{3t}, X_{3t}, A_3) + W_t L_t^W}{(1+r)^t} = \sum_{i=1}^{3} \sum_{t=0}^{2} \frac{P_{it} X_{it} + C_t}{(1+r)^t}. \tag{13}$$

The household can now be viewed as making consumption, human capital production, farm production and labor supply decisions in each of three periods by maximizing the objective function which is composed of Equation (8) subject to Equations (9), (11a)–(11c) and (13), including nonnegativity constraints. The Kuhn–Tucker first-order conditions are

$$\frac{\partial \xi}{\partial L_{1t}} = \frac{\partial U}{\partial L_{1t}} - \frac{\lambda_t}{(1+r)^t} = 0, \quad t = 0, 1, 2, \tag{14}$$

$$\frac{\partial \xi}{\partial X_{1t}} = \frac{\partial U}{\partial X_{1t}} - \lambda \frac{P_{1t}}{(1+r)^t} = 0, \quad t = 0, 1, 2, \tag{15}$$

$$\begin{aligned} &\frac{\partial \xi}{\partial L_{20}} = \lambda\left[PV_{Z_{20}}^{0} MP_{L_{20}}^{Z_2} - \lambda_0\right] \leqslant 0, \qquad L_{20} \geqslant 0, \\ &\quad L_{20}\left(PV_{Z_{20}}^{0} MP_{L_{20}}^{Z_2} - \lambda_0\right) = 0, \end{aligned} \tag{16}$$

[1] These allocations could be viewed as a share of the total amount of human capital services available in each time period.

where

$$PV^0_{Z_{20}} = \frac{W_1\alpha\lambda}{(1+r)} + \frac{W_2\alpha\lambda(1-\delta)}{(1+r)^2} \quad \text{and} \quad MP^{Z_2}_{L_{2t}} = \frac{\partial Z_{2t}}{\partial L_{2t}},$$
$$MP^{Z_2}_{X_{2t}} = \frac{\partial Z_{2t}}{\partial X_{2t}}, \quad t = 0, 1, 2,$$

$$\frac{\partial \xi}{\partial X_{20}} = \lambda\left[PV^0_{Z_{20}} MP^{Z_2}_{X_{20}} - P_{20}\right] \leqslant 0, \qquad X_{20} \geqslant 0,$$
$$X_{20} PV^0_{Z_{20}}\left(MP^{Z_2}_{X_{20}} - P_0\right) = 0, \tag{17}$$

$$\frac{\partial \xi}{\partial L_{21}} = \lambda\left[PV^0_{Z_{21}} MP^{Z_2}_{L_{21}} - \frac{\lambda_1}{1+r}\right] \leqslant 0, \qquad L_{21} \geqslant 0,$$
$$L_{21}\left(PV^0_{Z_{21}} MP^{Z_2}_{L_{21}} - \frac{\lambda_1}{1+r}\right) = 0, \tag{18}$$

$$\frac{\partial \xi}{\partial X_{21}} = \lambda\left[PV^0_{Z_{21}} MP^{Z_2}_{X_{21}} - \frac{P_{21}}{1+r}\right] \leqslant 0, \qquad X_{21} \geqslant 0,$$
$$X_{21}\left(PV^0_{Z_{21}} MP^{Z_2}_{X_{21}} - \frac{P_{21}}{1+r}\right) = 0, \tag{19}$$

where

$$PV^0_{Z_{21}} = \frac{W_2\alpha\gamma}{(1+r)^2},$$

$$\frac{\partial \xi}{\partial L_{22}} = \lambda\left[PV^0_{Z_{22}} MP^{Z_2}_{L_{22}} - \frac{\lambda_2}{(1+r)^2}\right] \leqslant 0, \qquad L_{22} \geqslant 0,$$
$$L_{22}\left(PV^0_{Z_{22}} MP^{Z_2}_{L_{22}} - \frac{\lambda_2}{(1+r)^2}\right) = 0, \tag{20}$$

$$\frac{\partial \xi}{\partial X_{22}} = \lambda\left[PV^0_{Z_{22}} MP^{Z_2}_{X_{22}} - \frac{P_{22}}{(1+r)^2}\right] \leqslant 0, \qquad X_{22} \geqslant 0,$$
$$X_{22}\left(PV^0_{Z_{22}} MP^{Z_2}_{X_{22}} - \frac{P_{22}}{(1+r)^2}\right) = 0, \tag{21}$$

$$\frac{\partial \xi}{\partial L_{3t}} = \lambda\left[\frac{P_{3t}}{(1+r)^t} MP^{Z_3}_{L_{3t}} - \frac{\lambda_t}{(1+r)^t}\right] = 0, \quad t = 0, 1, 2, \tag{22}$$
$$\frac{\partial \xi}{\partial X_{3t}} = \lambda\left[\frac{P_{3t}}{(1+r)^t} MP^{Z_3}_{X_{3t}} - \frac{P_{3t}}{(1+r)^t}\right] = 0, \quad t = 0, 1, 2, \tag{23}$$

$$\frac{\partial \xi}{\partial L_t^W} = (-\lambda_t + W_t)/(1+r)^t \leqslant 0, \qquad L_t^W \geqslant 0,$$

$$L_t^W(-\lambda_t + W_t)/(1+r)^t = 0, \quad t = 0, 1, 2. \tag{24}$$

In these equations, λ is the marginal utility of cash income, and λ_t, $t = 0, 1, 2$, is the marginal utility of human capital services in each of the time periods. Equation (24) implies that only if it is optimal for the household/individual to participate in off-farm wage work is $\lambda_t = W_t$, otherwise $\lambda_t > W_t$.

Equations (16)–(21) imply that the household minimizes cost in the production of human capital. If the constraints are satisfied, optimal human capital satisfies the conditions

$$PV^0_{Z_{2t}} = MC_{Z_{2t}} = \frac{\lambda_t}{MP^{Z_2}_{L_{2t}}} = \frac{P_{2t}}{MP^{Z_2}_{X_{2t}}}. \tag{25}$$

Implications The following important results follow from the three-period model of optimal behavior. First, the optimal size of the human capital investment in each period is the quantity or rate at which the present value of the marginal return from a unit of human capital service equals the present value of its marginal cost. Increases in the borrowing interest rate will cause the household to lower its current investments in human capital.

Second, insights about the tendency for investing in skill to weaken or strengthen ties to farming are obtained by examining the present value of the marginal return to investment in human capital. There are two effects – the change in the present value of the additional farm production that results from allocating part of an incremental unit of human capital services to this activity and the change in the present value of the additional labor market earnings that results from allocating the remaining part of an increment of human capital services to nonfarm wage work.

The allocation of an increment of human capital services between farm production and off-farm work is quite sensitive to the relative impact of human capital on the marginal product of labor in farm and nonfarm work or to the elasticity of demand faced by the individual for human capital services. If the marginal product of human capital services is low in farm production but relatively large in nonfarm wage work, and it is optimal to invest in human capital, then an agricultural household will increase the share of employed human capital services allocated to nonfarm wage work.

Third, given the three-period lifetime, a comparison of the present value of the marginal return to an investment in period $t = 0$ versus 1 shows that delaying the investment by one period significantly reduces the present value of the marginal return. Hence, it is optimal for agricultural households to make large human capital investments early in an individual's life rather than later. Furthermore, it is never optimal in this model for a household to invest resources in human capital production in the final period (period 2), because there is cost but no benefit (see Figure 1).

Fourth, because the marginal cost of human capital production is increasing, it will frequently be optimal for an agricultural household to spread its human capital invest-

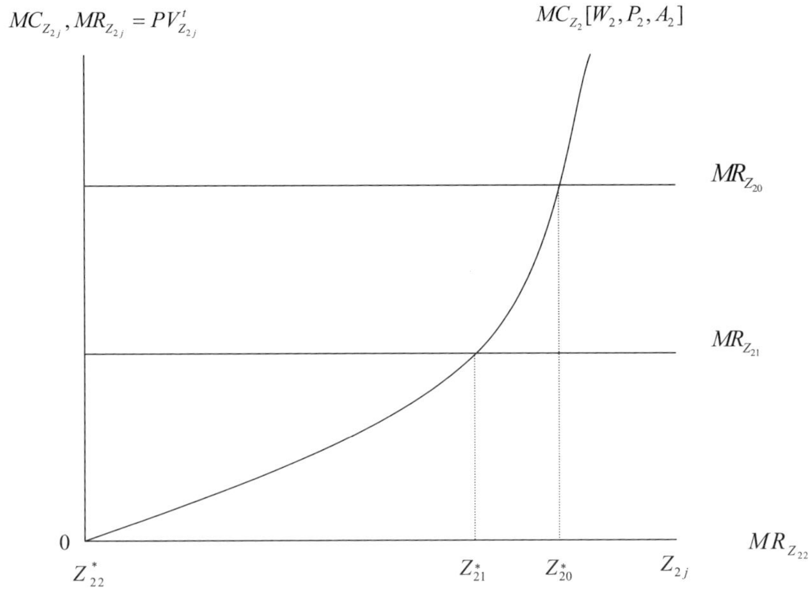

Figure 1. Optimal household decision making: production and investment in human capital over a three-period lifetime.

ment in an individual over two periods rather than a single period, even with finite life of three periods and associated reduction in the present value of the marginal return due to delaying the investment. Spreading the investment over time is a good decision when the cost saving exceeds the reduction in returns due to postponement (see Figure 1).

Fifth, if the length of life were to be extended to four periods (e.g., due to better public health measures), this would increase the household's demand for human capital, and other things being equal, increase life-time human capital (e.g., schooling) investment per individual.

2.2.2. *Turning to practical implications*

Schooling and learning-by-doing, human capital may be productive or unproductive in agriculture depending on economic conditions, but in economies with freely mobile resources, agriculture must compete with other sectors for skilled (and unskilled) labor. The wage for similarly skilled labor need not be equal across sectors, but in equilibrium the marginal compensation, including monetary value of nonmonetary attributes of the farm and nonfarm work, will be equal. Recently the U.S. farm–nonfarm compensating differential has been small [Huffman (1996)]. Although technical change in agriculture is frequently at least as large as in the nonfarm sector, the opportunities for raising labor productivity in agriculture through task specialization and coordination may be modest compared with the nonfarm sector. On a farm, the skilled individual may face a more inelastic demand for his/her services than in a large nonfarm business. Also, due to poor infrastructure and institutions, the agricultural sector may in some cases face small market size and high coordination costs that put it at a disadvantage. Mechanization of agricultural tasks in the U.S. has, however, created highly capital intensive agriculture.

In some agricultural environments, informal learning rather than schooling is the most important form of human capital, but in other environments where information processing about new technologies is important, schooling may have high payoffs [Schultz (1964); Huffman (1985, 1991); Becker (1993, pp. 1–13); Johnson (2000)]. For example, in a traditional environment that is static in technology and relative prices as exists in some low-income countries, accumulated experience is a better investment than schooling. Information accumulated informally does not depreciate when the decision-making environment is static. However, in a market economy where the political and economic environments are changing and new technologies are regularly becoming available, skills obtained from formal schooling provide an important foundation for later informal post-school learning. Most new agricultural technologies are geoclimatic and (or) land-specific, and changing technologies cause rapid depreciation in land-specific human capital. Being able to make good decisions on information acquisition and technology adoption is a valuable skill. Hence, a changing agricultural environment increases the expected return to formal schooling through allocative efficiency effects, which seem likely to be more important than technical efficiency effects.

3. A summary of 2000 years of world economic growth experiences

For many centuries most of the population lived near subsistence. Periodically, positive technological shocks would occur that permitted temporary increases in per capita income and population. However, most of the growth spawned by these shocks were transitory. Before the 1700s, the world's population grew at a very low steady state with standards of living near the subsistence level.

3.1. Early evidence

Maddison (2001) compiled data on the world population and per capita gross domestic product from year 0 to 1998 AD. As shown in Figure 2, for the first 1700 years of the series, world population hardly grew at all. From a base of approximately 231 million world inhabitants in year 0, the time of the first Roman Census, the world population grew to 603 million by 1700 – an average net increase of only 0.06% per year. Over the next 300 years, the world population increased ten-fold, rising at an average rate of 0.8% per year. Most of the growth has occurred during the 20th century when world population growth averaged 1.4% per year. The population may have grown even more rapidly were it not for the two world wars. Since World War II, world population growth averaged 1.8% per year.

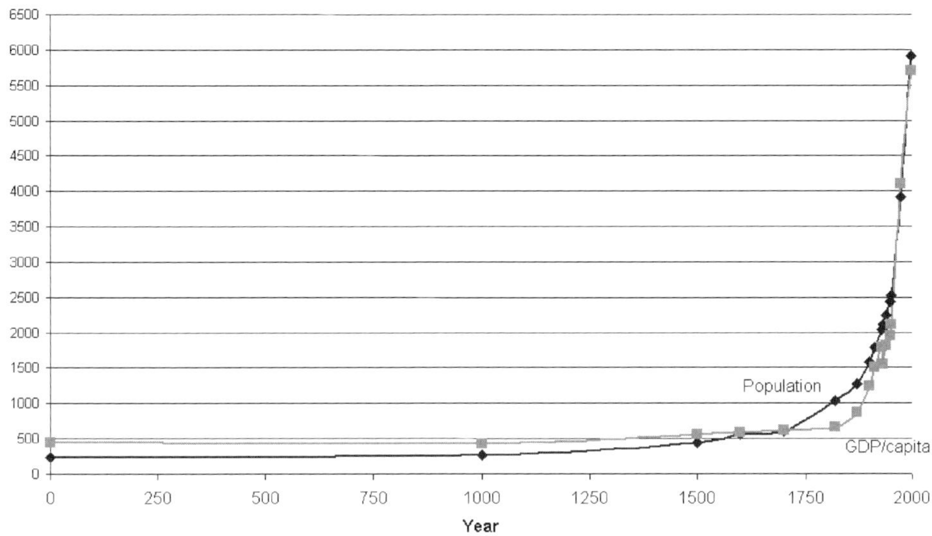

Figure 2. World population and GPD per capita, 0–1998; population in millions, income in 1990 U.S. dollars.

The very slow growth of the world population for the first 1700 years coincided with an even slower growth of GDP per capita that averaged 0.02% per year. Using GDP per capita as a rough indicator of labor productivity, it appears that workers in 1700 were no more than 1.4 times more productive than workers in year 0. Over the next 300 years, GDP per capita grew at 0.8% per year – roughly comparable to the population growth rate. However, the timing of the population and per capita GDP growth after 1700 differed. Before 1900, the world population grew faster than GDP per capita (0.50% vs 0.35%) but slower thereafter (1.4% vs 1.6%). Nevertheless, the consistency between the growth of population and the growth of output per capita suggests that the two series are structurally interrelated.

In 1750, over 90% of the world's labor force was engaged in agriculture. By 1830, France and Germany had reduced their labor force in agriculture to about 50%. The U.K., the most advanced industrial country at the time, had less than 25% of its labor force in agriculture. At the same time, the U.S. had about 65% of its labor force in agriculture [Grübler (1994)], a share that is equal to that of India and China in 2000 [World Bank (2000)].

The great wealth of today's industrialized nations and remarkable improvement in the wellbeing of people in developed countries have been made possible by farm people aided by organizational, institutional, and scientific advances. Farmers and farm people have played a central role in this transformation process. Change was possible because farmers could produce a surplus over and above their own consumption, and the surplus could be exported to the cities. Advances in the science of agriculture also contributed greatly to the early economic development of the currently developed countries [Johnson (2000)].

Fogel (1994) and Johnson (2000) have laid out the reasons why improvements in agricultural productivity were a necessary precondition for early economic growth to occur. Fogel begins his explanation with an examination of time-series data on death rates between 1550 and 1975. The rise in population growth rates after 1700 corresponded to a secular decline in death rates observed in European church records. Before 1700, the Malthusian prediction that the population expanded to consume any available increases in food production was essentially correct, as evidenced by the absence of appreciable growth in per capita output. For 1700 years, the economic conditions for the average person in the world hardly improved. What is less apparent is that the average income or food production level over that period was too low to energize the labor force for hard work, meaning that the low levels of per capita income also led to persistently low levels of labor productivity.[2]

Depending on the weather and on individual size, gender, and age, we can estimate the minimum caloric intake necessary to support productive labor over a full working day. Given the average stature of men and women in Europe in the 1700s, Fogel estimated that at least 2000 calories per person would be necessary to support productive work. In England and France in the latter part of the 18th century, food production was high enough to meet this target on average, but food was not equally distributed in the population. About 40% of the French males and 20% of the British males did not attain even this minimal level of nutrition, meaning that they were too undernourished to perform a full day of work.[3] Moreover, even those who attained the minimal level of nutrition on average, "were so stunted and wasted that they were at substantially higher risk of incurring chronic health conditions and of premature mortality".

Before 1700 in Europe, land-holdings were under a feudal system. Livestock mingled together as they grazed the "common pastures" and were shepherded by individuals, a time-intensive production method. These animals were called "common stock" and the genetic potential of these animals was low and static. Enclosure of the "commons" in the U.K. started about 1700 with the transition to private property. Enclosures – fencing private property – made controlled mating of farm animals possible, which was essential for improving livestock genetically. Enclosures also eliminated the need for labor to shepherd livestock, reducing the demand for labor in livestock production. With the spread of the enclosure system in the U.K., farmers started using nitrogen-fixing legumes in their crop rotations. This helped to boost crop yields. Crop productivity was also increased because crop farmers were largely free of the damage caused by wandering livestock herds [Huffman and Evenson (1993)]. These were important early changes

[2] This observation leads to models of efficiency wages which will be covered in a later section.

[3] Fogel (1994, p. 374) suggests that the very high proportion of beggars in cities (perhaps as high as 20% of the population) was related to the fact that the lowest fifth of the population would have caloric intake that was too low to support even a few hours of strolling per day.

that increased the productivity of agriculture in Europe in the 18th century and provided the nutrient base for the economic growth that followed.

With private ownership, farmers for the first time took an interest in farm-animal improvement, and improved strains began to appear. Farmers had known that some animals were better adapted to a particular environment than others, and in 1760, Bakewell, an English farmer, is credited with first establishing the pattern of modern animal breeding. He established purified lines, emphasizing selection for visual traits, and began the process of developing purebred animals [Huffman and Evenson (1993)]. New breeds were generally selected so that they were adapted to local geoclimatic conditions.

Before the Industrial Revolution, craftsmen operating small shops with a minimum of wage labor were the main producers of nonfarm goods. With increased agricultural productivity in the mid-18th century, the U.K. was able to initiate an Industrial Revolution that built on standardization and specialization of activity in the nonfarm sector [Grübler (1994)]. This industrialization first occurred in textile and iron production. By the early 19th century, Germany was making major technical advances through the application of science in laboratory chemistry. This provided the foundation of a new chemical industry and for further scientific advances to support agriculture. For example, during the 19th century, the U.S. and other countries sent students to Germany for training at the first agricultural chemistry laboratory – one established by Liebig at Giessen. He published his famous agricultural chemistry book, *Organic Chemistry in Its Relation to Agriculture*, in 1840. The early attempts to apply science to agriculture in the United States drew upon the German example for their model of institutional organization and the education of agricultural scientists [Huffman and Evenson (1993)].

Although nutrient intake data are not easily available, estimates of per capita GDP are widely available over countries and time. Pritchett (1997) created a conversion between caloric intake and per capita GDP, which allows a rough translation between the two measures of average welfare. He suggests that nutritional subsistence of 1600 calories per day requires an income of about $306 per person in 1990 dollars. Using Maddison's (2001) estimates, therefore, GDP per capita in the world was barely 150 calories per day above minimal subsistence in 1700, suggesting that much of the world's population was too malnourished to perform significant work.

Europe began to grow in 1700, at first slowly and then at an accelerated pace (Figure 3). The growth which occurred in Europe in the 18th and 19th century was made possible by improvements in agricultural productivity – increased crop yields and higher agricultural labor productivity. In addition, gains in labor productivity in agriculture freed up labor that could migrate to the nascent urban industrial sector. Furthermore, the improvements in agricultural productivity were large enough to improve the nutritional status of the growing urban population, although a shrinking share of the labor force was devoted to agricultural production.

Today, 50% of the world's labor force is engaged in farming, and many developing countries are still at levels of per capita GDP prevailing in Europe in 1700 (Figure 3). In particular, average GDP in Africa in 1998 was only modestly above the 18th cen-

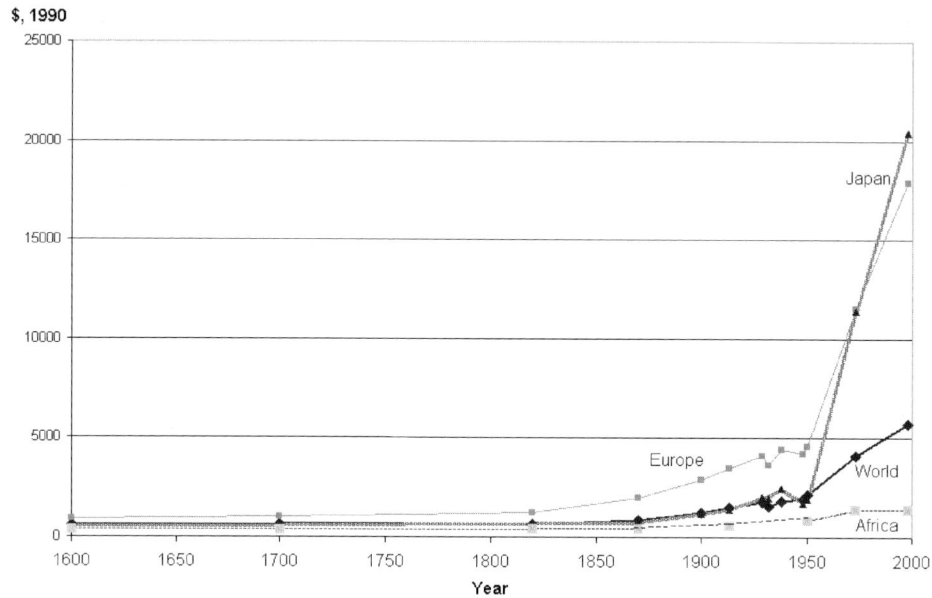

Figure 3. GDP per capita for various regions, 1600–1998, in 1990 U.S. dollars.

tury European average, and many countries in Africa have not yet attained that level. For example, in Ethiopia and Uganda, per capita calorie intake is approximately the same as for the U.K. three hundred years ago [Pritchett (1997)]. Hence, the population is stunted and wasted and does not receive enough calories to be able to undertake much work. Increased food availability and nutrition are potential sources of increased short- and long-term labor productivity in Africa and some other areas, but these countries are still constrained by 18th century agricultural technologies. For most of these countries, significant advances in per capita GDP will not occur until their agricultural sector undergoes a major transformation that raises labor productivity. Lacking the resources required to purchase significant quantities of food in the world market, these countries cannot circumvent the need to raise their own labor productivity in agriculture.

During the past century, the most fundamental and pervasive factors affecting the interaction of farm and nonfarm labor markets have been economic growth and science-based technological change. Referring back to the two sector model, the per capita income elasticity of demand for farm products is (and has been) positive but less than one, so the income elasticity of demand for nonfarm products is larger than one. Thus, domestic growth of real per capita income has caused a more rapid rate of growth of domestic demand for nonfarm products than for farm products. Furthermore, differences in rates of growth of domestic demand have increased as the rate of population growth has slowed. If the supply curve for domestic farm products shifts at least as fast as the supply for nonfarm products, the farm output share of national income and the relative farm output price must fall, provided that foreign demand growth for U.S. products is not offsetting. Even if agricultural technical change is factor neutral rather than biased toward capital, and even if technology advanced at the same rate in farm and nonfarm sectors, the differences in income elasticity between the nonfarm and farm sectors imply that growth of labor demand in the nonfarm sector must exceed that in the farm sector. If both sectors initially have equal rates of population and labor supply growth, the relative farm-labor wage rate must fall.

With the relative price of food falling, real income increases in the nonfarm population. With food quality and nutrition being luxury goods, this means that the nutritional status will improve for the nonfarm population. For labor to be fully employed and farm labor to earn its opportunity return under these conditions, the institutional structure of a country must be such that it permits the migration of labor to the nonfarm sector. With intersector mobility, wage rates become more equal, and they may approach equality when allowance is made for living-cost differences and an opportunity return on moving costs. The geographical distance of agricultural production from most other industries means that relative population density will fall in rural (farming) areas and rise in urban (nonfarm areas).

The transfer of labor occurs in three major ways. First, families may sever farm sector ties by quitting farm jobs, selling any farm capital they own, and taking a nonfarm job, perhaps moving to a city. Second, as children come of age, they may leave agriculture and take nonfarm jobs, but their parents remain in farming until retirement. Third, workers and their families may stay on farms, but some family members take full-time nonfarm jobs while others continue full-time work. Or some members may reduce their hours of farm work, take a non-farm job, and become multiple jobholders. The relative attractiveness of these alternatives depends on the location of nonfarm job opportunities, the types of skills of the people, and the costs of commuting or migrating to nonfarm jobs. All have important costs.

If the nonfarm real wage rises with economic growth, and if outmigration from the farm sector causes the real wage rate to be bid up there as well, new labor-saving technology may be induced for agriculture. Hayami and Ruttan (1985) suggest that induced innovation has occurred in U.S. agriculture and that it has been labor and land saving. Also the population and labor supply growth rates may initially be larger in the farm than in the non-farm sector as a result of the higher birthrates of farmwomen. Both of these changes increase migration rates needed to equalize earnings between sectors while maintaining full employment. This, however, provides labor to support a growing nonfarm sector. If, however, rural people do not have equal access to public schools, roads, and communication systems, this can be a major barrier to increasing labor productivity and inter-sector resource adjustment. Hence, it is critical that farm people have an opportunity to obtain a basic education that will facilitate making good decisions if they remain in the rural area or to support their occupational and geographical mobility to the nonfarm sector.

3.2. Jump-starting economic growth with an agricultural transformation

Before international trade was a reliable enterprise, gains from specialization and trade were limited. Each country had to produce the food and other goods consumed by its own people. Furthermore, each locality had to rely upon its own domesticated plants (and animals) for food. Major food sources were concentrated, generally lacking in available calories, essential proteins, vitamins, and mineral, and variety. Wheat was native to Egypt and the Middle East, rice was native to India and Southeast Asia, corn/maize was native to Mexico, the potato was native to South American and citrus was native to the subtropics. Chronic malnutrition was prevalent and life expectancy was short [Fogel (1994, 2004)].

With the development of transoceanic shipping, three things happened that were important with respect to the local food supply and human health. First, people migrated from densely populated and resource depleted areas to low-population and resource abundant areas where food could be more cheaply produced in the long run; e.g., from Western Europe to North and South America and Australia. Second, seeds, plants, and animals were collected from centers of origin and dispersed around the world. Although all crops and some animals are sensitive to local geoclimate, some crops and animals can be moved to new locations and grown successfully under the care of farmers. This is an example of how the diffusion of existing technology can be used to raise agricultural productivity in some locations. Third, transoceanic shipping made it possible for areas to specialize in production according to comparative advantage and to experience gains from trade. Some areas could produce non-perishable grains and nuts and trade them for nonagricultural products. Now a country could in principle specialize in manufactured goods and import food stuffs.

All 20 OECD countries jump-started economic growth by first having technical change in the agricultural sector [Hayami and Ruttan (1971, pp. 74–81), (1985, pp. 125–133); Hayami and Yamada (1975, pp. 4–6)]. In all cases, the source of increased agricultural production was from higher crop yields, based on advances in knowledge and technology, rather than expanding sown area. The economies of these countries have changed in response to lower prices of modern biological, chemical and mechanical inputs relative to the prices of land and labor and relative to the prices of agriculture products. These changes were associated with the transformation from traditional to modern agriculture [Schultz (1964); Ruttan (2001, pp. 611–614)].

Also, all OECD countries have undergone important demographic and economic transitions but not at the same time. The demographic transition occurs when a country goes from high birth and death rates to low birth and death rates and slow population growth [Schultz (1964); Ruttan (2001, pp. 611–614)], which facilitates growth in per capita food availability and incomes. An economic transition involves the movement of people from rural to urban areas and much of the labor force moving from agriculture to the nonagricultural sector, e.g., manufacturing, trades, and services. In most cases, if the economic transition does not follow the transformation of the agricultural sector, there is insufficient food and labor to support the development of urban centers and the nonagricultural sector [Johnson (2000)].

Among the currently poor countries such as those in Sub-Saharan Africa, most have not yet successfully completed any of the important transitions – agricultural, demographic, or economic – required for successful economic growth and development. They are all heavily agricultural and their agriculture remains heavily dependent on traditional technologies [Avila and Evenson (forthcoming)]. An important issue is whether these LDCs must undergo an agricultural transformation as a precursor to jump-starting sustainable economic growth.

The newly industrialized countries (NICs) of South Korea, Taiwan, Hong Kong, and Singapore have attracted considerable attention because of their high economic growth rates over the past four decades. For these NICs, per capita real GDP increased at slightly more than 7% per year over the 1960–1996 period [Heston, Summers and Aten (2002)].[4] As a result per capita incomes rose 10-fold since 1960. All four have moved into high income country status in under four decades.

Early growth in Korea and Taiwan was facilitated by an agricultural transformation aided by proximity to Japan. Both countries have a temperate climate and a rice culture. Hayami and Ruttan (1985, pp. 3304–3309) describe how in the early 20th century, these countries needed irrigation infrastructure to provide the water needed for higher rice yields. This irrigation infrastructure supported the adoption of higher yielding rice varieties imported from temperate Japan. Rice yields started to take off in the 1920s and 1930s. Gains in rice yields continued as the irrigated area expanded and early Japanese rice varieties were planted. As the irrigation systems improved, early Japanese rice varieties were replaced by newer, fertilizer responsive Japanese rice varieties.

Hence, jump starting economic growth in Korea and Taiwan built upon an agricultural transformation where agricultural technology was imported from the more advanced Japan. Later Korea and Taiwan were able to benefit from Green Revolution crop varieties starting in the 1960s [Evenson and Gollin (2003)]. This required building the intellectual capacity for incorporating improved rice germplasm obtained from IRRI into local varieties. Both countries had sufficient intellectual capacity to engage in adaptive research. Furthermore, the significant rise in rice yields freed labor from agriculture in Korea and Taiwan for work in the non-agricultural sectors. Nevertheless, by 1960, 60% of the labor force still worked in the agricultural sector. During the 1960–1975 period, investment rates in these countries shot up from 10% to 20–30% of GDP, becoming a major source of economic growth [Heston, Summers and Aten (2002); Jones (2002, p. 44)]. By 2000, the share of the labor force in agriculture had fallen to only 10%.

In contrast to Korea and Taiwan, Hong Kong and Singapore have been able to jump-start economic growth *without* an agricultural (or economic) transformation. Hong Kong is roughly 6 times the size of Washington, DC. It was originally part of China, but after a brief occupation by the United Kingdom in 1841, it became a U.K. colony.

[4] Growth did slow temporarily at the beginning of the 21st century.

In 1997, Hong Kong became a special administrative area of China with autonomy in all matters except foreign affairs and defense. Singapore, roughly 3.5 times the size of Washington, DC, was founded as a British Trading colony in 1814. It briefly joined the Malaysian Federation (1963–1965) and became independent in 1965 [CIA (2004)].

Hong Kong and Singapore do not have significant agricultural land or forests, and hence, had to look to other sectors for jumpstarting growth. In order to obtain food stuffs, they were forced to trade. Food stuffs rank in the top four types of commodities imported in both countries [CIA (2004)]. This meant that both needed to develop a very different set of institutions than those countries that jumpstarted growth with an agricultural transformation. Given their location and new institutions, they were able to flourish as regional trade centers over the past four decades. In particular, they have created an attractive environment for foreign direct investment associated with manufacturing and trade, and the trade intensity rate for these economies is quite high; the sum of exports plus imports divided by GDP is in excess of 150% [de Ferranti et al. (2003)].

In 1960, the service sector share of employment was already 62% in Hong Kong and 78% in Singapore. Manufacturing accounted for another 25% of the labor force in Hong Kong but only 12% in Singapore [World Bank (1981)]. Unusually high investment rates exceeding 40% of GDP for Singapore and 25% for Hong Kong helped fuel the growth processes of these two countries over four decades [Heston, Summers and Aten (2002); Jones (2002, p. 44)]. Technology needed to raise labor productivity in the nonagricultural sector was acquired largely through technology transfer associated with direct foreign investment by multinational companies. While adult literacy rates of primary and secondary attainment have risen rapidly over the period [Barro and Lee (2000)], neither Hong Kong nor Singapore had high proportions of college or post-college educated workers. Lacking the intellectual capacity to develop new technology themselves through basic and applied research, they were able to acquire it from abroad by creating a business environment that was attractive to technically advanced multinational companies. Moreover, Xu (2000) shows that an LDC can expect to attract technology from multinational enterprises only if it has an adult population that meets a threshold level of education of roughly 10 years of completed schooling. Hong Kong and Singapore are close to that threshold.[5]

Hong Kong and Singapore have been able to jump-start the economic growth process without an agricultural or economic transition. They were able to take advantage of improvements in agricultural productivity elsewhere through trade. By investing heavily in improving the schooling of their work forces and by establishing a political and economic environment that could attract foreign direct investment, they were able to generate a comparative advantage in exportable services and manufactured goods. This unconventional approach to economic growth by Hong Kong and Singapore has been sustained for four decades.

[5] Recently Singapore has begun to invest in the development side of R&D, although it is not undertaking much basic or applied research. See Ruttan (2001) and de Ferranti et al. (2003) for additional details on the economic growth process in Hong Kong and Singapore.

The Hong Kong and Singapore model of being a trade center is not easily replicable and is not an option for most LDCs. As trading centers, they exported nonagricultural goods and imported food, effectively importing the technology needed for an agricultural transformation. The Hong Kong and Singapore model of jump-starting economic growth through trade alone does not fit the countries of Sub-Saharan Africa or South Asia. These LDCs will need to follow the path of the OECD countries and of Taiwan and Korea, and jump start growth with a domestic agricultural transformation.

All of our discussion thus far has presumed that institutional structures do not limit labor mobility, that land markets and property rights are well-established, and that households are free to make optimal choices regarding human capital investments. Sokoloff and Engerman (2000) argue convincingly that countries that perpetuated unequal access to schooling, property rights, political power and occupational mobility tended not to grow in comparison to countries that fostered social and economic mobility. Acemoglu, Johnson and Robinson (2001, 2002) show strong negative effects on growth of institutions that fostered forced labor or economic immobility for the many while fostering the perpetuation of elite status for the few. Thus, underlying our simple two-sector model is a presumption that institutions exist to facilitate the actions of optimizing individuals, farmers, and households. In the countries where growth has failed to materialize, it is plausible that weak institutions have cut the linkage between improvements in agricultural technologies and broad-based economic growth and development.

3.3. Contemporary cross-sectional comparisons

In Section 2, we showed that investment in human capital is an important factor affecting wage rates and earnings. There is no consensus regarding how best to summarize schooling capital invested in people. One widely used proxy variable for education-based human capital is the number of years of formal schooling completed.[6] However, it takes approximately four years of formal schooling to attain permanent literacy. Hence, in many poor countries children are not completing enough schooling to attain permanent literacy. In these countries, education is best proxied by the adult literacy rate. Once the average years of schooling completed rises significantly above four years, then years of schooling completed is the best proxy. A remaining issue is whether the impact of education is always proportional to the change in years of schooling completed. See, for example, the discussion in Welch (1970).

We show the cross-sectional relationship between schooling and GDP per capita across countries in Figure 4. These plots use average years of education for women aged 15 and over as a measure of the level of human capital in the population, but the patterns would look similar if we were to use average education levels of males. As a further aid to illustrating the stylized facts concerning per capita output and human capital, we superimpose the results of a log-linear regression of GDP per capita on average levels of female and male schooling and a quadratic term in female schooling.

[6] See Greene (2003, p. 87) for a discussion of proxy variables.

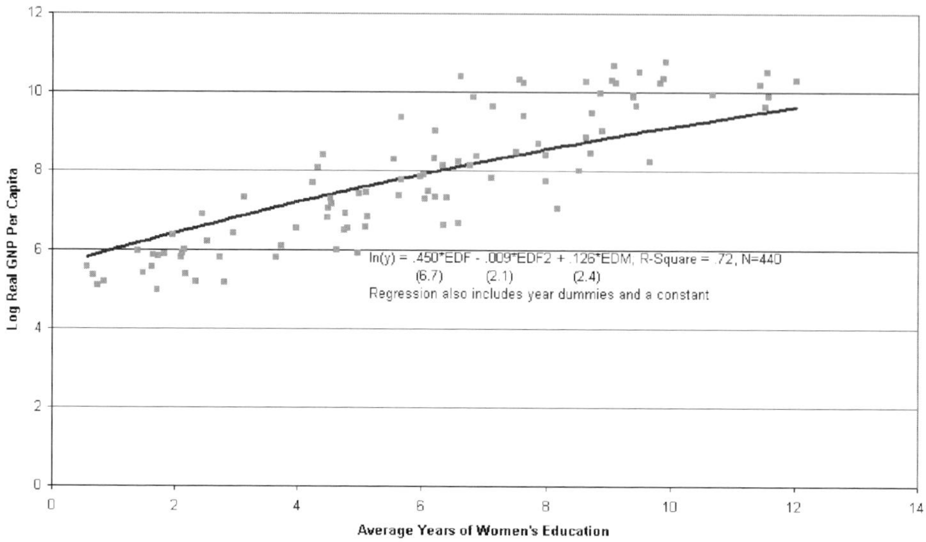

Figure 4. Log real GNP per capita and average years of women's education. Predicted and actual values 1998.

The specification is chosen to correspond to that used commonly in microeconomic analysis of labor earnings. These cross-country regressions cover the years 1960, 1970, 1980, 1990 and the most recent available data period. The regressions also include a complete set of year dummies whose coefficients are not reported. The original specification also included a quadratic term in men's schooling, but the quadratic terms in men's and women's schooling were so highly correlated (0.97) that we only included the one for women. The results illustrate a strong positive correlation between average years of schooling in the population and output per capita.[7]

[7] If households have a head and all individuals are part of some household, then the schooling level of the head is a key variable in explaining household and individual behavior as it impacts income or health. If there is no effective household head or if males have more than one wife, then the choice of schooling variables become more challenging. Several studies in low or middle income countries [e.g., King and Hill (1993); Smith and Haddad (2000); Klasen (2002)] have hypothesized that differential bargaining power exists between men and women, and that difference is well proxied by adult male–female schooling differences. Their models include a measure of average education in the country and also the difference in education levels between women and men. These models are subject to misleading interpretations. For example, suppose that the "true model" is $g = \beta_0 + \beta_1\text{EDF} + \beta_2\text{EDM} + \gamma Z + e$, so that both women's (mother's) education, EDM, and men's (father's) education, EDF, affect growth, g, and that both affect growth positively ($\beta_1 > 0$ and $\beta_2 > 0$). Z is a vector of other factors. If the estimating equation includes average adult education in the country plus the female–male education gap, the estimating equation will be $g = \alpha_0 + \frac{1}{2}\alpha_1(\text{EDF} + \text{EDM}) + \alpha_2(\text{EDF} - \text{EDM}) + \gamma' Z + e$. Some of the papers mentioned above claimed that a positive estimate for α_2 implied that a gap in education levels favoring men relative to women retarded growth. That is incorrect. It is straightforward to show that $\alpha_1 = \beta_1 + \beta_2$ so that the coefficient on average adult education will be the sum of the effects of men's and women's education on growth. Furthermore, $\alpha_2 = \frac{1}{2}(\beta_1 - \beta_2)$ where the sign

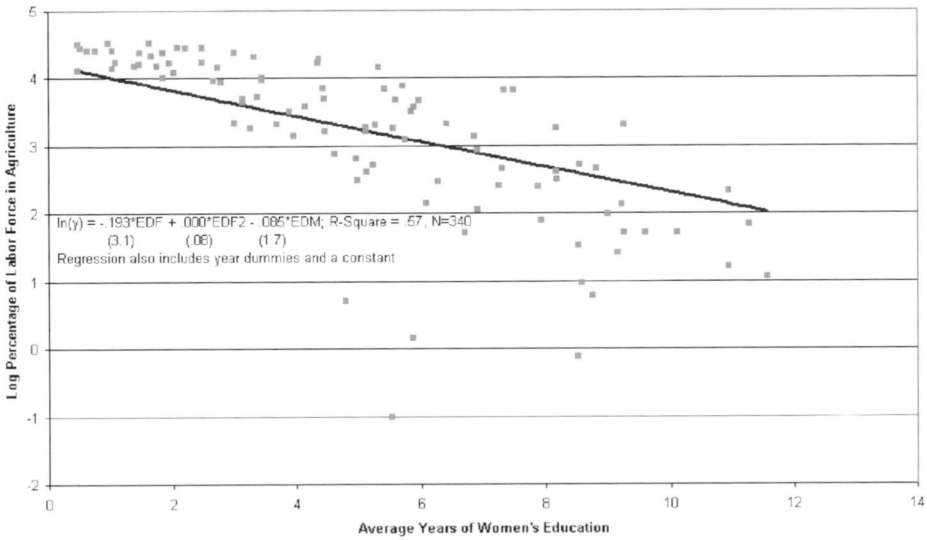

Figure 5. Log percentage labor force in agriculture and average years of women's education. Predicted and actual values 1990.

We conduct a parallel exercise illustrating the relationship between human capital and the proportion of the labor force engaged in agriculture. As discussed in the previous section, modern economic growth has been tied to a decline in the share of the labor force in agriculture for most countries of the world. We can also see that rising levels of education can also be tied to declining agriculture share of the labor force.

Although the agricultural transformation ultimately requires schooling of the masses, the early part of the transition involves more basic sources of improved labor productivity: learning-by-doing through apprenticeships or work experience; a larger stature that increases physical strength and ability to do work; and human migration. As the society becomes more advanced, formal training of teachers and schooling of children become cost effective, but this usually covers only elementary schooling. Furthermore, advancements are required before investing in high school teachers and devoting child

will depend on the relative size of the women's and men's education on growth. A positive estimated α_2 only implies that the marginal effect of women's education on growth is larger than that of men and not that the gap in education has an impact on growth *per se*.

Another specification contains the following form and reasoning: $g = \alpha_0 + \alpha_1 \mathrm{EDF} + \alpha_2(\mathrm{EDF} - \mathrm{EDM}) + \gamma' Z + e$. In this case, $\alpha_1 = \beta_1 + \beta_2$ so that the coefficient on women's education will be the sum of the male and female effects and $\alpha_2 = -\beta_2$ which will be negative if male education has a positive effect on growth. It will be particularly unwise to conclude from such a model that a redistribution of education away from men and toward women will improve growth. It seems safer to include separate measures of education for men and women and then to interpret the coefficients directly, as we do in this paper.

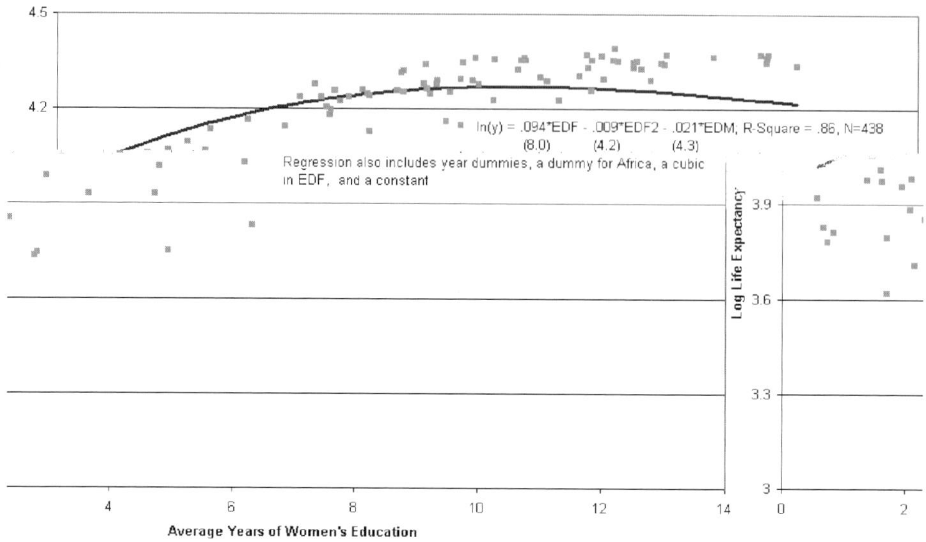

Figure 6. Log life expectancy and average years of women's education. Predicted and actual values 1999.

time to high school generate sufficient returns. Insights for all of these forms of investments can be obtained from the stylized Ben Porath human capital investment model in Section 2.

What are some of the reasons that the agricultural transition leads to increased human capital investment? First, as the nutritional status of the population rises, expected length of life at birth increases, which would increase human capital investment in our stylized Ben Porath model. In the Middle Ages the expected length of life at birth was about 24 years, roughly the expected length of life associated with subsistence levels of per capita GDP [Maddison (2001)]. By 1820, life expectancy at birth had risen to approximately 36 years in Europe, 39 years in the United States, and 34 years in Japan, but life expectancy remained near 25 years everywhere else. Since then, countries that raised GDP per capita have experienced increases in life expectancy. As predicted by the three-period human capital investment model, the near tripling of the world average length of life at birth has greatly strengthened the incentive to invest in skills early in life. For the average world resident in the Middle Ages, this incentive did not exist.

The present-day cross-country relationship between life expectancy and human capital is illustrated in Figure 6. Currently, life expectancy at birth stands at about 78 in the OECD countries. It averages only 52 years in Africa, and is actually declining in some African countries. As shown, a strong correlation exists between higher levels of human capital investment and greater life expectancy.

Second, as nutritional status improves, both labor productivity per hour and the number of hours per day an individual could potentially work productively or enjoy leisure increases. Hence, total allocatable time per year rises. As Fogel (1994) shows, during the agricultural transition in the U.K. and France, a disproportionate share of the increased available productive time was allocated to leisure and educational activities. In contrast, in earlier periods when much of the population was malnourished and the intensity of work was reduced, they actually allocated a large share of allocatable time to work.

Third, all the adult population in 1700 were stunted. Besides making them vulnerable to early onset of chronic diseases, their physical ability to do work was reduced by their small size. As schooling is considered a normal good, its demand rose with the rising incomes attributable to improved health of the population.

Fourth, as labor shifted out of agriculture, the need for child labor declined. This freed up children's time for larger investments in schooling. Current statistics show that the incidence of child labor is much higher in rural than in urban areas, with nine of every ten rural working children engaged in agriculture. As the rise of agricultural productivity helped to support the rise of cities and the shift of labor out of agriculture, it also supported the transfer of children from work to school.

Finally, a large literature exists showing that malnourishment at an early age retards brain development. Studies of the impact of nutrition on cognitive achievement have shown that schooling outcomes improve with nutritional sufficiency. In short, better-fed children do better in school, so the agricultural transformation has a direct impact on the returns to attending school. For all of these reasons, we argue that the agricultural transformation has contributed to the human capital transformation which has played such a prominent role in the theoretical and empirical literature on economic growth.

4. An overview of econometric evidence that schooling contributes to economic growth

We review both micro- and macro-evidence of the contributions of schooling to economic growth.

4.1. *Micro-evidence for schooling and economic growth*

One of the most widely investigated empirical relationships has been between schooling and earnings. Mincer (1974) showed that if the cost of schooling is the opportunity cost of time, and if the proportional return per year of schooling is constant over time, then an individual's wage will be well-explained by a function of the form

$$\ln(y_{it}) = \beta_0 + \beta_1 S_{it} + \beta_2 Z_{it} + \xi_i + e_{it}, \tag{26}$$

where $\ln(y_{it})$ is the natural logarithm of labor earnings per unit of work time of ith individual in period t, S_{it} is a measure of years of schooling of the ith individual, Z_{it} is a vector of other productive human capital attributes of the ith individual such as work experience and job tenure, and the β are associated regression coefficients. The coefficient on years of schooling, β_1, is commonly interpreted as the proportional change in labor

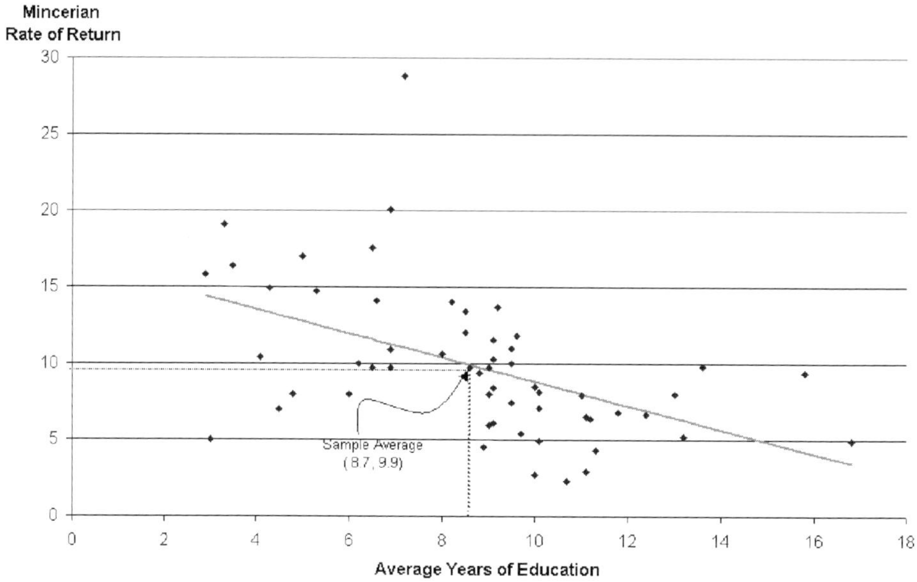

Figure 7. Plot of Psacharopoulos's estimates of Mincerian returns to schooling across 57 countries.

earnings from a 1 year increase in an individual's schooling attainment.[8] The last two terms are an individual-specific and time-invariant random effect ξ_i, which is known to the individual and might be ability; and a random disturbance term across individuals i and time periods t, e_{it}, that has a zero mean. If ξ_i is correlated with S_{it} and/or Z_{it}, then direct estimates of the coefficients in Equation (26) will be biased.

Mincerian earnings functions have been estimated using data on individuals in many different countries. The most recent extensive review of estimated private returns to schooling in developing countries is by Psacharopolous (1994). His results are illustrated in Figure 7. Estimated private returns to schooling are always positive. Furthermore, schooling appears to be subject to diminishing marginal returns, consistent with the Ben Porath model. At the lowest schooling completion levels the rate of return is highest and it declines for incremental increases in years of schooling completed. For the 57 countries surveyed by Psacharopoulos,[9] he also found that on average, private returns to girls' schooling exceeded returns to boys' schooling.

[8] Interestingly, specification tests conducted by Heckman and Polachek (1974) showed that this log-linear specification dominated all other alternatives. More recently, Welch (1999) has shown that a more complex spline-regression performs better when sample sizes are extremely large.

[9] Lam and Schoeni (1993) conducted a detailed examination of how the rate of return to schooling changed as years of school attainment rose in Brazil. They found nearly linear rates of return after controlling for family background variables, but the highest returns were in the first four years of schooling.

Jamison and Lau (1982) summarize empirical evidence from a set of 18 World Bank studies of the impact of farmers' schooling on agricultural productivity in 13 low and middle income countries. They report that 6 of the marginal products were negative, 13 were positive but not significantly different from zero, and 18 were positive and statistically significant. However, the Mincerian evidence that the returns to schooling decline as the amount of schooling increases was used by the World Bank and national governments to justify re-directing educational expenditures to elementary schooling investments. This had an unfortunate consequence of reducing public support for college education and training of local agricultural scientists in these developing countries. For growth in agriculture to occur, countries need to follow a schooling investment policy where all individuals have access to elementary schooling and some have the opportunity to attend high school and college. Countries cannot expect to successfully borrow agricultural technologies from more advanced countries over the long run without the local intellectual capacity to adapt technologies to local needs. That means that the country has access to local scientists with graduate-level training. This adaptive research produces national public goods that will require public financial backing, and so large developing countries, e.g., China, India, Brazil, have an advantage in establishing public research groups. Countries with small populations, say only a few million as in Sub-Saharan Africa, are at a great disadvantage in funding public research.

A large literature exists which explores the various sources of biases in estimated returns to schooling. Card (1999) provides a comprehensive review of the topic, so we will touch on it only briefly here. First, S_{it} might be endogenous and jointly determined with $\ln(y_{it})$.[10] Second, reported years of schooling may contain measurement error, e.g., at low years of completed schooling individuals regularly exaggerate years completed, which introduces measurement errors. Data on identical twins has been used to correct for unmeasured abilities.[11] Information on school availability or proximity, truancy laws, and school building projects has been used to correct for measurement error and/or self-selection in school choice.

Card (1999) reports that for industrialized economies, little difference exists between ordinary least-squares estimates and the more econometrically sophisticated estimates, suggesting that estimation bias in naïve models appears to be small or that various biases offset one another. In developing country settings, there is more variability in school attainment and consequently more potential for self-sorting to occur. Nevertheless, Krueger and Lindahl (2001), concluded that ability bias is approximately offset by measurement error in reported years of schooling for developing countries as well. Instrumental variable estimates are similar to those obtained from ordinary least squares

[10] If an individual chooses how much school to obtain based upon ξ_i, e.g., ability, then observed years of schooling (S_{it}) will almost certainly be correlated with ξ_i, so the least-squares estimate of Equation (8) will yield biased coefficients. Furthermore, correlation of ξ_i with Z_{it} creates a similar problem.

[11] Recent research has shown that "identical" twins are not genetically identical because the expression of certain genes is affected by the environment in which the individual finds him or herself. This weakens much of the economic evidence using identical twins.

[Psacharopolous (1994); Duflo (2001)]. Where researchers have found differences [Bedi and Gaston (1999); Bedi and Edwards (2002)], OLS estimates of returns to schooling appear to be biased downward. Thus, one might want to view the estimates in Figure 7 as a lower bound.

There is a small but important literature that examines whether it is years of schooling per se or the learning that occurs in school that matters. Glewwe (2002) reviews the few studies that have examined this issue in developing countries and concludes that it is cognitive skills (typically measured by standardized test scores) and not years of schooling per se that matter for income generation. Because cognitive tests are still only rarely available for data sets that also include earnings, most researchers will still be limited to years of formal schooling as a proxy for education. As discussed above, measures of literacy will be more useful in countries with the poorest schooling levels,[12] but years of schooling is an adequate measure for all but the poorest developing countries. Even in countries with higher average levels of schooling, literacy may be a reasonable proxy for school quality.

4.2. Macro-evidence

Given the virtually universal demonstrated success of education in generating private returns that meet or exceed returns on alternative investments, it seems clear that investments in education make good economic sense from an individual perspective. However, every country subsidizes education, meaning that the cost of education to society exceeds the marginal cost borne by the individual. For these public investments to make economic sense, there must be an external benefit from schooling other than the private return to individuals and their families.[13] To address this question, studies have typically used macroeconomic data that can capture spillover benefits and costs.

Returning to our regression estimates reported in Figure 4, the cross-country relationship reveals a strong positive correlation between average years of male schooling and GDP per capita, averaging 13% growth in per capita GDP for every year of added male schooling attainment. The relationship between female education and GDP is almost 4 times larger than that for male education. Consistent with the Psacharopoulos findings for private returns, the rates of return fall as the level of schooling rises.

If GDP per capita is interpreted as average income in the country, these rates of return can be interpreted as the social return as opposed to the private return from schooling. These returns from investment in female education are far higher than estimates of private returns, consistent with the view that education of girls generates greater positive

[12] Even in countries with higher average levels of schooling, literacy may be a reasonable proxy for school quality in the absence of other measures.

[13] Psacharopolous (1994) reports estimates of private and social rates of return averaged for country groups. Estimated social returns are uniformly lower than private rates of return, but this result is largely due to the construction of the estimates. Public costs are added that depress returns, but measures of external benefits from education are not. This would create a downward bias in his estimates of social returns.

externalities to the society than for men. In contrast, estimated social returns for male schooling are only marginally larger than typical estimates of private returns. Similar findings elsewhere [King and Hill (1993); Schultz (2002)] have led the World Bank and other international funding agencies to emphasize investments in girls' education as opposed to children's education generally, as a critical development tool [World Bank (2001, Ch. 2)].

The empirically oriented growth literature has concentrated on a first-difference variant of Equation (26):

$$\Delta \ln(y_{it}) = \tilde{\beta}_{0t} + \beta_t S_{it} - \beta_{t-1} S_{i\,t-1} + \beta_2 \Delta Z_{it} + e_{it}. \tag{27}$$

If returns to schooling are constant so that $\beta_1 = \beta_t = \beta_{t-1}$, then the impact of schooling can be captured by $\beta_t S_{it} - \beta_{i\,t-1} S_{it} = \beta_1 \Delta S_{it}$. The vector of regressors ΔZ_{it} is now used to represent per worker changes in physical and other human capital, and the constant term in (27) captures time-specific factors that have common effects on per capita income across countries.[14] Growth can also be linked to the Solow neoclassical growth model where changes in technology and physical and human capital are sources of growth [see Jones (2002, pp. 54–62)]. These factors fit under the ΔZ_{it} term in Equation (27).

By adding and subtracting $\beta_t S_{i\,t-1}$ to the right hand side of Equation (27), we obtain

$$\Delta \ln(y_{it}) = \tilde{\beta}_0 + \beta_t \Delta S_{it} + \Delta\beta_t S_{i\,t-1} + \beta_2 \Delta Z_{it} + e_{it}. \tag{28}$$

The coefficient on ΔS_{it} is interpreted as the average return to schooling across countries over the sample period, and the coefficient on $S_{i\,t-1}$ gives the change in the return to schooling over the sample period.

Equation (28) typically is estimated using average annual rates of change in per worker income over 5, 10 or 20-year intervals. For example, Benhabib and Spiegel (1994) estimated a human-capital model similar to (28) and found that the change in schooling had virtually no effect on changes in GDP per capita, but that the beginning period or initial level of schooling has a positive and significant effect. They justify this outcome by explaining that higher levels of education in the workforce lead to more rapid assimilation of existing technologies as well as more rapid innovations of new technologies. Topel (1999) argued that the Benhabib and Spiegel results were biased because they used logarithmic measures of schooling rather than the levels as suggested by the Mincerian specification. Krueger and Lindahl (2001) argue further that measurement errors in the international schooling data bias the coefficients. Correcting for these specification and measurement errors, Krueger and Lindahl found that a one-year increase in average schooling raised annualized growth in GDP per capita by as much as 30% over a twenty year period. This is consistent with the average of the male and female returns reported in Figure 4. However, using a similar regression specification but

[14] In the differencing process, the random individual-specific effect (ξ_i) is difference out.

different measures of education and physical capital, Pritchett (2001) finds negligible returns to schooling.

Researchers have found larger positive returns to schooling when longer time horizons are used in the averaging process, e.g., 10 year averages versus 5 year averages. Moreover, these estimated returns are higher than the private returns, suggesting that education generates positive external benefits to the economy as a whole.

Three issues could affect the interpretation of these results. First, there is the question of causality, i.e., does education cause income growth or does income growth make it possible to finance schooling either publicly or privately [Jones (2002)]. Bils and Klenow (2000) attempt to address the issue of causation between income growth and school enrollment rates. They used a calibrated version of the Mincerian relationship over a set of 85 countries and concluded that the impact of schooling on growth is less than one-third of that implied by the estimated cross-country growth coefficient. They also found that the size of the reverse causal effect from growth to schooling can be large enough to explain all of the cross-sectional effect. Both Krueger and Lindahl (2001) and Pritchett instrumented the schooling growth variable, but obtained opposite results. The true macroeconomic impact of schooling on growth remains elusive, and the linkage between macroeconomic and microeconomic estimates of the impact of schooling on labor productivity is open to further research.

Second, measurement errors in schooling exist in both the level and change form and they have implication for estimation of Equation (28). Recall that the coefficient on $S_{i\,t-1}$ is $\Delta\beta_t$, so if returns are constant over time, then $\Delta\beta_t = 0$. Measurement errors in schooling will bias these estimates toward zero. Furthermore, Krueger and Lindahl found that the estimate of $\Delta\beta_t$ was sensitive to the inclusion or exclusion of physical nonhuman capital in the vector of other variables ΔZ_{it}. Some estimates were negative, while others were so implausibly large as to imply that schooling levels are responsible for all growth in GDP per capita.[15] Nevertheless, it seems likely that the average level of schooling does affect the rate of growth, even if the effect is not precisely estimated. The reason is that rates of growth in capitalist markets with the highest levels of schooling have consistently outpaced rates of growth of the countries with the lowest levels of schooling, leading to a steady widening of the gap in income between the richest and poorest countries [Pritchett (1997)]. While other explanations can be advanced for this result, the role of schooling levels in raising long run growth rates has a strong theoretical appeal [e.g., Romer (1990)] that awaits a more definitive empirical test.

Finally, years of schooling that may be a satisfactory indication of relative education within a country may be fraught with error in a cross-country specification of (28). Variation in school quality across countries means that the average years of schooling will be a noisy measure of the average human capital stock. Hanushek and Kimko (2000)

[15] The lower bound of Krueger and Lindahl's positive estimates is about 0.003 log points of growth for every year of average educational attainment which would translate to roughly 2.6% growth in per capita GDP per year when evaluated at average world education levels. Average growth over the last 50 years was 2.2%.

undertake an examination of the impact of school expenditures per child and student cognitive performance on various international tests of academic achievement in mathematics and science. Consistent with the microeconomic evidence reviewed by Glewwe (2002), they show using data for about 80 countries that school expenditures per student is a relatively weak predictor of real income growth but that measures of school quality that can be associated with improvements in cognitive test scores are extremely important to later growth. Furthermore, the link between labor force quality and economic growth holds even when various subsets of East Asian countries are excluded. Their results suggest that a promising avenue for linking the macroeconomic and microeconomic studies is to use measures that are more closely tied to cognitive attainment in the growth analyses.

The presumed existence of positive externalities from schooling is a major justification for public subsidies for education. Returning to Figure 5, we find that as average educational attainment for women rises by one year, the proportion of women engaged in agriculture declines by 19%. Increases in male education also lower agriculture's share of employment, but the effect is half as large. The relationship is nearly linear, so the proportional decline in labor or out-migration from agriculture is constant as levels of education rise. Some have considered this outmigration from agriculture to be a form of "brain drain" from rural areas. This seems to be a misnomer. The education levels of those remaining in agriculture rise as well, but the process of development appears to raise returns to human capital in cities faster than it raises returns in the countryside, a theme to which we will return later.

With improvements in human capital and the shift of labor out of agriculture comes a change in how men and women allocate their time. Much of the academic literature has concentrated on changes in women's time allocation, but it is clear that there are dramatic changes in how men allocated their time as well, in terms of occupational and educational choices, residential choices, and time spent in work versus leisure over the lifetime. Nevertheless, the process of development does not affect male labor force participation rates at prime ages, which is not true for women.

The cross-sectional relationship between labor supply behavior and women's education is illustrated in Figure 8. Several scholars [Sinha (1967); Durand (1975); Psacharopoulos and Tzannatos (1989); Goldin (1995); Mammen and Paxson (2000)] have identified a U-shaped pattern in women's labor supply behavior as economic development progresses. The story behind the U-shape is that early in the development process, labor market opportunities expand off-farm rapidly. These opportunities disproportionately raise the value of time of men, either because men are more likely to engage in physically demanding factory work or because male education levels rise faster than that for women. Rising male wages combined with constant value of time for women results in an income effect away from women's work and toward nonmarket activities such as child or home care. Later in the progression of development, women's education also begins to rise, raising their opportunity cost of time. The rise of white-collar jobs as the economy develops and the opening of occupations to women appear also to be related to the movement of women into the labor market [Goldin (1995)].

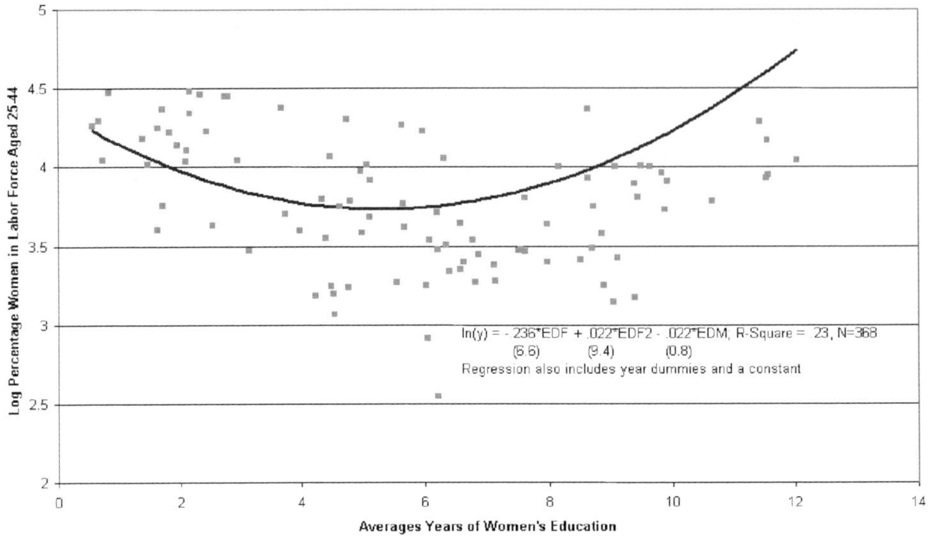

Figure 8. Log percentage women in labor force aged 25–44 and average years of women's education. Predicted and actual values 1998.

This story hinges on two presumptions that may or may not be correct: that educational and off-farm opportunities rise more rapidly for men than women early in development and that for women, income effects for leisure are dominated by substitution effects away from leisure as women's wages start to rise. These assumptions may hold in some countries and not others. In addition, variation in women's labor supply behavior out of the home across countries can be strongly influenced by local tastes and customs. Nevertheless, the hypothesized U-shape is supported by the simple data plots. Taken literally, the regression indicates that women's labor supply behavior declines until school attainment reaches 5 or 6 years and then rises thereafter.

An alternative interpretation consistent with the stylized Ben Porath model is that there may be little change in the value of market time as years of schooling completed increases in the range of one to five years. By 5 years of schooling, permanent literacy is usually attained,[16] and beyond that, years of schooling appear to have a rising effect on the value of market time. The value of time for women who have less than five years of schooling will reflect largely the ability perform physical work, and the value of this type of work may be greater in the rural household than in the labor market. However, as schooling levels rise above five years, women will have enough human capital to begin to engage in some skilled work whose value outside the household may dominate that in the household. Higher levels of schooling open up even a wider range

[16] This rule of thumb may hide considerable variation. Lloyd (2005, Figure 3-3) showed that across seven African countries, literacy rates for women with 5 years of schooling varied from 15 to 90%.

of skilled jobs to women in open societies. Consequently, the pattern of rising female labor supply beyond five years may just reflect the relative value of human capital at or beyond literacy attainment in off-farm versus on-farm activities.

The U-shape is more apparent when following labor supply patterns over time within a country. In the United States, for example, Goldin reports that labor force participation rates for married white women rose from under 15% in 1890 to 50% in 1980, but the change was not smooth or steady over the time period, e.g., major increases occurred during World War I and II which were followed by a post-war decline.

The cross-sectional relationship is also clouded by measurement problems in distinguishing between home production and labor supply in family farm enterprises [Schultz (1990)]. For example, women's time in home and farm production activities may be highly complementary and subject to joint products. A woman engaged in tending a family plot may also be tending children at the same time. It is difficult to distinguish between such jointly productive activities, and so there may be considerable error in characterizing women's work. Nevertheless, it is clear that the type of work that women do changes as development progresses. In the poorest countries, female labor force participation approaches 90%. This is well beyond the highest levels reported for industrialized economies. Women in these poorest countries are mainly engaged in unpaid physical labor for family enterprises. In the industrialized economies, most women are engaged in skilled paid work away from home.

Part of the public return from schooling is in the reallocation of male and female labor across sectors, and in the diversification of the economy which allows a greater degree of specialization according to comparative advantage. These gains occur in the formal labor market, but significant gains occur in home production as well. One of the avenues by which education can generate external benefits was illustrated in Figure 6 and two others in Figures 9 and 10. These gains are related to increases in life expectancy and reductions in infant mortality and fertility. Much of the literature has concentrated on the role of women's education in generating these welfare gains. In fact, development policies have concentrated on stimulating education for girls on the presumption that girls' education generates more externalities than boys' education, an assumption consistent with the higher estimated impact of girls' education on GDP per capita. What has been less commonly discussed is that married men and women should have similar objectives with respect to the number and health of their children, so that male education levels may have similar effects (in sign if not in magnitude) to those of women's education on these welfare indicators.[17] Also, in countries where individuals "choose"

[17] As an example, Johnson and Skinner (1986) examined whether divorce causes female labor supply or female labor supply causes divorce. Following married couples longitudinally, they found that when both the husband and the wife worked continuously after entering marriage, likelihood of divorce was not affected. Presumably, the husband's and wife's anticipated labor supply behavior was already incorporated into the marriage contract, so the wife's labor supply should not affect the probability of marital dissolution. Similarly, one would expect that number of children and interest in child care would be subjects agreed upon before entering marriage. It should be noted that these marriage contracts are not based on full information. Johnson

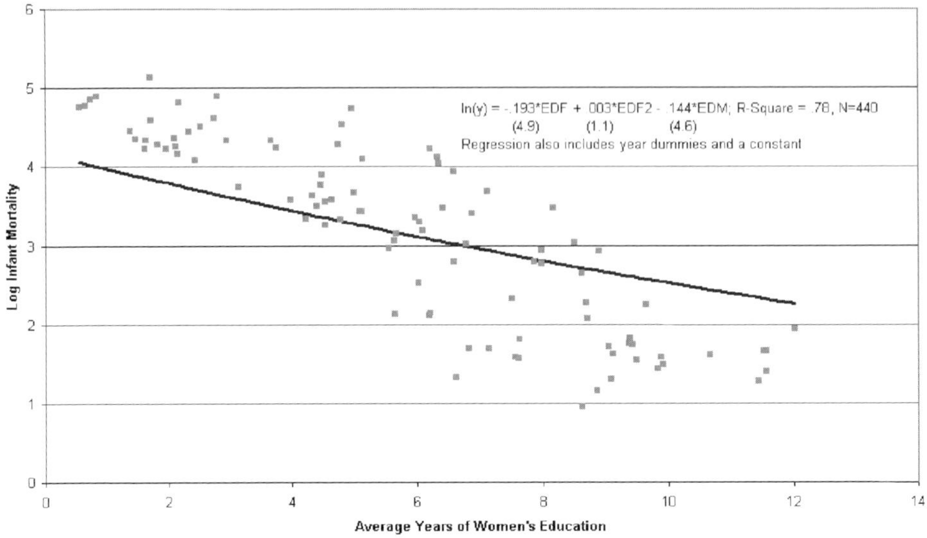

Figure 9. Log infant mortality and average years of women's education. Predicted and actual values 1998.

their own spouse, matching seems most likely to occur on education. This means that these societies will be quite open to educating girls.

As shown in Figure 6, increases in male and female education levels both raise life expectancy, but the effect is 3 times larger for women's education than for men. Taken literally, the gains in life expectancy from women's education dissipate after 14 years of schooling, but dissipate at 5 years of men's schooling. Both male and female education lower infant mortality. The impact is one-third larger for women's education than for men's education, suggesting a 19% or a 14% decline in infant mortality for every additional year of schooling for adult women and men, respectively.

The decline in mortality that accompanies the agricultural transition and the improvements in human capital will temporarily raise the population rate of growth. Without a change in the birth rate, reductions in the death rate would lead to rapid population growth that would threaten to reverse the initial gains in the country's development. Thus, it is critically important that fertility rates decline as a country develops. An additional year of average schooling for women lowers the fertility rate by 11% in Figure 10. The effect is concave and dissipates at about 15 years of schooling. Equally important is that fertility declines with improvements in male education levels, although the effect is about one-third smaller than for women's education.

and Skinner found that women who ultimately divorce were more likely to have entered the labor force 2–3 years before the divorce, suggesting that as the probability of divorce increases, women are more likely to work.

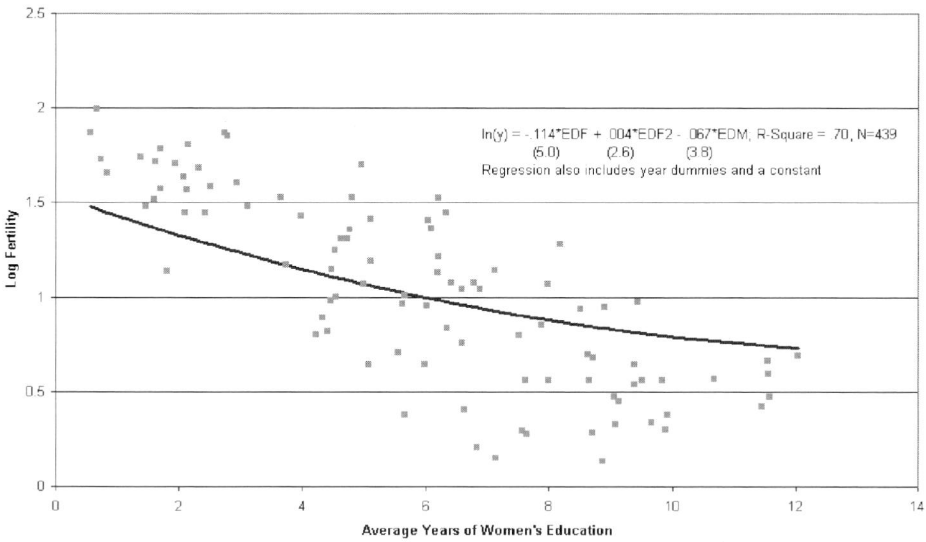

Figure 10. Log fertility and average years of women's education. Predicted and actual values 1998.

A common finding in all of these studies is that the measure of external health benefits from women's education is larger than that for men's education, but that the effects are in the same direction. This has been used to argue that there is underinvestment in female education relative to male education [Schultz (2002)]. Nevertheless, positive external benefits apparently arise from both boys and girls education.

5. Production of health, nutritional inputs, and work

Fogel (2004) has forged more strongly the links between physiological capital, nutritional intake of individuals, and economic growth. Physiological capital is a part of human capital broadly defined and related to health. Human capital was developed to explain differences in earnings between occupations, industries, regions, and life stages, using differences across individuals in education/schooling and on-the-job training. The concept has expanded to include health and information. The health capital concept was developed to explain the demand for goods and services that offset the depreciation rate on the initial endowment of health of an individual over a life cycle. Although the theory behind health capital takes for granted physiological capital, it does not deal with it explicitly. Health capital takes as exogenous an individual's health stock at birth and considers how later investments in health care can reduce the health stock's rate of depreciation. It does not address why some individuals are born with a greater stock of physiological capital than others, and it does not recognize the relationship between the size of the initial stock of physiological capital and rate of depreciation of physiological capital. Nor does it encompass the effect of an individual's date or country of birth on his or her initial stock or rate of depreciation of physiological capital. Furthermore, the theory of health capital does not confront the issue of how the average initial stock of physiological capital changes from one generation to another or why it differs across countries.

Improvements in physiological capital are reflected in larger stature and improved body-mass index (BMI) of populations over time [Fogel (2004)]. Variations in height and weight are associated with variations in the chemical composition of the tissues that make up vital organs, in the quality of the electrical transmission across membranes, and in the functioning of the endocrine system and other vital systems. Nutritional status, as reflected in mature height and weight for height, are critical links connecting improvement in technology to improvements in human physiology. The early onset of the degenerative diseases of old age has been linked to inadequate cellular development early in life, including intrauterine development. Fogel's theory of technophysio-evolution [Fogel (2004)] implies that health endowments in a given population change (on average improve) over time, and that they differ across countries that are at very different stages of development.

5.1. *Micro-evidence*

The microeconomic evidence of impacts of malnutrition on health and labor productivity is reviewed. First, consider the theoretical issues associated with estimating a person's or household's health production function. Let H be an individual's true health status, N be a vector of human nutritional intakes, and E be a vector of time investments in health such as exercise and acquisition of health-related knowledge. Z_H is a vector of observable individual and household attributes, μ is an unobservable individual health endowment, and e_H is a random measurement error. An epidemiological health production function can then be written as

$$H = H(N, E; Z_H, \mu, e_H). \tag{29}$$

Many studies have collected data on individual attributes such as sex, race, marital status and education as the elements of Z_H and information on the elements of N and E to estimate the effects of nutritional and exercise choices on measures of health. If information is missing on the health endowment, this will bias the estimated coefficients of the included variables, and often, lead to perverse results.

To see why, we must take explicit account of the individual's decision-making process. Following the pioneering work by Grossman (1972a, 1972b) and the human capital model of Section 2, we embed the health production function into an individual's utility function

$$U = U(H, L, X), \tag{30}$$

where L is leisure and X is a composite good that is purchased in the market and does not affect an individual's health. His or her time constraint is $T = L + E + h$, where h

is hours of work for pay. The individual is assumed to maximize utility, subject to the budget constraint

$$V + W(T - L - E) = \sum_{i=1}^{k} P_i N_i + P_x X, \tag{31}$$

where V is nonlabor income, W is an individual's hourly wage (assumed here to be unaffected by health) and P_i and P_x are the prices of nutrients and other goods, respectively. The reduced-form demand equations for nutrients, time investments in health, hours of work for pay will be of the form

$$\begin{aligned} N &= N(P_1, P_2, \ldots, P_k, P_x, V, W, Z_H, \mu), \\ E &= E(P_1, P_2, \ldots, P_k, P_x, V, W, Z_H, \mu), \\ h &= h(P_1, P_2, \ldots, P_k, P_x, V, W, Z_H, \mu). \end{aligned} \tag{32}$$

Without information on the unobservable health endowment, it is now clear why direct estimation of (29) is problematic. The endogenous variables N and E depend on μ. If μ is excluded from (29), the error term will include μ which will be correlated with the observed health inputs. Consequently, the estimated coefficients from the epidemiological production function will be biased.

Missing information on the health endowment can lead to bizarre findings in cross-sectional estimation of health outcomes. For example in a sample of young adults, Miller (1986) found that smokers tended to have larger than average lung capacity relative to nonsmokers. The reason is that asthmatics and others with poor pulmonary health endowments never started smoking. Conversely, those individuals who started smoking at a young age typically had stronger lung capacity when they first start smoking. Holding the initial health endowment fixed, the adverse consequences of smoking are more apparent in longitudinal studies that show that lung capacity declines with every additional year of cigarette smoking.

Similarly, epidemiological studies have frequently failed to find a positive impact of early prenatal care on the health of newborn babies. Rosenzweig and Schultz (1983) found that this odd result was due to the fact that pregnant women who were healthier were more likely to delay visiting the doctor, while women who had poor health endowments used doctors more intensively. When endogeneity of a woman's doctor visits is taken econometrically into account, the expected positive effect of doctor's visits on the baby's birth weight, a measure of infant health, occurs.

In the United States, empirical evidence shows that when farmers produce hogs using confined housing for farrowing and finishing, they have increased incidence of short-term respiratory problems. However, it is difficult to find evidence of longer-term loss of pulmonary function [Hurley, Kliebenstein and Orazem (2000)]. The reason is that those producers who have the lowest ability to adapt to the environmental hazards associated with hog production either never enter the sector or else exit once adverse health outcomes are experienced. Hence, when individuals self-sort into occupations based on their initial health status, farmers who select confined hog production disproportionately come from the tail of the health endowment distribution that can best accommodate the adverse consequences without becoming seriously ill.

Randomized experiments in which nutrient intake and/or time investments in health are exogenously varied could generate unbiased estimates of the impact of N and E on H. However, it can be difficult to extrapolate from such laboratory studies to actual behavioral outcomes. The primary reason is that the unconstrained choices of consumers in the market may not reflect the constrained choices dictated by the experimental design. Constraining the choices of consumers by, for example, limiting consumption of red meat will generally lead to increased consumption of other foods, which may have their own negative health consequences.

In developing countries, one method by which the level of N can be varied is by the deployment of government programs aimed at influencing health outcomes. Such deployments are partial rather than complete and are common because a government cannot afford to make universal implementation, or because the deployment is conducted in stages. Such programs include the installation of public health clinics, sanitation systems, tube wells or other improvements to water supplies, nutritional supplements, vaccination programs, and health educational programs. By locating these programs in some locations but not others, one can estimate the impact of the government intervention by comparing health outcomes in areas receiving the program against health outcomes in places in which the program is not yet (or never to be) deployed.

These quasi-experimental designs are rarely randomized because the government naturally wants the program to be deployed where it will have the largest effects [Rosenzweig and Wolpin (1986)]. Alternatively, the households that would benefit greatly from the program may relocate to take advantage of the new program [Rosenzweig and Wolpin (1988)]. Either of these problems makes the occurrence of the health intervention conditional on unobservable health endowments at either the individual household or community level, recreating the endogeneity problem we were trying to sidestep in the first place.

The problem of human migration from the control to the treatment areas can be overcome, but it requires that researchers create a good baseline estimate of health outcomes in the target population before the policies are announced and implemented. Then, one can difference-out the unobserved health endowment effect by examining changes in the health outcome from data before and after project's implementation date. The problem of strategic choice of treatment areas is more difficult to correct. However, if the criteria used for selecting an area for program implementation are known, they can be used to correct for the nonrandom selection of the treatment areas for program implementation.

The most ambitious experimental application of health interventions is the Progresa program recently implemented in Mexico. This program combines a health and nutrition program with a targeted income transfer program that is conditional on children being in school, not working, and attending a health clinic. The enhanced income improves child nutrition. Preliminary empirical findings suggest that children in the program have improved health outcomes and increased schooling, but the linkage between the two cannot be established.

An explicit randomized experiment that resolved the selection problem was the treatment of school-aged children for intestinal worms in Kenya [Miguel and Kremer (2004)]. Children who were scheduled to receive the treatment earlier had absentee rates that were 7 percentage points lower than children who were randomly assigned to a group scheduled to receive the treatment at a later time. Children who were treated grew faster and had fewer intestinal diseases, so one could relate the improved schooling to improved nutritional health. No associated increase between treatment and test scores occurred, although it may be that the time frame was too short to observe significant changes in cognitive skills associated with the increased time in school.

A nonexperimental alternative to the endogenous treatment problem is to find instruments that affect N or E but not H directly. The reduced-form demand equation (32) provides several plausible instruments, namely V, W, P_i, and P_x. The role of an individual's income and wage data is somewhat problematic. A strong positive correlation exists between income and caloric intake. Ray's (1998) survey of 26 studies covering 15 countries found universally positive effects of household income on caloric intake. However, the proper instrument should not be actual household income, which depends on the endogenous hours of work decision. In addition, an added problem of potential simultaneity exists when income is measured by aggregate household expenditures in households where a large fraction of total expenditures are for food.[18]

Better instruments are food prices, which are exogenous to individual household behavior and have no direct impact on human health. Strauss (1986), Sahn and Alderman (1988), and Chen et al. (2002) have used food prices to successfully identify nutritional inputs. Alternatively, time costs such as distance to health-service providers have been used sometimes to identify the use of health clinics [Behrman (1996)], and Chen et al. (2002) used wages to identify time spent in exercise. Another identifier is child's health status at an early age because any health investments in children cannot be made on the basis of innate ability which cannot be observed until a child reaches age 2 [Glewwe, Jacoby and King (2001)].

An alternative way to illustrate the importance of the unobserved health endowment is to derive a proxy measure of the endowment, μ, and then to estimate its impact on the choice of health inputs. Rosenzweig and Schultz (1983) proposed deriving an estimate of μ from the residuals of the reduced-form epidemiological health production function

$$H = H(P_1, P_2, \ldots, P_k, P_x, V, W, Z_H, Z_U; \mu), \tag{33}$$

where μ will be in the error term. If the instruments are truly exogenous, they will be uncorrelated with μ.[19] The error term will include μ as well as random measurement

[18] When instrumental variable procedures are used, standard errors should be adjusted for heteroscedasticity caused by individual time-invariant effects by using White standard errors [Wooldridge (2002)].

[19] If the wage and nonlabor income are a function of health, they will need to be excluded.

error in H that are orthogonal to the regressors in (33). This noisy estimate of μ can then be inserted into an estimable form of Equation (33) to derive an estimate of the impact of μ on N or E. Naturally, the random error in the estimate of μ will bias its impact toward zero, so this procedure is more suggestive than definitive. Nevertheless, they do yield reasonable implications. In the original Rosenzweig and Schultz study, mothers who had better health endowments do not consult doctors as early in their pregnancies, have more children, and have more children at an older age than did mothers with lower estimated values of μ. Chen et al. (2002) found that individuals who had better health endowments (as measured by the error term from a reduced form equation explaining low blood pressure) also engaged in more exercise and purchased fewer medicines.

5.2. *Effects of nutrition on physical and mental development*

Dasgupta (1993, Chs. 14, 15) and Fogel (2004) review the epidemiological evidence relating early malnutrition to health. Malnutrition in the first three years of life and illnesses associated with malnutrition can permanently affect physical stature at maturity. Mild malnutrition causes children to play less and sleep more. Over time, persistent undernourishment will retard development of motor skills. There is evidence that nutritional rehabilitation can reverse the adverse effects of malnutrition in mild cases, although the process is slow. In more severe cases, the damage may be irreversible. For example, severe malnutrition leads to wastage of heart muscle mass and it can retard brain development.

Few studies of early child nutrition have explicitly confronted the endogeneity of food choices. Nutritionally stressed households must choose how to allocate calories across the household members. In such households, some children may be favored at the expense of others. For example, in parts of South and West Asia, survival rates are higher for boys than girls [Rosenzweig and Schultz (1982)]. This is in contrast to the typical case that girls are less frail than boys. The enhanced relative survival rates for boys in these areas have been tied to differential access to food, medicine and parental care between boys and girls [Schultz (2001)]. The survival advantage for boys decreases in regions where a stronger labor market exists for females.

For older children, malnutrition has been associated with poorer schooling outcomes [Behrman (1996)]. In theory, child health and child schooling are jointly determined by parents, so the ordinary-least-squares (OLS) estimate of the effect of health or nutrition on school achievement will be subject to simultaneity bias. In addition, measured health indicators such as height-for-age or body mass are invariably subject to the measurement error, e_H, which will also cause bias in estimated regression coefficients of measured health on schooling. The direction of these two biases is unclear, but instrumental-variables methods can be used to correct for both the endogeneity and measurement-error problems [Greene (2003)]. In practice, the instrumental-variables method has proven to have large effects on the magnitude of the health or nutrition coefficients on cognitive development.

Glewwe and Jacoby (1995) found that malnourished children in Ghana were more likely to delay entry into school, but that instrumenting nutrition with food prices reduced the effect by 40%. In their study of child achievement in the Philippines, Glewwe and King (2001) found that a similar price-based correction for endogeneity raised the impact five-fold. Another study of Philippine data which used early child health as an instrument similarly found large effects of health on schooling outcomes [Glewwe, Jacoby and King (2001)]. Alderman et al. (2001) found that in Pakistan, correcting for endogeneity raised the effect of nutrition on enrollment three-fold.

While there have been too few studies to determine the direction of these biases, correcting for endogeneity and/or measurement error has not reversed the sign of the estimated OLS results. The general conclusion from both OLS, instrumental-variables estimation, and experimental methods is that enhanced nutritional status increases investment in and (often) returns to schooling. Nevertheless, the large change in estimated coefficients associated with instrumenting for education or health suggests that researchers must take these biases into account.

5.3. Nutrition and labor productivity: Micro-evidence

In the previous section, we assumed that wage rates did not depend on a worker's health or nutrient intake. However, if improved nutrition raises human physical and (or) mental capacity, then individual marginal product will increase. Numerous empirical studies have investigated this presumption, and Strauss and Thomas (1998) and Behrman (1999) offer detailed reviews of this literature. We summarize their findings and refer the interested reader to those papers for the details.

Studies differ by choice of dependent variable; some use a direct measure of production (output, profit, net revenue), and others use the market wage as a measure of marginal product. Studies also differ in choice of the primary regressor of interest; it might be a direct measure of health (H) or health inputs (N). Because an individual's income is causally related to his or her nutrition and health, a regression of income or wages on health will be subject to simultaneity problems. In addition, measurement-error problems are associated with the use of H or N. Once again, it seems that instrumental-variable methods should be used to derive valid inferences regarding the impact of nutrition or health on labor productivity and standard errors should be adjusted for unobserved heterogeneity [Wooldridge (2002)].

Height at maturity has been used as a summary indicator of long-term health status. Height, however, is positively correlated with educational attainment. Both health and education are human capital investments that are positively affected by a low rate of time preference for intertemporal consumption choices as shown in the conceptual model of Section 2. Furthermore, better nutrition while young leads to both better physical and better mental development. Concentrating on empirical studies where results were statistically significant, the effect of an individual's height on individual wage or productivity is positive in almost every study surveyed by Behrman (1999). The effect of an individual's height goes beyond its impact on the physical strength to do work.

Strauss and Thomas report that the steepest relationships between individuals' wage and height are for the more educated groups in Brazil and the United States. Similar results were obtained when body-mass index (BMI) is substituted for height as the summary measure of good health except that the BMI impact appears to be nonlinear.[20] Consequently, it is possible to obtain negative as well as positive wage elasticities with respect to BMI.

A major advantage of height and weight measures is that they are relatively easy to take, requiring simple measurement device and only a small amount of training to obtain high quality data. Hence, they have relatively small measurement errors.[21] Caloric intake is a measure of nutrition that is typically collected using a respondent's recollection of the last one-to-three days of food consumption. This method requires conversion of food availability into calorie levels. Given the heterogeneity of intake across days of the week and months of the year, caloric intake data have relatively large measurement errors. Furthermore, the elasticity of an individual's wage or output with respect to caloric intake tends to be small compared to elasticities with respect to height or BMI. Nevertheless, they are usually positive, supporting the conjecture that improved nutrition makes workers more productive.

If employers take the impact of higher wages on nutrition (and thus productivity) into account in setting pay, it is possible that wages will be set above the market clearing level. Leibenstein (1957) raised this possibility first for developing countries, but it came into prominence as a rationale for Keynesian fixed-wage and unemployment in developed country contexts of the 1980s.[22] To make the story more precise, let a worker's effort or energy be given by $e(w)$, where the worker's consumption of nutrients, $P_N N$, is a positive function of the wage rate. The presumption is that the added energy associated with a higher wage comes from the physics of work [Fogel (1994)] – added physical strength is fueled by larger nutritional intake.

The firm's short-run revenue function can be written $p \cdot q(\ell \cdot e(w))$ where p is the output price and ℓ is the number of workers. Assume that workers have a value of time outside the firm, or opportunity wage, equal to ν. The firm chooses ℓ and w so as to maximize profit $\pi = p \cdot q - w \cdot \ell$. The first order conditions are

$$e(w)pq' - w = 0, \tag{34a}$$

$$\ell pq'e'(w) - \ell = 0. \tag{34b}$$

By solving for pq' in (34a) and substituting into (34b), the optimum choice is characterized by $e'(w) = \frac{e(w)}{w}$ so that the marginal product of the wage is set equal to the average product of the wage. This condition does not depend on ℓ, p or q', which implies that if $w \geqslant \nu$, the wage is set independent of current product demand for q and

[20] BMI is defined as weight (in kilograms) divided by squared height (in meters).

[21] In the U.S., women uniformly under-report their weight. Men, however, at less than 220 pounds over-report, while those over 220 pounds tend to under report [Lakdawalla and Philipson (2002)].

[22] See reviews by Akerlof and Yellen (1986) and Stiglitz (1987).

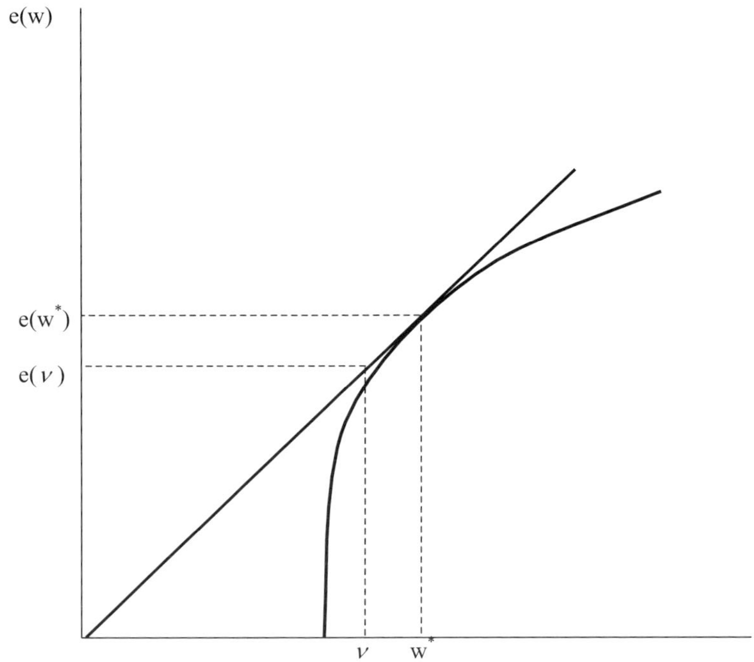

Figure 11. Illustration of the optimum wage, w^*, in the efficiency wage model.

fluctuations in the output price will not affect the optimum wage. More importantly, even if the wage is above the outside value of time ν, the firm will not lower the wage rate. This means that the wage may be set above the market clearing level, with the balance of the population earning ν or being unemployed.

Furthermore, this outcome can only occur if a nonconcave region exists in the work effort function, as illustrated in Figure 11. If work effort is concave in w at all wage rates, the first-order conditions will be satisfied at $w = 0$, and the firm will set the wage at ν, the opportunity wage of the worker. Although many variants on this theme exist, all have the feature that wages will be rigid in the face of persistent rural unemployment. As Rosenzweig's (1991) review illustrates, no convincing evidence exists of sticky wages in rural areas. Furthermore, Swamy (1997) argues that although actual wages paid to day laborers in rural India are higher than subsistence (i.e., $w \geqslant \nu$), they are simply too high relative to the marginal product of caloric intake to be consistent with the profit maximizing behavior (i.e., $w > w^*$). Hence, he concludes that wages are not being set with regard to their impact on nutrition.

Why do these models fail to fit well? One reason is that the firm may be owned by the workers, e.g., self-employed and unpaid family labor. High levels of self-employment and family enterprise characterize developing countries [Gollin (2000)]. It is plausible that Pareto-superior resource allocations would be fostered in household enterprises because the unemployed that would prefer to work would surely be allowed to do so.

5.4. Nutritional and labor productivity: Macro-evidence

Following Fogel (1994, 2004), the labor productivity of poor countries may be related to nutrient intake or availability. For example, FAO provides data on DES, dietary energy supply available from a country's domestic food production. Increases in DES are an outcome of the Green Revolution. DES is not perfect because it ignores the impact of net food imports and net food inventory change on currently available human energy from food.

Consider the following modified version of Equation (28):

$$\Delta \ln(y_t) = \beta_0 + \beta_1 \Delta S_t + \gamma_1 \Delta \ln N_t + \Delta\gamma_1 \ln N_{t-1} + \beta_2 \Delta \ln(K/L)_t + e_t, \quad (35)$$

where N_t is average nutrient availability of workers in time period t and K/L is the average capital–labor ratio. Although Equation (35) permits the impact of nutrition on growth to change between t and $t-1$, this seems only to be inviting estimation problems, given the quality of the available data. In our reported estimates, $\Delta\gamma_1$ is set equal to zero.[23]

We compiled data from the World Bank and FAO on 43 countries that the World Bank classified as low- or middle-income in 1970. There are 29 low-income countries and 14 middle income countries [Abdulla (2004)]. These are countries that are at a level of development where human energy availability could be expected to affect growth [Fogel (1994)]. Dietary energy supply (DES) is derived by FAO through the food balance sheet approach [FAO (1996, p. 40)]. Total food supply is based on information relating to domestic food production and net exports, food wastage from farm to retail, inventory changes, and nonfood use of food products.[24] Nutrient availability is defined as dietary energy supply (DES) per worker. Rates of growth (or change) of a variable is expressed as decade average rates of growth (or change) over 1961–1969, 1970–1979, 1980–1989, and 1990–1999.

Restricting $\Delta\gamma_1$ to zero and fitting Equation (35) to these country aggregate data, we obtain the results reported in Table 1. Regressions (1)–(3) are various OLS specifications. In regressions (1), the change in the literacy rate (schooling) has an estimated coefficient that is significantly positive. However, in regression (2), the coefficients of the log change in the literacy rate is not significantly different from zero. We conclude that the specification of literacy used in regression (1) is better. When change in dietary energy supply per worker is included but the change in the literacy rate is excluded

[23] Several reasons exist. First, the estimate of $\Delta\gamma_1$ will reflect the change in the returns to nutrient availability over time, so if the return is constant, the estimate of this change should be zero. Second, if energy availability is measured with error, the estimate will be biased toward zero.

[24] FAO acknowledges that data on domestic food production and net trade are frequently subject to significant error, but an average over several years is better than a single-year measure.

Table 1
Model explaining decade average rates of growth of real GDP per worker, 42 low- and middle-income countries, by decade 1960–1999

Regression/ equation	Estimated coefficients					R^2	N	Estimation method
	Intercept	d(literate)	d ln(DES/worker)	d ln(capital/worker)	d ln(literate)			
1. d ln y	0.030	0.007	0.003	0.111	–	0.136	168	OLS
	(8.86)	(1.86)	(1.64)	(4.03)				
2. d ln y	0.032	–	0.004	0.108	0.121	0.123	168	OLS
	(10.29)		(1.81)	(3.92)	(0.96)			
3. d ln y	0.035	–	0.004	0.106	–	0.118	168	OLS
	(16.6)		(1.85)	(3.84)				
4. d ln y	0.029	0.007	0.003	0.110	–	0.136	168	Random effects GLS by country
	(8.85)	(1.85)	(1.64)	(4.03)				
5. d ln y	0.030	0.008	–	0.120	–	0.122	168	Random effects GLS by country
	(9.00)	(2.05)		(4.48)				
6. d ln y	0.029	0.007	0.006	0.102	–	–	168	IV [instrument for d ln(DES/worker)]
	(8.45)	(1.61)	(1.18)	(3.24)				
7. d ln y	0.029	0.006	0.008	0.117	–	–	168	IV [for d ln(DES/worker)] and random effects GLS by country
	(8.42)	(1.49)	(1.47)	(4.35)				

(regression (3)), the estimated coefficient of energy from food is 0.004, and it is significantly positive. Investment in capital per worker is shown to contribute positively and significantly to per worker income growth in all three specifications.

The 43 countries in our sample seem unlikely to be impacted equally by country-specific random effects associated with political, social, and economic conditions. Regression (4) and (5) report models where country-specific random effects are included. The statistical significance of the changes in the literacy rate and DES are unchanged relative to regression equation (1). In regression (5) where the change in DES is excluded, the estimated coefficient of the change in literacy is only slightly different from and relative to its coefficient in regression (4) and (1).

Another issue is that current measures of DES almost certainly contain significant measurement errors that will lead to bias in the coefficients. Furthermore, the measures of DES are heavily based on agricultural production which is also a component of the numerator in the labor productivity measure that is used as the dependent variable. Consequently, the DES measures are endogenous by construction. One method for dealing with these problems is to instrument the change in DES [Greene (2003, pp. 86, 88–90)]. To test this hypothesis, we instrument food energy availability and refit Equation (34).[25] These results are reported in regression equation (5). Although the size of the estimated coefficient of nutrient availability doubles, it remains statistically weak.[26] Finally, we combine country-specific random effects and instrument food energy and report the results in regression equation (7). In this equation, neither the estimated coefficient on food energy nor that on the literacy rate is significantly positive at the 5% level.[27]

Smith and Haddad (2000) also conducted a cross-country investigation of the determinants of improved child nutrition status across countries and across time. Our interpretation of their results (see Footnote 6) is that roughly half of the improvement can be attributed to increased parental education, a quarter to improved food availability, and about one-fifth to improved water and sanitation conditions.

[25] As instruments to predict energy availability we use the production of corn, rice, wheat and milk per worker at the beginning of each decade. See Abdulla (2004).

[26] Avila and Evenson (2004) report a positive and significant effect of the change in DES per person in agriculture on the change in agricultural sector TFP for a set of 77 poor and middle income countries over 1961–1980 and 1981–2000. It makes little difference whether DES is instrumented in their results.

[27] Papers by Arcand (2001) and Wang and Taniguchi (2004) report per capita income growth equations including initial period nutrient availability and per capita GDP. Their specifications are equivalent to setting $\gamma_1 = 0$ and allowing $\Delta\gamma_1 \neq 0$ in (17). These studies seem to miss the dominant result predicted by Fogel (1994, 2004) in an enhanced Solow neoclassical growth model (including human capital and technical change), which is that change in real income per worker should be related to the change in nutrient availability per worker. Arcand and Wang and Taniguchi seem to be fitting some type of empirical growth convergence model [Jones (2002)], but their work fails to address the more important issue of the contribution of nutrient availability to steady state growth and some of their regressions include low, middle, and high income countries, which have very different histories of physiological capital development. Furthermore, nutrient availability in a given year can be measured with large error, but the average change in nutrient availability over a decade can be quite accurately measured. Hence, some caution should be used in interpreting the Arcand (2001) and Wang and Taniguchi (2004) results.

We conclude from these results that for poor and middle income countries, education measured as the literacy rate is a more important determinant of growth than dietary energy supply (DES). In fact, if we delete DES from Equation (35), and use regression (5) in Table 1, a one percentage point increase in the literacy rate of the low and middle income countries increases the rate of growth of labor productivity by 0.8% per year, an increase of 22% over the average labor productivity growth of 3.5% per year. This is not too large an effect – the average increase in literacy rates over a ten year period was less than one percentage point, so it takes a long time for a sample country to raise its adult literacy rate. Furthermore, The 95% confidence interval is quite wide (0.02 to 1.6), and so the true productivity payoff to improved literacy is subject to considerable uncertainty.

There is even less certainty about the payoff to improved DES. The estimated impact of nutrient/energy availability is sensitive to model specification. Hence, one should be cautious in drawing inferences about the contribution of dietary energy supply to growth even for relatively poor countries. It is possible that the impact of dietary energy supply would be stronger if better data were available for DES. Finding such information over a long enough period of time to assess its impact on a growth would be a tall order.

5.5. Obesity (over-nourished)

In some developed countries, e.g., the U.S., Great Britain, Greece, and Australia, average caloric intake has been rising while energy expended in work at home and market and transportation has been declining steadily [Huffman (2006)]. The net result of long-term energy imbalance is human weight gain. When an individual's body mass index is over 30, he or she is considered to be "obese". For the U.S., the obesity rate for adults was 31% in 2001. The rate has risen 15 percentage points over the past two decades. In contrast, in Japan and Norway, the obesity rate is about 6% [OECD (2005)].

Obesity translates with a time lag into future human health problems, including morbidity, mortality and increased demand for health care. Obesity is a major risk factor for diabetes, heart disease and some cancers. Other causes of heart disease include high cholesterol, high blood pressure and smoking cigarettes. In developed countries, food, especially high-fat and high-calorie convenience foods, has become relatively cheap and obesity rates are highest among the poor, low educated, and minority populations [Cutler, Glaeser and Shapiro (2003)].[28] Furthermore, at least 50% of the health-related obesity costs in developed countries are shifted to the public sector through public medical care and social insurance programs for the poor, disabled and elderly [Finkelstein, Fiebelkorn and Wang (2003)]. Thus, obese individuals are imposing negative externalities on society.

We showed earlier that under the most likely scenario, the relative price of food falls and real incomes rise as a country develops. Also, the demand for human energy falls

[28] As opposed to poor countries, the economic conditions in high income countries are such that poor people can purchase foods in large enough quantities to make obesity a problem.

with technical change in the household, market, and transportation sectors [Huffman (2006)]. These events may happen so rapidly that individuals fail to adjust properly to the new economic environment in which they are living. Hence, successful long term agricultural productivity growth seems to be the source of a new set of health-related problems associated with "over-nutrition". Economic research on obesity and associated issues is in its infancy, and we must wait for further analysis to judge whether future technical change in agriculture will improve the welfare of society in currently developed countries. Also, rising obesity rates may become a serious problem in developing countries that grow very rapidly and in population groups that emigrate from poor to rich countries [Mendez and Popkin (2004)]. Through evolution with selection, individuals today have genes and habits to gain weight in good times to build up energy reserve so that they can survive a famine. In rich countries, however, famines are absent. Hence, steady weight gain becomes a burden to good health.

5.6. *Poverty traps*

Dasgupta (1997) argued that it was less important that the efficiency wage model was literally true but rather that it highlighted the mechanism by which households could be trapped in poverty over several generations. In fact, substantial evidence exists of intergenerational transmission of poverty. Carter and May (2001) showed that 18% of the South African population were poor in both 1993 and 1998, and the bulk of these were structurally trapped into poverty. An additional 25% of the population were not poor in 1993 but were poor in 1998. Of those who fell into poverty, 85% were considered permanently trapped.

There are several plausible mechanisms that would cause intergenerational immobility. Emerson and de Souza (2005) found that parents who worked as children were more likely to have their children work, other things equal. Jacoby and Skoufias (1997) found that adverse income shocks caused parents to send their children to work. Poorer households are more prone to such adverse shocks than are wealthier households or households with higher levels of education [Glewwe and Hall (1998); McPeak and Barrett (2001)]. Underlying these findings is an explicit or implicit liquidity constraint on the poorest households, which prevents them from using short-term borrowing to smooth income shocks. Alternatively, income shocks prevent the household from repaying past debts, forcing it to devote all its personnel, children and adult, to current income generation rather than human capital investment [Basu (1997)]. Also, poorer households seem to apply higher discount rates to future versus current consumption related to poverty and also face higher borrowing costs. This lowers the incentive to invest in schooling and further intergenerational transmission of poverty.

This is the context in which malnutrition can have permanent adverse consequences. If a household is unable to feed its adults adequately, household earning capacity suffers. If it cannot feed its children, the next generation's earnings capacity will also suffer.

6. Schooling in agriculture

We first consider the impact of child labor on schooling choices and then turn to the impact of skilled/schooled labor on technical and allocative efficiency in agriculture. Finally, we consider the importance of schooling to off-farm work decisions of farmers.

6.1. Child labor and schooling

We indicated above that the incidence of child labor is much higher in rural than in urban areas, and agriculture is the primary employer of children in rural areas. To the extent that agricultural employment opportunities improve and child wages rise, it will be more likely that rural children work more and attend school less. It is clear that child labor and schooling are not mutually exclusive outcomes – most working children are also enrolled in school [Ravallion and Wodon (2000)]. Consequently, modest agricultural demand for child labor may not lower time spent in school. However, numerous studies have shown that increases in child wages or returns to child time in agriculture lower the probability of a child being in school [Rosenzweig and Evenson (1977); Levy (1985); King, Orazem and Paterno (2002); Orazem (1987)]. Furthermore, even if children are enrolled in school, child labor may reduce the amount of time they attend school, study, or learn per year. Very few studies exist of the impact of child labor on school achievement, except at the secondary level

A recent exception [Sanchez, Orazem and Gunnarsson (2005)] examined how a child's working affected student performance on 3rd and 4th grade tests of mathematics and language in 11 Latin American countries. In all cases, child labor lowered school performance, with the adverse effect increasing in magnitude for children who worked longer hours. If parents are more likely to send their children to work when they are performing poorly in school, then child labor and schooling are jointly determined. Hence, these results need to be interpreted cautiously. However, Ilahi, Orazem and Sedlacek (2005) found corroborating evidence that child labor lowers the production of human capital in schools. They found that Brazilian adults who worked as children received returns to a year of schooling that were 15–20% lower than adults who did not work as children. While it seems likely that child labor will lower human capital production, confidence in the exact magnitude await better data and more sophisticated modeling.

6.2. The choice of where to work: Rural–urban population shift and brain drain

Worldwide, about one-half of the labor force works in agriculture [World Bank (2000)]. A large majority are unpaid farm workers – the farmers who make decisions and work, and other farm family members who work generally without direct compensation – and a minority are hired (nonfarm family) workers. Hired workers are generally of two types: regular full time and seasonal. Seasonal labor demand variation arises largely from the definite seasonal pattern to biological events in plants, which creates unusually large labor demand at planting, weeding, and/or harvest time. The supply of seasonal agricultural labor frequently has a local component and a migratory component.

Over the long term, the share of the labor force employed in agriculture has declined dramatically in what are now developed countries, but slowly or not at all in low-income or developing countries [Grübler (1994); OECD (1995); Johnson (2000)]. Decisions on schooling by families and communities are an important factor determining whether individuals work in agriculture or elsewhere. Even in developed countries where farmers are relatively well educated, hired farm workers generally have significantly less education.

Whether to work in agriculture or in another industry is an important decision worldwide. In India and China, which account for about 40% of the world's population, about 65% of the labor force in 1990 was employed in agriculture. In Western Europe, less than 10% of the labor force was employed in agriculture, and in the United States the share was only 3%. In noncentrally planned countries, individuals make a choice of an occupation/industry for work. Schooling decisions affect later occupational choice decisions.

As economic conditions change in interconnected labor markets, workers in free societies invest in migration to improve their future economic welfare (see the three-period model in Section 2), which tends to reduce or eliminate intermarket wage differences. This complicates the problem of explaining migration, because individuals are acting on anticipated wage rate differences rather than the ex post values. Young adults have the longest time-period over which to obtain benefits from migration investment, and hence, they have the highest mobility rates (also see model in Section 2). Schooling also plays a significant role in these adjustments or reallocations because of its effect on the costs and returns to migration.

Although farmers tend to be tied to the land and to be geographically immobile, off-farm work of farmers is a relatively common international phenomenon. Since the 1950s and 1960s, aggregate demand for operator and family farm labor in all of the developed countries has declined [see OECD (1995)], the demand for housework in farm households has generally declined as family sizes have declined and labor-saving household technologies have been adopted, and the real nonfarm wage has generally increased. Faced with needing to make adjustments in labor allocation, farm households in developed countries have frequently chosen to continue in farming but also to supply labor of some of its members to the nonfarm sector [e.g., OECD (1994); Huffman (1980)].

The simple model of trade across agricultural and nonagricultural sectors demonstrated why even neutral technological change across the two sectors could lead to rising relative marginal revenue products in the nonagricultural sector. Consequently, the process of development will be accompanied by a shift of the population out of agriculture into other sectors and from rural to urban areas. Migration has been the subject of several recent reviews [Greenwood (1997); Lucas (1997); Taylor and Martin (2001)]. We touch only briefly on the topic here and refer interested readers to those other papers for more extensive reviews.

It has generally been observed that in both developed countries [Greenwood (1997)] and in developing countries [Schultz (1982); Williamson (1988)], more educated people are more likely to migrate from rural to urban areas. This process has been labeled the "brain drain". What's more, migrants tend to be younger, so the average age of city dwellers falls as the average age of rural dwellers rises. As Williamson (1988) demonstrates, this same phenomenon took place in England in the 18th and 19th centuries, and it is taking place in developing countries today. The incentives for younger people to migrate are well understood in the context of the human capital investment model – younger people have more years in which to obtain returns on their migration investment and they have less specific human capital invested in the place of origin. It is harder to explain why the relative returns from migration would be higher for more educated individuals. Nevertheless, this appears to be so, as demonstrated by recent studies in the United States [Mills and Hazarika (2001); Huang, Orazem and Wohlgemuth (2002)].

An additional year of schooling may raise worker productivity at off-farm work by more than at on-farm work. Numerous arguments explain why higher returns to human capital exist in cities than in dispersed populations. Human and physical capital may be complements, so if cities are concentrations of physical capital, they will enhance the returns to schooling. Specialized human capital in different areas may be complementary, so that educated labor is more productive when employed where other workers also have education [Becker and Murphy (1993)]. Cities may also lower the cost of information flows, making educated labor more productive. To the extent that cities are agglomerations of consumers, it is easier for labor to specialize according to comparative advantage, and so cities offer greater scope for specialization for skilled workers. By agglomerating jobs, cities also lower the costs of job search. If one job disappears, it is relatively easy to switch to another sector, which lowers the individual's risk of specializing. These and other arguments are presented by Glaeser (1998) and Quigley (1998).

However, the process of agricultural transformation will also change the input shares for educated labor in the countryside. Agriculture appears to be subject to constant returns to land and capital. Rising wages in the cities require that an educated farmer be paired with increasing levels of other inputs in order to generate sufficient income to match his or her opportunity costs in the city. Limitations on land in the face of this need for larger farms further accelerate the shift of the population out of agriculture. Thus, Kislev and Petersen (1982) and Barkley (1990) found that rising urban wages have been a driving force in raising farm size and lowering the number of farmers.

6.3. Technology adoption and information acquisition

The decision to adopt new technologies is an investment decision, because significant costs are incurred in obtaining information and learning about the performance characteristics of one or more new technologies, and the benefits are distributed over time. Huffman and Evenson (1993) summarize how public and private agricultural research has developed new crop varieties for U.S. farmers, and Evenson and Gollin (2003) summarize how public sector and CGIAR research efforts have developed new Green Revolution crop varieties for developing country farmers. For any given farmer only a small share of the new technologies that become available will be profitable to adopt. This means that there is a large amount of uncertainty facing farmers, and additional schooling may help them make better adoption decisions and increase farm profitability. Because schooling of farmers affects ability to acquire and process information, it ca have long term impacts. For optimal schooling decision making, the three-period model of Section 2 provides a useful guide.

When technology is new and widely profitable, farmers' schooling has been shown to be positively related to the probability of adoption. When a technology has been available for an extended period (e.g., several years) or it is not widely profitable, farmers' schooling is generally unrelated to adoption/use of the technology. Schooling has been shown to affect choice of information channels about new technologies. The most recent adoption literature is applying hazard models, and added farmer education has been shown to increase the hazard rate for hybrid cow technology in Tanzania [Sunding and Zilberman (2001); Abdulai and Huffman (2005)].

Although successful adoption of innovations clearly requires information, few studies have considered the important joint decisions of information acquisition and new technology adoption. This seems to be a fruitful area for new research. When several information sources exist, early adopters might prefer sources that facilitate faster learning about the innovation. The information channels for early adopters might also be different from those for late adopters.

Wozniak (1993) is an exception in that he examined farmers' joint decisions on information acquisition and technology adoption. He considered the adoption of two technologies, one new and one mature, and four channels of information, one active and one for both extension and private sector information providers. In this study, he found that farmers' education significantly increased the probability of adopting new and mature technologies; of acquiring information from extension by talking with extension personnel (passive); and of attending extension demonstrations or meetings (active) about the use of new products or procedures. Farmers' education did not have a statistically significant effect on a farmer's acquiring information by talking with private industry personnel or attending demonstrations or meetings on the use of new products or procedures sponsored by private companies. Farmers were more likely to be early adopters if they acquired information actively or passively from private industry than if they acquired information from extension. For both new and mature innovations, positive and significant interaction effects existed between farmers' acquisition of information from public and private sources, i.e., public and private information acquisition seems to be complementary.

Overall, the review of the literature [Huffman (2001); Hussain and Byerlee (1994)] shows that additional schooling of farmers increases the rate of early adoption of useful agricultural technologies in developed and developing countries. A surprisingly small amount of research has examined farmers' joint decisions on information acquisition and technology adoption, and this is an area for much needed new research.

6.4. Agricultural production

Education of farmers and other farm labor has the potential for contributing to agricultural production as reflected in gross output/transformation functions, and in value-added or profit functions. These effects are frequently referenced as technical efficiency effects, allocative efficiency effects, or economic efficiency effects of education. When the effects of schooling on production are considered in a gross output–complete input specification, the marginal product of education, a measure of technical efficiency, is limited by the other things that are held constant. A value-added or profit function representation of production accommodates a much broader set of effects of farmers' education associated with allocative efficiency. The effects include adoption of new inputs in a profitable manner, the allocation of land (and other quasi-fixed inputs) efficiently among alternative uses, the allocation of variable inputs efficiently, and the efficient choice of an output mix. The empirical evidence has shown that the productivity of farmers' education is enhanced by a wider range of choices, and Welch (1970) is generally given credit for delineating these substantive differences.

Overall, in developing and developed countries, the review of the literature [Schultz (1975); Huffman (2001)] shows that farmers' schooling has value under special but certainly not all environments. For schooling to be valuable to farmers, they must be in an environment where markets are in place and functioning for inputs and outputs, new technologies must be being made available by the nonfarm sector, and they must have access to credit. In this environment, allocative efficiency, associated with adjusting to disequilibria, has been shown to be valuable. The education for agricultural workers does not seem to enhance technical efficiency. For example, farmers' schooling has infrequently been shown to increase crop yields or gross farm output, because technical-efficiency gains from skills provided by farmers' schooling seem generally to be small. However, as summarized in Section 3, Jamison and Lau (1982) provide empirical evidence from 13 World Bank studies showing that in 18 of 37 reported estimates, farmers' schooling has a significantly positive impact on agricultural output. Also, farmers' education is valuable when they have the option of working off-farm at skilled jobs.

7. Conclusions and implications

We have provided a review and critique of five issues. First, we have shown that in a two-sector closed economy model technology shocks in agriculture can create growth and incentives for migration of labor out of agriculture to a growing nonagricultural sector. In poor countries, this labor is almost certainly unskilled. After a country reaches a certain level of development, additional formal schooling leads to permanent literacy that has value. A three-period dynamic model of optimal household decision making is used to show that with finite life and constraints on production, optimal human capital investments are generally largest early in life. If length of life is extended, optimal human capital investments in these early periods and over all periods taken together will increase.

Second, we have shown that over most of the history of the world, per capita income has been close to the subsistence level and population growth has been very slow. Most of the growth of per capita income and population has occurred over the past 250 years and is associated with major technical advances in the most advanced countries. Some of the technological successes in agriculture and industry were adapted to conditions in less advanced countries. Successful adaptation frequently required a minimum level of human capital in the receiving country. Almost all countries that have entered into modern economic growth started with sustained growth of agriculture. The few exceptions in which growth was not predicated on an agricultural transformation, e.g., Hong Kong and Singapore, are not useful examples for current low income countries in Sub-Saharan Africa. After sufficient advances have been made in institutions and technology, schooling of the general population is an important source of growth.

Third, considerable micro- and macro-economic evidence exists that investments in schooling contribute to economic growth of middle and high income countries. However, cognitive skill may be a better measure of the dimension of education that matters for growth than years of formal schooling completed. Furthermore, we found that for low- and middle-income countries schooling measured as the literacy rate is a stronger factor for explaining economic growth than is dietary energy supply (DES). Hence, we suggest that claims that human energy availability is an important source of grow is not supported by current data.

Fourth, it is now widely accepted that physiological capital is another important form of human capital. In addition, a nutritionally balanced diet and safe water are important factors in the production of physiological capital. With larger investments in physiological capital, the human organ and immune systems are better developed and stronger, and this superior physiological capital retards the on-set of diseases of old age – diabetes, cardiovascular diseases, and cancer.

Fifth, schooling for farmers and farm laborers has been shown to be productive in some but not all conditions. New evidence suggests that children who work regularly on farms have reduced achievements at school. This does not matter when average schooling levels are low, but when average schooling levels reach four or more years, missing school to participate in farm work detracts from later labor productivity. Parents may not have the necessary information to make optimal decisions on their children's schooling, and they may not even have the long-term interest of their children in mind. Educated farm labor seldom contributes to technical efficiency in agriculture, but farmers' education does frequently contribute to on farm allocative efficiency. The latter efficiency is important when the agricultural sector is in a type of disequilibrium due to the introduction of new, productive technologies. Also, farmers' schooling is frequently valuable when they work off-farm at nonfarm jobs.

The poorest countries in the subtropics have found it difficult to adapt agricultural technologies developed by the high income temperate zone countries. But, they also do not have the human capital necessary to successfully adapt these agricultural technologies to their local conditions. In addition, international experiment stations that have made major strides in improving corn, wheat and rice varieties have not been able to make similar improvements to crops that are most naturally suited to the subtropics, e.g., cassava, chick pea.

The poorest countries of the world continue to have a large share of their labor force employed in agriculture. GDP per capita is low and stagnant, birth rates and infant mortality are high and life expectancy at birth is low. Real economic growth cannot occur in these countries until they experience a positive agricultural productivity shock followed by steady productivity growth. With an increase in agricultural productivity, per capita incomes can rise and workers can be released from agriculture to work in other occupations. With luck, small-scale nonfarm industry can grow. Moreover, investments in schooling, health, information, and migration are important for the long-term welfare of farm people in these countries, and the local institutional structure must be such that it easily permits those activities to take place. These investments will help lower the birth rate, increase life expectancy, increase labor productivity, and facilitate successful rural to urban mobility and nonfarm employment that is necessary for long-term economic growth.

Many of the recent studies of economic growth have relied on cross-sectional data sets composed of a large number of countries but with relatively short time horizons, say one, two or three decades. It is important to ask whether such data can yield accurate inferences about the long-run economic growth process. If economies are subject to convergence in growth rates, growth over a given decade may be only weakly correlated with growth in prior or subsequent decades. All of the countries, for which we have a long time series of reliable economic data, are relatively advanced, and so we do not have a strong test of whether or not economic growth in current LDCs is subject to convergence to anything other than a subsistence level. Consequently, inferences about the economic-growth process drawn from cross-sectional data sets of relatively short duration must be viewed with a healthy degree of skepticism.

Although the Green Revolution is responsible for increased rice, wheat, and maize yields in Latin America and Asia, we do not find convincing evidence that increasing dietary energy supply (DES) is a source of economic growth. Schooling appears to be more important than DES in spurring economic growth.

Some of the current high income countries are experiencing over-nutrition or obesity, and its causes are still to be identified. As poor countries undergo development or their citizens migrate to developed countries, obesity, rather than malnutrition, promises to become a new and important problem in the future.

References

Abdulai, A., Huffman, W.E. (2005). "The diffusion of new agricultural technologies: The case of crossbreeding technology in Tanzania". American Journal of Agricultural Economics 87, 645–659.

Abdulla, M. (2004). "The impact of improved nutrition on labor productivity and GDP growth rates in low- and middle-income countries", Masters thesis. Department of Economics, Iowa State University.
Acemoglu, D., Johnson, S., Robinson, J.A. (2001). "The colonial origins of comparative development: An empirical investigation". American Economic Review 91, 1369–1401.
Acemoglu, D., Johnson, S., Robinson, J.A. (2002). "Reversal of fortune: Geography and institutions in the making of the modern world income distribution". Quarterly Journal of Economics 117, 1231–1294.
Akerlof, G.A., Yellen, J.L. (1986). Efficiency Wage Models of the Labor Market. Cambridge University Press, Cambridge.
Alderman, H., Behrman, J.R., Lavy, V., Menon, R. (2001). "Child health and school enrollment: A longitudinal analysis". Journal of Human Resources 36, 185–205.
Arcand, J.L. (2001). "Undernourishment and economic growth: The efficiency cost of hunger". FAO, http://www.fao.org/News/2001/stockholm/arcand.htm.
Avila, A.F., Evenson, R.E. (2004). Total Factor Productivity Growth in Agriculture: The Role of Technological Capital. Economic Growth Center, Yale University.
Avila, A.F., Evenson, R.E. (forthcoming). "FAO data-based TFP measures". In: Evenson, R.E., Pingali, P. (Eds.), Handbook of Agricultural Economics, vol. 4. Elsevier, Amsterdam. In press.
Barkley, A.P. (1990). "The determinants of the migration of labor out of agriculture in the United States, 1940–1985". American Journal of Agricultural Economics 72, 567–573.
Barro, R.J., Lee, J.-W. (2000). "International data on educational attainment: Updates and implications", Working Paper No. 42. Center for International Development, Harvard University.
Basu, K. (1997). Analytical Development Economics: The Less Developed Economy Revisited. MIT Press, Cambridge, MA.
Becker, G.S. (1993). Human Capital, third ed. University of Chicago Press, Chicago, IL, pp. 1–13.
Becker, G.S., Murphy, K.M. (1993). "The division of labor, coordination costs, and knowledge". In: Becker, G.S. (Ed.), Human Capital. third ed. University of Chicago Press, Chicago, IL, pp. 299–322.
Behrman, J.R. (1996). "Impact of health and nutrition on education". World Bank Research Observer 11, 23–37.
Behrman, J.R. (1999). "Labor markets in developing countries". In: Ashenfelter, O., Card, D. (Eds.), Handbook of Labor Economics, vol. 3B. Elsevier Science, Amsterdam.
Bedi, A.S., Edwards, J.H.Y. (2002). "The impact of school quality on earnings and educational returns – evidence from a low-income country". Journal of Development Economics 68, 157–185.
Bedi, A.S., Gaston, N. (1999). "Using variation in schooling availability to estimate educational returns for Honduras". Economics of Education Review 18, 107–116. Journal of the American Statistical Association 69 (1999) 350–354.
Benhabib, J., Spiegel, M.M. (1994). "The role of human capital in economic development: Evidence from aggregate cross-country data". Journal of Monetary Economics 34, 143–174.
Ben-Porath, Y. (1967). "The production of human capital and the lifecycle of earnings". Journal of Political Economy 75, 352–365.
Bils, M., Klenow, P.J. (2000). "Does schooling cause growth?". American Economic Review 90, 1160–1183.
Card, D. (1999). "The causal effect of education on earnings". In: Ashenfelter, O., Card, D. (Eds.), Handbook of Labor Economics, vol. 3A. Elsevier Science, Amsterdam.
Carter, M.R., May, J. (2001). "One kind of freedom: The dynamics of poverty in post-apartheid South Africa". World Development 29, 391–409.
Chen, S.N., Shogren, J.F., Orazem, P.F., Crocker, T.D. (2002). "Prices and health: Identifying the effects of nutrition, exercise and medication choices on blood pressure". American Journal of Agricultural Economics 84, 990–1002.
CIA (2004). The World Fact Book. http://www.cia.gov/cia/publications/factbook/geos.
Cutler, D., Glaeser, E., Shapiro, J. (2003). "Why have Americans become more obese?". Journal of Economic Perspectives 17, 93–118.
Dasgupta, P. (1993). An Inquiry into Well-Being and Destitution. Oxford University Press, Oxford.
Dasgupta, P. (1997). "Nutritional status, the capacity for work, and poverty traps". Journal of Econometrics 77, 5–37.

de Ferranti, D., Perry, G.E., Gill, I., Guasch, J.L., Maloney, W.F., Sanchez-Paramo, C., Schady, N. (2003). Closing the Gap in Education and Technology. The World Bank, Washington, DC.
Duflo, E. (2001). "Schooling and labor market consequences of school construction in Indonesia: Evidence from an unusual policy experiment". American Economic Review 91, 795–813.
Durand, J. (1975). The Labor Force in Economic Development. Princeton University Press, Princeton, NJ.
Emerson, P.M., de Souza, A.P. (2005). "The intergenerational persistence of child labor". In: Orazem, P.F., Sedlacek, G., Tzannatos, Z. (Eds.), Child Labor and Education in Latin America. In press.
Evenson, R.E., Gollin, D. (2003). "Assessing the impact of the Green Revolution, 1960 to 2000". Science 300, 758–762.
FAO (1996). The Sixth World Food Survey. FAO of the United Nations, Rome, Italy.
Finkelstein, E., Fiebelkorn, I., Wang, G. (2003). "National medical spending attributable to overweight and obesity: How much, and who's paying?". Health Affairs 22.
Floyd, J. (1967). "The effects of farm price supports on the returns to land and labor in agriculture". Journal of Farm Economics, 148–158.
Fogel, R.W. (1994). "Economic growth, population theory, and physiology: The bearing of long-term processes on the making of economic policy". American Economic Review 84, 369–395.
Fogel, R.W. (2004). "Health, nutrition, and economic growth". Economic Development and Cultural Change 52, 643–658.
Glaeser, E.L. (1998). "Are cities dying?". Journal of Economic Perspectives 12, 139–160.
Glewwe, P. (2002). "Schools and skills in developing countries: Education policies and socioeconomic outcomes". Journal of Economic Literature 40, 436–483.
Glewwe, P., Hall, G. (1998). "Are some groups more vulnerable to macroeconomic shocks than others? Hypothesis tests based on panel data from Peru". Journal of Development Economics 56, 181–206.
Glewwe, P., Jacoby, H. (1995). "An economic analysis of delayed primary school enrollment and child malnutrition in a low income country". Review of Economics and Statistics 77, 156–169.
Glewwe, P., Jacoby, H., King, E. (2001). "Early childhood nutrition and academic achievement: A longitudinal analysis". Journal of Public Economics 81, 345–368.
Glewwe, P., King, E.M. (2001). "The impact of early childhood nutritional status on cognitive development: Does the timing of malnutrition matter?". World Bank Economic Review 15, 81–113.
Goldin, C. (1995). "The U-shaped female labor force function in economic development and economic history". In: Schultz, T.P. (Ed.), Investment in Women's Human Capital. University of Chicago Press, Chicago, IL.
Gollin, D. (2000). "Nobody's business but my own: Self-employment and small enterprise in economic development". Department of Economics, Williams College.
Gollin, D., Parente, S., Rogerson, R. (2002). "The role of agriculture in development". American Economic Review, 160–164.
Grossman, M. (1972a). "On the concept of health capital and the demand for health". Journal of Political Economy 80, 233–255.
Grossman, M. (1972b). The Demand for Health: A Theoretical and Empirical Investigation. Columbia University Press, New York.
Greene, W.H. (2003). Econometric Analysis, fifth ed. Prentice Hall, Upper Saddle River.
Greenwood, M.J. (1997). "Internal migration in developed countries". In: Rosenzweig, M.R., Stark, O. (Eds.), Handbook of Population and Family Economics. Elsevier Science, Amsterdam.
Grübler, A. (1994). "Industrialization: A developing country perspective". In: Socolow, R., Andrews, C., Berkhart, F., Thomes, V. (Eds.), Industrial Ecology and Global Change. Cambridge University Press, New York, pp. 43–68.
Hanushek, E.A., Kimko, D.D. (2000). "Schooling labor-force quality, and the growth of nations". American Economic Review 90, 1184–1208.
Hayami, Y., Ruttan, V.W. (1971). Agricultural Development: An International Perspective. Johns Hopkins University Press, Baltimore, MD.
Hayami, Y., Ruttan, V.W. (1985). Agricultural Development: An International Perspective, revised and expanded edition. Johns Hopkins University Press, Baltimore, MD.

Hayami, Y., Yamada, S. (1975). "Agricultural research organization in economic development: A review of the Japanese experience". In: Reynolds, L.G. (Ed.), Agriculture in Development Theory. Yale University Press, New Haven, CT, pp. 224–249.
Heckman, J., Polachek, S.W. (1974). "Empirical evidence on the functional form of the earnings–schooling relationship". Journal of the American Statistical Association 69, 350–354.
Heston, A., Summers, R., Aten, B. (2002). Penn World Table Version 6.1. Center for International Comparisons at the University of Pennsylvania (CICUP), October, http://www.bized.ac.uk/dataserv/penndat/pennhome.htm.
Huang, T.L., Orazem, P.F., Wohlgemuth, D. (2002). "Rural population growth, 1950–1990: The roles of human capital, industry structure and government policy". American Journal of Agricultural Economics 83, 615–627.
Huffman, W.E. (1977). "Interaction between farm and nonfarm labor markets". American Journal of Agricultural Economics 59, 1054–1061.
Huffman, W.E. (1980). "Farm and off-farm work decisions: The role of human capital". Review of Economics and Statistics 62, 14–23.
Huffman, W.E. (1985). "Human capital, adaptive ability, and the distributional implications of agricultural policy". American Journal of Agricultural Economics 67, 429–434.
Huffman, W.E. (1991). "Human capital for future economic growth". In: Johnson, G.L., Bonnen, J.T. (Eds.), Social Science Agricultural Agendas and Strategies. Part III. Michigan State University Press, East Lansing, MI, pp. 61–67.
Huffman, W.E. (1996). "Labor markets, human capital, and the human agent's share of production". In: Antle, J.M., Sumner, D.A. (Eds.), The Economics of Agriculture, vol. 2 (Papers in Honor of D. Gale Johnson). University of Chicago Press, Chicago, IL, pp. 55–79.
Huffman, W.E. (2001). "Human capital: Education and agriculture". In: Gardner, B., Rausser, G. (Eds.), Handbook of Agricultural Economics. Elsevier Science, Amsterdam, pp. 334–381.
Huffman, W.E. (2006). "Understanding post-war changes in U.S. household production: A full-income demand-system approach", Working Paper No. 06036. Department of Economics, Iowa State University.
Huffman, W.E., Evenson, R. (1993). Science for Agriculture: A Long-Term Perspective. Iowa State Press, Ames, IA.
Huffman, W.E., Evenson, R. (2006). Science for Agriculture: A Long-Term Perspective, second ed. Blackwell Publishing, Ames, IA.
Hurley, T.M., Kliebenstein, J.B., Orazem, P.F. (2000). "An analysis of occupational health in pork production". American Journal of Agricultural Economics 82, 323–333.
Hussain, S.S., Byerlee D. (1994). "Education and farm productivity in post-green revolution agriculture in Asia", Invited paper, XXII International Conference of Agricultural Economists, Harare, Zimbabwe.
Ilahi, N., Orazem, P.F., Sedlacek, G. (2005). "How does working as a child affect wages, income and poverty as an adult?". In: Orazem, P.F., Sedlacek, G., Tzannatos, Z. (Eds.), Child Labor and Education in Latin America. In press.
Jacoby, H.G., Skoufias, E. (1997). "Risk, financial markets, and human capital in a developing country". Review of Economic Studies 64, 311–335.
Jamison, D.T., Lau, L.J. (1982). Farmer Education and Farm Efficiency. Johns Hopkins University Press, Baltimore, MD.
Johnson, D.G. (2000). "Population, food, and knowledge". American Economic Review 90, 1–14.
Johnson, W.R., Skinner (1986). "Labor supply and marital separations". American Economic Review 76, 455–469.
Jones, C.I. (2002). Introduction to Economic Growth, second ed. W.W. Norton & Co, New York.
Jorgenson, D.W. (1965). "The role of agriculture in economic development: Classical versus neoclassical models of growth". In: Wharton, C.R. (Ed.), Subsistence Agriculture and Economic Development. Aldine Publishing, Chicago, IL, pp. 320–348.
Jorgenson, D.W., Stiroh, K.J. (2000). "U.S. economic growth at the industry level". American Economic Review 90, 161–167.

King, E.M., Hill, M.A. (1993). Women's Education in Developing Countries. Johns Hopkins University Press, Baltimore, MD.
King, E.M., Orazem P.F., Paterno, E.M. (2002). "Promotion with and without learning: Effects on student dropout", Mimeo. Iowa State University.
Kislev, Y., Petersen, W. (1982). "Prices, technology and farm size". Journal of Political Economy 90, 578–595.
Klasen, S. (2002). "Lower schooling for girls, slower growth for all? Cross-country evidence on the effect of gender inequality in education on economic development". World Bank Economic Review 16, 345–373.
Krueger, A.B., Lindahl, M. (2001). "Education for growth: Why and for whom?". Journal of Economic Literature 39, 1101–1136.
Lakdawalla, D., Philipson, T. (2002). "The growth of obesity and technical change: A theoretical and empirical examination", NBER Working Paper No. 8946.
Lam, D., Schoeni, R.F. (1993). "Effects of family background on earnings and returns to schooling: Evidence from Brazil". Journal of Political Economy 101, 710–740.
Leibenstein, H. (1957). Economic Backwardness and Economic Growth. John Wiley & Sons, New York.
Levy, V. (1985). "Cropping pattern, mechanization, child labor, and fertility behavior in a farming economy: Rural Egypt". Economic Development and Cultural Change 33, 777–791.
Lloyd, C. (Ed.) (2005). Going up Global: The Changing Transitions to Adulthood in Developing Countries. National Academies Press, Washington, DC.
Lucas, R.E.B. (1997). "Internal migration in developing countries". In: Rosenzweig, M.R., Stark, O. (Eds.), Handbook of Population and Family Economics. Elsevier Science, Amsterdam.
Maddison, A. (2001). The World Economy: A Millennial Perspective. OECD, Paris.
Mammen, K., Paxson, C. (2000). "Women's work and economic development". Journal of Economic Perspectives 14, 141–164.
Matsuyama, K. (1992). "Agricultural productivity, comparative advantage, and economic growth". Journal of Economic Theory 58, 317–334.
McPeak, J.G., Barrett, C.B. (2001). "Differential risk exposure and stochastic poverty traps among East African pastoralists". American Journal of Agricultural Economics 82, 323–333.
Mendez, M.A., Popkin, B. (2004). "Globalization, urbanization, and nutritional change in the developing world". Journal of Agricultural Development Economics 1, 220–241.
Miguel, E., Kremer, M. (2004). "Worms: Identifying impacts on education and health in the presence of treatment externalities". Econometrica 72, 159–217.
Mincer, J. (1974). Schooling, Experience and Earnings. Columbia University Press, New York.
Miller, A. (1986). Pulmonary Function Tests in Clinical and Occupational Lung Disease. Grune and Startton, Orlando, FL.
Mills, B., Hazarika, G. (2001). "Migration of young adults from nonmetropolitan counties". American Journal of Agricultural Economics 83, 329–340.
OECD (1994). Farm Employment and Economic Adjustment in the OECD. OECD, Paris.
OECD (1995). Technological Change and Structural Change in OECD Agriculture. OECD, Paris.
OECD (2005). Health Data. OECD, Paris.
Orazem, P.F. (1987). "Black–white difference in schooling investment and human capital production in segregated schools". American Economic Review 77, 714–723.
Pritchett, L. (1997). "Divergence, big time". Journal of Economic Perspectives 11, 3–17.
Pritchett, L. (2001). "Where has all the education gone?". World Bank Economic Review 15, 367–391.
Psacharopolous, G. (1994). "Returns to investment in education: A global update". World Development 22, 1325–1343.
Psacharopoulos, G., Tzannatos, Z. (1989). "Female labor force participation: An international perspective". World Bank Research Observer 4, 187–201.
Quigley, J.M. (1998). "Urban diversity and economic growth". Journal of Economic Perspectives 12, 127–138.
Ravallion, M., Wodon, Q. (2000). "Does child labor displace schooling? Evidence on behavioral responses to an enrollment subsidy". Economic Journal 110, 158–175.

Ray, D. (1998). Development Economics. Princeton University Press, Princeton, NJ.
Romer, P. (1990). "Endogenous technological change". Journal of Political Economy 98, S71–S102.
Rosenzweig, M.R. (1991). "Labor markets in low income countries". In: Chenery, H., Srinivasan, T.N. (Eds.), Handbook of Development Economics, vol. I. Elsevier Science, Amsterdam.
Rosenzweig, M.R., Evenson, R. (1977). "Fertility, schooling, and the economic contribution of children in rural India". Econometrica 45, 1065–1079.
Rosenzweig, M.R., Schultz, T.P. (1982). "Market opportunities, genetic endowments, and intrafamily resource distribution: Child survival in rural India". American Economic Review 72, 803–815.
Rosenzweig, M.R., Schultz, T.P. (1983). "Estimating a household production function: Heterogeneity, the demand for health inputs and their effects on birth weight". Journal of Political Economy 91, 723–746.
Rosenzweig, M.R., Wolpin, K.J. (1986). "Evaluating the effects of optimally distributed public programs". American Economic Review 76, 470–487.
Rosenzweig, M.R., Wolpin, K.J. (1988). "Migration selectivity and the effects of public programs". Journal of Public Economics 37, 265–289.
Ruttan, V.W. (2001). Technology, Growth, and Development: An Induced Innovation Perspective. Oxford University Press, New York.
Sahn, D.E., Alderman, H. (1988). "The effect of human capital on wages, and the determinants of labor supply in a developing country". Journal of Development Economics 29, 157–184.
Sanchez, M.A., Orazem, P.F., Gunnarsson, V. (2005). "The effect of child labor on mathematics and language achievement in Latin America". In: Orazem, P.F., Sedlacek, G., Tzannatos, Z. (Eds.), Child Labor and Education in Latin America. In press.
Schultz, T.P. (1982). "Lifetime migration within educational strata in Venezuela: Estimates of a logistic model". Economic Development and Cultural Change 30, 559–593.
Schultz, T.P. (1990). "Women's changing participation in the labor force: A world perspective". Economic Development and Cultural Change 38, 457–488.
Schultz, T.P. (2001). "Women's roles in the agricultural household: Bargaining and human capital investments". In: Gardner, B., Rausser, G. (Eds.), Handbook of Agricultural Economics. Elsevier Science, Amsterdam.
Schultz, T.P. (2002). "Why governments should invest more to educate girls". World Development 30, 207–225.
Schultz, T.W. (1964). Transforming Traditional Agriculture. Yale University Press, New Haven, CT. Reprint edition: Arno Press, New York, 1976.
Schultz, T.W. (1975). "The value of the ability to deal with disequilibria". Journal of Economic Literature 13, 827–846.
Singh, I., Squire, L., Strauss, J. (Eds.) (1986). Agricultural Household Models. Johns Hopkins University Press, Baltimore, MD.
Sinha, J.N. (1967). "Dynamics of female participation in economic activity in a developing economy". In: Proceedings of the World Population Conference, Belgrade, 1965, vol. 4. United Nations, New York.
Smith, L.C., Haddad, L. (2000). Explaining Child Malnutrition in Developing Countries: A Cross-Country Analysis. International Food Policy Research Institute, Washington, DC.
Sokoloff, K.L., Engerman, S.L. (2000). "Institutions, factor endowments, and paths of development in the New World". Journal of Economic Perspectives 14, 217–232.
Stiglitz, J.E. (1987). "The causes and consequences of the dependence of quality on price". Journal of Economic Literature 25, 1–48.
Strauss, J. (1986). "Does better nutrition raise productivity?". Journal of Political Economy 94, 297–320.
Strauss, J., Thomas, D. (1998). "Health, nutrition and economic development". Journal of Economic Literature 36, 766–817.
Sunding, D., Zilberman, D. (2001). "The agricultural innovation process: Research and technology adoption in a changing agricultural sector". In: Gardner, B.L., Rausser, G. (Eds.), Handbook of Agricultural Economics, vol. 1A. Elsevier Science, New York, pp. 208–261.
Swamy, A.V. (1997). "A simple test of the nutrition-based efficiency wage model". Journal of Development Economics 53, 85–98.

Taylor, J.E., Martin, P.L. (2001). "Human capital: Migration and rural population change". In: Gardner, B., Rausser, G. (Eds.), Handbook of Agricultural Economics. Elsevier Science, Amsterdam.
Topel, R. (1999). "Labor markets and economic growth". In: Ashenfelter, O., Card, D. (Eds.), Handbook of Labor Economics, vol. 3C. Elsevier Science, Amsterdam.
Valdes, A.M., Andrew, T., Garnder, J.P., Klimura, M., Oelsner, E., Cherkas, L.F., Aviv, A., Spector, T.D. (2005). "Obesity, cigarette smoking, and telomere length in women". Lancet (June 14), 1–3. Available at: http://www.thelancet.com.
Wang, X., Taniguchi, K. (2004). "Does better nutrition enhance economic growth?". In: Taniguchi, K., Wang, X. (Eds.), Nutrition Intake and Economic Growth: Studies on the Cost of Hunger. Food and Agriculture Organization (FAO) of the United Nations, Rome, Italy.
Welch, F. (1970). "Education in production". Journal of Political Economy 78, 35–59.
Welch, F. (1999). "In defense of inequality". American Economic Review 89, 1–17.
Williamson, J.G. (1988). "Migration and urbanization". In: Chenery, H., Srinivasan, T.N. (Eds.), Handbook of Development Economics, vol. I. Elsevier Science, Amsterdam.
Wooldridge, J.M. (2002). Econometric Analyses of Cross Section and Panel Data. MIT Press, Cambridge, MA.
World Bank (1981). World Development Report, 1981. Oxford University Press, New York.
World Bank (2000). Entering the 21st Century. World Development Report 1999/2000. Oxford University Press, New York.
World Bank (2001). Engendering Development. Oxford University Press, Oxford.
Wozniak, G. (1993). "Joint information acquisition and new technology adoption: Later versus early adoption". Review of Economics and Statistics 75, 438–445.
Xu, B. (2000). "Multinational enterprises, technology diffusion, and host country productivity growth". Journal of Development Economics 62, 477–493.

Chapter 44

AGRICULTURAL EXTENSION*

JOCK R. ANDERSON and GERSHON FEDER

Agriculture and Rural Development Department, World Bank, Washington, DC

Contents

* The views expressed in this chapter are those of the authors, and do not necessarily reflect those of the World Bank. Needless to say, we have drawn unapologetically on much of the considerable World Bank experience with extension, including the work of many of our colleagues, most notably Gary Alex, Derek Byerlee, Ariel Dinar, David Nielson, Dina Umali-Deininger and Willem Zijp, but the interpretation is ours. In finalizing the chapter we have benefited from constructive comments by reviewers Carl Eicher and Wally Huffman, who we acknowledge gratefully with the usual disclaimer. Seniority of authorship is not assigned.

Handbook of Agricultural Economics, Volume 3
Edited by Robert Evenson and Prabhu Pingali

DOI: 10.1016/S1574-0072(06)03044-1

Abstract

In this chapter we analyze the considerations that lead policy makers to undertake extension investments as a key public responsibility, as well as the complex set of factors and intra-agency incentives that explain why different extension systems' performance varies. Accordingly, the chapter provides a conceptual framework outlining farmers' demand for information, the welfare economic characterizations of extension services, and the organizational and political attributes that govern the performance of extension systems. The framework is used to examine several extension modalities and to analyze their likely and actual effectiveness. Specifically, the modalities reviewed include "training and visit" extension, decentralized systems, "fee-for-service" and privatized extension, and farmer-field-schools. The chapter also provides a discussion of methodological issues pertaining to the assessment of extension outcomes, and a review of some of the recent empirical literature on extension impact. The chapter emphasizes the efficiency gains that can come from locally decentralized delivery systems with incentive structures based on largely private provision that in most countries will still be publicly funded. In wealthier countries, and for particular higher income farmer groups, extension systems will likely evolve into fee-for-service organizations.

Keywords

agricultural extension, training and visit system, fee-for-service system, extension impacts

JEL classification: H43, J43, O13, O47, Q16

1. Introduction

It is widely accepted that farmers' performance is affected by human capital, which encompasses both innate and learned skills, including the ability to process information [Jamison and Lau (1982)]. Extension services are an important element within the array of market and non-market entities and agents that provide human capital-enhancing inputs, as well as flows of information that can improve farmers' and other rural peoples' welfare; an importance long recognized in development dialog [e.g., Leonard (1977); Garforth (1982); Hazell and Anderson (1984); Jarrett (1985); Feder, Just and Zilberman (1986); Roberts (1989)]. The goals of extension include the transferring of knowledge from researchers to farmers, advising farmers in their decision making and educating farmers on how to make better decisions, enabling farmers to clarify their own goals and possibilities, and stimulating desirable agricultural developments [Van den Ban and Hawkins (1996)]. While extension agents often also provide services that are not directly related to farm activities (e.g., health, non-farm business management, home economics and nutrition), the focus of discussion in this chapter is on agricultural and farm management knowledge dissemination (which may include financial and marketing information).

The services provided by extension have significant public-good attributes. It is, therefore, not surprising that there are of the order of one-half million agricultural extension workers worldwide, and some 80% of the world's extension services are publicly-funded and delivered by civil servants [Feder, Willett and Zijp (2001)]. Universities, autonomous public organizations, and non-government organizations (NGOs) deliver about 12% of extension services, and the private sector delivers another 5%. There is a corresponding large volume of public budget allocated to extension activities (in 1988, for example, over six billion US dollars worldwide, likely rather less now in real terms). The regional distribution of workers in the developing countries and their budgets today is in need of updating but, based on Swanson, Farner and Bahal (1990, p. 56), it is something like what is depicted for public-sector employees in the un-numbered table that follows.

Developing region	Total public extension personnel ('000)
Latin America	28
Middle East–North Africa	34
Asia	277
Sub-Saharan Africa	57
Total developing countries	396

Source: Swanson, personal communication, 2005, with assistance of Kristen Davis (ISNAR Program of IPFRI), and adjustment for missing entries by Jock R. Anderson based on subjective judgment.

From a development-policy perspective, the investment in extension services or the facilitation of non-government extension, are potentially important tools for improving agricultural productivity and increasing farmers' incomes. More than 90% of the world's extension personnel are located in developing countries [Umali and Schwartz (1994)], where indeed the majority of the world's farmers is located. Yet, the record of extension impact on farm performance is, as we will review, rather mixed. The literature contains analyses indicating very high rates of return on extension investment, as well as documentation of cases of negligible achievements, implying a misallocation of public resources. Clearly, the format by which extension services are rendered, as well as the circumstances in which recipients of extension services operate, will affect the extent of the impact that is observed.

Productivity improvements are possible only if a differential exists between the actual productivity on the farms and what could potentially be produced with better know-how, subject as always, to farmers' preferences and resource constraints. In the past, rapid technological advances have created such a differential in many developing countries [e.g., Feder, Lau and Slade (1987)]. This productivity differential can be broadly classified into two types of "gaps": a technology gap and a management gap. The former might entail additional investment and higher recurring costs (e.g., for inputs such as seeds of improved cultivars or fertilizers) while the latter may offer the farmer a low-cost means of raising productivity by applying improved management practices [e.g., Byerlee (1988a, 1988b)]. These gaps are, in the first instance, a manifestation of the difference in the knowledge that farmers possess and the best-practice knowledge that exists at any point in time. Best practice is often, though not always, an embodiment of the latest science-based developments addressed to overcoming the limitations imposed by traditional technology and practices and thereby enhancing productivity. To realize their potential impact, however, the scientific advances must be aligned to the local agroecological and socioeconomic characteristics of the target areas.

Extension helps to reduce the differential between potential and actual yields in farmers' fields by accelerating technology transfer (i.e., to reduce the technology gap) and helping farmers become better farm managers (i.e., to reduce the management gap). It also has an important role to play in helping the research establishment tailor technology to the agroecological and resource circumstances of farmers. Extension thus has a dual function in bridging blocked channels between scientists and farmers: it facilitates both the adoption of technology and the adaptation of technology to local conditions. The first involves translating information from the store of knowledge and from new research to farmers, and the second by helping to articulate for research workers the problems and constraints faced by farmers.

The adoption of technology by farmers is inevitably affected by many factors [e.g., Feder, Just and Zilberman (1986); Sunding and Zilberman (2001)]. Adoption can be influenced by educating farmers about such things as improved varieties, cropping techniques, optimal input use, prices and market conditions, more efficient methods of production management, storage, nutrition, etc. To do so, extension agents must be capable of more than just communicating messages to farmers. They must be able to comprehend an often-complex situation, have the technical ability to spot and possibly diagnose problems, and possess insightful economic-management and risk-management [e.g., Anderson and Hazell (1994); Hardaker et al. (2004)] skills in order to advise on more efficient use of resources.

Effective extension involves adequate and timely access by farmers to relevant advice. However, while access to appropriate information is necessary to improve agricultural productivity, it is not sufficient. In general, farmers will adopt a particular technology if it suits their socioeconomic and agroecological circumstances. The availability of improved technology, access to "modern" inputs and resources, and profitability at an acceptable level of risk are among the critical factors in the adoption process. Further, farmers often get information from a number of sources. Public extension is one such source, but while it is not necessarily the most efficient, it is free of the conflict of interest that arises when private-sector suppliers of inputs are also the providers of agricultural information. Extension can increase the rate at which adoption occurs, but the extent and form that an extension service takes should be guided by considerations of cost-effectiveness and the nature of extension products. Thus, while extension, including that done in the public sector, can play an important role in improving the productive efficiency of the agricultural sector, the virtues and limitations of the alternative mechanisms need to be considered in assessing the cost-effectiveness of delivering information [e.g., Byerlee (1988a); Van den Ban (1999)]. These considerations are taken up in Sections 2 and 3 below.

While extension cannot be expected to be a single factor that can transform traditional agriculture, it usually has maximal impact at an early stage in the dissemination of, say, a new technology, when the informational disequilibrium (and the "productivity differential") is the greatest. At that stage, the perceived (necessarily subjective) risk of adopting new technology is high, as farmers do not have significant insights from others' experience [e.g., Feder and O'Mara (1982)]. Consequently, extension's role as decoder and transmitter of information from research is prominent, as noted by Huffman (1985). The decoding service provided by extension can substitute for farmers' education, and possibly also complement it [Huffman (1977, 1985, 2001a); Wozniak (1993); Evenson (2001)]. This view of extension has its roots in the insights of Schultz (1964, 1968, 1975) about traditional farmers being poor but efficient and their contribution to economic growth and their own escape from poverty largely coming from their being able to cope with disequilibria presented by the availability of new technology and new information. Over time, as increasing numbers of farmers become aware of a specific technological thrust, the impact of such extension diminishes, until the opportunity and need for more information-intensive technologies [Byerlee (1988b)] arise. The dynamic resolution of the information disequilibria associated with specific extension "messages" makes observing the impact of extension difficult. At the same time, the uneven flow of benefits from any particular extension message has significant implications from a policy and program design point of view [e.g., Simmonds (1988)]. The cost-effectiveness of information delivery at a given point in time should thus be established in the light of current and future benefits and costs in order to justify the marginal resources allocated to delivering the information. Aspects of these perspectives are pursued in Section 4.

Market distortions and infrastructural bottlenecks further affect the adoption of new technology and can help or hinder the effectiveness of extension services. Again, from an operational point of view, the cost-effectiveness of delivering messages must be considered within the prevailing policy and market environment. A restrictive environment has a high opportunity cost in terms of foregone benefits from extension advice, creating a divergence between potential and actual benefits. The prevailing policy regime thus has potentially important implications for an appropriate sequencing of policy interventions and program design.

The wider context of extension services, defined broadly as the rural knowledge and innovation system, was instructively overviewed by Alex et al. (2002), who argued that such services are key to informing and influencing rural household decisions. Unfortunately, rural areas usually lag behind urban areas in their access to information, and developing countries generally lag behind more developed countries in this regard. Such lags jeopardize the ability of rural people to realize their full potential and improve their economic, social and environmental conditions. Rural information services are, they argued, key to unleashing the potential of rural peoples and enabling them to change their living situations and bring about sustainable rural development.

We endeavor to analyze here the considerations that lead policy makers to undertake extension investments as a key public responsibility, as well as the complex set of factors and intra-agency incentives that explain why different extension systems' performance varies. The variation in extension outcomes is demonstrated in a review of the empirical results of studies focusing on extension effects. Accordingly, Section 2 provides a conceptual framework outlining farmers' demand for information, the welfare economic characterizations of extension services, and the organizational and political attributes that govern the performance of extension systems. Section 3 examines several extension modalities and analyzes their likely and actual effectiveness. This is followed in Section 4 by a discussion of methodological issues pertaining to the assessment of extension outcomes, and a review of the empirical literature on extension impacts. The final Section 5 highlights the conclusions.

2. Conceptual frameworks

2.1. *Information as an input to productivity growth: Demand for information*

Putting aside farming as a way of life, running a farm business can be thought of as deliberate management of diverse inputs – land, labor, physical capital of many types, and not to be forgotten, information – for producing outputs of value that can be consumed or traded to enhance the welfare of the dependent household. Extension as broadly conceptualized in this chapter is focused on the delivery of the information inputs to farmers. Information can be of many types, ranging from anticipated future prices for farm products, to new research products such as improved crop cultivars, to knowledge about techniques involved in using particular inputs, such as timing and intensity of use of fertilizer [e.g., Byerlee (1988b)]. As a productive input, farmers thus have a demand for information and, depending on how productive it is perceived to be, may be prepared to pay for it as they would for other purchased inputs [e.g., Dinar (1996)].

Yet information is a rather special type of input in many respects. Some information will have quite enduring value, such as when transferred managerial skills are encapsulated in the human capital of the farm manager, and such values are generally increasing over time as more complex and increasingly integrated managerial challenges are faced. At another extreme, some information may have quite ephemeral value, such as a forecast of tomorrow's wheat price in a local market. At an intermediate level, the value of input management information for a particular cultivar is likely as obsolescent as the cultivar itself. Clearly, different types of information can thus have many different inherent valuations to concerned farmers. In some cases, especially where the consequences of using the particular information include environmental outcomes, such as reduced soil erosion that might come with adoption of no-till farming [Pieri et al. (2002)], or with reduced overuse of fertilizer nitrogen [Byerlee (1988b)], the value of the information may go to many beneficiaries beyond the farm gate.

It is not surprising then that the delivery systems for supplying information can have diverse values to different client farmers, so getting a handle on the value of extension to farmers is not a trivial task, which may explain why it has so seldom been tackled. The task is made more challenging by the multitude of alternative suppliers of information; from friends and neighbors, to input supply firms and specialized consulting services, to media, to a government extension service. The complexity of the situation is instructively illustrated by Gautam (2000, p. 3) in his Figure 1, reproduced here with permission as Figure 1.

Taken together, the information delivery systems supporting farming should constitute something of a growth industry if, as is regularly argued by agricultural analysts, farming is becoming more information intensive [Byerlee (1988b)]. How suppliers meet the demands surely varies greatly around the world, depending on market and institutional conditions. Gautam (2000), for instance, concludes that there is a significant unmet demand in Kenya for general agricultural extension services. Just how different types of information are best delivered depends crucially on (a) the nature of the information concerned, a topic taken up in the following section, and (b) the type of farmer.

2.2. *Welfare economics contextualization*

The world of Adam Smith's perfect markets is seldom to be found in the environment in which most rural dwellers operate, especially those in the developing countries. The necessary conditions for such perfection include rivalry, excludability, appropriability, symmetric information, complete markets with no distortions or externalities, as is so effectively reviewed in the context of agricultural extension by Hanson and Just (2001). They appraise the extent of market failures along this spectrum for the case of farming in

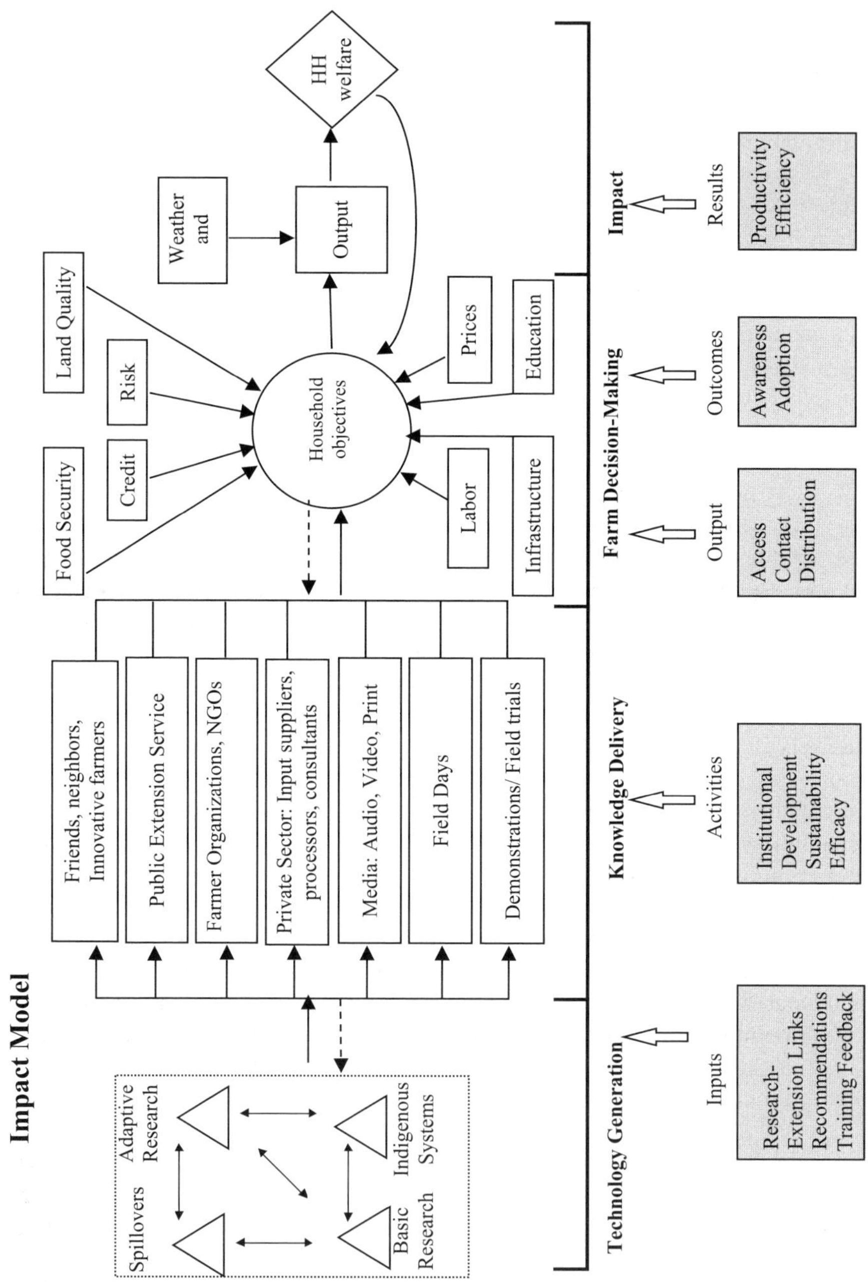

Figure 1. Scheme for an impact evaluation of extension [Gautam (2000)].

Table 1
Extension products by the nature of economic characteristics of information [based on Umali and Schwartz (1994, Figure 3.2, p. 24)]

Rivalry/ subtractability	Excludability	
	Low	High
Low	Public goods • Mass media information • Time insensitive production, marketing, and management information of wide applicability	Toll goods • Time-sensitive production, marketing, or management information
High	Common pool goods • Information embodied in locally available resources or inputs • Information on organizational development	Private goods • Information embodied in commercially available inputs • Client-specific information or advice

Maryland but their diagnosis of the prevalence of such failures surely applies to many if not most farming situations around the world. Several of the departures from perfection that they identify are returned to in Section 3 when we consider mechanisms that have been proposed for overcoming some of the problems of providing largely public-good extension products.

It has become almost standard to focus particularly on the first two elements of possible market failure in considering whether extension services are mainly public or mainly private goods based on a distinction using the principles of excludability and rivalry [e.g., Umali and Schwartz (1994)]. Excludability occurs when farmers who are not willing to pay for a service can be excluded from its benefits, such as tailor-made farm management advice. Rivalry occurs when one farmer, by using advice, reduces its availability to others, such as services embodied in commercial products. Rivalry and excludability are high for private goods and low for public goods. Other services are toll goods, characterized by high excludability and low rivalry, when some farmers can be excluded from access, even though their value to users is not diminished by use by others; or common pool goods, characterized by low excludability and high rivalry (Table 1). As noted in Section 2.1, the value of information may be influenced by time and place, as for example, market information that decreases in value as the information becomes more widely disseminated and markets adjust, or weather forecasts that have zero value after the event.

Knowledge delivered by extension may be information embodied in inputs or equipment (e.g., seed of improved cultivars or machinery) or more abstract, disembodied information on agricultural practice. Information embodied in inputs or equipment has high rivalry and tends to be a private good when the input or equipment must be purchased and a common pool good when the input is locally available. There are two broadly applicable types of disembodied agricultural information: general, non-excludable information (e.g., market information, cropping patterns, etc.), which tends to be a public good, and specialized, excludable information (e.g., fertilizer recommendations for a specific field or farm operation), which tends to be a toll good [Umali-Deininger (1996)].

The diverse types of knowledge and information can be provided by the public or private sector, or by NGOs, another often-important category of players in service provision. Different mechanisms are available for coordinating the supply of services – private-sector markets, public-sector hierarchies with state authority, and collective action by non-government groups [Picciotto (1995); Wolf and Zilberman (2001)]. The characteristics of an information service influence whether it is best supplied by the private, non-government (not-for-profit), or public sectors [Schwartz and Zijp (1994); Umali-Deininger (1996, 1997)]. Some implications of these observations drawn out by Picciotto and Anderson (1997) are that:

- Information closely associated with market goods (e.g., purchased inputs) is generally best left to the private sector;
- Information associated with toll goods can be effectively provided by combined public- and private-sector efforts;
- Information relative to management of common pool goods (forests, common grazing lands, water when it is not already subject to quota rules) is usually best provided by cooperative or other non-government institutions; and
- Only when market and participation failures are high should information provision be financed by the public sector and, even in these cases, the public sector might well finance private service delivery.

2.2.1. *Private extension services and cost recovery*

The private-good nature of many extension services has raised interest in privatizing extension services [e.g., Cary (1993, 1998); Lindner (1993)]. Indeed, as Vernon Ruttan has reminded us, this theme takes us back to the initial formal extension efforts in the US Mid-West, when the Farm Bureaus hired county extension agents to provide the information services they demanded. In reality now, most information services are provided outside of government, and farmers see public extension as only one option – perhaps even a last resort – in obtaining needed information services. The government has, however, a major role in establishing policies and programs to encourage development of private extension services, along with continued sustenance in some cases, and extension systems need to be designed with the understanding that they will be cost-effective only "if the public role is defined so as to complement what the private sector can and will fund and deliver" [Beynon et al. (1998, p. 135)].

Private consulting or advisory services generally address needs of commercial farmers. Developing private services for small-scale farmers often necessitates public investment to develop capacities of service providers and establish markets for services. Veterinarians and para-vets have pioneered private service provision in some countries

[Umali, Feder and de Haan (1994); de Haan et al. (2001)] and, in crop agriculture, pest control services present the same opportunities for private service delivery. Contracting schemes are another private-sector mechanism for providing services to small-scale farmers [Mullen, Vernon and Fishpool (2000); Rivera and Zijp (2002)]. The potential for conflict of interest in such arrangements may warrant a public regulatory and monitoring function backed up by public information, for quality checking on information supplied.

User financing mechanisms are a means of obtaining private financing to cover at least a portion of the cost of public extension services. Mechanisms include levies, direct user charges, or subsidies for services procured by users. Levies are most easily assessed on commercial crops with a highly centralized marketing system and a limited number of processors. User charges are more feasible for highly commercial operations, for more sophisticated producers, and for services that provide a clear and immediate benefit. Latin America has seen extensive experimentation with co-financing and private extension service provision [e.g., Keynan, Manuel and Dinar (1997); Dinar and Keynan (2001); Berdegué and Marchant (2002); Cox and Ortega (2004)], and small-scale farmers in various countries have indicated a willingness to pay for extension services that meet their needs [e.g., Gautam (2000); Holloway and Ehui (2001)]. A possible caveat to private user-pays extension is that, when farmers pay for extension information, they may be less willing to share that information freely with neighbors [Van den Ban (2000)]. This may significantly slow the spread of innovation. Producers may also want less intense service provision than is sometimes offered by public agencies [Gautam (2000)]. Practical issues that emerge in such changing private-public provision of services include an effective crowding out of public provision to the more remote clients when, by losing much of their traditional core business, such public providers incur diseconomies of size and scope (such as for training) for the provisioning task they are left with [Hanson and Just (2001)].

2.2.2. *Public financing of extension*

Public investment in extension is justified when the general public benefits more than the extension client, when government can provide services more cheaply or better, when extension services directly facilitate other programs, or when the private sector does not provide needed services [Van den Ban (2000)]. These conditions apply when there are positive externalities to innovation or market failure in service provision. Market failure is often due to: unorganized demand (small-scale farmers do not recognize potential benefits, have limited purchasing power, and are not organized to access services) or unorganized supply (few individuals or institutions are capable of providing technical services or there is limited opportunity for private firms to charge for provision of easily disseminated information). The most important externalities are: positive environmental [e.g., Byerlee (1988a); Mullen, Vernon and Fishpool (2000)] and health (human, livestock and crop) impacts of appropriate technology use; improvements in political stability and poverty reduction resulting from improved equity in access to information;

and improved national security, economic development and food security resulting from increased agricultural productivity, competitiveness and sustainability [e.g., Thirtle, Lin and Piesse (2003)]. Consumers often benefit more from increases in productivity than do farmers.

Despite the fact that public financing for extension services is often justifiable, the general trend toward fiscal restraint and a reduced role for the public sector has led to financial crises in many extension services. Two general options for improving financial sustainability of public extension involve scaling back public programs or improving cost-effectiveness [Beynon et al. (1998)]. Scaling back public programs might involve: reducing coverage to specific target farmer groups, reducing intensity of coverage (less frequent visits, fewer services), devolving service provision to private organizations or requiring cost sharing by users [Wilson (1991)]. State withdrawal from service provision might entail total abandonment of some programs or shifting of service responsibilities to others – requiring commercial farmers to arrange their own services; encouraging producer organizations to provide services; or promoting private extension by input suppliers (notwithstanding potential conflicts of interest in the content of advice), produce buyers, NGOs, environmental groups, or others. Improving cost-effectiveness can be achieved through improvements in program management, targeting and priority setting, and choice of appropriate extension delivery methods (e.g., greater use of mass media).

Sustainability of an extension service depends crucially on its ability to provide benefits and generate support from internal and external stakeholders [Gustafson (1994)]. Improving efficiency and quality of service provision and client involvement in setting priorities help to generate needed support. True farmer ownership of programs (often alluded to under the rubric of "empowerment" in contemporary development dialog) adds significantly to program sustainability [Scarborough et al. (1997)].

2.2.3. *Public–private partnerships*

There is growing recognition that, even where public financing of extension is justified, private service delivery is often more efficient in serving clients. This leads to strategies for contracting extension services – delinking funding from service delivery. Contracted extension strategies take many different approaches to division of responsibilities for financing, procurement, and delivery of services, but most reforms involve public funding for private service delivery [Rivera, Zijp and Alex (2000)]. Competitive contracting instills a private-sector mentality of cost-consciousness and results-orientation, even in public institutions too when they are forced to compete in providing services.

Contracted extension systems seek to reduce costs and improve cost-effectiveness of public extension services, but most current reforms go further and attempt to draw on private-sector funding to improve financial sustainability of extension. Table 2 illustrates the alternative arrangements possible in public and private financing and provision of extension services. These include the traditional public-sector extension services,

Table 2
Some alternatives for public–private financing and provision of extension services [according to Alex et al. (2002)]

Service provision	Finance provision: Public	Private (farmers)	Private (other)
Public	• Traditional extension	• Fee-for-service extension	• Contracts with public institutions
Private	• Subsidies to extension service providers • Publicly-financed contracts for extension services	• Commercial advisory services • Sale of newspapers, magazines	• Information provided with sale of inputs • Extension provided to contract growers • Advertising in newspapers, radio, television, magazines

fully private services, and public–private partnerships involving some type of contractual relationship.

The economic rationale for farmers to pay for extension services is generally clear and the trend toward such user payment is well established in OECD countries [e.g., Hone (1991); Marsh and Pannell (2000); Marsh, Pannell and Lindner (2003)]. In developing countries, many producers are unable or unwilling to pay for services, as they have not seen examples of effective, responsive extension.[1] Another constraint limiting private extension is that many countries have few extension service providers outside the public sector. Furthermore, few public institutions have incentives and institutional arrangements in place to encourage program cost-recovery. We return to these several issues in Section 3.3.

2.3. A conceptual framework for analyzing extension organizations

Earlier sections established the fact that many aspects of extension work entail strong public-good characteristics and other market failures that are not easy to overcome through taxes, subsidies and regulatory interventions. It is thus not surprising that public provision of extension services (whether by central or regional governments) has been common in most countries, at least at some stage of their history. While there have been some notable successes, it has also been observed, quite often, that public extension systems demonstrate weaknesses hampering their effectiveness. A worldwide review by Rivera, Qamar and Crowder (2001, p. 15) refers to extension systems as "failing" and "moribund", being in a state of "disarray or barely functioning at all". Others have made similar observations in the past [e.g., Kaimowitz (1991); Ameur (1994)]. It is conceivable that there are some generic and universal difficulties in the operation of public

[1] Chile's twenty five years of evolving privatized extension demonstrates a number of the challenges faced [Berdegué and Marchant (2002); Cox and Ortega (2004)].

extension systems, and in the typical bureaucratic-political environment within which they are budgeted and managed.

This hypothesis has been propounded by Feder, Willett and Zijp (2001), who suggest eight interrelated frequently encountered factors affecting the performance of public extension systems, causing the observed manifestations of deficient performance, low staff morale and financial stress. The identification of these factors enables an analysis of the conduct of different levels of extension personnel, which in turn determines the performance of the system as a whole. The approach also helps in analyzing the underpinnings of different organizational forms, as well as in predicting their likely performance. These characteristics of public extension systems are considered here under the eight headings that follow.

2.3.1. Scale and complexity

In countries where the farm sector comprises a large number of relatively small-scale farmers (as is common in most developing countries), the clients of extension services live in geographically dispersed communities, where the transport links are often of low quality, adding to the cost of reaching them. The incidence of illiteracy and the limited connections to electronic mass media can further limit the ability to reach clients via means that do not require face-to-face interaction (e.g., written materials, radio, television, Internet).

Thus, the number of clients who need to be covered by extension is large, and the cost of reaching them is high. The challenge is complicated further by the fact that farmers' information needs vary even within a given geographical area due to variations in soil, elevation, microclimate and farmers' means and capabilities. The large size of the clientele (all of whom are entitled to the public service in the common case of free extension) inevitably leads to a situation where only a limited number of farmers have direct interaction with extension agents. Since direct contacts are rationed, agents often exercise selectivity as to which farmers they interact with, and the selectivity often manifests preference for larger, better endowed, and more innovative farmers, who can provide some in-kind payment, as well as reflect better performance [Axinn (1988); Feder and Slade (1993)]. This sort of supply-side rationing is exacerbated by self-selection on the part of farmers, where those with a higher value (larger demand) for information tend to be large-scale farmers, with better opportunities to take advantage of information.

This selectivity of contacts has ramifications in terms of the likely extent of diffusion of information through farmer-to-farmer communications. As argued by Rogers (1995), information is more likely to flow from opinion leaders to followers if the former are not excessively superior in socioeconomic status to the would-be followers. Because those who tend to receive more extension contact are often not typical of the farming population, there is often a lesser inclination of other farmers to follow the example of contact farmers, or to seek advice from them (in spite of some contrary positive experiences, such as in Israel [Keynan, Manuel and Dinar (1997)]). This reluctance thus often diminishes the potential impact of extension services across the farm population. On the supply side, the reaction to the large clientele is the deployment of large numbers of agents, which presents a management challenge for national organizations or organizations dealing with large geographical-administrative units (e.g., states or provinces within a federal system). In organizations with a large number of field personnel, there is a tendency to adopt a hierarchical centralized management system, so as to facilitate the monitoring of the large and dispersed field-level labor force. The large and hierarchical bureaucracy is characterized by a top-down management style and is thus not conducive to participatory approaches to information delivery and priority setting [e.g., Waters-Beyer (1989); Fleischer, Waibel and Walter-Echols (2002)]. Furthermore, the many layers in the hierarchy distance the decision making from the field level and lead to suboptimal decisions.

2.3.2. *Dependence of extension on the broader policy environment*

The effectiveness of extension work is crucially dependent on complementary policy and institutional actions on which it has very limited influence. Thus, limiting factors such as credit, input and seed supplies, price incentives, marketing channels and human resource constraints determine the impact of the information that extension agents convey to farmers. While extension agents can adjust their advice, given the overall policy climate, the value of the information is diminished when the terms of trade are tilted against agriculture, rural infrastructure investment is inadequate, and farmers have irregular input supplies due to absent input markets [Axinn (1988)]. The coordination between agencies that influence these complementary factors and extension management is costly and difficult, and the degree of leverage that can be brought by extension is minor. The negative implications of this situation are particularly pronounced when one examines the poor record of linkages to the knowledge generation system, especially the national agricultural research system [e.g., Ewell (1989)], which is examined separately next, given its importance for extension performance.

2.3.3. *Interaction with knowledge generation*

In contrast to the situation in the US, where the cooperative extension service is embedded in the university system, the information on which extension advice is based in most developing countries is not generated within the extension organization itself but rather largely in separate systems (national agricultural research institutes and universities, and increasingly also private research firms), under separate management structures and subject to incentive systems where extension opinions and priorities often do not carry a significant weight. Because the performance indicators for research systems are often related primarily to recognition within the scientific community, the areas of priority are not necessarily aligned with what extension managers perceive as priorities, given their farm-level feedback [Kaimovitz (1991)]. Furthermore, many national agricultural research systems are weak, due to under-investment and distorted incentives, thus providing a poor foundation of technology knowledge for extension [Purcell and Anderson (1997); Swanson (2004); Avila and Evenson (2005)].

Public research and extension organizations often compete for budgets (as they are commonly located within the same ministry). Researchers typically enjoy a higher status (they are often better educated and have greater independence), and this produces tensions in the interactions between research managers and extension, which is not conducive to coordination and to a two-way feedback. The outcome is detrimental to extension effectiveness, as the information available to agents may not be specifically tailored to the problems faced by farmers, given their resource constraints [e.g., Mureithi and Anderson (2004) on the situation in Kenya]. A review in the World Bank of a large portfolio of extension projects [Purcell and Anderson (1997)] pointed out that research–extension linkages were generally weak, and neither research nor extension was sufficiently conscious of the need to understand the constraints and potentials of the different farming systems as a basis for determining relevant technology and technology development requirements. Consequently, the inadequate research–extension links and poor technology foundation led to adverse outcomes in a large proportion of the projects reviewed, and claims of insufficient relevant technology were frequently found. More recent World Bank operations have naturally built on the lessons of experience, so the contemporary landscape of extension-type interventions (including support for business development services assisting small and medium enterprise) differs greatly from that of earlier decades.

2.3.4. *Difficulty in tracing extension impact*

Because many factors affect the performance of agriculture in complex and contradictory ways, it is difficult to trace the relationship between extension inputs and their impact at the farm level. This difficulty, in turn, exacerbates other inherent problems related to political support, budget allocation, incentives of extension employees, and their accountability, both upward (to the managers) and downward (to their clients).

The evaluation of extension impact (Section 4) involves measuring the relationship between extension and farmers' knowledge, adoption of better practices, utilization of inputs, and ultimately farm productivity and profitability and the related improvement in farmers' welfare. But farmers' decisions and performance are influenced by many other systematic and random effects (prices, credit constraints, weather, other sources of information, etc.), and thus ascertaining of the impact of extension advice to farmers requires fairly sophisticated econometric and quasi-experimental methods. The decision makers who allocate funds, and even the direct extension managers, face great difficulties in assessing the impact of extension and in differentiating it from other contributing factors, or making allowances for the effects of counterveiling factors.

Given the difficulty in relating cause to effect, extension input indicators are often adopted as "performance" criteria, as they are cheaper and simpler to establish [Axinn (1988)]. Thus, the volume of contacts, numbers of agents, numbers of demonstration days, etc. are used to judge whether extension is "effective" or not. These, of course, are not necessarily indicative of the quality and relevance of the knowledge conveyed.

The inability to attribute impact and thus assess performance has adverse impact on the incentives of extension staff to exert themselves in outreach to farmers. The motivation to train and update knowledge is hampered too (as the improved performance that such training brings cannot be observed). Time is spent on collecting and reporting input indicators, as these are easier to obtain. There are some other perverse outcomes that result from the adverse impact on incentives, which are discussed below. All of these are likely to produce lower quantity, as well as deficient quality, in extension work.

2.3.5. Weak accountability

As in any public bureaucracy, extension personnel are accountable to the managerial cadres, but because the effectiveness of their activities cannot be easily established, their performance is measured in terms of input indicators that are easy to provide and confirm. The field staffs are thus practically not accountable for the quality of their extension work, and often even the quantity can be compromised with impunity. The higher level managers are nominally accountable for extension performance to the political level but, due to the same impact attribution problems, the extension system's performance is monitored in terms of budgets, staff levels, and other bureaucratic, rather than substantive, indicators. As is common in other large bureaucracies that are fully publicly funded, the accountability to the clientele (i.e., to the farmers) is only nominal, as typically there is neither a mechanism, nor incentives, to actually induce accountability to farmers [Howell (1986); Farrington et al. (2002)]. This is ironic, as the farmers are the only ones who can relatively easily observe the quality and effectiveness of the extension service they receive. In the absence of mechanisms to implement accountability to farmers (which would improve the effectiveness of extension), incentives are distorted. Non-extension activities, for which extra remuneration can be earned, such as promotion of certain inputs for which a commission can be secured, or intermediation in the acquisition of credit (e.g., assistance in filling forms), are undertaken by agents, as the amount of extension time diverted to these tasks cannot be easily detected. If such tasks are formally extension agents' responsibilities (as they are in some systems), they will tend to get higher priority than do information dissemination duties [Feder and Slade (1993)].

Earlier extension projects yielded evidence of accountability failures in many cases [e.g., Farrington et al. (2002)]. Little attention was given to the introduction of systematic participation by the farming community in problem definition, problem solving, and extension programming. In more than one-half of the projects reviewed in a World Bank retrospective, an "entrenched top-down" attitude by staff was noted, and, not surprisingly, three-quarters of failed extension projects were characterized by such conduct [Purcell and Anderson (1997)].

That this pattern of behavior has been so common in both more- and less-developed countries, and is derived from a common distorted incentive system, is evident from the comments of Hercus (1991, p. 25), characterizing the New Zealand extension service prior to its reforms, as an operation where the budget used was accounted for in terms of "activities, not results, and concerned almost exclusively with expenditure and hardly at all with outputs or efficiencies. The mandate of extension was derived by the ... service itself, and in the absence of any challenge or alternative definition by the taxpayers' representatives, the service regarded its charter as the right to exist on the prevailing terms and conditions."

2.3.6. Weak political commitment and support

Urban-bias and the weak standing of farmers in poor countries as an interest group has traditionally made agriculture a weaker contender for public investment resources in countries where agriculture is a large sector [e.g., Olson (1971, 1986); Binswanger and Deininger (1997)]. But even given this situation, extension tends to be a less powerful claimant for budgets. The review of extension operations assisted by the World Bank [Purcell and Anderson (1997)] pointed out that, in nearly one-half of the projects examined, lack of commitment and support by senior government officials adversely affected implementation and funding. Indeed, the failure to allocate funds is a key indicator of weak conviction by senior decision makers and, as reported by Umali-Deininger (1996), an overwhelming majority of extension projects in her review recorded inadequate operating funds. Feder, Willett and Zijp (2001) posit that a plausible reason for the lack of adequate support (and the resulting limited funding) by politicians and senior officials is the inability to derive political payoff that can be earned from a public outlay that has a visible impact (e.g., the double cropping that will follow from an irrigation investment, or the reduction in transport cost due to a bridge). Such a payoff cannot be obtained from an expenditure that has an unclear cause-effect nature, such as has sometimes been said of extension. In addition, it is possible that awareness of deficient accountability, and an overall impression of ineffectiveness, deter policy makers from allocating budgets to extension services.

2.3.7. Encumbrance with public duties in addition to knowledge transfer

Because the extension service typically has a large number of public servants functioning at the rural community level, governments are often inclined to utilize extension staff for other duties related to the farming population. Such duties include collecting statistics, administering loan paperwork, implementing special programs (e.g., erosion control), performing regulatory duties, and dealing with input distribution (for government-provided inputs) [Feder and Slade (1993)]. The assignment of input distribution functions to extension agents is quite common, and is often motivated by the absence, or poor functioning, of private input markets.

Many of the non-extension duties are easier to monitor by supervisors than the information dissemination function, as there are clear and quantifiable performance criteria (e.g., the number of loan applications returned or the submission of statistics reports). Consequently, extension workers naturally place greater attention on the accomplishment of these duties. Furthermore, there may be an extra monetary incentive in performing these other duties (such as input distribution) as some rents can be derived from handling services that have a clear cash value to the recipient farmer. The allocation of an inordinate amount of an extension agent's time to these duties, at the expense of time for technological information dissemination, can go undetected because the outcome of the core extension duty is so difficult to attribute, and because accountability to farmers is deficient. Swanson, Farner and Bahal (1990) estimated a diversion of as much as 25% of the education effort. Such patterns of behavior will tend to reduce the productivity impact of extension, and, over time, may exacerbate the image of ineffectiveness.[2]

2.3.8. *Fiscal sustainability*

Some of the preceding characterizations of public extension systems lead to persistent funding difficulties. The public-good nature of many extension services makes cost recovery at the individual beneficiary level difficult. The dependence on public funding, in turn, is problematic because weak political commitment implies lower budgets, relative to the large clientele that needs to be served. The image of ineffectiveness and of unenforceable accountability is possibly another reason for the reluctance to direct large budgets to extension. As pointed out by Howell (1985), a cyclical pattern may be observed, whereby, in years when budget is relatively large (such as when a foreign donor infuses funds for extension), large numbers of staff are recruited, imposing a large fixed cost on the extension service (public employees typically are tenured). When budgets dwindle, the fixed staff costs claim a large share of available funds, and field operations are curtailed (as they require funds for transport and living expenses), as well as other recurrent costs (vehicle maintenance, replacement of agents' modes of transport, etc.). The scaling down of field operations reduces not only the quantity of extension inputs, but also their quality, as the extent of feedback from farmers is reduced, and thus timely follow-up on farmers' issues is hampered.

References to fiscal inadequacy, and the consequent unsustainability of extension operations, are common in the extension literature [e.g., Howell (1985); Röling (1986); Ameur (1994); Feder, Willett and Zijp (2001); Hanson and Just (2001)]. Purcell and Anderson (1997) cited funding shortfalls as such a common phenomenon that over 70% of the extension projects in their sample of Bank-supported operations faced "unlikely" or "uncertain" sustainability. More recently this theme has come up for critical attention in the wider development literature [e.g., Kydd et al. (2001)].

[2] Nonetheless, there may be circumstances where a crisis situation, or some other high-priority assignment, could override the importance of extension duties. For example, the onset of the AIDS epidemic could have justified the diversion of extension staff into AIDS education activities in some countries.

3. Alternative extension modalities to overcome generic weaknesses

The discussion above provides a framework consistent with many "stylized facts" regarding the structure, operations, and performance of public extension systems. In this section we utilize this framework to analyze a number of specific formats of extension operations that have emerged in the past three decades. These newer approaches, which depart from the traditional public service models as described in Section 2.3, entail institutional innovations and reforms, often pluralistic [e.g., Anderson, Clément and Crowder (1999); Anderson (1999); FAO/WB (2000)], where specific design features reflect attempts to overcome some of the weaknesses inherent in the public extension systems of recent decades.

3.1. Training and visit (T&V) extension

The T&V model of extension organization was promoted by the World Bank between 1975 and 1995 as a national public extension system, with application in more than 70 countries [Anderson, Feder and Ganguly (2006)]. The system's designers stressed the following features [Benor and Harrison (1977)]: (i) a single line of command, with several levels of field and supervisory staff; (ii) in-house technical expertise, whereby subject matter specialists are to provide training to staff and tackle technical issues reported by field staff; (iii) exclusive dedication to information dissemination work; (iv) a strict and predetermined schedule of village visits within a two-week cycle where contacts are to be made with selected and identified "contact farmers"; (v) mandatory bi-weekly training emphasizing the key set of messages for the forthcoming two-week cycle; (vi) a seasonal workshop with research personnel; and (vii) improved remuneration to extension staff, and provision of transport (especially motorcycles and bicycles). It is evident that the T&V design attempts to tackle directly or indirectly some of the weaknesses highlighted above. But as we will argue, some of the modifications exacerbated other weaknesses, and the ultimate result was a widespread collapse of the structures introduced.

The problems of scale and complexity were tackled by heavy reliance on officially selected contact farmers within an identifiable farming group. By working with a small number of contact farmers (six to eight per group of about 100), agents were to maximize coverage. But the required staff–farmer ratios implied a significantly larger extension staff, and thus the costs of T&V extension systems were higher by some 25–40% than the systems they replaced [Feder and Slade (1993); Antholt (1994)]. This made T&V extension more dependent on public budget allocations. The design intended to tackle the accountability issue by improving management's ability to monitor staff activities, taking advantage of the strict visit schedule, the identifiable contact farmer, and the intensive hierarchy of supervisory staff. This would have indeed provided incentives for compliance with expectations regarding the quantity of service delivered. The monitorable daily activities schedule also eliminated much of the ability to divert time to activities other than information dissemination (which were formally removed from extension duties). But the quality of extension service was not practically monitorable and, ultimately, managers and policy makers could not observe the impact of extension. The lack of accountability to farmers was not resolved. The interaction with research was improved through the seasonal meetings but, in practice, little influence was gained regarding the setting of research priorities, and certainly the weakness of many national research systems could not be corrected through extension projects.

Several features of the design could not stand up to practical realities, however. The "contact farmer" approach was often replaced by a "contact group" approach because biases in the selection of contact farmers (universally observed due to extension agents' incentives) led to diminished diffusion. The strict bi-weekly visit schedule could not be maintained because often there were no important new messages that needed to be conveyed, and the farmers had limited interest in frequent visits. The consequences for extension impact were apparently negative. While a study by Feder, Lau and Slade (1987) showed a positive impact on yields in Haryana (India) three years after project initiation, studies in Pakistan [Hussain, Byerlee and Heisey (1994)] and in Kenya [Gautam (2000)] indicated no significant impact after a longer period.

Many observers, including early skeptics such as Moore (1984), agree that the single most crucial factor that eventually brought about the dismantling of the T&V extension system was the lack of financial sustainability, a generic problem made worse by the high cost of the system. As the ability to demonstrate impact was not improved, there was no significant change in the political commitment to support extension, and, in country after country, even in long-faithful India, once the World Bank ceased funding (assuming that the new system has been "mainstreamed"), the local budget process implied a return to the smaller funding levels of the past.[3] With lower funding, the T&V system could not be sustained and hard-pressed governments have struggled with downsizing options, in some cases supported directly by bilateral donors, inevitably coupled with other extension reforms [e.g., Sulaiman and Hall (2002)].

3.2. Decentralization

The decentralization of extension services retains the public delivery and public funding characteristics of traditional centralized extension, but transfers the responsibility for delivery to local governments (district, county, etc.). Several Latin American governments undertook this approach [Wilson (1991)] in the 1980s and 1990s, and it is being initiated in African countries such as Uganda [e.g., Crowder and Anderson (2002)]. The main expected advantage of the approach is in improving accountability, as agents become employees of local government, which (if democratically elected) is keen on receiving positive feedback on the service from the clientele-electorate [Farrington et al.

[3] World Bank willingness to continue promoting the T&V approach dissipated as well in the 1990s, as it became evident through reports of the Operations Evaluation Department that such projects will not be sustainable and no sustained productivity gains could be verified. The internal debate within the Bank started in the early 1990s but there was then only limited conclusive evidence on sustainability and impact.

(2002); Swanson (2004)]. This was expected to improve extension agents' incentives, and induce better service. Improved management capacity is another advantage, as the scale of the operation is reduced for each decision-making unit [Swanson (2004)]. Some advantages may also be realized in coordinating extension advice with activities of other agencies, as presumably the costs of coordination are lower for local agencies operating in a smaller geographical area. Political commitment may be stronger as well since the clientele is closer to the political leadership, and this can further lead to improved fiscal stability [Swanson (2004)].

But decentralized extension agencies also face a multitude of additional problems. There is greater potential for political interference and utilization of extension staff for other local government duties (including election campaign activities). Economies of scale in training and the updating of staff skills can be lost. Similarly, extension–research linkages are more difficult to organize. Analysis [Garfield, Guadagni and Moreau (1996)] of Colombia's experience with the decentralization of extension confirms these concerns, and documents a significant increase in the aggregate number of staff (and thus in aggregate costs). Issues of financial sustainability may, therefore, not have been resolved, but merely transferred to the local level.[4]

A related reform was the devolution of extension functions to farmers' associations, rather than to local governments, a strategy pursued in several West African countries, and where there have been some notable successes (e.g., Guinea). This approach is likely to have a greater impact on accountability, as the employer represents even more closely the clientele, and thus the incentives for higher quality of service are better. There is also a better potential for financial sustainability, as the farmers' association that provides the public good is better able to recover costs (say, as general membership fees) from its members, although typically government funding is also provided to the associations. Extension agents may be permanent employees of the associations, or contract employees from private entities, NGOs, or universities; conceptually, their incentives for better service are fairly similar regardless of their standing. The difficulties with maintaining agents' quality due to loss of economies of scale in training, and the problematic linkages with research that sometimes characterize decentralized systems, are likely to be present in this variant as well.

The fiscal burdens of extension can be mitigated to some extent if partnerships and complementarities with local NGOs' training activities can be exploited. These can entail cost sharing and allow expanded coverage. However, in many developing countries, NGOs do not have secure autonomous budgets, and thus the reliance on such partnerships over an extended period of time may not be generally feasible.

[4] The USA provides an example of decentralization of extension to the sub-state (county) level, within a partnership of federal-state-county authorities that provides for financial sustainability along with more efficient extension planning [Huffman and Evenson (2005)].

3.3. Fee-for-service and privatized extension

A format of fee for service for extension (where the provider may be a public entity or private firms or consultants) in developing countries usually still entails considerable public funding even if the provider is private [e.g., in the form of government-funded vouchers or other government funding, such as reported by Keynan, Manuel and Dinar (1997) and Dinar and Keynan (2001)], but it has the potential of reducing the fiscal burden of public extension services. Under such an arrangement, small groups of farmers typically contract extension services to address their specific information needs. The free-rider problems and non-rivalry in information use are resolved by defining the public good at the level of a small group, and having the whole group share in the cost. The difficulty of tracing extension impact is much less of a problem, although issues of asymmetric knowledge of the value of information and identifiability of benefits [Hanson and Just (2001)] will still be present and raise design challenges accordingly. Indeed, Chile's experience with privatized extension, where government-funded contracts were expected to be gradually reduced as farmers' cost sharing would increase, demonstrates that willingness-to-pay may be slow to materialize [Cox and Ortega (2004)].

With resolution of the accountability problem, the quality of service is expected to be higher. In fee-for-service modalities, farmers clearly determine the type of information that is of priority to them, and thus the impact of extension advice is expected to be higher. Practical problems of governance can lead to distortions, such as favoritism of well-connected (but not necessarily high-quality) providers, and illegal trade in government-issued vouchers for extension [Berdegué and Marchant (2002); Cox and Ortega (2004)]. Similarly, training and the update of skills will usually have to be undertaken by agents individually, with loss of economies of scale. These issues pose further design challenges. An important role for public extension and policy (such as has been supported by development agencies in Latin America) is to facilitate the development of private provision of extension services, so that the public system can withdraw as appropriate. A key drawback of fee-for-service modes of extension is that less commercial farmers (i.e., poorer farmers and those farming smaller and less favored areas), for whom the value of information is lower, may purchase fewer extension services, as the price of the service will tend to be market-determined (thus reflecting also the demand from farmers with higher value of information, to the extent that such farmers use these channels for their information). This may entail not only social considerations, but may be an inefficient outcome if the poor have a lesser ability to prejudge the value of information and tend to undervalue it. The resolution of this concern is the stratification of extension systems by types of clients within the country [e.g., Sulaiman and Sadamate (2000)]. That is, smaller scale and poorer farmers may be served by public extension or by formats of contract extension receiving larger shares of public funding (e.g., an association of smaller scale farmers receives a larger matching allocation to hire extension staff). In such ways, the particular needs of women farmers, for instance, may be addressed [e.g., Saito and Weidemann (1990)]. At the same time, commercial farmers are expected to pay a higher share of extension cost in a fee-for-service system [Wilson (1991); Dinar and Keynan (2001)]. Furthermore, as pointed out by Hanson and Just (2001), there may be several externalities (such as related to soil conservation) that imply inefficiency if a fully privatized extension system is introduced.

3.4. Farmer field schools

The farmer field school (FFS) was designed originally as a way to introduce knowledge on integrated pest management (IPM) to irrigated rice farmers in Asia. The Philippines and Indonesia were key areas in implementing this farmer training effort. Experiences with IPM-FFS in these two countries have since been documented and used to promote and expand FFS and FFS-type activities to other countries and to other crops. Currently, FFS activities are being implemented in many developing countries, although only a few operate FFS as a nationwide system.

A typical FFS educates farmer participants on agro-ecosystem analysis, including practical aspects of "...plant health, water management, weather, weed density, disease surveillance, plus observation and collection of insect pests and beneficials" [Indonesian National IPM Program Secretariat (1991, p. 5)]. The FFS approach relies on participatory training methods to convey knowledge to field school participants to make them into "...confident pest experts, self-teaching experimenters, and effective trainers of other farmers" [Wiebers (1993)].

A typical FFS entails some 9–12 half-day sessions of hands-on, farmer experimentation and non-formal training to a group of 20–25 farmers during a single crop-growing season. Initially, paid trainers lead this village-level program, delivering elements and practical solutions for overall good crop-management practices. Through group interactions, attendees sharpen their decision-making abilities and are empowered by learning leadership, communication and management skills [Van de Fliert (1993)]. Some of the participating farmers are selected to receive additional training so as to be qualified as farmer-trainers, who then take up training responsibilities (for some fee, possibly paid by their community) with official backup support such as training materials. While there is some debate on whether the FFS is an extension system or an informal adult education system, for purposes of our discussion, the distinction is not of much consequence, as the objectives of the FFS are similar to those of many extension systems. The approach whereby the training focuses more on decision making skills than on packaged messages is perceived by its proselytizers as superior to traditional extension methods.

The FFS seeks to rectify the problem of accountability. This aspect is addressed in two ways: (i) The official trainers who conduct the field school are bound by a strict timetable of sessions within a prespecified curriculum, which can be easily verified by supervisors; and (ii) continuous interaction with a cohesive group of trainees creates accountability to the group, which is enhanced by the participatory nature of the training methods. Later, when farmer-trainers who are members of the same community administer the training, accountability is presumed to be even greater. These features are thus expected to ensure the quality and relevance of the service (knowledge) provided to the farmers.

A key drawback of the farmer field school approach is its cost, which is likely to raise problems of financial sustainability. The intense training activities are expensive per farmer trained [Norton, Rajotte and Gapud (1999); Quizon, Feder and Murgai (2001a, 2001b); Thiele et al. (2001)], so the amount of service actually delivered (the number of farmers trained) on a national level would be small. Cost-effectiveness and financial sustainability could be improved if farmer-trainers were to become the main trainers, perhaps with significant community funding, and if informal farmer-to-farmer communications were effective in facilitating knowledge diffusion. In practice, however, farmer-trainers have been a minor factor in national FFS initiatives in Indonesia and the Philippines [Quizon, Feder and Murgai (2001a)].

A study in the Philippines documented improved knowledge among trained farmers, but little diffusion of knowledge from trained farmers to other farmers, presumably because the content of the training is difficult to transmit in casual, non-structured communications [Rola, Jamias and Quizon (2002)]. Similarly, recent analysis of FFSs in Indonesia found superior knowledge among field school graduates, but no significant diffusion of knowledge from trained to untrained farmers [Feder, Murgai and Quizon (2004b)]. A related study concluded that the training had no significant impact on yields and pesticide use by trained farmers or members of their communities [Feder, Murgai and Quizon (2004a)].[5] A study by Godtland et al. (2004) of potato growers in Peru reported on knowledge gains among trained farmers, but the study took place at an early stage of the program and could not analyze diffusion effects. Such findings suggest that both the curriculum and the training approach need to be modified so as to make information simpler and easier to diffuse, and to prioritize the content of the training in order to shorten the duration and reduce the cost.

4. The impact of extension

The extension operations of the past four decades may well be the largest institutional development effort the world has ever known. Evenson (2001) reviewed many of the impacts of such endeavor, and the present section is intended to complement his earlier review. As noted in our earlier section, the endeavor has been extensive; hundreds of thousands of technicians have been trained; and hundreds of millions of farmers have had contact with and likely benefited from extension services. As countries struggle with declining public budgets, a key question must be "How effective have these extension investments been and what impacts have they had?" Not all good questions have

[5] Limited diffusion of information from field school graduates to other farmers is also reported by Van de Fliert (1993, pp. 202, 230) and International Potato Center (2002). These studies, however, did not include a rigorous analysis of diffusion.

ready answers, however, in this case because of the many challenges of attribution and measurement that have been noted in earlier sections.

In principle, the economic analysis of extension projects is similar to that of any investment appraisal [see, e.g., Belli et al. (2001) for example], but inevitably challenges arise in appropriately valuing and attributing benefits. For projects that deliver agricultural knowledge products to producers effectiveness in enhancing productivity can be quantified by estimating the economic benefits to producers (or consumers) and computing a rate of return to the investment [e.g., Maredia, Byerlee and Anderson (2001)]. Rates of return can be estimated econometrically by relating productivity changes to investment in research and extension or by applying the economic surplus method, which builds benefits from the bottom up based on estimated productivity changes at the field level and adoption rates for each technology. With the data limitations that so frequently have plagued the econometric approach, the economic surplus approach has been much more widely applied in developing countries.

More comprehensive studies may also seek to trace wider economic benefits of research and extension through factor and product markets. Economic analysts are increasingly being asked to address other objectives beyond efficiency, such as equity improvements and poverty alleviation, environmental quality, food safety, and nutrition [e.g., Alston, Norton and Pardey (1995)]. The extent that research and extension organizations should depart from their traditional efficiency objective is much debated and there is yet no general resolution to guide, say, public policy analysts concerned with relevance and effectiveness of investment in research and extension. However, few studies have assessed extension achievements in their more comprehensive domains of ambition.

The econometric approach to impact assessment usually employs a production function, cost function, or a total factor productivity analysis to estimate the change in productivity due to investment in research and extension. The framework of, say, a production function incorporates conventional inputs (land, labor, etc.), non-conventional inputs (education, infrastructure, etc.), and the stock of technical knowledge (perhaps represented by some representation of investment in research and extension). Recent efforts have expanded the specification to include resource quality variables (e.g., soil erosion, nutrient status, etc.), and weather variables. The estimated coefficients on research and extension (measuring marginal products) are then used to calculate the value of additional output attributable to the respective expenditures (holding other inputs constant) and to derive marginal rates of return to the investments.[6]

There are many technical areas of debate and refinement in the literature on econometric methods, such as the length and shape of time-lag structures, the appropriate method of determining the rate of return from the estimations, the extent of selection

[6] An added complication, leading to a possible overestimate of extension impact, is the potential of a "Hawthorne Effect" [e.g., Freedman, Sears and Carlsmith (1981)], whereby farmers who have had an intensive interaction with extension change their performance temporarily simply because they perceive themselves to be under observation. Such a problem would be more likely in a cross-sectional study.

biases, and the quality of indices used as the dependent variable [Alston, Norton and Pardey (1995)]. The estimation of impact is made even more challenging by the fact that farmers' choices regarding technology, information acquisition, and risk-bearing methods are made simultaneously [Feder and Slade (1984)]. Most studies, however, deal with these decisions separately.[7] The main constraints on the wider application of econometric approaches in developing countries are data availability and quality. The econometric approach requires good-quality time-series data, which are difficult to obtain below the national or state level in most developing countries. Therefore, the approach is generally best for ex post evaluations of entire agricultural research and extension systems over a long period (say, 25–30 years), if the quantity and quality of data allow the use of statistical methods. Robert Evenson pioneered much of the work in this area in developing countries [e.g., various contributions in Evenson and Pray (1991)]. The approach is less relevant for individual research and extension organizations, since pertinent time-series data are rarely sufficiently long enough or complete enough or available at the needed level of disaggregation to allow useful estimation.

One good approach is to use panel data to capture both cross-sectional and time-series variability [e.g., Gautam (2000)]. Secondary data of a panel nature are increasingly available for many of the variables at the district level, especially production and input data, and some recent studies have even included district-wise data on resource quality. Maredia, Byerlee and Anderson (2001) offer a review of such studies, although the emphasis in them has been on the impact of research rather than extension. As panel data become more widely available, the use of econometric approaches to research and extension evaluation will expand.

Birkhaeuser, Evenson and Feder (1991) made an early review of studies of extension impact and found few studies of systematic comparison of costs and benefits with and without a project. Systematic social experiments comparing different methods of extension in similarly situated areas have yet to be carried out. Where extension programs have been evaluated by comparing outcomes in similar contiguous areas, the results have been nuanced. Thus, careful work by Feder and Slade [Feder and Slade (1986); Feder, Lau and Slade (1987)] comparing productivity differentials in Haryana and Uttar Pradesh suggested that T&V had no significant impact on rice production but yielded economic returns of at least 15 percent in wheat-growing areas. Similar work in Pakistan [Hussain, Byerlee and Heisey (1994)] found even smaller impacts in wheat areas, although the effect of T&V in increasing the quantity of extension contact was documented. Although evaluations of extension investments have criticized the observed low levels of efficiency and frequent lack of equity in service provision, they have in the past reported relatively high benefit/cost ratios [e.g., Perraton et al. (1983)].

More recent studies of extension impacts have also shown significant and positive effects [e.g., Bindlish, Evenson and Gbetibouo (1993) for Burkina Faso; Bindlish and

[7] Huffman (2001b) identified this issue as inadequately researched in the context of the impact of schooling on farmers' performance, but the same observation applies to extension.

Table 3
Estimated rates of return for economic impacts from extension in selected countries (number of countries)

Type of technological infrastructure in a country*	5–25%	26–50%	50%+
Traditional and emerging technology	0	1	9
Islands of modernization	1	1	4
Mastery of conventional technology	2	2	3
Newly industrialized	1	0	4
Industrialized	0	0	5

Source: Evenson (1997).
*Bangladesh, Botswana, Brazil, Burkina Faso, Côte d'Ivoire, Ethiopia, Indonesia, Japan, Malaysia, Nepal, Nigeria, Paraguay, Peru, Philippines, South Korea, United States and Thailand.

Evenson (1993) for Kenya; Bindlish and Evenson (1997)] and internal rates of return on extension investments in developing countries have reportedly ranged from 5% to over 50% (Table 3) [Evenson (1997)]. The overriding lesson from Evenson's review of 57 studies of the economic impact of agricultural extension is, however, that impacts vary widely – many programs have been highly effective, while others have not. Extension systems seem to have been most effective where research is effective and have the highest pay-off where farmers have had good access to schooling, although doubtless other factors also play key roles.

The most comprehensive review of impacts is found in a recent meta-study of 289 studies of economic returns to agricultural research and extension. This study found median rates of return of 58% for extension investments, 49% for research, and 36% for combined investments in research and extension [Alston et al. (2000)].[8] Similar success has been documented even for Sub-Saharan Africa alone [e.g., Oehmke, Anandajayasekeram and Masters (1997)]. Economic analysis has thus provided fairly strong justification for many past extension investments, but does not tell the full story.

Concern over data quality along with difficult methodological issues regarding causality and quantification of all benefits must be important qualifiers to the prevailing evidence of good economic returns from extension. In Kenya, perhaps [from Leonard (1977) to Gautam (2000)] the most closely studied case in developing countries, although previous evaluations had indicated remarkably high positive economic returns to extension investments, a comprehensive evaluation based on improved and new data revealed a disappointing performance of extension, with a finding of an ineffective, inefficient, and unsustainable T&V-based extension system and no measurable impact on farmer efficiency or crop productivity [Gautam (2000)]. Such findings bolster the skepticism of policy makers [reinforced by observations such as those of Hassan, Karanja

[8] The sample of studies reviewed in the meta-study was strongly oriented toward research, as only 18 out of 1128 estimates of rates of return were for "extension only". In contrast, 598 were for "research only" and 512 were for "research and extension combined".

and Mulamula (1998)] about getting returns to investment in public extension that are actually rather low, a skepticism that seems more than well justified. It is not our intention to end this survey on a note so salutary but evidently more evaluative work is called for to better assist policy insights and investment decisions.

5. Conclusion

Our review began by charting the important role that agricultural extension can play in development. We especially highlighted the public-good character of much actual and potential extension effort, as this underpins the extensive public investment in this domain.

We elaborated on the many administrative and design failures that have proved so problematic in public extension effort in the past, most notably those associated with: the scale and complexity of extension operations; the dependence of success in extension on the broader policy environment; the problems that stem from the less than ideal interaction of extension with the knowledge generation system; the difficulties inherent in tracing extension impact; the profound problems of accountability; the oftentimes weak political commitment and support for public extension; the frequent encumbrance with public duties in addition to those related to knowledge transfer; and the severe difficulties of fiscal unsustainability faced in many countries.

From our review of such problems, as well as due consideration of positive experience, we went on to reflect on the pros and cons of some specific formats of extension operations that have emerged in the past few decades, namely training and visit extension, decentralized mechanisms for delivery, fee-for-service and privatized extension, and farmer field schools. Naturally, specific situations will call for quite specific servicing methods but our review emphasizes the efficiency gains that can come from locally decentralized delivery with incentive structures based on largely private provision, much of which will inevitably remain largely publicly funded extension efforts, especially (and properly so) for impoverished developing countries.

Among these general problems of extension organization, the difficulty of attributing impact weakens political support, leading to small budgets and problems of fiscal sustainability. Ironically, this same difficulty may explain why international development agencies have heavily supported extension activities, financing some $10 billion in public extension projects over the past five decades. The economic justification for the investments is rarely based on solid ex ante cost–benefit analysis, since parameters are typically not available from past investments because of the difficulties of attributing impact. Attribution problems also imply that it will be difficult to establish failure once a project is completed (completion is the artificial point in time when donor funding is fully disbursed, but farming and extension activities continue).

Several other factors also account for the popularity of extension projects among donors. Extension projects are relatively easy to design, typically involving a small number of recipient government agencies, often just the ministry of agriculture. This reduces bureaucratic complexity. The activities funded by a project are well defined inputs; constructing and refurbishing extension offices, training agents, providing transport and budgets for field operations, and funding additional personnel. If a project is national in scale, it is easy for donors to build its budget to a significant size – a positive attribute for a development agency striving to maintain its own cost-effectiveness per dollar granted or lent.

There is thus some tension between domestic decision makers, who are reluctant to invest heavily in extension, and development agencies, which enthusiastically promote investment in it. The availability of external funds minimizes the need for trade-offs between investments in extension and investments in more politically rewarding undertakings, such as irrigation systems. But it also simply postpones the day of reckoning. Once the externally funded project is over, the lack of political support resurfaces and extension budgets are again cut. The more expensive features of the foreign-funded effort are abandoned, and the size of the extension service is cut way back [Purcell and Anderson (1997)].

There is clearly much yet to be done in bringing needed extension services to the poor around the world. Understanding of what works well in the diverse circumstances of the developing world is still far from complete and there is thus a clear need for continuing research effort to fill these gaps, as has been well articulated by observers such as Cox and Ortega (2004, p. 15). Meantime, investors need to be cautious in designing and adjusting public extension systems if they are not needlessly to re-learn the lessons of the past. Informed by these lessons governments should be able to increase the chance of reaping high returns to their investment and successfully assisting farmers to boost their productivity and income, and thereby contribute more strongly to economic growth.

References

Alex, G., Zijp, W., Byerlee D., with input from the AKIS Thematic Team (2002). "Rural extension and advisory services: New directions", Rural Development Strategy Background Paper #9, ARD. World Bank, Washington, DC.

Alston, J.M., Norton, G.W., Pardey, P.G. (1995). Science Under Scarcity: Principles and Practice for Agricultural Research Evaluation and Priority Setting. Cornell University Press and ISNAR, Ithaca. Republished in 1998 by CAB International, Wallingford.

Alston, J.M., Chan-Kang, C., Marra, M.C., Pardey, P.G., Wyatt, T.J. (2000). "A meta-analysis of rates of return to agricultural R&D: Ex Pede Herculem?", Research Report 113. IFPRI, Washington, DC.

Ameur C. (1994). "Agricultural extension: A step beyond the next step", World Bank Technical Paper 247. World Bank, Washington, DC.

Anderson, J.R. (1999). "Institutional reforms for getting an agricultural knowledge system to play its role in economic growth". In: Pakistan Society of Development Economics Papers and Proceedings. Pakistan Development Review 38, 333–350.

Anderson, J., Clément, J., Crowder, L.V. (1999). "Pluralism in sustainable forestry and rural development: An overview of concepts, approaches and future steps". In: Pluralism and Sustainable Forestry and Rural Development, Proceedings of an International Workshop. FAO, Rome, pp. 17–26.

Anderson, J.R., Feder, G., Ganguly, S. (2006). "The rise and fall of training and visit extension: An Asian mini-drama with an African epilogue". In: van den Ban, A.W., Samanta, R.K. (Eds.), Changing Roles of Agricultural Extension in Asian Nations. B.R. Publishing Corporation, New Delhi, pp. 149–174.

Anderson, J.R., Hazell, P.B.R. (1994). "Risk considerations in the design and transfer of agricultural technology". In: Anderson, J.R. (Ed.), Agricultural Technology: Policy Issues for the International Community. CAB International, Wallingford, pp. 321–339.

Antholt, C.H. (1994). "Getting ready for the twenty-first century: Technical change and institutional modernization in agriculture", World Bank Technical Paper 217. World Bank, Washington, DC.

Avila, A.F., Evenson, R.E. (2005). "TFP calculations from FAO production data". Yale Economic Growth Center, New Haven, personal communication from Robert Evenson.

Axinn, G. (1988). Guide on Alternative Extension Approaches. FAO, Rome.

Belli, P., Anderson, J.R., Barnum, H.N., Dixon, J.A., Tan, J.-P. (2001). Economic Analysis of Investment Operations: Analytical Tools and Practical Applications, WBI Development Studies. World Bank, Washington, DC.

Benor, D., Harrison, J. (1977). Agricultural Extension: The Training and Visit System. World Bank, Washington, DC.

Berdegué, J.A., Marchant, C. (2002). "Chile: The evolution of the agricultural advisory service for small farmers: 1978–2000". In: Rivera, W.M., Zijp, W. (Eds.), Contracting for Agricultural Extension: International Case Studies and Emerging Practices. CABI Publishing, Wallingford, pp. 21–27.

Beynon, J., Akroyd, S., Duncan, A., Jones, S. (1998). Financing the Future: Options for Agricultural Research and Extension in Sub-Saharan Africa. Oxford Policy Management, Oxford.

Bindlish, V., Evenson, R. (1993). "Evaluation of the performance of T&V extension in Kenya", World Bank Technical Paper No. 208. World Bank, Washington, DC.

Bindlish, V., Evenson, R.E. (1997). "The impact of T&V extension in Africa: The experience of Kenya and Burkina Faso". World Bank Research Observer 12, 183–201.

Bindlish, V., Evenson, R., Gbetibouo, M. (1993). "Evaluation of T&V-based extension in Burkina Faso", World Bank Technical Paper No. 226. World Bank, Washington, DC.

Binswanger, H.P., Deininger, K. (1997). "Explaining agricultural and agrarian policies in developing countries". Journal of Economic Literature 35, 1958–2001.

Birkhaeuser, D., Evenson, R.E., Feder, G. (1991). "The economic impact of agricultural extension: A review". Economic Development and Cultural Change 39, 607–650.

Byerlee, D. (1988a). "Agricultural extension and the development of farmers' management skills". In: Howell, J. (Ed.), Training and Visit Extension in Practice, Agricultural Advisory Unit Occasional Paper 8. ODI, London, pp. 8–27.

Byerlee, D. (1988b). "Knowledge-intensive crop management technologies: Concepts, impacts, and prospects in Asian agriculture". In: Pingali, P. (Ed.), Impacts of Rice Research. Thailand Development Research Institute and International Rice Research Institute, Bangkok and Los Baños, pp. 113–134.

Cary, J.W. (1993). "Changing foundations for government support of agricultural extension in economically developed countries". Sociologia Ruralis 33, 336–347.

Cary, J.W. (1998). "Issues in public and private technology transfer: The cases of Australia and New Zealand". In: Wolf, S.A. (Ed.), Privatization of Information and Agricultural Industrialization. CRC Press, Boca Raton, pp. 185–203.

Cox, M., Ortega, H. (2004). "Chile: Origin and evolution of a privatized extension system". In: Rivera, W., Alex, G. (Eds.), Extension Reform for Rural Development, vol. 2, Privatization of extension systems, Case Studies of International Initiatives Agriculture and Rural Development Discussion Paper 9. World Bank, Washington, DC, pp. 9–15. Accessed 5/5/5 at: http://lnweb18.worldbank.org/ESSD/ardext.nsf/11ByDocName/PublicationsExtensionReformforRuralDevelopment.

Crowder, L.V., Anderson, J. (2002). "Uganda: Private sector secondment of government extension agents". In: Rivera, W.M., Zijp, W. (Eds.), Contracting for Agricultural Extension: International Case Studies and Emerging Practice. CABI Publishing, Wallingford, pp. 155–162.

de Haan, C., Schillhorn van Veen, T., Brandenburg, B., Gauthier, J., Le Gall, F., Mearns, R., Siméon, M. (2001). Livestock Development: Implications for Rural Poverty, the Environment, and Global Food Security, Directions in Development. World Bank, Washington, DC.

Dinar, A. (1996). "Extension commercialization: How much to charge for extension services". American Journal of Agricultural Economics 78, 1–12.

Dinar, A., Keynan, G. (2001). "Economics of paid extension: Lesson from experience in Nicaragua". American Journal of Agricultural Economics 83, 769–776.

Ewell, P.T. (1989). Linkages Between On-Farm Research and Extension in Nine Countries, OFCOR Comparative Study No. 4. ISNAR, The Hague.

Evenson, R. (1997). "The economic contributions of agricultural extension to agricultural and rural development". In: Swanson, B., Bentz, R., Sofranko, A. (Eds.), Improving Agricultural Extension: A Reference Manual. FAO, Rome, pp. 27–36.

Evenson, R.E. (2001). "Economic impacts of agricultural research and extension". In: Gardner, B.L., Rausser, G.C. (Eds.), Handbook of Agricultural Economics, vol. 1A. Elsevier Science, Amsterdam, pp. 574–625.

Evenson, R.E., Pray, C. (Eds.) (1991). Research and Productivity in Asian Agriculture. Cornell University Press, Ithaca.

FAO/WB (2000). Agricultural Knowledge and Information System for Rural Development (AKIS/RD): Strategic Vision and Guiding Principles. Food and Agriculture Organization of the United Nations and the AKIS Thematic Group of the World Bank, Rome.

Farrington, J., Christoplos, I., Kidd, A., Beckman, M., Cromwell, E. (2002). "Creating a policy environment for pro-poor agricultural extension: The who? What? And how?", Natural Resource Perspectives 80. ODI, London.

Feder, G., Just, R.E., Zilberman, D. (1986). "Adoption of agricultural innovations in developing countries: A survey". Economic Development and Cultural Change 35, 255–298.

Feder, G., Lau, L.J., Slade, R.H. (1987). "Does agricultural extension pay: The training and visit system in northwest India". American Journal of Agricultural Economics 69, 677–686.

Feder, G., Murgai, R., Quizon, J.B. (2004a). "Sending farmers back to school: The impact of farmer field schools in Indonesia". Review of Agricultural Economics 26, 45–62.

Feder, G., Murgai, R., Quizon, J.B. (2004b). "The acquisition and diffusion of knowledge: The case of pest management training in farmer field schools, Indonesia". Journal of Agricultural Economics 55, 217–239.

Feder, G., O'Mara, G.T. (1982). "On information and innovation diffusion: A Bayesian approach". American Journal of Agricultural Economics 64, 141–145.

Feder, G., Slade, R. (1984). "The acquisition of information and the adoption of new technology". American Journal of Agricultural Economics 66, 311–320.

Feder, G., Slade, R.H. (1986). "Methodological issues in the evaluation of extension". In: Jones, G.E. (Ed.), Investing in Rural Extension: Strategies and Goals. Elsevier Applied Science, London, pp. 255–266.

Feder, G., Slade, R.H. (1993). "Institutional reform in India: The case of agricultural extension". In: Hoff, K., Braverman, A., Stiglitz, J.E. (Eds.), The Economics of Rural Organizations. Oxford University Press, New York, pp. 530–542.

Feder, G., Willett, A., Zijp, W. (2001). "Agricultural extension: Generic challenges and the ingredients for solutions". In: Wolf, S., Zilberman, D. (Eds.), Knowledge Generation and Technical Change: Institutional Innovation in Agriculture. Kluwer, Boston, pp. 313–356.

Fleischer, G., Waibel, H., Walter-Echols, G. (2002). "The costs of transforming public extension services towards participatory approaches". In: CIMMYT Impact Assessment Conference, San Jose, Costa Rica.

Freedman, J.L., Sears, D.O., Carlsmith, J.M. (1981). Social Psychology. Prentice-Hall, Englewood Cliffs.

Garfield, E., Guadagni, M., Moreau, D. (1996). "Decentralization of agricultural extension in Colombia". In: Proceedings of Extension Workshop, Alternative Mechanisms for Funding and Delivering Extension. World Bank, Washington, DC. Mimeo.

Garforth, C. (1982). "Reaching the rural poor: A review of extension strategies and methods". In: Jones, G.E., Rolls, M.J. (Eds.), Progress in Rural Extension and Community Development, vol. 1. Wiley, New York, pp. 43–69.

Gautam, M. (2000). Agricultural Extension: The Kenya Experience: An Impact Evaluation. Operations Evaluation Department, World Bank, Washington, DC.

Godtland, E.M., Sadoulet, E., de Janvry, A., Murgai, R., Ortiz, O. (2004). "The impact of farmer field schools on knowledge and productivity: A study of potato farmers in the Peruvian Andes". Economic Development and Cultural Change 53, 63–92.

Gustafson, D.J. (1994). "Developing sustainable institutions: Lessons from cross-case analysis of 24 agricultural extension programmes". Public Administration and Development 14, 121–134.

Hanson, J., Just, R. (2001). "The potential for transition to paid extension: Some guiding economic principles". American Journal of Agricultural Economics 83, 777–784.

Hardaker, J.B., Huirne, R.B.M., Anderson, J.R., Lien, G. (2004). Coping with Risk in Agriculture, second ed. CABI Publishing, Wallingford.

Hassan, R.M., Karanja, D., Mulamula, L. (1998). "Availability and effectiveness of agricultural extension services for maize farmers in Kenya". In: Hassan, R.M. (Ed.), Maize Technology Development and Transfer: A GIS Application for Research Planning in Kenya. CAB International, Wallingford, pp. 175–188.

Hazell, P.B.R., Anderson, J.R. (1984). "Public policy toward technical change in agriculture". Greek Economic Review 6, 453–482.

Hercus, J. (1991). "The commercialization of government agricultural extension services in New Zealand". In: Rivera, W.M., Gustafson, D.J. (Eds.), Agricultural Extension: Worldwide Institutional Innovation and Forces for Change. Elsevier, Amsterdam, pp. 23–30.

Holloway, G.J., Ehui, S.K. (2001). "Demand, supply and willingness-to-pay for extension services in an emerging market setting". American Journal of Agricultural Economics 83, 764–768.

Hone, P. (1991). "Charging for agricultural extension services". Review of Marketing and Agricultural Economics 59, 297–307.

Howell, J. (1985). Recurrent Costs and Agricultural Development. Overseas Development Institute, London.

Howell, J. (1986). "Accountability in extension work". In: Jones, G.E. (Ed.), Investing in Rural Extension: Strategies and Goals. Elsevier Applied Science, London, pp. 213–217.

Huffman, W.E. (1977). "Allocative efficiency: The role of human capital". Quarterly Journal of Economics 91, 59–80.

Huffman, W.E. (1985). "Human capital, adaptive ability and distributional implications of agricultural policy". American Journal of Agricultural Economics 67, 429–434.

Huffman, W.E. (2001a). "Human capital, education and agriculture". In: Peters, G.H., Pingali, P. (Eds.), Tomorrow's Agriculture: Incentives, Institutions, Infrastructure and Innovations, Proceedings of 23rd International Conference of Agricultural Economists. Ashgate Publishing, Aldershot, pp. 207–222.

Huffman, W.E. (2001b). "Human capital, education and agriculture". In: Gardner, B.L., Rausser, G.C. (Eds.), Handbook of Agricultural Economics, vol. 1A. Elsevier Science, Amsterdam, pp. 333–381.

Huffman, W.E., Evenson, R.E. (2005). Science for Agriculture: A Long-Term Perspective. Blackwell Publishing, Ames.

Hussain, S.S., Byerlee, D., Heisey, P.W. (1994). "Impacts of the training and visit extension system on farmers: Knowledge and adoption of technology: Evidence from Pakistan". Agricultural Economics 10, 39–47.

Indonesian National IPM Program Secretariat (1991). Farmers as Experts: The Indonesian National IPM Program. Indonesian National IPM Program, Jakarta.

International Potato Center (2002). Impact Evaluation of Participatory Development of Integrated Insect and Disease Management (IPM) for the Potato Crop in San Miguel, Peru. CIP, Lima.

Jamison, D., Lau, L. (1982). Farmer Education and Farm Efficiency. Johns Hopkins University Press, Baltimore.

Jarrett, F.G. (1985). "Sources and models of agricultural innovation in developed and developing countries". Agricultural Administration 18, 217–234.

Kaimovitz, D. (1991). "The evolution of links between research and extension in developing countries". In: Rivera, W.M., Gustafson, D.J. (Eds.), Agricultural Extension: Worldwide Institutional Innovation and Forces for Change. Elsevier, Amsterdam, pp. 101–112.

Keynan, G., Manuel, O., Dinar, A. (1997). "Co-financed public extension in Nicaragua". World Bank Research Observer 12, 225–248.

Kydd, J., Dorward, A., Morrison, J., Cadisch, G. (2001). "The role of agriculture in pro poor economic growth in Sub Saharan Africa". Imperial College, Wye. Mimeo produced for DFID.

Leonard, D.K. (1977). Reaching the Peasant Farmer: Organization Theory and Practice in Kenya. University of Chicago Press, Chicago.

Lindner, R.K. (1993). "Privatising the production of knowledge: Promise and pitfalls for agricultural research and extension". Australian Journal of Agricultural Economics 37, 205–225.
Maredia, M., Byerlee, D., Anderson, J.R. (2001). "Ex post evaluation of economic impacts of agricultural research programs: A tour of good practice". In: The Future of Impact Assessment in the CGIAR: Needs, Constraints and Options, Proceedings of a Workshop Organized by the Standing Panel on Impact Assessment (SPIA) of the Technical Advisory Committee (TAC), 3–5 May 2000. FAO, Rome. Annex 3, pp. 5–42. A version on the web accessed 5/5/05 at: http://www.cgiar.org/tac/spia/maredia.pdf.
Marsh, S.P., Pannell, D.J. (2000). "Agricultural extension policy in Australia: The good, the bad and the misguided". Australian Journal of Agricultural and Resource Economics 44, 605–627.
Marsh, S.P., Pannell, D.J., Lindner, R.K. (2003). "Does extension pay? A case study for a new crop, lupins, in Western Australia". Agricultural Economics 30, 17–30.
Moore, M. (1984). "Institutional development, the World Bank, and India's new agricultural extension program". Journal of Development Studies 20, 303–317.
Mullen, J.D., Vernon, D., Fishpool, K.I. (2000). "Agricultural extension policy in Australia: Public funding and market failure". Australian Journal of Agricultural and Resource Economics 44, 629–645.
Mureithi, J., Anderson, J.R. (2004). "Farmer–extension–research interfaces". In: Ndiritu, C.G., Lynam, J.K., Mbabu, A.N. (Eds.), Transformation of Agricultural Research Systems in Africa: Lessons from Kenya. Michigan State University Press, East Lansing, pp. 261–286.
Norton, G., Rajotte, E., Gapud, V. (1999). "Participatory research in integrated pest management". Agriculture and Human Values 16, 431–439.
Oehmke, J.F., Anandajayasekeram, P., Masters, W.A. (1997). Agricultural Technology Development and Transfer in Africa: Impacts Achieved and Lessons Learned. Office of Sustainable Development, Bureau for Africa, USAID, Washington, DC.
Olson, M. (1971). The Logic of Collective Action: Public Goods and the Theory of Groups, revised ed. Harvard University Press, Cambridge.
Olson, M. (1986). "Toward a more general theory of government structure". American Economic Review 76, 120–125.
Perraton, H.D., Jamison, D.T., Jenkins, J., Orivel, F., Wolff, L. (1983). "Basic education and agricultural extension: Costs, effects, and alternatives", Staff Working Paper No. 564. World Bank, Washington, DC.
Picciotto, R. (1995). "Putting institutional economics to work: From participation to governance", World Bank Discussion Paper No. 304. World Bank, Washington, DC.
Picciotto, R., Anderson, J.R. (1997). "Reconsidering agricultural extension". World Bank Research Observer 12, 249–259.
Pieri, C., Evers, G., Landers, J., O'Connell, P., Terry, E. (2002). "No-till farming for sustainable rural development", Agriculture and Rural Development Working Paper. World Bank, Washington, DC.
Purcell, D.L., Anderson, J.R. (1997). Agricultural Extension and Research: Achievements and Problems in National Systems, World Bank Operations Evaluation Study. World Bank, Washington, DC.
Quizon, J., Feder, G., Murgai, R. (2001a). "Fiscal sustainability of agricultural extension: The case of the farmer field school approach". Journal of International Agricultural and Extension Education 8, 13–24.
Quizon, J., Feder, G., Murgai, R. (2001b). "Fiscal sustainability of agricultural extension: The case of the farmer field school approach: Supplementary remarks". Journal of International Agricultural and Extension Education 8, 73–76.
Rivera, W.M., Qamar, M.K., Crowder, L.V. (2001). Agricultural and Rural Extension Worldwide: Options for Institutional Reform in Developing Countries. FAO, Rome.
Rivera, W.M., Zijp, W. (Eds.) (2002). Contracting for Agricultural Extension: International Case Studies and Emerging Practices. CABI Publishing, Wallingford.
Rivera, W., Zijp, W., Alex, G. (2000). Contracting for Extension: Review of Emerging Practice: AKIS Good Practice Note. Agricultural Knowledge and Information System Thematic Group, World Bank, Washington, DC.
Roberts, N. (Ed.) (1989). Agricultural Extension in Africa, A World Bank Symposium. World Bank, Washington, DC.

Rogers, E. (1995). Diffusion of Innovations, fourth ed. The Free Press, New York.
Rola, A.C., Jamias, S.B., Quizon, J.B. (2002). "Do farmer field school graduates retain and share what they learn?: An investigation in Iloilo, Philippines". Journal of International Agricultural and Extension Education 9, 65–76.
Röling, N. (1986). "Extension and the development of human resources: The other tradition in extension education". In: Jones, G.E. (Ed.), Investing in Rural Extension: Strategies and Goals. Elsevier Applied Science, London, pp. 51–64.
Saito K.A., Weidemann, C.J. (1990). "Agricultural extension for women farmers in Africa", Working Paper 398. World Bank, Washington, DC.
Schultz, T.W. (1964). Transforming Traditional Agriculture. Yale University Press, New Haven.
Schultz, T.W. (1968). Economic Growth and Agriculture. McGraw-Hill, New York.
Schultz, T.W. (1975). "The value of the ability to deal with disequilibrium". Journal of Economic Literature 13, 827–846.
Schwartz, L.A., Zijp, W. (1994). "Public and private roles in the delivery of extension services". In: Public and Private Sector Roles in the Provision of Agricultural Support Services, Proceedings of the International Symposium. IICA, Coronado, Costa Rica, pp. 95–126.
Scarborough, V., Killough, S., Johnson, D.A., Farrington, J. (Eds.) (1997). Farmer-Led Extension. Intermediate Technology Publications for ODI, London.
Simmonds, N.W. (1988). "Observations on induced diffusion of innovations as a component of tropical extension systems". Agricultural Administration & Extension 28, 207–216.
Sulaiman, R., Hall, A. (2002). "Beyond technology dissemination: Can Indian agricultural extension re-invent itself?", Policy Brief 16. National Centre for Agricultural Economics and Policy Research, Pusa, New Delhi.
Sulaiman, R., Sadamate, V.V. (2000). "Privatising agricultural extension in India", Policy Paper 10. National Centre for Agricultural Economics and Policy Research, Pusa, New Delhi.
Sunding, D., Zilberman, D. (2001). "The agricultural innovation process: Research and technology adoption in a changing agricultural sector". In: Gardner, B.L., Rausser, G.C. (Eds.), Handbook of Agricultural Economics, vol. 1A, Agricultural Production. Elsevier Science, Amsterdam, pp. 207–261.
Swanson, B.E. (2004). "Extension strategies for poverty alleviation in a global economy", Paper presented at the 2nd International Conference on Agricultural Education and Environment, Suwon, Republic of Korea, October 14. Mimeo.
Swanson, B.E., Farner, B.J., Bahal, R. (1990). "The current status of agricultural extension worldwide". In: Agricultural Education and Extension Service, Report of the Global Consultation on Agricultural Extension. FAO, Rome, pp. 43–76.
Thiele, G., Nelson, R., Ortiz, O.N., Sherwood, S. (2001). "Participatory research and training: Ten lessons from the farmer field schools (FFS) in the Andes". Currents – Swedish Academy of Agricultural Sciences 28, 4–11.
Thirtle, C., Lin, L., Piesse, J. (2003). "The impact of research-led agricultural productivity growth on poverty reduction in Africa, Asia and Latin America". World Development 32, 1959–1975.
Umali, D.L., Feder, G., de Haan, C. (1994). "Public and private sector roles in the delivery of livestock services". In: Public and Private Sector Roles in the Provision of Agricultural Support Services, Proceedings of the International Symposium. IICA, Coronado, Costa Rica, pp. 129–148.
Umali, D.L., Schwartz, L. (1994). "Public and private agricultural extension: Beyond traditional frontiers", World Bank Discussion Paper 236. World Bank, Washington, DC.
Umali-Deininger, D. (1996). "New approaches to an old problem: The public and private sector in extension". In: Extension Workshop, Alternative Mechanisms for Funding and Delivering Extension. World Bank, Washington, DC. Processed.
Umali-Deininger, D. (1997). "Public and private agricultural extension: Partners or rivals?". World Bank Research Observer 12, 203–224.
Van den Ban, A.W. (1999). "Agricultural development: Opportunities and threats for farmers and implications for extension organizations". Journal of Agricultural Education and Extension 6, 145–156.

Van den Ban, A.W. (2000). "Different ways of financing agricultural extension", Agricultural Research & Extension Network Paper 106b. Overseas Development Institute, London, pp. 8–19.
Van den Ban, A.W., Hawkins, H.S. (1996). Agricultural Extension, second ed. Blackwell, Oxford.
Van de Fliert, E. (1993). "Integrated pest management: Farmer field schools generate sustainable practices. A case study in central Java evaluating IPM training", Paper No. 93-3. Wageningen Agricultural University, the Netherlands.
Waters-Beyer, A. (1989). "Participatory technology development in ecologically-oriented agriculture: Some approaches and tools", Agricultural Administration (Research and Extension) Network Paper 7. ODI, London.
Wiebers, U.-C. (1993). "Integrated pest management and pesticide regulation in developing Asia", World Bank Technical Paper No. 211. World Bank, Washington, DC.
Wilson, M. (1991). "Reducing the costs of public extension services: Initiatives in Latin America". In: Rivera, W.M., Gustafson, D.J. (Eds.), Agricultural Extension: Worldwide Institutional Innovation and Forces for Changes. Elsevier, Amsterdam, pp. 13–21.
Wolf, S., Zilberman, D. (Eds.) (2001). Knowledge Generation and Technical Change: Institutional Innovation in Agriculture. Kluwer, Boston.
Wozniak, G. (1993). "Joint information acquisition and new technology adoption: Late versus early adopters". Review of Economics and Statistics 75, 438–445.

PART 3

INVENTION AND INNOVATION

Chapter 45

THE ROLE OF INTERNATIONAL AGRICULTURAL RESEARCH IN CONTRIBUTING TO GLOBAL FOOD SECURITY AND POVERTY ALLEVIATION: THE CASE OF THE CGIAR*

PRABHU PINGALI

Agricultural and Development Economics Division, Food and Agriculture Organization, Rome, Italy

TIM KELLEY

CGIAR Science Council Secretariat, FAO, Rome, Italy

Contents

* The views expressed in the paper are those of the authors and should not be attributed to FAO or the Science Council of the CGIAR.

Handbook of Agricultural Economics, Volume 3
Edited by Robert Evenson and Prabhu Pingali

DOI: 10.1016/S1574-0072(06)03045-3

Abstract

Considering the deep pessimism about the limits to growth that prevailed throughout much of the 60s and early 70s, the rapid growth in food crop productivity and food supplies triggered by the Green Revolution was a remarkable achievement. The driving force behind this success was the application of modern science for enhancing food crop productivity, particularly in the favorable production environments. The CGIAR played a crucial role in adapting scientific knowledge to the conditions of developing countries as well as in coordinating international efforts in transferring technologies across national boundaries. Implicit in the CGIAR mission was, and still remains today, a primary focus on the production of international public goods (IPGs), i.e., goods that are non-exclusive in access and non-rival in use, and that have widespread applicability, i.e., of potential use beyond national boundaries. This chapter focus on the origins, evolution and major accomplishments of the CGIAR and its partners in meeting global food security and poverty reduction goals, and highlights the challenges facing the CGIAR in the decades ahead. Particular attention is paid to the existing evidence on the diffusion and impacts of CGIAR products and to the evidence on the rates of return to international agricultural research investments. The broader impacts of the CGIAR on poverty and food security are discussed. The chapter ends with a discussion of the future need for and the challenges facing the CGIAR.

Keywords

technology change, adoption, diffusion, impact, Green Revolution

JEL classification: O13, O14, O31, O32, O38

In the period between 1960 and 2000, the world's population grew by 90%, mostly in developing countries. At the same time, using only 10% more cultivated land, world food production grew by 115%, resulting in a 25% increase in per capita availability of food. Food prices consequently fell by 40% in real terms over the four decade period. Considering the deep pessimism about the limits to growth that prevailed throughout much of the 60s and early 70s, these achievements were as unexpected as they were outstanding. The driving force behind this success was the application of modern science for enhancing food crop productivity, particularly in the favorable production environments.

Many factors contributed to the rapid improvements in developing country food supplies: government investment in agricultural R&D and infrastructure improvements, particularly for irrigation and fertilizers, accompanied by supportive policies and institutions. The technological breakthroughs in rice and wheat provided the early impetus for the concerted effort at raising agricultural productivity. CIMMYT in Mexico and IRRI in the Philippines both established in the early 1960s – funded in large part by the Rockefeller and Ford Foundations, were the forerunners to an international agricultural research system that specialized in the generation and promotion of research and technological spillovers across national boundaries. Scientists at these centers, drawing on breakthroughs in rice and wheat breeding undertaken in Japan, China, Taiwan, and Mexico, achieved remarkable early success in the development and release of new, high yielding varieties (HYVs) of wheat and rice. The new short-statured, fertilizer-responsive cultivars had significantly higher yield potential than traditional varieties and were quickly adopted in farmers' fields across many parts of Asia and Latin America.

Inspired by the early successes at CIMMYT and IRRI a special partnership within the global agricultural research community was formed to address the chronic food supply deficits in many developing countries through production-oriented research. This partnership came to be known as the Consultative Group for International Agricultural Research (CGIAR). Established in 1972, the CGIAR mission was to expand and coordinate international efforts in transferring and adapting scientific knowledge to the conditions of developing countries. Implicit in this mission was, and still remains today, a primary focus on the production of international public goods (IPGs), i.e., goods that are non-exclusive in access and non-rival in use, and that have widespread applicability, i.e., of potential use beyond national boundaries.[1] The comparative advantage of the CGIAR derives partly from the fact that private firms operating thorough markets have limited interest in public goods since they do not have the capacity to capture much of the benefit through proprietary claims. Socially desirable levels of investment in such

[1] Accordingly, CGIAR research is organized around major problems that are of international relevance. Indeed, production of IPGs is one of the four criteria used to assess opportunities for strategic choices in shaping the CGIAR future research agenda [TAC (2000)]. This extends to all types of CGIAR activities, including training and advisory services, though producing IPGs for these and other research-related activities are more difficult to achieve.

goods can only be elicited from the public sector. The CGIAR investment in developing IPGs complements investments by both the private sector and governments who would invest in national public goods, irrespective of potential spillovers. Although it contributes less than 5% of the total global agricultural research budget, the CGIAR has played a fundamental role in helping spur agricultural growth and poverty reduction in developing countries.

This chapter focus on the origins, evolution and major accomplishments of the CGIAR and its partners in meeting global food security and poverty reduction goals, and highlights the challenges facing the CGIAR in the decades ahead. Section 1 of the paper discusses the evolution and funding of the CGIAR over the past three decades. Section 2 assesses the existing evidence on the diffusion and impacts of CGIAR products and Section 3 reviews the evidence on the rates of return to international agricultural research investments. Section 4 addresses the broader impacts of the CGIAR on poverty and food security. The last section of the paper looks at the future need for and the challenges facing the CGIAR.

1. Evolution and funding of the CGIAR

The research focus of the CGIAR has evolved over time (Figure 1). IRRI and CIMMYT, before and shortly after the establishment of the CGIAR, had a clear mandate focused on improving rice, wheat and maize productivity through the development and release of improved varieties. The CGIAR commodity improvement mandate quickly expanded to include other key food crops, such as legumes (beans, cowpea, pigeonpea and chickpea),

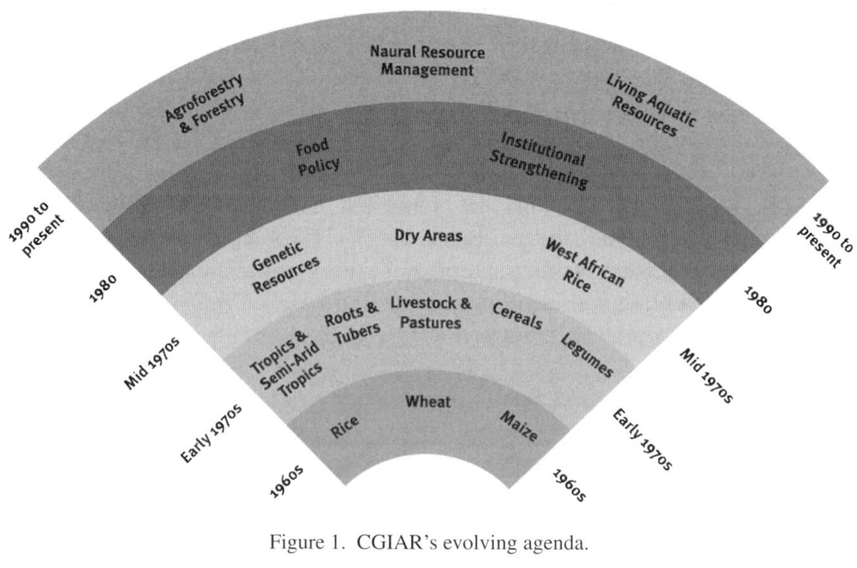

Figure 1. CGIAR's evolving agenda.

roots and tubers (cassava, yams, potato and sweet potato), other cereals (sorghum and millet) and to concentrate on better management of livestock and pastures.

By the beginning of the 1980s, the CGIAR through its network of 10 International Agricultural Research Centers (IARCs) was actively involved in varietal improvement and crop management research for most crops and cropping systems of major importance in developing countries. What had not been explicitly addressed by the CGIAR until then were constraints related to distorted food and agricultural policies and weak research institution capacity. Recognizing the critical role that good agricultural policies and strong national agricultural research systems (NARS) play in accelerating technological change and fostering agricultural development, two new CGIAR centers were established in the 1980s, one focused on agricultural policy (IFPRI) and the other on capacity building and institutional strengthening (ISNAR). As sustainability and conservation of the environment emerged as dominant themes, genetic resource conservation (IPGRI), forestry (CIFOR), agro-forestry (ICRAF), water (IWMI) and aquatic resource management (ICLARM, now WFC) were added to the CGIAR agenda by the early 1990s. At this point, the CGIAR mission moved well beyond simply raising food production in developing countries to explicitly embrace poverty reduction through sustainable agriculture, forestry and fisheries production.[2]

In its thirty-year history, the CGIAR saw its agenda broaden considerably as it moved from a strongly supply-side orientation to a considerably more demand (often donor) led one. At the same time it moved away from delivering concrete research products (e.g., improved crop varieties) to one of developing approaches, articulating problems and deriving common agendas and solutions with its partners. Its role as a moderator, initiator, facilitator, stimulator and a bridge to the broader stakeholder groups has become increasing important. Yet it is not clear that the success that the CGIAR has achieved in terms of helping raise the productivity of basic food crops can be replicated in the future as it attempts to address the broader development agenda it has now adopted.

1.1. Growth in funding base for the CGIAR

Member support for CGIAR research activities expanded dramatically during the first 15 years of its operation. From an initial funding base of US $19.5 million in 1972, support to the CGIAR broadened and deepened. By 1987, 35 CGIAR members were investing over $240 million each year to support the System's 12 IARCs.[3] Today, the aggregate annual contributions from the 64 members is just under US $440 million. This provides the financial base of support for the core research programs of 15 IARCs,

[2] Reflected most fully in the CGIAR's most recent mission statement: "to achieve sustainable food security and reduce poverty in developing countries through scientific research and research-related activities in the fields of agriculture, forestry, fisheries, policy, and the environment" [CGIAR (2000)].

[3] This total includes financial support for all Center and CGIAR System-level activities, including overheads.

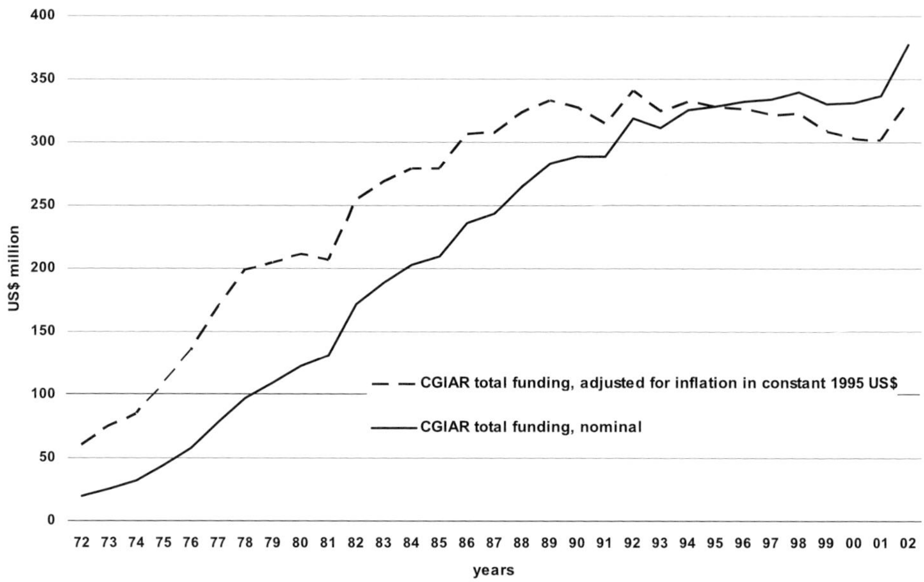

Figure 2. CGIAR total contributions to the agreed research agenda 1972–2002, nominal and real. *Source*: CGIAR annual reports 1986–1988/89 and CGIAR financial reports 1989–2002; and Executive Summary of the 2004 CGIAR Financial Results (May 2005).

16 Systemwide (inter-center) Programs and four global Challenge Programs (CPs),[4] in addition to supporting System-level governance and management functions. The pattern of growth of total investments to the CGIAR from 1972 to 2002, nominal and real,[5] is depicted in Figure 2. The most recent data show a continuing increase in donor investments in the CGIAR – now at US $450 million (in 2005).

Although the current pattern of investment still reflects the dominant position and contributions of a small group of donors, this is changing. In 1994, the top 10 donors accounted for over 78% of the total CGIAR investment; by 2004, their relative contribution accounted for only 64%, suggesting a broadening of the ownership within the System over the last 10 years. The developing countries themselves, who stand to gain the most today from the CGIAR efforts, contribute only 4% of the total budget currently, which has risen slightly over the last five years but still suggests a long way to

[4] At AGM '01, the CGIAR adopted the "CP" concept as a major pillar of change in the CGIAR reform process. This new programmatic approach was focused on large multi-institutional research programs addressing specific problems of regional or global significance using expertise of Center programs and expanded partnerships. There are currently four CPs in the CGIAR: HarvestPlus CP, Water and Food CP, Genomics CP, and Sub-Saharan Africa CP.

[5] Only a relatively small part of the growth can be attributed to inflation. Using 1995 as the base year, the annual investment in the CGIAR in 2005 is calculated at US $352 million, compared to US $60 million in 1972.

go before the CGIAR becomes an organization predominantly financed and managed by the developing countries themselves.

1.2. Trends in CGIAR investment by activity/undertaking

Up until 1991, activities of the CGIAR were classified into four simple activities: Research; Strengthening NARS, Research Support, and Management Research, with sub-activities within each. Subsequently,[6] as the mandate of the CGIAR evolved and donor interest in some key issues emerged, the activity classification system was revised to embrace five major 'Undertakings' of relevance to the CGIAR, with several sub-activities, as follows[7]:

- Increasing Productivity
  - Germplasm enhancement and breeding
  - Production systems development and management
- Protecting the Environment
- Saving Biodiversity
- Improving Policies
  - Socio-economic, policy and management research
- Strengthening National Research Programs

By making some simple assumptions about investments in the early years based on specific support to individual CGIAR centers and the correspondence between certain sub-activities in the each of the classification schemes, it is possible to reconstruct a time series of investment based on the latter classification system of major Undertakings. This is presented in Figure 3 where shares of total investment in the CGIAR by major undertakings are shown for five-year averages beginning in 1972–1976.

During the first five-year period (1972–1976), coinciding with the establishment and early activities of the CGIAR, almost 75% of the total investment went to IARCs for activities related primarily to "Increasing Productivity". A large percentage of this, particularly during the early years when the CGIAR's primary focus was on crop improvement, consisted of sub-activities related to "germplasm enhancement and breeding".[8] "Increasing Productivity" type activities received about twice as much importance as all other Undertakings combined up until the early 90s, with almost 65% of the total

[6] In fact, an interim system of CGIAR activity classification was used between 1992 and 1994. Although this further complicates constructing a consistent time series of activity investment within the CGIAR, for purposes here, simple assumptions were made in converting those to the more standard one used between 1995 and 2001.

[7] There is no clear dividing line between these categories. In some cases activities may overlap into several categories. Indeed, a number of CGIAR objectives cross-cut all these categories, such as the strengthening of NARS, and improvement of the sustainability of production systems.

[8] Beginning in the late 70s and early 80s, increasing emphasis was given to activities related to crop and farm management issues, e.g., farming systems research. Precise figures are lacking, but in 1992, the first year for which separate data became available, "germplasm enhancement and breeding" and "production systems development and management" each accounted for about half of the "Increasing Productivity" Undertaking.

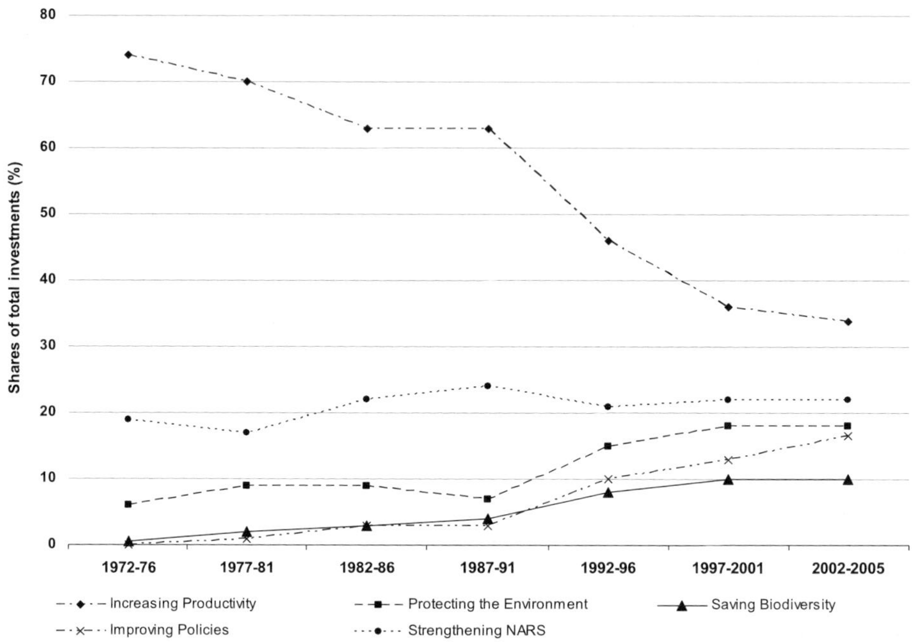

Figure 3. Expenditures in shares of CGIAR total investment, 1972–2001*, by undertakings. *Source*: CGIAR annual report 2001 for data from 1972 to 2001; Executive Summary of the 2004 CGIAR Financial Results (May 2005) for data from 2002 to 2005 (latter year projected). Note, after the CGIAR changed the categories from 'Undertakings' to 'Output' it was assumed that 'Protecting the Environment' remains 18% similar to the data from 2000 to 2002 and that 'Increasing Productivity' composed of the new allocations by the Outputs 'Sustainable Production' and 'Germplasm Improvement'.

CGIAR investment allocated to it. During the 90s, however, donors shifted their interest to several other Undertakings. In 2002, the last year for which actual Undertaking data are available, the share of CGIAR investment in Increasing Productivity had fallen to 34%, of which the sub-activity "germplasm enhancement and breeding" comprised 18% (down from 24% in 1992) and "production systems development and management" to 17% (down from 25%).[9] This trend in investment away from productivity-enhancing activities, for which there are proven impacts on poverty, raise questions about the current direction and focus of the CGIAR [World Bank (2003)].

The three CGIAR Undertakings that continued to expand, and most prominently since 1991, were "Protecting the Environment", "Improving Policies", and "Saving Biodiversity". From a negligible base in 1972, these three Undertakings emerged as

[9] The two largest components within the "production systems" sub-activity saw their investments shares fall the most, e.g., 'cropping systems' from 16% to 9% and 'livestock systems' from 6% to 4%. At the same time, investments in 'tree systems' fluctuated around 3% while investments in 'fish systems' actually rose.

dominant investment activities of the donors, such that by 2002–2005 they constituted 43% of the CGIAR agenda (18%, 15% and 10%, respectively). The fifth undertaking, "Strengthening NARS" grew slightly in relative importance over the 30-year period, from 19% to 22%. Thus, a major change in the CGIAR over the last 30 years has been a shift in focus away from increasing productivity activities toward those relating to protecting the environment, saving biodiversity and improving policies. The current system of investment accounting is based on targeted 'Outputs' and reflects a similar emphasis with 'germplasm improvement' accounting for only 16% of the total CGIAR investment, 'germplasm collection' for 11%, 'policy' for 18%, 'sustainable production' for 34% and 'enhancing NARS' for 21%. The policy and financial context under which drove this trend toward a stronger environmental focus was (a) the run up to the UNCED and its outcome (Agenda-21 goal of achieving sustainable agriculture) and (b) a financial crisis for the System in the early 1990s at a time when the major focus was on germplasm improvement. In 1995 a renewal program in the CGIAR was launched where commitments to the newer sustainability agenda were made, with the overall budget position of the CGIAR improving significantly but resulting in a diversion away from productivity-oriented activities in favor of a stronger environmental protection focus, a trend which continued until very recently.

1.3. Investments by center type

The shift in priority emphasis within the CGIAR can also be seen by examining the investments to the individual Centers over time. Table 1 shows the donor contributions to each of the Centers since 1972. Centers have been grouped into four categories according to the primary focus of their work: (i) commodity focus; (ii) ecoregional plus commodity focus; (iii) policy/institutional focus; and (iv) NRM focus.

Investments in the commodity-focused centers of the CGIAR, i.e., CIMMYT (maize and wheat), IRRI (rice), ILRI (livestock) and CIP (potatoes and sweet potatoes), registered most of their growth up until 1989, with total funding for these four centers rising from an initial \$8.5 million to \$107 million. Investments for this group declined thereafter and did not return to 1989 levels until 1997, and then only in nominal terms. The real value of investment actually declined, from \$125 million in 1989 to \$98 million in 2005 (using 1995 dollars). In relative terms, the predominantly commodity-focused centers now account for less than 30% of the total CGIAR investment (2004–2005 average) versus about 50% during the 1970s.

The ecoregional + commodity focused centers include ICRISAT, ICARDA, CIAT, IITA and WARDA.[10] These centers too experienced their biggest nominal growth up to 1990, rising from an initial \$11 million to \$107 million. Like the commodity-focused

[10] ICRISAT ecoregional mandate is global semi-arid tropics and commodity mandate is sorghum, millet, groundnut, pigeonpea and chickpea. ICARDA's ecoregional mandate is the Caucuses and West Asia–North Africa and commodity mandate for wheat (with CIMMYT), lentils, chickpea, etc.

Table 1
CGIAR budget, by center: 1972–2005

Commodity focused centers	1972–1975	1976–1979	1980–1983	1984–1987	1988–1991	1992–1995	1996–1999	2000–2003	2004–2005
CIMMYT	6.3	11.6	17.7	21.2	26.9	25.7	30.0	37.31	41.76
CIP	1.7	5.6	9.1	11.5	17.6	17.2	21.9	20.82	22.83
ILRI[a]	2.4	13.1	18.5	24.0	32.4	25.0	25.5	28.52	31.3
IRRI	5.2	12.0	18.2	22.5	28.2	27.6	31.2	32.51	33.0
Total center	**15.6**	**42.23**	**63.5**	**79.2**	**105.1**	**95.5**	**108.6**	**119.16**	**128.89**
% to total	**47.85**	**49.24**	**45.26**	**43.68**	**45.46**	**37.58**	**33.76**	**32.61**	**29.89**
Total contrib. to the research agenda[b]	32.6	85.9	140.3	181.3	231.2	254.1	321.7	365.36	431.14
Ecoregional/commodity-oriented centers									
ICRISAT	3.2	10.3	15.6	23.1	29.3	26.7	25.5	23.2	25.08
ICARDA		5.9	14.9	18.8	18.5	17.8	22.0	25.06	24.98
CIAT	5.5	10.2	17.9	22.7	27.1	27.0	30.9	31.5	35.68
IITA	6.9	12.7	17.4	20.6	22.0	22.2	27.1	34.47	41.11
WARDA	0.6	1.5	2.4	3.0	6.1	6.5	9.5	11.36	15.62
Total center	**16.2**	**40.56**	**68.2**	**88.2**	**103**	**100.2**	**115**	**125.59**	**142.47**
% to total	**49.69**	**47.26**	**48.61**	**48.65**	**44.55**	**39.43**	**35.75**	**34.37**	**33.04**
Total contrib. to the research agenda	32.6	85.9	140.3	181.3	231.2	254.1	321.7	365.36	431.14
NRM focused centers									
IWMI[c]						6.8	9.2	13.1	25.89
ICLARM						5.2	10.8	13.72	18.43
ICRAF						13.5	20.1	23.95	29.46
CIFOR						5.5	10.5	13.42	14.95
Total centers						**31**	**50.6**	**64.19**	**88.73**
% to total						**12.20**	**15.73**	**17.57**	**20.58**
Total contrib. to the research agenda	32.6	85.9	140.3	181.3	231.2	254.1	321.7	365.36	431.14
Policy/institution strengthening centers									
IFPRI	0.3	1.4	3.1	4.9	8.9	8.9	18.8	22.62	37.27
ISNAR			2.2	4.3	7.2	6.5	9.6	9.1	*[d]
IPGRI[e]	0.5	1.6	3.3	4.7	7.0	12.0	19.1	24.7	33.78
Total centers	**0.8**	**3.0**	**8.6**	**13.9**	**23.1**	**27.4**	**47.5**	**56.42**	**71.05**
% to total	**2.45**	**3.49**	**6.13**	**7.66**	**9.99**	**10.79**	**14.76**	**15.44**	**16.48**
Total contrib. to the research agenda	32.6	85.9	140.3	181.3	231.2	254.1	321.7	365.36	431.14

Source of data: CGIAR annual report 1997 for 1972–1997, financial reports 1998–2003. Source to the total research agenda: CGIAR annual report 1997 (1972–1981), 1986–1988/89 and financial reports 1989–2003, and Executive Summary of the 2004 CGIAR Financial Results (May 2005).

[a]Previously ILCA and ILRAD.

[b]Figure does not include CGIAR investments that supported System governance and management, e.g., CGIAR Secretariat, TAC Secretariat, etc.

[c]Previously IIMI.

[d]ISNAR subsumed into IFPRI in 2004.

[e]Previously IBPGR and INIBAP.

centers, investments here fell thereafter and did not recover to 1990 levels until 1996, and only in nominal terms. But by 2004–2005, the aggregate level of investment for these five centers had risen to $142 million, roughly equivalent to real resource levels enjoyed during the late 1980s and early 1990s. These five Centers currently account for a third of the total CGIAR funding, although during the early 1970s they accounted for about half.

The first two policy/institution related centers were established in 1975: IFPRI, with a mandate for research on food and agricultural policy, and IPGRI focusing on genetic resources policy. A third, ISNAR with a mandate for strengthening national agricultural research in developing countries, joined the CGIAR five years later. In 1980, the aggregate level of investment in these three centers was just over $5 million. Funding for this group of policy/institution centers grew rapidly during the 1980s and the 1990s. By 2004–2005, the total annual investment to this group was over $70 million and accounted for almost 17% of the total CGIAR budget, from a mere 3% during the 1970s.[11] This is a clear statement about the growing relative importance of agricultural and food policy, institution strengthening and genetic resources policy in the eyes of CGIAR members.

The NRM-focused centers are the newest group to be added to the CGIAR and consist of four centers: IWMI (water management), ICLARM now WorldFish Center (aquatic resources), ICRAF now World Agroforestry (agroforestry) and CIFOR[12] (forestry). These centers joined the CGIAR in 1992 with an initial aggregate funding of $25 million. Funding for this group grew steadily throughout the next 10 years and reached $89 million in 2004–2005. In terms of relative importance, the NRM-focused centers today account for 21% of the CGIAR budget, having risen from virtually nothing before 1992, a vivid statement about the importance of conservation and environmental related issues to CGIAR members. This growth, in relative and absolute terms, came at the expense of primarily germplasm enhancement and breeding and thus affected primarily the commodity-focused centers and to a lesser extent the commodity + ecoregional centers.

1.4. The changing nature of the investment: From unrestricted to restricted

One of the hallmarks of the CGIAR during its early years was the relatively large percentage of funds contributed by donors in an unrestricted manner, i.e., contributions were not earmarked for a particular project or type of activity. Over time this has changed, with an increasing amount of "special project" funds coming to the Centers, thereby restricting the freedom they have in selecting priorities and activities. Figure 4 shows the trend from 1988 to 2001 in absolute contributions to the CGIAR centers

[11] At the Annual General Meeting of the CGIAR in October, 2003, the CGIAR members decided to integrate ISNAR into IFPRI, as a distinct program. Hence, from 2004 there are only two centers in this group.

[12] While classified as a NRM institute, in fact, much of CIFOR's research focuses on policies.

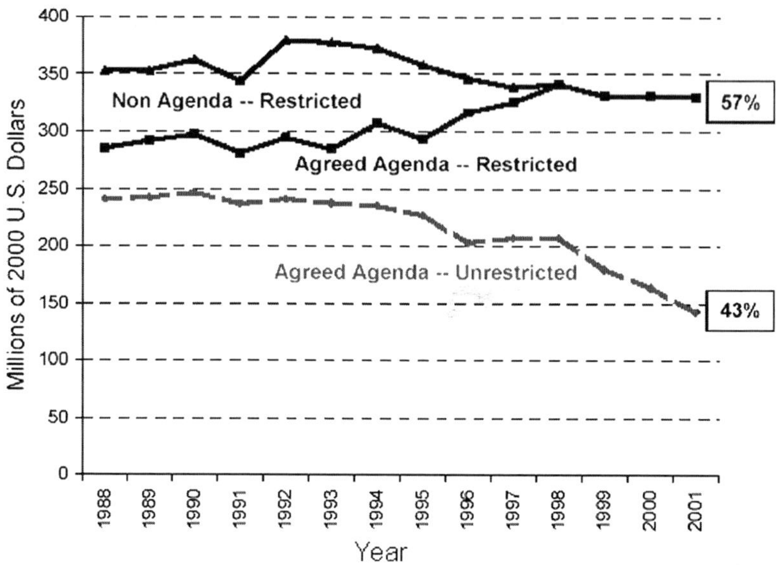

Figure 4. Unrestricted and restricted funding for the CGIAR 1988–2001. [Adapted from World Bank (2003).]

in unrestricted and restricted (targeted) forms. The level of restricted funding has increased by almost 2½ times since 1988. The relatively rapid rise of restricted funding, from 36% of the CGIAR budget in 1992 to 64% in 2005, continues to threaten the integrity of the System's research agenda, and puts at risk the essential functioning of the CGIAR as a mechanism for coordinated research and funding. Disproportionate levels of restricted funding could move the CGIAR emphasis away from the generation of International Public Goods toward the adaptive and development end of the research spectrum. Piecemeal project funding could over emphasize short term outputs at the cost of reduced Center emphasis on long term strategic research.

According to a recent World Bank evaluation, there is an overwhelming consensus now that the growing share of restricted funding is distorting research priorities, increasing transactions costs, and reducing the efficient use of resources at both the System and Center levels [World Bank (2003)]. While there is no tangible documented evidence for this as yet, there is growing awareness that this phenomenon has contributed to a "strategic drift" within the CGIAR. The recent priority setting exercise led by the Science Council focuses on a more coherent IPG research agenda with a strong commitment by donors to unrestricted funding. This is a step in the right direction, but it remains to be seen whether the System can develop an effective mechanism for channeling its resources to the avowed System priorities to maximize impacts, with as few restrictions as possible.

2. Diffusion and impact of CGIAR research and technology generation

The activities and outputs generated by the CGIAR have evolved over the last three decades, from an exclusive focus on crop genetic improvements, to the current diverse mix of products that includes agronomy and crop management technologies, new paradigms and knowledge on resource management and environmental conservation, policy analysis, human capacity building and germplasm and knowledge networks. Despite the current activity and output diversity, it is not clear that the CGIAR's ability to effectively disseminate its products and to have an impact on developing country agriculture has progressed beyond its success in promoting crop varieties and improved breeding lines. To-date, there is insufficient evaluation of the other CGIAR products to assess their diffusion and impact. This section presents the evidence that is currently available on the extent of diffusion and impact of CGIAR products.

2.1. Crop genetic improvements

Most development experts agree that the IARCs have played a pivotal role in the agricultural change and intensification process witnessed in developing countries over the last three decades. Some of the major achievements and impact derived from the CGIAR's crop germplasm improvement efforts have been documented by Evenson and Gollin (2003). In partnership with the national agricultural research systems (NARS), the CGIAR and its forerunner institutes were instrumental in facilitating and sustaining the impact of the "Green Revolution" – the development of improved, fertilizer-responsive high yielding varieties (HYVs) of wheat, rice, and a host of other important staple commodities, in the developing regions of the world. Evenson and Gollin (2003) estimates that the productivity gains from crop germplasm improvement alone averaged 1.0% per annum for wheat (across all regions), 0.8% for rice, 0.7% for maize, and 0.5% and 0.6% for sorghum and millets respectively.

Documentation of the extent of diffusion, adoption and farm-level impact of modern, HYVs of rice, wheat, and maize began in the early 1970s. Similar work for other crops, such as cassava, sorghum, millets, and potatoes, followed in the 1980s and thereafter. Dalrymple (1978), for example, documented that approximately 30 million hectares of wheat and 25 million hectares of rice in the developing world had been planted to HYVs by 1977. Dalrymple's work triggered numerous other efforts by researchers over the next two decades to study adoption and impacts of improved varieties of rice, wheat, maize, sorghum & millet, and potatoes [e.g., Herdt and Capule (1983); Byerlee and Moya (1993); López-Pereira and Morris (1994); Evenson and David (1993); Walker and Ryan (1991); Walker and Crissman (1996); Rohrbach et al. (1999)].

Evenson and Gollin (2003) report for all developing countries, that the adoption of HYVs during the first twenty years of the Green Revolution, if aggregated across the major crops, rose from 9% in 1970 to 29% by 1980 and reached 63% by 1998. As might be expected, varietal releases by the national programs were rising as well over this period. In the case of wheat, varietal releases doubled from an average of 40 per year between 1965 and 1970 to 80 per year between 1986 and 1990. Annual rice releases tripled during the same time period and maize releases increased five fold. The same pattern holds for sorghum, and even for crops that were relatively less researched such as millet, barley and lentils.

The spillover benefits of CGIAR breeding efforts can be measured by the parentage and pedigree of the varieties released by national programs. Evenson and Gollin (2003) report that the CGIAR content in modern varieties was high in most crops with 36% of all varietal releases based on crosses made in CGIAR centers. For Sub-Saharan Africa and the WANA region this figure was more than 50%. In addition, another 26% of the modern varieties released had a CGIAR cross as a parent or other ancestor. Evenson and Gollin (2003) conclude that the expanding pool of genetic resources and varieties made available to the national programs through the CGIAR helped avoid the diminishing returns to breeding efforts that would have occurred in the NARS programs had they been forced to work with the pool of genetic resources available to them at the beginning of the period.

International spillovers are likely highest for a commodity like wheat, which is grown in relatively homogeneous production environments, with little variability in local tastes and preferences for quality characteristics [Byerlee and Traxler (2001)]. Quality characteristics are a limiting factor in the direct transferability of varieties for some major commodities such as rice and maize. Consumer tastes may be so highly location specific in some cases, such as beans in Africa, to make it difficult even for country programs to develop widely accepted varieties [Sperling, Loevinsohn and Ntambovura (1993)]. NARS programs have generally used varieties or crosses made in the CGIAR centers as parents for the development of varieties that are more closely adapted to particular agro-ecological environments or specific taste preferences. In practice, a large proportion of varietal transfers take the form of adaptive transfers. The CGIAR has contributed significantly to the improvement of research efficiency and to the reduction of research costs by enabling such adaptive transfers.

It often takes a long time for knowledge to be developed through research and then adopted. Typically, ten years pass from the initiation of a research project to the dissemination of research results. By borrowing research results (e.g., plant lines or varieties) from other countries, a country can shorten its research time and contribute to increased returns to research investments [Alston, Norton and Pardey (1995)]. The CGIAR's numerous crop improvement networks allowed for the best breeding materials and knowledge to be widely and freely available across the developing world. Several attempts have been made to trace and quantify such spillovers at the level of individual countries. Wood and Pardey (1998) explicitly accounted for public goods cross-border technology spillovers in agricultural research priority setting in Latin America. Morris, Dubin and Pokhrel (1994) conducted a similar assessment for wheat research spillovers from India to Nepal. Maredia and Byerlee (1999) quantified spillover benefits for improved wheat germplasm across agro-ecological boundaries – in other words, they measured the transferability of wheat varieties developed for one production environment (e.g., an irrigated environment) to another (e.g., a rainfed environment). In general, large NARS's engage in adaptive transfers rather than direct use of CGIAR generated varieties and crosses [Byerlee and Traxler (2001)].

Developed country agricultural research systems also benefited from the IPG technology spillovers generated by the CGIAR. Brennan (1986) measured the benefits to Australian wheat breeding programs of access to CIMMYT breeding materials. Pardey et al. (1996) measured the benefits to US wheat and rice production from germplasm developed at CIMMYT and IRRI. Most of these spillovers have been adaptive transfers, but in the case of wheat in California, most varieties have originated directly from CIMMYT. The aggregate benefits of these spillovers have been valued in the billions of dollars.

2.2. Crop and resource management impacts

Traditional crop management research in the CGIAR tended to be more narrowly defined and included such agronomy-related themes such as soil and nutrient management, irrigation and land-cover management, pest management and water harvesting. It had a strong emphasis on increasing or maintaining resource productivity. The primary aim of the research was to complement the germplasm improvement research to exploit the benefits of new cultivars. More recently, there has been a growing interest in the CGIAR in integrated natural resource management (INRM) research. This is a broader research paradigm that emphasizes the nexus of productivity enhancement–environmental protection–human development as a multiple research objective across different time and spatial scales, from field plot to landscape levels [Sayer and Campbell (2001); Turkelboom et al. (2003)]. Invariably, INRM must concern itself with sociopolitical, economic, and ecological variables [Campbell et al. (2001)]. Clearly, this represents a significant departure from traditional crop management research that aimed to raise or simply maintain ('maintenance research') productivity of resource use in a sustainable manner, i.e., over the long term.

Unlike the case for crop genetic improvement, the documented evidence of the impact of NRM research in the CGIAR is virtually nil, at least when considering moderate to large scale effects [Kelley and Gregersen (2005)]. A review of the literature by Pingali (2001) found relatively few 'crop management and improved input use' and other NRM-related CGIAR impact studies to-date, a finding that corroborates an earlier review by Byerlee and Pingali (1994). Raitzer (2003) systematically reviewed and evaluated IA studies of economic benefits derived from CGIAR innovations (known 'success stories'), so as to produce a range of plausible and highly credible benefit-cost ratios for the entire investment in the CGIAR. Results show a notable absence of large-scale success stories for NRM, with notable the exception of biocontrol and integrated pest management (IPM) research [e.g., Zeddies et al. (2001); Bokonon-Ganta, DeGroote and Neuenschwander (2001)]. A comprehensive survey of rates of returns for all types of agricultural research, including large and small-scale studies, found few NRM-related studies among them, indeed less than 4% of the total studies reviewed [Alston et al. (2000)]. Unlike the case for crop germplasm improvement, for which large-scale adoption of yield-enhancing CGIAR-derived varieties has been documented for a range of CGIAR crops, there are as yet few examples of widely adopted CGIAR-generated improved NRM technologies for which demonstrable impact has been measured. The NRM IAs included in the Alston et al. study also showed significantly lower average rates of return than for crop germplasm improvement research.

The reason for there being so few documented success stories of NRM research in the CGIAR, i.e., studies that go beyond anecdotal evidence and selective small-scale case study results, is not yet clear. Some of the more plausible reasons discussed below are: (i) a lack of sustained critical mass of effort and investment; (ii) inappropriate methods for measuring NRM research impact; and, possibly, (iii) a lack of impact per se.

2.3. *Lack of sustained critical mass investment*

It can be argued that the lack of evidence to-date reflects, at least partly, an insufficient and sustained emphasis on NRM research over the last few decades. While time series data on specific categories of funding is difficult to reconstruct, most would concede that total CGIAR investments in crop germplasm improvement have been considerably larger than those for NRM research. Notwithstanding, the absolute levels of investments by the centers in NRM-related research and its earlier precedents, e.g., farming systems research, are still considerable. This applies not only to the NRM-focused centers, but also, to a larger extent, to the predominantly commodity-oriented centers and to the commodity + ecoregional centers as well.

Research in soil and water management and cropping/farming systems in general represented a significant component of many CGIAR centers' research agenda during the CGIAR's first two decades, and these were typically focused on productivity-enhancing aspects of NRM. Major investments were made in such areas as broadbed-and-furrow management, minimum and zero tillage systems, alley cropping, watershed management and other soil and water management related research. To-date, far too little of this has been assessed in terms of impact, whether measured in terms of improvements in resource productivity, or in enhancing the environment.

2.4. *Inappropriate methods for measuring NRM impact*

NRM research IA has lagged behind assessment of the impacts of research on germplasm improvement and certain technology developments. Approaches are needed that capture environmental services and other (non-crop yield) gains due to such NRM research as maintenance and loss reduction, risk reduction, quality improvement, reduction of negative environmental externalities, and compatibility with off-farm labor schedules. Certainly, lack of appropriate methods has constrained efforts to document impact from NRM research [Izac (1998)]. Economic surplus methods for measuring and attributing the impact for crop germplasm research may often not be appropriate in the case of NRM research. While this may apply to some of the current efforts in process-oriented integrated NRM (INRM), it does not adequately explain the lack of NRM impact assessments for research focused mainly on productivity improvements, the lion's share of NRM efforts before the mid-1990s and a significant portion of it afterward.

When addressing NRM research impacts, a wide range of other issues needs to be considered. Markets are largely missing for the environmental services provided. Different valuation methods exist, all of which are highly imperfect and tricky to use, and hence there is a need for a range of values reflecting different perspectives and valuation methods. Externalities are spread over different scales and hence difficult to capture as each level needs to be done with different tools. The time dimension is crucial and hence the choice of discounting key. There are also important problems of resilience and irreversibilities that need to be taken into account in constructing counterfactual scenarios. For these reasons, designing control groups for NRM treatments is particularly difficult because of the spatial and temporal dimensions involved.

The difficulty in measuring and attributing impact of NRM/INRM research is generally recognized to be of a significantly higher order than for crop germplasm research [Izac (1998)]. The issues relate particularly to complexity issues (in scale, in time), non-linearity (causality), the economic and non-economic dimensions, operation-indicator issues, higher costs, more disciplines involved, longer time lags, attribution problems, and difficulty in extrapolation. The recognition that some of the gains and impacts from crop improvement have been supported by improved crop and soil management derived through NRM research is not always apparent.[13] This is a measurement/allocation problem, but without some evidence it remains conjectural, or anecdotal at best. There is a need to develop means to measure and subsequently document the role improved resource management has played in realizing on-the-ground impacts.

Given the levels of investment to-date in NRM research in the CGIAR and that much of it has targeted productivity improvements, NRM productivity impacts using the conventional market model should not be dismissed *a priori*. Underpinning this is the core issue of efficiency of resource use. Virtually all sustainable paths to poverty alleviation are derived directly or indirectly through increased productivity. Thus, notwithstanding the present need for new methods and approaches to measure the more-complex and less-tangible effects of NRM research, it remains the case that even simple impact measures, such as adoption and use of NRM products and practices, are still scarce. Complexity itself may not be the primary reason for a lack of documented impact in NRM research in the CGIAR.

[13] Bell et al. (1995) attempted to measure the genetic and agronomic contributions to increasing wheat yields in Northwest Mexico. They estimated a 28% yield gain due to genetic factors, a 48% yield gain attributed to increasing N fertilizer and the remaining 24% was attributed to 'other factors' – possibly including increasing P fertilizer, among others. These results, while isolating the pure genetic contribution, do not in themselves establish a contribution from NRM research.

2.5. *Lack of impact per se*

It must be recognized that, as in the case of other types of research, some NRM research in the CGIAR has failed to generate the appropriate technologies or institutional arrangements that adequately address the needs of poor farmers and communities. CGIAR Center annual reports from the late 1970s through to the early 1990s attest to the range of NRM research-related activities in which the CGIAR has been involved including water harvesting, broadbed-and-furrow management, erosion control through contour bounding, zero tillage in Africa, use of green cover crops and mulching, ley farming, alley cropping, and better management of the crop stover. The analyses of the problems and the long-term basic and applied research undertaken to address them are, for the most part, highly commendable from a scientific point of view. What is missing is impact. There is not much evidence, even today that, over the long run, the work has generated sufficiently wide-scale adoption of improved resource management practices among farmers. Admittedly, in such cases where it is evident that adoption is lacking, there is little incentive to assess impact. Thus, lack of impact *per se* could be a major reason behind the lack of *evidence* of impact. This is not an indictment of the quality of research conducted – not all research can be expected to result in a proven, adopted technology – nor does it overlook the fact that some technologies have indeed been adopted by some farmers.

One hypothesis to explain why NRM research may not have had more impact is that the innovation generated through the research is not, in itself, sufficient to catalyze wide-scale adoption. Its use and adoption is contingent on a great many other pre-conditions including, in some cases, institutional reform. Relative to germplasm improvement, NRM improvements require many more actors to get impacts on the ground, such as extension, policy, institutions, organized farmers and communities. For example, lack of an effective delivery mechanism could explain low adoption, although this reason may be used more frequently than is justified. Also, because NRM information is often more location-specific, and the CGIAR has not yet developed adequate links with many of these actors at the local level, it is inherently difficult to generate these impacts.

To the extent that for either technical, economic, or social reasons, research-led innovations have not been adopted, it might be useful to distinguish between NRM research focused on individual farmer-based decision-making (more technology-focused) vs that focused on group/community-based decision-making (more rules/institution-focused, technology less important). With respect to the individual farmer, the attractiveness of a new management practice depends largely on expected profitability/risk levels and additional labor or other inputs required with the 'improved NRM-based technology'. Perhaps insufficiently appreciated is the fact that many farmers in developing countries are looking for innovations that allow them to reduce their labor input in agriculture, not increase it, or to have other opportunities that are more profitable or less risky, or that give them higher utility, e.g., investments in children's education. An opportunity-cost assessment approach is more relevant in this case. Some of ICRISAT's research on the non-use of fertilizers in southern Africa shows this to be the case [Rusike, Dimes and Twomlow (2003)].

With respect to NRM research focused on community-based decision making, the emphasis on key issues such as property rights and the need for community action has resulted in a number of promising pilot success stories, as brought out in, e.g., the Systemwide Collective Action and Property Rights (CAPRi) external review [Interim Science Council (2003)]. Here, the major constraint is scaling up, or scaling out and without that capacity, the investment cannot usually be shown to be cost-effective. Indeed, this was one of the major conclusions reached at the Agroforestry Dissemination Workshop held at ICRAF in September 1999: "The developing world has no shortage of successful 'pilot' schemes and projects that have sought to address the problems of poverty, food security and environmental degradation. There are too few cases where these successful pilots have led to widespread impact on a sustainable basis" [Cooper and Denning (2000)]. Exacerbating scaling problems is the fact that funding for extension has fallen significantly in recent years, and the greater the complexity of technological adoption, the greater the need for extension [Douthwaite, Keatinge and Park (2001)]. Thus, this lack of impact may not be attributable to research itself. The entire impact pathway needs to be considered, including dissemination and adoption processes. Acquiring a better understanding of how resource management practices change over time and under different sets of agricultural policies and economic and social environments is fundamental. This sets the stage for more effective targeting of technology and greater impact.

3. Rates of returns to IARC research investment

A large body of benefit–cost and rates of returns studies exists that document the efficacy of earlier investments in the CGIAR, both for specific types of research and for the organization as a whole. These studies show in a relatively consistent manner that rates of returns to agricultural research compare favorably with alternative public investments.

3.1. Returns to crop improvement research investment

When considering the broad spectrum of research and research-related activities undertaken by the CGIAR, rates of return studies for crop germplasm improvement (CGI) research are the most numerous and best documented. Studies by Scobie and Posada (1977) at CIAT and Flores-Moya, Evenson and Hayami (1978) at IRRI, focusing on rice improvement in Columbia and the Philippines, respectively, were some of the earliest studies to calculate rates of returns on CGIAR research investment. Although focused on relatively short periods of time, 1957–1964 and 1966–1975, the studies estimated relatively high rates of return, well above 50%. This marked the beginning of numerous other studies which followed, expanding analyses to cover other geographic regions, longer timeframes and other crops where the CGIAR was having impact. Taken together, studies estimating the rates of return to CGIAR commodity research investments have consistently shown the investment to be extremely profitable. Evenson (2001) and Alston et al. (2000) provide a detailed synthesis of studies conducted across crops and countries. Their reviews confirm the widespread evidence of high economic rates of return for crop improvement research in the CGIAR.

Gardner (2003) provides a comprehensive review and critique of the huge body of CGIAR impact and rates of return literature in a recent analysis done for the World Bank. This meta-analysis looks at a range of Center impact assessments as well as some major earlier meta-analyses. Drawing on a major piece of work by Anderson, Herdt and Scobie (1988) examining the benefits and costs of research on (but not only) wheat and rice breeding in the CGIAR, and making conservative assumptions about total economic gains and modest attribution to the CGIAR, Gardner estimates a b/c ratio of 6.7, just considering efforts on wheat and rice. Some other more recent studies reviewed by Gardner include one by Anderson and Dalrymple (1999) where the economic surplus generated by improved CGIAR-derived varieties of wheat and maize is estimated at $1.8 billion and $1.0 billion, respectively (up to 1997). A study by Jha and Kumar (1998) calculated internal rates of return from the joint IRRI – Indian national program between 32 and 74% across various states in India. Heisey, Lantican and Dubin (1999) on wheat and Morris (2001) on maize estimate rates of returns on a global scale that result in "phenomenal rates of return". These are a sample of the many studies that have sought to document the positive economic effects of CGIAR research.

Gardner highlights important caveats and qualifying statements for many of these studies. These include the degree to which adequate documentation is provided (poorly in some cases), the degree to which reasonable assumptions have been made (often not), the transparency provided (missing in several studies), biases in case study selections (the winners) and problems invariably associated with attribution effects and the development of the counterfactual. Notwithstanding these caveats and the different concerns expressed for different studies, taken together, the body of evidence is fairly robust and widespread and typically consistent (albeit with sometimes large degrees of error) for the high rates of return from the CGIAR commodity programs – particularly for wheat, rice and maize – and to a lesser extent for cassava, potato, sorghum, and other crops. Not surprisingly, Gardner also found that for CGIAR efforts in areas less directly related to agricultural productivity, the evidence available on adoption and impacts is very little indeed, if any.

Some have claimed, based largely on anecdotal evidence that rates of returns to agricultural research in general, and by implication rates of return to investments in the CGIAR, have been declining. Alston et al. (2000) in their assessment of trends and characteristics of the rates of return in agricultural research and development examined 292 case studies with 1900 estimated rates of returns. The median annual rate of return estimate from these studies fell within 40–60% – consistent with the broad literature. More importantly, they found no evidence that rates of return to agricultural research had declined over time. Evenson (2001) in his review of over 100 studies estimating rates of return to research came to a similar conclusion. Gardner (2003) concurs with that assessment based on his analysis but notes that even the most recent impact assessment reflects the product of research that occurred many years earlier and that there is essentially no evidence from these studies on returns to research conducted after 1990.

3.2. Non-commodity focused efforts

Pingali (2001) concluded from his review of impacts and rates of return literature that there were relatively few 'crop management and improved input use' and other NRM-related CGIAR impact studies to-date. The exception here are the large scale impacts from biocontrol or IPM research already alluded to. Using biocontrol to manage the cassava mealy bug during the late 1980s and thereafter delivered huge benefits to small producers virtually throughout SSA. Total economic returns, i.e., crop losses averted, were estimated between \$8 and \$20 billion [Zeddies et al. (2001)], depending on assumptions about the counterfactual. Biocontrol of the mango mealybug in Benin was another IITA success story generating high rates of returns [Bokonon-Ganta, DeGroote and Neuenschwander (2001)]. CIP's potato pest management impacts were summarized by Walker (2000) in ten case studies of the impact of improved potato varieties and pest control – generating rates of return ranging between 27 and 200%. In a review of the CGIAR's systemwide IPM efforts, Waibel (2000) estimated that the rate of return to investment in IPM has been in the order of 15 to 40%. As already alluded to, rates of return for investments in NRM research have averaged considerably lower than for crop improvement research [Alston et al. (2000)].

The degree of difficulty in measuring rates of economic return for some types of research is of a significantly higher order than that for commodity improvement. Measuring the rates of return on policy research and training, for example, is fraught with measurement difficulties. Nevertheless, IFPRI has made some impressive attempts to document – usually in non-economic terms, but sometimes in economic as well – the value of some of its policy research investments. Ryan (1999), for example, estimated that IFPRI's role in the changes in Vietnam's rice policy amounted to a gain of \$45 million during 1996–1997, generating a huge benefit–cost ratio of 45 to 1 (due to the low cost of research). But this is more the exception than the rule. The pathway from policy research to policy change to effects on agricultural productivity and, ultimately, to poverty impacts is long and complex and involves many actors. Attribution problems are probably insurmountable in many situations.

The CGIAR has made immense contributions to strengthening research capacity in the national research systems of developing countries. This has mainly occurred by building human capacity through training programs and by improving the exchange of information and technology through networks.

Building national capacity is believed by some to be the area where the CGIAR has had the largest impact. Indeed, strengthening NARS capacity has always been one of the key objectives of the CGIAR. However, unlike other forms of research which have more direct paths to poverty and for which economic and social rates of return can be estimated, investments in training and capacity building have thus far not been systematically assessed in terms of their impact.[14] Nevertheless, investments in this activity by the CGIAR have been major, typically accounting for 20 to 25% of the total CGIAR budget each year.

CGIAR germplasm and crop management networks have been essential to the rapid dissemination of knowledge and products. The networks allowed a global community of commodity scientists to be connected to each other and to exchange ideas and experiences. Yet almost no work has been done to formally assess the impact of CGIAR networks on productivity improvement for developing country agriculture. An exception is a study of CIMMYT Regional Maize Program in Central America (a maize research network) that found high returns to participation in the network, especially for small countries that could not afford a critical mass of crop research and development specialists [Gómez (1999)]. Substantial work is needed on the economic and social returns to network participation, since this mode of linking researchers in national programs and the CGIAR is expected to continue into the foreseeable future. Measuring the impacts of networks is very difficult, however, because of problems in clearly identifying inputs and outputs and attributing them to the participants in the network.

In his benefit–cost meta-analysis, Raitzer (2003) systematically reviewed and critically evaluated IA studies of economic benefits derived from CGIAR innovations (known 'success stories') to produce a range of (a) plausible and (b) highly-credible benefit–cost ratios for the entire investment in the CGIAR since 1972. Against an aggregate investment of 7120 million 1990 US dollars (6900 million of investment in the CGIAR, plus relevant pre-CGIAR costs) from 1960 through 2001, all scenarios produced benefit–cost ratios in substantial excess of one, using the benefits accruing from 1972–2001. Including only "significantly demonstrated" studies that empirically attribute CGIAR derived contributions to collaborative efforts results in a ratio of 1.94. When all "significantly demonstrated" studies are included with assumed attributive coefficients applied, the ratio rises to 3.77. The "plausible" scenario results in a ratio of 4.76, and when extrapolated to 2001 this rises to 9.00. If the latter is extrapolated through to 2011, the ratio rises to 17.26. Since costs are distributed over the benefit period, and many benefits peaked in the early 1990s, the discount rate applied only significantly affected generated ratios in the extrapolative scenarios.

Raitzer asserts that the true value of benefits arising from the CGIAR is probably in excess of even the upper bounds of the results demonstrated here, as only a small subset of System impacts have been assessed or can be measured easily in economic terms. Indeed, 98% of "significantly demonstrated" and 93% of "plausible" benefits were generated by just three research areas – cassava mealybug biocontrol, breeding of spring bread wheat and modern varieties of rice. Since these are not the only areas of CGIAR research success, so there is substantial scope for expanded impact coverage.

[14] A CGIAR Science Council externally commissioned study is currently underway to evaluate CGIAR training of NARS and, to the extent possible, its impact.

4. Impacts on poverty and food security

Productivity growth that resulted from agricultural R&D under the auspices of the CGIAR and its partners during the GR and post-GR period has had an enormous impact on food supplies and food prices, and consequent beneficial impacts on food security and poverty reduction. Rising productivity in the agricultural sector has also stimulated growth in the non-agricultural sectors and has acted as an engine of overall economic growth. While the benefits of agricultural productivity growth have been shared widely, some sections of society have undoubtedly gained relatively less than others have, such as landless labor, female headed households, and farm households in marginal environments. However, such statements must be qualified. For example, while the Green Revolution has been criticized for bypassing millions of resource-poor farm households living in marginal environments, this was not the case in all unfavorable environments. There is increasing evidence of agriculture success stories even within the agro-climatically stressed and geographically isolated environments [Kelley and Byerlee (2003)]. This involves not only technology transfer and capital investments but also the "software" of development, such as local institutions, property rights and social capital.

4.1. Food supplies and food prices

Widespread adoption of modern seed-fertilizer technology led to a significant shift in the food supply function, contributing to a fall in real food prices. The primary effect of agricultural research on the non-farm poor, as well as on the rural poor who are net purchasers of food, is through lower food prices:

> The effect of agricultural research on improving the purchasing power of the poor – both by raising their incomes and by lowering the prices of staple food products – is probably the major source of nutritional gains associated with agricultural research. Only the poor go hungry. Because a relatively high proportion of any income gains made by the poor is spent on food, the income effects of research-induced supply shifts can have major nutritional implications, particularly if those shifts result from technologies aimed at the poorest producers [Alston, Norton and Pardey (1995)].

Early efforts to document the impact of technological change and the consequent increase in food supplies on food prices and income distribution were made by Hayami and Herdt (1977) at IRRI, Pinstrup-Andersen, Ruiz de Londoño and Hoover (1976) and Scobie and Posada (1978) at the International Center for Tropical Agriculture (CIAT), and Binswanger (1980) at ICRISAT. Pinstrup-Andersen argued strongly that the primary nutritional impact for the poor came through the increased food supplies generated through technological change.

In examining some of the economic and social welfare effects of crop genetic improvement programs of the CGIAR and its partners, Evenson and Rosegrant (2003) estimate that without the CGIAR and national program crop germplasm improvement efforts, food production in developing countries would have been almost 20% lower (requiring another 15–20 million hectares of land under cultivation in addition to at least 5% higher food imports). World food and feed prices would have been 35 to 65% higher, and as a consequence, average caloric availability would have declined by 11 to 13% globally (more in some regions). Finally, child malnutrition would have gone up by 6–8% – affecting some 30 to 45 million more children than otherwise. Overall, these efforts benefited virtually *all* consumers in the world – and the poor relatively more so.

The profitability of modern farming systems has been maintained despite falling real food prices, owing to a steady decline in the unit cost of production. The point that producers have continued to benefit from technological change despite falling output prices has not been emphasized adequately in the literature, although empirical evidence does show quite clearly that unit cost of production has fallen significantly for modern varieties of crops, such as rice [e.g., see Pingali, Hossain and Gerpacio (1997) for several Asian locations and Hossain (1998) for Bangladesh] and wheat [e.g., see Sidhu and Byerlee (1992) for evidence from the Indian Punjab].

The proposition that agricultural growth acts as an engine of overall economic growth and poverty reduction has been empirically supported by several in-depth case studies in Green Revolution areas [Hazell and Haggblade (1993); Fan, Hazell and Thorat (1998)]. Hayami et al. (1978) provided for the Philippines a village-level illustration of the impacts of rapid growth in rice production on land and labor markets and the non-agricultural sector. Long term changes in village economy were traced through periodic revisits over three decades [Hayami and Kikuchi (2000)]. Hazell and Ramaswamy (1991) showed the development of backward and forward linkages from increased agricultural productivity growth in India. Delgado, Hopkins and Kelly (1998), found similar evidence for Africa – growth being stimulated in the non-agricultural sector by growth in agricultural productivity.

4.2. *Differential impact of technological change*

The impact and benefits of technological change have varied by ecological domain, socioeconomic factors (such as farm size), and gender. Many studies have addressed the differential impact of technological change in favorable and unfavorable production environments. David and Otsuka (1994) conducted a study on the differential impact of technological change across rice environments in Asia. Although the favorable, high-potential environments gained the most in terms of productivity growth, the less favorable environments benefited as well through technology spillovers and through labor migration to more productive environments. Wage equalization across favorable and unfavorable environments was one of the primary means of redistributing the gains of technological change. Renkow (1993) found similar results for wheat grown in high- and low-potential environments in Pakistan.

Indeed, there are many examples where people living in marginal environments did in fact benefit from new agricultural technologies and development efforts, with significant impacts [Kelley and Byerlee (2003); Lantican, Pingali and Rajaram (2003)]. Notwithstanding these important qualifiers, few would argue that rural producers in marginal areas have received benefits comparable to their counterparts in the better endowed areas, where irrigation and associated inputs are more readily available, and modern varieties have been widely adopted. For example, Byerlee and Morris (1993) confirmed that improved seed-fertilizer technologies for wheat were less widely adopted in marginal environments worldwide and had less of an impact there than in favored environments. Byerlee (1996), too, found that almost full adoption of wheat and rice HYVs had been achieved in irrigated environments by the mid-1980s, but very low adoption in environments with scarce rainfall, or poor water control (in the case of rice). Moreover, whereas HYVs of wheat provided yield gains of 40% in irrigated areas, with modest use of fertilizer, in dry areas gains were often no more than 10%. On the consumer side, the picture is more straightforward. Increased food supplies and lower food prices via GR effects benefited virtually *all* consumers in the world, and the poor relatively more so since they spend a larger share of their income on food. Indeed, many consumers in marginal areas have benefited from lower food prices, and even the majority of farmers in these environments are often net food purchasers [Renkow (1993)].

Poorly endowed environments, nevertheless, pose a tremendous challenge to researchers and policy makers alike, namely to identify new agricultural R&D opportunities and to facilitate adoption of technologies and appropriate institutions to meet the needs of the poor living there.

Income distribution effects across the various socioeconomic groups within a rural community have received some attention in the impact literature. In a detailed study of North Arcot District of Tamil Nadu, India, Hazell and Ramaswamy (1991) estimated the distribution of benefits of technological change across landless laborers, tenant farmers, and small and large landowners. David and Otsuka's (1994) study paid particular attention to effects on landless labor and tenant farmers, and found that the benefits were shared across the various farm size groups. The early criticism that the Green Revolution had benefited only large-scale farmers stood in sharp contrast to the findings of all of these studies. Empirical evidence indicates that small farms also benefited, albeit later, in terms of productivity and income growth.[15] Hazell and Ramaswamy found that initially the new technology was principally adopted by large-scale farmers, but as time

[15] One of the common arguments made for small farmer reluctance to adopt new technologies is that it is riskier than the ones they use currently. Anderson and Hazell's (1989) volume on variability in grain yields provided an important synthesis of evidence on production variability in agricultural systems that had recently switched to modern varieties. The volume covered most CGIAR crops and all continents where the CGIAR centers worked. The worry that modern varieties may be more risky and therefore less attractive to farmers does not seem to have held up in practice. Stochastic dominance tests of the distribution of returns from improved and traditional varieties typically show new varieties to be dominant. The following studies provided crop-specific results for sorghum and millets [Walker (1989); Witcombe (1989)] and rice [Flinn and Garrity (1989); Coffman and Hargrove (1989)]. More recent studies at CIMMYT [CIMMYT (1991)] for wheat and at ICRISAT for millet [Adesina (1988); Shapiro (1990)] have reported reduced coefficients of variation for yields over time.

passed smaller farms began adopting high-yielding varieties at similar rates of large-scale farmers. In North Arcot District of Tamilnadu, between 1973/74 and 1983/84 "small paddy farmers and landless laborers gained the largest proportional increases in family income". The authors point out that the results of these papers are quite different from many previous studies which evaluated the new technology too early in the adoption process.

The evidence on landless labor is less clear. Pinstrup-Andersen and Hazell (1985) argued that the landless labor did not adequately share in the benefits of the Green Revolution because of depressed wage rates attributable to migrants from less endowed regions. David and Otsuka (1994), however, found that migrants shared in the benefits of the Green Revolution through increased employment opportunities and wage income.

The differential impact literature has focused on identifying the distribution of benefits between men and women farmers and male- and female-headed households. The general finding across crops and continents is that women farmers and female-headed households have gained proportionally less than their male counterparts. It is not gender alone that determines whether an individual benefits from technological change, however, but rather the initial social and economic status of the individual [Paris (1998)]. Quisumbing, Haddad and Peña (1995) concluded from a ten-country study that among the very poor the economic welfare of male- and female-headed households differed very little. Differences emerged only where cultural or institutional factors prevented equal participation in the labor force, as in Bangladesh. For excellent recent reviews of the literature, see Doss (1999) on African maize farming systems and Paris (1998) on rice in Asia.

5. Challenges ahead

Some of the emerging trends in the global food economy are highlighted in Box 1. As should be obvious, the world today is dramatically different from when the CGIAR was first established. Trade liberalization and greater integration of global food markets is leading to more reliable food supplies and lower food prices in real terms. Income growth, urbanization and growing global inter-connectedness are leading to diet diversification and homogenization. The locus of technological innovation has shifted from the public to the multi-national private sector with a consequent rise in the transactions costs faced by developing countries in accessing technology. Access problems are particularly acute in the case of the emerging biotechnologies. While hunger and food insecurity continue to be an important focus of developing country food policy, the emphasis has shifted from a pre-occupation with increasing food supplies to one of enhancing access to food. Potential trade-offs between agricultural productivity growth and the conservation of the natural resource base receive substantially greater attention today than in earlier decades.

Our review of the existing evidence shows that the CGIAR has had a significant impact on increasing national as well as global food supplies, and thereby has helped in

Slowdown in population growth: The growth rate of 1.35% per annum in the second half of the 1990s is expected to decline to 1.1% in 2010–15 and to 0.5% by 2045–50 [UN Habitat (2001)].
Income growth and reductions in poverty: Per caput income growth in developing countries will increase from 2.4% per annum between 2001 and 2005 to 3.5% between 2006 and 2015. The incidence of poverty will be reduced from 23.2% in 1999 to 13.3% in 2015. These figures are for developing countries as a whole. It should be acknowledged that reductions in the incidence of poverty will be geographically uneven with the greatest progress being made in East Asia and the least progress in Sub-Saharan Africa [FAO (2002c)].
Urbanisation: Virtually all of the world's anticipated population growth between 2000 and 2030 will be concentrated in urban areas [UN Habitat (2001)]. At the present rate of urbanisation, urban population will equal rural population as early as 2007 and will exceed it from that point on.
Average food intake will increase but hunger will remain high: Daily per capita calorific intake in developing countries will increase from an average of 2681 kcal in 1997/9 to 2850 in 2015. Under 'business as usual', undernourishment will decline from 20% in 1992 to 11% in 2015, but reductions in absolute numbers of undernourished people will be modest – from 776 million in 1990/92 to 610 million in 2015, far from meeting the WFS target.
Diet transitions: The pace of dietary change, both qualitative and quantitative, accelerates as countries become richer and populations become increasingly urbanised, with a shift in diet structure towards a higher energy density diet in developing countries and a dramatic increase in the contribution to food calories from livestock products (meat, milk and eggs), vegetable oils, and, to a lesser extent, sugar.
Growing agricultural imports: Agricultural trade surpluses in developing countries are shrinking and, by 2030, will have become a deficit of about US$31 billion, with a rapid rise in imports of cereals and livestock products and a decline in surpluses in vegetable oils and sugar [FAO (2002a)]. In addition to the emergence and strengthening of international trade agreements, growing developing country food imports are also due to increased demand combined with low competitiveness of their domestic agriculture.
Changes in product composition: Between 1997 and 2015, wheat and rice production in developing countries will grow modestly (by 28 and 21%, respectively). However, significant increases are expected in coarse grains (45%), vegetable oils and oilseeds (61%), beef and veal (47%), mutton and lamb (51%), pig meat (41%), poultry meat (88%), milk and dairy production (58%) [FAO (2002a)].
Market structures: Food markets in developing countries are undergoing profound changes that are fueled by economic development, increase in per capita incomes, changing technology and urbanisation. Urbanisation increases the scope for economies of scale in food marketing and distribution, while reductions in transactions costs increase the size of the market for distributors and retailers. The result is an impressive increase in the volume of food handled by supermarkets [Reardon et al., this volume], but also substantial organisational and institutional changes throughout the food marketing chain [Dolan and Humphrey (2001)].
Towards larger farm sizes: With the increasing commercialization of production systems and the rising opportunity cost of labor, small family farm operations for subsistence production become increasingly unprofitable. Landless tenant farmers will gradually find their way to the urban industrial sector. Small landowners will likewise find it more profitable to sell or lease their holdings rather than to cultivate them.
Production growth based mostly on yield growth: Yield improvements will account for about 70% of production growth, land expansion for 20% and increased cropping intensity for the rest. Nevertheless FAO projections show that the arable area in developing countries will increase by almost 13% (120 million ha) and water withdrawals for irrigation by 14% by 2030. One in five developing countries will face water shortages [FAO (2002a)].
Growing environmental and social costs: Although food will, in general, be cheaper, failure to internalise the environmental costs of the expansion and intensification of agriculture will result in the price of food being lower than its social cost, holding back incentives for further research in yield improvements. Preserving biodiversity will be a formidable challenge. Also, where property rights are not clearly established, high value crop production in upland environments could lead to higher risks of soil erosion and land degradation. Finally, the absorption of the rural poor in the industrial and service sectors has significant costs in terms of learning new skills and family dislocations.
Changing locus of agricultural R&D: While the national agricultural research systems in some developing countries are getting stronger, the locus of agricultural research and technology development is moving towards the private sector, especially in the case of biotechnology. Accessing these emerging technologies and targetting them towards the needs of the poor is a major challenge faced by the developing country public research systems.

Box 1. Facts and trends in the emerging world food economy. [Adapted from DeHaen et al. (2004).]

making food supplies more accessible to the urban poor and to many rural poor as well. The circumstances are considerably different today, however. There now exists a vast number of alternative suppliers of research and development, both at the national and global level – new actors in the global agricultural research and knowledge system capable of generating new technologies and improved policies and facilitate in disseminating new ideas, new information and the resulting technology innovations. In the future, the CGIAR is unlikely to play the dominant role it once did, nor should it be expected to. Although its role is more complementary today than it was 30 years ago, and is likely to become increasingly so, its strategic contribution in the future may have as much, if not more, value than when it was the predominant driving force in the global agricultural research system.

The overriding goal of the CGIAR, to reduce poverty, hunger and malnutrition by sustainably increasing the productivity of resources in agriculture, forestry and fisheries, will continue to be relevant into the foreseeable future. Indeed, the work of the CGIAR aligns and will no doubt contribute directly to several of the MDGs related to hunger and environmental sustainability. But the CGIAR cannot possibly hope to achieve its goal independent of what others are doing. There are numerous other stakeholders with which it must remain (or become) actively engaged and with whom it must coordinate and synchronize its efforts. Why is this so necessary, and now perhaps more so than ever?

Firstly, confidence in the CGIAR succeeding in its goal rests on the belief that raising agricultural productivity will improve the welfare of many millions of poor producers, laborers and consumers. These productivity improvements may have direct effects on poor producers or indirect effects on poor consumers and poor laborers or combinations of both [de Janvry and Sadoulet (2003)]. The CGIAR, working in close partnerships with global, regional and national development partners (e.g., international development organizations, government ministries, development banks, regional fora, international NGOs, etc.) across a broad spectrum of development fields, has a critical role to play in helping identify the appropriate role of and therefore emphasis given to agricultural development in the context of a broader poverty-focused development strategy for different regions. In some cases, this may include taking on the role of advocacy.

Secondly, to the extent that raising agricultural productivity is considered a potentially cost-effective instrument in addressing poverty within the context of a broader development strategy, it must be recognized that making agriculture more productive can be achieved through a number of channels, e.g., via improved roads, markets, and communication and through investments in education, health and clean water supplies, among others. An important channel, and one that has, as discussed earlier, served well in the past, is through the generation, dissemination and adoption of improved agricultural technologies, policies and institutions. In cases where productivity constraints are not well understood, or where processes that might contribute to those solutions are not well understood, or require testing or validating, there is a strong justification and a clear role for agricultural research. Here again, it is essential that the CGIAR, working in close partnership with various stakeholders within the agricultural, forestry and fisheries R&D network, spell out the specific role of research, vis-à-vis other ag development initiatives, in addressing acute and long-term productivity constraints that affect the poor.

Finally, the specific role and function of CGIAR research efforts must be defined in the context of a much broader global agricultural research system than the one that existed thirty years ago. Ideally, this should occur through a highly consultative and well-coordinated priority setting process where CGIAR and partner priorities are discussed and determined in the context of systematic constraint (physical, technological, institutional and policy) analyses and identifying opportunities for raising the productivity of crops, trees, livestock and fish for target groups of poor for whom agriculture development offers the most likely pathway out of poverty. This encompasses, in addition to definition and validation of the most relevant *researchable* issues related to overcoming widespread productivity constraints, an assessment of each partner's comparative advantage in specific areas and types of research. Obviously, close consultation with a range of alternative suppliers of research and relevant R&D partners is essential, including with representatives of advanced research institutes, NGOs, farmer organizations, civil society, and the private sector in addition to NARS clients, including GFAR and the regional and sub-regional organizations.

This is fairly basic, but it is important to emphasize that the main intervention point for the CGIAR and its R&D partners in alleviating poverty is through raising productivity of resource use in agriculture, forestry and fisheries – a focus that must be retained at all costs, and that today, much more so than 30 or 40 years ago, this requires a broad coalition of effort across stakeholders and actors working together and complementing each others strengths. In summary, the major challenge for the CGIAR will be to carefully map out its strategic role within the broader context of (a) other development instruments with a focus on poverty alleviation, e.g., investments in education, health, infrastructure, and economic policy reform; and (b) the emerging demand for and supply of global agricultural research. With respect to the latter, a stronger focus on raising productivity (in the short *and* long term) to the factors in greatest demand in areas where the poor are concentrated will help provide the focus needed in the System at this time.

The material contributions of the CGIAR in the past have derived from research that generated new information that, ultimately, was embedded (by the CGIAR or its clients) in new seeds, management innovations, policies and institutions. Generating and disseminating new knowledge of relevance to clients and partners globally remains the *première raison d'être* for the CGIAR, and likely will remain so for the foreseeable future. As argued earlier in this paper, however, this is insufficient in itself to justify an international agricultural research system. The emphasis must be firmly kept on generating knowledge of both relevance and wide applicability, i.e., IPGs, rather than on more visible but narrowly focused products, despite the demand for the latter.

5.1. CGIAR public goods

There are five major types of public goods/roles that we believe the CGIAR can and should be providing/performing over the next 10 to 15 years. They are:

1. Maintaining and securing the vast collections of genetic resources held in trust and preserving and enhancing the information contained therein.
2. Generator and provider of new knowledge through applied biological and social sciences.
3. Assessing the biophysical and socioeconomic consequences of technical change.
4. Strategic leadership and integrator within the global agricultural research community.
5. Facilitator and an 'honest broker' in access to knowledge and technology.

5.1.1. CGIAR as global custodian of genetic resources

Modernization of agriculture and continued genetic improvement does not necessarily lead to reduced reliance on genetic diversity. Access to diverse sources of germplasm is of great importance to the success of public and private breeding programs. The continued advances in yield potential that are a necessary condition for alleviating hunger are thought to depend on increasingly complex combinations of genes and novel alleles [Pingali and Smale (2001)]. Landraces and wild relatives have served as repositories for resistance to biotic and abiotic stresses when these were absent in advanced breeding material. Moreover, even in areas where modern varieties are well established, idiosyncratic growing conditions and consumer preferences may provide economic incentives for the continued cultivation of traditional varieties and local land races. Hence, a strong case exists for continued investment in the collection and conservation of genetic resources.

The CGIAR has an exemplary record in collecting and conserving genetic resources for the cereals of major importance to the developing world. The 11 genebanks maintained by the CGIAR Centers conserve more than 666,000 accessions of staple crops and tree species [Koo, Pardey and Wright (2003)]. CGIAR collections account for roughly 30% of the unique entries in all genebank collections worldwide. Studies have indicated that the *ex situ* collections in the CGIAR gene banks have been cost effective and have had very high rates of return on the investment.[16] With respect to further collection activities for existing crops held in trust by the CGIAR or for the collection

[16] Koo, Pardey and Wright (2003) estimate that the annual cost of conserving and distributing the existing collections in the CGIAR is around 5.7 million US dollars. Moreover, the services provided by the gene banks can be ensured for perpetuity with a relatively modest endowment of 149 million US dollars invested at a real rate of interest of 4% per annum [Koo, Pardey and Wright (2003)]. The authors argue that 40% of the amount would underwrite the cost of maintaining the collections and 60% for sustaining the distribution activities. The Global Conservation Trust has been setup to create such an endowment and to create a mechanism to sustain the activities of the CGIAR gene banks.

of new or underutilized crops that have not been part of its collection, the CGIAR may not in the future be in a position to undertake such activities. Instead, it should focus on building capacity within the NARS for collection activities and to set protocols and standards for maintaining the collections. This is especially relevant given that many underutilized or "orphan" crops are more geographically isolated.

Extracting genetic and phenotypic information from the material that is available in its collections is an area where the CGIAR could advance much further than it has to-date. Use of modern molecular biology (genomics and bioinformatics) could and should allow not only CGIAR breeders but the global community to tap the vast pool of information residing in the CGIAR genebanks. Assessing and releasing such information into the public domain would have enormous benefits for the poor, in terms of identifying traits that are important for the poor and in terms of the speed with which varieties can be developed and released.

Some of the genetic diversity may best be preserved *in situ*. Accordingly, CGIAR and NARS efforts may increasingly be focused on *in situ* germplasm collections and characterization and understanding the future of local landraces and local genetic diversity in a rapidly changing rural economy. Incentive-based mechanisms for future preservation of such material should be explored.

5.1.2. From product development to knowledge generation though application of social and biological sciences

The number of alternate suppliers of agricultural technologies, specifically seed-based technologies, has expanded rapidly over the last two decades. Strong NARS and the private sector have become major players in the research, generation and release of new varieties. Even NGOs and civil society organizations are becoming active in developing community seed systems. While there are still a significant number of weak NARS that depend directly on CGIAR products – particularly improved seed, because of the alternative sources of supply, varietal development should not be the primary responsibility of the CGIAR. With a growing demand for "custom made" varieties that can address country specific taste and quality preferences, even where the production environments are homogeneous across continents, the CGIAR would not appear to have the resources or comparative advantage to carry out that function.

The CGIAR now needs to seriously move its breeding and germplasm enhancement efforts more 'upstream', interact and build strategic partnerships with private sector and focus its efforts on generating basic information, methodologies and producing intermediate breeding products that could be utilized by national programs in the development of finished products. The CGIAR has an extremely important role to play in the basic science behind germplasm improvement, i.e., functional genomics, and its application in the development of new plant ideo-types with traits of particular importance to the poor, e.g., drought tolerance. It must keep at the forefront of science and be adept in applying the latest molecular biology tools, such as molecular marker aided selection and transgenic methods, to develop new breeding lines of relevance to its clients in the national programs.

In the area of NRM, the CGIAR must concentrate its efforts on improving its understanding of tropical and sub-tropical agro-ecologies to facilitate innovation and new management practices that raise productivity of land, water and, particularly, labor. Understanding the processes that lead to sustainable soil fertility and pest management – essential conditions for ensuring long-term sustainable productivity improvement – should be the overriding objective of this priority area. The emphasis must be strategic knowledge generation rather than the development of location-specific techniques and products. While universities and advanced research institutes in the developed world may have significant capacity to conduct such strategic work, it is important to realize that their emphasis on tropical agro-ecosystems is limited. The CGIAR, with its expertise on tropical and sub-tropical agriculture and its familiarity across a broad range of crops and ecosystems, has a comparative advantage in this area. Although the challenge of keeping focused on generating IPGs in the area of NRM is perhaps more challenging than in other areas of research, there are specific resource management problems of an international nature for which the CGIAR could play a unique role, for example, in understanding the extent and magnitude and major causes of unsustainable agricultural practices, such as soil erosion, or excessive water use.

5.1.3. Assessing the consequences of technical change

This chapter has highlighted only a small portion of the immense volume of work undertaken, primarily by CGIAR economists, but not exclusively, in measuring the socioeconomic impacts and consequences of technical change due to CGIAR research. They leave no doubt about the far-reaching impact that modern technologies, particularly seed based technologies, have had on poverty reduction, food security improvement, and the enhancement of rural livelihoods. However, these studies have had two major shortcomings. First, the focus has typically been on successes with much less effort on providing a broader and more comprehensive assessment of the distributional effects, both positive and negative. Second, biophysical consequences of research-derived new technologies have typically been ignored. Hence, impact assessment as a tool for strategic learning and as input for making mid-course correction with current research has not been adequately exploited.

Dramatic changes in farming practices have occurred on millions of hectares of land across the tropics over the last fifty years, driven by research-induced opportunities for agricultural intensification. Yet, little systematic work has been conducted to understand the biophysical consequences of these changes, in particular, its impact on long-term sustainability of these systems. While it cannot reasonably be argued that research *per se* has been the sole driver of this intensification process, the CGIAR is well placed to assess these changes in terms of developing methods and undertaking comparative analysis of the biophysical consequences. In terms of socioeconomic consequences, the emphasis on "what works" – which has been reasonably addressed, has not been sufficiently augmented by assessments of "what has not worked" or what has worked less well. Understanding factors that affect adoption, i.e., the complex of environmental, technological, socio-economic, institutional and political, still remains inadequately investigated despite the vast amount of work to-date by economists within and outside the CGIAR.

Successes and failures are often predicated on the existing policy and institutional environment in which technological change is being promoted. Understanding and documenting the interface between technology and policy would be an important IPG generated by the CGIAR. Developing appropriate frameworks for assessing and understanding the relative importance of different types of productivity constraints (technological, policy, institutional) in the context of different target groups of poor, is an area that needs considerably more emphasis, and would build on a vast amount of CGIAR experience. Such understanding could lead to better *ex ante* assessments of where sustainable productivity improvement can be anticipated.

5.1.4. Strategic leadership and integrator within the global agricultural research community

With its pool of technical expertise and accumulated experience in developing country agriculture, the CGIAR has played an important role, in association with others such as the World Bank and FAO, in shaping the vision for the food and agricultural sector, and in targeting research investments toward future needs. But this role can be enhanced even further. Indeed, the advisory body to the CGIAR, the Science Council, embraces this challenge specifically, through its standing panel on mobilizing global science. It plans to release a report reviewing the state of agricultural science and technology later in the year. The neutrality of the CGIAR is an additional asset in the international acceptance of its strategic vision. The CGIAR can also become an advocate, promoting attention to the problems of the poor, facilitating South–South and North–South partnerships, and drawing attention to the key role of investments in agriculture development and sound policies have in fostering growth and alleviating poverty.

Experience with IFPRI's 2020 demand/supply projections indicates that the CGIAR's vision on the food and agriculture sector has become a standard against which other scenarios are compared. IFPRI's projections have also played an important role in commodity specific strategic planning, priority setting and research investments, at other CGIAR centers as well as in various national programs. While 'visions of the future' have traditionally been the domain of economists within the CGIAR, important contributions can be made by biological scientists too, in contributing to existing exercises and in projecting detailed scenarios of likely 'technology futures'.

In addition to remaining active in research and contributing to the generation of IPGs, the CGIAR must strengthen its role as a catalyst, integrator and disseminator of knowledge within the overall global agricultural research system. The need for intensifying its efforts in such functions was highlighted in the CGIAR Vision and Strategy document [TAC (2000)]. The major focus should be on issues important to NARS and on the changing external international environment. These include issues related to genetic resources conservation and characterization, bioinformatics, IP, ICT and knowledge management. The CGIAR Centers may also facilitate greater linkages between the NARS and the research institutions in industrial countries concerned with international agricultural research, such as those in Europe, North America, Japan and Australia. In this way, the CGIAR's investment in research could be combined with those of others to support the development of a global system for international agricultural research.

5.1.5. Honest broker in access to knowledge and technology

The final area where the CGIAR should continue to provide an important public service to the developing country agricultural science community is in helping reduce transactions costs in the acquisition of knowledge and technology. This chapter documented the contribution that the CGIAR germplasm networks played in the rapid dissemination of improved varieties and breeding materials across the developing world. The continuing momentum of the Green Revolution can in part be explained by the extensive cooperation and collaboration between the IARC's and the NARS's in the development and exchange of improved germplasm [Traxler and Pingali (1998)]. As the CGIAR moves upstream these networks will continue to be relevant, although they would need to be transformed into 'knowledge networks', even as they continue to provide advanced breeding lines and finished varieties for some clients.

The changing locus of agricultural research from the public to the private transnational sector, with particular reference to biotechnology, poses important challenges and opportunities for the CGIAR. The CGIAR could become an important conduit for the flow of modern biotechnology knowledge, tools, methods and products to developing country scientists. In order to do so, it would first be necessary to invest significantly more in strengthening its own capacity in molecular biology research and, thereafter, helping build NARS capacity. The CGIAR could also play a role in designing institutional mechanisms and policy measures needed to promote the sharing of private intellectual property for public goods research.

References

Adesina A.A. (1988). "Farmer behavior and new agricultural technologies in the rainfed agriculture of Southern Niger: A stochastic programming analysis", Ph.D. thesis. Purdue University, Indiana.

Alston, J.M., Norton, G.W., Pardey, P.G. (1995). Science under Scarcity: Principles and Practices for Agricultural Research Evaluation and Priority Setting. Cornell University, Ithaca.

Alston, J.M., Marra, M.C., Pardey, P.G., Wyatt, T.J. (2000). "Research returns *redux*: A meta-analysis of the returns to agricultural R&D". Australian Journal of Agricultural Resource Economics 44 (2), 185–215.

Anderson, J.R., Dalrymple, D.G. (1999). The World Bank, the Grant Program, and the CGIAR: A Retrospective Review. Operations Evaluation Department, World Bank, Washington, DC. 46 pp.

Anderson, J.R., Hazell, P.B.R. (1989). Variability in Grain Yields: Implications for Agricultural Research and Policy in Developing Countries. Johns Hopkins University Press, Baltimore.

Anderson, J.R., Herdt, R.W., Scobie, G.M. (1988). Science and Food: The CGIAR and Its Partners. World Bank, Washington, DC.
Bell, M.A., Fischer, R.A., Byerlee, D., Sayre, K. (1995). "Genetic and agronomic contributions to yield gains: A case study for wheat". Field Crops Research 44, 55–65.
Binswanger, H.P. (1980). "Income distributional effects of technical change: Some analytical issues". Southeast Asian Economic Review 1 (3), 179–218.
Bokonon-Ganta, A.H., DeGroote, A.H., Neuenschwander, P. (2001). "Socioeconomic impact of biological control of mango mealybug in Benin". Agriculture, Ecosystems and Environment 1905, 1–12.
Brennan, J.P. (1986). "Impact of wheat varieties from CIMMYT on Australian wheat production", Agricultural Economics Bulletin No. 5. Department of Agriculture New South Wales, Sydney.
Byerlee, D., Morris, M. (1993). "Research for marginal environments: Are we underinvested?". Food Policy 18 (5), 381–393.
Byerlee, D., Moya, P. (1993). Impacts of International Wheat Breeding Research in the Developing World 1966–1990. International Maize and Wheat Improvement Center (CIMMYT), Mexico, DF.
Byerlee, D., Pingali, P. (1994). "Rice research in Asia: Fulfillments and frustrations", Presented at the XXII Conference of the International Association of Agricultural Economists in Harare, Zimbabwe, 22–29 August 1994 (mimeo).
Byerlee, D., Traxler, G. (2001). "The role of technology spillovers and economies of size in the efficient design of agricultural research systems". In: Alston, J., Pardey, P., Taylor, M. (Eds.), Agricultural Science Policy: Changing Global Agendas. Johns Hopkins University Press, Baltimore.
Campbell, B., Sayer, J.A., Frost, P., Vermeulen, S., Ruiz Pérez, M., Cunningham, A., Prabhu, R. (2001). "Assessing the performance of natural resource systems". Conservation Ecology 5 (2), 22.
CGIAR (2000). Charting the CGIAR's Future: A New Vision for 2010. CGIAR Secretariat, Washington, DC.
CIMMYT (1991). "World wheat facts and trends 1990–1991: Wheat and barley production in the rainfed marginal environments of the developing world".
Coffman, W.R., Hargrove, T.R. (1989). "Modern rice varieties as a possible factor in production variability". In: Anderson, J.R., Hazell, P.B.R. (Eds.), Variability in Grain Yields: Implications for Agricultural Research Policy in Developing Countries. Johns Hopkins University Press, Baltimore.
Cooper, P.J.M., Denning, G.L. (2000). "Scaling up the impact of agroforestry research: Report of the agroforestry dissemination workshop". World Agroforestry Centre (ICRAF), Nairobi, Kenya, 14–15 September 1999. 43 pp.
Dalrymple, D.G. (1978). "Development and spread of high yielding varieties of wheat and rice in the less developed nations", USDA Foreign Agricultural Economics Report No. 95. United States Department of Agriculture, Washington, DC.
David, C., Otsuka, K. (Eds.) (1994). Modern Rice Technology Income Distribution in Asia. Lynne Rienner, Boulder.
DeHaen, H., Stamoulis, K., Shetty, P., Pingali, P. (2004). "The world food economy in the twenty-first century: Challenges for international co-operation". In: Maxwell, S., Slater, R. (Eds.), Food Policy – Old and New. Blackwell Publishing, Oxford, UK.
de Janvry, A., Sadoulet, E. (2003). "Achieving success in rural development: Toward implementation of an integral approach", Paper presented at the IAAE Conference in Durbin, S.A., August 2003.
Delgado, L.C., Hopkins, J., Kelly, V.A. (1998). "Agricultural growth linkages in Sub-Saharan Africa", IFPRI Research Report No. 107. International Food Policy Research Institute (IFPRI), Washington, DC.
Doss, C.R. (1999). "Twenty-five years of research on women farmers in Africa: Lessons and implications for agricultural research institutions with an annotated bibliography", CIMMYT Economics Program Paper No. 99-02. International Maize and Wheat Improvement Center (CIMMYT), Mexico, DF.
Douthwaite, B., Keatinge, J.D.H., Park, J.R. (2001). "Why promising technologies fail: The neglected role of user innovation during adoption". Research Policy 30, 819–836.
Evenson, R.E. (2001). "Economic impacts of agricultural research and extension". In: Gardner, B., Rausser, G. (Eds.), Handbook of Agricultural Economics, vol. 1. Elsevier Science, Amsterdam. Ch. 11.
Evenson, R., David, C. (1993). Adjustment and Technology: The Case of Rice. Development Centre Studies, Organisation for Economic Cooperation and Development (OECD), Paris.

Evenson, R.E., Gollin, D. (Eds.) (2003). Crop Genetic Improvement and Agricultural Development. CABI.
Evenson, R., Rosegrant, M. (2003). "The economic consequences of crop genetic improvement programmes". In: Evenson, R.E., Gollin, D. (Eds.), Crop Genetic Improvement and Agricultural Development. CABI.
Fan, S., Hazell, P., Thorat, S. (1998). "Government spending, growth, and poverty: An analysis of interlinkages in rural India", EPTD Discussion Paper No. 33. International Food Policy Research Institute (IFPRI), Washington, DC.
Flinn, J.C., Garrity, D.P. (1989). "Yield stability and modern rice technology". In: Anderson, J.R., Hazell, P.B.R. (Eds.), Variability in Grain Yields: Implications for Agricultural Research and Policy in Developing Countries. Johns Hopkins University Press, Baltimore.
Flores-Moya, P., Evenson, R.E., Hayami, Y. (1978). "Social returns to rice research in the Philippines: Domestic benefits and foreign spillover". Economic Development and Cultural Change 26 (3), 591–607.
Gardner, B. (2003). The CGIAR at 31: An Independent Meta-Evaluation of the Consultative Group on International Agricultural Research. Operations Evaluation Department, World Bank, Washington, DC. 46 pp.
Gómez, M.I. (1999). "Economic benefits of research cooperation: The case of the regional maize program for Central America and the Caribbean", Ph.D. thesis. University of Illinois at Urbana–Champaign.
Hayami, Y., Herdt, R.W. (1977). "Market price effects of technological change on income distribution in semi-subsistence agriculture". American Journal of Agricultural Economics 59 (2), 245–256.
Hayami, Y., Kikuchi, M. (2000). A Rice Village Saga: Three Decades of Green Revolution in the Philippines. Barnes and Noble, New York.
Hayami, Y., Kikuchi, M., Moya, P.F., Bambo, L.M., Marciano, E.B. (1978). Anatomy of a Peasant Economy: A Rice Village in the Philippines. International Rice Research Institute (IRRI), Los Baños.
Hazell, P., Haggblade, S. (1993). "Farm–nonfarm growth linkages and the welfare of the poor". In: Lipton, M., van de Gaag, J. (Eds.), Including the Poor. The World Bank, Washington, DC.
Hazell, P.B.R., Ramaswamy, C. (1991). Green Revolution Reconsidered: The Impact of the High-Yielding Rice Varieties in South-India. Johns Hopkins University Press, Baltimore.
Heisey, P., Lantican, M., Dubin, H.J. (1999). "Assessing the benefits of international wheat breeding research in the developing world: The global impacts study, 1996–1997". In: Pingali, P.L. (Ed.), CIMMYT 1998–99 World Wheat Facts and Trends. Global Wheat Research in a Changing World: Challenges and Achievements. CIMMYT, Mexico, DF.
Herdt, R.W., Capule, C. (1983). Adoption, Spread, and Production Impact of Modern Rice Varieties in Asia. International Rice Research Institute (IRRI), Los Baños.
Hossain, M. (1998). "Rice research, technological progress, and the impact on the rural economy: The Bangladesh case". In: Pingali, P.L., Hossain, M. (Eds.), Impact of Rice Research. Proceedings of the International Conference on the Impact of Rice Research, 3–5 June 1996, Bangkok, Thailand. Thailand Development Research Institute (TDRI) and International Rice Research Institute (IRRI), Bangkok and Los Baños.
Interim Science Council (2003). Report of the First External Review of the Systemwide Program on Collective Action and Property Rights (CAPRi). Science Council Secretariat, FAO, Rome, Italy.
Izac, A.-M.N. (1998). "Assessing the impact of research in natural resources management. Synthesis of a workshop". World Agroforestry Centre (ICRAF), Nairobi, Kenya, 27–29 April 1998. 38 pp.
Jha, D., Kumar, P. (1998). "Rice production and impact of rice research in India". In: Pingali, P., Hossain, M. (Eds.), Impact of Rice Research. Thailand Development Research Institute and International Rice Research Institute, pp. 279–291.
Kelley, T.G., Byerlee, D. (2003). "Surviving on the margins: Agricultural research and development for poverty reduction in marginal areas", Keynote paper presented at the 14th International Farm Management Congress in Perth, Western Australia, 10–14 August, 2003.
Kelley, T.G., Gregersen, H. (2005). "Lessons from CGIAR impact assessment research". In: Shiferaw, B., Freeman, H.A., Swinton, S. (Eds.), Natural Resource Management in Agriculture: Methods for Assessing Economic Environmental Impacts. CABI Publishing, pp. 341–360.
Koo, B., Pardey, P.G., Wright, B. (2003). Endowing Future Harvests: The Long Term Costs of Conserving Genetic Resources. IFPRI, Washington, DC.

Lantican, M.A., Pingali, P.L., Rajaram, S. (2003). "Is research on marginal lands catching up? The case of unfavourable wheat growing environments". Agricultural Economics 29 (3), 353–361.
López-Pereira, M.A., Morris, M.L. (1994). Impacts of International Maize Breeding Research in the Developing World, 1966–1990. International Maize and Wheat Improvement Center (CIMMYT), Mexico, DF.
Maredia, M.K., Byerlee, D. (Eds.) (1999). The Global Wheat Improvement System: Prospects for Enhancing Efficiency in the Presence of Spillovers, CIMMYT Research Report No. 5. International Maize and Wheat Improvement Center (CIMMYT), Mexico, DF.
Morris, M.L. (2001). "Impacts of international maize breeding research in developing countries, 1966–1998", Mimeo draft. CIMMYT, Mexico, DF.
Morris, M.L., Dubin, H.J., Pokhrel, T. (1994). "Returns to wheat breeding research in Nepal". Agricultural Economics 10 (3), 269–282.
Pardey, P.G., Alston, J.M., Christian, J.E., Fan, S. (1996). Hidden Harvest: U.S. Benefits from International Research Aid, Food Policy Report. International Food Policy Research Institute (IFPRI), Washington, DC.
Paris, T.R. (1998). "The impact of technologies on women in Asian rice farming". In: Pingali, P.L., Hossain, M. (Eds.), Impact of Rice Research, Proceedings of the International Conference on the Impact of Rice Research, 3–5 June 1996, Bangkok, Thailand. Thailand Development Research Institute (TDRI) and International Rice Research Institute (IRRI), Bangkok and Los Baños.
Pingali, P.L. (2001). Milestones in Impact Assessment Research in the CGIAR, 1970–1999, with an Annotated Bibliography of Impact Assessment Studies Conducted in the CGIAR, 1970–1999. Standing Panel on Impact Assessment, Technical Advisory Committee (TAC) of the CGIAR, World Bank, Washington, DC.
Pingali, P.L., Hossain, M., Gerpacio, R.V. (1997). Asian Rice Bowls: The Returning Crisis? CAB International, Wallingford.
Pingali, P.L., Smale, M. (2001). "Agriculture industrialized". In: The Encyclopedia of Biodiversity, vol. I. Academic Press.
Pinstrup-Andersen, P., Hazell, P.B.R. (1985). "The impact of the Green Revolution and prospects for the future". Food Review International 1 (1), 1–25.
Pinstrup-Andersen, P., Ruiz de Londoño, N., Hoover, E. (1976). "The impact of increasing food supply on human nutrition: Implications for commodity priorities in agricultural research". American Journal of Agricultural Economics 58 (2), 131–142.
Quisumbing, A.R., Haddad, L., Peña, C. (1995). "Gender and poverty: New evidence from 10 developing countries", Food Consumption and Nutrition Division (FCND) Discussion Paper No. 9. International Food Policy Research Institute (IFPRI), Washington, DC.
Raitzer, D. (2003). "Benefit–cost meta-analysis of investment in the Consultative Group on International Agricultural Research (first draft)", Report prepared on behalf of CGIAR Standing Panel on Impact Assessment. Science Council Secretariat, Food and Agriculture Organization of the United Nations (FAO), Rome, Italy. 45 pp.
Renkow, M. (1993). "Differential technology adoption and income distribution in Pakistan: Implications for research resource allocation". American Journal of Agricultural Economics 75 (1), 33–43.
Rohrbach, D., Lechner, W., Ipinge, W., Monyo, S. (1999). "Impact from investments in crop breeding: The case of Okashana-1 in Nimbia". In: ICRISAT Impact Series, vol. 4. ICRISAT, Patancheru, Andhra Pradesh, India.
Rusike, J., Dimes, J., Twomlow, S. (2003). "Risk-return tradeoffs of smallholder investments in improved soil fertility management technologies in the semiarid areas of Zimbabwe", Presented at the International Association of Agricultural Economists (IAAE) Mini-Symposium on Soil Fertility and Food Security for the Poor in Southern Africa: Technical, Policy and Institutional Challenge, Durban, South Africa, August 16–22, 2003. Unpublished.
Ryan, J. (1999). "Assessing the impact of rice policy changes in Viet Nam and the contribution of policy research", Impact Assessment Discussion Paper No. 8. International Food Policy Research Institute (IFPRI), Washington, DC.

Sayer, J.A., Campbell, B. (2001). "Research to integrate productivity enhancement, environmental protection, and human development". Conservation Ecology 5 (2), 32.
Scobie, G.M., Posada, R.T. (1977). The Impact of High-Yielding Rice Varieties in Latin America – with Special Emphasis on Colombia. Series JE-01. Centro Internacional de Agricultura Tropical (CIAT), Cali.
Scobie, G.M., Posada, R.T. (1978). "The impact of technical change on income distribution: The case of rice in Colombia". American Journal of Agricultural Economics 60 (1), 85–92.
Shapiro, B.I. (1990). "Potential constraints, policy, and new technologies in the Niamey Region of Niger", Ph.D. thesis. Purdue University, Indiana.
Sidhu, D.S., Byerlee, D. (1992) "Technical change and wheat productivity in the Indian Punjab in the post-Green Revolution period", CIMMYT Economics Working Paper No. 92-02. International Maize and Wheat Improvement Center (CIMMYT), Mexico, DF.
Sperling, L., Loevinsohn, M., Ntambovura, B. (1993). "Rethinking the farmers' role in plant breeding: Local bean experts and on-station selection in Rwanda". Experimental Agriculture 29, 509–519.
TAC (2000). A Food Secure World for All: Towards a New Vision and Strategy for the CGIAR. TAC Secretariat, FAO, Rome, Italy.
Traxler, G., Pingali, P.L. (1998). "Enhancing the diversity of modern germplasm through the international coordination of research roles". In: Smale, M. (Ed.), Farmers, Gene Banks, and Crop Breeding: Economic Analyses of Diversity in Wheat, Maize, and Rice. Kluwer Academic Publishers, Boston.
Turkelboom, F., La Rovere, R., Hagmann, J., El-Khatib, R., Jazeh, K. (2003). "Putting INRM into action: Workshop documentation". 4th INRM Workshop, International Center for Agricultural Research in the Dryland Areas (ICARDA), Aleppo, Syria, 16–19 September, 2002.
Waibel, H. (2000). Impact of Integrated Pest Management Research in the CGIAR. Impact Assessment and Evaluation Group, Technical Advisory Committee of the CGIAR, Rome.
Walker, T.S. (1989). "High-yielding varieties and variability in sorghum and pearl millet production in India". In: Anderson, J.R., Hazell, P.B.R. (Eds.), Variability in Grain Yields: Implications for Agricultural Research and Policy in Developing Countries. Johns Hopkins University Press, Baltimore.
Walker, T.S. (2000). "Reasonable expectations on the prospects for documenting the impact of agricultural research on poverty in ex-post case studies". Food Policy 25, 515–530.
Walker, T., Crissman, C.C. (1996). Case Studies of the Economic Impact of CIP-related Technologies. International Potato Center (CIP), Lima.
Walker, T., Ryan, J. (1991). Village and Household Economies in India's Semi-Arid Tropics. Johns Hopkins University Press, Baltimore.
Witcombe, J.R. (1989). "Variability in the yield of pearl millet varieties and hybrids in India and Pakistan". In: Anderson, J.R., Hazell, P.B.R. (Eds.), Variability in Grain Yields: Implications for Agricultural Research and Policy in Developing Countries. Johns Hopkins University Press, Baltimore.
Wood, S., Pardey, P.G. (1998). "Agroecological aspects of evaluating agricultural R&D". Agricultural Systems 57 (1), 13–41.
World Bank (2003). The CGIAR at 31: An Independent Meta-Evaluation of the CGIAR, vol. 1, Overview Report. Operations Evaluation Department (OED), World Bank, Washington, DC. 46 pp.
Zeddies, J., Schaab, R.P., Neuenschwander, P., Herren, H.R. (2001). "Economics of biological control of cassava mealybug in Africa". Agricultural Economics 24, 209–219.

Chapter 46

CONTRIBUTIONS OF NATIONAL AGRICULTURAL RESEARCH SYSTEMS TO CROP PRODUCTIVITY

ROBERT E. EVENSON

Economic Growth Center, Yale University, New Haven, CT

DOUGLAS GOLLIN

Department of Economics, Williams College, Williamstown, MA

Contents

Handbook of Agricultural Economics, Volume 3
Edited by Robert Evenson and Prabhu Pingali

DOI: 10.1016/S1574-0072(06)03046-5

Abstract

This chapter describes the impact of national agricultural research systems on the unfolding of the Green Revolution in four regions: Asia, Latin America, the Middle East and North Africa, and Sub-Saharan Africa. Although international institutions contributed much of the research that led to the Green Revolution, national programs also proved important in the development and diffusion of modern varieties. This chapter documents the Green Revolutions that occurred in 11 food crops – wheat, rice, maize, barley, sorghum, millets, lentils, groundnuts, beans, cassava and potatoes. The chapter traces the overall role of national institutions and the growing importance of national agricultural research systems in the developing world. It also discusses the interaction of National Agricultural Research Systems and International Agricultural Research Centers, which have largely played complementary roles. Finally, it discusses the political economy of support for national agricultural research.

Keywords

the Green Revolution, Green Revolution modern varieties, the Gene Revolution, Gene–Green Revolution congruity

JEL classification: O12, O13, O30, Q12

1. Introduction

National agricultural research systems (NARS) play a central role in the development of new agricultural technologies for developing countries. As Byerlee and Alex (2003) note, NARS programs include "not just ... public organizations ... [but] all organizations that generate, share, import, and utilize agriculture-related knowledge and information". In this sense, NARS include public sector agencies, universities, non-profit organizations, cooperatives, and (in principle) private sector actors.

For most of the developing world, however, the private sector has until now provided little in the way of improved agricultural technology. In crop agriculture, private sector research has been essentially limited to the development of hybrid varieties of maize, sorghum, and millet, although recent biotechnology advances have also encouraged the use of privately marketed varieties of cotton and soybeans. Pardey and Beintema (2001) estimate that the private sector accounted for only 5.5% of research expenditures around 1995, for the developing world as a whole. By contrast, in the developed countries, private expenditures on agricultural R&D were equal to public expenditures.

This chapter will thus focus largely on the role of public sector institutions in developing countries, although we will refer in places to the role of private sector actors. The private sector is addressed in greater detail in Chapters 49 and 50 in this volume. In this chapter, we also focus primarily on the major government agencies that account for the bulk of NARS research expenditures in developing countries. According to Pardey and Beintema (2001), government agencies – rather than universities or non-governmental organizations – do most of the public spending on agricultural research in developing countries. By contrast, universities play a much more central role in many high income countries, including notably the United States.

To limit our scope further, this chapter will primarily deal with crop agriculture. (The following chapter addresses the role of international and national programs in livestock production.) Within crops, we focus on a set of major food crops that are consumed in developing countries. We include the most important grain and root crops (rice, wheat, maize, sorghum, millet, barley, cassava, potato, as well as lentils, beans, and groundnuts); we will not consider in depth soybeans, bananas and plantain, fruits and vegetables, coffee, tea, cocoa, sugar, cotton, and other crops that are primarily produced for export markets. (See Chapter 73 in Volume 4 for coverage of these crops.)

Finally, this study will pay particular attention to the relationship between NARS and the international agricultural research centers (IARCs). Although the IARCs are a crucial source of technologies for developing countries, they typically work in partnership with NARS programs, providing improved germplasm, training, and support.

Section 2 of this chapter describes the history and background of national agricultural research programs in developing countries. Section 3 discusses patterns of investment in these institutions and the responsiveness of investment to cost–benefit information. Section 4 describes the performance of NARS in generating improved crop varieties, focusing on the period before and during the Green Revolution (beginning in 1964). We consider the diversity in rates of modern variety (MV) production and adoption by crop and region, and we note that these patterns are related to the research investments made before the Green Revolution. Section 5 describes the adoption of MVs and reports estimates of the productivity impacts of the Green Revolution. Section 6 addresses the IARC-NARS role in the Gene Revolution. Section 7 discusses the fact that we have "mass poverty" in many developing countries in a "world awash with grain".

2. Background

National Agricultural Research Systems (NARS) were established in many developing countries in the late 19th century, often by colonial governments. An explicit goal of many NARS was to generate crop improvement in commodities that were imported by the colonial powers. Thus, early NARS research efforts were focused on crops such as sugarcane, rubber, coffee, tea, bananas, and cotton. Some NARS programs were remarkably successful in their production of new varieties, and the benefits of this research were accrued primarily by consumers in the colonial powers. In some developing countries, NARS programs also focused on food crops, chiefly rice and wheat, with a goal of increasing food security, thereby reinforcing the power and legitimacy of governments (or colonial administrations).

After World War II, NARS systems were further expanded and strengthened in many countries. This expansion proceeded at different rates in different countries and at different rates for different crops. This was partly related to the dates of independence from colonial regimes, but it was also affected by geo-climate or agro-economic zone (AEZ) conditions. Latin American countries were generally most advanced in NARS development, while Sub-Saharan African countries, most of whom did not achieve independence until after 1960, were least advanced.

By the mid-1950s, the demographic changes associated with declines in infant and child mortality – and death rates more generally – were recognized. Virtually every developing country was then entering a demographic transition phase where mortality had fallen and birth rates had not yet declined. This implied rapid population growth and major changes in the demographic structure of populations. Today, in retrospect, we know that the demographic transitions realized in developing countries were actually quite short, because birth rates did fall relatively quickly in most countries, with the most rapid transition being realized in the countries achieving rapid economic growth. But we also know in retrospect that developing countries did realize major population expansions in the second half of the 20th century. The population of developing countries (including China) increased from 1.67 billion in 1950 to 4.75 billion in 2000.

In recognition of the food production demands associated with this population "explosion", the international community (broadly defined) responded by creating an international network of scientific institutions to bring concerted effort to the agricultural problems of the developing world. From initial efforts by the Ford and Rockefeller Foundations, along with the Food and Agriculture Organization (FAO) of the United

Nations, a network of international agricultural research centers (IARCs) eventually emerged.

The combined IARC-NARS systems essentially produced the Green Revolution in developing country agriculture. The Green Revolution was based on "conventional" plant breeding techniques, and the initial successes of the Green Revolution drew on relatively simple "first generation" benefits from moving germplasm across countries and regions. Semi-dwarf genes, originating in varieties from Japan and China, were incorporated into rice and wheat varieties suitable for tropical and subtropical conditions.

In the 1980s and 1990s, a second scientific revolution, the "Gene Revolution", was initiated. This revolution is based on "recombinant DNA" (rDNA) or genetic engineering techniques. The Gene Revolution is still in its infancy. Most of the Gene Revolution products available to farmers in developing countries have been produced by private sector "multinational corporations" (MNCs). But the IARC-NARS systems will also have a major role to play in the realization of production gains from the Gene Revolution. During the Green Revolution, we know that those countries with strong NARS capacity in conventional breeding were able to benefit the most from new technologies developed in the IARCs. A similar phenomenon is likely to occur with respect to the Gene Revolution; some countries are likely to be left behind as new technologies arrive, with potentially serious harm for producers and consumers.

3. NARS institutions: Investment patterns and characteristics

NARS programs vary considerably in size, funding, and effectiveness. The strongest of the NARS programs in the developing world rank among world leaders in generating substantial amounts of new technology. By contrast, some weaker programs exist in name only and have accomplished little in the way of technology generation.

3.1. NARS expenditures

Table 1 from Pardey and Beintema (2001) and Boyce and Evenson (1975) provides data on expenditure in public agricultural research systems around the world. Although developing countries now spend more collectively on public agricultural research than rich countries – $11.5 billion compared to $10.2 billion – many countries have NARS systems that are funded and staffed at perfunctory levels. The largest of the NARS in developing countries are found in China, India, and Brazil, which together accounted for half the developing world expenditure in the mid-1990s.

Many other developing countries lag far behind, whether measured by total expenditures or staffing, or by various measures of research intensity (e.g., spending per hectare of agricultural land, spending per person employed in agriculture, spending as a fraction of agricultural GDP). Table 2 from Pardey and Beintema (2001) summarizes research intensity measures. More detailed data have recently been made available by the Consultative Group on International Agricultural Research (CGIAR). The CGIAR's

Table 1
Global expenditures on agricultural research (millions 2001 US dollars)

	1965	1976	1985	1995
Public sector				
Developed countries	6532	8270	10192	11900
Developing countries				
China	377	709	1396	2063
Other Asia	441	1321	2453	4619
Middle East–North Africa	360	582	981	1521
Latin America & Caribbean	562	1087	1583	1947
Sub-Saharan Africa	472	993	1181	1270
International Agric. Research Centers	12	163	315	400
Private sector				
Developed countries				10829
Developing countries				672

Source: Pardey and Beintema (2001) and Boyce and Evenson (1975).

Table 2
Public agricultural research intensities

	Expenditure as a share of agricultural GDP			Expenditures per capita		
	1976	1985	1995	1976	1985	1995
Developed countries	1.53	2.13	2.64	9.6	11.0	12.0
Developing countries	0.44	0.53	0.62	1.5	2.0	2.5
China	0.41	0.42	0.43	0.7	1.3	1.7
Other Asia	0.31	0.44	0.63	1.1	1.7	2.6
Latin America and Caribbean	0.55	0.72	0.98	3.4	4.0	4.6
Sub-Saharan Africa	0.91	0.95	0.85	3.5	3.0	2.0

Source: Pardey and Beintema (2001) (Evenson estimates for SSA).

Agricultural Science and Technology Indicators (ASTI) offer recent data on NARS staffing and expenditure for a large number of countries around the world. These data show that many developing countries have very limited NARS research efforts. For example, the 44 countries in Sub-Saharan Africa together accounted for \$1.27 billion in public agricultural research expenditure, with a large number spending purely notional amounts (under \$20 million annually). Measuring research inputs by the number of scientists (adjusted for degree level), we find that many NARS carry fewer than 50 research scientist equivalents.[1] These programs are unlikely to be able to conduct original research – particularly if the research effort is spread (as is typical) across a range of crops and animal species.

On a positive note, Byerlee and Alex (2003) point out that developing countries were, by the mid-1990s, spending more on public sector agricultural research than were developed countries, and their expenditure was growing at a faster rate. Pardey and Beintema (2001) document that the average annual growth rate of public agricultural research expenditure for 1976–1995 was 4.5% per year for developing countries and only 1.9% in developed countries.

The training and educational background of scientists employed in NARS programs has also increased over time. More researchers have postgraduate degrees than in previous time periods. In some regions, there has also been a shift in the national background of researchers.

Unfortunately, for many NARS programs, expenditures have not kept pace with increases in staffing, leading to declines in spending per scientist – at least in some countries and regions. This pattern appears to hold outside of Africa, and it remains a concern for policy makers [Byerlee and Alex (2003)]. A related concern is the decline in growth rates of spending on agricultural research in the late 1990s, described by Byerlee and Alex (2003) and documented to some extent in the ASTI data (at least for Sub-Saharan Africa, for which the data are relatively recent).

In part, the declines in spending by many public sector agricultural research agencies reflect a worldwide phenomenon. As Heisey, Srinivasan and Thirtle (2001) note, "public sector agricultural research in general, and public plant breeding research in particular, is in trouble in both industrialized and developing nations". Research budgets for public institutions in rich countries have leveled off or fallen in real terms, funding of international research (the IARCs) has fallen, and foreign aid for agricultural research has also declined. Heisey, Srinivasan and Thirtle (2001) examine some of the reasons for this secular trend, including the emergence of private sector research in industrial countries.

Issues relating to private sector involvement in agricultural research and technology creation are discussed in greater detail in Chapters 49 and 50. For now, we simply note that crop improvement in developing countries – with the important exception of maize – remains largely a public sector activity. Both international and national programs are involved in crop improvement, and we next consider their past performance.

3.2. The political economy of support for NARS programs

The expansion of funding for NARS programs has been heavily dependent on international aid. Figure 1 shows support from World Bank loans and from USAID. Both of these funding sources have declined in magnitude. This is particularly alarming for

[1] Byerlee and Alex (2003) estimate, based on these data, that "perhaps three quarters of NARSs in developing countries employ fewer than 200 researchers".

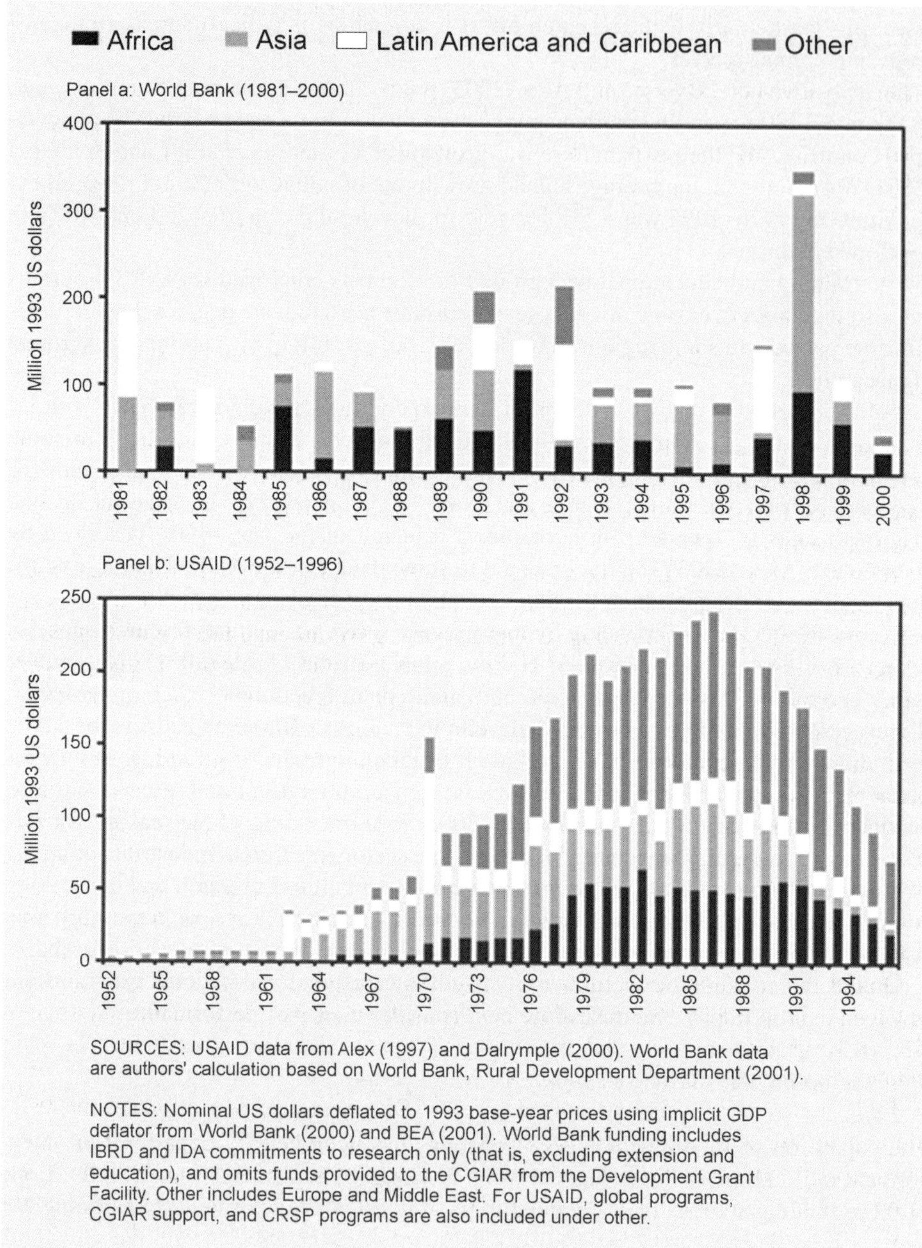

Figure 1. World Bank and USAID expenditures on agricultural research, from Pardey and Beintema (2001).

Table 3
Internal rate of return estimates: NARS agricultural programs

Studies from:	Number reported	Distribution by IRR						Approximate median IRR
		0–20	21–40	41–60	61–80	81–100	100+	
Asia	120	0.08	0.18	0.21	0.15	0.11	0.26	67
Latin America	80	0.15	0.29	0.29	0.15	0.07	0.06	47
Africa	44	0.27	0.27	0.18	0.11	0.11	0.05	37

Source: Evenson (2001).

many Sub-Saharan African countries, where domestic support bases have not been developed.[2]

Table 3 reports a summary of estimated "internal rates of return" (IRRs) for NARS research programs [Evenson (2001)]. The central feature of the rates of return reported here is that they have a wide dispersion, ranging from low IRRs to very high ones. But median IRRs are high and show clearly that successful research programs produce economic growth.

Yet few NARS programs have managed to establish solid support bases in their countries. Many are still dependent on international aid agencies, with little or no domestic constituency for research funding. Often, national governments, faced with limited resources and difficulty in mobilizing revenues, treat agricultural research as a dispensable item. Because its payoffs are long-term, and because other needs are more pressing, agricultural research is an easy target for budget cuts and reallocations.

In those developing countries that have been able to provide support for agricultural research, it is often a response to widespread recognition of food security problems.

The rate of return studies summarized in Table 3 measured returns to investments by taxpayers in NARS programs. These returns varied across studies, but it is noteworthy that the median rates of return calculated from individual studies were consistently higher than estimates from the Green Revolution studies for NARS programs. (See Section 5.)

One of the variables that could explain some of these differences is the research intensity of the project. The individual studies seldom report the research intensity of the NARS programs reviewed. But it is possible to use the scientists/cropland ratio for countries as a proxy for research intensity.

Table 4 reports a statistical analysis of the IRR estimates. It shows that the IRRs are indeed related to the scientists/cropland ratio. The range of the scientists/cropland ratio across countries is from 0.01 to 0.20. The estimate indicates that at the upper limit, IRRs are quite modest. The coefficient implies that at the median scientist/cropland ratio (0.05), IRRs should be adjusted downward by 7 percentage points.

[2] As the Green Revolution section of this chapter will note, Sub-Saharan Africa did not realize the same type of Green Revolution that was experienced in Asia.

Table 4
Determinants of internal rate of return estimates

Dependent variable: IRR estimate	
Independent variables: (*t*-ratios in parentheses)	
Scientist/cropland	−148.01
	(3.49)
Indicator variable:	
Africa study	−24.26
	(1.49)
Latin America study	−18.66
	(3.72)
Ex ante study	−3.85
	(0.50)
Yield study	7.26
	(1.32)
Project evaluation study	9.38
	(1.33)
Production function study	39.80
	(1.59)
Time shape estimates	21.99
	(2.55)
Geographic spill-ins	−18.35
	(2.16)
Post-1979 study	5.75
	(1.05)
Constant	
R^2	0.235
F	7.82
N	265

4. NARS contributions to crop improvement: The Green Revolution

Prior to the development of "formal" agricultural experiment station-based plant breeding programs around 1870, crop genetic improvement (CGI) and animal genetic improvement (AGI) was the province of farmers and livestock breeders. Farmer breeding of crops was achieved by seed selection for the next crop. Some farmers, however, specialized in seed selection and production. As populations expanded and moved to new areas, new "types" of the cultivated species were selected. Many of these types, known as "landraces", were collected in *ex situ* collections by early agricultural scientists, and many remain the basic genetic resources used by modern plant breeders.[3]

[3] Modern "gene banks" have been established for all major cultivated crop species. These *ex situ* collections include most "landraces" in the species as well as accessions of closely related of "wild" (uncultivated) species.

The settlement of the New World, i.e. the Americas and Oceania, was associated with a major expansion in crop landraces in the 18th and 19th centuries. These New World landraces have been particularly valuable to plant breeders.[4]

Some of the earliest plant breeding programs actually originated in developing countries. Tropical crops with "mother country" export potential (sugarcane, cotton, rubber, coffee, tea, spices) were among the first to be given attention in "scientific" plant breeding programs. In 1878, sugarcane breeders in Java and in Barbados independently discovered techniques to induce flowering in sugarcane species. This opened the door to the basic methods of plant breeding; strategic crossing of parental materials and "selection under pressure" of resultant progeny over several generations to achieve "stability" in varietal performance.

A major advance in plant breeding was achieved around 1920, when inter-specific hybridization was achieved in sugarcane in Java and India. Inter-specific hybridization is the sexual crossing of closely related species (usually species in the same genus) to incorporate traits from a non-cultivated (wild) species into a cultivated species. Inter-specific hybridization has been achieved in almost all cultivated species and has been an important part of Green Revolution breeding.[5]

Another major advance was achieved with the development of "heterosis" breeding methods creating "hybrid" varieties. The first "heterosis" hybrids were for maize varieties. This development began at Harvard and Yale around 1900 and was developed more fully at the Connecticut State Experiment Station in New Haven, CT. Major impetus to hybrid maize varietal development was achieved when private sector firms (Pioneer, Funks) began breeding hybrid maize varieties in the U.S. Griliches (1957) describes the development of hybrid maize. Today, as noted below, Green Revolution heterosis hybrids have been important in maize, sorghum, millets and rice.

While the "mother country" crops did receive most plant breeding attention in developing countries, some food crops received attention as well. Many Latin American countries had plant breeding programs by the 1930s and strengthened them in the post WWII period. In Asia, plant breeding programs for wheat, rice and vegetables were well developed before the Green Revolution. In Africa, except for Kenya, South Africa and Rhodesia, few major plant breeding programs were built until after independence from colonial regimes (and most African countries did not achieve independence until after 1960).

Thus, at the beginning of the Green Revolution period, a great deal of diversity by crop and region in terms of the proportions of crops still in "landrace" form existed.

In general, we can view landrace agriculture as "traditional" agriculture, in the terminology of T.W. Schultz.[6] While landrace crop yields do vary by region, being highest in

[4] Gollin and Evenson (1997) show that for rice genetic resources, the United States is the major net exporter of rice landrace genetic resources to other countries. Many authors ignore the landraces created by the settlement of the New World. But these landraces have been particularly valuable to plant breeders.

[5] See Evenson and Kislev (1975) for an account of early sugarcane breeding.

[6] Schultz (1964) in *Transforming Traditional Agriculture* argued that traditional agriculture was "poor but efficient" and "efficient but poor".

temperate regions, they nonetheless are consistently lower than yields associated with pre-Green Revolution breeding. Farmers selected landraces under conditions of limited availability of modern crop inputs; fertilizer, insecticides and herbicides. This selection process produced cultivars with low responsiveness to fertilizer and high natural resistance to diseases and insect pests.

Pre-Green Revolution MVs were generally "selections" from improved landraces, with a few varieties developed from parental crosses of landraces.[7] Typically, breeders were pursuing higher yields; in selecting for higher yields, breeders sometimes had to sacrifice disease resistance and insect pest resistance features.

4.1. The development of Green Revolution MVs: An overview

In keeping with the literature, we use the term "Green Revolution" to describe the production, using modern scientific principles, of "high yielding" or "modern" crop varieties (MVs) for developing countries.[8] We also understand the term to apply to the subsequent diffusion of those varieties into farmers' fields. Most studies of the Green Revolution date its beginning to 1964–1965 when both rice and wheat MVs were made available to farmers in Asia and Latin America, although earlier efforts had developed "proto-MVs" by the late 1950s.[9] Many accounts of the Green Revolution treat it as a phenomenon confined to wheat and rice, and some suggest that the Green Revolution produced a "one-time" increase in production of wheat and rice that was effectively completed by 1985.

This "narrow" perspective on the Green Revolution is in part due to limited data on MV adoption. Dana Dalrymple, in a series of important studies, documented MV adoption in many countries for rice and wheat MVs for the period 1965 to 1984.[10] The Dalrymple studies were very important both for documenting the production and diffusion of MVs and for identifying the international and national plant breeding programs responsible for them.[11]

[7] For example, in rice, India's agricultural system had developed varieties like GEB24 that were superior to landrace varieties under favorable growing conditions, but they apparently did not outperform landraces in more marginal environments.

[8] By "modern scientific principles", we mean that the breeders understood fully the process of obtaining new genotypes (i.e., varieties) by combining the genes of existing varieties through "crossing". In some cases, they also used techniques to induce mutations (e.g., through radiation). These breeders also understood the process of selecting the offspring of new varieties through a number of generations to obtain essentially stable (i.e., homozygous) lines.

[9] For example, an FAO program in Asia developed a number of improved *indica* rice varieties, such as Mahsuri, that remain widely planted. Similarly, the Rockefeller Foundation wheat program in Mexico, under the direction of Norman Borlaug, had made significant strides in developing new varieties usable in Latin America.

[10] See Dalrymple (1986a, 1986b). Dalrymple played an important role in bringing attention to the Green Revolution.

[11] Wheat and rice Green Revolution MVs were first adopted in the 1964 and 1965. It was several years before Green Revolution MVs were produced for other crops.

But this narrow perspective is very incomplete and misleading. The Green Revolution described in this paper offers a much broader perspective.[12] This broader perspective is based on data for 11 food crops included in the mandates of seven International Agricultural Research Centers (IARCs).[13] More than 500 National Agricultural Research System (NARS) plant breeding programs were involved in the Green Revolution. We consider data on the production of approximately 9000 Green Revolution Modern Varieties (GRMVs) in 11 crops over the 1965 to 2000 period.[14]

This broader perspective supports a different interpretation of the Green Revolution experience. The narrow perspective suggested that GRMVs were adopted only in "favorable" production environments (i.e., with high-quality soil, water control and climate conditions) and that GRMV adoption was accompanied by high levels of "chemical" use.

The broader perspective is considerably more nuanced. GRMVs have been produced and adopted for all IARC mandate crops, including crops that are naturally suited to unfavorable environments (e.g., crops produced in semi-arid and dryland environments). The broader perspective also shows that "generations" of GRMVs have been developed for most Green Revolution crops and that each new generation of GRMVs has both displaced earlier generations of GRMVs and extended the "margin" of GRMV adoption. This generational feature of GRMV production creates productivity gains that accumulate over time.

The broader perspective also notes that the "complementarity" between GRMVs and other inputs is not uniform for different crops and is not uniform for different regions. For example, consider the use of agricultural chemicals, such as fertilizers, herbicides and insecticides. The adoption of these is not uniform across crops and regions. Low-wage economies generally do not use herbicides for weed control. Insecticide use varies by crop, but in all crops "host plant resistance" to insect pests has been an important breeding objective in the Green Revolution. Most successful GRMVs are complementary with fertilizer (for sound economic reasons) but actual fertilizer use differs from country to country because market efficiencies (transaction costs) differ.

This broader perspective of the Green Revolution does not allow one to conclude that the Green Revolution was "ideal" in the sense that it benefited all countries and all regions in an equitable way. The production increases enabled by the Green Revolution in the aggregate in developing countries constitute a "global" success, but for a number of countries, the Green Revolution represents a "local" failure. The population of developing countries increased from 1.67 billion in 1950 to 4.75 billion in 2000. Pakistan,

[12] This perspective is described below.

[13] These include IRRI and CIMMYT, the International Center for Tropical Agriculture (CIAT), the International Potato Center (CIP), the International Center for Agricultural Research in the Dry Areas (ICARDA), the International Crops Research Institute for the Semi-Arid Tropics (ICRISAT) and the West African Rice Development Association (WARDA).

[14] Our analysis is based on work carried out by a large number of collaborators and presented in more detail in Evenson and Gollin (2003a).

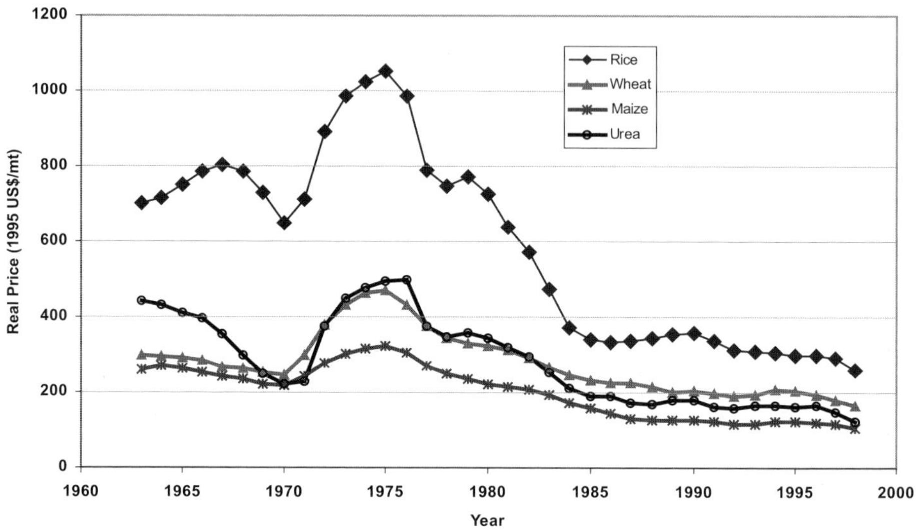

Figure 2. Real world prices of rice, wheat, maize and urea (1961–2000, 5-yr moving average). *Source*: IFPRI.

for example, experienced a tripling of population over this period but actually increased food production per capita. This accomplishment came despite the fact that Pakistan was regarded as having already utilized most of its cultivable land. In the absence of the Green Revolution (i.e., had farmers been constrained to use varieties available in 1965), millions of children would not have survived childhood and millions more of those survivors would have suffered from malnutrition.[15]

The Green Revolution, however, did not reach all farmers. Indeed, some 15 developing countries with populations over a million were effectively excluded from the Green Revolution. Many of these countries are in Sub-Saharan Africa. But for farmers reached by the Green Revolution, supply increases were large. During the 1961–1981 period, food grain increases in South and Southeast Asia were more than 3% per year (see below). The actual MV contributions to production were around 1% per year, but increased fertilizer use and irrigation expansion led to larger supply increases. As a consequence, world prices for basic food grains and other crops declined even though many farmers did not realize Green Revolution gains at all. Real world grain prices in 2000 were less than 40% of their levels in 1950. (See Figure 2.)

Consumers benefit from lower food prices and the Green Revolution brought lower food prices to virtually everyone in the world. Since low-income consumers spend disproportionately on food, this had favorable distributional consequences.[16]

[15] See Evenson and Rosegrant (2003) for malnutrition estimates.

[16] See Avila and Evenson (forthcoming) for estimates of cost reductions in developing countries. Avila and Evenson show that countries not achieving significant GRMV adoption did not achieve cost reductions to match price reductions.

Farmers, however, benefited where access to Green Revolution technology enabled their costs of production to decline more than prices declined, but they lost income when they were delivered lower prices but did not have access to Green Revolution technology. (See Section 6.)

4.2. *GRMV production and adoption*[17]

4.2.1. *Wheat*

Wheat is the world's most extensively cultivated crop, with roughly 225 million ha under cultivation. Approximately 100 million ha are under cultivation in developing countries: 8 million ha in Latin America, 1.2 million ha in Sub-Saharan Africa, 25 million ha in the Middle East and North Africa, and 65 million ha in Asia (of which 29 million ha are in China). Several wheat types are cultivated. Most are bread wheats, but significant quantities of durum wheats suited to pasta products are also grown. Wheats are classified as spring type or winter type, with northern temperate regions (i.e., Canada and the northern United States) producing traditional spring types. Winter types are produced in temperate and sub-tropical regions with mild winters, as in the southern growing regions of the U.S. In tropical climates with relatively warm winters, spring types are planted, but in the fall, as are winter types.

Because of extensive investments in wheat research programs in Europe and North America, the temperate zone spring and winter types had been considerably improved relative to the tropical spring types by the 1960s. The Rockefeller Foundation supported a wheat breeding program in Mexico in the 1940s and 1950s under the direction of Norman Borlaug. This program eventually was transformed into the wheat program at CIMMYT, where, after 20 years of dedicated breeding work, the Green Revolution semi-dwarf bread wheats were adapted for widespread use in Asia, beginning in the mid-1960s.

CIMMYT's program maintains a staff of 35 senior scientists (70 scientists) in a number of locations in the 1990s, with an annual budget of about $12 million [Heisey, Lantican and Dubin (2003)]. Approximately 1700 scientists in NARS programs were working on wheat improvement in 1997, with an annual budget of roughly $100 million in the 1990s. More than 2900 GRMV wheat varieties have been produced by these programs, with roughly 80 varieties released each year from 1980 to 1997. (See Table 5.)

Since CIMMYT does not release varieties directly to farmers, all of these released varieties involve some degree of NARS collaboration. However, according to Heisey, Lantican and Dubin (2003), in the late 1960s, about one-third of all the wheat varieties

[17] This section draws heavily on chapters by various authors and collaborators collected in Evenson and Gollin (2003a). We provide citations to the individual chapters where appropriate, but we note here that the book chapters were in turn based on more detailed work by chapter authors, and we encourage those interested in the subject to pursue additional work that may be available on the web sites of individual international agricultural research centers.

Table 5
Wheat Green Revolution Modern Varieties (GRMVs)

	Latin America & Caribbean	Asia	Middle East & North Africa	Sub-Saharan Africa
		Average annual releases		
1960s	17.6	8.6	4.4	10.2
1970s	23.6	17.2	9.1	8.7
1980s	32.0	25.6	13.0	8.1
1990s	26.3	22.4	21.0	9.6
		GRMV adoption (%)		
1970	11	19	5	5
1980	46	48	25	27
1990	82	74	42	52
2000	90	86	66	66

Source: Based on Heisey, Lantican and Dubin (2003).

released by NARS in developing countries had actually been crossed at CIMMYT, and another 15% had one or more CIMMYT parent. Thus, NARS were primarily releasing material developed at CIMMYT (presumably after screening and testing) or else using CIMMYT materials in their own breeding programs, perhaps to cross with local varieties.

The same pattern holds today, with NARS working extensively with CIMMYT material. In the 1990s, Heisey, Lantican and Dubin (2003) report that released varieties were approximately half CIMMYT crosses and another quarter had one or more CIMMYT parent. For the period 1991–1997, fully 90% of spring wheat bread varieties released in developing countries had CIMMYT ancestry. However, almost half of the varieties were based on some breeding work taking place in NARS.

These modern varieties were widely used by farmers. Table 5 shows the fraction of wheat area under MV cultivation and, for recent years, the fraction planted to varieties with CIMMYT content. By the late 1990s, over 80% of world wheat area was planted to MVs, including 20% of world wheat area planted to MVs developed without CIMMYT breeding materials. MV adoption was slower and later in Africa and the Middle East than in other regions.

4.2.2. *Rice*

Rice is arguably the most important crop in developing countries. Asian countries dominate production with 133 million ha. (India, with 43 million ha, and China, with 33 million ha, are the leading countries.) Latin America and African countries each produce on roughly 8 million ha. Developed countries, including Japan, produce on only 5 million ha. Rice is produced in several different environments. The dominant production environments are irrigated and rainfed "paddy" environments. Rice is also produced in "upland" and "deepwater" environments. Most upland production is in Africa and Latin America; most deepwater production is in Asia.

Three different IARCs have been involved in rice MV development, along with a large group of NARS. For Asia, the International Rice Research Institute (IRRI) has played a major role in producing important rice varieties. In Latin America and the Caribbean, the International Center for Tropical Agriculture (CIAT) has played a central role, while in West Africa, the West African Rice Development Association (WARDA) has been the lead institution.

The relationship between IRRI and Asian NARS is in some sense the most "mature" of such relationships. After the 1970s, IRRI's role was increasingly that of a germplasm supplier to national programs, rather than a direct producer of varieties for farmers' use. IRRI in recent years has primarily produced parent material and other breeding lines for NARS breeders, except for some countries with relatively weak NARS, where IRRI has supplied more finished material. IRRI's shift in roles has been facilitated by an international network for germplasm exchange that provides NARS breeders with ready access to breeding materials.

IRRI's success was first concentrated in irrigated rice environments and then extended to favorable rain-fed environments. This success has not yet extended in any significant degree to upland rice environments nor to deepwater environments. The early IRRI Asian rice varieties were also not particularly well adapted to Latin America or to Africa.

CIAT, in Colombia, established a rice breeding program a number of years after the IRRI program was established. It was this program that undertook adaptive breeding required to bring the high-yielding semi-dwarf varieties from Asia to Latin America. In contrast to the release pattern in Asia, where releases leveled off in the 1980s and 1990s, annual releases have continued to increase in Latin America.

For African production conditions, neither IRRI nor CIAT had much success in generating varieties that met with widespread adoption. The West African Rice Development Association (WARDA), the regional rice development center, experienced considerable instability in the 1960s and 1970s, and was not effective until it was established as a center capable of doing its own breeding, and not fully effective until it moved from an urban location in Liberia to an experiment station in Côte d'Ivoire. By the mid-1990s, WARDA's program was beginning to show effectiveness.

With respect to the more advanced Asian NARS, IRRI's role shifted to that of a germplasm supplier in the 1980s and early 1990s. However, in the later 1990s, the political opening of Cambodia, Vietnam and Laos, where national programs were relatively undeveloped, put IRRI back in the position of breeding varieties for direct release. Outside of Asia, the extension of the Green Revolution in rice to Latin America was greatly assisted by CIAT's program. And with a delay, WARDA now is assisting in the expansion of the rice Green Revolution to Africa. (However, WARDA has been forced to relocate once again by violence in Côte d'Ivoire. This disruption in research has had tragic consequences for African rice producers and consumers.)

Table 6
Rice Green Revolution Modern Varieties (GRMVs)

	Latin America & Caribbean	Asia	Sub-Saharan Africa
		Average annual releases	
1960s	5.5	17.5	–
1970s	3.2	17.5	4.8
1980s	6.4	25.9	8.2
1990s	6.4	24.5	6.7
		GRMV adoption (%)	
1970	2	10	0
1980	22	35	2
1990	52	55	20
2000	65	74	40

Source: Based on Hossain et al. (2003) and Dalton and Guei (2003).

IARC and NARS programs have produced more than 2500 rice GRMVs. Although it is difficult to get comparable figures for different regions, it appears that more than 2000 varieties have been developed for Asia, with around 300 varieties each in West Africa and the Latin America/Caribbean region. Table 6 shows the patterns of varietal release and diffusion for these regions. Relatively few varieties have been released for West Asia and North Africa, reflecting in part the fact that rice is a minor crop in that region.

By the late 1990s, MV diffusion was at 74% for Asia as a whole, and about 65% for Latin America. Within those regions, however, there were broad differences across countries, with some Asian countries (e.g., China, Korea, Malaysia, Philippines, Sri Lanka) at 90% or above, and other countries (e.g., Cambodia, Laos) with 10% or less of the area planted to modern varieties. In Latin America, too, there is wide variation across countries. Most of these country-specific differences reflect differences in growing conditions and agro-ecological zones. To some extent, however, they also reflect variation in the strength of national programs. IRRI's scientific staff in the late 1990s totaled around 200 scientists (master's level or above), while NARS research staff in South and Southeast Asia numbered around 1700 (excluding China). Numbers of scientists working in Latin America and Africa were quite small, by comparison. Expenditures per scientist were greater at IRRI than in NARS, resulting in an overall level of spending of about \$35 million each at IRRI and in Asian NARS (again excluding China), as of the late 1990s.

4.2.3. *Maize*

Maize is grown in both temperate (mostly developed country) regions and in tropical and sub-tropical regions (almost entirely in developing countries – 23 million ha in

Africa, 27 million ha in Latin America, and 42 million ha in Asia). Major advances in maize varieties were achieved in temperate zone regions based on the "heterosis hybrid" technique of breeding in the first part of the 20th century. Most of these hybrids were used for livestock feed. The transfer of heterosis-based hybrid maize technology to the tropics was very limited as of 1960. CIMMYT, the main international center mandated with maize production improvement, chose to pursue improvements in both hybrids and so-called "open pollinated varieties" (OPVs), which farmers can grow from saved seeds.[18] In its early years of operation, however, CIMMYT focused primarily on open pollinated varieties, which implied a substantial change in direction relative to research conducted on maize in North America and Europe. Another dimension in which CIMMYT's maize research differed from previous programs was that it focused on varieties to be consumed directly by humans.

Most of the public sector NARS varietal releases in maize have been open-pollinated varieties, although in the 1990s hybrids have become more important. Almost all private sector varietal releases by contrast have been hybrids. By the 1990s private sector programs were developing more varieties than public sector programs in Latin America. They are also becoming important in Asia and Sub-Saharan Africa.

NARS research investments for maize in developing countries were not quite as high as for rice or wheat. In the public sector, about 900 full-time equivalent researchers were working on maize in the late 1990s, with an additional 400 or so working in private sector firms (including 240 in multinational firms).[19] In dollar terms, CIMMYT's research expenditures on maize have been roughly \$10–\$20 million annually, depending on how the expenditure is categorized and measured. [Morris, Mekuria and Gerpacio (2003) do not offer a comparable figure for private sector expenditures.]

Public sector IARC and NARS programs had produced more than 1200 maize GRMV varieties by the late 1990s (Table 7). Another 700 GRMV hybrid varieties have been produced by private firms. Many of these private sector programs drew heavily on high-quality inbred lines developed in CIMMYT and NARS programs, following the pattern established in North America and Europe, where public sector breeding programs generated many of the raw materials used in hybrid seed programs.

By the late 1990s, improved maize varieties were planted on about 62.4% of the maize area in developing countries [Table 7 and Morris, Mekuria and Gerpacio (2003), based on CIMMYT Global Maize Impacts Survey]. The rate in tropical and other non-temperate zones was lower, at 47.2% of the maize area. Adoption was highest in East,

[18] Heterosis hybrids take advantage of "hybrid vigor" – a productivity advantage affecting, for partly unexplained reasons, the first-generation progeny of two inbred parent lines are crossed. To benefit from the heterosis effect, farmers must purchase seeds of first-generation progeny (F1 seeds) each growing season. Seeds saved from these F1 plants (known, in the second generation, as F2 seeds) do not perform as well as F1 seeds, and in fact may perform worse than the parent lines. By contrast, open-pollinated varieties do not need to be purchased fresh each year.

[19] The public sector figures, from Morris, Mekuria and Gerpacio (2003), do not include about 1500 Chinese breeders working in temperate zone maize production in northern and central China.

Table 7
Maize Green Revolution Modern Varieties (GRMVs)

	Latin America & Caribbean	Asia	Sub-Saharan Africa
	Average annual releases (private sector MVs in parentheses)		
1960s	12.4	(?)	1.0
1970s	17.6	(?)	1.5
1980s	25.1 (18)	(?)	5.4 (3)
1990s	24.5 (51)	(?)	14.3 (5)
	GRMV adoption (%)		
1970	10	10	1
1980	20	35	4
1990	30	45	15
2000	45	82	52

Source: Based on Morris, Mekuria and Gerpacio (2003) and Manyong et al. (2003).

South, and Southeast Asia, although this includes a large amount of temperate maize grown in China. Adoption rates were lowest in West and Central Africa and in Eastern and Southern Africa other than South Africa, where about 36% of maize area was planted to modern varieties. Table 7 shows adoption rates in the late 1990s across different regions. Note that area planted to MVs is difficult to measure for maize, given that recycled seeds of both hybrid varieties and OPVs can exhibit rapid genetic drift, since maize outcrosses fairly promiscuously. Nonetheless, the data shown here represent CIMMYT's best estimates of area planted.

4.2.4. *Sorghum and pearl millet*

Sorghum and pearl millet are grown extensively in semi-arid regions of Asia (chiefly India) and Sub-Saharan Africa. Of the 45 million ha planted to sorghum, 23 million ha are in Africa and 14 million in Asia. For pearl millet, 38 million ha are planted worldwide, of which 20 million are in Africa and 16 million in Asia. The International Crops Research Institute for the Semi-Arid Tropics (ICRISAT) has developed research programs for both crops in both India and Africa. ICRISAT maintains germplasm collections and provides germplasm to approximately 750 sorghum scientists and 300 pearl millet scientists in NARS programs.

India maintained breeding programs for both sorghum and pearl millet prior to the ICRISAT program and did release a number of modern varieties of sorghum and pearl millet in the 1960s and 1970s. Annual varietal releases in both Asia (India) and Africa show upward trends, with annual releases in the 1990s being roughly double the releases in the 1970s (Tables 8 and 9). More than 400 sorghum GRMV varieties and more than 100 pearl millet GRMV varieties have been released. During the 1990s, private firms began to develop and release hybrid varieties of sorghum. The relevant feature of this

Table 8
Sorghum Green Revolution Modern Varieties (GRMVs)

	Asia	Sub-Saharan Africa
	Average annual releases	
1960s	3.4	3.0
1970s	6.5	1.9
1980s	7.8	3.6
1990s	11.6	4.4
	GRMV adoption (%)	
1970	4	0
1980	20	8
1990	54	15
2000	70	26

Source: Based on Deb and Bantilan (2003).

private sector development is that ICRISAT material contributes a significant part of the genetic content of private varieties (as CIMMYT does for maize). Thus, private sector firms are building programs on public sector "platforms". It should be noted that a number of the public sector varieties are also hybrids, requiring the seed production methods assorted with hybrids [Deb and Bantilan (2003)].

As of the late 1990s (Table 8), adoption of improved sorghum varieties ranged from over 90% in China and about 70% in India to very low levels (e.g., 10–30%) in some other countries, including much of the African growing region [Deb and Bantilan (2003)]. For millet (Table 9), Deb and Bantilan (2003) report that adoption of improved varieties varies substantially across countries, with the highest levels of adoption in India (65%), Namibia (50%) and Zambia, Mali, Zimbabwe, and Botswana (20–30%).

4.2.5. Barley

Barley is a significant food source in many arid regions and marginal areas of the developing world, where it may be the only crop that can tolerate agro-ecological conditions. It is also one of the oldest food grains to have been domesticated. In modern times, barley is used as an important animal feed in West Asia and North Africa, as well as in Ethiopia and Eritrea and several Andean countries. FAO data show that in 2000, about 14.5 million ha in developing countries were planted to barley, with about 8 million ha in the Middle East and North Africa. The largest areas planted in the developing world are in Morocco, Kazakhstan, Syria, Iran, Ethiopia, China, and India. Yield levels vary considerably but are quite low in many producing countries, with average annual yields below 1.0 t/ha in many countries. Yields are also quite variable, since barley tends to be planted on marginal lands.

Intensive research on barley in developing countries has a comparatively recent starting date. International research began in 1977, with the establishment of the In-

Table 9
Pearl millet Green Revolution Modern Varieties (GRMVs)

	Asia	Sub-Saharan Africa
		Average annual releases
1960s	0.8	0.0
1970s	1.1	0.0
1980s	3.7	1.2
1990s	4.9	3.0
		GRMV adoption (%)
1970	na	0
1980	na	0
1990	60	0
2000	78	19

Source: Based on Deb and Bantilan (2003).

ternational Center for Agricultural Research in the Dry Areas (ICARDA), which has a mandate for research on barley in all developing countries. ICARDA's average annual research expenditures were around US$1 million during the 1990s, with about 3.5 scientist years of investment. National programs in nine countries having close collaborative relationships with ICARDA employed an additional 25 full-time equivalent scientists, with about half trained at the Ph.D. level [Aw Hassan and Shideed (2003)]. Thus, investments in barley research have been quite low globally.

In spite of the low investments in research, the production of new varieties has proceeded at a moderate pace, with over 100 GRMVs developed in the period from 1980 to 2000. As shown in Table 10, about 5–6 new varieties per year have been released across 23 developing countries [Aw Hassan and Shideed (2003)]. About three quarters of these varieties were developed on the basis of ICARDA-derived materials, with more than half of them being crossed at ICARDA [Aw Hassan and Shideed (2003)].

Modern barley varieties have been planted on 50% or more of the barley area in Egypt and Jordan, with high levels also reported in Tunisia and Ecuador. Across the major growing countries of West Asia and North Africa, however, total adoption rates remained around 15% by the late 1990s. (See Table 10.)

4.2.6. *Beans*

Dry beans are planted on 26 million hectares worldwide, with approximately 6 million ha in Latin America and 35 million ha in Sub-Saharan Africa. (China and India are also major producers.) CIAT initiated a bean program in South America in 1973, with its mandate extending to Central America in 1979 and to Africa in 1983. Seventeen Latin American NARS had bean programs in the 1970s. Only two programs were active in Africa in the 1970s, but by 1998, twelve programs in Africa were operating.

Table 10
Barley Green Revolution Modern Varieties (GRMVs)

	Middle East & North Africa
	Average annual releases
1960s	0.0
1970s	0.0
1980s	5.5
1990s	6.5
	GRMV adoption (%)
1970	2
1980	7
1990	17
2000	49

Source: Based on Aw Hassan and Shideed (2003).

CIAT in Colombia has been supporting bean improvement research for a number of years. Beans are an important source of protein in the diets of many consumers in Latin America, especially Brazil, and in Africa. Because of limited genetic improvements, beans have effectively been "crowded out" of productive areas by crops with greater genetic improvement, especially corn and soybeans.

CIAT's investments in bean improvement averaged around US$10 million per year through the 1980s and 1990s (measured in real 1990 dollars), with 20 or more scientists working on beans. National program investments in research grew substantially over the time period, with as many as 80 breeders in Latin America and 40 in Africa at various moments in time [Johnson et al. (2003)].

Table 11 reports varietal release data showing steady increases in releases in both Latin American and Africa with high CIAT content (especially CIAT crosses) in both regions. In the early years of CIAT's research activities, fewer than 10 varieties per year were released in Latin America, and essentially none in Africa. By the late 1980s and early 1990s, more than 20 varieties were being released annually in these regions, with 10–20 in Latin America and 5–10 in Africa. These varieties were adopted fairly extensively; by the late 1990s, CIAT-related varieties were grown on about 20% of the bean area in Latin America and about 15% of the area in Africa [Johnson et al. (2003)]. In many cases, these varieties were developed on the basis of work done in NARS as well as at CIAT, and the respective contributions are difficult to disentangle.

4.2.7. Lentils

Lentils are a major food crop in a number of parts of the world, including the Middle East and North Africa, which together with Ethiopia account for about one third of world area and production. The largest producers are Turkey, Iran, and Syria, with the

Table 11
Beans Green Revolution Modern Varieties (GRMVs)

	Latin America & Caribbean	Sub-Saharan Africa	Middle East & North Africa
	Average annual releases (private sector MVs in parentheses)		
1960s	4.0	0.0	0.0
1970s	5.7	2.0	0.0
1980s	13.2	4.5	9.3
1990s	24.3	10.0	3.9
	GRMV adoption (%)		
1970	1	0	0
1980	2	0	0
1990	15	5	5
2000	20	15	23

Source: Based on Johnson et al. (2003).

Middle East and North Africa together producing lentils on about 1.2 million ha, comprising one third of the world total. Yields are generally quite low – less than 1.0 t/ha in most of the developing world. Nonetheless, the value of world lentil production in the late 1990s was around US$1.4 billion.

Investments in lentil research, however, have been extremely low. Along with technical issues making lentil breeding quite difficult, this led to a very modest pace of varietal improvement through the 1970s and 1980s. Until now, international research expenditure on lentils has been less than US$1.0 million annually, with 2.5 to 3.0 scientist full-time equivalents devoted to lentil research. In NARS programs, only 13 Ph.D. scientists worked on NARS in the six largest breeding programs, as of the late 1990s, with 34 scientists total working on the crop. NARS expenditures in 1997 were estimated at US$1.7 million across the seven largest breeding programs [Aw Hassan and Shideed (2003)].

The production of modern varieties has correspondingly been low. Table 12 presents details on varietal release and adoption. Relatively few GRMV varieties were generated until the 1990s, at which point breeding at ICARDA began to generate significant payoffs. By 2000, as much as 23% of the area in the Middle East and North Africa was planted to modern varieties. Yields from improved varieties average 29% higher than for traditional varieties [Aw Hassan and Shideed (2003)], but it is not clear whether this effect is entirely driven by genetic improvement or whether it also reflects the fact that improved varieties are grown under better conditions and using higher levels of inputs.

4.2.8. *Groundnuts*

Groundnuts are produced on 24 million hectares, over 95% in developing countries. India, China, Nigeria, and Sudan are leading countries for production. Several other

Table 12
Lentil Green Revolution Modern Varieties (GRMVs)

	Middle East & North Africa
	Average annual releases
1960s	0.0
1970s	0.0
1980s	1.8
1990s	3.9
	GRMV adoption (%)
1970	0
1980	0
1990	5
2000	23

Source: Based on Aw Hassan and Shideed (2003).

semi-arid countries produce significant amounts of groundnuts, which represent an important export crop in some countries.

Research efforts have largely been based in the public sector, with some limited private sector efforts in high-income countries such as the US. In many developing countries, one or two scientists work on groundnut research. India has the largest program, with as many as 150 scientists engaged in groundnut research [Bantilan, Deb and Nigam (2003)]. International research, based at ICRISAT, has been supported with 10–12 scientists per year. Breeding has focused on resistance to various insects, diseases, and fungi, as well as for adaptation to specific environments. Varietal releases, described in Table 13, have risen to around 10 per year, mostly in India, where over 40 varieties had been released by the end of 1999.

Few modern groundnut varieties suited to developing countries were released prior to 1980. Since then, varietal production in both Asia and Africa has increased steadily. By the late 1990s, adoption rates for modern varieties were very high in China, and high in India and parts of Africa. The number of groundnut GRMVs produced has been modest.

4.2.9. Potatoes

Potatoes are produced over a wide range of climate systems. Globally, 38 million ha are planted. Roughly one quarter of this acreage is in developing countries (Asia has 6 million ha, Africa 1 million, Latin America 1.5 million), and this acreage is growing as new varieties are developed. Many potato landraces (farmers' varieties) continue to be planted for local preference reasons. More than 30 developing countries have potato-breeding programs with more than 300 breeders and related scientists working on

Table 13
Groundnut Green Revolution Modern Varieties (GRMVs)

	Asia	Sub-Saharan Africa
	Average annual releases	
1960s	0.0	0.0
1970s	0.0	0.0
1980s	8.5	0.6
1990s	9.5	1.1
	GRMV adoption (%)	
1970	0	0
1980	0	0
1990	0	0
2000	5	49

Source: Based on Deb and Bantilan (2003).

genetic improvement in potatoes. CIP provides genetic resource and breeding support in major regions. Roughly 700 potato GRMVs have been produced.

Research on potatoes for developing countries is relatively recent, with the International Potato Center (CIP) in Peru having been founded only in 1971. Previously, potatoes were primarily seen as a crop of high-income countries, plus a few countries of the Andean region. In most other places, including essentially all of the warm tropics and subtropics, potatoes were not widely cultivated. Nonetheless, the past three decades have seen a rapid increase in potato production. Potatoes are now grown in many countries where wheat is produced, with China and India having emerged as large producers [Walkerm et al. (2003)].

Research takes place on a significant basis in at least 30 developing country NARS, where almost 1000 scientist full-time equivalents were working on potatoes in the late 1990s [Walkerm et al. (2003)]. Of these, 400 were based in China.

Walkerm et al. (2003) report that from 1957 to 1998, about 500 varieties were released, but the pace of varietal production increased substantially following the 1950s. By the 1980s and 1990s, varietal releases were about 16 per year, with about 3–4 of those in Sub-Saharan Africa. About half the varieties were developed in NARS exclusively, with no contribution from CIP; about one quarter were developed in collaboration between CIP and NARS, and about one quarter were developed in high-income countries and then released by NARS.

Adoption of modern varieties was quite high by the 1990s, as shown in Table 14. NARS-developed varieties accounted for the largest fraction of the area, with developed country varieties also important. CIP was not a major contributor of varieties except in Africa [Walkerm et al. (2003)].

Table 14
Potato Green Revolution Modern Varieties (GRMVs)

	Latin America and Caribbean	Asia	Sub-Saharan Africa
		Average annual releases	
1960s	2.0	0.0	0.0
1970s	5.6	4.5	1.6
1980s	6.5	7.1	3.7
1990s	6.5	8.4	4.7
		GRMV adoption (%)	
1970	25	30	0
1980	54	50	25
1990	69	70	50
2000	84	90	78

Source: Based on Walkerm et al. (2003).

4.2.10. Cassava

Cassava is a major food crop in Africa, where 11 million of the world's 16 million ha are planted. Latin America plants 2.4 million ha and Asia 3.3 million ha. Cassava is thus produced almost exclusively in tropical developing countries. Two IARCs, CIAT in Latin America and IITA in Africa, support NARS programs. There are important differences in cassava consumption across regions. Essentially all of the cassava produced in Africa is used for human consumption, but in Asia most is industrial (for starch), and in Latin America, significant amounts are used for industry and for animal feed.

Very little cassava research had been done prior to the establishment of CIAT in 1969. In contrast to most other tropical food crops, there were essentially no improved varieties that had been developed in high-income countries. Thus, research progressed very slowly. CIAT, for example, spent five years collecting farmers' varieties of cassava, and the first breeder was appointed in 1974 [Johnson et al. (2003)]. CIAT's spending on cassava research reached US$5–6 million in the 1980s and 1990s, with as many as 10–15 principal staff members devoted to cassava improvement. NARS programs employed about 90 cassava breeders by 1999, but it is difficult to know what resources they had available to them.

The production of new varieties has been limited, although the years following 1980 have seen substantial increases in the rate of variety production. (See Table 15.) About 130 varieties had been released using CIAT and/or IITA germplasm, as of 1999 [Johnson et al. (2003)], with new releases averaging about 2 per year each in Asia and Latin America; African countries released 5–6 varieties per year in this period. IARC content is high in releases, reflecting the fact that the IARCs have dominated the genetic improvement in this crop.

Table 15
Cassava Green Revolution Modern Varieties (GRMVs)

	Latin America and Caribbean	Asia	Sub-Saharan Africa
		Average annual releases	
1960s	0.0	0.0	0.0
1970s	0.0	0.0	1.5
1980s	1.5	6.5	6.0
1990s	2.4	3.0	7.5
		GRMV adoption (%)	
1970	0	0	0
1980	1	0	0
1990	1	2	2
2000	5	10	16

Source: Based on Johnson et al. (2003).

4.3. GRMV production: A summary

To summarize, Table 16 reports a summary of GRMV production by crop and for all crops, by region by five-year periods. This table shows some of the dimensions of the Green Revolutions. Some crops, notably pearl millets, barley, lentils and cassava did not have significant GRMV production until after 1980. By contrast, wheat, rice and maize GRMVs were produced at high rates in the 1960s. This was to a considerable extent, related to the existence of pre-Green Revolution MVs. This table also shows that the production of GRMVs is rising over time.

Crops for which little research was available in the early years have, in most cases, experienced rapid and significant progress in more recent years. Notably, the Green Revolution is a phenomenon that has cut across crops and across regions, contrary to the popular perception that it was limited to rice and wheat in Asia and Latin America.

5. Impacts of the Green Revolution

5.1. GRMV production and adoption

Tables 5–15, above, show MV adoption rates by decade and region for GRMVs. In the Sub-Saharan Africa region, significant GRMV production in the 1960s (mostly of IARC-crossed MVs) led to little GRMV adoption. This reflected the geographic limitations of even the strongest IARC programs and the failure of "Technology Transfer" programs to facilitate broad international transfer.

In the 1960s and 1970s, several IARCs, IRRI (rice) and CIMMYT (wheat and maize), and later CIAT (beans) and ICRISAT (sorghum and millets), expected germplasm that

Table 16
Average annual varietal releases by crop and region, 1965–2000

Crop	Average annual releases						
	1965–1970	1971–1975	1976–1980	1981–1985	1986–1990	1991–1995	1996–2000
Wheat	40.8	54.2	58.0	75.6	81.2	79.3	80.0
Rice	19.2	35.2	43.8	50.8	57.8	54.8	58.5
Maize	13.4	16.6	21.6	43.4	52.7	108.3	71.3
Sorghum	6.9	7.2	9.6	10.6	12.2	17.6	14.3
Millets	0.8	0.4	1.8	5.0	4.8	6.0	9.7
Barley	0.0	0.0	0.0	2.8	8.2	5.6	7.3
Lentils	0.0	0.0	0.0	1.8	1.8	3.9	5
Beans	4.0	7.0	12.0	18.5	18.0	43.0	45
Cassava	0.0	1.0	2.0	15.8	9.8	13.6	15
Potatoes	2.0	10.4	13.0	15.9	18.9	19.6	20
All crops							
Latin America	37.8	55.9	65.9	92.5	116.2	177.3	150
Asia	27.2	59.6	66.8	86.3	76.7	81.2	90
Middle East North Africa	4.4	8.0	10.2	12.2	28.4	30.5	82
Sub-Saharan Africa	17.7	18.0	23.0	43.2	46.2	50.1	57
All regions	87.1	132.0	161.8	240.2	265.8	351.7	379

was successfully transferred to countries in Asia and Latin America, to be transferable to Sub-Saharan Africa. With the exception of wheat GRMVs adopted in North Africa, this simply did not happen. Many GRMVs were released to farmers, but were not adopted, even after significant extension activities. As a result of these failures on the part of IARC programs, Sub-Saharan Africa was delivered MVs at least 20 years later than was the case for Latin America and Asia. IARCs in Africa (IITA and ILRI) were also less effective than their counterparts in Asia and Latin America.

5.2. Productivity effects of GRMVs

Evenson and Gollin (2003b) estimated GRMV contributions to yield from experimental evidence (including evidence from farmers' fields) and from three country studies for India, China and Brazil. The objective was to determine the production gains from converting acreage from pre-Green Revolution varieties to post-Green Revolution varieties; i.e., to compute the yield contribution of genetic improvements. These contributions were not entirely independent of the use of other inputs, but the methodology attempted to isolate these genetic gains, holding input use constant.

Table 17
Growth rates of food production, area, yield, and yield components, by region and period

	Early Green Revolution 1961–1980	Late Green Revolution 1981–2000
Latin America		
Production	3.083	1.631
Area	1.473	−0.512
Yield	1.587	2.154
MV contributions to yield	0.463	0.772
Other input/ha	1.124	1.382
Asia		
Production	3.649	2.107
Area	0.513	0.020
Yield	3.120	2.087
MV contributions to yield	0.682	0.968
Other input/ha	2.439	1.119
Middle East–North Africa		
Production	2.529	2.121
Area	0.953	0.607
Yield	1.561	1.505
MV contributions to yield	0.173	0.783
Other input/ha	1.389	0.722
Sub-Saharan Africa		
Production	1.697	3.189
Area	0.524	2.818
Yield	1.166	0.361
MV contributions to yield	0.097	0.471
Other input/ha	1.069	−0.110

Source: Evenson and Gollin (2003b), based on FAO data and authors' estimates.

Table 17 reports the results of an exercise relating growth in production to growth in cropped area and yields.[20] The growth in yields was further decomposed into genetic MV contributions and the contributions of other inputs (fertilizer and irrigation). This was done for two periods, the early Green Revolution, 1961 to 1980, and the late Green Revolution, 1981 to 2000. The calculations are for the 11 Green Revolution crops in the study.

A striking feature of Table 17 is that for both periods, production growth exceeded population growth, except for Sub-Saharan Africa in the early period (recall that this period includes the Nigerian food production decline associated with higher oil prices).

[20] Since production is area times yield, the growth rate of production can be decomposed into the growth rate in area plus the growth rate in yield.

A second feature of note is that the MV contribution to production was higher in the late Green Revolution than in the early Green Revolution in all regions. The Green Revolution has not run its course by any means. A further feature of Table 17 is the disparity in MV contribution between regions. Asia and Latin America realized large MV gains over both periods. The Middle East–North Africa region realized significant gains in the late period. But Sub-Saharan Africa realized minuscule gains in the early period and only modest gains in the late period.

The area contributions to production are also of interest. By the late period, Latin American countries were reducing area planted to these food crops, as were all developed OECD countries. Asia had virtually ceased expanding cropped area in the late period. The Middle East and North African regions reduced the area contribution, but it remains high. Sub-Saharan Africa, by contrast, realized most of its production growth from expanded crop area. Furthermore, the Green Revolution MVs in Sub-Saharan Africa in the late period were not accompanied by increased input use.

Thus, even though these data show that Sub-Saharan Africa is finally realizing some Green Revolution gains, the nature of these gains is disquieting. Most production gains over the late period were from high rates of cropland expansion. This source of growth will soon be exhausted. Perhaps more relevantly, the absence of increased fertilizer use in the Sub-Saharan Africa Green Revolution suggests that the MVs in Africa were not bred to be responsive to cheaper fertilize or that the markets for fertilizer are very inefficient with high transaction costs.

5.3. Returns to investment in IARC and NARS programs

In the classic work of Griliches (1958) on hybrid maize, a benefit/cost analysis was performed. This requires a cost series $\{c_t\}$ over time and a benefit series $\{b_t\}$ over time. It is possible to construct a series for each region from ISNAR data on research expenditures and estimates of the CGI share of the expenditures. This cost series can be constructed for the 1950–2000 time period.

The data on GRMV adoption and impact can be used to construct the benefits series, $\{b_t\}$.

The cost and benefit series can then be utilized to calculate the following:

PVB: The present value of the benefits stream computed at a specific interest rate (we use 6%).

PVC: The present value of the costs stream computed at the same specific interest rate.

$B/C = PVB/PVC$, the benefit–cost ratio.

IRR: The rate of interest at which $PVB = PVC$.

Table 18 reports the IRRs for both NARS crop improvement programs and IARC crop improvement programs, by region. Note that these estimates include long periods of costs where few benefits are achieved. For example, for Sub-Saharan Africa, benefits exceeded costs almost 15 years later than was the case for Latin America and Asia.

Table 18
Estimated internal rates of return from Green Revolution contributions

Region	NARS B/C	NARS IRRs	IARC B/C	IARC IRRs
Latin America	56	31	34	39
Asia	115	33	104	115
West Asia–North Africa	54	22	147	165
Sub-Saharan Africa	4	9	57	68

Source: Evenson calculations reported in Evenson and Gollin (2003b).

The review of the rates of return for agricultural research [Evenson (2001)] summarized in Table 3 reported regional 'median' IRRs. As noted above, these IRRs, from individual studies for NARS programs, are higher than those reported in Table 18. This is primarily because individual studies tend to ignore the research costs required to reach the stage where benefits are produced. Some of this research is 'unproductive' but a considerable part of it is required to build the germplasm stocks and to enter the staging area where MV1s can be produced.

The IARC program IRRs are very high. These high IRRs appear to be very real and they reflect the 'leveraging' associated with the high production of IARC crosses and the high volume of IARC germplasm. Table 4 reports a statistical analysis of IRRs showing that the investment ratio did affect IRRs. But IRRs for NARS programs in Sub-Saharan Africa in the Green Revolution study are quite low.

6. NARS and the "Gene Revolution"

The section above argues that NARS programs, working in conjunction with IARCs, generated large and important returns through conventional breeding in the period from 1960 to 2000. What are the prospects for NARS programs in the emerging "Gene Revolution"?

The Gene Revolution is distinguished from the Green Revolution in several dimensions. First, the Gene Revolution is based on techniques of molecular manipulation, based on the enormous advances in molecular biology that took place in the second half of the 20th century. Some use "recombinant DNA" (rDNA) techniques, often referred to as genetic engineering techniques, to insert desirable segments into a plant cell. This is usually described as "transformation". Several techniques for this insertion (biolistic, agrobacterium) have been developed for this process. Note that a sexual cross between parents is not required as in the case of conventional breeding. Conventional breeding techniques do include wide-crossing and wide hybridization techniques but these require a sexual cross. Genetic engineering techniques, however, did not evolve from wide-crossing techniques. They emerged as a by-product of advances in the science of molecular biology. Nor are genetic engineering techniques separable from conventional breeding; once a transformed plant is developed, it can typically be used in a sexual cross. The "trait" associated with transformation can then be expressed in the progeny of the sexual cross through simple selection.[21]

Early claims for biotechnology methods implied that the new molecular techniques held out the promise of a "Gene Revolution" to complement the Green Revolution. Specifically, it was suggested that genetically modified (GM) crops had great promise for production environments in poor countries that had not yet been reached by conventional breeding techniques. These are the "disadvantaged" environments that continue to be dominated by landrace varieties ("farmers' varieties"). Claims about the potential for GM crops were often put forth by private firms with little or no experience in dealing with disadvantaged environments (but with vested interests in securing regulatory approvals and public acceptance of GM crops).

The reality is that the private sector has hitherto had little interest in developing breeding programs for poor countries, except for hybrid varieties – for the sensible reason that there is limited potential for them to recover the research costs that would be involved. Since plant breeders typically cannot retain proprietary control of most varietal technologies (hybrid varieties representing an important exception), they have little incentive to invest in breeding.[22] Yet, in spite of this, several GM crops have been adopted on significant areas in developing countries, and the potential for further application is great. This is because some GM crop products are "transportable", even though crop varieties are not. GM products can be installed on many different crop varieties.

It should be noted that at present, most commercial GM products are essentially "qualitative trait" products – i.e., specific attributes that plants either possess or do not possess. Such traits endow plants with specific cost advantages that vary from environment to environment but are "static" in nature. That is, the cost advantages conferred by these traits are essentially constant over time.[23] It is possible to "stack" more than one GM product in a crop variety, but stacking does not necessarily produce cumulative gains.

[21] It is also worth pointing out that molecular techniques have been put to extensive use within conventional breeding programs, even where transformation is not the goal. For example, marker-assisted selection, DNA fingerprinting, and mapping of resistance genes are all tools that have been widely used by breeders.

[22] An apt, though imperfect, analogy is to think of crop varieties as being like computer programs that can be readily duplicated and disseminated from user to user. In countries where users are numerous, poor, and effectively beyond the reach of enforceable contracts, this kind of piracy cannot readily be stopped. Although piracy may not prevent producers from writing computer programs that have a large market among corporate users in rich countries, it may tend to limit their enthusiasm for making products that would be targeted to the needs of poor people in developing countries. This may not be important in the software business; but for crop varieties, it should be clear that private firms have little incentive to develop varieties that are specifically targeted at the needs of poor farmers in developing countries. Where the varieties developed for rich countries can be used without significant modification, they may be pleased to distribute them internationally. Similarly, for public relations purposes, private firms may release crop technologies that have use in poor countries. But there is little reason to expect that private sector firms will develop improved plant types for marginal environments in developing countries.

[23] They may diminish as resistance erodes.

Although GM crops have begun to spread widely within the developing world, it is a mistake to imagine that the current mechanisms for generating a Gene Revolution will displace the need for a continuation of the conventional Green Revolution. The current mechanisms do not support the use of GM technologies to generate new variety platforms – new plant types adapted to marginal growing environments. It is sometimes said that the Gene revolution will replace the Green Revolution. To date, GM products can only complement conventional Green Revolution breeding by installing single-trait or multiple-trait GM products on the varietal platforms produced by conventional Green Revolution methods.

6.1. Gene–Green Revolution congruity

The first stage of the Green Revolution can be characterized as the "MV1 stage", in which first-generation modern varieties were delivered to producers. The delivery of MV1 varieties varied by crop and region. Much of this variation was dictated by preconditions. In many cases, where landrace agriculture has predominated MV1 varieties have not been successfully produced, even today.

Beginning a few years later, scientists began to develop "second generation" modern varieties (MV2s), based primarily on the insertion of useful "qualitative traits" into the MV1 varieties. A recurring theme was that MV2s added disease and pest resistance to MV1s, which were frequently susceptible to major biotic stresses. Furthermore, the susceptibility to specific diseases and insect pests was not easily predictable. The MV1 varieties were largely based on the relatively quick identification of new plant types (e.g., semi-dwarfs). By contrast, the MV2 varieties were based on broader evaluation of genetic resources for resistance traits. In the case of rice varieties, the MV1 generation (typified by IR8) was susceptible to brown planthopper, green leafhopper, and several diseases carried by these pests (e.g., rice tungro virus). The Genetic Evaluation Unit (GEU) program at IRRI eventually produced a number of disease-resistant MV2 varieties, most notably IR26 and IR36.[24]

The lesson for the Gene Revolution is clear. Current molecular methods are useful for transferring traits across varieties, but they must have robust and locally adapted platform varieties on which these traits can be placed. These varieties have historically come from IARCs working with NARS; the private sector has little incentive to develop the platforms, which are essentially public goods. Moreover, as with MV2 varieties, there will be many locally important traits to incorporate into platform varieties. The location-specificity of these traits means that much of the needed work will be done in national programs, rather than international ones. We believe that ultimately, national public sector programs will play a large role in making the Gene Revolution a reality for most developing countries.

[24] These MV2 varieties were developed very quickly at IRRI. IR-36 became one of the most widely-planted MVs ever developed as a result.

Table 19
Adoption of GM crops: 2003

Country	Percentage of crop in GM				
	Soybeans	Maize	Cotton	Canola	Total GM hectares (million)
US	81	44	59	✓	42.8
Argentina	99	✓	✓		13.9
Canada	✓	✓		75	4.4
Brazil	30				3.0
China			58		2.8
South Africa		3	20		0.4
Australia			✓		0.1
India			10		0.1
Romania	✓				>0.5
Uruguay	✓	✓			>0.5
Spain		6			<0.5
Mexico	✓		20		<0.5
Philippines		5			<0.5
Colombia			✓		<0.5
Bulgaria		✓			<0.5
Honduras		✓			<0.5
Germany		✓			<0.5
Indonesia			✓		<0.5

Source: ISAAA.

6.2. GM crop coverage to 2003

For now, the GM crop varieties available to developing countries are few in number and are limited to a few crop species. These are species and varieties where private firms have essentially been able to take advantage of direct research spillovers; varieties developed for North American markets have proven to be well adapted to other countries. Nonetheless, the impact to date has been substantial. Data from the International Service for the Acquisition of Agro-Biotech Applications (ISAAA) indicate that for the year 2003, the global area planted to GM crops is 67.5 million hectares grown by more than 7 million farmers in 18 countries (Table 19).

For each of these crops, a check mark indicates that the crop is approved for adoption by farmers. All of these countries plus many more have conducted field trials in preparation for approval to sell GM crops to farmers. More than 80 countries have approved GM field trials in 30 to 40 crops.

In addition to the four major crops, some adoption in the US is reported for GM potato, GM squash, GM sugar beets, GM tomato, and GM tobacco. The leading GM trait has been herbicide tolerance. Of the global acreage of 67.7 million hectares in these four crops, 73% were herbicide tolerant (all four crops utilized herbicide tolerant traits).

Insect resistance (*Bt*) was important in cotton and maize and 18% of the 67.7 million hectares had this trait. The area planted to "stacked" insect resistant and herbicide tolerant crops was 8% of the global total. Virus resistance was the third most important trait.

Several advanced developing countries (Argentina, Brazil, China, South Africa, and India) have significant area planted to GM crops. Several other developing countries have introduced GM crops but still do not have large areas planted to them.

Note that there are, as yet, no significant areas planted to GM crops in those species that are uniquely important in poor countries (e.g., cassava, lentil, sorghum). This suggests that the resource allocation in GM research is not, as of now, optimal for developing countries.

6.3. IARC and NARS "failures" in the Gene Revolution

The discussion of similarities between the Green Revolution and the Gene Revolution shows that IARC programs were successful in producing MV1 varieties through conventional methods. However, rDNA techniques are not particularly advantageous for producing MV1s, which involve major shifts in plant type. They are, however, well suited to MV2 production, which requires the incorporation of particular traits. Furthermore, they are ideally suited to germplasm conversion where GM transgenic products can be incorporated into breeding lines that can then enter conventional breeding programs.

IARC programs were successful in using conventional breeding techniques (including wide-crossing) to produce the host plant resistance traits in MV2 varieties. MV2 varieties in turn, made NARS programs more productive. And, on balance, IARC MV2 programs led to increased NARS investments in plant breeding (although not for small countries with low population densities).

Why then, are IARC programs not providing the same kind of leadership in the development of rDNA techniques, given the high degree of congruity with conventional breeding techniques? IARC spending on biotechnology is estimated to be $25 million, only 7% of IARC expenditures.

There are two possible explanations for the "failure" of IARC programs to provide leadership in the Gene Revolution comparable to the leadership that they provided in the Green Revolution. The first is essentially a political explanation. Bluntly stated, individual IARC programs are inhibited by the potential loss of donor support, given the political controversy surrounding GM crops (particularly in Europe). A second explanation is more systemic. The IARC programs in the Green Revolution did not use "recently developed" techniques; the conventional breeding methods used in the 1960s, 1970s, and 1980s were already well established. Expertise was readily available, and the IARCs could easily follow the well-blazed trails for conventional breeding. Arguably, Gene Revolution techniques are "recently developed" and given the limited spending in IARC programs on biotech programs, it is difficult to attain mastery of these technologies.

Both problems are even more acute for NARS. Since many NARS are heavily dependent on donor funding, there may be explicit or implicit pressures for NARS programs to hold off on investments in rDNA capacity. Moreover, NARS programs have even less capacity than IARCs to pursue frontier research techniques. Historically, most developing country NARS programs (with the exception of a few large actors) have had difficulty operating state-of-the-art technologies.

7. Economic impact of NARS programs

In addition to the impacts of NARS programs described above, a number of additional areas of impact are important. We note briefly the role that agricultural research (including NARS research) has played in conserving land for nature; in mitigating the effects of changes in world food prices; and in alleviating mass poverty.

7.1. Yield–cropland tradeoffs (land for nature)

In many countries, the land available for crop agriculture is limited, and expansion of cropped area comes at the expense of natural areas – valuable as habitat for wildlife and for ecosystem services. In most countries, intensification of agriculture on existing farmland is seen as a substitute for extensive expansion onto new areas. The data support this view: those countries that have experienced growth in crop yields have seen relatively slower growth in cropped area.

Table 20 shows the results of a regression illustrating this relationship. In general, NARS programs recognize this relationship and seek to increase crop yields. We can understand Table 20 as a simple measure of their success in reducing growth in cropland through increases in crop yields. We present the reverse relationship as well.

In either case, there is a strong relationship, indicating that NARS-produced yield increases are related to slower cropland growth. The relationship shown in Table 20 indicates that yield growth reduced cropland area growth substantially.

7.2. Prices of agricultural commodities and mass poverty

World prices of cereal grains have declined in real terms for decades, as shown in Figure 2. Why is this so, when many countries are not realizing high rates of cost reduction? And how do these prices affect farm incomes in different countries? (See Figure 2 for prices.)

World prices, of course, are determined by global supply and global demand. Global supply is generally considered to be relatively price inelastic for aggregate agricultural production. But supply does respond to cost reductions. Most OECD countries have had relatively high rates of growth in agricultural total factor productivity (TFP). In fact most OECD countries have realized higher rates of TFP growth (this is equivalent to unit cost reduction) in agriculture than in the rest of the economy.

Table 20
Yield–cropland relationships

Independent variable	Dependent variables	
	Growth in cropland	Growth in yield
Constant	1.486	0.137
	(8.93)	(0.59)
Period 2	0.198	−0.087
	(1.48)	(0.58)
Growth in yield	−0.609	
	(−11.00)	
Growth in cropland		−0.752
		(−11.00)
R^2	0.466	0.46
F	62.52	60.81
N	146	146

Regressions weighted by value of agricultural production.

For developing countries, the TFP experience is mixed. Most countries realizing a Green Revolution had TFP growth rates similar to or greater than rates of price reduction. In other words, these countries were able to reduce their costs of production faster than prices were falling. Typically these countries were also able to achieve increases in the supply of food that were more rapid than population increases. Thus, countries realizing Green Revolution gains generally saw increasing availability of food per capita.

Many countries, however, had relatively modest rates of TFP growth. And many of these countries had relatively high rates of population growth in countries. Some were able to meet their food needs by expanding the area under cultivation (with predictable environmental effects). Others were not able to meet their food needs through domestic production.

At the global level, with the low price and income elasticities of demand, and with population-related supply effects, it is very plausible that the real prices of cereal grains in world markets should decline even when significant parts of the cereal-producing world do not realize high rates of TFP-induced cost reductions. A major factor is that most of the OECD countries are currently in situations of "overproduction" in agriculture. Many engage in costly subsidization of agriculture. In a sense, their overproduction problems are the result of "uneven" delivery of cost-reducing technology in regional markets. With increasingly globalized markets, this overproduction has become global. Thus, the OECD countries and some other large producers (e.g., China) have delivered low food prices to their own populations – but also to the rest of the world.

For developing countries that have experienced Green Revolutions, farm incomes have declined. But many of the world's poorest farmers are trapped in a situation where real prices are declining faster than real costs. Without dramatic changes in their TFP levels – and perhaps also in the structure of world agricultural markets – farmers in these countries will continue to be impoverished by declining world prices. Farmers in Sub-Saharan Africa and some other regions of the developing world need access to improved technologies if they are to cope with the long-term decline in world prices.

8. Policy issues for NARS and IARC programs

The dominant theme in both the Green Revolution and the Gene Revolution is that of "uneven delivery". The development of GRMVs was uneven by crop and by region. Landrace varieties dominated the agriculture of many countries in 1960. But in 2004, crop production in many countries continues to be based on landrace varieties. And for a number of countries, landrace agriculture is the dominant form of agriculture.

The Gene Revolution products have been delivered to developing countries in a still more uneven fashion. Their introduction is clearly related to the institutional setting in developing countries. In many cases, a country with adequate institutions can realize significant cost reductions. (*Bt* cotton offers an example.)

The IARCs provided real leadership in the Green Revolution era, except for those IARCs serving Sub-Saharan Africa, which had a more mixed experience. The IARCs initiated genetic resource evaluations programs and developed elite breeding material for the express purpose of raising the productivity of NARS breeders.

NARS programs in Latin America and Asian countries effectively responded to IARC programs in the Green Revolution. NARS programs in Sub-Saharan Africa have been much less successful. This lack of success is partly related to country size. Small countries (measured either in area or economic activity) cannot develop the kinds of programs that larger countries can develop. The lack of success is also related to geo-climate conditions. Productivity gains in mountainous regions and in regions with soil and water limitations are difficult to achieve.

The early experience with the Gene Revolution is that most IARCs are not (yet) providing leadership in the Gene Revolution. Most IARCs have invested little in agricultural biotechnology research. NARS program investments are also very diverse. Several countries, notably China, India and Brazil, have invested more in NARS-based biotechnology capacity. These countries are providing Gene Revolution leadership. Several other countries have made significant investments. But many countries are simple allowing the Gene Revolution to pass them by.

The difficulties of extending the Green Revolution to more countries and more agro-ecologies are real. The Green Revolution experience to date has changed the landscape of development. Most countries in Latin America, except for a number of Caribbean countries, have demonstrated that they can produce agricultural productivity growth. Many Latin American countries have had a tendency to "shoot themselves in the foot" by pursuing erratic macroeconomic policy, but most have the institutions and technological capacity to achieve economic growth. Most Asian countries also have the institutions and technological capacity to achieve economic growth. There are exceptions (Afghanistan, Cambodia, Bhutan and Nepal), but these are associated with weak governments. The countries of the Middle East and North Africa also have growth capacity, although at this writing, their domestic political situations and global geopolitical situations make growth difficult.

At the end of the Green Revolution, then, we see growth capacity everywhere, except in parts of Sub-Saharan Africa and in individual countries in other regions. Problems are typically due to local conflicts, resource capacity, legal systems, property rights systems and other inadequate institutions.

The agricultural research system – IARCs and NARS alike – has also failed Sub-Saharan Africa. Farmers in Sub-Saharan Africa have paid a high price for these failures. Farmers in the rest of the world have received significant cost reductions over the past 50 years. With price inelastic demand in globalized markets, prices have declined, generally at the maximum rate of cost reduction in the rest of the world. But farmers in many Sub-Saharan African countries have received little in the way of cost reductions. They have been delivered price reductions without cost reductions.

The failure of the IARCs with respect to Sub-Saharan Africa is an issue for their individual boards and for their governing entity, the CGIAR. The performance of NARS programs, in contrast, is ultimately a national government responsibility. International agencies cannot assume this responsibility. The bottom line in both the Green and Gene Revolutions is that national governments determine success or failure.

References

Avila, A.F., Evenson, R.E. (forthcoming). In: Evenson, R.E., Pingali, P. (Eds.), Handbook of Agricultural Economics, vol. 4. Elsevier Science, Amsterdam. Ch. 73.

Aw Hassan, A., Shideed, K. (2003). "The impact of international and national investment in barley germplasm improvement in the developing countries". In: Evenson, R.E., Gollin, D. (Eds.), Crop Variety Improvement and Its Effect on Productivity: The Impact of International Research. CAB International, Wallingford, UK, Ch. 11, pp. 241–256.

Bantilan, M.C.S., Deb, U.K., Nigam, S.N. (2003). "Impacts of genetic improvement in groundnuts". In: Evenson, R.E., Gollin, D. (Eds.), Crop Variety Improvement and Its Effect on Productivity: The Impact of International Research. CAB International, Wallingford, UK, Ch. 14, pp. 293–314.

Boyce, J.K., Evenson, R.E. (1975). Agricultural Research and Extension Programs. Agricultural Development Council, New York.

Byerlee, D., Alex, G. (2003). "National agricultural research systems: Recent developments and key challenges", Mimeo. Consultative Group on International Agricultural Research, Washington, DC.

Dalrymple, D.G. (1986a). Development and Spread of High Yielding Rice Varieties in Developing Countries. Bureau for Science and Technology, Agency for International Development, Washington, DC.

Dalrymple, D.G. (1986b). Development and Spread of High Yielding Wheat Varieties in Developing Countries. Bureau for Science and Technology, Agency for International Development, Washington, DC.

Dalton, T.J., Guei, R.G. (2003). "Ecological diversity and rice varietal improvement in West Africa". In: Evenson, R.E., Gollin, D. (Eds.), Crop Variety Improvement and Its Effect on Productivity: The Impact of International Research. CABI Publishing, Wallingford, UK, Ch. 6, pp. 109–134.

Deb, U.K., Bantilan, M.C.S. (2003). "Impacts of genetic improvement in sorghum". In: Evenson, R.E., Gollin, D. (Eds.), Crop Variety Improvement and Its Effect on Productivity: The Impact of International Research. CAB International, Wallingford, UK, Ch. 9, pp. 183–214.

Evenson, R.E. (2001). "Economic impacts of agricultural research and extension". In: Gardner, B., Rausser, G. (Eds.), Handbook of Agricultural Economics, vol. 1A. Elsevier Science, Amsterdam, Ch. 11, pp. 574–616.
Evenson, R.E., Gollin, D. (Eds.) (2003a). Crop Variety Improvement and Its Effect on Productivity: The Impact of International Agricultural Research. CAB International, Wallingford, UK.
Evenson, R.E., Gollin, D. (2003b). "Objectives and methodology: Crop genetic improvement studies". In: Evenson, R.E., Gollin, D. (Eds.), Crop Variety Improvement and Its Effect on Productivity: The Impact of International Research. CABI Publishing, Wallingford, UK, Ch. 3, pp. 7–38.
Evenson, R.E., Kislev, Y. (1975). Agricultural Research and Productivity. Yale University Press, New Haven, CT.
Evenson, R.E., Rosegrant, M. (2003). "The economic consequences of crop genetic improvement programmes". In: Crop Variety Improvement and Its Effect on Productivity: The Impact of International Research. CABI Publishing, Wallingford, UK. Ch. 21.
Gollin, D., Evenson, R.E. (1997). "Genetic resources, international organizations, and rice varietal improvement". Economic Development and Cultural Change 45 (3), 471–500.
Griliches, Z. (1957). "Hybrid corn: An exploration in the economics of technological change". Econometrica 25 (4), 501–522.
Griliches, Z. (1958). "Research costs and social returns: Hybrid corn and related innovations". Journal of Political Economy 66, 419–431.
Heisey, P.W., Lantican, M.A., Dubin, H.J. (2003). "Wheat". In: Evenson, R.E., Gollin, D. (Eds.), Crop Variety Improvement and Its Effect on Productivity: The Impact of International Research. CAB International, Wallingford, UK. Ch. 4.
Heisey, P.W., Srinivasan, C.S., Thirtle, C.G. (2001). "Public sector plant breeding in a privatizing world", ERS Agriculture Information Bulletin No. 772. U.S. Department of Agriculture Economic Research Service, Washington, DC.
Hossain, M., Gollin, D., Cabanilla, V., Cabrera, E., Johnson, N., Khush, G.S., McLaren, G. (2003). "International research and genetic improvement in rice: Evidence from Asia and Latin America". In: Evenson, R.E., Gollin, D. (Eds.), Crop Variety Improvement and Its Effect on Productivity: The Impact of International Research. CAB International, Wallingford, UK. Ch. 5.
Johnson, N.L., Manyong, V.M., Dixon, A.G.O., Pachico, D. (2003). "The impact of IARC genetic improvement programs on cassava". In: Evenson, R.E., Gollin, D. (Eds.), Crop Variety Improvement and Its Effect on Productivity: The Impact of International Research. CAB International, Wallingford, UK, Ch. 16, pp. 337–356.
Manyong, V.M., Kling, J.G., Makinde, K.O., Ajala, S.O., Menkir, A. (2003). "Impact of IITA germplasm improvement on maize production in West and Central Africa". In: Evenson, R.E., Gollin, D. (Eds.), Crop Variety Improvement and Its Effect on Productivity: The Impact of International Research. CAB International, Wallingford, UK, Ch. 8, pp. 158–182.
Morris, M., Mekuria, M., Gerpacio, R. (2003). "Impacts of CIMMYT maize breeding research". In: Evenson, R.E., Gollin, D. (Eds.), Crop Variety Improvement and its Effect on Productivity: The Impact of International Agricultural Research. CAB International, Wallingford, UK, Ch. 7, pp. 135–158.
Pardey, P.G., Beintema, N.M. (2001). "Slow magic: Agricultural R&D a century after Mendel", IFPRI Food Policy Report. International Food Policy Research Institute, Washington, DC.
Schultz, T.W. (1964). Transforming Traditional Agriculture. Yale University Press, New Haven, CT.
Walker, T., Bi, Y.P., Li, J.H., Gaur, P.C., Grange, E. (2003). "Potato genetic improvement in developing countries and CIP's role in varietal change". In: Evenson, R.E., Gollin, D. (Eds.), Crop Variety Improvement and Its Effect on Productivity: The Impact of International Research. CAB International, Wallingford, UK, Ch. 15, pp. 315–336.

Chapter 47

LIVESTOCK PRODUCTIVITY IN DEVELOPING COUNTRIES: AN ASSESSMENT

ALEJANDRO NIN

International Food Policy Research Institute (IFPRI), Washington, DC

SIMEON EHUI

The World Bank, Washington, DC

SAMUEL BENIN

International Food Policy Research Institute (IFPRI), Washington, DC

Contents

Handbook of Agricultural Economics, Volume 3
Edited by Robert Evenson and Prabhu Pingali

DOI: 10.1016/S1574-0072(06)03047-7

Abstract

This chapter assesses livestock productivity in the developing world in the context of growing demand for livestock products and substantial changes in the structure of food demand, commercialization of production, and growing importance of international markets. Issues related to the roles of the various partners in livestock research are also considered and modalities to ensure more effective linkage between research and development agencies to improve technology transfer and impact are discussed.

The chapter starts discussing the factors affecting livestock development and how these factors interact at different stages in the process of commercialization and development of the livestock sector. This serves as the conceptual framework to analyze trends in livestock demand; structural characteristics of livestock production in developing countries; global trends in productivity, and the evolution of research and development in developing regions. Past trends show that changes in global livestock production are explained by demand and supply changes associated to economic growth mainly in East and Southeast Asia while on the other hand productivity in Sub-Saharan Africa has remained stagnated for most of the past 40 years. National research systems in developing countries have focused mostly on research in ruminants, while research in poultry production has been dominated by the private sector. Despite these research priorities in developing countries, productivity differences in poultry and pig production between high income and some of the fast growing developing countries have been reducing in the past years, while large differences in productivity still persist in the case of milk and beef production with no evidence of developing countries catching-up with most productive countries. Major improvements in livestock productivity are possible and could contribute to economic growth in developing countries, but increased investment in livestock research and a framework for international action to support livestock research and development in developing countries are needed.

Keywords

livestock productivity, livestock demand, Livestock Revolution, research and development, livestock

JEL classification: Q1, Q18, Q55, Q11

1. Introduction

It is estimated that livestock and livestock products make up over half of the total value of agricultural gross output in industrial countries, and about a third of the total in developing countries, but this latter share is rising rapidly principally due to rapidly growing demand for livestock products [Bruinsma (2003)]. The global importance of livestock and their products is increasing as consumer demand in developing countries expands with population growth, rising incomes, and urbanization. This rapid worldwide growth in demand for food of animal origin has been called "Livestock Revolution" [Delgado et al. (1999)]. Although the global growth rate of livestock production is currently just over 2% per year and is declining over time, this masks the large regional disparities (see Table 1). While production growth rates in industrial countries, where people already enjoy adequate supplies of animal protein, have remained at just over 1% for the past 30 years, growth rates in developing countries as a whole have been high and generally accelerating to match the rapid growth in demand. The trends in East Asia are highest, with livestock product growth rates of over 7% a year in the past 30 years, albeit from a low base. South Asia and the Middle East and North Africa have maintained long-term growth in livestock product output of over 3% per year. Sub-Saharan Africa is the only region of the world that has lagged behind in livestock production. Standing at about 2% per annum, per capita livestock output has hardly increased at all in the past 30 years [Ehui et al. (2002)].

To date, overall growth in livestock production has been sufficient to meet increases in demand without significant price increases, and relative to the long-term downward trend in prices for cereals, oils, and fats, the prices for livestock products have remained relatively stable. However, developing countries, as a group, have become net importers of livestock products from industrial countries, revealing structural constraints for a sustained expansion of livestock production. Between 1990 and 2000, net imports of meat and milk to developing countries grew by more than 6% a year, while net imports of

Table 1
Global and regional growth in livestock output (percent per year)

Regions	1969–1999	1979–1999	1989–1999
Sub-Saharan Africa	2.4	2.0	2.1
Near East/North Africa	3.4	3.4	3.4
LAC	3.1	3.0	3.7
South Asia	4.2	4.5	4.1
East Asia	7.2	8.0	8.2
Developing	4.6	5.0	5.5
Industrial	1.2	1.0	1.2
Transition	−0.1	−1.8	−5.7
World	2.2	2.1	2.0

Source: Bruinsma (2003).

eggs declined by a little over 16%. Projections of future demand and supply of livestock products in developing countries show that demand for all products will grow faster than supply and that among regions, the gaps are greatest in Sub-Saharan Africa (Table 2), which indicates that developing countries will face R&D challenges (among others) if livestock production is to keep pace with demand. How will these constraints affect future livestock production in developing countries? How will productivity growth contribute to growth in supply of livestock products? What modalities can be put in place to ensure more effective linkages between livestock research and development?

This chapter assesses livestock productivity in the developing world in the context of growing demand for livestock products and substantial changes in the structure of food demand, commercialization of production, and growing importance of international markets. The chapter also addresses issues related to the roles of the various partners in livestock research and modalities to ensure more effective linkage between research and development agencies to improve technology transfer and impact. In order to do this, the next section presents a conceptual framework discussing the factors affecting livestock development and how these factors interact at different stages in the process of commercialization and development of the livestock sector. In the context of this framework, Section 3 discusses the trends in livestock demand as well as the factors fueling consumer demand and Section 4 presents trends and structural characteristics of livestock production in developing countries. Sections 5 and 6 focus on the analysis and comparison of partial and total factor productivity growth in developing regions respectively, while Section 7 discusses the evolution of research and development in these regions. The summary and conclusions are discussed in Section 8.

2. Factors affecting livestock development

Livestock production structure and organization evolves with economic growth and development. Income growth and urbanization impact traditional patterns of consumption increasing demand for value added food and livestock products. It is assumed for analytical purposes that development of the livestock sector is the transformation process that the sector follows from household's backyard production for self-consumption to commercialization and specialization [see Pingali and Rosegrant (1995)]. Figure 1 presents a diagram showing the main relationships and determinants of the process of commercialization and development of the livestock sector. During this process, structural changes occur in consumption patterns, transport and transaction costs and in the relative availability of resources, all of which result in output and input market development and the transformation of institutions.

The diagram in Figure 1 puts consumer preferences and demand as the main forces driving the process of commercialization of the livestock sector through output markets. Changes in consumption patterns are mainly associated with income growth and urbanization but several other factors affecting market development and supply of livestock products would determine the possibilities of the sector to keep pace with demand.

Table 2
Recent productivity (1982–1994) and projected (1993–2020) demand growth rates (% p.a.)

	Sheep & goat meat		Beef		Pork		Poultry		Milk	
	Productivity growth[a]	Demand growth[b]	Productivity growth	Demand growth	Productivity growth	Demand growth	Productivity growth	Demand growth	Productivity growth	Demand growth
East Asia	n.a[c]	0	3.8	3.2	1.5	2.8	2.4	3.6	0	2.8
South Asia	n.a.	4.2	1.5	3.7	0	18.6[d]	0.8	21.7[d]	3.9	4.0
Southeast Asia	n.a.	0	0.9	4.2	0.8	3.2	−0.3	2.6	2.4	3.0
Latin America & Caribbean	n.a.	18.6[d]	0.2	2.2	0.5	2.6	1.1	2.6	0.7	1.9
Central Asia, West Asia and North Africa	n.a.	2.6	2.7	3.5	0.1	0	0.2	2.6	1.5	3.0
Sub-Saharan Africa	n.a.	2.6	−0.5	4.2	0.1	2.6	−0.1	2.6	0.6	3.0
Developing countries	n.a.	2.6	0.5	2.9	1.2	2.7	0.8	3.2	1.9	3.2
Developed countries	n.a.	0	0.9	0.4	0.4	0.3	0.8	1.0	1.3	0.3

Source: Derived from Delgado et al. (1999).
[a]Productivity growth is on a per animal basis.
[b]Demand growth is expressed on a total basis, not per capita.
[c]n.a., not available.
[d]These growth rates were from a very low initial base which inflate the figures.

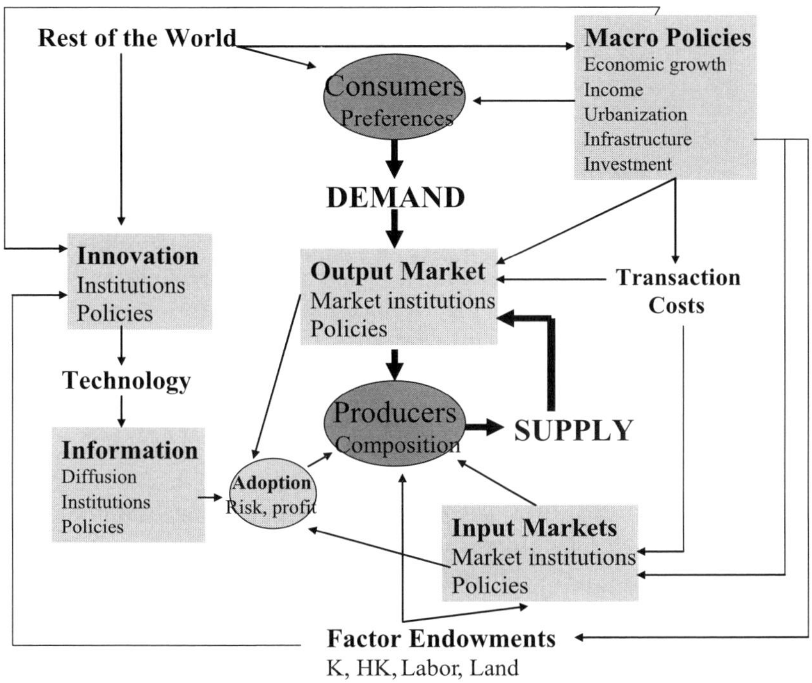

Figure 1. Factors affecting development of the livestock sector.

These factors are: transaction costs, institutions, policies, input and labor markets, and technology availability for improving productivity and disease control.

Transaction costs, institutions and policies affect development of output and input markets. Transaction costs are among the main factors determining maturity of input (e.g., feed) and output markets and can be affected by local infrastructure conditions, including distances to the nearest grain market, road conditions, transportation, frequency of local farm fairs. Local policies affect market maturity as well. At the regional level, regions with better infrastructure commonly enjoy advantageous terms of trade shifts while producers in rural hinterlands often lose out [Reardon and Barret (2000)]. Markets are also affected by macro policies including trade policy, which regulates the impact of international prices on domestic markets. Policies and government interventions can also facilitate or retard the transition to new stages of development by favoring certain institutions and agents. They can contribute to define winners and losers of structural changes in the sector, enhance market participation of smallholders, employment generation or development of large-scale operations.

In the long run, changes in consumption patterns, economic growth, infrastructure and macroeconomic changes will impact on factor and resource availability and transaction costs, affecting relative prices in input markets. As changes in feed and labor markets play a key role in the development of the livestock sector, increases in crop production and productivity could contribute to the development of the feed industry and of feed markets, reducing prices of feed relative to other inputs. With respect to labor markets, growth of off-farm employment reduces the advantages of backyard livestock production and promotes specialization and commercialization of livestock production. This implies further demand for new relatively more capital-intensive technologies and an expansion of the commercial sector. Availability of higher quality and cheaper feed and increasing labor costs would eventually results in increasing demand for new technologies like genetically improved animals and animal health technologies which would allow producers to take advantage of the development of feed markets and cheap quality feed.

2.1. The process of livestock development

We compare stylized and most distinctive facts of the livestock sector in a low income country with those of the livestock sector in middle or high income countries in order to illustrate the expected changes during the process of development or commercialization of the livestock sector.

At the farm level, the two extreme contrasting cases are the non-specialized household backyard producer (traditional model) and the specialized producer (commercial model). Roughly associated with these stylized extreme forms of livestock production are contrasting models for the processing and marketing of livestock products. The traditional livestock product market systems can be characterized by atomized market structure, consisting of many small scale market agents; artisan processing, labor intensive handling and transport methods; low value added products, generally limited in diversity; great diversity in market behavior and roles, with different types of market agents, which is associated with a lack of vertical integration.

The commercial livestock market systems on the other hand has normally a concentrated market structure, with relatively few large scale, vertically integrated market agents; industrial processing is based on capital intensive technologies at all market levels; large scale collection and distribution associated with long distance market chains; value added and diverse products; little diversity in market enterprise types, with production concentrated in specialized commercial farms.

The path between the traditional and commercial models is a gradual process. Generalizing from farm level situations in China, Chen and Rozelle (2001) distinguish 3 phases of the income–herd relationship in smallholder producers, which coincide with the process of commercialization of the livestock sector: emergence, expansion and contraction. Poor farmers raise few livestock because there are factors that constrain household's livestock production (low crop production and residues availability). At this stage, livestock plays multiple roles in the livelihood strategy of rural households, providing livestock products but also, store of wealth, insurance, draught power and inputs to crops.

As economic growth begins to take place, poor rural households gradually expand their livestock holdings and production of livestock products grows in importance as the main function of livestock holding due to increased food availability and development of feed markets, reduction of transaction costs and access to output, credit and insurance markets. The herd size and production specialization gradually expands with further development, but there is a point in many rural economies after which most farmers choose to stop raising livestock (higher opportunity cost of labor and job opportunities outside the farm). After a certain income level, livestock production for most household falls and only a few specialized households evolve toward larger-scale commercial operation with marketing and processing activities evolving concomitantly.

For Sub-Saharan Africa, McIntire et al. (1992) distinguish four stages in the process of animal agricultural intensification. In the first stage, at low population density and abundance of land, crop and livestock production activities are extensive and specialized, with few interactions between cultivators and pastoralists. In the second stage, agriculture intensifies due to population growth and changes in markets. There are more crop–livestock interactions with crop farmers using more animal power and manure, and with pastoralists using more crop residues for feed. Higher competition for land between cultivators and pastoralists also sets in. In the third stage, as population pressure increases further, cropland expands, while fallows and natural pasture contract. Further intensification increases the demand for more crop inputs in livestock production and for more livestock inputs in crop production. Increasing impediments to obtaining inputs in markets or through contracts promote closer integration of crop and livestock activities or mixed farms. Such impediments create economic incentives to provide inputs directly on farm, thus encouraging crop–livestock integration. In the fourth stage, if markets and exogenous technologies accompany further population growth, purchased inputs can replace crop–livestock integration, thereby developing specialized enterprises.

Empirical evidence presented and discussed by Chen and Rozelle (2001) and McIntire et al. (1992) confirms the different phases of the livestock commercialization and industrialization process as presented above, indicating the presence of an inverted-U relationship between livestock holdings by households and income level.

2.2. Vertical coordination in the livestock sector

As the process of commercialization results in an increasing importance of vertical coordination in livestock production chains and as technology also flows between high income and developing countries, it is important to understand how these changes affect livestock production and how they relate to technical change and production of different species. This section looks at some of the evidence of vertical coordination in livestock sectors in high income countries and the main factors explaining it, mainly based on work by Martinez (2002) and Ward (1997).

Evolution of the livestock sector in the U.S. shows the importance of vertical coordination in livestock production but at the same time shows important differences in the organization of the food chain between the egg, broiler, pork, milk and beef sectors. While vertical integration is not extensive in U.S. agriculture, with less than 8% of volume produced by farm operations owned by processors or input supply businesses in 1998, contracts are common among all types of farms, accounting for 35% of total production.

The egg industry is the only livestock sector where vertical integration is important due to transaction costs and significant site specificities. Technological breakthroughs in the 1960s led to high-speed, in-line grading, in which eggs are conveyed directly from laying cages to grading and packing machines making on-farm egg processing the norm [Hayenga et al. (2000)]. Over two thirds of contract volume is in marketing contracts, and one-third in production contracts. Production contracts dominate the broiler sector with significant use also in eggs and hogs. Spot markets are still predominant in cattle exchanges. Poultry and dairy contracts represented 45% of total contract value in U.S. agriculture in 1998. Within each sector, 95% of commodity produced in the poultry industry was produced under contracts, 43% in hogs, 57 in dairy and 25 in cattle [Hayenga et al. (2000)].

Generally, poultry and hog productions are more coordinated than beef. This is because of [according to Martinez (2002)] asset specificity together with uncertainty (technological changes, unpredictable changes in consumer preferences), and information asymmetry among trading partners regarding product value and producer effort.

Asset specificity is the degree to which assets are specifically designed or located for a particular use or user. "As assets become more specialized, the investing party will expend more resources to specify more contract contingencies because there are greater benefits from "holding-up" the asset owner. Consequently, vertical integration, which eliminates the exchange relationship, becomes more prevalent as asset specificity and the potential benefits to reneging on contracts increase" [Martinez (2002)].

Three main biological characteristics determine the incentives to vertical integrate or coordinate the industry, affecting the degree of asset, space and time specificity of the three species [Ward (1997)]: biological production cycle, industry stages, and geographic concentration in production. According to Ward (1997), the main determinant of vertical integration in livestock production is the speed with which biological changes such as genetic improvements can be made. There is more incentive to vertically integrate in an industry which has a shorter biological process and in which genetic changes can be made more quickly. Making quicker genetic changes also affects efforts to reduce production costs and increase consistency of products for consumers. Ward (1997) compares the genetic base for poultry with that of hogs and cattle in order to explain differences in production coordination between species. While poultry's genetic base is relatively narrow (only a few breeds or genetic lines, i.e., fewer than ten, are used), the genetic base for hogs has narrowed considerably in recent years, with specialized firms providing breeding stock for larger hog operations. With respect to the beef industry, Ward points out that in contrast with developments in the poultry and hog industries, the genetic base is widening instead of narrowing while many cattlemen are attempting to create new breeds, resulting in further amalgamation of the genetic base.

According to Ward (1997), the biological process, the number of stages in the production process and the spatial organization of the beef chain are major disincentives for vertical coordination in the industry. The biological process is a serious constraint to quickly changing the genetic base in beef production given that a cow produces only one calf per year and it takes about 24 months to learn whether or not the breeding process resulted in beef with more or less desirable characteristics. Additionally, the production process for cattle consists of three stages: cow-calf, stocker or growing, and feeding. Thus, the beef industry has a third production stage,[1] and each stage also has different resources and management needs and thus increases the difficulty in managing a vertically integrated beef production unit. Finally, the geographic concentration in cattle production is different from that in poultry and pigs. The significant land and forage base required for cattle production is what determines, according to Ward, that cattle stocker or growing operations are diverse and usually not concentrated in the same geographic regions as cow-calf production. Geographic dispersion combined with an added production stage, results in significant costs for moving animals from dispersed cow-calf operations to more concentrated stocker or growing areas and to still more concentrated cattle feeding areas.

3. Demand for livestock products

The magnitude and significance of the projected increases in demand for livestock products in developing countries over the next 15 years has prompted the introduction of the term 'Livestock Revolution' to describe this process [see Delgado et al. (1999)]. The implications, opportunities and challenges represented by the Livestock Revolution are considered by some to be just as great as those that accompanied the Green Revolution of the 1970s. There is, however, a major difference between the two. The Green Revolution occurred as a result of action taken by the international community to address the widespread threat of famine in developing countries, whereas the Livestock Revolution is driven by economic growth and increased demand for livestock products.

There is currently a marked disparity between consumption of livestock products in the developed and developing countries. People in developed countries typically consume up to four times as much meat and six times as much milk as those in developing countries This generally low level of consumption in developing countries signals the enormous potential for growth in production of foods of animal origin in these regions.

While according to Delgado et al. (1999) demand for meat in the developed countries is projected to grow only marginally over the next 15 years, demand in developing countries is expected to grow at 2.8% per year. This will increase the annual demand for meat in developing countries from 89 million tons in 1993 to 188 million tons by

[1] The poultry industry has two primary production stages, hatching and growing, apart from the processing and distribution stages, which are common to beef, pork, and poultry. The pork industry also has two primary production stages, farrowing and finishing.

Table 3
Regional projections of demand/consumption of livestock products (million metric tonnes)

Region	Sheep meat		Beef		Pork		Poultry		Milk	
	1993	2020	1993	2020	1993	2020	1993	2020	1993	2020
East Asia	2	2	3	7	31	65	7	18	9	19
South Asia	1	3	3	8	0	1	0	2	69	201
Southeast Asia	0	0	1	3	3	7	3	6	5	11
Latin America & Caribbean	0	1	10	18	3	6	7	14	46	77
Central Asia, West Asia and North Africa	2	4	2	5	0	0	3	6	23	51
Sub-Saharan Africa	1	2	2	6	1	2	1	2	14	31
Developing	6	12	22	47	39	81	21	49	168	391
Developed	4	4	32	36	38	41	26	34	245	263

Source: Delgado et al. (1999).

2020 – that means an additional 100 million metric tons of meat will be required every year to meet this demand. Two-thirds of the increased demand will be for pork and poultry meat but again there will be important regional differences, partly due to cultural factors (see Table 3). The relative importance of the different livestock species varies within regions: cattle are generally most important in Latin America and the Caribbean, small ruminants in Sub-Saharan Africa, small ruminants and buffalo in South Asia, pigs in East Asia and poultry in East Asia and Latin America.

Given the dramatic changes that demand of livestock products is experiencing in developing countries and the consequences that these changes are having in agricultural production patterns in the developing world, this section presents a brief characterization of global consumption patterns of livestock products and historical trends of changes in demand experienced by developing regions.

3.1. Consumption structure

Table 4 shows the level of consumption of livestock products in high income and developing regions for 1961–1980 and 1981–1999 expressed in kilograms of meat equivalent per capita. Consumption per capita in high income countries was 85 kilograms per person on average for the period 1961–1980, and increased to 93 kilograms during 1981–1999. Consumption in developing regions is well below this figure although there are significant variations between regions. Latin America is the region with the highest consumption among developing regions with almost 50 kilograms of meat equivalent per capita, less than high income countries but two to four times the levels in other developing regions. Africa shows also very low levels of consumption, around 15 kilograms in Sub-Saharan Africa and 20 kilograms in North Africa.

Table 4
Per capita consumption of livestock products (kilograms of meat equivalent)

Region	1961–1980	1981–1999
Australia–N. Zealand	164	134
EU	85	94
North America	109	111
Japan	22	45
High income	**85**	**93**
Central Africa	6	7
East Africa	21	19
South Africa	46	40
Rest of Southern Africa	11	10
Nigeria	7	9
Rest of West Africa	12	11
Sub-Saharan Africa	**16**	**15**
North Africa	**16**	**22**
Africa	**16**	**16**
China	6	19
East & Southeast Asia	8	12
India	8	11
Rest of South Asia	12	15
West Asia	30	31
Asia	**8**	**16**
Mexico	32	44
Central America & Caribbean	26	25
Andean	32	36
Brazil	35	54
Southern Cone	95	90
Latin America	**41**	**49**

Source: Authors based on FAO (2003).

Table 5 shows the distribution of consumption of livestock products between regions, compared with the distribution of population. High income countries with 16% of total population consume 43% of the total amount of livestock products produced worldwide. Latin America also has a larger share in total livestock production than in total population. Asia as a whole is still consuming less of the total livestock products than her share in total population. Within Asia, China and West Asia are at present consuming a share of global livestock similar to their share in total population. Other regions in Asia are consuming well below the consumption level that corresponds to their share in total population.

With respect to the share of different products in total consumption of livestock products, Table 6 shows that developing regions can be divided in two main areas: regions where ruminant products are the main consumed products and regions where consumers

Table 5
Population and demand for livestock products (average 1995–1999)

Region	Demand	%	Population 000'	Population %
Australia–N. Zealand	7276	1.6	22081	0.4
EU	88340	19.1	373746	7.2
North America	86592	18.7	301944	5.8
Japan	17202	3.7	126013	2.4
High income	**199411**	**43.1**	**823784**	**15.9**
Central Africa	1393	0.3	82319	1.6
East Africa	9067	2.0	192719	3.7
South Africa	3541	0.8	38723	0.7
Souther Africa	1436	0.3	63172	1.2
Nigeria	2284	0.5	103923	2.0
West Africa	2545	0.5	95830	1.8
Sub-Saharan Africa	**20266**	**4.4**	**576686**	**11.1**
North Africa	**8026**	**1.7**	**135472**	**2.6**
Africa	**28292**	**6.1**	**712158**	**13.7**
China	93847	20.3	1250903	24.1
East & Southeast Asia	16455	3.6	472851	9.1
India	31294	6.8	966027	18.6
South Asia	13982	3.0	307484	5.9
West Asia	12303	2.7	168705	3.3
Asia	**167880**	**36.2**	**3165970**	**61.0**
Mexico	11355	2.5	94268	1.8
Central America & Caribbean	3722	0.8	60550	1.2
Andean	10332	2.2	106910	2.1
Brazil	28408	6.1	163684	3.2
Southern Cone	13784	3.0	58647	1.1
Latin America	**67601**	**14.6**	**484059**	**9.3**
Developing	263773	56.9	4362187	84.1
Total	463184	100.0	5185971	100.0

Source: Authors based on FAO (2003).

have preference for non-ruminant products. Ruminant products (beef, milk and meat of shoats) constitute almost 90% of consumption of livestock products in South Asia, 80% in Africa, and 70% in Latin America. Among these regions, milk is the main livestock product consumed in India and in the rest of South Asia, North Africa and West Asia, while beef is the preferred product in Sub-Saharan Africa and Latin America. North Africa and West Asia have similar shares of beef and meat of sheep and goats as the preferred livestock products.

Chinese and Southeast Asian consumers prefer pork meat to other livestock products, representing respectively 53 and 35% of total consumption of livestock products. In

Table 6
Structure of consumption of livestock products by region (%)

Region	Bovine	Eggs	Milk	Pig meat	Poultry	Shoats	Ruminants	Non-ruminants
Central Africa	44	2	16	11	12	15	75	25
East Africa	40	2	35	4	4	15	90	10
South Africa	39	5	21	5	17	13	73	27
Southern Africa	44	6	19	8	12	11	74	26
Nigeria	34	11	15	5	8	27	76	24
West Africa	34	7	21	5	15	19	73	27
Sub-Saharan Africa	39	5	21	6	11	17	77	23
North Africa	25	6	31	0	16	22	78	22
Africa	32	6	26	3	13	19	77	23
China	12	14	3	53	12	6	22	78
East & Southeast Asia	35	6	8	35	12	4	47	53
India	21	3	66	2	2	6	92	8
South Asia	27	5	45	0	7	15	87	13
West Asia	17	6	37	0	16	24	78	22
Asia	22	7	32	18	10	11	65	35
Mexico	35	8	26	12	17	2	63	37
Central America & Caribbean	35	7	28	10	19	1	65	35
Andean	37	5	27	8	19	3	68	32
Brazil	51	3	21	7	16	1	73	27
Southern Cone	54	2	24	8	9	3	81	19
Latin America	42	5	25	9	16	2	70	30

Source: Authors based on FAO (2003).

other regions, poultry is preferred to pork representing 16% of total consumption of livestock products in Latin America (the share of pork is 9%); also 16% in North Africa and 11% in Sub-Saharan Africa (the share of pork is 6%).

3.2. Trends in consumption

Table 7 shows growth rates of demand for livestock products in high income and developing countries and the decomposition of total demand growth into its components: population and consumption per capita. Growth rates are estimated by regressing the logarithms of the FAO data series on a time trend and a constant term for two periods, 1961–1980 and 1981–2000:

$$\ln(y_i) = a^y + b_i^y T,$$

where y_i are quantities of commodity i, T is a time trend and b_i^y is commodity's i growth rates.

Annual growth rates in high income countries are low: 1.8% during 1961–1980 and decreasing to only 0.88% in the last 20 years. In contrast with the low growth rates in

Table 7
Demand growth rate and decomposition in population and per capita growth rate (%)

Region	Demand 1961–1980	Population 1961–1980	Cons. per capita 1961–1980	Demand 1981–1999	Population 1981–1999	Cons. per capita 1981–1999
Australia–N. Zealand	0.62	1.67	−1.04	1.06	1.28	−0.23
EU	1.81	0.58	1.23	0.19	0.28	−0.09
North America	1.43	1.09	0.34	1.29	0.98	0.31
Japan	8.59	1.09	7.49	2.96	0.42	2.54
High income	**1.80**	**0.85**	**0.95**	**0.88**	**0.58**	**0.30**
Central Africa	3.06	2.70	0.36	2.96	2.98	−0.02
East Africa	2.46	2.87	−0.40	2.11	2.77	−0.66
South Africa	2.02	2.31	−0.29	0.85	1.96	−1.12
Souther Africa	3.22	2.65	0.57	1.33	2.67	−1.34
Nigeria	6.55	2.82	3.72	1.56	2.71	−1.15
West Africa	3.82	3.51	0.31	2.03	2.87	−0.84
Sub-Saharan Africa	**2.92**	**2.86**	**0.06**	**1.73**	**2.74**	**−1.01**
North Africa	**3.79**	**2.50**	**1.30**	**3.81**	**2.30**	**1.51**
Africa	**3.09**	**2.78**	**0.31**	**2.25**	**2.65**	**−0.40**
China	9.00	2.13	6.87	8.53	1.26	7.27
East & Southeast Asia	3.64	2.43	1.22	4.26	1.88	2.38
India	2.29	2.24	0.05	4.15	1.97	2.18
South Asia	2.62	2.64	−0.02	4.62	2.45	2.17
West Asia	3.11	2.86	0.25	1.94	2.61	−0.67
Asia	**4.29**	**2.28**	**2.01**	**6.01**	**1.74**	**4.27**
Mexico	5.94	3.06	2.88	2.93	1.94	0.99
Central America & Caribbean	3.46	2.47	0.99	1.62	2.02	−0.40
Andean	3.63	2.85	0.78	2.60	2.15	0.45
Brazil	4.53	2.58	1.94	4.27	1.71	2.55
Southern Cone	1.76	1.64	0.12	1.44	1.52	−0.08
Latin America	**3.56**	**2.58**	**0.98**	**2.87**	**1.86**	**1.01**

Source: Authors based on FAO (2003).

high income countries, consumption of livestock products has been growing steadily in all developing regions: 4.29, 3.56 and 3.09% in Asia, Latin America and Africa, respectively.

Within developing regions, the spectacular growth in consumption that occurred in Asia is reflected in the figures for China where consumption of livestock products increased at rates above 8% in 1961–1999. Growth in Southeast Asia, India and rest of South Asia did not reach the levels of growth in China but consumption increased above 4% per year on average in the last 20 years. In Latin America, Brazil appears as the country with the most dynamic demand, with growth rates above 4% in the past 40 years. In contrast with the evolution of demand in Asia and Brazil, Sub-Saharan Africa shows a dismal growth in demand, especially during 1981–1999. An annual growth rate of 1.73% for this period is even below the growth rate shown by high income countries in 1961–1980, in a region where consumption of livestock products is almost 5 times smaller than in high income countries. Demand growth was higher during 1961–1980 in Africa and Latin America, while Southeast Asia, India and the rest of South Asia have seen demand growth accelerated in the last 20 years.

Table 7 also shows the decomposition of total growth in demand of livestock product into population growth and consumption per capita. There is a qualitative difference in demand growth between Sub-Saharan Africa and other regions. In the case of the latter, demand growth is explained exclusively by population growth. During 1981–1999, total demand in Sub-Saharan Africa increased at an annual rate of 1.73% while population increased at 2.74%. This implies that consumption per capita has been decreasing at an annual rate of 1.01% for the past 20 years. This reduction in consumption per capita occurred in all sub-regions within Sub-Saharan Africa. On the other hand, demand growth in Asia is mainly explained by growth in consumption per capita rather than population growth (4.27 and 1.74, respectively). Latin America and North Africa are in an intermediate position: while population growth makes the major contribution to demand growth, consumption per capita is still growing at rates above 1% per year on average.

This qualitative difference in demand growth has implications in terms of development of the livestock sector. Growth in consumption per capita should have a positive effect on development of the livestock sector as it is probably related to increased demand of higher value added products and better opportunities to develop the livestock food chain. In the case of Sub-Saharan Africa, rapid growth of an impoverished population is not even allowing the sustainability of previous consumption levels, which were very low even compared with other poor developing regions.

In order to better understand the main determinants of demand growth in developing countries, all countries in our sample of 79 countries are sorted according to their growth rate of demand of livestock products and clustered in 3 groups. Group 1 includes those countries with the highest demand growth and group 3 includes countries with the lowest demand growth. Countries within group 1 are mainly from Southeast and South Asia (37%), 25% are Latin American countries and 21% are countries from Sub-Saharan Africa. On the other hand, countries within group 3 are mainly from Sub-Saharan Africa (58%), 23% are countries from Latin America and only 8% are from East Asia (Mongolia and North Korea).

Table 8 shows average values of different variables for fast growing (group 1) and slow growing (group 3) livestock demand and an intermediate growth group (group 2). On average, demand in group 1 countries grew at an annual rate of 5.24% while demand growth rate in group 3 countries was only 0.86%. Growth in demand appears to be associated with growth in consumption per capita and less so to population growth given the small differences between population growth rates in the three groups. Also, growth in consumption per capita appears to be related to growth in income (GDP per capita) and less so to growth in urbanization.

Table 8
Demand, population, income, urbanization and composition of consumption for 3 groups of countries clustered according to demand growth

	Group 1, high demand growth	Group 2, intermediate demand growth	Group 3, low demand growth
Demand	5.24	2.93	0.86
Population	2.48	2.45	2.41
Cons. per capita	2.77	0.48	−1.55
GDP growth	4.52	3.45	1.46
GDP per capita	1433	1262	1291
GDP per capita growth	2.00	0.99	−1.01
% Urban	35.61	32.58	37.80
Urban growth	4.16	4.19	3.95
Share in consumption			
Bovine	31.43	36.53	33.39
Eggs	5.37	5.02	5.56
Milk	25.23	25.73	25.12
Pig meat	15.18	6.52	8.34
Poultry	14.45	13.59	11.95
Shoats	8.34	12.62	15.64
Average annual growth rates			
Bovine	7.40	2.45	1.46
Eggs	5.17	3.94	2.84
Milk	4.98	3.44	1.56
Pig meat	6.02	4.41	2.49
Poultry	8.18	6.78	4.79
Shoats	4.87	0.86	3.14

Source: Authors based on FAO (2003).

Consumption growth rates of different livestock products in different developing regions are presented in Table 9. In spite of the great variability between regions and products, demand for poultry meat in all regions increased faster than demand for other livestock products, followed by demand for pork. Table 10 distributes total change in consumption of different livestock products among regions. Most of the action in demand change of livestock products is in Asia while Africa contributes only marginally to these changes. Demand growth for pork in China and Southeast Asia accounts for 95% of total demand growth in developing countries. Growth in demand for beef and poultry is concentrated in two countries: China and Brazil. They account for 60 and 58% of total demand growth for beef and poultry, respectively. South Asia explains 57% of total demand growth for milk. The only livestock product where Africa has a significant contribution is meat of sheep and goats.

Table 9
Consumption growth rates of livestock products in different developing regions (%)

Region	Bovine	Eggs	Milk	Pig meat	Poultry	Shoats
Central Africa	5.65	2.66	4.51	4.99	5.79	3.70
East Africa	1.96	2.77	2.50	5.01	3.10	2.09
South Africa	0.00	4.09	0.14	2.24	4.82	0.97
Southern Africa	5.07	4.10	0.89	4.21	7.49	3.82
Nigeria	−0.73	5.17	6.54	8.68	1.45	5.69
West Africa	2.23	4.99	2.62	2.87	6.00	2.79
Sub-Saharan Africa	2.36	3.96	2.87	4.67	4.77	3.18
North Africa	2.86	5.72	2.96	9.04	7.09	3.64
Africa	2.61	4.84	2.91	6.85	5.93	3.41
China	16.57	11.53	6.98	6.83	11.70	9.61
East & Southeast Asia	5.53	3.62	2.77	14.23	7.41	2.68
India	2.63	5.95	4.79	4.15	8.96	2.26
South Asia	2.25	5.35	3.71	1.04	6.03	5.65
West Asia	7.92	5.93	2.01	−9.74	5.97	2.40
Asia	6.98	6.48	4.05	3.30	8.02	4.52
Mexico	4.85	5.23	1.25	−0.13	8.68	3.74
Central America & Caribbean	1.60	2.55	1.91	3.53	8.14	16.67
Andean	2.26	3.49	2.51	2.71	8.79	1.92
Brazil	4.17	3.45	3.59	3.32	7.83	4.90
Southern Cone	2.26	1.93	3.17	2.31	6.81	−0.71
Latin America	3.03	3.33	2.48	2.35	8.05	5.30
Average	3.60	3.94	3.27	4.28	6.53	2.83

Source: Authors based on FAO (2003).

3.3. Elasticities

In order to understand past and future trends of demand for livestock products in different regions we need to consider income and price elasticities. Information on income elasticities from Delgado et al. (1999) is presented in Figure 2. It is clear from the figure that income elasticities for livestock products are higher for low income countries, meaning that income growth in poor countries (or income growth of low income households within a country) would result in higher impacts on meat and milk consumption than income growth in rich countries (or in developing countries with low prices for livestock, e.g., Argentina). Elasticity values for different commodities show that preferences for additional milk and beef decreases marginally when moving from poorer to richer developing countries. Preferences for poultry are stable across wealth groups and preferences for pork and mutton rises with income. This could be related with quality changes in consumption as income increases [Delgado et al. (1999, pp. 11–12)].

Table 11, extracted from Coyle et al. (1998), shows income elasticities for livestock products and grains, for different regions and two years (1980 and 1995). Coyle et al. (1998) highlight two main conclusions from the figures in Table 11. First, expenditure

Table 10
Contribution of different regions to global demand growth of demand for livestock products (%)

Regions change	Bovine	Eggs	Milk	Pig meat	Poultry	Shoats	Livestock
Central Africa	0.92	0.04	0.16	0.12	0.29	1.24	0.38
East Africa	2.57	0.32	4.10	0.32	0.47	5.47	1.86
South Africa	−0.22	0.63	0.07	0.12	1.35	0.25	0.29
Souther Africa	0.34	0.17	−0.21	0.12	0.39	0.50	0.17
Nigeria	−0.40	0.75	0.31	0.11	0.17	4.40	0.39
West Africa	0.72	0.32	0.39	0.05	0.82	1.80	0.51
Sub-Saharan Africa	3.93	2.23	4.82	0.84	3.48	13.66	3.60
North Africa	3.95	1.34	3.55	0.01	3.55	7.22	2.64
Africa	7.88	3.57	8.37	0.85	7.03	20.89	6.24
China	36.46	73.12	7.60	87.98	43.41	52.52	51.27
East & Southeast Asia	4.27	5.95	2.71	7.40	9.89	2.39	5.76
India	8.15	4.45	40.32	0.90	1.93	6.70	10.91
South Asia	4.74	1.39	17.60	0.02	1.73	9.11	5.37
West Asia	2.91	3.04	2.27	0.00	3.60	6.51	2.31
Asia	56.54	87.94	70.50	96.29	60.55	77.23	75.62
Mexico	7.26	3.51	1.93	−0.57	6.58	1.17	3.14
Central America & Caribbean	0.13	0.48	0.58	0.45	1.93	0.08	0.61
Andean	3.19	1.31	4.27	0.58	5.38	0.10	2.61
Brazil	24.03	2.86	9.77	1.85	14.72	1.70	9.98
Southern Cone	0.98	0.33	4.57	0.56	3.80	−1.16	1.79
Latin America	35.58	8.49	21.13	2.85	32.41	1.88	18.14
Total	100.00	100.00	100.00	100.00	100.00	100.00	100.00

Source: Authors based on FAO (2003).

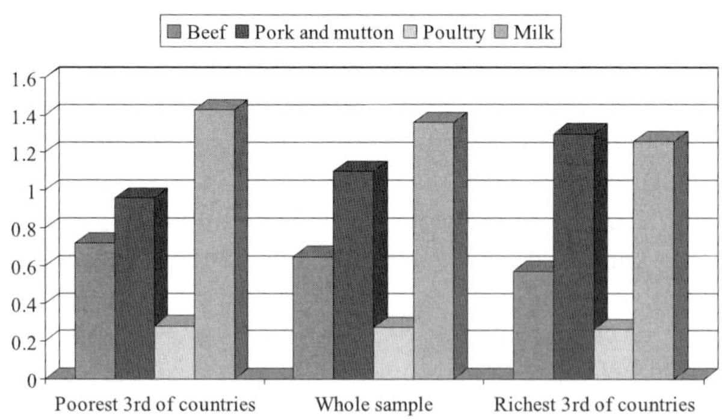

Figure 2. Demand elasticities for major food production of animal origin from a cross-country, systems estimation, 1970–1995, developing regions. *Source*: Delgado et al. (1999).

elasticities of demand for livestock products fall over the period for the lower income countries while in high income countries expenditure elasticity is marginally higher in

Table 11
Income elasticities for livestock and cereals[a]

	Livestock		Cereals	
	1980	1995	1980	1995
Australia	0.696	0.76	0.103	0.057
Japan	0.652	0.727	0.082	0.032
Asian NICs	0.685	0.663	0.439	0.084
ASEAN	0.769	0.677	0.597	0.307
China	1.079	0.959	0.938	0.757
Canada	0.716	0.776	0.048	0.025
United States	0.754	0.814	0.022	0.01
Mexico	0.662	0.68	0.184	0.142
MERCOSUR	0.633	0.651	0.19	0.122
Western Europe	0.694	0.738	0.098	0.065
Economies in transition	0.686	0.685	0.337	0.335
Middle East & North Africa	0.714	0.704	0.439	0.404
Rest of the world	0.982	0.936	0.812	0.739

Source: Coyle et al. (1998).
[a]Estimated using an AIDADS demand system.

1995 than in 1980. Second, given this pattern of change, the impact of global economic growth on the incremental demand for livestock products is ambiguous, and depends on the composition of economic growth worldwide. However, economic growth will have a greater relative stimulus on livestock products.

Cranfield et al. (1998) obtained income elasticities for food estimating an AIDADS demand system and projecting food consumption to 2020 for countries at different income levels. Table 12 summarizes their findings. Changes in budget shares evaluated at fixed prices show that grain budget shares decline in all countries with a smaller reduction projected in Ethiopia and a large reduction in Korea. However, when per capita expenditure increases to the projected 1995 and 2020 levels, livestock budget shares decline in all countries except Ethiopia. This means that Engel's law holds for all countries, however, at low per capita expenditure levels, the average budget share is increasing. Above a certain income level (income level of Kenya in 1985), the average livestock budget share would start to decrease with further increases in expenditure. In sum, information on income elasticities shows that low income countries have the highest grain budget shares. With income growth, there is an expansion of livestock budget shares which grows beyond the share of other food products while grain has the smallest budget share of the food goods [Cranfield et al. (1998)].

3.4. Transformation of the livestock sector and consequences for international trade

There are important linkages between livestock development in high income and developing countries as technological outputs resulting from research in developed countries

Table 12
Budget shares and shares' annualized growth rates for food, grain and livestock evaluated at each country's price levels

	Budget share			Annual growth rate	
	1985	1995	2020	1985–1995	1995–2020
Food					
Ethiopia	0.52	0.52	0.51	−0.06	−0.01
Pakistan	0.42	0.4	0.37	−0.61	−0.29
Senegal	0.41	0.4	0.37	−0.3	−0.26
Korea	0.31	0.2	0.09	−4.41	−2.9
France	0.16	0.14	0.09	−1.22	−1.62
United States	0.11	0.1	0.07	−1.09	−1.19
Grain					
Ethiopia	0.22	0.21	0.21	−0.52	−0.07
Pakistan	0.12	0.11	0.1	−0.89	−0.41
Senegal	0.13	0.12	0.11	−0.45	−0.4
Korea	0.08	0.04	0.02	−6.01	−3.38
France	0.03	0.03	0.02	−1.51	−1.86
United States	0.02	0.02	0.01	−1.3	−1.29
Livestock					
Ethiopia	0.12	0.13	0.13	0.62	0.07
Pakistan	0.14	0.14	0.13	−0.43	−0.19
Senegal	0.14	0.14	0.13	−0.21	−0.16
Korea	0.12	0.09	0.07	−2.72	−0.97
France	0.08	0.08	0.07	−0.45	−0.45
United States	0.07	0.07	0.07	−0.29	−0.24

Source: Cranfield et al. (1998).

flow to the other countries, and as inputs and output are traded. Growth in consumption in developing countries will certainly drive output in these countries, but it will also have strong effects on developed countries, as grains and meat and milk products are imported. When should developing countries import grain and produce livestock as opposed to importing livestock?

The growing importance of China as a market for livestock products motivated a series of studies discussing the future role of China in the international market for livestock products and its impact on production and trade from developed countries. Several of these studies [e.g., Hayes (1999); Hayes and Clemens (1997); Fang and Fabiosa (2002)] argues that Chinese agricultural resources favor labor intensive crops and that China is in "... the middle of a transition from Iowa-type corn prices to Japanese-type corn prices, a transition that will make the world's largest pork industry uncompetitive with imported products". When China becomes an importer of feed grains, Chinese grain prices will increase reflecting international grain transportation costs. As Chinese consumer's tastes are complement to US tastes in that Chinese consumer will pay a premium for cuts that US consumers dislike, it will be cheaper to transport boneless boxed pork than the feed grain equivalent [Hayes and Clemens (1997)].

From a different perspective, Tuan and Peng (2001) estimate domestic resource costs, net social profitability and effective rates of protection to demonstrate that China's production of hogs, beef cattle and poultry has been competitive in international markets. However, these authors also highlight the fact that hog production has received positive protection and that a policy change will reduce its competitiveness in international markets. Also, China's exports of livestock products will be negatively affected by sanitary standards.

In order to assess the likely consequences of future changes in livestock productivity on international trade in livestock products, Nin Pratt (2001) uses a general equilibrium model to make projections to the year 2010. Although an acceleration of supply is projected, this acceleration appears to be insufficient to satisfy the emerging demand for meats and developing countries are projected to increase their share of global meat imports by the end of the present decade. Findings with respect to China's future in livestock trade show that this country will increase net imports of non-ruminant products becoming a net importer of these products by 2010. These results, however, are sensitive to economic growth and productivity changes. If livestock productivity growth is at the high end of possible outcomes, and if there is a slow-down in the rest of the economy, China could still be a competitor in export markets by 2010. On the other hand, slower than expected diffusion and adoption of livestock technology coupled with a rapidly growing macro-economy could transform China in a major future market for meat exports.

Some insight of the role that different developing countries could play in international markets for livestock products can be gained by referring to Table 13, which shows the evolution of an index of revealed comparative advantage (RCA) for selected regions and for the period 1961–1998. It should be noted that the RCA estimates used here referred to historic data and that changes in population, per capita income, economic growth and changing demand, as discussed above in the case of China, can have a strong impact on the position of developing countries in international livestock markets. So one has to exercise caution when using these figures to predict direction of trade.

The original RCA index was developed by Balassa (1965) in order to analyze comparative advantage of countries for different commodities. Given the difficulties of directly measuring this comparative advantage, Balassa argued that advantages can be "revealed" through actual trade patterns given that trade reflects relative costs and differences in factors. The index used here is from Vollrath (1991):

$$\mathrm{RCA}_i = \frac{s_i^x - s_i^m}{S_i^w},$$

Table 13
Index of revealed comparative advantage for livestock products and for selected regions

Country	Beef		Pig & poultry		Milk	
	1965–1980	1981–1998	1965–1980	1981–1998	1965–1980	1981–1998
Tanzania	–	−1.40	–	−0.05	–	−0.90
Uganda	0.00	−0.46	−0.08	−0.95	−1.04	−1.67
Malawi	0.00	−0.07	−1.05	−0.94	−1.48	−1.40
Mozambique	0.22	−0.75	−0.47	−0.47	−2.03	−1.60
Zambia	−0.54	−0.03	−0.33	−0.14	−1.35	−0.44
Zimbabwe	10.94	4.21	0.30	0.19	−0.32	0.13
Rest of South African Customs Union	1.59	0.39	−0.11	−0.35	−0.24	−0.19
Other Southern Africa	−0.67	−2.55	−0.77	−2.32	−3.58	−4.10
Rest of Sub-Saharan Africa	−0.11	−0.64	−0.49	−0.78	−1.90	−2.25
Morocco	0.40	−0.36	−0.37	−0.07	−2.07	−1.44
Rest of North Africa	−0.84	−1.81	−0.88	−0.66	−2.18	−3.80
Colombia	1.61	0.00	−1.32	−0.60	−0.67	−0.33
Peru	−0.88	−1.38	−0.35	0.68	−2.49	−2.81
Venezuela	−0.35	−0.46	−0.68	−0.26	−1.78	−2.36
Rest of Andean Pact	0.06	0.10	−1.70	−0.35	−1.11	−0.75
Mexico	1.21	−1.08	−0.92	−0.94	−1.30	−1.40
Central America and the Caribbean	1.96	1.06	−1.24	−1.24	−1.71	−1.88
Argentina	16.06	8.28	3.37	1.91	0.53	0.91
Brazil	1.04	0.59	0.43	3.18	−0.38	−1.06
Chile	−1.44	−1.47	−0.71	0.46	−2.06	−0.55
Uruguay	39.39	33.00	2.98	2.67	1.02	5.43
Bangladesh	0.00	−0.28	0.44	−0.38	−3.78	−2.90
India	0.33	1.01	−0.22	−0.02	−1.57	−0.38
Sri Lanka	0.00	−0.14	−0.22	−0.21	−3.34	−3.27
Rest of South Asia	0.00	−0.29	−1.13	−0.63	−1.13	−0.62
China	0.27	−0.02	2.84	0.90	−0.03	−0.10
Indonesia	−0.03	−0.17	0.02	0.02	−1.03	−0.71
Malaysia	−0.34	−0.41	1.53	0.78	−1.68	−0.76
Philippines	−0.28	−0.44	−0.17	−0.03	−2.56	−2.26
Singapore	−0.24	−0.16	−0.58	−0.44	−0.38	−0.25
Thailand	−0.02	−0.03	0.18	1.93	−1.29	−0.73
Vietnam	0.81	0.03	−0.10	0.25	−5.50	−1.11
Turkey	0.95	0.51	−0.14	0.07	−0.17	−0.07
Rest of Middle East	−0.91	−1.68	−0.91	−1.10	−1.32	−1.72
France	−0.59	−0.65	−1.03	0.15	1.85	1.52
Germany	−0.35	−0.06	−1.48	−1.35	0.10	0.25
United Kingdom	−1.48	−0.12	−2.38	−1.32	−1.68	−0.35
Canada	−0.22	−0.06	0.42	0.64	0.13	0.11
United States	−1.10	0.46	1.30	0.77	−0.07	−0.02
Australia	12.88	12.53	1.99	0.77	2.46	2.65
New Zealand	42.95	36.40	5.01	2.34	24.44	25.87
Japan	−1.04	−2.12	−1.21	−2.56	−0.29	−0.42
Korea, Rep.	−0.10	−0.87	−0.70	−0.17	−0.08	−0.12
Taiwan	−0.15	−0.62	0.32	1.54	−0.79	−0.60

Source: Authors based on Dimaranan and McDougall (2002).

where s_i^x and s_i^m are respectively the share of commodity i's exports and imports in a country's total exports and imports, while S_i^w is the share of commodity i's trade in world's total trade. A positive value of the index results when the export share of commodity i is greater than the import share of the same commodity for that country. In that case, a comparative advantage for commodity i is "revealed" for that country.

Looking at Table 13 and focusing first in East and Southeast Asia, the region where demand for livestock products is expanding faster, we observe that the region has a comparative disadvantage in milk production. Countries with the bigger disadvantages are Philippines, Malaysia, Vietnam and Indonesia. In general, most countries (except China) have improved their RCA indices in the past years. The region also shows disadvantages for livestock production although on average these are smaller than those in dairy production. On the other hand, most countries have advantages for pig and poultry production, which had been reduced in the past years, except for Thailand a competitive poultry exporter. Similar patterns are observed in the cases of Japan, Taiwan and South Korea, countries with higher income, which followed a similar path in development of the livestock sector. These figures suggest that the advantages for these regions to compete in ruminant's production are low given land constraints and competition for scarce labor and capital, combined with environmental constraints.

Latin America shows diversity in terms of RCA results. Argentina and Uruguay are among the regions with larger comparative advantages for livestock production in the world. Brazil also shows RCA in beef and non-ruminants (poultry) and Central America has advantages in beef production. All countries except Argentina and Uruguay have comparative disadvantages in milk production.

Disaggregated data for all regions in Africa are not available but in general, Sub-Saharan Africa shows significant disadvantages in dairy production, moderate disadvantages in poultry and pig production and some advantages for beef and small ruminants meat production, although this advantage has been eroding in the past years.

4. Livestock production

In this section we describe the livestock production structure in developing countries, recent trends in production patterns, and main characteristics of livestock systems.

4.1. Production structure

Table 14 compares livestock production in different regions and for two periods (1961–1980 and 1981–1999). For the group of 80 countries considered here, the average value of livestock production (meat, milk and eggs) during 1961–1980 ascended to a total of \$225 billion of 1980. Of this, only 36% was produced by developing countries where almost 80% of total population was located. This distribution of output and population results in per capita production in developing countries of only 12 kilograms compared

Table 14
Livestock production (million $ 1980)

Region	1961–1980					1981–2000				
	Output	%	Population	%	Production per capita	Output	%	Population	%	Production per capita
Australia–N. Zealand	13002	5.8	15512	0.5	327	17341	4.5	20235	0.4	335
EU	66882	29.8	340298	10.5	77	88603	23.1	365565	7.9	95
North America	58819	26.2	232323	7.2	99	81690	21.3	282133	6.1	113
Japan	5533	2.5	75493	2.3	29	9361	2.4	95249	2.1	38
High income	144237	64.2	663626	20.6	85	196995	51.3	763182	16.5	101
Central Africa	738	0.3	38962	1.2	7	1189	0.3	68111	1.5	7
East Africa	4177	1.9	92609	2.9	18	6725	1.8	161053	3.5	16
South Africa	2564	1.1	22523	0.7	44	3256	0.8	34064	0.7	37
Southern Africa	1038	0.5	31268	1.0	13	1359	0.4	53007	1.1	10
Nigeria	787	0.4	50910	1.6	6	1692	0.4	87457	1.9	8
West Africa	1257	0.6	46920	1.5	10	1893	0.5	80149	1.7	9
Sub-Saharan Africa	10561	4.7	283192	8.8	15	16114	4.2	483841	10.5	13
North Africa	2436	1.1	72876	2.3	13	5232	1.4	117430	2.5	17
Africa	12997	5.8	356069	11.0	14	21347	5.6	601271	13.0	14
China	14112	6.3	841592	26.1	7	61789	16.1	1152985	24.9	21
East & Southeast Asia	5142	2.3	277326	8.6	7	13194	3.4	419074	9.1	12
India	10487	4.7	564792	17.5	7	22933	6.0	850783	18.4	11
South Asia	4125	1.8	159911	5.0	10	8752	2.3	264080	5.7	13
West Asia	4985	2.2	83257	2.6	23	8745	2.3	144518	3.1	24
Asia	38850	17.3	1926878	59.7	8	115413	30.1	2831440	61.2	16
Mexico	4250	1.9	51991	1.6	32	8653	2.3	83281	1.8	41
Central America & Caribbean	2082	0.9	34814	1.1	23	3039	0.8	52930	1.1	22
Andean	4178	1.9	57947	1.8	28	7712	2.0	93101	2.0	32
Brazil	7279	3.2	97603	3.0	29	17616	4.6	147230	3.2	47
Southern Cone	10690	4.8	39069	1.2	107	12942	3.4	53011	1.1	95
Latin America	28478	12.7	281423	8.7	40	49961	13.0	429553	9.3	45
Developing	80325	35.8	2564370	79.4	12	186721	48.7	3862264	83.5	19
Total	224561	100.0	3227995	100.0	27	383716	100.0	4625446	100.0	32

Source: Authors based on FAO (2003).

with 85 kilograms in high-income countries. During this period, 50% of output in developing countries was produced in Asia (30% in China and India alone), 35 in Latin America (23% in Brazil and the Southern Cone) and 15% in Africa.

Twenty years later, total livestock production increased by 72% and this growth is mainly explained by growth in developing countries, which increased to almost 50% of global livestock production. This changes in livestock production patterns are explained by a phenomenal growth of livestock production in Asia, which produces 30% of global livestock output. The major transformation occurred in China where production per capita increased from 7 kilograms of meat in 1961–1980 to 21 in 1981–1999. Similarly, East and Southeast Asia and India and rest of South Asia also increased per capita production of livestock, more than doubling total output in absolute values. Growth in Latin America was 75% during this 20-year period, slightly increasing its share in global livestock production. This is explained by growth in Brazil, Mexico and the Andean countries. Livestock production in Africa increased 64% losing share in global production. Within Africa, production in North Africa showed high dynamism compared with Sub-Saharan Africa. While the latter only increased production by 52%, North Africa more than doubled production levels. Sub-Saharan Africa was the only region reducing total production per capita producing at present at similar levels than Asian countries, which were far behind Africa's production during 1961–1980.

4.2. *Output growth*

Output and input growth rates are estimated by regressing the logarithms of the production data series from FAOSTAT (2003) on a time trend and a constant term following the same procedure used for the estimation of growth rates in demand. Outputs included in the analysis are: beef, sheep and goat meat, cow and buffalo milk, poultry meat, eggs and pig meat.

Figure 3 and Table 15 show average growth rates of livestock production for different regions and for two periods: 1961–1980 and 1981–2000. Asian countries in East, Southeast and South Asia show the fastest growth in livestock production for the period. On the other extreme, the growth performance of the livestock sector in Sub-Saharan Africa in the last four decades has been poor. Output grew at an average annual rate of 2%, with East and Central Africa showing above the average growth and West and Southern Africa showing output growth rates below 2%. In contrast with Sub-Saharan Africa, the livestock sector in North Africa has showed a dynamic behavior, growing above 4% per year on average in the past 20 years. Tunisia, Egypt and Libya show the higher growth rates in the region.

Growth rates of livestock production in Latin America are between those in Asia and Sub-Saharan Africa: 2.85% during 1961–2000. Brazil has been the most important source of growth in the region increasing production at 4.4% on average for 40 years.

The output growth rate in Sub-Saharan Africa (SSA) was below the level of the population growth rate for the period resulting in a reduction of output per capita between 1961 and 2000 (Figure 4). Output per capita in Sub-Saharan Africa in the early 1960s

Table 15
Livestock output growth rate by country 1961–2000

	1961–2000	1961–1980	1981–2000
Central Africa			
Cameroon	3.71	3.88	3.51
Central African Republic	5.54	4.79	4.73
Chad	0.82	0.25	2.56
Congo, Dem.	1.19	0.41	1.53
Congo, Rep.	2.75	2.75	2.32
Gabon	1.41	1.41	1.81
Rwanda	3.06	4.87	0.86
East Africa			
Ethiopia	0.91	0.32	1.40
Kenya	3.27	2.97	2.92
Madagascar	1.39	1.36	1.35
Somalia	1.59	2.58	0.26
Sudan	3.31	3.40	3.24
Tanzania	2.82	2.48	2.85
Uganda	2.46	3.14	2.82
North Africa			
Algeria	4.95	5.12	3.67
Egypt	3.35	2.70	4.42
Libya	5.41	7.24	4.28
Morocco	3.21	2.14	3.60
Tunisia	3.92	3.47	4.49
Southern Africa			
Angola	2.18	2.67	2.25
Botswana	1.04	1.55	0.17
Malawi	3.54	5.24	1.48
Mozambique	1.54	2.52	0.77
Namibia	0.26	1.76	1.40
South Africa	1.18	1.57	0.59
Zambia	2.52	4.06	2.04
Zimbabwe	1.47	3.95	0.60
West Africa			
Benin	3.62	4.25	2.29
Burkina Faso	2.71	0.12	4.54
Côte d'Ivoire	3.81	4.37	2.87
Gambia	1.14	2.75	0.78
Ghana	2.78	3.89	1.26
Guinea	1.11	1.51	2.55
Liberia	2.71	3.78	0.96

(*continued on next page*)

Table 15
(*continued*)

	1961–2000	1961–1980	1981–2000
Mali	1.92	1.95	0.66
Mauritania	0.86	0.35	0.95
Niger	0.86	0.94	1.60
Nigeria	3.81	4.89	1.51
Senegal	3.29	2.46	4.50
Sierra Leone	2.28	2.64	1.77
Andean countries			
Bolivia	3.71	4.71	3.00
Colombia	2.86	2.81	3.23
Ecuador	3.60	3.80	4.38
Peru	2.76	2.79	3.20
Venezuela	3.63	5.61	1.28
Central America & the Caribbean			
Cuba	−0.01	2.25	−3.49
Dominican Republic	3.97	4.44	3.96
Haiti	1.52	2.75	1.16
Costa Rica	3.96	5.74	3.16
El Salvador	2.33	3.65	2.54
Guatemala	2.56	3.17	3.00
Honduras	3.26	3.73	4.21
Mexico	3.66	4.75	2.34
Nicaragua	0.81	5.39	1.69
Panama	3.37	3.98	2.46
South America			
Argentina	0.72	1.24	0.89
Brazil	4.40	4.28	4.51
Chile	2.77	1.93	4.00
Paraguay	2.77	1.25	4.39
Uruguay	0.97	−0.01	1.49
East & Southeast Asia			
Cambodia	3.58	−1.43	7.02
China	7.11	5.82	8.48
Indonesia	4.71	2.90	4.39
Korea dem.	7.63	8.80	6.20
Laos	3.89	0.63	5.59
Malaysia	6.96	6.32	7.57
Mongolia	0.80	1.58	0.30
Philippines	3.84	4.02	4.83
Thailand	3.75	3.66	4.09
Vietnam	4.22	0.80	5.66

(*continued on next page*)

Table 15
(*continued*)

	1961–2000	1961–1980	1981–2000
South Asia			
Bangladesh	2.02	2.08	3.80
India	3.70	2.56	4.01
Nepal	2.56	2.58	2.32
Pakistan	4.05	2.70	5.36
Sri Lanka	1.66	1.79	2.18
West Asia			
Iran	3.66	3.15	4.29
Iraq	0.34	1.66	−3.18
Jordan	5.58	3.44	6.58
Syria	4.22	3.71	2.44
Turkey	1.99	2.09	1.18

Source: Authors based on FAO (2003).

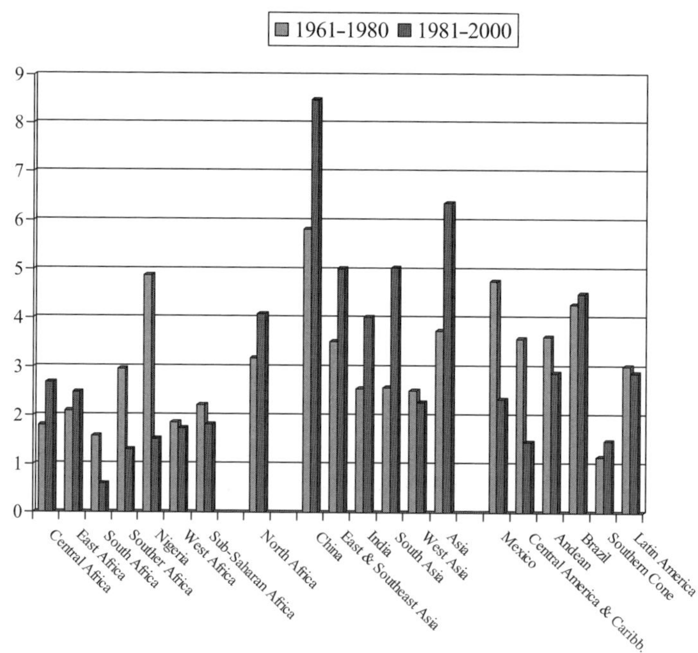

Figure 3. Average growth rates of livestock production for different regions and for two periods: 1961–1980 and 1981–2000. *Source*: Authors based on FAO (2003).

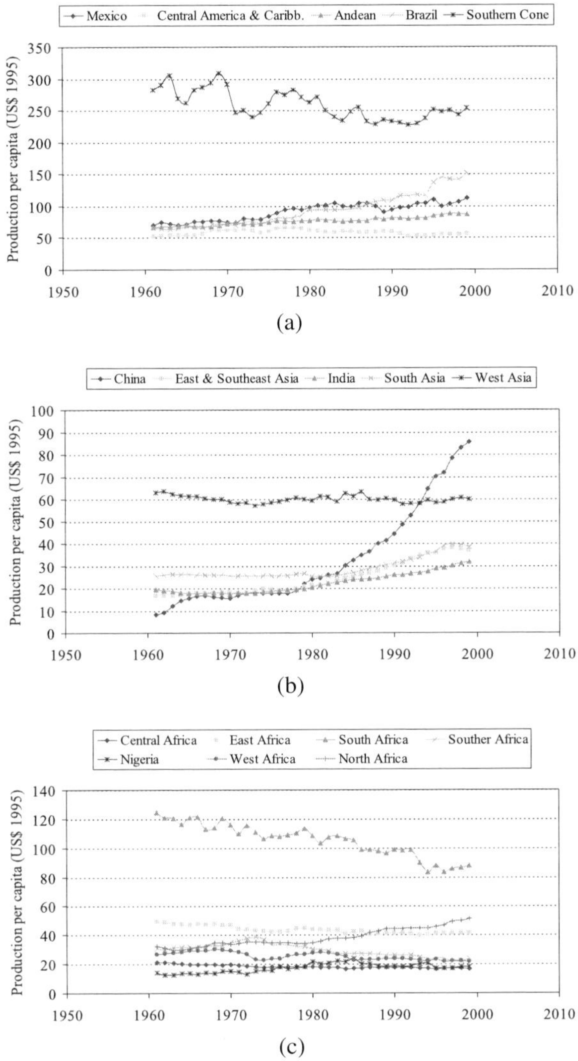

Figure 4. Livestock output per capita in different regions: (a) Latin America; (b) Asia; (c) Africa. *Source*: Authors based on FAO (2003).

was close to that in North Africa and it was twice that in Southeast Asia, South Asia, and China. By the late 1990s China was producing between two and three times more output per capita than Sub-Saharan Africa (SSA), and the trend shows that SSA will become the region with the lowest livestock output per capita in the coming years.

In Latin America, the Southern Cone experienced a reduction in output per capita from a high level of $300 in the 1960s to around $250 during the late 1990s. Brazil appears to be catching-up with the Southern Cone in terms of production per capita although it is still $100 below Southern Cone's level in 1999.

4.3. Output growth in different sectors

Table 16 shows the contribution of the different livestock sub-sectors to growth of livestock production. Coinciding with the shares of different sectors in production, in one extreme, Africa's growth have been driven by growth in ruminant production, which explains more than 50% of growth in all regions between 1961 and 1980. The slowdown of beef production in the last 20 years reduced its contribution to total livestock growth in Africa, while increasing the contribution of milk. The contribution of poultry and pig meat remained almost unchanged.

On the other extreme, East and Southeast Asian countries growth is explained by growth in pig and poultry production, with increasing contribution of pigs in Southeast Asia and of poultry and beef in China in the past two decades. In India and South Asia, growth in livestock production is based on the expansion of milk production In Latin America, the importance of the contribution of the different sectors to growth varies by region and period. Central America and the Andean countries based their growth on beef and milk production during 1961–1980 with an increasing share of poultry contribution in the last two decades. In the Southern Cone milk and poultry contributed to 75% of livestock growth in the past 20 years. On the other hand, mainly beef followed by poultry and milk explains Brazil's growth in livestock production.

5. Partial factor productivity and resource use

This section assesses the role of livestock in the process of agricultural growth under the perspective of induced technical change. We argue that this role depends on the initial resource endowments, which determine the initial share of livestock in agricultural production. According to the induced innovation perspective, we would expect that countries with low land/labor ratio would have a lower initial stock of livestock, lower animal stock/labor ratios, lower labor productivity and higher land productivity than countries with high land/labor ratios (except in highly forested zones where diseases prevent livestock rearing). Expansion of livestock production in countries with low land/labor ratios would increase output per hectare of land in agriculture by technologies that increase the number of animals per hectare and/or increase output per head of animal stock through a more intensive use of labor and feed, increasing land productivity. In the case of countries with high land/labor ratios, expansion of the livestock sector would occur through technical change allowing further substitution of land and capital for labor, increasing labor productivity. In some cases the new technology (embodied in improved pastures, new equipment, or new production practices) may not always be a substitute per se for labor as explained by Ruttan (1997). Rather, it is a catalyst that facilitates the substitution of relatively abundant factors.

Table 16
Contribution of different regions to total growth in livestock output (%)

	1961–1980						1981–2000					
	Beef	Shoat meat	Pig meat	Poultry meat	Milk	Total	Beef	Shoat meat	Pig meat	Poultry meat	Milk	Total
Central Africa	54.01	21.16	9.61	4.45	10.77	100	53.95	20.74	9.34	5.39	10.57	100
East Africa	50.06	22.94	1.02	3.57	22.41	100	44.88	20.50	2.09	3.87	28.67	100
South Africa	45.99	16.91	4.18	4.78	28.15	100	47.07	13.44	5.00	12.06	22.43	100
Souther Africa	59.82	9.55	5.60	4.14	20.90	100	58.69	9.09	6.81	6.97	18.45	100
Nigeria	66.77	11.93	5.24	8.18	7.88	100	47.39	29.89	5.95	10.57	6.20	100
West Africa	46.22	28.59	6.88	5.71	12.60	100	41.10	28.07	8.87	11.01	10.95	100
Sub-Saharan Africa	50.99	19.90	3.65	4.55	20.91	100	47.03	19.85	4.71	7.37	21.04	100
North Africa	31.32	27.56	0.19	8.59	32.34	100	26.67	26.37	0.12	17.81	29.04	100
Africa	47.61	21.21	3.06	5.24	22.88	100	42.60	21.24	3.70	9.69	22.77	100
China	3.64	5.51	76.01	10.00	4.84	100	7.69	6.49	71.99	10.21	3.63	100
East & Southeast Asia	30.74	10.32	40.17	16.07	2.70	100	21.94	7.70	41.81	25.63	2.92	100
India	17.50	11.84	2.75	0.91	67.00	100	15.05	8.43	2.84	1.62	72.06	100
South Asia	22.97	15.89	0.19	2.18	58.78	100	16.38	20.31	0.20	4.09	59.02	100
West Asia	14.02	34.82	0.05	5.66	45.46	100	17.99	28.02	0.01	13.55	40.43	100
Asia	14.40	13.40	31.55	6.40	34.25	100	12.66	10.32	40.81	9.72	26.50	100
Mexico	37.00	2.99	22.38	6.66	30.98	100	39.66	2.31	18.47	12.87	26.69	100
Central America & Caribbean	57.64	1.37	6.44	5.70	28.86	100	46.59	1.11	6.44	16.19	29.67	100
Andean	51.03	5.24	7.87	6.35	29.50	100	44.83	2.75	8.10	15.37	28.95	100
Brazil	56.00	1.81	11.63	5.13	25.43	100	55.65	1.31	7.90	14.05	21.09	100
Southern Cone	70.66	6.19	4.04	2.33	16.78	100	65.58	3.42	4.58	6.07	20.34	100
Latin America	58.44	4.09	9.25	4.51	23.71	100	53.86	2.22	8.57	12.11	23.24	100
Developing	37.14	11.01	18.21	5.46	28.19	100	29.45	9.04	26.17	10.46	24.88	100

Source: Authors based on FAO (2003).

In what follows, we present different livestock PFP measures and input relationships for aggregated regions and PFP growth and trends in input use. Information for individual countries is also provided.

5.1. *PFP and input use in different developing regions and countries*

Table 17 presents the labor/land ratio associated to different PFP measures and input ratios. The relative intensity in the use of inputs explains the differences in PFP between regions. Latin America produces livestock using land intensively and saving labor and feed compared to Asia. The intensity in the use of labor in Asia can be verified by looking at the number of animals per worker, less than 2.2 in all cases compared to 17 on average for Latin America (and as high as 44 in the Southern Cone). Livestock production in West Asia and North Africa is an intermediate case between the farthest cases of Latin America and East, Southeast and South Asia.

Livestock production in Sub-Saharan Africa shows differences between regions. Southern Africa's labor/land ratio is similar to that in South America, while other regions show higher values than North Africa and West Asia. Output per hectare is lower in Southern Africa than in other regions as expected, but it is much lower than output per hectare in South Africa, a country with labor/land ratio similar to that of Southern Africa, reflecting in part the low quality of land and natural resources in this region.

With respect to other regions in Sub-Saharan Africa, PFP values do not correspond to the actual figures of the labor/land ratio. Comparing Sub-Saharan Africa to Central America, a region with similar labor/land ratio, we verify that output per hectare in Central America almost doubles output per hectare in Nigeria and is several times larger than land productivity in West, Central and East Africa. Similar differences can be found in output per worker. Differences in output per head of animal stock are smaller, although Sub-Saharan Africa still shows lower values than other regions. The use of inputs reveals that Sub-Saharan Africa's livestock production is based on the intensive use of labor and land. The figures of heads of animal stock per worker are closer to those in Asia, while the number of animals per hectare is almost half that in Latin America. South Africa differs clearly from other regions in Sub-Saharan Africa, showing patterns of input use and PFP similar to those in South America. Nigeria appears to perform better than countries in West, East and Central Africa presenting higher values of land and labor productivity.

In order to better understand the differences in PFP and input use between regions, Table 18 presents growth rates of PFP measures and input relationships. The trend in Asia is to substitute labor for land, which results in 80% growth rate of output per hectare of land compared to 55% growth of output per workers between 1961–1980 and 1981–2000. These changes were achieved differently in different regions. Southeast Asia significantly increased the number of animals and the number of workers per hectare. China and India increased labor per head of animal stock, resulting in higher growth of output per stock than in Southeast Asia.

Table 17
Labor/land ratio, partial factor productivity measures and input relationships for different regions (average 1981–2000)

Country	Labor/land	PFP feed	PFP land	PFP labor	PFP stock	Feed/labor	Feed/land	Feed/stock	Stock/land	Stock/labor
Central Africa	0.24	2.00	75	319	122	157	36	60	0.60	2.65
East Africa	0.22	2.28	67	308	111	138	29	49	0.59	2.83
South Africa	0.02	0.58	77	3933	184	6497	129	310	0.42	21.02
Souther Africa	0.09	0.96	24	273	124	281	23	126	0.18	2.23
Nigeria	0.24	0.62	184	778	138	1267	303	220	1.26	5.27
West Africa	0.18	1.40	70	385	118	266	48	83	0.56	3.20
Sub-Saharan Africa	0.16	1.07	67	414	126	373	59	115	0.51	3.24
North Africa	0.17	0.39	183	1089	440	2590	431	1087	0.39	2.37
West Asia	0.17	0.40	289	1653	313	4082	696	778	0.89	5.25
China	0.95	0.46	415	437	293	976	920	660	1.40	1.49
East & Southeast Asia	1.55	0.43	913	585	282	1296	1976	639	3.06	2.02
India	1.35	1.04	611	448	203	420	559	192	2.90	2.20
South Asia	1.66	1.32	800	479	288	349	566	211	2.69	1.67
Mexico	0.08	0.48	191	2279	262	4540	380	533	0.72	8.56
Central America & Caribbean	0.25	0.57	442	1749	173	3116	787	304	2.61	10.32
Andean	0.07	0.89	140	2013	153	2173	149	168	0.88	12.90
Brazil	0.07	0.55	233	3565	130	6090	418	240	1.74	25.13
Southern Cone	0.01	1.12	130	8952	195	7606	109	170	0.64	44.64
Latin America	0.06	0.66	187	3029	165	4366	271	248	1.09	17.52

Notes: Feed is measured in kilograms of maize equivalents; labor represents total economically active population in agriculture; stock is the number of animals measured as cow equivalents; PFP is US$ 1980.
Source: Authors based on FAO (2003).

Table 18
Total growth in PFP and input relationships between 1981 and 2000 (in percentage)

Country	PFP land	PFP labor	PFP stock	Stock/ha	Stock/labor	Labor/ha
Central Africa	46	3	1	44	5	42
East Africa	43	−8	14	22	−19	57
South Africa	33	68	20	13	45	−22
Souther Africa	24	−16	4	17	−19	49
Nigeria	85	83	2	68	64	2
West Africa	55	7	16	28	−11	45
Sub-Saharan Africa	49	5	11	29	−7	41
North Africa	77	63	46	16	7	8
Africa	54	12	18	25	−7	37
China	86	78	128	−20	−23	4
East & Southeast Asia	80	42	25	39	11	27
India	80	36	74	−1	−24	31
South Asia	74	36	57	8	−13	27
West Asia	89	65	40	35	21	14
Asia	80	55	85	−5	−17	16
Mexico	65	42	43	16	−3	18
Central America & Caribbean	9	14	30	−15	−10	−5
Andean	55	34	30	16	2	15
Brazil	59	117	35	16	50	−25
Southern Cone	44	35	11	26	19	6
Latin America	53	55	22	23	23	−1

Source: Authors based on FAO (2003).

In Latin America, productivity growth per worker and per hectare was similar, growing more than twice as fast than output per head of animal stock. Within the region Brazil shows a clear contrast with Mexico and the Andean countries. While the latter increased land productivity faster than labor productivity, Brazil appears as the region with higher growth in labor productivity. The Southern Cone shows low growth in all PFP measures.

PFP growth in Sub-Saharan Africa shows the lower values for all regions. Growth in output per worker and per head of animal stock remained almost unchanged (5 and 11% growth in 20 years) while output per hectare increased by 49%, and this is explained mainly by growth in the number of workers per hectare. This pattern of growth can be verified in all major regions in Sub-Saharan Africa, with the exceptions of the two countries shown separately: South Africa and Nigeria. South Africa's growth patterns resemble those of Brazil or the Southern Cone, substituting labor for land and capital. Nigeria, with almost no change in the number of workers per hectare of land during the period, increased land and labor productivity above 80% in 20 years by increasing the number of animals per hectare and per worker.

Figure 5 compares the share of different livestock species and products between groups of countries with different labor/land ratios. Countries with high labor/land

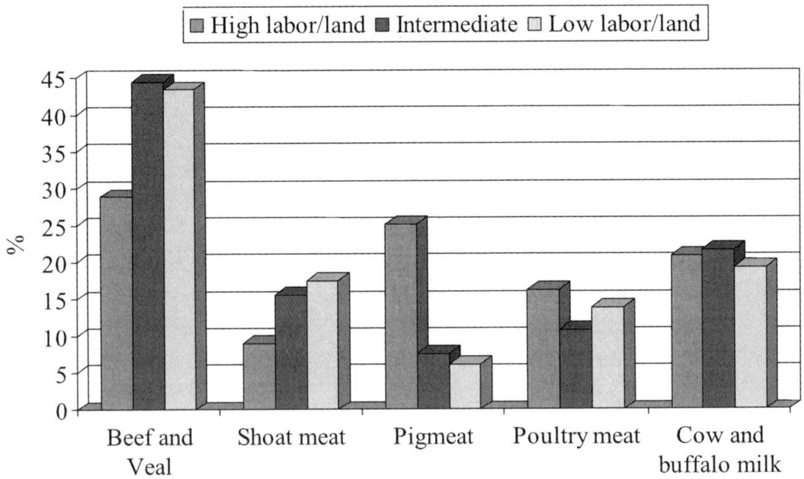

Figure 5. Share on total livestock production of different livestock species and labor land ratios. *Source*: Authors based on FAO (2003).

ratios produce less beef and shoat meat and more pig meat than countries with low labor/land ratios. There are no differences between groups in the share of milk production and the intermediate group produces less poultry than the high and low groups, which shows that poultry production cannot be associated to differences in resource endowments.

5.2. *PFP of livestock sub-sectors*

In general, all regions experienced high growth in poultry meat and egg production (Table 19). Almost half of growth in the poultry meat sector in Latin America, East and South Asia is explained by growth in yields. Sub-Saharan Africa also experienced high growth in poultry meat production but most of this growth is explained by growth in South Africa during the 1960s and 1970s, which occurred also with significant contributions of yield growth. The data also show that accelerated growth in poultry production occurred in Latin America, South Africa and Southeast Asia in the first 2 decades of the period considered here, while growth in East and South Asia followed in the last 20 years. During the period of high growth and development of the poultry industry, total output is boosted by increases in stocks and substantial yield growth. After this period, output continued to grow at relatively high rates showing increases in animal stock but a slowdown in yield growth.

The beef and milk sectors show a very different pattern. Output growth occurs at a slower pace compared with growth in poultry production, and Asia appears as the most dynamic region. High growth of beef and milk production in East Asia is explained by China's ruminant sector, a relatively small sector which is rapidly expanding with

Table 19
Average annual growth rate of PFP and stock of different species, 1961–1980 and 1981–2000 (%)

	Beef		Shoats		Milk		Pigs		Chicken		Eggs	
	Yield	Stock	Yield	Stock	Yield	Stock	Yield	Stock	Yield	Stock	Yield	Stock
1961–1980												
Central Africa	0.13	2.12	0.48	2.13	0.91	2.53	−1.43	5.37	0.28	5.09	0.02	4.33
East Africa	0.33	1.69	0.04	1.59	0.44	2.16	−1.05	5.15	0.80	2.77	0.49	2.57
North Africa	1.93	0.47	−0.20	1.65	0.88	3.08	3.24	−3.69	1.04	3.95	0.57	4.04
South Africa	0.83	1.60	1.76	−0.19	−0.68	1.13	−0.25	2.91	5.29	4.51	0.27	5.30
West Africa	1.58	1.81	0.44	4.07	−0.20	2.35	−0.33	3.54	2.28	3.94	1.07	3.90
Sub-Saharan Africa	0.76	1.69	0.39	1.74	−0.13	2.15	−0.97	4.11	2.29	3.76	0.85	3.70
Andean	1.16	2.51	0.34	−0.04	0.90	1.83	0.32	2.64	3.65	5.35	1.36	4.83
Central America	1.36	2.37	2.26	0.36	0.89	4.31	1.03	4.53	1.69	5.36	4.16	3.00
Caribbean	−1.76	2.87	−0.11	0.76	3.71	1.45	2.08	0.36	0.99	6.25	1.34	5.02
Southern Cone	−0.40	2.75	−0.75	−0.87	−0.57	3.62	1.67	1.35	6.16	6.03	0.47	4.26
South America	−0.16	2.70	−0.53	−0.66	−0.27	3.24	1.39	1.61	5.40	5.86	0.72	4.36
Central America & Caribbean	0.72	2.42	1.88	0.41	1.35	3.84	1.67	3.64	1.57	5.53	3.68	3.34
East Asia	5.00	0.35	4.19	2.34	2.02	1.29	3.58	7.28	1.81	2.83	1.70	1.91
Middle East	3.43	−0.05	0.67	0.95	0.33	2.46	−4.05	−4.26	2.79	6.62	0.95	6.23
South Asia	2.04	0.21	0.31	2.00	0.79	1.26	−0.13	2.97	0.57	3.65	2.26	4.56
Southeast Asia	3.06	0.20	1.81	1.07	1.16	2.87	1.42	1.50	3.69	2.97	1.40	5.14

(continued on next page)

Table 19
(continued)

	Beef		Shoats		Milk		Pigs		Chicken		Eggs	
	Yield	Stock	Yield	Stock	Yield	Stock	Yield	Stock	Yield	Stock	Yield	Stock
1981–2000												
Central Africa	1.66	1.68	0.87	2.30	0.22	1.47	0.14	2.01	−0.58	4.10	−0.04	1.79
East Africa	−0.01	1.52	0.11	1.84	0.51	2.73	2.62	1.96	0.54	1.83	−0.07	2.04
North Africa	3.81	−0.13	3.19	1.14	2.40	1.64	−0.56	1.48	0.97	6.19	0.35	4.70
South Africa	0.40	0.04	−0.06	0.17	−0.29	0.32	−0.08	1.82	0.50	2.85	0.78	2.66
West Africa	−0.98	1.31	0.61	2.62	0.24	1.46	−0.72	5.22	0.54	2.81	−0.05	3.96
Sub-Saharan Africa	0.18	1.19	0.85	1.73	0.36	2.11	−0.26	3.34	0.95	3.48	0.37	3.34
Andean	0.45	0.91	0.16	0.64	0.34	3.05	0.77	1.68	0.69	5.63	0.66	2.39
Central America	2.43	−0.02	1.66	−0.14	0.73	1.16	−0.06	−0.76	2.87	4.90	1.15	3.32
Caribbean	−2.00	1.23	0.03	1.39	−1.07	−0.40	1.80	1.41	0.82	3.15	0.24	0.70
Southern Cone	1.17	1.31	2.20	−1.81	3.13	0.00	3.24	−0.42	2.92	4.14	−0.62	3.10
South America	1.08	1.23	1.52	−1.03	2.47	0.70	2.61	0.13	2.36	4.54	−0.34	2.98
Central America & Caribbean	1.83	0.12	1.42	0.09	0.51	0.98	−0.12	−0.37	2.36	4.59	1.16	2.90
East Asia	10.42	3.22	6.04	2.06	1.23	7.69	4.96	1.46	3.22	7.00	4.26	6.33
Middle East	4.42	−1.20	1.49	0.41	1.54	0.96	8.14	−7.67	0.24	6.03	0.84	4.48
South Asia	1.89	0.67	0.73	2.12	2.78	1.71	0.73	3.12	3.53	4.52	1.90	3.70
Southeast Asia	−0.35	3.18	−0.52	3.31	3.44	5.25	2.20	3.41	−0.23	7.04	0.44	4.82

Source: Authors based on FAO (2003).

changes in consumption patterns and income growth. The successful transformation of the dairy sector in India and Pakistan in the last 20 years and the expansion of beef production explain output growth in South Asia.

In the case of Latin America, milk production grew during the 1960s and 1970s due to increases in the stock of milking cows (4.3% in Central America, 3.6% in the Southern Cone). In the last 20 years, most of growth in milk and beef production is explained by yield growth.

Sub-Saharan Africa is the region showing the lowest growth in ruminant production, which could be explained by the fact that this growth is based mostly in an expansion of the animal stock. During the first half of the period (1961–1980), Sub-Saharan Africa was able to increase beef production at an average rate of almost 2.5% by increasing stock. Because of the constraints imposed by natural resource availability and degradation, this growth rate decreased to 1% during the 1981–2000 period. However, the substitution of beef cattle for milking cows is still possible and that could explain growth in milk production mainly based on increasing stocks at a rate above 2% a year during between 1961 and 2000, mainly in East Africa.

6. Total factor productivity measures

We use the Malmquist index approach (see Appendix A) to estimate total factor productivity and the product-specific measure for a group of developing and developed countries. Data are from FAOSTAT (2003) and comprise 115 regions (92 developing and 23 high income regions) considering two outputs (livestock and crops), and seven inputs (feed, animal stock, pasture, land under crops, fertilizer, tractors and labor). The specific definition of these variables is given in Appendix A.

Comparisons of livestock TFP in different regions are presented in Figure 6. Two main results need to be highlighted. Firstly, we verify that the second half of the period analyzed (1981–1999) shows the higher dynamism in TFP growth. Most regions show negative TFP growth during 1961–1980 but they resumed growth in the last two decades. Secondly, Asia is the region showing the fastest expansion in TFP followed by Latin America, while Sub-Saharan Africa has remained stagnated for almost the whole period of our sample.

Within regions, China is the country with the fastest expansion of livestock TFP, growing at an average of 6% during the past 2 decades. South Asia has also shown high dynamism during the 1980s and 1990s. Figure 7a shows the cumulative TFP growth for different regions in Asia between 1965 and 1999. The figure clearly shows how China's TFP growth took off after the economic reforms in the late 1970s and how TFP decreased during the period of the "Cultural Revolution" starting in 1966. In the case of India, TFP growth started in the 1970s coinciding with the expansion of the dairy sector (operation flood). Countries in South and Southeast Asia performed poorly during the 1960s and 1970s and started sustained growth in the 1980s. Similar trends are shown by West Asia.

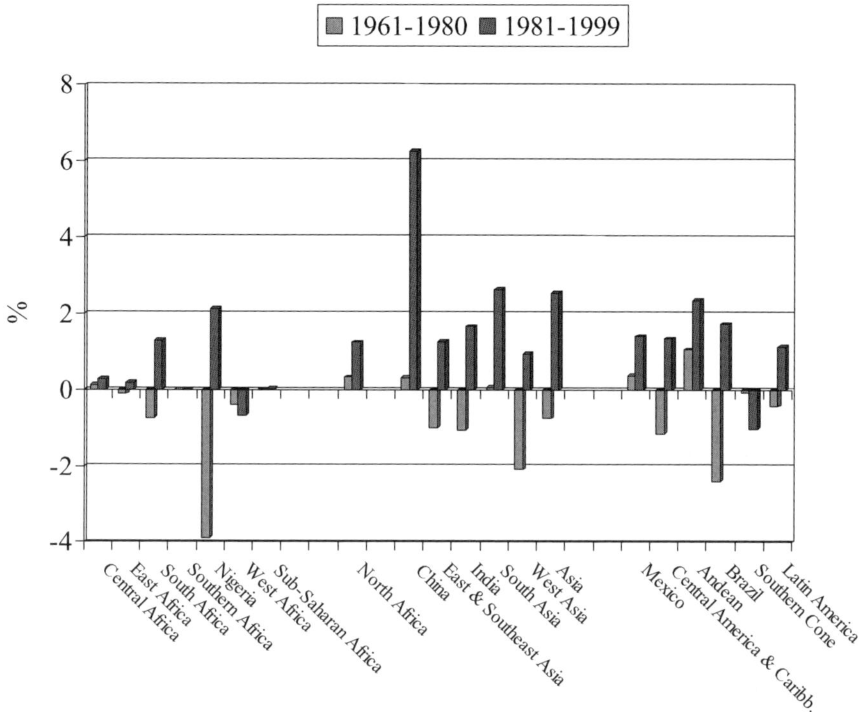

Figure 6. Annual average TFP growth in different regions. *Source*: Estimated by authors using FAO (2003) data.

In Latin America, the Andean countries show the highest rate of TFP growth (2.34%). There are differences in terms of growth during the 1960s and 1970s as shown by Figure 7b. While Mexico and the Andean countries show some growth, Central America and specially Brazil experienced negative growth for most of the period. The Southern Cone appears to experience modest growth during 1965–1980 and negative growth in the 1980s, showing a recovery during the 1990s.

As can be seen in Figure 7c, there is no significant growth of livestock TFP during 1961–2000 in Sub-Saharan Africa, showing a growth rate in livestock TFP close to zero (0.02%). Within the region, East and Central Africa show almost no changes in TFP in 40 years, with a decreasing trend in the first half of the period and a recovery during the past 20 years. The final balance after 40 years is only a 2–3% increase in TFP compared to the levels in the mid 1960s. The same trend is shown by South Africa, although in this case growth in the past 20 years reached an average annual rate of 1.3%. West Africa on the other hand, has performed poorly according to our TFP estimates with negative, though small in absolute value, growth rates. Southern Africa shows zero growth in TFP, even with less variability than that shown by East and Central Africa.

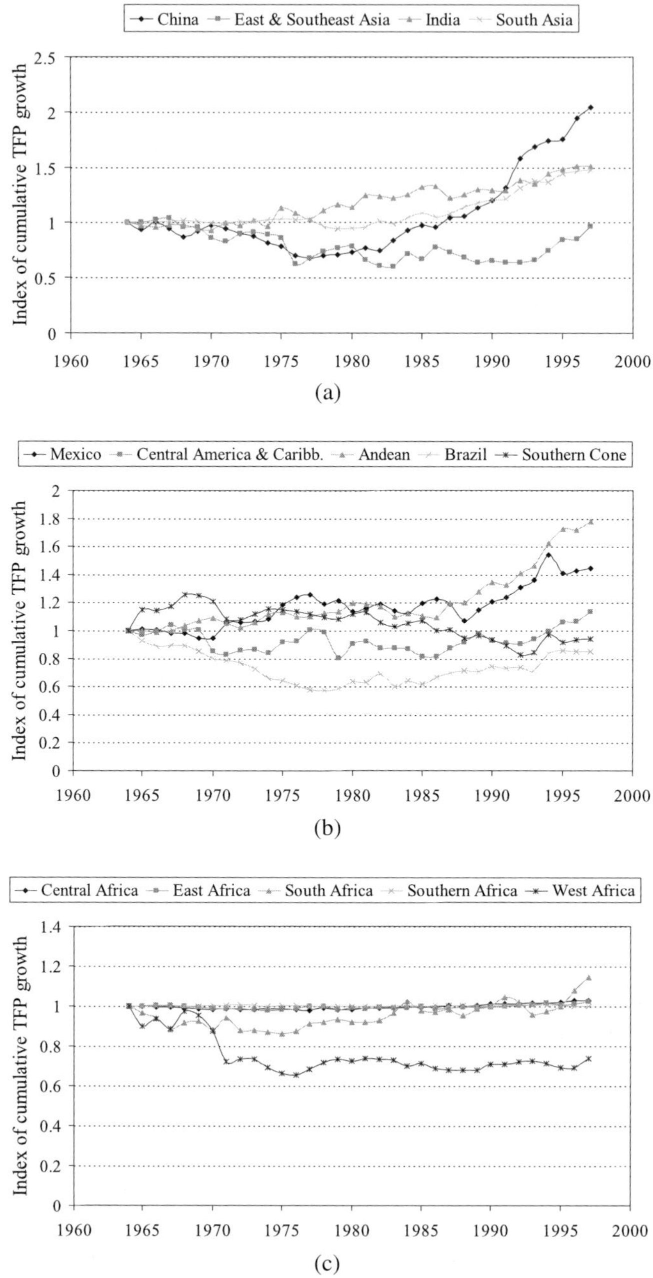

Figure 7. Cumulative livestock TFP growth: (a) Asia; (b) Latin America; (c) Africa. *Source*: Authors using FAO (2003) data.

Table 20
Livestock total factor productivity growth and components (technical change and efficiency)

	1961–1980			1981–1999		
	Total factor productivity	Technical change	Efficiency	Total factor productivity	Technical change	Efficiency
Central Africa	0.13	0.20	−0.07	0.28	0.34	−0.06
East Africa	−0.11	0.17	−0.28	0.19	0.05	0.14
South Africa	−0.73	0.67	−1.39	1.30	0.74	0.56
Southern Africa	0.00	0.08	−0.07	0.00	0.03	−0.03
Nigeria	−3.89	0.01	−3.89	2.14	0.15	1.98
West Africa	−0.40	1.27	−1.65	−0.67	0.43	−1.09
Sub-Saharan Africa	0.01	0.01	0.00	0.02	0.03	0.00
North Africa	0.33	0.60	−0.27	1.25	0.38	0.86
Africa	0.17	0.31	−0.14	0.63	0.21	0.43
China	0.32	1.04	−0.72	6.27	0.82	5.40
East & Southeast Asia	−0.99	2.25	−3.17	1.28	6.08	−4.53
India	−1.05	0.64	−1.67	1.66	1.62	0.05
South Asia	0.06	0.89	−0.83	2.65	2.85	−0.20
West Asia	−2.09	0.75	−2.81	0.96	0.38	0.58
Asia	−0.75	1.11	−1.85	2.54	2.33	0.21
Mexico	0.39	0.77	−0.38	1.41	0.64	0.76
Central America & Caribbean	−1.14	1.52	−2.62	1.35	3.01	−1.61
Andean	1.06	1.03	0.03	2.34	2.40	−0.06
Brazil	−2.38	1.10	−3.44	1.73	1.07	0.66
Southern Cone	−0.08	0.61	−0.68	−1.03	0.66	−1.68
Latin America	−0.44	1.01	−1.43	1.15	1.55	−0.39

Source: Authors using FAO (2003) data.

The decomposition of TFP growth in efficiency and technical change in Table 20 shows that growth in technical change follows TFP patterns described above, with Asia showing higher technical change growth, followed by Latin America, North Africa and with Sub-Saharan Africa showing very low growth. Results in Table 20 also show that the poor performance in TFP growth during 1961–1980 is explained mainly by efficiency losses and a relatively slow expansion of the technical frontier. Acceleration of TFP growth in the 1980s and 1990s is mainly the result of an increased rate of expansion of the technical frontier.

Different regions in Asia, Latin America and Sub-Saharan Africa, show similar patterns to those discussed at an aggregated level but there is variability in terms of the speed in the expansion of the frontier and changes in efficiency. TFP growth in China appears to be mostly the result of efficiency gains according to our estimates. Technical change occurred at a high rate in the case of Southeast Asia, but the region still experiences negative changes in efficiency, reducing the overall impact of the expansion of the frontier in TFP. In Latin America, growth rate of technical change was higher for

Table 21
Annual productivity growth rate for aggregate and disaggregate livestock TFP measures (1961–2001)

Country/region	Aggregate			Disaggregate livestock			
	Agriculture	Crops	Livestock	Agriculture	Crops	Ruminants	Non-ruminants
Brazil	0.68	0.65	1.04	0.81	0.79	1.20	4.33
China	0.80	0.71	3.66	1.00	0.75	2.87	3.39
World	0.76	0.93	0.78	0.75	1.11	−0.07	1.81
Developed countries	1.45	2.64	1.51	1.04	2.57	0.93	2.11
Developing countries	0.66	0.43	1.81	0.57	0.51	0.38	2.38
Least developed countries	0.19	−0.03	0.15	0.54	0.14	0.40	1.24
Africa	0.98	0.75	0.82	0.65	0.70	0.20	1.54
Sub-Saharan Africa	0.38	0.21	0.78	0.57	0.24	0.59	0.80
Asia	0.44	0.03	1.90	0.44	−0.09	0.63	1.96
Asia developing	0.71	0.19	2.55	0.96	0.25	0.50	2.71
East and South East Asia	0.37	0.17	1.49	0.44	0.03	−0.28	1.58
North Africa and Middle East	−0.37	−0.52	−0.30	0.42	−0.06	−0.05	1.54
Western Europe	1.56	3.50	2.16	0.93	3.33	0.96	2.47
Eastern Europe	1.27	1.05	1.62	0.93	2.03	1.18	2.10
North & Central America	1.10	1.78	1.78	1.11	2.07	1.73	1.63
Latin America & Caribbean	0.83	0.72	0.82	0.71	0.98	0.10	2.59
Caribbean	0.39	−2.19	1.16	−0.30	−2.03	−0.96	1.02
South America	0.85	0.85	0.45	0.73	1.23	0.29	3.05

Source: Ludena et al. (2005).

Central America and the Andean countries and low for Mexico and the Southern Cone, with Brazil showing intermediate values. Brazil and Mexico appear to be improving also in terms of efficiency while other regions show negative efficiency growth. Technical change growth rate for Sub-Saharan Africa is low compared with that in Latin America and Southeast Asia.

TFP values are estimated also for ruminants and non-ruminants within the livestock sector (Table 21). Most countries in the sample display largeraverage productivity growth gains in non-ruminant production, compared to ruminant productivity gains. For developing countries, non-ruminant productivity grew at a rate of 2.38%, while productivity of ruminant production at an annual average rate of 0.38%. Most regions show small (less than 1%) or negative average growth rates for ruminants.

In order to better understand TFP growth patterns in different regions we compare the evolution of TFP growth to growth in PFP measures for three countries with different labor/land ratios and different growth performance. Figure 8 compares the path followed

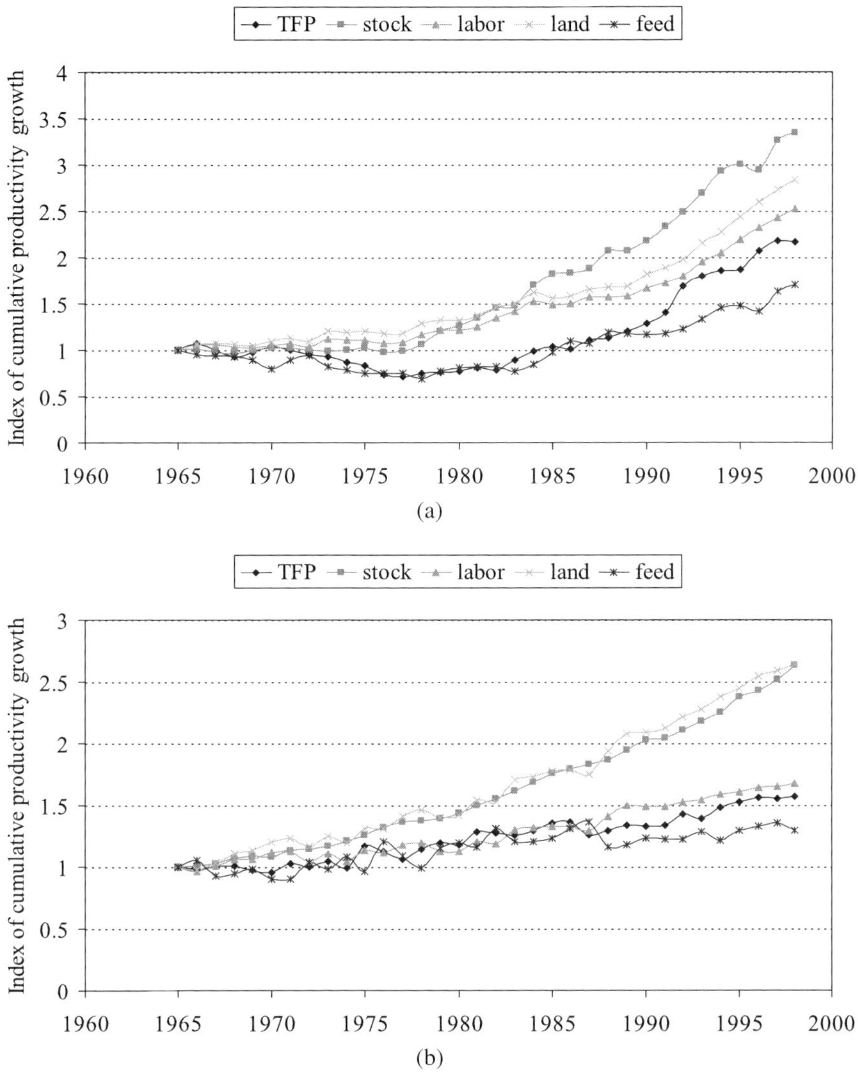

Figure 8. Cumulative TFP and PFP growth: (a) China; (b) India; (c) Brazil; (d) Kenya. *Source*: Authors using FAO (2003) data.

by the TFP measure and PFP estimates for different inputs between 1965 and 1998 in China, India, Brazil and Kenya. We verify that in all cases, TFP growth is below growth in land, labor and stock PFP. Comparisons between China and India and Brazil also reveal the role of different inputs in TFP growth in countries with high and low labor/land ratio. In China and India, growth in land productivity together with output per head of animal stock drives TFP growth. In the case of Brazil the most significant

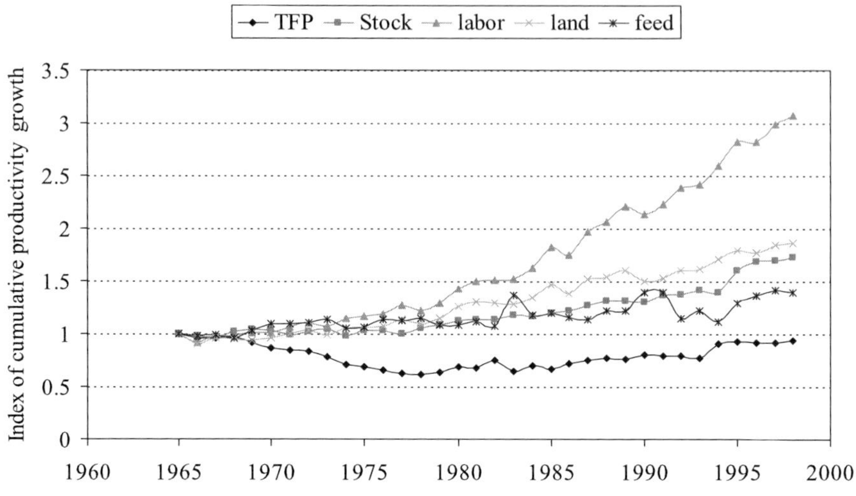

(c)

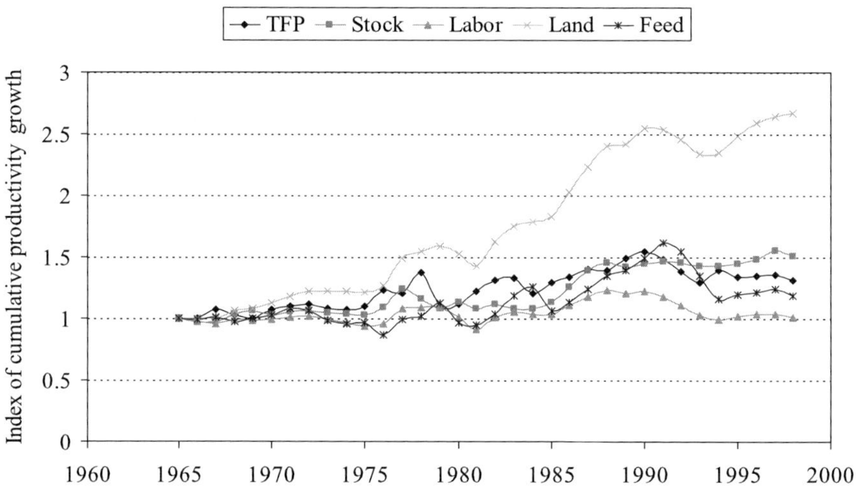

(d)

Figure 8. *(continued)*

contribution to TFP growth comes from increases in labor productivity. In the case of Kenya, with labor/land ratios between those in India and Brazil, growth is explained mainly by increases in land productivity. In contrast with India, output per head of animal stock, feed and labor PFP did not follow and TFP remained stagnated. It is only between 1985 and 1993 that growth in PFP of animal stock and feed occurred, positively affecting TFP. Labor productivity also increased during this period but less than productivity in other factors.

The livestock productivity performance just discussed relies primarily on the performance of the research and development systems. We discuss this subject in Section 7 below.

7. Livestock R&D systems

Improving the productivity of livestock in the developing countries will require significant investment in livestock research by the public and the private sectors. However, livestock research in the developing countries faces two major challenges [Delgado et al. (1999)]. First, appropriate existing and new technologies and production systems have to be adapted and disseminated to the developing world to eliminate low productivity. Second, the limits of livestock production technology and systems have to be extended to include the environmental and public health problems that have appeared in high-intensity livestock production.

National Agricultural Research Systems (NARS) in developing countries have evolved rapidly in the past three decades. According to Byerlee and Alex (1998), they employ over 100,000 scientists and manage an annual investment of over $8 billion, a higher level than that in high income countries but still low in comparison with agricultural gross domestic product (0.5% in developing countries and 2.4% in high income countries). Several studies have shown the important contribution of agricultural research to overall productivity growth, increasing crop yields and production and the high rate of return to investment in research [see, for example, Evenson, Pray and Rosegrant (1999)].

In crop research, much of the benefit to date has been generated through varietal development. Livestock research, on the other hand, is slower, more costly, and more difficult than crop research. The nature of these complexities is well summarized by Jarvis (1986): "Individual animals are dramatically more expensive than individual seeds or plants. For animals, several years elapse between conception and maturity, and substantial time is required before the impact of new technologies can be evaluated. Experimental control is difficult because animals move about and animal personality affects the results. Interactions with management variables are also complex. Livestock research is essential, but technological advances are piecemeal and slow; governments must be prepared to provide funds over lengthy periods without expecting quick, dramatic breakthroughs."

In what follows we briefly describe R&D efforts in different developing regions, summarizing disperse information on livestock research and quantifying the efforts made by developing countries in livestock R&D.

7.1. Livestock R&D in Sub-Saharan Africa

With political independence, which started in the late 1950s and early 1960s, African countries faced a complex transition which will influence the future paths and changes the system experienced in different countries. Domestic financial support for research increased in the 1960s and early 1970s peaked in 1981 and declined throughout the 1980s, while real research expenditures stalled. Low salaries and poor incentives resulted in reduced quality of human capital in the system. Because of reduced expenditures, NARS became increasingly reliant on donor-sourced funds. Donor funding currently accounts for 61% of total agricultural research expenditures in Francophone countries and 26% in Anglophone countries.

Nigeria and South Africa accounted for about 25% of the region's total investment in agricultural R&D in 1991 and Pardey, Roseboom and Beintema (1997) mention them as following contrasting paths during the period. The South African system grew comparatively slowly but steadily with the rate of growth of expenditure keeping pace with its growth in research staff. Nigeria on the other hand, followed an explosive growth due to a boom in public revenues from oil exports, during the 1980s research spending declined dramatically while the number or research staff stayed constant, a common trend with other African countries.

According to Rukuni, Blackie and Eicher (1998), "Today most African countries have weak and ineffective public research systems that are not serving the majority of farmers. Although foreign aid has assisted in human capital development and increasing the size of many NARS in Africa, there is increasing concern that the size of most NARS is not going to be sustainable after foreign aid is withdrawn."

Trying to explain the decline of NARS in Africa, Eicher (2001) mention four factors that have contributed to this: (a) lack of African political support for research; (b) project design problems such as preparing projects that are too large and too complex for too short a time span; (c) institutional coordination problems, that is, the coordination and sequencing of investments in research, extension and higher education; (d) the fourth problem is the financial sustainability problem and the tendency for donors to contribute to the expansion of a NARS beyond its capacity to be financed after donor aid is phased out.

Focusing specifically on livestock research in Sub-Saharan Africa, Winrock International (WI) states that NARS as a group have not generated sufficient new technology to adequately fuel animal agricultural development. They argue that poor overall performance of African NARS prominently includes problems in research planning and organization, research environment, training, funding scientific communication and career opportunities. In particular, funding for animal research programs is insufficient to serve the needs of animal agriculture considering the complexity and scale of development needs.

Analyzing the impact of animal agriculture development project in the past, WI argues that range–livestock projects have been the most disappointing. Range–livestock systems, designed to replace traditional systems with new production forms and improved technology like reseeding and improved grazing systems, failed completely. In crop–livestock systems where more options are available, projects have performed better. In an assessment conducted by the World Bank of 125 animal agriculture projects implemented in Sub-Saharan Africa from 1967 to 1983, it was found that crop–livestock projects and other livestock components projects were more successful than pure livestock projects.

Similar conclusions were reached by an USAID evaluation of 104 livestock-related projects implemented between 1954 and 1981 [WI (1992, p. 54)]. The success stories in smallholder dairy production in East Africa and animal traction in West Africa are concentrated in the crop–livestock systems. In East Africa, smallholder dairy development, which started about 1955, has been one of the successful stories. Returns from milk and forage production have been consistently higher than the returns to crops like beans and maize, a key factor explaining this success [WI (1992, p. 53)]. In semi-arid zones of West Africa, where the introduction of animal traction began in the 1940s, number of oxen was almost doubled between 1979 and 1981–1983. This was made possible by the existence of profitable cash crops (cotton and groundnuts) and effective input supply, credit and extension services for cotton production, especially in the francophone area [WI (1992, p. 54)]. A third success story is the introduction of animal health technology in many regions and production systems reducing the threats of diseases such as rinderpest and contagious bovine pleuropneumonia [WI (1992, p. 54)].

The priority livestock research areas in the region include feed/nutrition, animal health, genetics/biodiversity, natural resource management and mixed crop–livestock systems. Seasonal variations in the quality and quantity of feed supply continue to be the major constraints to productivity improvement. For example, continued research on evaluation of indigenous and novel feed resources and their integration into the farming system is needed. In the area of animal health research on the epidemiology of diseases and development of control measures as well as the impact and risk assessment to identify key areas for animal health intervention is essential. In the area of genetics/biodiversity it will be important to continue to support research on identification and use of resistant genotypes in livestock breeding programs. Sustained attention to the sustainability and environmental and human impacts of livestock production should continue. Today global safety concerns regarding human and animal disease associated with livestock products are increasing.

7.2. Livestock R&D in East, South and Southeast Asia

In contrast with Africa, there were no fundamental structural and organizational changes in most Asian NARS after independence other than an increase in the resources committed to agricultural research. Raising demands from different groups led to an increase in the number of commodity research institutes financed by taxes on producers, contributing to fragment national agricultural research efforts. This fragmentation, especially in the larger countries like India and Indonesia resulted in "inefficiencies, duplication of effort, lack of coordination and the like, while at the same time domestic food shortages continued to grow in many countries" [Pardey, Roseboom and Anderson (1991a, p. 239)].

Within the region, the cases of China and India deserve special attention. China's investment in agricultural research in relation to agricultural GDP in the early 1960s was high compared with investment in other developing regions and dropped during the 1980s and 1990s not because of reductions in the amount of research expenditure but as a result of high growth rates in agricultural GDP mainly as the result of institutional reforms [Pardey, Roseboom and Anderson (1991a)]. India has one of the largest and more complex agricultural research systems in the world, which, associated with the education and extension systems has greatly contributed to the rapid growth of agricultural production since independence. Long-term growth in public agricultural research expenditure has increased rapidly until 1968, very fast from 1968 to 1980, going to a phase of slow growth after 1980.

Focusing on the case of livestock research in the region, Devendra (1996) argues that research in animal agriculture has not been a priority in comparison to research in crops in terms of efforts and funding supports. In general, animal systems were not linked through crop–animal systems to enhance agricultural production.

Livestock research in Asia has tended to focus on ruminant's breed improvement and feeding and nutrition with variable success. The emphasis on ruminants is based on its wide distribution across agro-ecological zones; association with poor smallholders; need to produce more animal protein; ability to utilize crop residues and comparatively more advanced and self-sufficient non-ruminant sectors [Devendra (1996)].

In the case of India, projects are mainly directed toward: (a) evolving high yielding strains of livestock; (b) evaluating the utilization of conventional and non-conventional feed resources and developing new feed resources; (c) reproduction and adaptation; (d) developing immuno-prophylactic and laboratory diagnostic methods; and (e) improving techniques for processing milk, meat and animal fiber [Devendra (1995)].

In an analysis of major constraints and priorities for research in Asia, Devendra (1995) highlights the need of a more complete utilization of animal genetic resources, increased investment for research in rain-fed areas and animal diseases, and increased intensification and efficiency in the use of available feeds given that this is the principal constraint among the non-genetic factors, which affects productivity.

Referring specifically to India's livestock research, the World Bank (1996) stresses that livestock research is skewed toward bovines and mainly focuses on cattle. Research in buffalo, which supply half the national milk output and small ruminants, which supply most of the national ruminant meat production, does not correspond to the importance of these sectors. The ICAR livestock research institutes tend to focus their efforts on genetic improvement, with considerable emphasis on the import of exotic breeds and biotechnology. Given the sector's technical and socioeconomic constraints and institutional weaknesses, research efforts should be reoriented, according to the World Bank, to focus on:

- Farming systems research;
- Feeds and fodder production, including basic research using modern biotechnology techniques to improve nutritive value of low-quality forage and crop residues, and also applied research on feed and fodder production, especially in the more arid zones;
- Buffalo and small ruminants and feeding systems;

- Genetic improvement, shifting from the current emphasis on introducing exotic breeds to one that identifies simple breeding schemes to improve local breeds;
- Epidemiological research and development of cost-effective control strategies;
- Free range chicken production, with emphasis on backyard production systems and disease control (Newcastle disease).

With some differences with the diagnosis and proposed focus by the World Bank report, Devendra (1995) thinks that research emphasis needs to be on genetic studies of crossbred cattle with a view to develop selection criteria, determine genetic and phenotypic trends and generate superior crossbred germplasm for milk production. On the other hand, and agreeing with the World Bank report, he argues that similar studies need to be carried out on draft, dual-purpose and milk breeds of indigenous cattle and buffaloes. There is also the need to develop, according to Devendra, a package of practices for herd–health management to maintain the achieved level of production including development of a system for animal disease monitoring, surveillance and forecasting.

7.3. Livestock R&D in North Africa and West Asia

The region is characterized by a fast reduction of agricultural population reflecting the rapid expansion of the region's oil industry. Jobs were created in the high-income oil-exporting countries and labor-surplus countries have profited by exporting their labor to the oil-exporting countries [Pardey, Roseboom and Anderson (1991a)]. Agricultural research in the region has been led by Egypt, which had nearly one-half of the region's agricultural researchers in 1981–1985, but represented only 10% of total expenditure in agricultural R&D.

Natural resources in the region impose serious constraints to livestock development. Only 37% of the land is suitable for agriculture of which 8% is arable, 23% rangelands, 7% woodlands, and the remaining area is desert or semi desert. The main constraints to livestock production are: insufficient and poor quality food; low genetic potential; poor animal health and poor management practices [Sidahmed (1995)]. In spite of a long history of research attempts at technology transfer, adoption of technologies for enhancing fodder production for improved grazing management systems has been very poor. This is explained mainly by the nature and complexity of technologies, lack of adequate quantities for seed of pasture and forage legume species including the lack of appropriate equipment [Sidahmed (1995)]. Research as suggested by Sidahmed (1995) should focus on development of sustainable livestock-based system based in crop–livestock production aimed to conserve the natural resource base in the rangelands.

7.4. Livestock R&D in Latin America

The history of modern agricultural research in Latin America started with the consolidation of the system of national research institutes (INIAs) in the late 1950s [see Pardey, Roseboom and Anderson (1991a)]. According to Echeverria (1998), there are in total about 100 public and private agricultural R&D organizations in the region, employing in

1995 close to 10,000 professionals with a total annual budget of over US$900 million including activities of the private sector, producer associations, research foundations and universities. Private investment still accounts for only 10% of total investment. The main problems facing NARS in Latin America according to Echeverria (1998) are related with funding and organization and management of these organizations. Public funding for agriculture stagnated and even declined in many countries of the region during the 1990s after two decades of rapid growth.

Jarvis concludes that livestock productivity in Latin America is well below its potential and technological progress has been relatively slow in recent decades [Jarvis (1986)]. In his extensive study of Latin America's livestock sector, Jarvis highlights research priorities to improve performance of the livestock sector:

- Given that culling and selection possibilities on Latin American farms are very limited due to lack of records, low calving and weaning rates, little selection is permitted. Improve reproductive performance is required to increase production growth rates and allow for progressive animal selection.
- Pasture improvement, and improved management techniques are required to increase calving and weaning rates and expand possibilities for culling and selection.
- Farming system research is needed to promote and develop smallholder livestock producers. Dual purpose cattle make a large contribution to total milk production offering scope for increasing the welfare of small farmers.
- Research in technologies that emphasize the complementarities of crops and livestock are needed. Ruminants produce meat at lower cost than poultry or swine only if they are fed mostly on pasture and crop residues with a low opportunity cost. Introduce of improved grass and legume species could increase livestock output in areas with abundant pastureland that is not suited for crops (low quality pasture).

7.5. *International agricultural research centers*

International agricultural research centers (IARCs) are relatively new institutional innovations, created to assist developing countries in generating technology to increase agricultural production and in building research capacity. The Consultative Group on International Agricultural Research (CGIAR), established in 1971, is an informal association of more than 40 countries, international and regional organizations, private foundations, and representatives from national research systems in the less-developed world, formed to guide and support a system of international research centers [Gryseels and Anderson (1991); Alston, Pardey and Roseboom (1998)]. CGIAR research activities cover a broad spectrum of crop and livestock production plant breeding, farming systems, natural resource conservation and management, animal diseases and production, plant protection, post-harvest technology and food policy. The CGIAR has provided the predominant institutional model for development research.

Livestock research was included as one of the priorities for the CGIAR system since its early years given that high priority was given to improving the quality of diets through research on food legumes and ruminant livestock. The specialized GCIAR center for livestock research is the International Livestock Research Institute (ILRI), which was created by the merger in 1994 of the International Laboratory for Research on Animal Disease (ILRAD) based in Kenya and the International Livestock Center for Africa (ILCA) in Ethiopia. ILRI made significant contribution to the advancement of science and knowledge in livestock and understanding and solving problems faced by the smallholders. Some of the contribution of ILRI (among many) resulting from combined work of ILCA and ILRAD include: (1) a comprehensive survey of the state of African cattle genetic resources which revealed that at least 21 breeds/strains have become extinct in the last century and, out of the remaining 144 breeds, about 45 are currently at risk of extinction; (2) breeding cattle for trypanotolerance; (3) factors determination immune responses to trypanosomosis; (4) techniques for producing immunity to East Coast fever; (5) developing and field testing of recombinant East Coast fever vaccine; (6) research to isolate rumen microbes to better utilize tropical feeds; (7) developing models for predicting environmental impact of combinations of crop, livestock and natural resource management interventions; (8) research leading to significant productivity increases as a direct result of using forages in Africa and Latin America; (9) development of models for determining priorities for future research; (10) state of the art poverty mapping using models and GIS data; (11) policy research for identifying factors affecting access for smallholder livestock producers; (12) long term research on the dynamics of Sahel rangelands.

7.6. R&D and investment in livestock research

We proceed now to quantify the effort different developing regions and countries have put in livestock R&D. The information on expenditure and researchers in livestock R&D presented in this section was built using detailed data on percentage of total researchers in livestock (full time equivalents) for 32 countries in Africa, Latin America and Asia extracted from several statistical briefs on the national agricultural research systems from ISNAR's Indicator Series Project. These percentages are applied to figures of total researchers in agriculture extracted from Pardey, Roseboom and Anderson (1991b) assuming that no changes in the proportion of researchers in livestock occurred through time. Expenditure in livestock R&D was derived from the number of researchers using average values of expenditure per researcher for each country. Values for countries for which there was no information available were estimated using figures for similar countries in their region adjusted by the share of livestock in agricultural output.

Expenditure in livestock R&D and number of livestock researchers for aggregated regions are presented in Table 22. According to our estimates, the total number of researchers in livestock R&D in developing countries averaged 7525 over 1961–1980 and increased to 16711 in 1981–1985. Total expenditure in livestock R&D by developing countries increased from \$416 million to \$751 million in the same period. Number of researchers increased rapidly in West Africa, China, Southeast and West Asia, Mexico, Central America and Brazil. South Africa, the Southern Cone and the Andean region

Table 22
Expenditure in livestock R&D and number of livestock researchers for aggregated regions

	Researchers in livestock			Expenditure in livestock		
	1961–1980	1981–1985	Annual growth	1961–1980	1981–1985	Annual growth
Central Africa	39	72	4.92	3410	5363	3.69
East Africa	127	250	5.57	13093	16578	1.91
South Africa	244	326	2.34	30530	38664	1.91
Southern Africa	58	102	4.54	6216	8535	2.57
Nigeria	95	221	6.97	12353	17622	2.88
West Africa	61	216	10.61	6391	12859	5.75
Sub-Saharan Africa	625	1185	5.25	71994	99620	2.63
North Africa	288	663	6.89	12152	21102	4.51
Africa	913	1848	5.80	84146	120723	2.93
China	3289	8745	8.14	118213	253401	6.29
East & Southeast Asia	391	953	7.38	31870	55684	4.57
India	1433	2433	4.33	67686	130500	5.39
South Asia	398	699	4.61	10068	26511	8.05
West Asia	176	374	6.19	18706	30736	4.05
Asia	5687	13204	6.97	246543	496832	5.77
Mexico	75	190	7.71	6278	23220	11.03
Central America & Caribbean	34	81	7.19	2049	3516	4.41
Andean	236	326	2.61	24557	28071	1.08
Brazil	313	683	6.44	29970	52614	4.61
Southern Cone	267	379	2.86	22115	25600	1.18
Latin America	925	1659	4.79	84969	133021	3.65
Developing	7525	16711	6.59	415658	750575	4.84

Source: Authors based on Pardey and Roseboom (1989).

show the slowest growth in number of researchers. India, rest of South Asia and Mexico are the only regions where expenditure per researchers increased during this period.

Total number of researchers in livestock R&D for the same regions compared to the share of livestock production in total production of developing countries is presented in Table 23. Asia accounts for more than 75% of total researchers in developing countries even though its share in total livestock output among developing countries is 48% in 1961–1980 and increased to 62% in 1981–1985. China and India alone accounted for 62% of all researchers in developing countries in 1961–1980 increasing to 67% in the early 1980s. Sub-Saharan Africa's share in total number of researchers is 8%, similar to its share in output. Latin America's share of total researchers reduced from 12 during 1961–1980 to 10% in the 1980s, which is well below its share in output.

The congruency model is a commonly used method of assessing the allocation of research resources. The model assumes that an additional dollar spent on research would

Table 23
Total number of researchers in livestock R&D compared to the share of livestock production in total agricultural output

Region	1961–1980		1981–2000	
	Output	Researchers	Output	Researchers
Central Africa	0.92	0.52	0.64	0.43
East Africa	5.20	1.69	3.60	1.50
South Africa	3.19	3.24	1.74	1.95
Souther Africa	1.29	0.78	0.73	0.61
Nigeria	0.98	1.26	0.91	1.32
West Africa	1.56	0.81	1.01	1.29
Sub-Saharan Africa	13.15	8.31	8.63	7.09
North Africa	3.03	3.83	2.80	3.96
Africa	16.18	12.13	11.43	11.06
China	17.57	43.71	33.09	52.33
East & Southeast Asia	6.40	5.20	7.07	5.70
India	13.06	19.04	12.28	14.56
South Asia	5.13	5.29	4.69	4.18
West Asia	6.21	2.35	4.68	2.24
Asia	48.37	75.58	61.81	79.01
Mexico	5.29	1.00	4.63	1.14
Central America & Caribbean	2.59	0.45	1.63	0.48
Andean	5.20	3.13	4.13	1.95
Brazil	9.06	4.16	9.43	4.09
Southern Cone	13.31	3.55	6.93	2.27
Latin America	35.45	12.29	26.76	9.93
Developing	100	100	100	100

Source: Authors based on Pardey and Roseboom (1989).

yield a higher return if spent in areas with a relatively low ratio of research funding to output ratio [Beintema, Dias Avila and Pardey (2001)]. This implies that funds should flow toward programs with relatively low research intensities. To be congruent with the corresponding value of output, the share of livestock research spending in agricultural research should equal livestock share in total agricultural output. A congruence test between the share of livestock research in agricultural research and livestock share in total agricultural output is shown in Figure 9. Congruence assumes that in Latin America, East Africa, South Africa, and North Africa, South (excluding India) and West Asia and in a lesser degree West Africa, the share of livestock research is much smaller than might be expected given its share in agriculture. On the contrary, livestock research in Nigeria and India accounts for a larger than congruent share of total agricultural research. Other regions in Asia show congruent shares of livestock research.

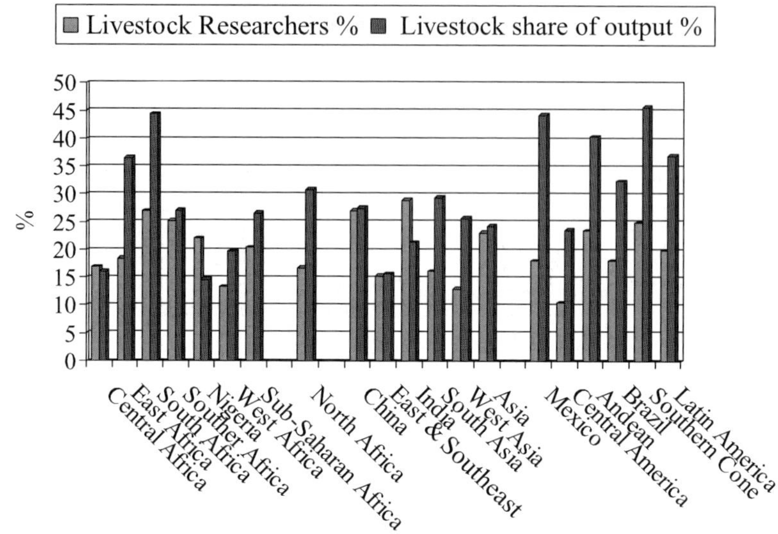

Figure 9. Congruence test: percentage of livestock researchers compared with livestock share of output. *Source*: Authors based on Pardey and Roseboom (1989).

Considering livestock research as an investment activity it is possible to determine research productivity and factor-intensity ratios as discussed in Pardey, Roseboom and Anderson (1991b). As these authors point out, it is important to bear in mind, when analyzing these indicators, the problems of interpreting agricultural output compared to contemporaneous agricultural research expenditures given that investment in research generates a stream of benefits over several years and that there are lags in the research process itself.

Figure 10 measures the level of livestock output per unit of contemporaneous research expenditure and per full-time equivalent researcher. The figure shows that the most productive investment in livestock research occurs in Latin America, especially in Mexico, Central America and the Southern Cone. On the other hand, China, South and Southeast Asia show the lowest values of research productivity. Regions in Sub-Saharan Africa reveal intermediate values. This contrasts with results presented by Pardey, Roseboom and Anderson (1991b) for total agricultural research. When considering agriculture as a whole, Asia and Pacific records the highest output per unit of research expenditure and personnel while Latin America and Caribbean and China show the lowest output per researcher and expenditure.

Assuming that diminishing returns to livestock research prevail we should expect lower factor-intensity ratios for livestock research in Latin America than in Asia. As shown in Table 24, research intensities are higher in Asia than in Latin America in the case of intensity ratios per head of animal stock and per unit of land. Expenditure per million animal head in Latin America is below \$0.5 while in Asia it is above \$1.5

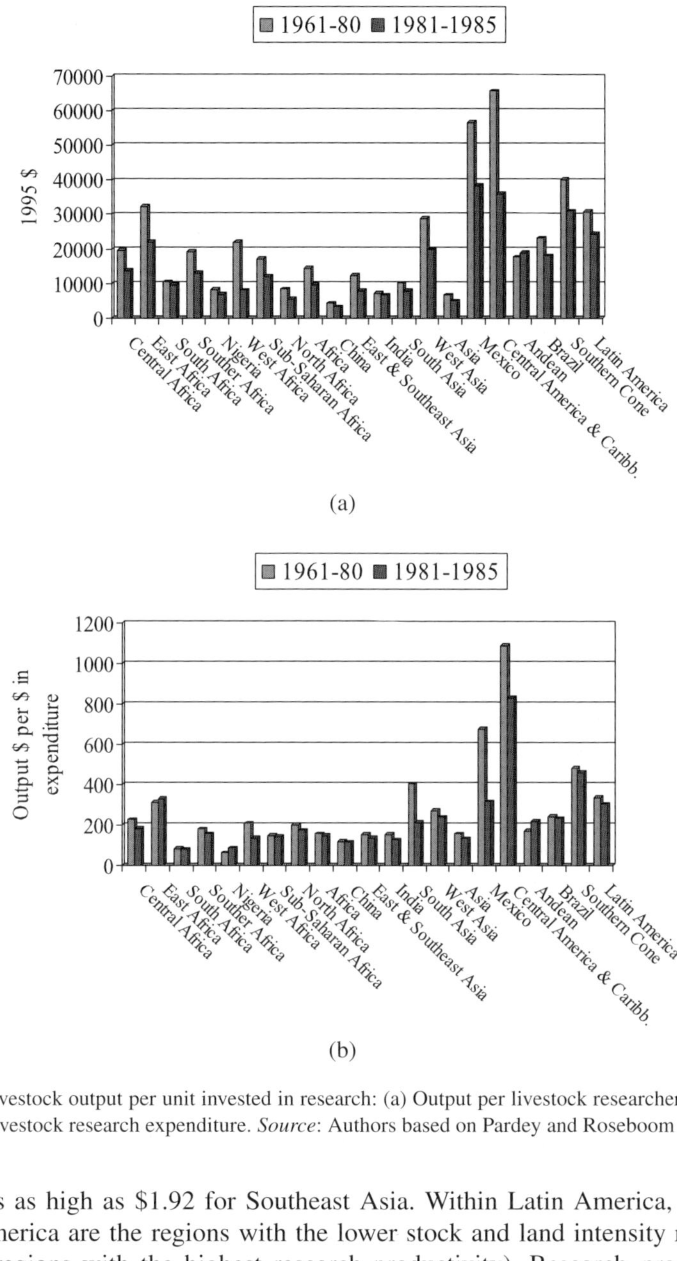

Figure 10. Livestock output per unit invested in research: (a) Output per livestock researcher; (b) output per $ in livestock research expenditure. *Source*: Authors based on Pardey and Roseboom (1989).

with values as high as $1.92 for Southeast Asia. Within Latin America, Mexico and Central America are the regions with the lower stock and land intensity ratios (while being the regions with the highest research productivity). Research–productivity ratios declined much more rapidly in the regions with higher productivity ratios, which is consistent with the fact that incremental gains to research occur at diminishing rates.

Table 24
Research intensity ratios for land, worker and animal stock

	$1980/million ha		$1980/worker		$1980/stock	
	1961–1980	1981–1985	1961–1980	1981–1985	1961–1980	1981–1985
Central Africa	34.2	53.1	0.2	0.3	0.5	0.6
East Africa	50.0	62.6	0.4	0.3	0.3	0.3
South Africa	318.7	412.4	12.7	20.1	1.8	2.2
Souther Africa	27.1	37.0	0.5	0.5	0.7	0.8
Nigeria	183.8	258.5	0.8	1.1	2.1	1.9
West Africa	44.3	88.2	0.4	0.6	0.5	0.8
Sub-Saharan Africa	80.1	110.2	0.7	0.8	0.8	0.9
North Africa	133.3	231.6	0.9	1.4	1.5	2.2
Africa	85.0	121.3	0.7	0.8	0.8	1.0
China	311.7	546.8	0.3	0.6	1.1	1.8
East & Southeast Asia	456.4	746.1	0.4	0.5	1.5	1.9
India	389.5	745.5	0.4	0.6	0.8	1.3
South Asia	291.3	741.0	0.2	0.5	0.5	1.1
West Asia	157.8	262.5	1.0	1.6	0.8	1.1
Asia	317.7	573.7	0.4	0.6	0.9	1.5
Mexico	64.9	237.5	0.9	2.8	0.3	0.7
Central America & Caribbean	79.7	119.8	0.3	0.5	0.1	0.2
Andean	193.6	208.0	3.2	3.2	0.7	0.6
Brazil	166.9	243.4	1.9	3.1	0.4	0.5
Southern Cone	103.6	117.4	7.6	8.6	0.4	0.4
Latin America	132.3	191.1	2.1	3.0	0.4	0.5

Source: Authors based on Pardey and Roseboom (1989) and FAOSTAT (2003).

Labor intensity ratios show contrasting results with those for land and animal stock. Latin America reveals higher intensity in research per worker than Asia and Africa. Labor intensities in the Southern Cone are more than 16 times larger than those in Southeast Asia and China implying that livestock research productivity and diminishing returns are determined by land and stock intensity.

Commodity focus for the aggregate regions was determined in a similar way than the information on expenditure and researchers in livestock R&D, using detailed data on percentage of total researchers in different commodities using information available for a group countries in Africa, Latin America and Asia extracted from several statistical briefs on the national agricultural research systems from ISNAR's Indicator Series Project. As explained above, values of countries for which there was no information available, were estimated using figures for similar countries in their region adjusted by the share of livestock in agricultural production.

Figure 11 shows aggregate numbers of percentage of researchers and expenditure in different commodities for our group of 80 developing countries. One-third of all

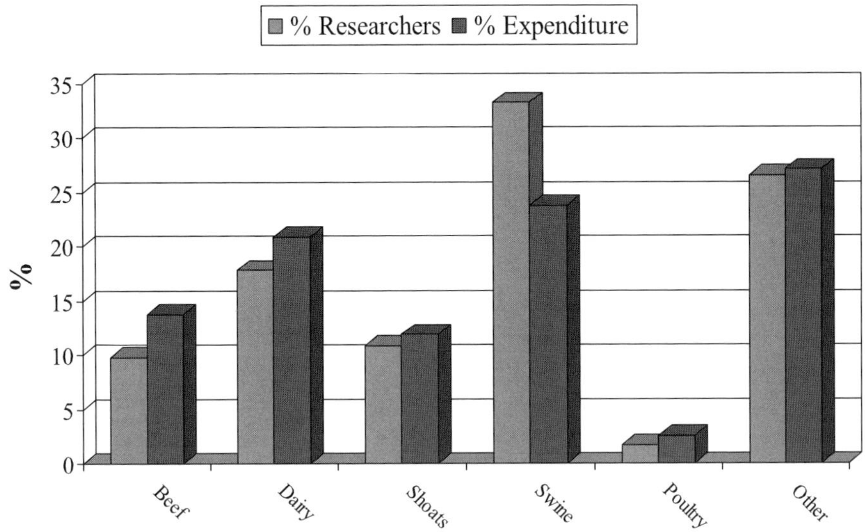

Figure 11. Proportion of researchers and expenditures in different commodities and programs. *Source*: Authors based on Pardey and Roseboom (1989).

researchers and one-fourth of total expenditure in developing countries are allocated to swine, which is explained by the importance of this commodity in China and Southeast Asia and the share of this region in total research. More than one-fourth of resources in developing countries go to non-specific commodity research (production systems, forages, feed and nutrition, health programs, etc.). Around 20% of research investment is allocated to dairy production and less than 15% each to beef and shoats. The share of researchers and expenditure allocated to poultry is only 2 and 3%, respectively.

7.7. Strategies for improving impact of livestock research

Clearly where investment has been inadequate, such as in Africa, there has been a marked lack of success in livestock development and transfer of new technologies. To function, NARS need not only trained personnel but also adequate funds to cover salaries, and fixed and operating costs. Government funding for NARSs in Africa for example, has rarely been sufficient and has often been used ineffectively. National resources are mainly used for maintaining staff salaries, which often account for 90% of the total budget, and infrastructure. Thus, in spite of donor aid that makes it possible to cover the marginal costs of experiments, many NARS are still performing badly. The human capacity/institution building model must replace long-term technical assistance in order to develop national research capacity. Building national systems capable of doing adaptive as well as applied research will require continued investment in human capacity though higher education in the agricultural sciences. However, it is questionable if whether the necessary human and institutional development can take place unless national scientists are more adequately remunerated given their education level, expertise and contribution to livestock and agricultural development in general.

International agricultural research centers such as ILRI can play an enhanced role in NARS' capacity building through the training provided in collaborative research. Short term training courses are good. But they need to continue to be supplemented with increased collaborator research training in the context of projects in cross-center initiatives (e.g., the challenge programs sponsored by the CGIAR). A research environment increasingly dominated by short-term project-related research funding also creates obstacles to development of national research capacity. Short term funding is even a greater problem for livestock research given the length of the biological cycle in animals. Thus, effective human capacity/institution building strategies and adequate long-term funding for livestock research projects need to be critical components of efforts to support livestock development.

What research strategies can promote livestock development goals? Until recently the IARCs had global or continental objectives for specific commodities. The NARS' role was seen as selecting and adapting what was most useful for their own environments from IARC results and doing location-specific production system research [Lynam and Blackie (1994) cited in Ehui and Shapiro (2004)]. NARS have taken over more of the breeding functions as their breeding capacity has developed and they have fought for a larger share of international resources as they have become scarcer. The emerging regional research institutions (RRIs), meanwhile, are achieving critical mass to attempt to solve problems common to more than one developing country. These factors are causing the IARCs to re-define their roles and seek to move upstream in the technology development process. At the same time, the IARC must ensure that their work results in impact at the farm level. To accomplish this there is need for greater cooperation and closer collaboration between IARCs and NARS so that more impact is achieved with limited available resources. Among modes of functioning are the global and system wide research programs initiated by the CGIAR. ILRI has been charged with managing in the system-wide livestock program that involves closer collaboration with sister CGIAR institutions concerned with livestock research such the International center for Tropical Agriculture (CIAT), the International Center for agricultural Research in the Dry Areas (ICARDA), and the International Food Policy Research Institute (IFPRI). A continuing challenge facing ILRI and the CGIAR is effective integration of NARS in the global initiatives to help them develop their human and institutional capacities. In these new modes of functioning, to ensure efficient use of scare resources, the comparative advantage of the various partners in research needs to be taken into consideration in defining their respective roles. Research can be categorized as basic/strategic, applied and adaptive (Table 25). These categories form a continuum in the research spectrum and all has had implications for development. Basic/strategic research is scientific investigation that advances the knowledge of feasible biological processes, but may have immediate application in farming practices. In basic/strategic research the problem definition is more general, the degree of predictability of results is moderate and time of

Table 25
Characteristics of types of research and comparative advantage of research institutes that make up the global research system

Characteristics	Types of research		
	Adaptive	Applied	Strategic
Definition of problem	Specific	Specific	Specific
Predictability of results	High	Moderate	High
Likelihood of achieving impact	High	Moderate	Low–high
Applicability for impact	Narrow	Moderate	Broad
Time to impact	Short	Intermediate	Long
Types of research institutes	Comparative advantage		
	Adaptive	Applied	Strategic
NARS	+++	++	+
RRIs	++	+++	+
IARs	+	++	+++
Advanced institutes	n.a.	++	+++

Source: Ehui and Shapiro (1995).

impact are broad and long. Applied research is oriented toward achieving a practical objective, such as the genetic resistance of animals to parasites. Adaptive research refers to adjustment of technology to a particular set of farming conditions, an example being the selection of certain forage species for use as feed in a specific agro-ecological zone or region. Problem definition in adaptive research is very specific, the predictability of results is very high and the extent time of impact is narrow and short.

In the expanded global research system, NARS and RRIs have a comparative advantage in adaptive and applied research, whereas the IARCs such as ILRI and advanced institutes are better adapted to basic/strategic and applied research. This does not imply that NARS should not engage in basic/strategic research. But it must be recognized that basic research requires expensive equipment and staff skills that few SSA countries possess. IARCs are well positioned to assist NARS and RRIs with transfer of basic research results from specialized institutes in developed countries. IARCs and RRIs have comparative advantage for doing research from which results "spillover" to similar agro-ecological and socio-economic conditions across national boundaries.

One of the challenges of future research efforts will be to define strategies to increase cooperation with advanced institutes in the donor countries. Some of these already contribute to the development of the livestock sector in developing countries. Involvement of these institutions in the new modes of functioning will help determine the extent of support provided by their governments. Also effective collaboration with RRIs offers opportunities to accomplish more with the scarce resources available. Take the case of Sub-Saharan Africa where regional organizations (e.g., CIRDES in Burkina Faso and ITC in the Gambia) have been formed in an attempt to use resources more efficiently while tackling problems of a regional nature. Such institutes can enable NARS to apply a critical mass of scientist resources to a problem when this would not be possible for individual NARS due to limited manpower. Regional programs can complement the functions of NARS and IARCs and serve as mechanisms for NARS to pool resources and rationalize responsibilities in the accomplishment of individual and collective objectives. Like NARS, however, they frequently lack sufficient funds and adequate funding support will be crucial. Closer collaboration and coordination among NARS could result in greater impact and higher returns from scarce investment and human resources.

Networks, another mechanism to enhance cooperation, are maturing rapidly as an effective means of allocating and utilizing scarce research resources. Networking allows collaborating NARS partners to pool and coordinate scientific efforts, do more effective research on problems of mutual interest and avoid duplication of effort. National scientists are increasingly well trained, but there are few in the same discipline in one institute or even in one country. Multi-locational projects managed through networks provide opportunities for enhancing research efficiency and allow the introduction of standardized methodologies that lead to more significant conclusions than can be obtained from isolated, location-specific experiments.

IARCs already play a major role in networks as partners in collaborative research, providing training opportunities to network participants, disseminating research methods and results, and facilitating exchange of information. IARCs also assist with network support functions which include helping to attract donor funding, organization of meetings for setting up network steering committees, sponsoring meetings of participating scientists and providing services in areas such as data analysis, documentation and publishing. A future challenge for the expanded global livestock research system will be how to maintain and expand networks to include new NARS and RRI partners, to be able to tackle new problems and research important aspects of existing problems that have not been adequately addressed.

How to link research and technology transfer? As the number of partners and stakeholders expands the effective linkage of livestock research and technology transfer is becoming more complicated. Greater coordination and synergy between research and technology transfer will be required if impact is to be achieved [Ehui and Shapiro (1995)]. The expanding global research system will need greater interaction with the development agencies, including multilateral organization such as the World Bank, FAO, IFAD, the United Nations Development Program (UNDP), bilateral government agencies, and NGOs. The recently established multi-stakeholder platform for African Livestock Development (Alive), coordinated at the World Bank provides an excellent framework for enhancing interaction among researchers, policy makers and development practitioners. Within developing countries, governments will also have an increasingly greater say in the research and development activities that take place within their borders.

Where will the resource come from? As economies in the developing countries improve they need to play a greater role in funding research, but continued developed country contributions will clearly be required for some time. This need not be seen solely as a philanthropic or humanitarian activity as funding research in developing countries can benefit agriculture in the developed countries. This is true, for instance, in the resistance to endoparasites in some African sheep and goat breeds. Embryo transfer has allowed five African goat and sheep breeds to be introduced to Australia where these breeds will be reared for live export. Live animal exports from Africa have not been possible before due to disease considerations. The transfer of this technology occurred due to the public goods nature of research generated technologies.

There is a current debate in the donor countries on whether the private or the public sector should finance the development of agriculture in developing countries. Its origin lies in concerns for liberalization and privatization arising from the process of structural adjustment being undertaken in the developing world. The belief is that in many countries the economic situation can only improve if the public sector disengages from economic activity. This argument neglects, however, the role that government can play to correct for market failures arising from the existence of public goods.

An example of this counter argument is the significant role that research can play in economic development. There is now broad consensus that a large proportion of agricultural research must be treated as a public good and requires funding by the public sector even in countries pursuing free market philosophies. According to Ellis (1992) the reasons are: (1) most agricultural innovation (including livestock) (e.g., disease resistant animal breeds) are in the public domain after release and cannot be protected by patents or copyright laws; (2) private enterprise usually restricts itself to applied research that lend itself to copyright protection but this is a small fraction of the research needed to achieve the long-run output, equity and food security goals of society; (3) smallholders, who are often the main beneficiaries from research in developing countries, cannot easily organize and finance the scale of research required for widespread advances in technology; (4) consumers, who are the other main category of research beneficiaries, would not organize and finance agricultural research of their own volition. According to Schultz (1984), "... the only meaningful approach to modern agricultural research is to conceptualize most of its contribution as public goods. As such they must be paid for on public account, which does not exclude private gifts to be used to produce public goods." Public investment in livestock research can also be very profitable judged by the high rates of return to research of greater than 50% obtained in other parts of the world [Alston et al. (2000)].

8. Summary and conclusions

Major improvements in livestock productivity are possible and needed to assist economic growth in developing countries. Research can provide technologies to help achieve productivity increases, but transfer of technology has to take place to achieve impact. In this chapter we assess livestock productivity growth in the developing countries and consider issues related to the roles of the various partners in livestock research and modalities to ensure more effective linkage between research and development agencies to improve technology transfer and impact.

We begin the chapter by characterizing livestock development in different phases, each of those phases with different demands in terms of production technologies, policies and institutions. The first phase is a food self-sufficiency phase where households produce low levels of grain and deficiencies in grain, and labor and credit markets are severe. Rural people are unable to participate fully in livestock output and input markets because of low production values and high transaction costs. The second phase develops with increases in crop productivity. There is a boost in income for the rural household population. As households produce enough grain and accumulate enough wealth, they have an incentive to expand livestock production. There is increased crop–livestock interactions and integration. The third phase also called the industrialization phase occurs with income growth. Higher incomes trigger increases in demand for high-quality, high value-added diverse products and generate incentives for vertical coordination and integration of the livestock food chain. In general several factors combine together to impact on livestock development.

A key factor driving the commercialization process of livestock production is demand changes related to income growth and urbanization. Associated to economic growth mainly in East and Southeast Asia, there has been a rapid increase in demand for livestock products over the past two decades. During that period the food sector in many developing countries has experienced dramatic structural change with respect to livestock production, consumption and trade. Consumers in these rapidly growing economies tend to prefer an increasingly diverse diet, and expenditures on items such as meats, beverages, and fruit tend to grow faster than food staples such as cereals and legumes.

Since demand drives the process of livestock development, development of input markets and market institutions and technology availability are necessary to allow the supply response. Consequently livestock production patterns changed to match demand.

Livestock production in the developing countries increased from 36% during 1961–1981 to almost 50% of global livestock production during 1981–1999. These changes in livestock production patterns are explained by a phenomenal growth of livestock production in Asia. On the other extreme, the growth performance of the livestock sector in Sub-Saharan Africa in the last four decades has been poor.

With regard to the product structure of growth, three broad groupings of trends can be distinguished. Poultry meat production has expanded almost nine fold between 1961 and 2001. This quite remarkable growth in output has been achieved through rapid expansion of industrial ("landless") chicken rearing and processing facilities located in peri-urban areas throughout the world. These enterprises in turn depend on supplies of quality grain-based feedstuffs from national or international markets.

While growth in the poultry meat sector has been relatively consistent since the early 1960s, the output of eggs and pork was slower both in its takeoff and in its subsequent growth, with higher and more sustained growth starting only in the early 1980s. Pig and poultry meat each now account for about a third of all meat produced worldwide, and more than one half of total pig production is in China.

Growth in milk (cattle and buffalo), beef, and mutton and goat meat production has, on the whole, kept pace with population growth rates, and average per capita global production has stayed relatively constant over the last 40 years. Milk production has risen faster in developing than in industrial countries, but still lies far below the 264 kilograms per capita per year of industrial countries. Annual per capita production of beef increased in developing countries while in industrial countries, despite the large-scale switch to poultry meat, annual per capita beef production edged up from 19.6 kilograms in 1961 to 22.4 in 2001. Almost the entire expansion in output from poultry and pigs, globally, and from beef and milk cattle in industrial countries, has taken place in intensive, industrial production systems.

The contribution of technical change and productivity to these changes has varied across regions, species and production processes. Biological differences and different dependence on natural resources result in constraints and differential possibilities to develop technological packages for different species and regions. The period between 1981 and 1999 shows the highest dynamism in TFP growth, while Asia is the region showing the fastest expansion in TFP, followed by Latin America, while Sub-Saharan Africa has remained stagnated for most of the past 40 years. Most countries show larger average productivity growth gains in non-ruminant production, compared to ruminant productivity gains. In countries with high labor/land ratio like China and India, changes in land productivity together with changes in output per head of animal stock drives TFP growth. In regions with low labor/land ratios like Brazil, the most significant contribution to TFP growth comes from increases in labor productivity.

The complexity of livestock production systems, and the resources involved in this process imply higher research costs and investments, and uncertain results compared to research in crops. Related to this, NARS in developing countries have focused mainly on research on ruminants, which appears to be more complex than research in other species due to biological characteristics of these species. Research in poultry production on the other hand has been mainly dominated by the private sector.

As a result livestock productivity (output per head of livestock) continued to be higher in industrial than in developing countries, but while productivity differences in poultry and pig production between high income and some of the fast growing developing countries have been reducing in the past years, large differences in productivity still persist in the case of milk and beef production with no evidence of developing countries catching-up with most productive countries. In general, Sub-Saharan Africa and South Asia have the lowest output per animal compared with other parts of the world. In Sub-Saharan Africa, milk production per animal has been declining since 1961, and in 2001, while production of beef per animal was about 65% of the world average; production of milk per animal was only 14% of the world average.

Assuming that most efforts in livestock research by NARS in developing countries are going to ruminant production, the results of these efforts have been mixed. These can be explained on the one hand by the complexities of ruminant's production processes as discussed above; and on the other hand, by the interaction of these complexities with the weaknesses of research systems and institutions in developing countries. The total number of researchers in livestock R&D in developing countries more than doubled between 1961–1980 and 1981–1985. Total expenditure in livestock R&D by developing countries increased from $416 million to $751 million in the same period, but while in Sub-Saharan Africa investment per researcher was reduced substantially, reflecting institutional problems in the region, other regions like South Asia, and Mexico showed expenditure per researchers increased during this period.

The effort in livestock R&D is also different between regions. Asia accounts for more than 75% of total researchers in developing countries even though its share in total livestock output among developing countries is 48% in 1961–1980 and increased to 62% in 1981–1985. China and India alone accounted for 67% of all researchers in developing countries during the early 1980s. Sub-Saharan Africa's share in total number of researchers is 8%, similar to its share in output, while Latin America's share of total researchers was 10% in the 1980s, which is well below its share in output.

Increasing investment in livestock research is needed. Given the expanded number of partners and stake holders involved in livestock research and development, as well as the fact that available resources may be more limited, a framework for international action to support livestock research and development is required. The goal of such a framework would be to help achieve increased and sustainable food production from animal origin and regenerate more income for improved food security in low-income countries. This goal fits within the global mandate of ILRI, the CGIAR center in charge of livestock research, which although its main function is research, cannot ignore the need to link its research to technology transfer efforts if it is to make an impact. Such a framework would ensure overall coordination of all the partners and stakeholders involved by rationalizing their roles and more efficiently organizing their efforts. This framework for action will need to include research and technology transfer policies, as well as strategies for action. It will require measures to promote an increase in livestock research investment within the framework of a human capacity and institution-building model to replace the technical assistance model that has been dominant up to now. An effective mix of the various types of research and the strengths of all the partners in the global research and development system need to be included in the strategies chosen to promote livestock development goals. Models such as the CGIAR system-wide livestock and the World Bank 'Alive' initiatives provide useful examples.

Appendix A: Estimating the Malmquist index

International comparisons of livestock Total Factor Productivity (TFP) measures are difficult because of data constraints and technical dependencies that limit our ability to allocate inputs across sectors and hence our ability to measure sectoral productivity. It is in the face of these limitations that we use an alternative approach to measure commodity-specific efficiency and productivity [see Nin et al. (2003)]. This approach adapts a directional efficiency measure to focus on a single commodity at a time and does not require allocation of *all* the inputs to specific outputs. On the other hand, where such allocations are available (e.g., feed usage, crop land, etc.), we can take advantage of them. Specifically, we construct non-parametric, output-oriented distance functions that evaluate output efficiency for a specific commodity with respect to allocatable and unallocable inputs, given output levels for all other commodities in the sector. These distance functions are used to estimate a Malmquist index to measure productivity growth in an output-specific direction.

The production technology $\boldsymbol{S}$ describes the possible transformations of inputs $\boldsymbol{x} \in R_+^N$ into outputs $\boldsymbol{y} \in R_+^M$: $\boldsymbol{S} = \{(\boldsymbol{x}, \boldsymbol{y})$ such that $\boldsymbol{x}$ can produce $\boldsymbol{y}\}$. For all $\boldsymbol{x} \in R_+^N$, define the output possibility set as the set of output vectors that can be produced with input vector $\boldsymbol{x}$:

$$P(\boldsymbol{x}) = \left\{\boldsymbol{y} \in R_+^M \mid (\boldsymbol{y}, \boldsymbol{x}) \in \boldsymbol{S}\right\}. \tag{A.1}$$

The frontier of the output possibilities for a given input vector is defined as the output vector that cannot be increased by a scalar multiple without leaving the set. Shephard's output distance function is defined as the reciprocal of the maximum proportional expansion of output vector $\boldsymbol{y}$ given input $\boldsymbol{x}$. That is,

$$D_{\mathrm{O}}(\boldsymbol{x}, \boldsymbol{y}) = \left(\sup\left\{\theta\colon (\boldsymbol{x}, \theta\boldsymbol{y}) \in \boldsymbol{S}\right\}\right)^{-1}, \tag{A.2}$$

where θ is the coefficient dividing $\boldsymbol{y}$ to get a frontier production vector given $\boldsymbol{x}$. This function offers a complete characterization of the technology, because $\boldsymbol{y} \in P(\boldsymbol{x}) \Leftrightarrow D_{\mathrm{O}}(\boldsymbol{x}, \boldsymbol{y}) \leqslant 1$ as shown by Färe and Primont (1995).

In contrast to Shephard's output distance function, which seeks to increase all outputs simultaneously, the directional distance function allows output expansion in a specified direction [Chambers, Chung and Färe (1998, 1996); Chung, Färe and Grosskopf (1997); Färe and Grosskopf (2000)].

We use the directional distance function to define an output-oriented, product-specific directional Malmquist index. This index can also be decomposed into an efficiency component and a technical change component. The Efficiency Index measures the change in the gap between observed production and maximum potential production between period t and $t + 1$, and the Technical Change Index captures the shift of technology between the two periods. A value of the efficiency component of the Malmquist index greater than one means that the production unit is closer to the frontier in period $t + 1$ than it was in period t; the production unit is catching-up to the frontier. A value less than one, indicates efficiency regress. The same range of values is valid for the technical change component of total productivity growth. Technical progress is observed when the value is greater than one and technical regress when the index is less than one.

As with the conventional Malmquist index, the DM index measure indicates productivity improvements if their values are greater than one and decreases in productivity if the values are less than one.

As indicated in the text data are from FAOSTAT 2003, the database from the Food and Agriculture Organization of the United Nations. The data are for 115 regions (92 developing and 23 high income regions) considering two outputs (livestock and crops), and seven inputs (feed, animal stock, pasture, land under crops, fertilizer, tractors and labor).

Table A.1
Comparison of estimated share based TFP indexes with Malmquist measures available

Country	Malmquist				Shared based[a]	
	1961–1999	1961–1980	1981–1999	1961–2001	1961/80	1981/01
Algeria	−0.0075	−0.0045	−0.0106	0.0329	0.0408	0.0249
Angola	0.0024	−0.0128	0.0186	−0.0057	−0.0005	−0.0108
Bolivia	0.0146	0.0009	0.0292	0.0210	0.0281	0.0139
Botswana	0.0027	0.0099	−0.0047	0.0072	0.0078	0.0065
Brazil	−0.0045	−0.0238	0.0162	0.0216	0.0071	0.0361
Burkina	−0.0138	−0.0167	−0.0107	0.0130	−0.0089	0.0349
Central Africa	0.0351	−0.0038	0.0777	0.0267	0.0298	0.0236
Chad	−0.0071	−0.0095	−0.0046	0.0166	0.0084	0.0248
Chile	−0.0030	0.0054	−0.0117	0.0106	0.0024	0.0187
China	0.0290	0.0032	0.0568	0.0459	0.0258	0.0659
Colombia	0.0280	0.0166	0.0401	0.0137	0.0049	0.0224
El Salvador	0.0027	−0.0008	0.0065	0.0150	0.0199	0.0100
Ethiopia	−0.0024	−0.0085	0.0041	0.0019	−0.0037	0.0074
Guatemala	0.0033	0.0010	0.0057	0.0031	0.0090	−0.0028
Guinea	−0.0156	−0.0227	−0.0080	0.0184	0.0105	0.0263
Honduras	0.0030	0.0007	0.0054	0.0199	0.0207	0.0191
India	0.0030	−0.0105	0.0174	0.0265	0.0263	0.0266
Iran	−0.0292	−0.0274	−0.0311	0.0369	0.0237	0.0500
Iraq	−0.0338	−0.0246	−0.0433	−0.0228	0.0125	−0.0581
Kenya	0.0029	−0.0035	0.0096	0.0137	0.0164	0.0109
Libya	−0.0031	0.0023	−0.0087	0.0139	0.0315	−0.0038
Madagascar	0.0042	0.0023	0.0062	0.0061	0.0062	0.0059
Malawi	−0.0080	0.0131	−0.0298	−0.0090	−0.0029	−0.0150
Mali	−0.0076	0.0163	−0.0322	0.0175	0.0314	0.0035
Mauritania	0.0120	0.0202	0.0034	0.0101	0.0069	0.0133
Mexico	0.0034	0.0039	0.0028	0.0233	0.0302	0.0163
Mongolia	0.0153	0.0114	0.0194	0.0054	0.0109	−0.0002
Morocco	−0.0069	0.0056	−0.0198	0.0096	0.0036	0.0156
Mozambique	−0.0122	0.0053	−0.0303	0.0247	0.0407	0.0087
Namibia	0.0107	0.0393	−0.0185	0.0301	0.0381	0.0221
Niger	0.0057	0.0217	−0.0109	0.0118	0.0073	0.0162
Pakistan	0.0193	0.0099	0.0294	0.0258	0.0117	0.0398
Panama	−0.0021	−0.0203	0.0173	0.0155	0.0161	0.0149
South Africa	0.0069	−0.0073	0.0220	0.0248	0.0305	0.0191
Sierra Leone	0.0125	−0.0034	0.0295	0.0248	0.0137	0.0358
Sudan	0.0058	−0.0054	0.0178	0.0128	0.0131	0.0124
Syria	−0.0222	0.0043	−0.0493	0.0165	0.0262	0.0067
Tanzania	0.0010	−0.0103	0.0130	−0.0089	−0.0055	−0.0123
Tunisia	−0.0066	0.0021	−0.0158	0.0275	0.0229	0.0321
Turkey	−0.0262	−0.0233	−0.0293	0.0068	0.0143	−0.0007
Uruguay	0.0084	0.0104	0.0062	0.0011	−0.0032	0.0053
Venezuela	0.0077	0.0155	−0.0006	0.0224	0.0341	0.0107
Zambia	−0.0005	0.0050	−0.0064	−0.0092	−0.0042	−0.0141
Zimbabwe	−0.0074	−0.0020	−0.0131	−0.0040	0.0040	−0.0119

[a]From Evenson and Dias Avila.

The specific definition of these variables is given as follows: the quantity of livestock production is in millions of 1979–1981 international dollars. FAO's livestock production index estimated for each country is scaled using the value of livestock output for 1980. The quantity of crop production is in millions of 1979–1981 international dollars. FAO's crop production index is used as in the case of livestock. The quantity of feed is in metric tons of total protein supplied to animals per year. Amounts of edible commodities (cereals, bran, oilseeds, oilcakes, fruits, vegetables, roots and tubers, pulses, molasses, animal fat, fish, meat meal, whey, milk and other animal products from FAOSTAT food balance sheets) fed to livestock during the reference period, whether domestically produced or imported, are transformed into protein quantities using information of feed protein content. The quantity of animal stock is in number of cattle, sheep, goats, pigs and chicken in cow-equivalent units. Given the variability of body sizes of the main animal species across geographical regions, animal units are standardized for comparisons across the world following Seré and Steinfeld (1995). The quantity of pasture is in hectares of land used permanently (five years or more) for herbaceous forage crops, either cultivated or growing wild (wild prairie or grazing land). The quantity of land under crops is in hectares of land under temporary crops (double-cropped areas are counted only once), temporary meadows for mowing or pasture, land under market and kitchen gardens, land temporarily fallow (less than five years), land cultivated with permanent crops such as cocoa, coffee, rubber and fruit trees. The quantity of fertilizer is in metric tons of plant nutrient consumed in agriculture. The quantity of tractors is total number of wheel and crawler tractors (excluding garden tractors) used in agriculture. The quantity of labor is the total economically active population in agriculture, engaged in or seeking work in agriculture, hunting, fishing or forestry.

Table A.1 compares measures of TFP growth obtained using the Malmquist index with share based estimates. Correlation between measures for the period 1961–1999/2001 is 0.301, significant at the 5% level. Estimates could vary significantly in comparisons at the individual country level.

References

Alston, J.M., Pardey, P.G., Roseboom, J. (1998). "Financing agricultural research: International investment patterns and policy perspectives". World Development 26 (6), 1057–1071.

Alston, J.M., Chan-Kang, C., Marra, M.C., Pardey, P.G., Wyatt, T.J. (2000). "A meta-analysis of rates of return to agricultural R&D. Ex pede herculem?". International Food Policy Research Institute, Washington, DC.

Balassa, B. (1965). "Trade liberalization and 'revealed' comparative advantage". The Manchester School of Economic and Social Studies 33, 99–123.

Beintema, N.M., Dias Avila, A.F., Pardey, P.G. (2001). "Agricultural R&D in Brazil. Policy, investments, and institutional profile". FONTAGRO-EMBRAPA-IFPRI, Washington, DC.

Bruinsma, J.E. (2003). World Agriculture: Towards 2015/2030. An FAO Perspective. Earthscan Publications, London.

Byerlee, D., Alex, G.E. (1998). "Strengthening national agricultural research systems. Policy issues and good practice". In: Environmentally and Socially Sustainable Development Series. Rural Development. World Bank, Washington, DC.

Chambers, R.G., Chung, Y.H., Färe, R. (1996). "Benefit and distance functions". Journal of Economic Theory 70, 407–419.

Chambers, R.G., Chung, Y.H., Färe, R. (1998). "Profit, directional distance functions, and Nerlovian efficiency". Journal of Optimization Theory and Applications 98, 351–364.

Chen, J., Rozelle, S. (2001). "Economic development and the evolution of backyard livestock production: A case study of hog production in China", Working paper. Department of Agricultural and Resource Economics, University of California, Davis.

Chung, Y.H., Färe, R., Grosskopf, S. (1997). "Productivity and undesirable outputs: A directional distance function approach". Journal of Environmental Management 51, 229–240.

Coyle, W., Gehlhar, M., Hertel, T., Wang, Z., Yu, W. (1998). "Understanding the determinants of structural change in world food markets". American Journal of Agricultural Economics 80 (December), 1052–1062.

Cranfield, J., Eales, J., Hertel, T.W., Preckel, P. (1998). "Changes in the structure of global food demand", GTAP Working Paper No. 01.

Delgado, C., Rosegrant, M., Steinfeld, H., Ehui, S., Courbois, C. (1999). "Livestock to 2020: The next Food Revolution", IFPRI/ILRI/FAO joint publication in IFPRI's 2020 Vision Discussion Paper Series.

Devendra, C. (1995). "Animal production systems in South-East and East Asia: Potential and challenges for research". In: Global Agenda for Livestock Research. Proceedings of a Consultation. ILRI, Nairobi, Kenya.

Devendra, C. (1996). "Present status of research and development and institutional considerations for animal production system in the Asia–Pacific region". In: Proceedings of the 8th AAAP Animal Science Congress, vol. 1. Japanese Society of Zootechnical Science.

Dimaranan, B.V., McDougall, R.A. (2002). Global Trade, Assistance, and Production: The GTAP 5 Data Base. Center for Global Trade Analysis, Purdue University.

Echeverria, R. (1998). "Agricultural research policy issues in Latin America: An overview". World Development 26 (6), 1103–1111.

Ehui, S., Shapiro, B.I. (1995). "Research and technology transfer for livestock development". In: Wilson, R.T., Ehui, S., Mack, S. (Eds.), Livestock Development Strategies for Low Income Countries. Proceedings of the Joint FAO/ILRI Roundtable on Livestock Development Strategies for Low Income Countries, ILRI, Addis Ababa, Ethiopia, 27 February–02 March 1995. Food and Agriculture Organization/International Livestock Research Institute, Nairobi, Kenya.

Ehui, S.K., Shapiro, B.I. (2004). "Improving livestock research and technology transfer in developing countries". In: Ehui, Lynam, Okike (Eds.), Adapting Social Science to the Changing Focus of International Agricultural Research. International Livestock Research Institute, Addis Ababa, Ethiopia.

Ehui, S., Benin, S., Williams, T., Meijer, S. (2002). "Food security in Sub-Saharan African to 2020", Socio-Economic and Policy Working Paper No. 49. ILRI (International Livestock Research Institute), Addis Ababa, Ethiopia.

Eicher, C.K. (2001). "Africa's unfinished business: Building sustainable agricultural research systems", Staff Paper No. 2001-10. Department of Agricultural Economics, Michigan State University.

Ellis, F. (1992). Agricultural Policies in Developing Countries. Cambridge University Press, Cambridge.

Evenson, R., Pray, C., Rosegrant, M. (1999). "Agricultural research and productivity growth in India", Research Report No. 109. International Food Policy Research Institute, Washington, DC.

Fang, C., Fabiosa, J. (2002). "Does the US Midwest has a cost advantage over China in producing corn, soybeans, and hogs?", MATRIC Research Paper No. 02-MRP 4.

FAO (2003). FAOSTAT data. http://faostat.fao.org.

Färe, R., Grosskopf, S. (2000). "Theory and application of directional distance functions". Journal of Productivity Analysis 13, 93–103.

Färe, R., Primont, D. (1995). Multi-Output Production and Duality: Theory and Applications. Kluwer Academic Publishers, Boston, MA.

Gryseels, G., Anderson, J.R. (1991). "International agricultural research". In: Pardey, P.G., Roseboom, J., Anderson, J.R. (Eds.), Agricultural Research Policy. International Quantitative Perspective. ISNAR, Cambridge University Press.

Hayenga, M.L., Schroeder, T., Lawrence, J., Hayes, D., Vukina, T., Ward, C., Purcell, W. (2000). Meat Packer Vertical Integration and Contract Linkages in the Beef and Pork Industries: An Economic Perspective. American Meat Institute, Arlington, VA.

Hayes, D.J. (1999). "China's role in world livestock and feed grain markets". CHOICES (first quarter).

Hayes, D.J., Clemens, R. (1997). "Chinese market for U.S. pork exports", Center for Agricultural and Rural Development (CARD) Publication 97-bp14. Center for Agricultural and Rural Development (CARD), Iowa State University.

Jarvis, L.S. (1986). Livestock Development in Latin America. The World Bank, Washington, DC.

Ludena, C.E., Hertel, T.W., Preckel, P.V., Foster, K., Nin Pratt, A. (2005). "Disaggregate productivity growth in livestock: A directional Malmquist Index approach", Presented at the American Agricultural Economics Association Annual Meeting, Providence, RI, July.

Martinez, S.W. (2002). "Vertical coordination of marketing systems: Lessons from the poultry, egg, and pork industries", Agricultural Economic Report No. 807. USDA.

McIntire, J., Bourzat, D., Pingali, P. (1992). "Crop–livestock interaction in Sub-Saharan Africa". In: Regional and Sectoral Studies Series. World Bank, Washington, DC.

Nin, A., Arndt, C., Hertel, T.W., Preckel, P.V. (2003). "Bridging the gap between partial and total factor productivity measures using directional distance functions". American Journal of Agricultural Economics 85, 928–942.

Nin Pratt, A. (2001). "Agricultural productivity: Measurement, forecasting and implications for trade", Ph.D. dissertation. Purdue University.

Pardey, P.G., Roseboom, J. (Eds.) (1989). ISNAR Agricultural Research Indicator Series: A Global Database on National Agricultural Research Systems. Cambridge University Press.

Pardey, P.G., Roseboom, J., Anderson, J.R. (1991a). "Regional perspectives on national agricultural research". In: Pardey, P.G., Roseboom, J., Anderson, J.R. (Eds.), Agricultural Research Policy. International Quantitative Perspective. ISNAR, Cambridge University Press.

Pardey, P.G., Roseboom, J., Anderson, J.R. (1991b). "Patterns of agricultural growth and economic development". In: Pardey, P.G., Roseboom, J., Anderson, J.R. (Eds.), Agricultural Research Policy. International Quantitative Perspective. ISNAR, Cambridge University Press.

Pardey, P.G., Roseboom, J., Beintema, N.M. (1997). "Investments in African agricultural research". World Development 25 (3), 409–423.

Pingali, P.L., Rosegrant, M.W. (1995). "Agricultural commercialization and diversification: Processes and policies". Food Policy 20 (3), 171–185.

Reardon, T., Barret, C.B. (2000). "Agroindustrialization, globalization and international development. An overview of issues, patterns and determinants". Agricultural Economics 23, 195–205.

Rukuni, M., Blackie, M.J., Eicher, C.K. (1998). "Crafting smallholder-driven agricultural research systems in Southern Africa". World Development 26 (6), 1073–1087.

Ruttan, V.W. (1997). "Induced innovation, evolutionary theory and path dependence: Sources of technical change". The Economic Journal 107 (444), 1520.

Schultz, T.W. (1984). "The economics of agricultural research". In: Eicher, C.K., Staatz, J.M. (Eds.), Agricultural Development in the Third World.

Seré, C., Steinfeld, H. (1995), in collaboration with J. Groenewold, "World livestock production systems: Current status, issues and trends". In: Global Agenda for Livestock Research. Proceedings of a Consultation. ILRI, Nairobi, Kenya.

Sidahmed, A.E. (1995). "Livestock and feed development and improvement research needs in West Asia and North Africa". In: Global Agenda for Livestock Research. Proceedings of a Consultation. ILRI, Nairobi, Kenya.

Tuan, F.C., Peng, T. (2001). "Structural changes in China's livestock and feed production: Trade implications", Paper presented in the winter IATRC Meetings, Auckland, New Zealand, January 18–19.

Vollrath, T.L. (1991). "A theoretical evaluation of alternative trade intensity measures of revealed comparative advantage". Weltwirtschaftliches Archiv 130, 265–279.

Ward, C.E. (1997). "Vertical integration comparison: Beef, pork, and poultry", Selected paper submitted to the Western Agricultural Economics Association.

Winrock International (1992). Assessment of Animal Agriculture in Sub-Saharan Africa. Winrock International Institute for Agricultural Development, USA.

World Bank (1996). "India livestock sector review: Enhancing growth and development". Agriculture and Water Operations Division, Country Department II, South Asia Region, World Bank.

Chapter 48

AGRICULTURAL INNOVATION: INVESTMENTS AND INCENTIVES*

BRIAN D. WRIGHT

Department of Agricultural and Resource Economics, University of California, Berkeley, CA

PHILIP G. PARDEY

Department of Applied Economics, University of Minnesota, St. Paul, MN

CAROL NOTTENBURG

Cougar Patent Law, Seattle, WA

BONWOO KOO

Department of Management Sciences, University of Waterloo, Ontario, Canada

Contents

* The authors thank the Universities of Minnesota and California, Berkeley and the International Food Policy Research Institute for financial support, and Steve Dehmer, Jennifer Drew, Doug Ashton, Jade Sharples, Eduardo Castelo-Magalhães, and Hannah Wright for their help in preparing this chapter.

Handbook of Agricultural Economics, Volume 3
Edited by Robert Evenson and Prabhu Pingali

DOI: 10.1016/S1574-0072(06)03048-9

Abstract

Investments in R&D and agricultural innovations have been fundamental to long-term economic growth worldwide. But global resource allocation has been uneven, with some developing countries closing in on developed-world scientific capacities, others regaining ground lost over the past decade or so, but, unfortunately, many others either stalled or slipping behind. Recently, substantial shifts in the scientific basis for much of biological research have created a new and promising set of opportunities for innovation in agricultural biotechnologies. Moreover, the institutional, regulatory and, especially, intellectual property regimes that affect agricultural R&D are also undergoing rapid change, providing enhanced private incentives for applications of biotechnology in agriculture. But the nature of these developments is raising real concerns about the extent to which developing-country agriculture will be able to partake in the benefits offered by the advances in biotechnology.

In this chapter we present an overview of the globally evolving structure of funding and performance of agricultural research. This establishes the special nature of agricultural research as a truly joint endeavor of North and South, and of the public and private sectors. We then sketch the economics of intellectual property protection – highlighting static and dynamic models and the effects of competition – and alternative incentives to innovate. A brief list of means that have been used or proposed for protecting intellectual property in agriculture is presented, followed by a discussion of the ongoing global evolution of the intellectual property environment and relevant institutions. Then we discuss instruments that are available for transacting intellectual property rights (IPRs), provide a quantitative view of trends in agricultural intellectual property protection worldwide, and end with a short conclusion.

Keywords

agricultural research and development, intellectual property rights, patents, prizes, research contracts, plant breeders rights, open source, innovation, technology licenses

JEL classification: O16, O19, O31, O32, O33, O34, O38

1. Introduction

When Parson Malthus wrote that unconstrained population growth would outrun agricultural productivity in the long run, he had history on his side. In 1798, the publication date of his *Essay on the Principle of Population*, the practice of agriculture relied on local labor and natural resources, including land, seed, water and organic fertilizer. Within the constraints of locally available knowledge and technologies, farmers had over the generations come close to optimizing their production in their own agro-ecologies. This meant little scope for further productivity advances, absent relaxation of constraints originating beyond the farm [Schultz (1964)]. Agricultural research beyond localized farmer experimentation, and collection and dispersal of foreign plants or animals, was essentially unknown. Such production increases as were still being achieved were primarily the consequences of expansion in arable land. The rate of improvement in yield, where it occurred, was slow.

Starting a century or so after Malthus' *Essay*, yield growth began to accelerate in Europe, America and Japan, and through the latter half of the 20th century spread around most of the less-developed world. Dire Malthusian predictions of famines and starvation in the 1960s by Ehrlich (1968) and Paddock and Paddock (1967), among others, proved misguided. Currently, world food oversupply and low prices are widely viewed as pressing policy problems. In an ironic reversal, some modern sages now appear to believe that supply shortages can never be a prime cause of famines, which they see primarily as a distributional phenomenon.

The widespread productivity increase of agriculture in the last century was neither anticipated nor inevitable, as is no doubt clear from other entries in this Handbook. The currently favorable dynamic balance between food supply and demand is the result of sustained interactions between farmers, input suppliers, and an overwhelmingly publicly-supported research and extension system that generated and disseminated innovations and public knowledge for free. Sharing of knowledge and innovation has been international in scope, but its local effectiveness generally has depended on local adaptive investments.

The record of research in agriculture over the last century is unlike that of research in any other sector. First, it has been globally dispersed across regions and levels of development. The enterprise has been characterized by regional interdependence and widespread international exchange of genetic resources, education and training, and by multilateral sharing of discoveries and technologies. Second, this activity has sustained a rate of return that is the envy of other sectors, though an embarrassment to econometricians who time and again find rates of return that seem too high to be true [Alston et al. (2000); Pardey et al. (2004)].

In the new millennium, the recent history of technical change in agriculture must again be viewed with skepticism as a guide to the future. The sustained uptrends in farm productivity in most regions, and in world agricultural output, and the downtrends in prices, make the case for continued public support less than obvious to the citizens who are asked to fund continued public research. As a result, public funding of research institutions is softening just as important new opportunities and challenges are emerging in the agricultural research environment, and substantial investments in maintenance research are required simply to sustain past productivity gains [Adusie and Norton (1990)]. In particular, changes in intellectual property rights and important new opportunities for innovations in agricultural biotechnology have emerged in the developed North, particularly in the United States, and quickly attained global significance.

In 1980, the sea change in the intellectual property protection of agriculture was heralded in the United States by the award of a fundamental patent, a key Supreme Court Decision and an Act of Congress. In that year the U.S. Supreme Court held that the class of patentable subject matter included living organisms in the landmark *Diamond v. Chakrabarty* decision.[1] In addition, the 1980 Bayh–Dole Act established the general right of grant recipients (e.g., universities) to apply for patents on most federally-funded research.[2] Finally, Cohen and Boyer, after almost a decade of examination and continuances, were awarded U.S. patent (number 4,237,224) in 1980 on their recombinant DNA technology. Their federal funding contract happened to allow patenting by the researchers, pre-Bayh–Dole, but their work appears to have been unaffected by any patent incentives; only at the last minute were they persuaded to apply for a patent for their innovations.

In the same decade the U.S. federal court structure was reformed. The newly established Court of Appeals of the Federal Circuit was given jurisdiction over all patent appeals [Jaffe (2000)], and the result was litigation outcomes more favorable, on average, to patentees. Later in the decade the patentability of both plants and animals was explicitly confirmed.

Thus, by coincidence, the patentability of life forms, including plants and animals, was confirmed in the United States just as detection of infringement of life form patents was becoming technologically more feasible, and the prospects for usefully applying biotechnology were rapidly expanding. Until the 1980s, patents on specific genetic characteristics of plants, animals or other life forms would have been problematic if not impossible to enforce due to the difficulty of verifying that the genetic material of the life form in question was within the scope of the patent claims.

Subsequently, biotechnology itself began producing a new stream of potentially valuable innovations, fueled by a surge in health research funding, via the U.S. National Institutes of Health, that had widespread political support. Large investments were also made in the private sector, often founded on technology initiated in universities with

[1] In the early 1970s General Electric brought a powerful test case on the patentability of a living organism (specifically an oil eating bacterium). In 1980, the U.S. Supreme Court (in the case of Diamond v. Chakrabarty) ruled in GE's favor, and although the bacterium was never commercialized, a legal basis for patenting life forms had been established.

[2] Some of the pressures for strengthening U.S. patent law and expanding its scope originated outside of agriculture. They included the pessimistic perception that the United States had lost its technological edge in the 1970s and that other countries were insufficiently compensating the United States for past innovations. See Landes and Posner (2003) for more perspectives.

public funding. Participants included startups funded by venture capitalists, who typically demanded evidence of patenting as a condition for investment, and existing pharmaceutical and chemical companies. The majority was interested in health applications, though agricultural applications, initially thought to be a complementary component of a "life sciences" approach, also attracted considerable interest.

Thus the changes in the 1980s in patent law and in biotechnology were complementary in their effects on subsequent agricultural research incentives in the United States. Other developed nations followed, to various degrees, the United States' lead in allowing patents on life forms. In the 1990s, the Trade-Related Aspects of Intellectual Property Rights (TRIPS) agreement of the World Trade Organization (WTO) ensured that adoption of intellectual property protection relevant to agricultural research would quickly become a global phenomenon.[3]

In this chapter we present an overview of the globally evolving structure of funding and performance of agricultural research. We also review means of intellectual property protection relevant to agriculture, and identify some effects of the dynamic co-evolution of the legal and scientific environments of agricultural research. This research environment is unique in its global nature, in its continuing strong public and nonprofit involvement, and in its history of sustained high research productivity.

At the outset it is important to establish clearly the limits of the scope of this chapter. Handbook chapters typically include very useful, exhaustive literature reviews. We do not attempt this here, even with respect to the topics we address. Nor do we discuss important issues such as the politics and economics of research support and the determinants of the direction and biases of innovation, the diffusion of innovation [see Sunding and Zilberman (2001)], techniques of measurement of research benefits and their distributional effects [see Alston, Norton and Pardey (1995) and references therein], the interactions between public and private support [for example, Ulrich, Furtan and Schmitz (1986)], public/private collaborations [Bush (1945); Rausser (1999)], and the effects of market regulation on innovation [for example, Ulrich, Furtan and Schmitz (1987); Constantine, Alston and Smith (1994)]. In addition, our focus is narrowly on the supply of food and fiber. We have not been asked to address important environmental, ecological, industrial, nutritional, and ethical considerations.

The chapter is structured as follows. We begin with an overview of the structure and levels of funding of agricultural research worldwide. This establishes the special nature of agricultural research as a truly joint endeavor of North and South, and of the public and private sectors. We then sketch the economics of intellectual property protection – highlighting static and dynamic models and the effects of competition – and alternative incentives to innovate. After presenting a brief list of means that have been used or proposed for protecting intellectual property in agriculture, we discuss the

[3] Building on ideas broached by Boyle (2003), Runge (2004) draws conceptual and practical parallels between the English land enclosure movements of the seventeenth and eighteenth centuries, and the late twentieth century efforts to enclose the genomic and related intellectual properties used in agriculture.

ongoing global evolution of the intellectual property environment and relevant institutions. Then we discuss instruments that are available for transacting intellectual property rights (IPRs), provide a quantitative view of trends in agricultural intellectual property protection worldwide, and end with a short conclusion.

2. Research investments and the structure of funding

The public and private roles in agricultural science are evolving, reflecting changes in economic conditions in the broader economy as well as in agriculture, in intellectual property rights, and in public attitudes and perceptions. While many elements of these changes reflect global influences, there have been some important divergences among countries as well.

2.1. International trends[4]

Looking back over the past three decades, worldwide, public investments in agricultural research nearly doubled in inflation-adjusted terms, from an estimated \$11.8 billion (1993 international dollars) in 1976 to nearly \$21.7 billion in 1995 (Figure 1).[5] These data reveal a modern historical first: during the 1990s, developing countries as a group undertook more of the world's public agricultural research than the developed countries.

What the regional totals fail to reveal is that the public spending was concentrated in only a handful of countries. Just four countries – the United States, Japan, France, and Germany – accounted for two-thirds of the \$10.2 billion of public research done by rich countries in 1995, about the same as two decades before. Similarly, just four of the developing countries – China, India, Brazil, and South Africa – spent 44% of the developing world's public agricultural research money in 1995, up from 35% in the mid-1970s.

Despite a pattern of strong longer-term growth in spending since the 1970s, for many parts of the world the rapid and quite pervasive growth in spending during the 1970s and early 1980s gave way to a dramatic slowdown in the first half of the 1990s. In the rich countries, public agricultural research investment grew just 0.2% annually between 1991 and 1996, compared with 2.2% per year during the 1980s. In Africa, there was no growth at all – the continuation of a longer-run trend, which began after rapid growth in spending in the 1960s gradually was halted by debt crises in the 1980s and curbs on government spending and waning donor support in the 1990s. Preliminary results from new surveys of African countries suggest no substantial recovery in the latter part of the 1990s, with spending totals shrinking for the region during the decade ending 2001 if Nigeria and South Africa are excluded [Beintema and Stads (2004)]. Spending in Asia

[4] This section draws heavily on Pardey and Beintema (2001).

[5] All these data involved conversions from local currency units to U.S. dollar equivalents using purchasing power parities to account for cross-country price differentials, rather than using market exchange rates to do the conversions. See Pardey, Roseboom and Craig (1992) for details.

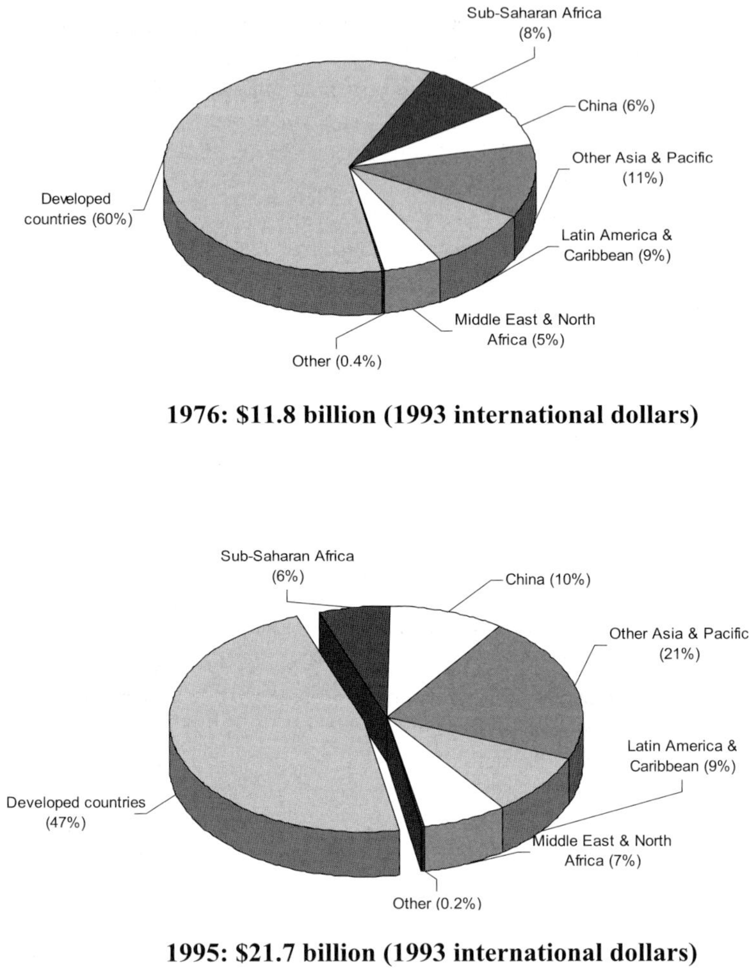

Figure 1. Global agricultural research expenditures, 1976 and 1995. *Source*: Adapted from Pardey and Beintema (2001). *Note*: Agricultural research intensities represent research spending as a percentage of corresponding agricultural gross domestic product.

grew by an average of 4.4% per year during 1991–1996, compared with 7.5% annually during the previous decade. Growth slowed in the Middle East and North Africa as well.

China is an exception. Growth in spending during the first half of the 1990s rebounded from a period of stagnation during the last half of the 1980s.[6] Things look a little better in Latin America, too, with growth in spending of 2.5% per year from 1991

[6] China has developed its agricultural biotechnology capacity as well [see, for example, Huang et al. (2002)].

to 1996, following little or no growth during the previous decade. But the recovery in Latin America seems fragile, and is not shared widely throughout the region. Public research in countries like Brazil (with public spending approaching a billion dollars a year, a considerably larger commitment than any of the developed countries besides the United States and Japan) and Colombia did better in the early 1990s but suffered cutbacks in the later part of the decade. Many of the poorer (and smaller) countries have failed to experience any sustained growth in funding for the past several decades.

Geography is one way to group countries; another is to group them according to per capita income. Spending by low-income countries grew fastest, so their combined share of the global total increased from 19% in 1976 to 28% in the mid-1990s. Note that this trend reflects the comparatively rapid growth of India and China, two large countries whose expenditures dominate the group average. In fact, the other low-income countries, as a group, lost a little ground. Their share of global agricultural R&D spending dropped from 8.7% in 1976 to 8.3% in 1996.

2.2. *Research intensities*

Turning now from absolute to relative measures of R&D investments, developed countries as a group spent $2.64 on public agricultural R&D for every $100 of agricultural output in 1995, a sizable increase over the $1.53 they spent per $100 of output two decades earlier. Since 1975, research intensities rose for the developing countries as a group, but unevenly. Despite having gained a greater absolute share of the developing world's total agricultural research spending, China's agricultural research intensity in the mid-1990s was no greater than in the mid-1980s. In other words, China's research spending grew, but its agricultural sector grew just as quickly. Although public research throughout the rest of Asia and Latin America appears to have grown in intensity during the last decade of our data, Africa lost considerable ground, with research intensities now lower than in the 1970s (Figure 2).

Other research intensity ratios are also revealing (Table 1). Rich countries spent over $590 per agricultural worker, more than double the corresponding 1976 ratio. Poor countries spent just $8.50 per agricultural worker in 1995, less than double the 1976 figure. These rich–poor country differences are, perhaps, not too surprising. A much smaller share of the rich-country workforce is employed in agriculture, and the absolute number of agricultural workers declined more rapidly in rich countries than it did in the poor ones. Agricultural research spending per capita rose, too, by an average of 25% for developed countries (from $9.60 per capita in 1975 to $12.00 in 1995) and 79% in developing countries (from $1.50 per capita in 1975 to $2.50 in 1995). Notably, per capita research spending (in terms of both total population and agricultural workers) declined in Africa, the only region of the world where this occurred.[7]

[7] Roe and Pardey (1991) provide a political economy perspective on these various research intensity ratios. See also Alston and Pardey (1993) and references therein.

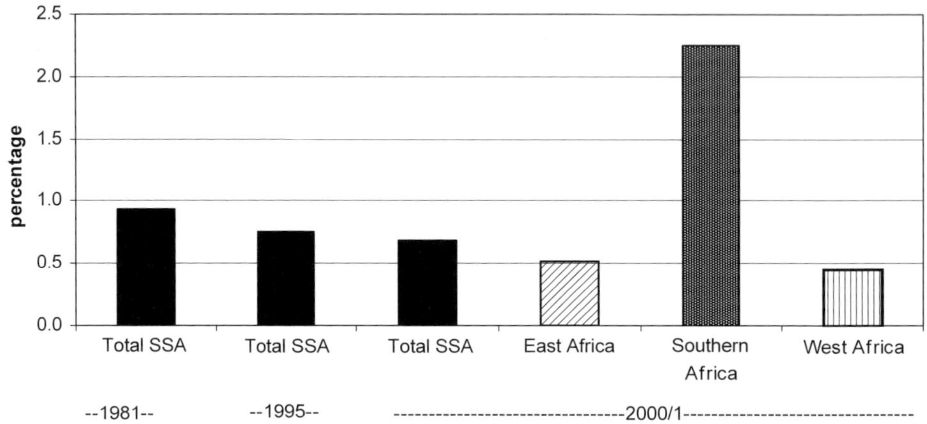

Figure 2. Public agricultural R&D intensity ratios. *Source*: Pardey and Beintema (2001) and Beintema and Stads (2004). *Note*: Data for West Africa, with the exception of Nigeria, are for 2001.

Table 1
Alternative agricultural research intensity ratios, 1976–1995

Region/country	Expenditures as a share of AgGDP (%)			Expenditures per capita (1993 international dollars)			Expenditures per economically active agricultural population (1993 international dollars)		
	1976	1985[a]	1995[a]	1976	1985	1995	1976	1985	1995
Developing countries	0.44	0.53	0.62	1.5	2.0	2.5	4.6	6.5	8.5
Sub-Saharan Africa	0.91	0.95	0.85	3.5	3.0	2.4	11.3	10.6	9.4
China	0.41	0.42	0.43	0.7	1.3	1.7	1.8	3.1	4.1
Other Asia	0.31	0.44	0.63	1.1	1.7	2.6	3.8	6.1	10.2
Latin America	0.55	0.72	0.98	3.4	4.0	4.6	26.0	36.0	45.9
Developed countries	1.53	2.13	2.64	9.6	11.0	12.0	238.5	371.0	594.1
Total	**0.83**	**0.95**	**1.04**	**3.3**	**3.8**	**4.2**	**12.9**	**15.3**	**17.7**

Sources: Pardey and Beintema (2001). Agricultural GDP data are from World Bank (2002); total and economically active agricultural population data are from FAO (2000).
Notes: Data are based on provisional estimates of public agricultural R&D expenditures and exclude Eastern European and countries of the former Soviet Union. The developing-country total includes Greece, designated as a middle-income country in 1996 by World Bank (1996) criteria used here to group the countries. Developed countries include only high-income countries specified by the World Bank in 1996, the latest year of the data series. 1985 and 1995 expenditure data are three-year averages for 1984–1986 and 1994–1996, respectively.

Table 2
Private and public agricultural R&D investments, circa 1995

Region	Expenditures (million 1993 international dollars)			Shares (%)		
	Public	Private	Total	Public	Private	Total
Developing countries	11,469	672	12,141	94.5	5.5	100
Developed countries	10,215	10,829	21,044	48.5	51.5	100
Total	**21,692**	**11,511**	**33,204**	**65.3**	**34.7**	**100**

Source: Pardey and Beintema (2001).
Note: Combining estimates from various sources resulted in unavoidable discrepancies in the categorization of "private" and "public" research. For example, data for private spending in Asia included nonprofit producer organizations, whereas we included research done by nonprofit agencies under public research, when possible, in Latin America and elsewhere.

2.3. Private and public research roles

By the mid-1990s, about one-third of the $33 billion total investment in agricultural research worldwide was contributed by private firms, including those involved in providing farm inputs and processing farm products (Table 2). But little of this research took place in developing countries. The overwhelming majority ($10.8 billion, or 94% of the global total) was conducted in developed countries. In the less-developed countries, the private share of research was just 5.5%. Public funds account for about half of the total support in rich countries. Agricultural research is one area where the public sector, worldwide, dominates the private sector as a source of support. More than one-half of the world's public agricultural R&D dollars is spent in developing countries, while only one-third of the public plus private research spending occurs in that part of the world. These facts help put the scope of the role of private research in agriculture into proper perspective.

In addition, the research intensity gap between rich and poor countries is wide and growing. As we saw, in 1995, public research intensities were four times higher in rich countries than they were in poor ones; if total (that is, private and public) spending is considered, the gap grows to more than eightfold, with rich countries spending about $5.40 on agricultural R&D per $100 of agricultural GDP.

2.4. Research knowledge stocks

The eightfold difference in total research intensities is an indication of the present gap between rich and poor countries in the flow of funding of research. However, it is the size of the accumulated stock of knowledge – not merely the amount of investment in current research and innovative activity – that provides a more meaningful measure of a country's technological capacity and a better account of cross-country differences in

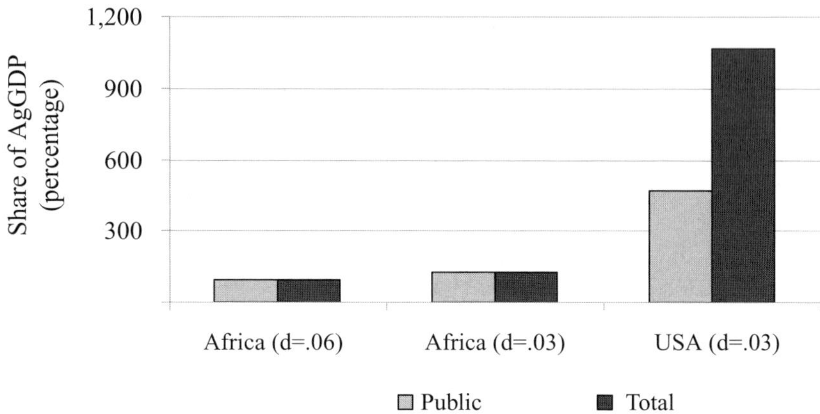

Figure 3. African and American stocks of research knowledge, 1995. *Source*: Pardey and Beintema (2001). *Note*: The lag length relating innovations, I_t, to present and past research expenditures, R_{t-s}, was taken to be 10 years for both regions, so the stock of knowledge for year t, K_t, was formed as $K_t = (1-d)K_{t-1} + I_t$, where d is the rate of knowledge depreciation and $I_t = \sum_{j=0}^{10} R_{t-j}$.

agricultural productivity. As noted above, agricultural scientific research tends to be a cumulative endeavor. Innovations beget new ideas and further rounds of innovation and additions to the cumulative stock of knowledge.[8]

The current stock of knowledge and the contribution of past research spending to that stock are sensitive to the types of science being done, the institutional structures surrounding the science, and the economic context that affects the use of this stock. Some science spending makes persistent and even perpetual contributions to the changing stock of locally produced knowledge: the same spending in societies continually ravaged by wars, institutional instability, and outright collapse may have a much more ephemeral effect.[9]

Figure 3 represents money measures of the stock of scientific knowledge based on research performed in the United States (assuming a baseline rate of depreciation of the knowledge stock of 3% per annum) and Africa. Knowledge stocks in 1995 – representing a discounted accumulation of research spending from 1850 for the United States and 1900 for Africa – are expressed as percentages of 1995 agricultural GDP to

[8] The sequential and cumulative nature of scientific progress and knowledge is starkly illustrated by crop improvement. It typically takes 7 to 10 years of breeding to develop a uniform, stable, and superior crop variety, and in doing so, today's breeders build on an accumulation of knowledge and improved germplasm built up by their predecessors.

[9] There can be catastrophic losses, too, tied to the political instability that is also a cause of hunger. Civil strife and wars cause an exodus of scientific staff, or at least a flight from practicing science. Much of Uganda's scientific facilities, for example, was in shreds when its civil war ended in the early 1980s. It is hard to imagine that today's Congo once had one of the most sophisticated scientific infrastructures in colonial Africa, comparable to the facilities and quality of staff found in most developed countries at the time.

normalize for differences in the sizes of the respective agricultural sectors. Assuming the 3% baseline depreciation rate, this accumulated stock of knowledge in the United States was about 11 times the amount of agricultural output produced in 1995. In other words, every \$100 of agricultural output was supported by a stock of knowledge of around \$1100. In Africa, the stock of knowledge in 1995 was actually less than the value of African agricultural output that year. The ratio of the U.S. knowledge stock relative to U.S. agricultural output in 1995 was nearly 12 times higher than the corresponding amount for Africa. If a depreciation rate of 6% instead of 3% is used, the gap in American and African ratios is more than 14-fold.

2.5. International dimensions of agricultural R&D

2.5.1. Conduct of research

United States Secretary of Agriculture Henry Wallace was sent on a diplomatic visit to Mexico by President Roosevelt in late 1940, just prior to his inauguration as Vice President in the Roosevelt administration. Secretary of Agriculture Henry Wallace (founder of Pioneer Hi-Bred International) traveled for one month around Mexico. He returned very concerned with the condition of the farmers and their lack of access to modern agricultural technologies, and brought this issue to the attention of the Rockefeller Foundation. Subsequently, donor initiatives to strengthen agricultural R&D in developing countries resulted in the foundation of international agricultural research centers (IARCs). In 1971 the supporting donors formed the Consultative Group on International Agricultural Research (CGIAR) (Table 3). The CGIAR is an informal partnership that now includes developing and developed countries, private foundations, and regional and international organizations co-sponsored by the Food and Agriculture Organization of the United Nations, the International Fund for Agricultural Development, the United Nations Development Program, and the World Bank. It provides oversight to a system of international research centers, a mechanism for collectively funding those centers, and a forum to discuss and affirm overall research policy objectives [Alston, Dehmer and Pardey (2005)].

The CG system began modestly. Between 1960 and 1964, of the institutes that would become the CG, only IRRI was operating as such. After an initial expenditure of \$7.4 million in 1960, total spending was \$1.3 million per year in 1965. By 1970, the four founding centers – IRRI, CIMMYT, IITA, and CIAT – were allocated a total of \$14.8 million annually. The progressive expansion of the total number of centers, and the funding per center, during the next decade involved a tenfold increase in nominal spending, to \$141 million in 1980. During the 1980s, spending continued to grow, more than doubling in nominal terms to reach \$305 million in 1990. The rate of growth had slowed but was still impressive. In the 1990s, however, although the number of centers grew – from 13 to 18 at one point, but now 15 – funding did not grow enough to maintain the level of spending per center, let alone the growth rates.

Table 3
CGIAR supported centers

Center	First year of CGIAR support	Foundation date	Location of headquarters	Main areas of focus: Commodity	Activities	Region	2004 expenditures (million US$)
International Rice Research Institute (IRRI)	1971	1960	Los Baños, Philippines	Rice		World	32.9
					Rice-based ecosystems	Asia	
International Maize and Wheat Improvement Center (CIMMYT)	1971	1966	El Batan, Mexico	Wheat and maize		World	41.1
International Center for Tropical Agriculture (CIAT)	1971	1967	Cali, Colombia	Phaseolus bean and cassava		World	36.7
					Tropical pastures	Lowland tropics	
				Rice		Latin America	
International Institute of Tropical Agriculture (IITA)	1971	1967	Ibadan, Nigeria	Cassava, cowpea, maize, plantain and banana, soybean, and yam		World	42.6
					Smallholder cropping and postharvest systems	Dry, moist and mid-altitude savannas, and humid forest	
International Crops Research Institute for the Semi-Arid Tropics (ICRISAT)	1972	1972	Patancheru, India	Sorghum, millet, pigeonpeas, chickpeas, groundnuts		World	26.8
					Farming systems	Semi-arid tropics (Asia, Africa)	
International Potato Center (CIP)	1973	1970	Lima, Peru	Potato, sweet potato, and other root crops		World	21.5
International Laboratory for Research on Animal Diseases (ILRAD)	1973	1973	Nairobi, Kenya		See ILRI		n.a.
International Livestock Center for Africa (ILCA)	1974	1974	Addis Ababa, Ethiopia		See ILRI		n.a.

(continued on next page)

Table 3
(*continued*)

Center	First year of CGIAR support	Foundation date	Location of headquarters	Main areas of focus			2004 expenditures (million US$)
				Commodity	Activities	Region	
International Plant Genetic Resources Institute (IPGRI)[a]	1974	1974	Rome, Italy		Promote activities to further collection, conservation, evolution, and utilization of germplasm	World	32.0
Africa Rice Center (WARDA)[b]	1975	1971	Bouaké, Côte d'Ivoire	Rice		Sub-Saharan Africa	10.1
International Center for Agricultural Research in the Dry Areas (ICARDA)	1977	1977	Aleppo, Syria	Barley, lentils, faba beans		North Africa and Near East	24.6
					Farming systems	North Africa and Near East, World	
				Wheat, kabali chickpeas		North Africa and Near East	
International Service for National Agricultural Research (ISNAR)[c]	1980	1979	The Hague, Netherlands		Strengthen national agricultural research systems	World	2.4
International Food Policy Research Institute (IFPRI)	1980	1975	Washington, DC		Identify and analyze policies for sustainably meeting the food needs of the developing world	World, with primary emphasis on low-income countries and groups	31.4
World Agroforestry Center[d]	1991	1977	Nairobi, Kenya		Agroforestry, multipurpose trees	World	28.5
International Water Management Institute (IWMI)[e]	1991	1984	Colombo, Sri Lanka		Water and irrigation management	World	23.1

(*continued on next page*)

Table 3
(*continued*)

Center	First year of CGIAR support	Foundation date	Location of headquarters	Main areas of focus			2004 expenditures (million US$)
				Commodity	Activities	Region	
WorldFish Center[f]	1992	1977	Batu Maung, Malaysia		Sustainable aquatic resource management	World	14.1
Center for International Forestry Research (CIFOR)	1993	1993	Bogor, Indonesia		Sustainable forestry management	World	15.1
ILRI, International Livestock Research Institute[g]	1995	1995	Nairobi, Kenya, and Addis Ababa, Ethiopia		Livestock production and animal health	World	31.7

Sources: Alston, Dehmer and Pardey (2005).

Note: n.a. indicates not applicable.

[a]First established in 1974 as the International Board of Plant Genetic Resources (IBPGR). The Board was funded as a CG center but operated under the administration of the FAO, located at FAO headquarters in Rome. In 1993, IBPGR changed its name to IPGRI and was established as a self-administering CG center in its own headquarters in Rome. The International Network for the Improvement of Banana and Plantain (INIBAP) was established in Montpellier, France, in 1984. In 1992, INIBAP became a CG sponsored center, but in 1994 INIBAP's functions were placed under the administration of IPGRI, INIBAP, however, continues to maintain its own board.

[b]Formerly known as the West Africa Rice Development Association. Originally located in Liberia, moved to Bouaké, Côte d'Ivoire in 1987, and has been working out of the IITA station in Cotonou, Benin since January 2005 because of civil unrest in Côte d'Ivoire.

[c]Ceased operations in March 2004, and reconstituted as an "ISNAR Program" within IFPRI in April 2004.

[d]Known as the International Center for Research in Agroforestry (ICRAF) until 2002.

[e]Known as the International Irrigation Management Institute (IIMI) until 1998.

[f]Formerly known as the International Center for Living Aquatic Resource Management (ICLARM), its headquarters were relocated from Metro Manila to Batu Maung, Malaysia in 2001.

[g]ILRI became operational in January 1995 through a merger of the International Laboratory for Research on Animal Diseases (ILRAD) and the International Livestock Center for Africa (ILCA). ILRAD research focused on livestock diseases (world) and tickbone disease and typsanomiasis (Sub-Saharan Africa). ILCA did research on animal feed and production systems for cattle, sheep, and goats for Sub-Saharan Africa.

Stagnant total CG spending through to the early 2000s (with signs of a resumption of growth beginning in 2002) was accompanied by a shift in spending away from performing research – intramurally or jointly with others – toward other activities. These other activities include hosting and managing research networks that facilitate research performed by others, some in conjunction with CG centers [Plucknett, Smith and Ozgediz (1990)]; rehabilitating seed stocks in war-ravaged countries like Rwanda and Afghanistan, and the rice research system of Cambodia; promoting zero-till systems in the wheat systems of the Indo-Ganges Plains; and efforts to develop smallholder milk supply systems in Africa. Some of these initiatives entail technology transfer activities that complement CG research; others involve a move into development efforts less directly related to research.

2.5.2. *Sources of support*

Agriculture, and with it agricultural R&D, no longer command the attention international donor and aid agencies once gave them. Precise data are hard to come by, but the evidence suggests that after several decades of strong support, international funding for agriculture and agricultural research began to decline around the mid-1980s as support for economic infrastructure as well as health, education, and other social services began to grow. Africa, where agricultural R&D in many countries was reliant on donors for more than 40% of its total funding in the early 1990s, was particularly hard hit.

The following gives some quantitative highlights of the decline in international aid for agriculture and the research that directly supports the sector:

- Even though the European Community (EC) increased overall aid to developing countries during the period 1987–1998, aid to agriculture declined substantially. Agriculture accounted for 12% of total EC contributions in the late 1980s, but only 4% during 1996–1998.
- Over the past two decades World Bank lending to the rural sector has been erratic, but after adjusting for inflation the general trend has been downward. Agriculture's share of total lending has also declined (from an average of 26% during the first half of the 1980s to only 10% in 2000).
- There is no discernible pattern in the amount of World Bank lending authorized for agricultural R&D, other than a temporary increase in loan approvals in the late 1980s, early 1990s, and an exceptionally large amount of lending in 1998, resulting mostly from loans with large research components approved for India ($136 million, current prices), China ($68 million), and Ethiopia ($60 million) (Figure 4, Panel a). The size of the loans has been highly variable, ranging from $0.1 million for Argentina in 1992 and Niger in 1997, to $136 million for India in 1998.
- The amount of funding that USAID directed toward agricultural research in LDCs declined by 70% from the mid-1980s to 1997 (Figure 4, Panel b). Asian countries suffered the largest losses, from around $42 million (in 1999 prices) in the mid-1980s to only $1.1 million in 1997. Support to Africa and Latin America and the

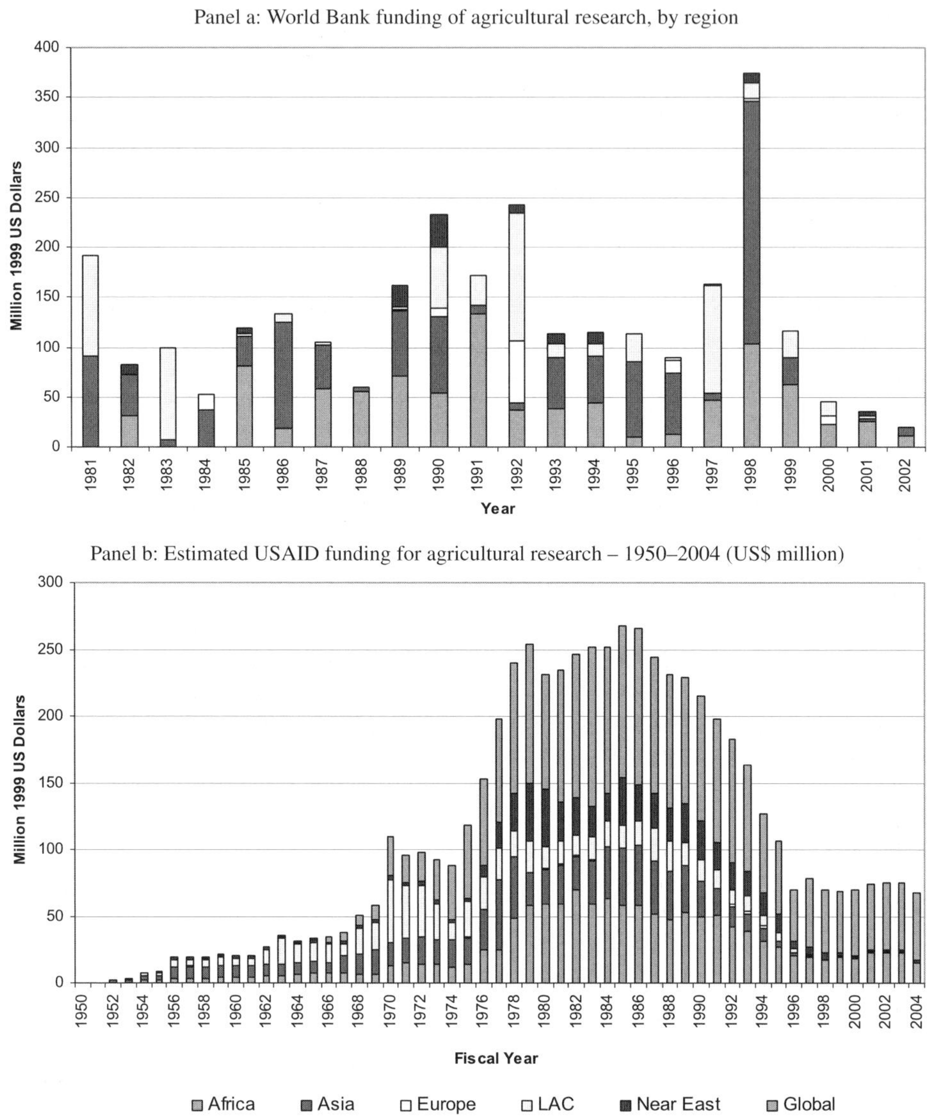

Figure 4. World Bank and USAID funding of agricultural research by region. *Source*: World Bank (2004) for Panel a, and Alex (2004) for Panel b.

Caribbean was also cut severely – by 1997 funding had fallen to only 34% of mid-1980s levels for Africa and 7% for Latin America and the Caribbean. Since 1997, USAID funding for agriculture has failed to regain the ground it lost, and funding for agricultural R&D has fared no better.

3. The economics of innovation incentives – in brief

Innovations are a fundamental source of change, economic growth and improvement in well-being. Societies relying exclusively on private market incentives are likely to under-invest in research and support a mix of R&D that is sub-optimal. Indeed it was only after leading nations found ways to intervene in providing, funding, or offering incentives for agricultural innovations that agricultural productivity began to outrun population growth on a sustained basis. Introduction of intellectual property rights such as patents, plant breeders' rights and copyrights are among the more prominent public policy measures intended to stimulate the creation and dissemination of inventions. However, other mechanisms (such as research prizes, contracts or open-source initiatives) may also be effective in generating new innovations in certain circumstances. Here we provide a brief introduction to the economics of intellectual property protection relevant to agricultural innovation.[10]

The output of agricultural innovation activities is, to a large extent, information which is non-excludable and non-rivalrous absent public intervention.[11] Often a valuable innovation can be easily copied and used by others who had no role in its production; this use does not diminish the availability of the innovation to other users. Nonexcludability and nonrivalry increase the social value of an innovation by increasing the speed and reducing the cost of diffusion to potential users, and reducing the price of the products of innovation to consumers. But lack of excludability means there is insufficient incentive for the private sector to produce the innovation in the first place. Without some form of public intervention, the appropriable returns to agricultural innovators are far less than the social benefits. Historically, there has been insufficient incentive for agricultural innovation, and indeed the pace of technical progress was slow, absent public intervention.

From ancient times, rulers have supported expeditions in search of new plants and animals [Juma (1989)] and members of the U.S. Congress in the nineteenth century distributed packages of seeds of newly introduced crop varieties gratis to their constituents [Huffman and Evenson (1993)]. In the nineteenth century, Germany, followed by several other European countries and the United States, introduced institutional innovations in the form of publicly funded agricultural research institutions. These included agricultural research stations and the United States Land Grant universities [Ruttan (1982)]. Thus the public sector became the means by which the interests of diffuse, individual beneficiaries of agricultural innovation could act collectively to increase the amount of resources devoted to research, funded by tax revenues. Over the past century, this institutional innovation spread across the globe, and today, as noted above, about two thirds of research resources worldwide are allocated by the public sector. Most of this

[10] For an excellent overview of the economics of innovation generally see Scotchmer (2004b); see Langinier and Moschini (2002) for a good introductory review in an agricultural context.
[11] Alston and Pardey (1999, Ch. 2) expand on these ideas in the context of agricultural R&D.

research has been performed by employees in the public and nonprofit sectors. As the many economic evaluation studies cited by Alston et al. (2000) established, overall, the rate of return on this investment in innovation has been extraordinarily high, worldwide, relative to other public or private investments.

Nevertheless the political climate has shifted in developed countries in favor of greater involvement of the private sector in agricultural research related to on-farm technologies. Traditionally, private-sector innovation has been largely focused on agricultural inputs, such as farm machinery, fertilizers and crop protection chemicals, or on post-harvest technologies and marketing, where market power and intellectual property rights (such as patents, copyrights and trademarks) have long been available to increase the appropriability of the benefits of innovation. In the 1980s, the confirmation of the patentability of life forms in the United States and the concurrent development of the biotechnology necessary to prove infringement enabled the emergence of utility patents as new high-powered incentives for research on on-farm agricultural biotechnology. In addition, the Bayh–Dole Act of 1980 encouraged the spread of patenting and utilization of public-funded research [Mowery et al. (2001)].

In recent years, the adoption of patent systems in developing countries worldwide has been accelerated by the need to comply with the TRIPS agreement – a necessary condition for membership of the World Trade Organization. In addition, many countries have strengthened their patent systems as part of domestic initiatives to upgrade their national innovation systems [Mowery (1998)], or to comply with post-TRIPS bilateral or multilateral agreements. Some observers might infer that these developments reflect the increasing influence of economics on intellectual property policy. Today, many and perhaps most economists support the inclusion of a patent system as part of a modern system of innovation and economic development. However, a substantial minority has a different view [see, for example, Boldrin and Levine (2002)], and indeed economists have traditionally been more ambivalent with respect to the inclusion of patents as a part of public policy [Plant (1934); Machlup and Penrose (1950); Machlup (1958)].[12] The cost of restriction of the use of the patented product or process for the duration of the patent monopoly has long been recognized as a downside effect of a patent system, while the incentive effects have been less widely accepted.

3.1. Static patent theory

The modern economic theory of patents has its roots in the fundamental contribution of Arrow (1962) and has been strongly influenced by the subsequent model of Nordhaus (1969). Nordhaus derived an optimal finite patent life which balances the gains from the

[12] Adam Smith wrote in *The Wealth of Nations* that monopolies should be temporary. If maintained, the result would be to tax individuals "... very absurdly in two different ways: first, by the high price of goods, which, in the case of a free trade, they could buy much cheaper; and secondly, by their total exclusion from a branch of business which it might be both convenient and profitable for many of them to carry on" [Smith, Book III, p. 339, cited in Boyle (2003, p. 55, note 87)].

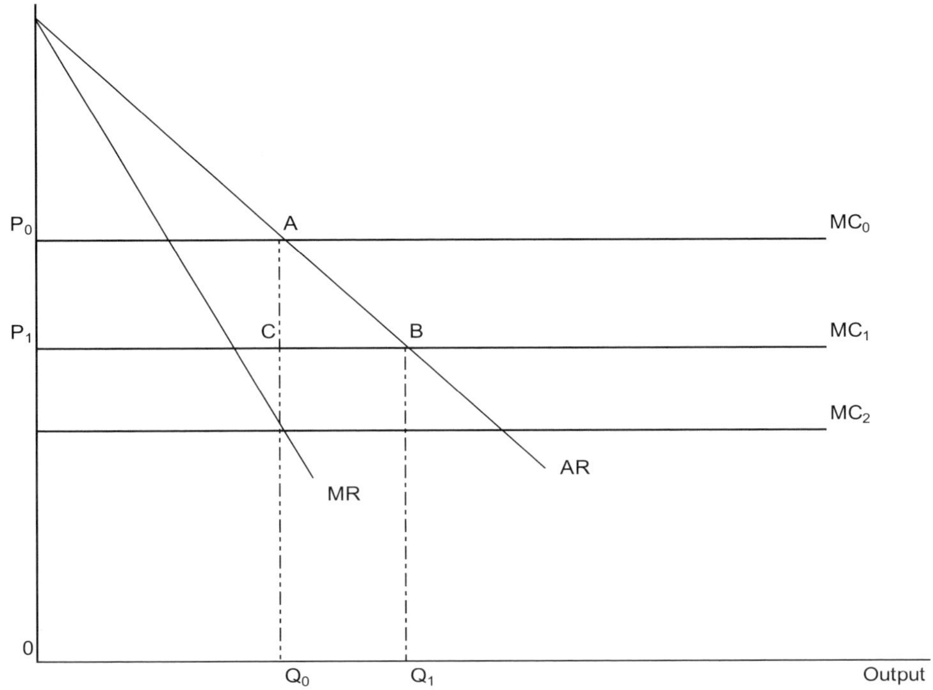

Figure 5. The static welfare implication of a patent. *Source*: Adopted by authors from Scherer (1972).

cost reductions due to the extra innovative effort encouraged by longer patent life against the deadweight loss associated with greater duration of the single-price monopoly. Like Arrow, he considered the case of a discrete, cost-reducing invention, produced with certainty by a single firm and sold at an undifferentiated price. Assuming patent value is a concave function of research effort, the aggregate surplus of the innovator and consumers is optimized by a patent life around those typically observed in modern patent systems.

The tradeoff in the Nordhaus model is illustrated in Figure 5, which shows a product demand curve AR with a corresponding marginal revenue curve of MR. With the initial cost of production at MC_0, the initial equilibrium under competition is (P_0, Q_0). Now, consider the innovation of a new process which reduces the marginal cost of production from MC_0 to MC_1. If the new cost-reducing innovation is freely available, the equilibrium price and output are P_1 and Q_1, respectively, and the increased annual consumer surplus from the innovation is P_0ABP_1. If a patent bestows a monopoly on the cost-reducing process, then the monopolist's revenue is maximized by keeping the output at the pre-innovation level of Q_0.[13] The additional annual revenue accruing to the monop-

[13] In Figure 5, we highlight the static welfare implications for a "run-of-the-mill" innovation (with comparatively small reductions in production costs) such that the monopolist's price and output before and after the

olist is equivalent to P_0ACP_1, while the annual deadweight loss of the patent monopoly is represented by the triangle ABC.

The significant virtue of a patent system is the administrative economy achieved by fixing a uniform patent life, regardless of technology field, cost of research, or value of the innovation, all of which might be difficult and costly to identify. In the Nordhaus model, this uniformity of patent life comes at modest social cost. If the marginal cost of achieving reductions in production costs (i.e., of lowering MC_1) is increasing, the effects of symmetric deviations from optimal patent life in the Nordhaus model are asymmetric; too long is less costly than too short. But as long as the patent life is longer than some minimum level, social welfare is relatively insensitive to patent life.

Patent life is only one of many policy instruments that policy makers can choose to increase social welfare in a patent system. Renewal fees charged periodically during the life of the patent encourage early abandonment for innovations of modest value, by making effective patent life shorter [Cornelli and Schankerman (1999)]. The scope (or breadth) of protection awarded by the patent determines the extent to which competitors might be motivated to "invent around" a patent by developing another innovation. The importance of patent scope along with patent life was emphasized in the literature during the 1980s and 1990s [for example, Gilbert and Shapiro (1990); Klemperer (1990); Gallini (1992)]. However, as used by economists (as distinct from lawyers) this concept of scope (or breadth) is vague. It has been defined variously as the size of available profit (Gilbert and Shapiro), the distance in characteristics space between a patented product and others (Klemperer), the cost of imitation (Gallini), or the number of possible applications [Matutes, Regibeau and Rockett (1996)]. Depending in part on the definition of scope and assumptions about industry structure, the social welfare maximizing mix of patent life and scope tends to vary from model to model. Some models imply that the optimal policy should have an infinite patent life with finite scope [Gilbert and Shapiro (1990)], while others imply that finite patent life with infinite scope can be optimal [Klemperer (1990); Gallini (1992)].

The models discussed above are static, in that they focus on the situation where a discrete innovation is made by a single firm at a single point of time with no relation to previous and future innovations and no competition for the innovation. In the Nordhaus model, the innovating firm at the outset holds, for some reason, a monopoly over an "innovation prospect" in which it can invest to produce, with certainty, an innovation that it can monopolize for a specified period via a patent. Thus, the patent extends the monopoly in the prospect to a monopoly of the innovation. The value of the patent is taken to be the surplus directly generated by its use. The model ignores any information, useful for further research, that might be disclosed in the published patent. The

innovation are unchanged. If it were a "drastic" invention reducing marginal cost below MC_2, then a revenue maximizing monopolist would increase production beyond Q_0, lowering the price it charges below P_0 [see Nordhaus (1969, pp. 71–73) and Scherer (1972) for more details]. Scherer points out that the extent to which the monopoly power conferred by the patent is exercised depends on various market structure issues. Moschini and Lapan (1997) extend the model to consider vertically integrated agricultural markets in which innovations in one stage of the market (e.g., input suppliers) have consequences for other stages (e.g., farmers).

latter can be important, and indeed has been viewed, especially by lawyers, as the main consideration received by society in exchange for grant of the patent monopoly. Thus the innovation is doubly static: the monopoly of the prospect comes from nowhere, and the innovation has no implications for further innovation.

3.2. The implications of competition in research

Where the innovation prospect is known and accessible to all, there would be competition to be the first to obtain the patent. Recognizing the stochastic nature of innovation and the "winner take all" nature of the award offered by the patent system, the marginal incentive perceived by each firm is the marginal increase in the firm's probability of winning in response to marginal effort, multiplied by the expected total value of the returns (from direct exploitation or from licensing). With perfect competition in the research industry, each competitor takes the overall probability of success as given and so the expected return equals the average expected return – intuitively, the "return to a ticket in the (large) innovation lottery". This exceeds the marginal expected return, assuming the latter is diminishing. Thus the social value of the patent will be diminished by competitive "rent seeking" in the race to be the first successful innovator, if there are no other sources of market failure. This rent dissipation problem was noted by Usher (1964), and is akin to the problem of overexploitation of an unregulated fishery identified by Gordon (1954) and often dubbed the "common pool problem".

If the marginal cost of research is constant (that is, the short-run industry supply of research effort is perfectly elastic), and the innovation is non-drastic, the entire value of the patent is dissipated in excess competitive research effort if patent life is infinite. In this case, the annual cost of rent-seeking is represented in Figure 5 by the rectangle P_0ACP_1, and the conventional deadweight loss is represented by the triangle ABC. A shorter patent life, which reduces the rent dissipation problem, is optimal in this model. The social value of a prospect is quite sensitive to deviations from optimal patent life in either direction, in contrast to Nordhaus' analysis of a monopoly of the prospect where social value is insensitive to patent life beyond an optimal patent life [Wright (1984)]. If the supply of research is less elastic, this problem is less serious. Paradoxically, the patent incentive is more robustly useful when research is not too responsive to that incentive. This rent dissipation inefficiency "rectangle" under competition is of a larger order of magnitude than the simple monopoly deadweight loss "triangle" typically identified as the welfare cost of a patent [Wright (1983)]. In this situation private research in response to patent incentives is not necessarily more efficient than publicly controlled research, even if the public sector is less informed than the private sector about relevant market parameters.[14]

[14] On the other hand, factors that diminish the private value of patents, some of which are mentioned below, can reduce this rent dissipation and make patents more socially attractive as instruments for encouraging research.

3.3. Innovation dynamics and the role of patents

So far we have ignored the fact that much modern innovation is cumulative. Technological change frequently arises from, or embodies, a sequence of prior innovations [Merges and Nelson (1990)]. An excellent example is the crop improvement process where a modern high-yield crop variety represents the latest achievement in a long cumulative process of incremental improvements in the genetic resources responsible for current yield levels, and in techniques of crop breeding. When innovation is cumulative, a patent system may create dynamic distortions of incentives for subsequent innovations (i.e., dynamic inefficiencies), in addition to the static deadweight loss discussed above. A strong patent on an initial innovation might stimulate the earlier-than-otherwise development of the innovation but reduce the incentive for subsequent innovations, while a weak patent may not even induce the initial innovation thereby stymieing subsequent innovations. This intertemporal, dynamic distortion of incentives can be more serious than the static inefficiency of the monopoly loss because the entire research sequence can easily be blocked if incentives at any stage are inappropriate.

The design and performance of patents when innovation is cumulative have received considerable attention recently. Green and Scotchmer (1995) argue that first-generation innovators should be given strong protection so as to overcome the intertemporal externality that arises when second-generation improvements can be obtained by others [see also Chang (1995); Scotchmer (1996); Matutes, Regibeau and Rockett (1996)]. In some cases, patents on follow-on innovations can reduce the incentive for the first-stage research, and Scotchmer (1996), in an informal argument, concludes the second-stage innovation should not be patentable. On the other hand, as emphasized by Merges and Nelson (1990) and Heller and Eisenberg (1998), broad patent protection can stifle subsequent stages of innovations [see also O'Donoghue (1998); O'Donoghue, Scotchmer and Thisse (1998); Denicolo (2000); and Hopenhayn and Mitchell (2001)]. When patents are first introduced into a given industry, the newly privatized incentives for innovation are high-powered at the first stage since there are no existing claims of IPR and research inputs are freely available. But after a first round of innovation protected by IPRs, claims on inputs act as disincentives for follow-on innovations, reducing the overall incentive effect, as shown by Koo (1998) for a linear innovation sequence.

A special case of cumulative innovation involves the development of a research tool, that is, a product or process whose only value is as an input to follow-on innovations. In agricultural biotechnology, a research tool could for example be a patent on a DNA sequence modified to enhance the expression of a trait such as insect-resistance, while the follow-on innovation may be a new transgenic variety of cotton.

The dynamic inefficiency problem is particularly serious when the second innovation prospect is monopolized, and the second innovation is not highly profitable. Koo and Wright (2005) show that in this situation, bilateral *ex post* bargaining can result, depending on the profitability of the second innovation, in either immediate follow-on research after the first patent, or delay until the first patent expires. The result can be very sensitive to patent life, in contrast to the implications of the one-stage Nordhaus model. The *ex ante* negotiation emphasized by Green and Scotchmer (1995) is important if it can prevent such delay. Koo and Wright also show that a competitive race for follow-on innovation makes the patent system more robust to differences in profitability of different innovations, given uniform patent life and *ex post* negotiation. Entry of follow-on innovators during the life of the first patent reduces the delay for innovations that are too costly to be profitable if royalties must be paid during the entire life of the second patent. Furthermore, dissipation of the rents on more profitable second-round innovations can be reduced by royalties paid for the license on the first patent, assuming no rent seeking in the search for the first innovation.

In agricultural biotechnology, there may be multiple independent claims on multiple, mutually blocking inputs, as in the well-publicized case of Golden Rice technology [Kryder, Kowalski and Krattiger (2000)].[15] Decentralized ownership of blocking claims, in the presence of significant transaction costs, introduces the possibility of an "anti-commons" phenomenon – the underutilization of innovations subject to multiple, fragmented (perhaps uncertain, or at least legally untested) property rights [Heller and Eisenberg (1998)].

3.4. Is the anti-commons impeding agricultural research?

Concerns with access to research tools have been expressed for some time in biotechnology in the health sector [National Academy of Sciences (1996)]. Quantitative evidence on the question is however scarce. In a survey of 70 IP attorneys, business managers, scientists and technology transfer officers from the private and public sectors related to pharmaceuticals and biotechnology, Walsh, Arora and Cohen (2003) found that changes in intellectual protection involving increased patenting of research tools in biotechnology have not substantially impeded drug discovery or research in the health area, at least partly because researchers assumed a *de facto* research exemption or simply ignored intellectual property claims. This survey was conducted before the landmark *Madey v. Duke* decision by the U.S. Court of Appeals of the Federal Circuit in 2002, which made clear to researchers what many lawyers claim to have known all along, that any research exemption in United States law was so narrow as to rule out virtually all meaningful biotechnology research. A second, larger survey by Walsh, Cho and Cohen (2005) of scientists actively engaged in genomics, proteomics and related fields[16] appears largely to confirm the key finding of the first survey: public sector and nonprofit scientists usually ignore intellectual property claims in accessing, making and using their research tools. *Duke v. Madey* notwithstanding, they consider themselves to have a *de facto* research exemption. From their viewpoint, they encounter "freedom to operate" as a problem when they cannot get their hands on a research tool, held by another,

[15] Note, however, that the valid claims in the major rice-producing and rice-importing nations were few if any [Kryder, Kowalski and Krattiger (2000); Pardey et al. (2003)].

[16] The survey was commissioned by the U.S. National Academies Committee on Intellectual Property Rights in Genomics and Protein Related Inventions, formed at the request of the National Institutes of Health to address continuing concerns related to the effects of IPR in this specific area.

as quickly and easily as usual. In such cases the perceived problem arises not only from intellectual property claims, but also from more traditional academic concerns with priority of publication and competition for support.[17]

Preliminary results of a survey of 90 plant biologists at four public universities (UC Berkeley, Davis and Riverside and the University of Arizona) by Lei, Juneja and Wright (2006) indicate that plant biologists have attitudes to proprietary technology similar to those of biologists in the medical field. When scientists send research tools to colleagues at other universities and in industry, the interaction continues to be more frequently informal than formal. Respondents rarely try to check the IPR status of the tools they use, though more than one third of them have made invention disclosures. Consistent with the results of Walsh and Cohen, problems with freedom to operate materialize as lack of easy and quick access to materials held by others, whether associated with patents, secrecy, or problems with material transfer agreements.

Over the past 5 years, more than one third of the surveyed scientists experienced delays in obtaining access to research tools, with a mean of 2 delays, and a mean duration of over 8 months. More than one quarter of the respondents reported one or more cases in which difficulties in obtaining research tools affected their research projects in ways other than causing delays. In one third of these cases, alternative tools of equivalent effectiveness were used, but in about 40% of cases less effective tools were substituted. More seriously, in about one quarter of the problem cases a project or line of research that was part of a project had to be abandoned, or not initiated, due to lack of access to research tools. There does not appear to be strong evidence distinguishing academia from industry as the major source of these problems.

The majority of all respondents, and of the subset who have made invention disclosures, believe that IPRs on research tools are, overall, having a negative effect on research in their area. Thus claims that scientists working on biotechnologies for crop breeding in the United States perceive that IPRs are adversely affecting their research can no longer be dismissed by economists as merely "anecdotal".

Recent papers measuring citation behavior of biologists give additional quantitative evidence of a possible anti-commons effect of research tool patents on research.[18] How-

[17] The exception is diagnostic tools which are used by universities in revenue-generating treatment of patients, which jointly contributes to ongoing research. Concerns with commercial application have led some patentees to vigorously pursue infringers, generating a great deal of controversy.

[18] Murray and Stern (2004) consider the 340 peer-reviewed scientific articles published in *Nature Biotechnology* between 1997 and 1999, and were able to link about half to a United States patent. In a "difference in differences" of the annual citations received by papers with and without a paired patent, they found that the patent grant reduced the subsequent citation rate relative to the trend in citations for papers reporting unpatented research. In addition, citations of papers declined when a matched patent was granted. The overall post-grant decline was estimated to be between 9 and 17%, and was more pronounced for papers with public-sector authors. A subsequent paper by Sampat (2004) analyzed the genomic patents assigned to the top 15 recipients of NIH funding over the period 1990–2000, and identified 499 patent–paper matches. He uses the variation in the lag to patent grant in his identification strategy, and finds that the grant of a patent causes a decline of 14% in US citations for post-1990 articles on gene sequences. Non-sequence patents ("techniques")

ever, scientists seem to be minimizing the adverse effects by ignoring IPR where they can, even after *Madey v. Duke*.

Where public agricultural biotechnology research is directed at providing upstream input to private-sector development, as in human health (with the prominent exception of research complimentary with clinical use of diagnostic tools), the public scientists can leave the achievement of freedom to commercialize to private developers, often large oligopolistic firms, who have achieved freedom to operate principally through consolidation via mergers and acquisitions [Graff et al. (2001)], rather than by licensing.

But as discussed above, the public agricultural research (especially in developing countries) not only provides upstream inputs to the private sector, but has the responsibility to see research through to commercialization in all but the few lucrative markets that attract the bulk of private sector attention. Accordingly, effects of IPR on nonprofit "commercialization" of innovations have been a key issue for the agricultural sector. Prevented from avoiding transactions costs by merger or acquisition, they look to licensing for an alternative. Where feasible, *ex ante* licensing could solve anti-commons problems as pointed out by Scotchmer (1999). But in agricultural biotechnology such *ex ante* licensing presents many challenges [see Merges and Nelson (1990)] and is not prevalent. Even *ex post* licensing of input technologies for modern plant breeding for commercial application has proven to be expensive, labor-intensive and difficult. Difficulties with licensing appear to generalize to other applications of biotechnology. As Jaffe and Lerner (2004) note, the broad cross-licensing typical of the semiconductor industry has not emerged in biotechnology.

For agricultural biotechnology research, the key issue is the effect of IPR on freedom to operate of the integrated enterprise of public sector commercialization of biotechnology, in applications beyond those attractive to the private sector. But any direct effect on commercialization is hard to measure objectively. The law of large numbers can be brought to bear on samples of thousands of citations, or hundreds of scientist person-years. But thousands of citations and scientist years might relate, in the end, to only a handful of commercialized products.

The set of biotechnology innovations that have been, or but for IPR problems might have been, developed to successful commercialization by public and nonprofit agriculture, is necessarily small. Thus a statistically valid quantitative analysis of effects of IPR on agricultural biotechnology innovation is impossible at this time. To discuss the issue we are forced to resort to case studies, anecdotes and the opinions of industry participants.

In the United States and some other developed countries, there is some evidence that university research projects designed to produce new crops with modern biotechnology have been shut down because of refusal of IPR-holders to permit commercialization of varieties incorporating their intellectual property. These typically involve intellectual property such as genes, markers or promoters incorporated in the final product, or

showed no such effect. As the authors clearly recognize, confirming any link between these findings and the effects on research productivity is a significant challenge.

products protected as compositions of matter generated via a novel process, such as a means of transformation, where the means is verifiable from the regulatory record. And the problem of freedom to operate generally arises after the development succeeds sufficiently to become a candidate for commercialization.

In one instructive example [described by Wright (1998)], UC researchers, with producer support, successfully created a tomato variety genetically engineered to express the university's endoglucanase gene to retard softening and improve shelf-life characteristics. However, the promoter they used was one for which a patent application surfaced during the development of the new cultivar. The patentee, a private corporation, refused to negotiate terms for the use of its embodied technology for commercialization of the cultivar. (Had the proprietary technology been a disembodied process, pre-patent application would not have been infringing.) The research and development effort came to naught, shattering the confidence of the producers in the capacity of the university to successfully breed and commercialize new transgenic cultivars. A few other examples of holdups of this type are available in the literature [Wright (1998); Erbisch and Maredia (2004)].

Economists unfamiliar with public-sector cultivar development tend to react skeptically to examples of holdups. They reasonably assume that when there are gains to be made from a trade, the trade will occur. But economists increasingly recognize the importance of the problem that trades can fail if the transactions costs are too high. Perhaps the public sector negotiators had unrealistic expectations regarding private sector largesse. The owner of the key IPR might have been concerned with protecting itself from liability or from damage to its reputation due to misuse or bad publicity beyond its control. In some cases the expected financial gains, given the size of the market, might have been less than the cost in time and money to the IPR owner (public or private) of making and enforcing an agreement. Or perhaps the patent holder saw no reason to help out a potential competitor, for little financial return, in a market that could one day be of financial interest to the patentee.

In our experience, scientists who have experience with managing startups and public-sector development in this area generally strongly support the view that IPR thickets and holdups can be serious impediments to successful development to commercialization of transgenic cultivars for agriculture and horticulture in the United States. If this is true, prudent research managers will avoid the lines of research that could furnish sufficient new examples to provide a large enough sample to be amenable to statistical analysis.[19] Policy, North and South, must and will be made, and strategies committed, in the absence of direct statistical rejection of the hypothesis that IPR is not a serious impediment to public sector commercialization of applications of modern biotechnology.

[19] Taylor and Cayford (2004), Commission on Intellectual Property Rights (2002, Ch. 3), and Correa (2000) present some developing-country perspectives regarding the access to proprietary technologies and intellectual property rights.

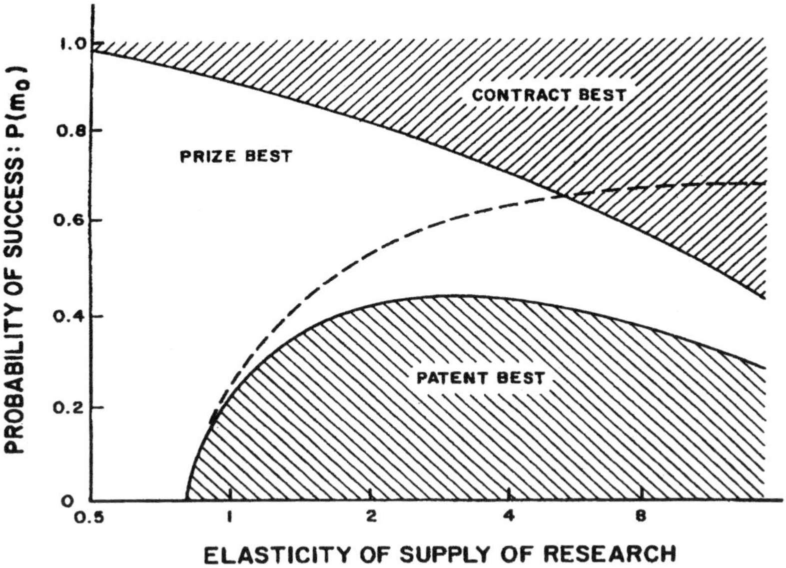

Figure 6. Choice between patents, prizes and contracts. *Source*: Wright (1983, p. 703).

3.5. Alternative inducements to innovate

Thus far, we have focused on patents as incentives for innovation. But is the patent system the only economically justifiable means of inducing innovation? If, as implicitly assumed thus far, there are no information asymmetries between government and researchers, the answer is clearly negative. If the government can collect research funds from an efficient tax system, distribute them to researchers through research contracts, and make the final research output freely available, it can avoid monopoly pricing without distorting the innovation incentive. Similarly, if government awards a prize to the one who first achieves a research output, innovations could pass immediately into the public domain. Setting the appropriate amount of prize or contract support is considered the main problem of this type of policy when government lacks relevant information held by the researchers. An extension to the case of research of Weitzman's (1974) prices versus quantities analysis of regulation of externalities reveals that if researchers have superior *ex ante* information about the cost of research or the value of the innovation then patents can be superior to contracts, assuming the allocative improvement in response to this information outweighs the cost of the excess burden of the patent [Wright (1983)]. But contracts are not the only alternative. Prizes have also long been used to encourage innovation.

The choice between patents, prizes and contracts is illustrated in Figure 6. The vertical axis shows the probability of success at the social planner's optimum, $P(m_0)$, were the benefit of success common knowledge and equal to its mean. If the marginal research incentive equaled the marginal social value of research so that there is no common pool problem, then patents would be dominant below the dashed line, and prizes above it if researches have private information about the costs and benefits of research. Given the possibility of competitive rent dissipation, however, patents are best only in the shaded area at the bottom of the figure, even if, as assumed here, patent lives (and the values of prizes) are optimally adjusted in line with accurate knowledge of research supply elasticity.[20] When the research supply elasticity is low, the deadweight loss from patents dominates the advantage of using inventors' information about value, and prizes dominate patents. As the research supply elasticity increases, the common pool problem increasingly favors contracts. In general, asymmetry of information is necessary to justify moving away from contracts in allocating research resources in a Nordhaus-type model, but patents dominate contracts and prizes only in a subset of the possible cases. When competition in research is recognized, this subset shrinks, because the common pool problem distorts the market signals that are observed by private researchers.[21]

Thus patents would not be justified if the research process did not involve informational asymmetries: i.e., if each party has complete information about the cost of research, the probability of success, and the value of a successful innovation. For staple food crops where the aim of research is to lower the price of food or to improve public health and nutrition, the benefits of success might well be better perceived by the policy maker than by potential researchers. But the latter might be far better informed regarding their own costs of research. In this situation, Wright (1983) shows that a prize awarded to the first successful innovator could ensure that the policy-maker's information on value, and the researchers' information on cost, would both be brought to bear in determining resource allocation.

Only if there is an asymmetry in the information about the benefits (or value) of a successful innovation, to the disadvantage of the policymaker, can patents in theory dominate prizes or contracts. Even in this case, the situation depends on the value of relevant parameters, including the supply response of research. The Nordhaus analysis assumes the innovator is a monopolist of research, essentially the owner of the unique research prospect. Paradoxically, however, patents appear particularly undesirable if research supply is competitive and very responsive to incentives, given there is a sufficient

[20] In the example in Figure 6, the coefficient of variation of the costs and benefits of a non-drastic, cost-reducing innovation is 0.25, and the deadweight loss is 2.5% of its social value given no patent monopoly. At an equilibrium with the probability of success, $P(m) = 1 - (1 - \pi)^m$ of 0.2, wherein π is the probability of success of any one project and m is the number of researchers, the patents and prizes are optimally reduced in value by about 10%, while at a probability of 0.4, the reduction to optimally reduce the common pool problem is about 25%. For the technical details of the model, see Wright (1983).

[21] This discussion might in fact overstate the case for patents. Scotchmer (1999) argues that one could use an efficient mechanism to virtually costlessly extract the information held by the private sector, rendering patents dominated in all cases in this type of model. We are unaware that any such mechanism has been successfully implemented in this context.

degree of irresolvable uncertainty about the benefits of success. Contracts can avoid this competitive misallocation of resources in response to private information, not shared with the government, that is distorted by the common pool problem. When there is no informational asymmetry regarding the probability of success or the ability of researchers, contracts are in general the dominant choice.

However, even if contracts or prizes dominate patents as a means of resource allocation in a closed economy, they are problematic when the world is divided into multiple nations and the benefits of research spill over across national boundaries. The temptation to "free-ride" can lead to a "prisoner's dilemma" equilibrium where too little research is funded [Scotchmer (1999)]. In this context, patents enable the innovator to capture some of the international benefits, reducing the free-rider problem. An international agreement on amounts of national research funding, or joint funding of international research organizations (observed currently on a small scale in the example of the CGIAR), can, when feasible, be a better alternative.[22]

When the limits of the advantages of patents over other alternative means of research resource allocation are recognized, it is easier to understand the success of the innovation systems that have driven a high rate of technical change in agriculture over the past century, before intellectual property rights were a significant force in agricultural biotechnology. When the overall objective (more food at lower prices or improved nutrition and public health, for example) was clear and the information gaps between funders and researchers not too broad, contracts (supplemented by "prizes", including professional recognition, academic tenure and salary enhancements) have certainly supported research with a very high average rate of return, relative to other investments, for more than a century. Whether this performance can be sustained in the future when relevant prior innovations are protected by patents or other intellectual property rights is an important issue.

Research on alternative means of encouraging innovation was further developed by de Laat (1996) who considers the choice of patents and prizes in a non-competitive R&D environment, and by Shavell and van Ypersele (2001) who show that a reward system (such as a prize) together with a patent system can be superior to the patent system alone.[23] Kremer (1998) proposes a patent buyout as a mechanism to increase the incentive for innovation without creating monopoly price distortion, again ignoring the rent dissipation problem [see also Maurer (2005)]. In the area of agriculture, Kremer and Zwane (2005) present a system of awards for pre-specified agricultural innovations, and Masters (2003) has advocated a system of prizes, calculated as a percentage of the surplus generated, to be awarded for innovations adopted in Africa in the agricultural

[22] Currently, the United States is encountering considerable difficulties in persuading other OECD nations to carry their "fair share" of the cost of funding modern pharmaceutical research, when drug prices are subject to public control.

[23] Their analysis relies on a crucial assumption that the government can identify *ex ante* those proposed projects that will fail to at least break even, a nontrivial condition.

sector. In India, the HoneyBee Network gives individual and collective awards for small-scale innovations, many of them related to agriculture.[24]

3.6. Intangible incentives – open source initiatives

At first blush, economics cannot by itself provide a full explanation of innovative performance. This is made obvious by the fact that economists have not made much progress in explaining the key roles of farmers and local blacksmiths in fundamental mechanical innovation in agriculture in the nineteenth century in the United States documented by Huffman and Evenson (1993), or the crop improvement efforts in the same era, described by Olmstead and Rhode (2002). Nor does it explain, for example, the leadership of farmers in the development of "no-till" agriculture, one of the crucial innovations of the late twentieth century. Farmers, in short, in attacking their local problems, produce many innovations with no prospects of reaping a significant fraction of their full monetary value.

The record of small-scale agricultural mechanical innovations in the last few centuries seems no more nor less puzzling than the recent "open source" development in software production that has attracted so much attention as a "new" collaborative innovation phenomenon [Benkler (2004a, 2004b)].[25] Explanations for the latter include "career concerns" of participants who expect to gain indirectly from the reputational effects of involvement in open source [Lerner and Tirole (2002)], and the efficiency of decentralized debugging of a system with millions of potential configurations [Bessen (2004)], as well as the intrinsic motivation of delight in solving an intellectual challenge and the reward of recognition by one's peers. The existence of such phenomena is obviously related to the availability of communication systems that allow dispersed individuals to participate in, or benefit from, the decentralized research processes.

The recent Biological Innovation for Open Society (BIOS) initiative arising out of CAMBIA is an attempt to initiate open-source development of key enabling technologies for agricultural biotechnology using licensing strategies inspired by the open source movement in software [Nature (2004)]. In addition, the Public Intellectual Property Resource for Agriculture (PIPRA) initiative is an attempt by public and nonprofit researchers to give each other access to their proprietary technologies, and to make such technologies available to third world researchers, without relinquishing the option to license for royalties to private-sector entities in developed countries [Graff et al. (2001); Atkinson et al. (2003); Delmer et al. (2003)]. These initiatives respond to the high cost of transactions in agricultural biotechnologies associated with multiple intellectual property claims on key inputs, tools and processes. They are being assessed with great interest by the wider human biotechnology community as innovations that might provide models for addressing freedom to operate issues within that community.

[24] See http://www.nifindia.org/secondaward/hbn_background.html for more details.

[25] A similar point is made by von Hippel (2002) with respect to innovation networks more generally.

4. Means of protecting innovations relevant to the agricultural biosciences

4.1. Intellectual property rights

4.1.1. Plant patents

The practice of applying "intellectual" property rights to claims to life forms appears to have originated in the United States' Plant Patent Act enacted in 1930.[26] It provides protection rights to breeders who invent or discover new and distinct varieties of asexually propagated plants, other than a tuber propagated plant or a plant found in an uncultivated state (35 U.S.C. §161). The criteria for patentability are similar to those of a utility patent (i.e., novelty, non-obviousness and utility), as is the duration of protection (20 years).[27] During the protection period, the patent holder (or breeder) may exclude others from asexually reproducing, selling, or using the plant so reproduced without the breeder's prior authority.

One of the limitations of the original 1930 plant patent was that the law did not apply to "parts of plants". Therefore, cut flowers, fruit and other plant parts or products are outside the scope of protection, and these products were often grown outside the United States using the protected varieties and imported and sold domestically without violating the U.S. patent rights. The Plant Patent Amendments Act of 1998 dealt with this limitation by including plant parts under the scope of its protection, and the patentee can now exclude others from importing protected plants or any of their parts.[28]

This form of protection was useful, principally in horticulture, well before the advent of modern biotechnology (e.g., genetic engineering). In contrast to utility patents, plant patents do not protect against independent achievement of a given invention. Plants with similar characteristics that were developed by means other than clonal propagation from a patented plant do not infringe a plant patent. In many other countries, protection of

[26] While personal or corporate intellectual property rights for plant biotechnology are recent phenomena within most countries, attempts at asserting national property rights over breeding materials internationally are nothing new [Boettiger et al. (2004)]. Monopolization of valuable markets has long been accomplished by nation-states prohibiting access to breeding materials. Examples include the Dutch monopolization of the European tea supply [Juma (1989)], the Italian Piedmont's prohibition on rice seed export famously violated by Thomas Jefferson [Fowler (1994); Root and De Rochemont (1976)], Australia's twentieth century ban on the export of merino sheep [Quilkey (1970)], and more recently Ethiopia's ban on the export of some coffee tree varieties [Fowler and Mooney (1990)]. These cases are, however, atypical; in general, traders, collectors, and breeders have had relatively free access to landraces and farmers' varieties from around the world.

[27] Any plant patent application filed after June 7, 1995 has, if granted, a protection life of 20 years from the date of application, compared with earlier applications which were granted up to 17 years of protection from the date of grant. Given a few years of grant lag, the net impact of this change in patent life may be insignificant.

[28] Now the law states that "in the case of a plant patent, the grant shall include the right to exclude others from asexually reproducing the plant, and from using, offering for sale, or selling the plant so reproduced, or any of its parts, throughout the United States, or *from importing the plant so reproduced, or any parts thereof*, into the United States" (35 U.S.C. §163) [italics added by authors].

clones became available much later via plant breeders' rights [Koo, Nottenburg and Pardey (2004)].

4.1.2. Plant breeders' rights

The makeup of plant breeders' rights (PBRs) varies from nation to nation. Many countries base their PBR legislation on a model PBR system, called the International Convention for the Protection of New Varieties of Plants (commonly known by its French acronym UPOV) established by a group of Western European countries in Paris in 1961, and revised in 1972, 1978 and 1991. The only international intellectual property convention focused directly on agriculture, it is directed primarily to the interests of commercial plant breeders and originally aimed to offer them an alternative to utility patents for protection of plant varieties, including both sexually and asexually propagated varieties.

PBRs consistent with the UPOV system protect varieties that are deemed new, uniform, distinct and stable from unauthorized acts in furtherance of commercialization.[29] The novelty criterion is less demanding than for patents, though it has been tightened over time. A variety can be novel if it has not been previously commercialized for more than four years in member states, or for one year in the country of application. Distinctness means that the variety be clearly distinguishable from others commonly known to exist at the time of filing. Stability and uniformity mandate that the plant variety satisfy its original description and retain its relevant characteristics under repeated reproduction or propagation. These criteria do not focus on the value or significance of distinctive characteristics, but on the need to identify reliably a protected plant prior to the availability of the tools of modern biotechnology. For example, a United States soybean cultivar distinguished only by its flower color was awarded Plant Variety Protection.

Repeated reproduction of a protected cultivar for commercial sale constitutes infringement, but use of germplasm for research or as breeding stock for producing new cultivars does not. Thus breeders are protected against reproduction of protected varieties by competitors. In addition, inbred parent lines of hybrid corn have been protected under UPOV from use by competitors in commercial hybrid production. But breeders remain free to use protected varieties for their own breeding, and to protect newly developed varieties bred from the protected varieties, while, under the farmer's exemption, farmers can save and re-use germplasm for propagation and reproduction. Indeed, the original UPOV model did not require that inter-farmer exchange or sale of seed be prohibited by member countries [Pardey, Koo and Nottenburg (2004)].

The 1991 UPOV Act significantly increased the rights conferred to breeders; it introduced the notion that a variety "essentially derived" from a protected parent is covered

[29] Protection confers the right to exclude others from: (1) producing or reproducing, (2) propagating, (3) offering for sale, (4) selling or other marketing, (5) exporting, (6) importing or (7) stocking for any of the above purposes the protected variety [Article 14(1) of the Convention].

by the protection of the parent. What constitutes an essentially derived variety is set forth in the UPOV Act in Article 14(5)(b) and (c).[30] Moreover, and not without controversy, an essentially derived variety may be obtained by transformation, among other methods. Thus, for example, if a breeder starts with a protected corn variety and transforms the variety with a genetic construct (e.g., containing a gene for insect resistance), the resulting corn is an "essentially derived variety" and protected by virtue of its parentage. While the 1978 UPOV Act conferred rights to exclude others from production for purposes of commercial marketing, offering for sale and marketing of the protected variety, these rights were significantly expanded in the 1991 Act. Rights now additionally encompass conditioning, exporting, importing and stocking of the protected variety. The farmers' exemption was also weakened in the 1991 UPOV Act by limiting the exchange of protected seeds.

Like patents, plant breeders' rights pertain only to the jurisdiction in which they are awarded. When seeking a broader geographical scope of protection, some experienced practitioners have reported that the requirements for evaluation in each country, including trial plantings, and the need for local legal representation and perhaps translation of documents, can render achievement of broad international protection via plant breeders' rights comparable in cost to international utility patent protection [Margeules (2002)].[31]

In the United States, protection for sexually propagated varieties is conferred by award of a Plant Variety Protection Certificate (PVPC) administered by the United States Department of Agriculture (USDA) and based on the 1991 UPOV Convention. The fact that germplasm of protected varieties could legally be saved, reproduced and even sold to other farmers might explain the failure of efforts to develop a private market for wheat seed in the United States [Knudson and Ruttan (1988)]. Essentially, the problem for self-pollinating crop seeds is that in the year after introduction of a cultivar, the entry of general harvest into the commerce stream competes with the seed producer's output; only the first year of sales is free from this competition if the seed is

[30] In brief, an essentially derived variety is defined as a variety that is predominantly derived from another variety, while retaining the expression of the essential characteristics of the initial variety; is clearly distinguishable from it; and except for differences that result from the act of derivation, conforms to the initial variety in the expression of the essential characteristics of the initial variety.

[31] Koo et al. (2006) report that the total administrative cost of establishing and maintaining a varietal right in China for 20 years is \$8679 at market exchange rates (or \$39,704 per variety at purchasing power parity rates) compared with \$3652 per variety in the United States (as of February 2003), and up to €8100 for applications lodged and maintained to term with the Community Plant Variety Office, CPVO (as of June 2004). A U.S. plant patent held to expiration costs \$1760 in administrative fees (without maintenance fee) and a utility patent \$9400, including maintenance fees (as of December 2004). Legal fees for filing a U.S. plant patent are typically a few thousand dollars; between \$12,000 and \$15,000 using outside counsel to file for a biotechnology patent of average complexity (typically less for a varietal-related utility patent), and another \$10,000 to obtain a patent, exclusive of the USPTO maintenance fees required to renew the patent. The costs of litigation to defend and exercise one's rights can run from \$50,000 to \$500,000 per claim.

easily stored by farmers in a way that maintains its viability [Perrin, Kunnings and Ihnen (1983)].[32] If profits from those first year sales were sufficient to induce an efficient level of breeding, this would not be a problem. Although in some circumstances early sales could conceivably provide sufficient incentives to innovate [Boldrin and Levine (2002)], this does not appear to be the case for seed breeding.

Some other countries have enacted quite different versions of plant breeders' rights. India and Thailand, for example, allow protection of varieties with traditional knowledge or varieties that have been traditionally cultured and also include clauses for "benefit sharing" with local communities when landraces are used in commercial breeding. The Indian Protection of Plant Varieties and Farmers' Rights Act (PPVFR) of 2001 appears to allow for unrestricted sale by farmers of seeds from protected varieties, as long as they are sold in "brown bags", unidentified by variety or registration [Brahmi, Saxena and Dhillon (2004)]. The PPVFR has some additional and atypical requirements for obtaining protection, beyond the UPOV criteria: for example, the applicant must provide information about the origin of the genetic material and declare that the variety does not incorporate Genetic Use Restriction Technologies (v-GURTs).[33] While the Indian Act sets an interesting precedent in mandating identification of the origin of genetic materials, it might offer insufficient incentives for private breeders [Koo, Nottenburg and Pardey (2004)].

It remains to be seen whether the PBR systems such as legislated in India will qualify as adequate *sui generis* systems under TRIPS. Although TRIPS does not define requirements for a *sui generis* system, most commentators believe that a UPOV-approved scheme will pass under TRIPS. This may have motivated India to apply for membership of UPOV. So far, UPOV has not given approval, and, in the view of the authors, India's 2001 PPVFR Act does not appear to conform to the 1991 UPOV Act.

Plant breeders' rights afford freedom to operate for breeders through a research exemption, weakening its incentive effect for breeding new elite material for use by commercial breeders. Most empirical studies [including Perrin, Kunnings and Ihnen (1983); Knudson and Pray (1991); and Alston and Venner (2002)] found weak or indeterminate empirical evidence to suggest that plant breeders' rights are effective in stimulating investments in varietal-improvement research. Nor do plant breeders' rights generate substantial *ex post* licensing and enforcement activity [Janis and Kesan (2002)].[34]

[32] An end-point royalty (in contrast to a point-of-seed-sale) scheme, wherein revenue is derived as levies imposed on the first sale of harvested material, was introduced in Australia in 1999 to explicitly address this issue [see Kingwell and Watson (1998) and Kingwell (2001)]. This might make sustained private-sector innovation in crops such as wheat feasible. [See Lindner (2004) for a discussion of related issues.]

[33] GURTs, discussed in more detail below, are genetic use restriction technologies [Jefferson et al. (1999)]. The idea behind a v-GURT (varietal-GURT) is to engineer a gene or gene(s) that inhibit the development of viable seed. The most notorious system, dubbed "terminator technology" by a creative non-governmental organization, is disclosed in the United States Patent No. 5,723,765.

[34] See also studies by Godden (1998) for Australia, Diez (2002) for Spain, and CFIA (2002) for Canada.

4.1.3. Utility patents

Utility patents are considered the strongest form of intellectual property rights that apply to inventions of both processes and products (compositions of matter) that are embodied in tangible things. In contrast to trade secrecy, the patent award is conditional on disclosure of the invention in a manner that would enable a person with ordinary skill in the relevant art to produce the invention. This disclosure usually becomes public eighteen months after application.[35] Thus one important benefit of the patent system is that it facilitates the flow of information regarding innovations. Patents have a limited term, generally 20 years from the date of filing, and the scope of the property rights is defined by the claims made in the patent which, in the event of litigation, may be subject to interpretation by a court of law.

Issuance of a patent does not mean that the patent office will directly defend the patented innovation from infringement. Rather, a patent confers on its owner a legally enforceable right to *exclude others* from practicing the invention as described and claimed in the patent document.[36] Thus the value and strength of a patent system relies directly on the effective operation of the legal system in preventing infringement and penalizing infringers.

Although international treaties and conventions govern key aspects of patenting, patents are awarded by national governments and are valid only in the specific national jurisdiction that awards them. To protect an innovation in any given country, a patent must be obtained in that country. Multi-country applications are facilitated by the Patent Cooperation Treaty (PCT), administered by the World Intellectual Property Organization (WIPO). Applications in multiple European countries are facilitated by application through the European Patent Office.[37] English-speaking African nations that are members of the African Regional Industrial Property Office (ARIPO) can file patents through that office subject to confirmation in each member country, and a similar system applies for Francophone countries through the Organisation Africaine de la Propriété Intellectuelle (OAPI). The cost of obtaining a patent varies from country to country; the cost of obtaining protection in all important markets can be very substantial – hundreds of thousands of U.S. dollars. Components of this cost include the attorney fees for drafting the original patent document and negotiating with the patent office over claim language, filing fees for each country, as well as translation and local legal fees.

[35] The United States patent law allows an applicant to opt-out of publication upon petition, if the patent application has not been filed in any other patent office that publishes patent applications (37 CFR §1.213).

[36] The classic articles by Coase (1937, 1960) concern the boundary conditions of liability and exposure created by property rights, and as Runge (2004) observed: "Read together, it is evident that they pertain to the economic advantages of rights to exclude and be included, respectively" (p. 808).

[37] Following grant of a patent, the applicant must register and, where required, renew the patent in each designated European country that is desired, paying the national patent office fees and, if necessary, filing a translation into the official language.

For countries that belong to the WTO and have implemented the TRIPS agreement (discussed below), the minimal criteria for patentability do not differ from nation to nation. To be patentable an invention must be novel, involve an innovative step (called non-obviousness in the United States), and have industrial applicability. In addition, the invention must be adequately described and disclosed in a manner that would enable the making and using of the invention by a person ordinarily skilled in the relevant arts. While the United States Patent and Trademark Office contends that the standards for each requirement are uniformly applied across different classes of innovation, not all practitioners or commentators believe that the Patent Office achieves this result. In particular, for innovations in biotechnology, many observers believe the U.S. standards of novelty and non-obviousness are weak [Barton (2003)], whereas the standards for written description and enablement are very high. On the other hand, the utility requirements for DNA sequences are at present generally regarded as atypically stringent.

Implementation of the criteria for patentability can vary among countries. To begin with, TRIPS does not define the term "invention", allowing countries great latitude in deciding what discoveries qualify for patents. Many countries have deemed that higher life forms are not patentable. For agriculture this means that plants and inventions directed to plants or plant products (e.g., seeds) might not be eligible for a patent grant. In practice, at least some patent protection for plants and plant products is offered by the United States as well as member countries of the European Patent Convention.

The United States has a particularly expansive patent system, allowing utility patents on many different kinds of innovations including research tools, transformation processes, vectors, components of vectors such as markers, promoters, and genes and proteins, as well as on organisms and their parts. Following the 1980 landmark decision (*Diamond v. Chakrabarty* 447 US 303, 1980), plants and plant parts, including seeds and tissue cultures, were explicitly held to be patentable (*Ex parte Hibberd* 227 USPQ 433 (Bd. Pat. App. & Int. 1985)).[38] Moreover, patents may be granted for plant groups, individual plants and their descendants, plant parts, transgenic plants, particular plant traits, plant components (e.g., specific genes or chromosomes), plant products (e.g., fruit, oils, pharmaceuticals), plant material used in industrial processes (e.g., cell lines used in cultivation methods, bio-fermentation or bio-remedial applications), reproductive material (e.g., seeds or cuttings), plant culture cells, plant breeding methodologies, and vectors and processes involved in the production of transgenic plants. The United States also permits dual protection for plants, namely both a utility patent and a PVPC, or a utility patent and a plant patent (*J.E.M. AG Supply v. Pioneer Hi-Bred International* 122 S. Ct. 593, 2001).

[38] Later it was established that patentable subject matter under U.S. patent law included multicellular organisms (polyploid oysters, *Ex parte Allen*, 2 USPQ2d 1425 (Bd. Pat. App. & Interf. 1987), *aff'd mem.*, 846 F.2d 77 (Fed. Cir. 1988)) and mammals (a cancer prone mouse – the "onco-mouse" of Harvard University, U.S. Patent No. 4,736,866).

In Europe, the European Patent Office[39] has ruled that plant varieties are not patentable, yet the transgenic methods and plants produced using these methods are not *per se* unpatentable. This means that an applicant can effectively claim plant groupings other than varieties (e.g., corn, rice, cereals).

The Andean Community, a sub-regional organization with Bolivia, Colombia, Ecuador, Peru and Venezuela as member countries, has a common intellectual property regime that is embodied in Decision 486, which entered into effect in 2000. Article 20(c) of Decision 486 expressly prohibits patents on "plants, animals, and essentially biological processes for the production of plants or animals other than non-biological or microbiological processes". These prohibitions are the extent of what TRIPS allows (Article 27).

Some countries do not allow patents on DNA and amino acid sequences corresponding to the proteins produced by a naturally occurring organism. Presumably these countries, including Brazil, Cameroon, Colombia, Cuba, Guatemala, and Uzbekistan [Thambisetty (2002)], are relying on the flexibility of TRIPS to define isolated products from organisms as "natural" products which are not new; as a consequence, they are unpatentable.[40]

The fact that patent law prohibits protections for plants (and other higher organisms) does not necessarily leave technology owners entirely at risk. If a plant contains patented genetic material, would the protection of the gene "extend" protection to the whole plant? Although Canadian patent law does not allow patenting of 'higher life forms' (e.g., plants and animals), the courts have held that growing transgenic canola plants that contained a patented herbicide-tolerance gene infringed the patent on the gene (*Monsanto Canada Inc. v. Schmeiser*, 2004 SCC 34). This ruling clearly favors the economic interests of the patent holder [Nottenburg (2004)]. If other countries follow suit, it could alter interpretation of the treatment of plants and other life forms in the TRIPS agreement, which is discussed below.

The values of patents are highly skewed. While, for example, the widely-licensed Cohen–Boyer patent earned more than $200 million in royalties, most patents generate revenues ranging from zero to just tens of thousands dollars (meaning that the costs far outweigh revenues). It is not surprising, then, that most inventions are patented in just one or a few developed countries with large markets. Even highly relevant biotechnology innovations have been protected in few if any developing countries, including those where patenting of the relevant type of technology is available. For example, none of

[39] The European Patent Office (EPO) is a regional organization that examines patent applications for its member countries. An applicant that obtains a grant from the EPO must register the patent in each country where she wants it to have effect. Despite recent efforts to implement a Europe-wide patent, patents are still national in scope.

[40] The reasoning behind the policy that allows isolated DNA and protein products to be patentable in the United States, for example, keys on the fact that these products do not occur in isolation in nature. To isolate them requires human intervention, which propels them into the realm of patentable subject matter in the United States.

a set of key *Agrobacterium* technologies is patented, or has a patent pending, in more than four jurisdictions outside Europe, while the very popular CaMV 35S promoter, used widely in plant transformation, is patented only in European countries, Japan, and the United States [Roa-Rodríguez (2003); Pardey et al. (2003)].

4.1.4. Trade secrets

Proprietary information (e.g., customer lists, business plans, manufacturing processes, inbred lines required for hybrid seed production) that has commercial value can be protected as a trade secret if the holder conceals it from its competitors to prevent others from duplicating or using it [Friedman, Landes and Posner (1991)]. Information protected under trade secrecy can include new genetic material, notes of experiments, the information in an employee's head regarding proprietary processes or procedures, and generally inventions for which patent protection will be sought prior to the publication of the patent application (or the patent itself for inventions patented only in the United States). The secret (or information) is not a property right *per se*, but is recognized as a personal right by common law.[41] This right, in contrast to patents and copyrights, is not registered and is of unlimited duration, as long as the holder of the trade secret makes reasonable efforts to maintain secrecy. Identification by private code of fields of inbred parent lines of corn has been deemed, in the 1994 Iowa case *Pioneer Hi-Bred v. Holden Foundation Seeds*, sufficient to constitute a reasonable effort to maintain secrecy, even though the corn was grown outdoors and subject to misappropriation by informed "flashlight breeders". Furthermore, submission of sample seeds to the Plant Variety Protection Office as part of an application for a PVPC does not constitute a disclosure that prevents subsequent patentability.[42]

If a trade secret is infringed, the rights holder has recourse through the courts for remedies, including injunction and damages. A product or process made using the trade secret is not infringing, however, if the secret information is revealed by reverse engineering, independent discovery, or obtained by other legal means (for example, from

[41] Common law is that body of law made by judges and not by legislation. Countries with legal systems based on English law have common law.

[42] In *Advanta USA Inc. v. Pioneer Hi-Bred International Inc.* W.D. Wis., No. 04-C-238-S, 10/27/04, Advanta (now owned by Syngenta) admitted that they selected parental lines from seeds included in bags of hybrid corn seeds. Pioneer had a patent on the hybrid corn lines, but not on the parental lines, which were however registered under the PVPA. Of course, registration requires some amount of disclosure, which would seem at first blush to extinguish any claim of misappropriation. The Court found otherwise. The issuance of a PVPC does *not* per se extinguish the secrecy necessary to maintain a trade secret. In particular, the PVPA does not require that information about registered plant varieties be made accessible to the public. Moreover, the disclosure requirement for obtaining a PVPC does not need to be enabling, such as is required for a patent. Therefore, there is no inherent conflict between obtaining plant variety protection and maintaining the protected variety as a trade secret.

a published paper or in a seminar).[43] Trade secret law is applicable to biotechnology, although may not be widely used due to ease of reverse engineering. On the other hand, the revolution in analysis of genetic material has created genetic fingerprinting technologies well suited to detection of unauthorized reproduction or breeding, so that trade secrets might be used more frequently in the future.

Trade secrecy is in fundamental conflict with academic ideals of open communication and sharing of knowledge. Yet most, if not all, academics utilize secrecy to their advantage to at least some extent, when they withhold some relevant information about research plans, approaches, preliminary findings, and similar matters, prior to publication, for at least some period of time. But academics have strong incentives to publish research achievements. On the other hand, attempts by firms to maintain trade secrecy over the long term can prevent new discoveries from being presented at conferences or published and can result in restrictions on the inter-firm mobility of scientists. It is noteworthy that surveys indicate that in most industries, trade secrecy outranks patents as a means of protecting innovations [Levin et al. (1987); Cohen, Nelson and Walsh (2000)]. Though in principle of indefinite term, in practice trade secrecy is frequently used for short-term protection, allowing innovators to be "first to market", or to safeguard their innovation until a utility patent is awarded and published.

4.1.5. Trademarks

The identity of a product, as perceived by the customer, has important bearing on the value of that product. Thus, a registered trademark under which a plant variety or genetic trait is sold can be a significant means of protecting the value of genetics embodied in the product. Even under conditions where copying of an underlying innovation is extensive, in order to assure quality or performance purchasers will often pay more for a brand name version with a good reputation as compared with a generic or knock-off version, especially when a crucial quality or trait is not directly observable, as is the case with herbicide tolerance or pest resistance. Horticultural producers will often pay more for a trademarked variety in order to be able to label their product with the trademarked name, if the latter is recognized and valued by consumers. Trademarks have an additional advantage in the duration of protection. Registration is renewable for a modest fee as long as the mark is being used. As well, obtaining trademark registration in multiple countries is becoming easier under the Madrid Protocol.

4.1.6. Geographical indications

Geographical indications (GIs), like trademarks, have been used to affirm the quality, reputation and other aspects of a good. TRIPS defines geographical indications as

[43] In the headlines of computer news in mid-2004 was the report that Real Networks reverse-engineered Apple Computers' music encoding used on iPods. Although Apple publicly tried to discredit Real Networks, Apple had no legal recourse, and Real Networks still sells music for iPods through e-commerce.

"indications which identify a good as originating in the territory of a Member, or a region or locality in that territory, where a given quality, reputation or other characteristic of the good is essentially attributable to its geographical origin".

In order to qualify for GI protection then, there has to be a link between the characteristics of the product and its place of origin. Unlike other intellectual property rights including trademarks, geographical indications do not reside in a single owner but collectively belong to all producers in the geographical area.

Long an important protection for wines and spirits (e.g., Champagne, Scotch), the issue of expansion to other products has attracted global attention. TRIPS limits protection to "goods", meaning that services do not qualify. Under TRIPS, wines and spirits have a higher level of protection under Article 23. For other goods, Article 22 provides a protection scheme. Two main requirements that WTO Members must fulfill are the implementation of legal mechanisms for rights holders (i) to exclude others from presenting a good that misleads the public as to its geographical origin, and (ii) to prevent use of the geographical indications that constitutes an act of unfair competition. As with other Articles in TRIPS, key terms are undefined, including the definition of "public" and "misleading".[44]

4.2. *Alternatives to intellectual property rights*

4.2.1. *Hybrid plants*

Until the advent of biotechnology, the principal effective strategy for protecting inventions embodied in plants was the production of hybrids for commercial sale. Since hybrids do not breed true, replanting of their seeds yields an inferior crop and thus farmers are induced to purchase new seed for each planting season, rather than saving their own seed. The case of corn in the United States is the most prominent agricultural (as distinct from horticultural) example. Protection via hybridization was strong enough to foster the establishment of a profitable private seed industry in the 1930s [Griliches (1957)] in which future Vice-President Henry Wallace was a prominent participant, well before the effective strengthening of legal intellectual property protection of plants.[45]

In developing countries such as China and India, hybridization offers weaker protection because parent lines cannot be effectively protected for more than a few years from falling into the hands of competitors. Until the 1980s, appropriation of parent lines by competitors was also a problem for leaders in the hybrid corn industry in the United States, as confirmed in informal reports of comparative analysis of proprietary germplasm. Identification of parent lines has become more feasible with the advance

[44] For additional perspective on geographical indications see Maskus (2000), Taubman (2001) and Josling (2005).

[45] See also Knudson and Ruttan (1988) on the history of the commercially unsuccessful efforts to develop hybrid wheats in the United States.

in the tools of biotechnology. Nevertheless misappropriation has still occurred, as indicated in the recent payment by Cargill to Pioneer Hi-Bred in settlement of a charge of misappropriation of corn germplasm by a breeder who had moved from Pioneer to Cargill.

Some critics have interpreted hybrid breeding strategies as motivated by capitalistic designs to monopolize germplasm commercially [e.g., Kloppenberg (1988)]. Note, however, that the hybrid cultivars developed under communism constitute large fractions of China's rice and maize crops.[46]

4.2.2. Genetic use restriction technologies (GURTs)

Despite the expanding scope of intellectual property protection, enforceability and the cost thereof are still major issues. This is especially true at the farm level. Even in the mature institutional environment of the United States, it may not be cost-effective to directly sue farmers in court for IPR infringement because the sums at stake and the limits on infringing farmers' assets are usually less than the cost of an average lawsuit. Only their deterrent effects on others' behavior can justify such actions. As a result, scientists have been researching new biological means of restricting the copying of germplasm or proprietary traits, and thus appropriating the returns to investment. A diverse set of technologies is being considered in the general class characterized as Genetic Use Restriction Technologies (GURTs) [Jefferson et al. (1999)].

GURTs come in two broad types, variety-level (V-GURT) and trait-specific (T-GURT). In 1998, a U.S. patent (5,723,765) was granted jointly to the USDA and Delta & Pine Land Company, the largest U.S. supplier of cotton seed, on a V-GURT technology [Jefferson (2001)]. This technology was designed so that a seed producer could inoculate the seed with a specific regulator that renders the plant infertile before delivery to the farmer, thus making seed saving infeasible. Well before this type of physical protection could be developed into a marketable technology, the prospect of V-GURT cultivars generated substantial opposition from farmers' groups and other non-governmental organizations. Although V-GURTs had the potential to allay oft-expressed concerns of these groups regarding transgenic flows to second-generation seeds (of the same crop or a cross fertilized weedy relative), critics dubbed them "terminator technologies", emphasizing the possibility that they might sterilize neighboring non-transgenic crops by drifting pollen, or bankrupt farmers who unknowingly received terminator seed by mistake.

On the other hand, T-GURTs do not terminate reproduction of the plant. Generations grown from saved seed will be fertile. But to induce expression of a protected trait in a given year, an activator, for example a chemical, must be applied [Jefferson (2001)]. Shoemaker (2001) suggests that a T-GURT might be activated by spraying a "standing

[46] Fan et al. (2003, p. 8) estimate that hybrid varieties accounted for 61% of total Chinese rice production in 2001.

crop" with a highly specific and proprietary compound. The fee for using the gene could be collected as part of the purchase price of the activator. If such a technology proves feasible, the farmer might be able to wait until the trait, such as disease resistance, is clearly needed before purchasing and applying the activator, in essentially the same manner as a chemical biocide. In this case the trait offers "self-protection" rather than "self-insurance", in the terminology of risk management. Such a targeted strategy could help reduce selective pressure and the buildup of resistance in the target pest population.

We are not aware that any of these technologies has moved (or is even close) to commercial implementation; however they have generated much controversy and concern. In response, varieties using V-GURTs have been disqualified from registration in India [Pendleton (2004)], and V-GURTs have been rejected by the Rockefeller Foundation [Rockefeller Foundation (1999)] and by the Consultative Group on International Agricultural Research (CGIAR), but not by the USDA, co-developer of the original cotton V-GURT technology.[47] A UPOV memorandum critical of GURTs was issued in 2003, but under political pressure from the United States UPOV withdrew its critique in a subsequent position paper [Pendleton (2004)]. As the important distinctions between V-GURTs and T-GURTs become better understood, it is possible that attractive applications of the latter (and perhaps of the former, in preventing pollen drift problems for pharmaceutical-producing plants, for example) might be developed and adopted with scientifically credible safeguards.

4.2.3. Contractually-defined rights over tangible property

Finally, it should be noted that the exercise of rights over physical property can be invoked via contracts to prevent dissemination of genetic or intellectual constructs embedded within something claimed as physical property. This can be relevant in the research setting, where a particular vector or other biological material created by researchers may be treated as their own or their employer's physical property [Kryder, Kowalski and Krattiger (2000)]. For example, breeding pairs of mice with specific genetic characteristics have been successfully commercialized using material transfer agreements, without invoking patent protection. University offices of technology transfer are becoming increasingly attracted to this alternative, where feasible, due to its simplicity and the avoidance of the high costs, in time and money, of obtaining patents and negotiating licenses. This alternative is of relevance for the farm production of high-value proprietary products such as modified-oil canola, where a supplier of germplasm may lease or lend plants to the farmer under contract, and directly claim the harvested

[47] On October 30, 1998, the CGIAR eschewed the use of terminator gene technology (or more specifically, "any genetic systems designed to prevent seed germination") based on the "recognition of concerns over potential risks of its inadvertent or unintended spread through pollen; the possibility of the sale or exchange of nonviable seed for planting; importance of farm-saved seed, particularly to resource-poor farmers; potential negative impacts on genetic diversity; and the importance of farmer selection and breeding for sustainable agriculture" [CGIAR (1998, p. 52)].

product, maintaining identity preservation and retaining legal physical ownership and thus a right to reclaim the property at the end of the contract or if the contract is breeched [Margeules (2002)].

4.2.4. Mergers and acquisitions

The problems posed by the multiplicity of proprietary claims to key research inputs have led firms to bring the claimants "in-house" by engaging in strategies of merger and acquisition to obtain commercial freedom to operate in plant breeding. Thus since 1995 there has been a significant increase in concentration in the agricultural input firms producing agricultural biotechnologies, seeds, and crop protection chemicals. For example, de Janvry et al. (2000) report that in 1994 individuals and independent biotechnology firms held 77% of U.S. patents related to *Bt*, whereas years later, in 1999, six integrated firms held 67% of *Bt* patents, of which 77% were obtained by acquisition of smaller biotech and seed firms.

5. The globally evolving institutional environment for agricultural IPR

The current international environment for intellectual property relevant to agriculture is the result of continuing complex interactions between many entities representing many different interests, in many different settings. We consider, chronologically, several quite distinct, and in some ways conflicting, international agreements. UPOV focuses on protecting the rights of seed producers, TRIPS is directed to strengthening the rights of inventors, while the International Undertaking, CBD, and the International Treaty are oriented toward protecting the rights of farmers who use landraces, or at least of the nations in which the farmers reside.

5.1. International Union for the Protection of New Varieties of Plants (UPOV)

Until the 1960s, plant genetic resources were considered to be products of nature, and the work of breeders received very little institutional protection. A few countries (the United States and several European countries) gave limited rights to plant breeders prior to the 1960s. As the size of plant breeding industry increased, commercial plant breeders and seed sellers lobbied for international standards for plant breeders' rights. The International Union for the Protection of New Varieties of Plants (UPOV) was founded in 1961 to create a set of uniform and clearly defined principles on protecting new varieties, which are mostly developed by plant breeders. Originally founded by 5 countries (Belgium, France, Germany, Italy, and Netherlands) in 1961, membership in UPOV has gradually increased over time to 58 states as of November 2004.

The UPOV Convention sets out exclusive property rights in the form of plant breeders' rights (PBRs) that cover new plant varieties satisfying the criteria of distinctness, uniformity, and stability. Over time, the UPOV Convention has undergone several revisions, notably in 1978 and in 1991, in the direction to expanding breeders' rights (discussed above). In 2002, the UPOV Council explicitly adopted a concept of novelty that includes discoveries such as selections from pre-existing populations. But the new variety must not be a matter of common knowledge, anywhere in the world, and thus a landrace is not eligible for protection.[48] Evidence on the effect of PBRs on the rate of innovation suggests that PBRs offer at best only modest incentives for innovative breeding, if revenue collection relies on initial sales of seed.[49] They have important uses in protecting horticultural varieties distributed as clones, and in protecting inbred parent lines of hybrids from utilization for hybrid production by competing breeders. In Australia, preliminary evidence from the wheat market indicates that protection under UPOV can be more effective in ensuring a return to breeders' investment if collection of a royalty on crop output is feasible.

5.2. *International undertaking on plant genetic resources*

The structure of plant breeders' rights afforded by UPOV strikes a balance between the interests of plant breeders who desire open access to breeding materials including farmers' landraces, and their interests in protecting their newly developed modern varieties. Critiques of the UPOV Convention and intellectual property protection mechanisms in general focused on the perceived inequality of patterns of germplasm protection and exchange between developed and developing countries. In response to these concerns, the Food and Agriculture Organization (FAO) of the United Nations drafted the International Undertaking on Plant Genetic Resources (the "International Undertaking") in 1983. As originally framed, the principal aim of the International Undertaking was to define the common heritage of mankind to encompass all plant genetic resources, including improved modern varieties developed (and protected) by plant breeders. In addition, the 1983 International Undertaking emphasized that free exchange of genetic resources was vital for food and agriculture. However, strong opposition by plant breeders in developed countries to free access to their modern varieties led to a negotiated compromise: the simultaneous and parallel international recognition of the plant breeders' rights and farmers' rights. This recognition is embodied in FAO conference Resolutions 4/89, 5/89, and 3/91.[50]

[48] See "Revision of Document C(EXR.)/19/2 The Notion of Breeder and Common Knowledge", UPOV, August 9, 2002, especially paragraphs 14,15, 22, 23, and 24.

[49] The studies are listed in footnote 34 above, and the associated text.

[50] Resolution 4/89 acknowledges that the rights of plant breeders are not incompatible with farmers' rights and allows for legal protection of patented seed varieties. Resolution 3/91 deals with financial and institutional aspects of the implementation of farmers' rights. It also restricts the free flow of germplasm by stating that though the common heritage principle is still recognized, it is subordinated "to the sovereignty of the states over their plant genetic resources" (International Undertaking on Plant Genetic Resources, Annex III, Resolution 3/91, available at http://www.fao.org/ag/AGP/AGPS/pgrfa/pdf/iu.pdf).

The concept of farmers rights stated in Resolution 5/89 was intended to form the basis of a formal recognition and reward system to encourage and enhance the continued role of farmers and rural communities in the conservation and use of plant genetic resources [Esquinas-Alcázar (1996)]. Its adoption constituted path-breaking international recognition of the role of farmers in conserving and developing plant genetic resources. The agreement was not legally binding, however. It merely represented a moral commitment on the part of member countries to recognize and reward "the enormous contribution that farmers of all regions have made to the conservation and development of plant genetic resources" (Resolution 4/89).

5.3. Convention on Biological Diversity (CBD)

The Convention on Biological Diversity (CBD) entered into force in 1993, and a total of 188 parties have signed (though not necessarily ratified) this agreement as of December 2004. It began as an international initiative, with United States support, to conserve flora and fauna *in situ*, in response to claims of an alarmingly high rate of species extinction. The three main objectives of the CBD are the conservation of biological diversity, the sustainable use of its components, and the fair and equitable sharing of benefits arising out of the utilization of genetic resources (Article 1). Although the CBD does not expressly refer to any international IPR agreements, it contains provisions relating to IPRs in the context of access to and transfer of technology (Article 16).

One of the most noteworthy aspects of the CBD is the recognition that nation states have the sovereign right to exploit their own resources as well as the authority to determine the conditions of access to them (Articles 3 and 15). This ownership clause of genetic resources is in stark contrast to the "heritage of mankind" concept regarding genetic resources envisioned previously in the International Undertaking. Inspired by generalization to agriculture of overly optimistic expectations of the marginal value of *in situ* biological resources for the development and marketing of pharmaceutical products, the CBD emphasized the intellectual property protection of genetic resources and technology transfer as a basis for benefit sharing [Dutfield (2002); Boettiger et al. (2004)]. Were such rights granted, contracts could exist between the nations holding the resource rights (or indigenous populations, if their nations give them the rights) and those standing to benefit from their exploitation (biotechnology or pharmaceutical firms in developed countries).[51] Note that the CBD emphasizes the sovereign rights of states, not the rights of farmers within those states. Nor does it describe mechanisms for sharing the benefits arising from the use of genetic resources.

Several access frameworks for genetic resources have been developed at the national and regional levels since the CBD. Legislations regarding access and benefit sharing to the country's genetic resources was introduced or is being introduced in several

[51] For one creative initiative involving community maintenance of trade secrecy regarding its genetic resources, see Vogel (1994).

countries including Philippines, Costa Rica, the Andean Pact countries, and the Organization of African Unity (OAU) representing 53 African nations [Diaz (2000)]. In practice, however, access and benefit sharing provisions vary greatly among countries. Their strength appears to be empirically related to the significance given to environmental protection within a nation's governmental structure [Carrizosa et al. (2004)]. The main practical effect to this point has been to slow or stop international bio-prospecting activities for both scientific and business purposes in many countries, beginning with pharmaceutical bio-prospecting, and extending to agricultural genetic resources [Dalton (2004)]. It is now evident that the financial value of bioprospecting for pharmaceuticals or for germplasm is generally insufficient to justify conservation of *in situ* resources at the margin, a result predicted by a controversial paper by Simpson, Sedjo and Reid (1996).[52] Indeed the value might often be insufficient to outweigh the current levels of transactions costs associated with access and benefit sharing rules designed to implement the CBD. Thus, access and benefit sharing legislation has become an important influence on the transfer of genetic resources between countries, but not perhaps as intended by the CBD.

5.4. *Trade-Related Aspects of Intellectual Property Rights (TRIPS)*

Currently, the incentive and property rights structures of agricultural research are undergoing a global transformation unprecedented in its scope and speed. The dominant influence on this transformation is an agreement known as TRIPS (Trade-Related Aspects of Intellectual Property Rights), a multilateral agreement among the members of the World Trade Organization (WTO), including 148 countries as of October 2004. TRIPS was negotiated from 1986 to 1994 as a part of the Uruguay Round of the General Agreement on Tariffs and Trade (GATT).

The TRIPS agreement requires member countries to satisfy setting minimum standards for all major types of intellectual property rights (copyright, trademarks, geographical indications, industrial designs, patents, topographies of integrated circuits, and trade secrets). Further, it details how countries should enforce these rights and how disputes are to be resolved. When the agreements took effect in January 1995, developed countries were given one year to comply, developing countries and transition economies five years,[53] and least-developed countries 11 years (which is now extended to 2016 for pharmaceutical patents).

Perhaps most importantly, Article 27.3(b) of TRIPS states that all plants may be excluded from patentability, provided the member country adopts alternative intellectual

[52] If the species-area relation were linear, the result might be different [see Rausser and Small (2000) and the comment by Costello and Ward (2003)]. However, ecological studies indicate that the species–area relation is concave.

[53] If a developing country did not provide product patent protection in a particular area of technology (e.g., pharmaceutical and agricultural chemical products) when the agreement came into force, it has up to 10 years to introduce the protection.

property legislation referred to as an "effective *sui generis*" system to protect plant varieties. *Sui generis* in Latin means "of its own kind", and in TRIPS the phrase is used to indicate a flexibility whereby member countries can individually design a system of plant variety protection that works for their country. It is important to note that *all plants* can be excluded from patentability while the alternative form of protection (or in combination with patents) is required to cover only new *plant varieties*. A gap is left whereby a plant, which is not a variety, may be unprotected.[54] Complicating the issue, neither of the terms "effective" nor "plant variety" is defined within the TRIPS document, and as discussed above, the interpretation of these requirements in the field of biotechnology differs among countries. Ultimately, the interpretation of what TRIPS mandates as patentable in the field of plant biotechnology will require resolution in the dispute settlement processes of the WTO.

The huge global effort to apply minimum standards of IPR to agriculture is posing significant challenges to less developed economies.[55] It is also evident that the net benefits of these minimum standards with respect to agriculture, for individual countries or the world as a whole, were never seriously addressed prior to adoption.[56] In general, a nation with a large market has a greater interest in domestic protection of innovations while a nation with a small domestic market derives greater relative benefit from international protection of its innovations.[57] After the multilateral adoption of TRIPS, the European Union and the United States have pursued a program of bilateral and multilateral negotiations with small or weak states, in which stronger protection than provided for in Article 27.3(b) feature prominently in the agreements.[58] The significance of these agreements, and indeed of TRIPS itself, depends on implementation involving domestic legal enforcement.

Economic analysis of the effects of trade liberalization and intellectual property rights has been a key element of the worldwide movement toward trade liberalization, at least since Adam Smith. Although there is a significant amount of literature on patenting in an international context, it is generally based on models with assumptions inappropriate to the question of protection of agricultural innovation. Almost all

[54] The same provision, Article 27.3(b), also allows "animals" to be excluded from patentability.

[55] See Boettiger et al. (2004) and references therein for a discussion of the history and motivation for the TRIPS negotiations. For an initiative to increase access of the poor to some pharmaceuticals, see Lanjouw (2003).

[56] Samuelson (2004) provides an informative treatment of the effect that foreign laws on IPR can have on the operation of domestic IPR.

[57] Scotchmer (2004a) argues that small countries have a greater interest in international harmonization with strong IPR than do large countries. Of course, they could find a regime of strong protection in other countries with national treatment of large markets, but weak protection in small markets, even more attractive. See also the discussion of national IP options in The Crucible II Group (2001).

[58] As of October 2004, it has been reported [GRAIN (2004)] that eight of twelve completed bilateral "TRIPS+" agreements involving the United States include mandatory patentability of plants and animals, well beyond the requirements of Article 27.3(b). In an analysis of the trade implications of IPRs, Smith (2002) showed that stronger foreign patent rights enhance the market power of U.S. drug exports.

models assume that there are only two types of countries, the North and the South, and most assume that only the aggregate North can innovate [Chin and Grossman (1990); Diwan and Rodrik (1991); Deardorff (1992); Helpman (1993)]. In more recent work, Grossman and Lai (2004) and Lai and Qiu (2003) assume innovation is possible in the South, but, like all the foregoing, assume that without patents there is no innovation and that the South is homogeneous. It seems that the stylized models available so far are not appropriate for economic analysis of the globalization of agricultural intellectual property rights. In reality, agricultural research capacity worldwide is concentrated primarily in nations with some of the largest land areas and markets for agricultural products, nations who can hope to internalize domestically a substantial part of research gains on agricultural staples via lower food prices, and secondarily among cash-crop exporters whose farmers will gain from locally-specialized cost reductions or quality improvements. North versus South is just the wrong match-up for understanding the impact of IPR on global agriculture, especially in the long run [Yang (1998)].

For the future, there are plans afoot to implement a "Substantive Patent Law Treaty" under WIPO. This would, unlike TRIPS, impose a global system of patent protection with centralized administration. The administrative economies could be particularly important for small or poor countries. But the further toughening of patent enforcement might prove to work against their interests.

5.5. *International treaty on plant genetic resources for food and agriculture*

The International Undertaking has been revised as a treaty, and brought into harmony with the CBD in the 1990s, especially with respect to access to plant genetic resources and equitable sharing of the benefits. By incorporating changing perspectives on the ownership and international exchange of germplasm, the International Treaty on Plant Genetic Resources for Food and Agriculture (in short, the "International Treaty"), was adopted in November 2001. The International Treaty came into force on June 29, 2004, after it had been ratified by 40 governments.

The International Treaty exists in a legal and political space between the CBD and the TRIPS agreement [Brush and Stabinsky (1996)]. It seeks to preserve aspects of a free exchange system for germplasm that is seen as fundamental to global food security, while achieving harmony with the CBD. Its focus is to create a multilateral system in which member states designate plant materials that can be shared and accessed by all other members. Exchange of these materials is to be governed by a common material transfer agreement (MTA) and without any further bilateral negotiations. The current agreement lists 40 crops that can be exchanged through a common MTA, including all of the major staple crops such as rice, wheat, maize, potato, and beans (with the notable exception of soybeans).

The provisions on access and benefit sharing are aimed to address equity concerns of contributing parties, including, prominently, groups of farmers in developing countries.[59] Four major types of benefit sharing are set out: (i) exchange of information; (ii) access to and transfer of technology; (iii) capacity building; and (iv) sharing of monetary and other benefits of commercialization. The financial aspect of benefit sharing is required only if a commercialized product is restricted for further research and breeding.[60] The practical mechanisms for tracking and documentation will be critical to whether or not the multilateral system of exchange and benefit sharing can be workable, keeping the costs of each transaction low enough to approximate the system of free exchange. For any material accessed from the system, the Treaty's governing body will have to develop a common MTA, and a database to ensure that all parties involved comply with the access agreements, and to track future utilization of material. The initial constructions of databases and legal structures are not difficult; however, the longer-term tasks of tracking all materials utilized in breeding future generations of crop cultivars and assessing royalty payments for varieties used in producing future commercial cultivars for which further uses will be restricted will be extremely challenging, to say the least.

If the agreement mandates complex and potentially costly systems to track material accessed through to any subsequent "derived product", the result might be that exchanged seeds will be limited to the point that little or no financial return is generated. The recent experience with access and benefit sharing under the CBD, discussed above, is instructive in this regard.

6. Instruments for transactions in agricultural IPR

An IPR has value to the extent there is an expectation that a product or service relying on the underlying technology might be profitably commercialized. In agriculture the ultimate commercializers are usually farmers, and in general it is not feasible for biotech innovators to integrate forward to such an extent that they monopolize production of a given crop. Therefore transactions that enable the use of the technology for further innovation or for production of intermediate inputs or of outputs are key elements of the transfer of agricultural biotechnology domestically or internationally. As countries follow the lead taken by the United States, in the Bayh–Dole Act of 1980 and a series of other legislative initiatives in encouraging the patenting of publicly supported research, transactions involving intellectual property rights are becoming increasingly important for public and nonprofit research organizations as well.

The transaction of an IPR is a formal or informal contract by which the rights holder pledges not to exert his right to exclude another party from practicing the specified invention or technology under specified conditions, in return for some form of compensation ("consideration", in legal jargon). In essence, it grants permission for, or waives

[59] Farmers' rights are mentioned in the text of the agreement, but are acknowledged to be within the purview of the member countries. Furthermore, farmers' rights are envisioned to be much broader than the right to save, use, and sell farm-saved seed.

[60] The amount of payment is set by a Governing Body and may be reviewed from time to time.

the right to prevent, the use of the innovation in return for an agreed amount of compensation or a specified royalty payment. Thus the ability to trade IPRs depends in large part on the ability of the rights-holder to detect violations and enforce those rights, which in turn hinges on the nature of the technology (process versus product), the strength of the legal system, and the cost of establishing and evaluating an allegation that someone has violated the IPR.[61] We have seen that biotechnologies such as genetic fingerprinting have helped prove IPR violations, and others such as GURTs could be used to prevent violations directly. The various instruments used in IPR transactions include:

6.1. Licenses

Licenses are formal contracts between a technology owner and a licensee extending permission to use technology covered by any form of property right according to the terms and conditions of the contract. Permission may be granted on an exclusive or non-exclusive basis, and the conditions of the contract may define or restrict the scope of use allowed by the patent monopoly. Payment for licenses can take many forms, but usually includes an up-front lump sum, milestone payments, or a running royalty, the latter dependent on the value or volume of production. A licensee cannot be committed to pay royalties beyond the expiration of the patent term, or after the patent has been declared invalid.

An owner of a patent on an enabling technology might offer a license in exchange for royalties on products generated from innovations achieved using the technology. Such "reach through" licenses are controversial. They give the enabling technology inventor some control over other inventions enabled by their technology, decreasing the incentives for follow-on innovative work. Yet in some circumstances the reach through provisions might be a reasonable means of sharing risk and reward from an uncertain research path. [See Arora and Fosfuri (2005) for an analysis using this insight.]

Patent licenses are often negotiated in combination with "know-how" contracts, generating royalties and fees that may be major sources of revenue flows for technology providers. In 2001, royalties and fees received by the United States, the largest technology-providing nation, were U.S. \$38.7 billion, while it paid U.S. \$16.4 billion to other countries, the bulk of this going to other developed nations. However, U.S. \$21.5 billion of those receipts came from foreign affiliates of U.S. firms and about half (U.S. \$8.6 billion) of U.S. payments went to foreign affiliates of U.S. firms [U.S. Census Bureau (2001)].[62] These figures might well be distorted because of transfer pricing schemes by the parent organization and its affiliates to minimize worldwide tax payments, taking advantage of differences in tax rates between countries. Assuming the reported figures are informative, it seems that a net U.S. \$9.4 billion of receipts came

[61] Although especially in the case of non-exclusive licenses, free-riding may be tolerated.

[62] Another source reports that worldwide about 70% of total royalty and license payments flow between trans-national corporations and their overseas affiliates [UNDP (1999)].

from arms-length transactions (predominantly non-agricultural). This is a large amount but, given the prominence of technology transfer issues in international negotiations, some perspective on the potential role of private technology transfer may be gained by noting that this net figure, in which agricultural technologies no doubt have only a minor share, is substantially less than the U.S. $11.8 billion spent by developing countries on public agricultural research alone in 1995 [Pardey and Beintema (2001)].

Licenses that allow use of patented technology in research, but do not permit use in a way that results in commercialization, are often cheaper and easier to get than commercial licenses. But such research licenses can be a dubious blessing: any innovations achieved under the research license may be blocked from subsequent commercialization by refusal of the rights holder to sign a commercial license. The innovator is left with sunk costs of research dependent on someone else's permission for commercialization and thus is placed in a very weak bargaining position. In some recent instances, biotechnology researchers have been blocked from application of the technology [Wright (1998)].

If full legal freedom to operate in research is desired, a license is in general necessary. Contrary to popular perceptions, few countries have liberal legal exemptions for use of patented technology in research, even research done by public or non-profit agencies [Nottenburg, Pardey and Wright (2002)]. In the United States, a recent judicial decision has emphasized the highly restrictive nature of the research exemption.[63] Even if available they generally allow research on, rather than research with, an innovation. Research exemptions have the same disadvantages as research licenses, discussed above, in cases that might lead to commercialization.

6.2. *Material transfer agreements (MTAs)*

An MTA is a contract for transfer and use of an input to the research process, either for basic research or for commercial use. The transferred material must be covered by some independent means of intellectual property protection (for example, a patent or trade secrecy) or be a defined piece of tangible property to provide a basis for preventing its appropriation by third parties. MTAs are often means of transferring material under trade secret protection, as embodied, for example, in the various state laws in the United States. They may restrict the user's rights to modify, improve, resell, or commercialize the biological material.

MTAs are used by the research centers of the Consultative Group on International Agricultural Research (CGIAR) to control the use of plant varieties held "in trust" on behalf of the countries of origin in their genebanks [Binenbaum et al. (2003)]. If access to the materials is not otherwise available, this protection may be effective in preserving the provider's rights over the germplasm. A major limitation of MTAs is that they generally confer no direct control over the material, should it be acquired by a third party.

63 See Eisenberg (2003) for discussion of a recent appeals court ruling stressing the narrow scope of research exemption in U.S. patents (*Madey v. Duke University*, 307 F.3d 1351, 2002).

6.3. Bag-label contracts

In selling seeds in developed countries, a contract is often printed on a seed bag label. Users of the product are presumed to have agreed to comply with the contract if they break the label to open the bag. Thus a "bag-label contract" seems analogous to the "shrink-wrap" contract common in software transactions. If use of the seed is found to violate such an implicit contract, then bag label contracts are an additional means of protection of intellectual property. The validity of bag labels as licenses has thus far been upheld by courts.[64]

6.4. Technology use agreements

In recent years in the United States a major producer of transgenic crops has adopted and enforced "technology use agreements", an innovation in property rights for agricultural technology. Their provisions typically include the right to plant a given seed type on a specific area of land for a certain period, and might also include restrictions on the use of proprietary traits in the creation of new varieties, or permission to the technology provider for access to the relevant property to check for violations.[65] The latter provision is not popular with farmers. These agreements have the efficiency advantage that royalty collection does not distort seeding rates; fixed running royalties per bag, on the other hand constitute an incentive to reduce seeding rates below their efficient levels.

7. IPR trends for plant varieties

International treaties like TRIPS and intergovernmental organizations like UPOV leave scope for much variation in the specifics of plant IP protection. Countries are exploiting these degrees of freedom, tailoring plant IP legislation to local circumstances. The variations include such fundamentals as the types of IP offered, species and genera encompassed, costs, and extent of farmers' rights. The long-run effects of these variations on the rate and direction of plant innovation are yet to be determined. Past productivity gains in agriculture were heavily reliant on the international spillovers of plant varieties. Although the geographical scope of protection is expanding, IP markets have to date been quite segmented – the preponderance of protection pertains to rich-country jurisdictions, leaving poor countries free to tap these technologies.

[64] *Monsanto Co. v. Ralph* (Fed. Cir. 2004) (03-1243); *Monsanto Co. v. McFarling*, 302 F.3d 1291 (Fed. Cir. 2002), cert. denied, 537 U.S. 1232 (2003); *Monsanto Co. v. McFarling*, 363 F.3d 1336 (Fed. Cir. 2004), cert. appl.

[65] For example, in 1998, a Mississippi farmer, Homan McFarling, purchased Roundup Ready soybean seed by signing the "technology agreement" which allows planting the seed only for one season and prohibits seed saving and selling. In 2000, he was sued by Monsanto for saving and planting the seed in the next season. The court decided that McFarling indeed breached the technology agreement and ordered him to pay Monsanto damages (Monsanto v. McFarling, Fed. Cir. 2004).

Table 4
Plant variety protection legislation worldwide (circa 2004)

Economies	Statutory protection	Legislation under consideration	Member UPOV	Member WTO
		(number of countries)		
High-income economies (54)[a]	29	4	23	37
OECD (24)	23		21	24
Non-OECD (30)	6	4	2	13
Upper-middle-income economies (37)	20	4	14	28
Lower-middle-income economies (56)	20	18	13	36
Low-income economies (61)	22	3	4	42
Total (208)	**91**	**29**	**54**	**143**

Source: Koo, Nottenburg and Pardey (2004) based on on-line searches of national intellectual property offices, http://www.upov.org/en/about/members/index.htm for UPOV data, and http://www.wto.org for WTO data.
Notes: Countries are classified into income classes according to World Bank (2004) criteria. Countries with 2003 per capita gross national incomes greater than $9386 are designated high income; $3036–9385 are upper-middle income; $766–3035 are lower-middle income; and less than $765 are low income.
[a]Bracketed figures indicate total number of countries in each income class according to World Bank (2004).

Putting policies and legal frameworks into place to protect plant varieties is one thing, seeking and maintaining varietal rights is something else entirely. Not least, these exclusionary IP rights are costly to obtain and to exercise, meaning economic choices based on the benefits versus costs of the rights are paramount. Agriculture in many developing countries involves subsistence or semi-subsistence cropping systems, with few commercial opportunities to market seed and consequently little incentive to seek varietal rights, even where they are a legal option. Here we review the IP rights on offer and provide new international evidence on the crop varietal rights in practice worldwide.

7.1. *Global trends in crop-related IPRs*

As of 2004, of a total of 191 countries, nearly half (91 countries) offered statutory intellectual property protection for crop varieties, while another 15% (29 countries) had legislation under consideration (Table 4). However, only a quarter of the countries (54 countries) were members of the UPOV Convention. Most of the Organization for Economic Cooperation and Development (OECD) countries have plant variety protection laws and are UPOV members, while the majority of the middle- and low-income countries do not have varietal protection legislation nor are UPOV member countries.

Figure 7 shows the changing pattern of applications for plant breeders rights (PBRs) for 37 countries grouped into four per-capita-income classes. More than 140,000 PBR

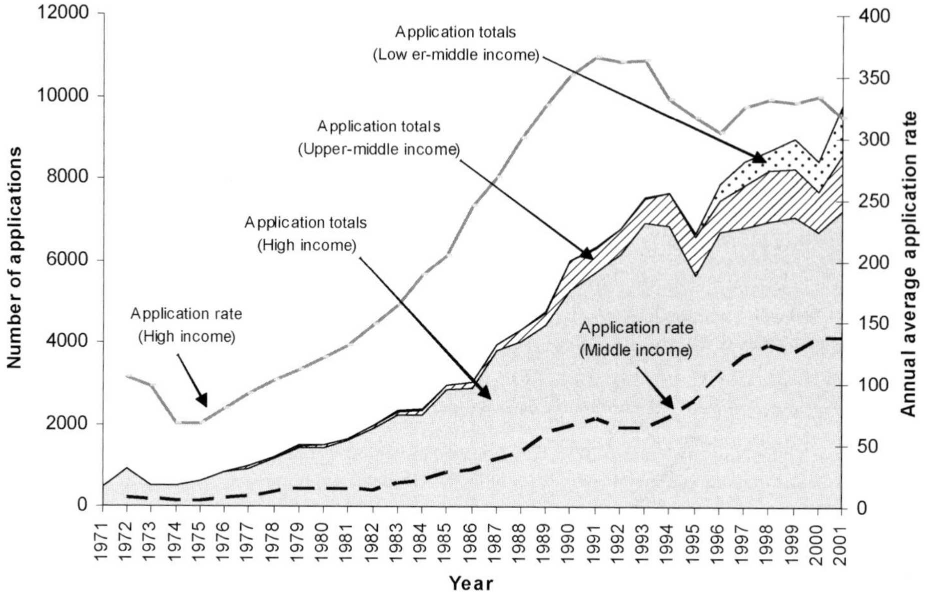

Figure 7. Plant breeders' rights applications for countries grouped by income, 1971–2001. *Source*: Koo, Nottenburg and Pardey (2004) based on UPOV (2003b) plus Koo et al. (2006) for China; Koo, Nottenburg and Pardey (2004) for Brazil; and USDA (2004) and USPTO (2004) for the United States. *Notes*: See notes to Table 1 for per capita income criteria used to classify countries. Application totals (LH axis) report total number of applications per year for countries grouped by income class. Application rates (RH axis) report annual rate of applications averaged across countries in each income class. Data for 2002 and 2003 were available to the authors but omitted due to likely underreporting stemming from lags in recording the rights claimed or granted.

applications have been lodged worldwide since 1971.[66] During the 1970s and 1980s, rich countries accounted for 92 to 96% of the total applications, declining throughout the 1990s to 77% in 2001–2002. In contrast, PBR applications filed in upper-middle-income countries – including Argentina, Chile, Czech Republic, Hungary, Poland, Slovakia, South Africa, and Uruguay – grew steadily since the early 1970s, while reported PBR applications in lower-middle-income countries – now including Brazil, Bulgaria, China, Colombia, Romania, the Russian Republic, and Ukraine – only began increasing a decade later.

For high-income countries, the growth in the total number of applications stems largely from an exceptionally rapid increase in the rate of applications per country per year through the 1970s and 1980s. Moreover, most high-income countries had PBR legislation in place for most of the period reported here. In contrast – and setting aside

[66] Some applications were lodged before 1970, but the number is small (3%) compared with the totals reported in Figure 1.

some initial "start-up" blips in PBR applications – the majority of middle-income countries have shown no general tendency to increase their rates of application over time.[67] In fact, application rates declined for some of these countries. The preponderance of growth for the middle-income group was due to an increase in the number of countries offering plant breeders rights (3 countries in 1971, 5 in 1985, 8 in 1990 and 14 in 2002).[68] For lower-middle-income countries there was a particularly marked jump from 131 applications during 1991–1995 to 2437 applications during 1996–2000. One would expect the number of applications to increase further over time as awareness of the existence and effectiveness of PBRs in a particular country increased and as the economic costs of applying for and evaluating applications declined with improved bureaucratic procedures.[69] Notably, the number of plant breeders rights sought in low-income countries is negligible. The principal proximate cause of this situation is the lack of rights on offer in poor countries. More fundamentally, it reflects a range of economic influences regarding the costs and benefits of securing breeders rights in a particular jurisdiction.

7.1.1. Foreign PBR applications

Overall, 34% (17,529 of a total of 51,258) of the applications filed in 50 UPOV member countries during 1998–2002 were lodged by foreigners (Table 5). This substantial fraction of foreign applications indicates extensive potential spillovers of varietal improvement research done in one locale on seed market and production developments elsewhere in the world. The *intensity* of foreign participation in domestic varietal rights markets differs markedly. Application by foreigners accounted for 31% of the total in

[67] Koo et al. (2006) describe an initial blip in applications in China after it began issuing PBRs in April 1999.

[68] Plant breeders' rights have been available in many rich countries for at least the past three decades. Germany, for example, has issued plant breeders rights since at least the 1950s and likewise for a few other European countries. The United States began issuing plant variety protection certificates (PVPCs) in 1971 for sexually reproduced plants: asexually reproduced plants (like grape vines, fruit trees, strawberries, and ornamentals that are propagated through cuttings and graftings) have had recourse to intellectual property protection since 1930 when the Plant Patent Act was passed. Many middle-income countries passed PVP legislation during the 1990s in compliance with their *sui generis* obligations to offer the intellectual property rights over plant varieties enshrined in Article 27(3)b of the 1995 Trade-Related Aspects of Intellectual Property (TRIPS) agreement in the World Trade Organization (WTO). An indication of the geographical extent of plant breeders' rights is the listing of member countries of the International Union for the Protection of New Varieties of Plants (UPOV). At its inception in 1961, UPOV had 5 member countries (Belgium, France, Germany, Italy, and Netherlands, all of them high-income countries), growing to 20 countries by the end of 1992, then increasing rapidly to 53 countries – 21 high-income, 27 middle-income and 5 low-income – as of September 2003. Notably, under the TRIPs agreement, the "least developed" countries (a WTO designation) were exempt from complying with Article 27(3)b until January 2006.

[69] In addition, some countries have expanded the scope of crops eligible for protection overtime. In China, for instance, a total of 10 species was eligible for protection in September 1999, growing to 30 species by March 2002 (including 5 major cereals, 2 oil crops, 2 roots and tubers, 10 vegetables and fruits and 11 flowers and grasses, but excluding cotton).

Table 5
Share of plant breeders' rights applications lodged by foreigners, 1998–2002

Economies	Total	Residents	Non-residents
		(number of applications)	
High-income economies (23)[a]	39,079	26,893	12,186
Upper-middle-income economies (11)	5,583	1,945	3,638
Lower-middle-income economies (12)	6,109	4,592	1,517
Low-income economies (4)	487	299	188
Total (50)	**51,258**	**33,729**	**17,529**

Source: Koo, Nottenburg and Pardey (2004) based on UPOV (2003a).
Notes: See Table 1 for country income classification criteria.
[a]Bracketed figures indicate number of countries included in the data.

high-income countries, 65% in upper-middle-income countries, 25% in lower-middle-income countries, and 38% in low-income countries. The country-by-country participation of foreigners is even more variable. For example, the foreign share of applications is 85% in Switzerland and Canada, 42% in the United States, 37% in the United Kingdom, 24% in Japan, 16% in the Netherlands and Germany, and 11% in France.

7.1.2. *European and United States trends*

Worldwide, seed sales are estimated to be $30 billion annually (ISF 2003). While the value of sales in seed markets within the European Union (about $5.2 billion in total) is a little less than U.S. seed sales ($5.7 billion), there have been three times more IPR applications for plant protection since 1971 lodged throughout Europe than related applications in the United States (Table 6). Much of the difference may stem from multiple applications for the same variety among national jurisdictions in Europe, whereas only one application is required per variety in the United States. Part of the difference may arise from the different forms of varietal protection on offer in Europe (plant breeders rights) versus the United States (plant patents and plant variety protection certificates, as well as utility patents).

Four countries – the Netherlands (35%), France (22%), Germany (16%) and the United Kingdom (8%) – account for most of the European applications, according to the applications lodged with the Community Plant Variety Office (CPVO).[70] The number

[70] Prior to April 27, 1995, when the Community Plant Variety Office (CPVO) was established, a breeder seeking protection for a variety throughout the European Union was required to submit an application to each of the member states. Now, with a single application to the CPVO, a breeder can be granted varietal protection rights throughout the European Union. This European-wide system – CPVO members currently include Austria, Belgium, Denmark, Finland, France, Germany, Greece, Ireland, Italy, Luxembourg, Netherlands, Portugal, Spain, Sweden, and the United Kingdom – operates in parallel with respective national systems,

Table 6
Extent of intellectual property rights on plants in the European Union and United States

Country/region	Before 1970	1971–75	1976–80	1981–85	1986–90	1991–95	1996–2000	2001–02	Total
					(counts)				
European Union[a]	**598**	**843**	**4,369**	**6,374**	**13,254**	**20,290**	**19,232**	**7,472**	**72,432**
Netherlands	140	213	518	1,369	4,252	6,838	4,278	1,386	18,994
France	–	–	2,151	2,046	3,206	3,395	2,326	686	13,810
Germany	212	244	436	1,007	2,275	3,042	1,306	472	8,994
UK	2	6	8	6	500	2,365	1,334	359	4,580
Others	244	380	1,256	1,946	3,021	4650	1344	188	12,029
CPVO[b]	–	–	–	–	–	–	8,644	4,381	13,025
United States[a]	**3,495**	**1,313**	**1,587**	**2,046**	**3,150**	**3,754**	**6,539**	**2,013**	**23,896**
Plant variety protection	–	600	614	934	1,228	1,505	1,943	562	7,386
Plant patents	3,495	713	973	1,105	1,883	2,089	3,666	1,346	15,270
Utility patents[c]				6	39	160	930	105	1,240

Source: Koo, Nottenburg and Pardey (2004) based on commissioned data obtained from the USPTO (2004) for U.S. plant and utility patents; the USDA (2004) for U.S. plant variety protection certificates; UPOV (2003b) for data of European Union countries; and CPVO (2003) for CPVO series.

[a]European Union aggregate includes applications for plant breeders' rights in 13 European countries. U.S. aggregate of plant and utility patents granted and plant variety protection certificate applications.

[b]CPVO (Community Plant Variety Office) members currently include Austria, Belgium, Denmark, Finland, France, Germany, Greece, Ireland, Italy, Luxembourg, Netherlands, Portugal, Spain, Sweden, and the United Kingdom. Individuals or companies from member states of UPOV, but not a member of the European Union, can also apply, provided that an agent domiciled in the Community has been nominated [see CPVO (2003) for further details]. Since it was first implemented in 1995, around 35% of these applications are lodged from the Netherlands, 16% from Germany, 14% from France, 19% from elsewhere in the European Union and 16% from outside the European Union.

[c]Preliminary tabulation by authors based on patent-by-patent scrutiny of 4080 patents from 1972–2003 assigned USPTO patent code 800 (which includes patents pertaining to living multi-cellular organisms except humans) to identify patents with claims that encompass plant varieties.

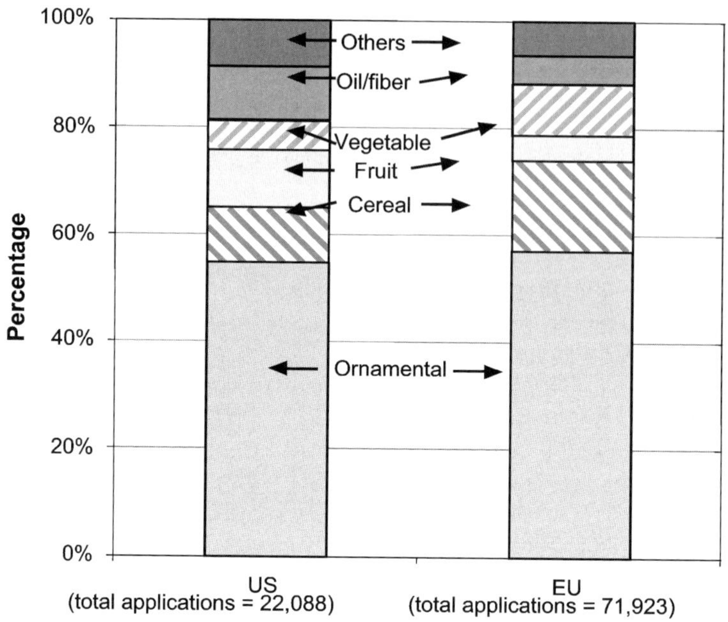

Figure 8. Plant breeders' rights stratified by crop categories. *Source*: Koo, Nottenburg and Pardey (2004) based on Table 6 sources. *Notes*: United States data includes total number of plant patents granted from 1930 to 2003 and plant variety protection applications from 1970 to 2003. Data for European Union includes plant breeders' rights applications to national plant variety offices beginning at or near their inception dates (1942 for Netherlands, 1955 for Germany, 1970s and 1980s for most other countries) to 2003 and applications to the CPVO from 1995 to 2003.

of PBR applications filed with the CPVO has increased over time, offsetting declines in the number of applications lodged with national protection offices. In 1996, there were 1385 applications lodged with the CPVO and a total of 2766 applications made to individual national systems. By 2000, almost equal numbers of PBR claims were filed with the CPVO and the respective national offices (about 2000 applications each), and in 2001 CPVO applications (2158) exceeded those filed with national offices (1864) [CPVO (2003)].

Ornamental crops account for more than half the total applications in both the United States and Europe (Figure 8). In the United States, cereal crops (such as wheat and corn), as well as oil and fibers, and fruit crops each make up more than 10% of the total number of applications since 1970. Ornamentals and fruits are mostly protected

although the owner of a variety cannot simultaneously exploit both a community plant variety right (CPVR) and a national plant breeders right in relation to that variety. Individuals or companies from member states of UPOV that are not members of the European Union can also apply, provided that an agent domiciled in the Community has been nominated. The duration of CPVR protection is 25 years for most crops, and 30 years for potato, vine and tree varieties.

by plant patents, while cereal, oil and fiber crops, and vegetables are usually protected by plant variety rights. About 5% of all the IP applications in this area in the United States are for utility patents, of which 55% pertain to corn and 40% to soybeans. In Europe, cereals account for more than a quarter of the total PBR applications, followed by vegetable (10%), oil and fiber crops (5%) and fruit (5%).

7.2. Biotechnology patenting patterns

An initial assessment by Pardey et al. (2004) pertains to the international dimensions of patent activity in biotechnology and specific sectors, such as agriculture and health, is presented in Figure 9. Numbers of patent applications submitted to the World Intellectual Property Organization (WIPO) under the Patent Cooperation Treaty (PCT) (Panel a) and patents granted by the European Patent Office (EPO) (Panel b) are plotted against the year published. For this analysis, patent documents were selected on the basis of the International Patent Classification (IPC) scheme used by the patent offices. Data were obtained for documents satisfying criteria for "biotechnology" and further sub-divided into "agricultural biotechnology" and "health biotechnology".[71] The numbers of the two sub-divisions add to more than for total biotechnology as some documents fit into both categories. While initially agricultural biotechnology patent documents exceeded health related documents both at EPO and WIPO, the situation reversed in 1999. Furthermore, the spectacular rise in patent filings in the late 1980s and through the 1990s appears to be leveling off.

The data presented here appear, at first glance, to contrast with recently reported analyses of Graff et al. (2003) who noted drops in patent grants in plant biotechnology at the EPO after peaking in 1994–1995. The differences are explained by differences in the definition of plant or agricultural biotechnology. The definition of Graff et al. (2003) focuses on biotechnology for plant breeding and comprises a description of the scope of technologies, such as genetic engineering of plants, plant genes, and plant breeding methods, covering a small subset of IPC codes and specific technology keywords. In contrast, the definition used by Pardey, Koo and Nottenburg (2004) encompasses broader aspects of plant biotechnology, including genetic modification of plants, biocides, organismal or enzymic-based methods for preservation of foods, microbiological treatment of water and soil, compositions containing micro-organisms or enzymes, and processes using micro-organisms or enzymes. The definitional differences are highlighted by the order of magnitude difference in the number of documents that satisfy the criteria. For example, in 2000, we obtained 8859 PCT patent filings and 5097 EP patent grants for inventions concerning agricultural biotechnology, compared with around 625

[71] As used here, "biotechnology" refers to "the application of science and technology to living organisms as well as parts, products and models thereof, to alter living or non-living materials for the production of knowledge, goods and services", a definition used by the OECD (see "Statistical definition of biotechnology" 12 June 2002 in the Biotechnology, Statistics section of http://www.oecd.org).

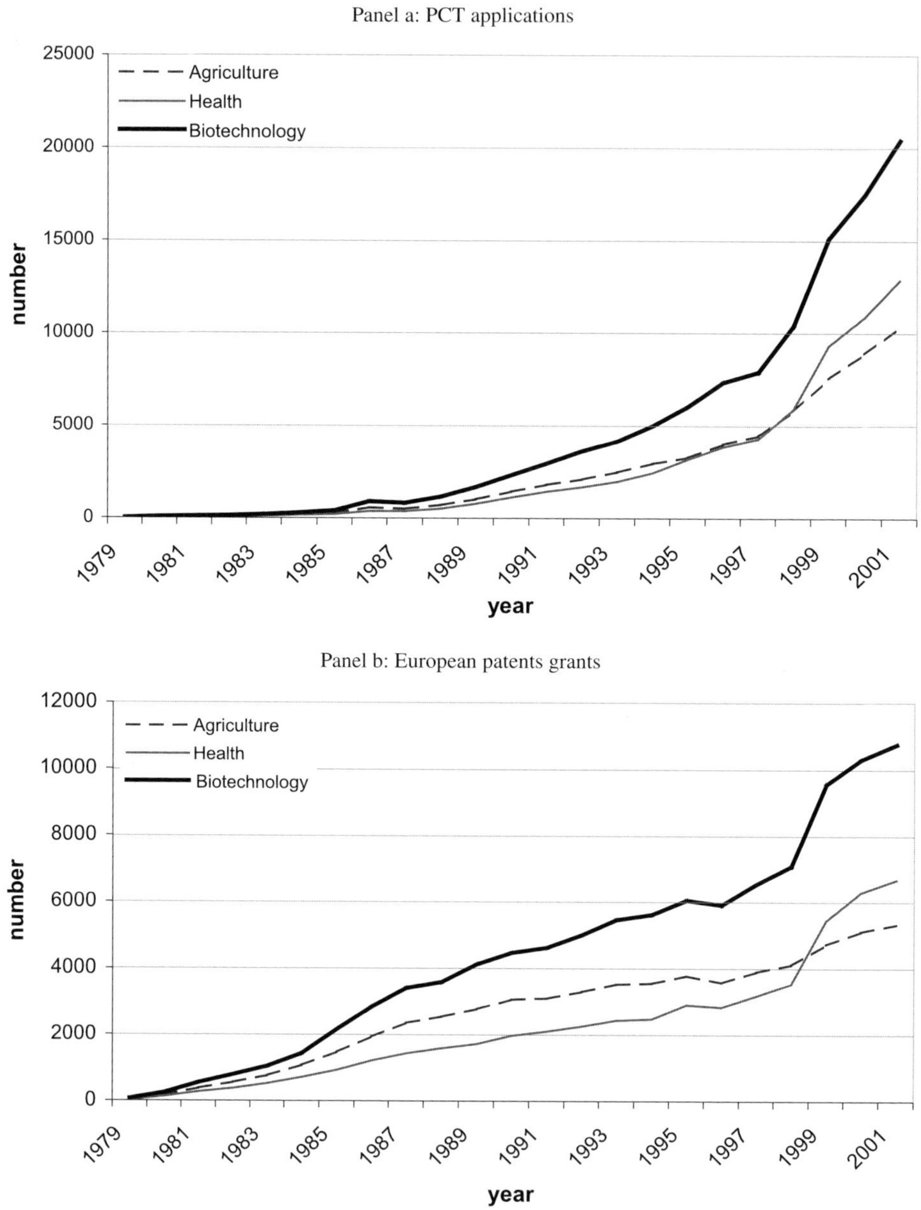

Figure 9. Biotechnology patents. *Source*: Pardey et al. (2004) based on data compiled from CAMBIA-IP resource database.

PCT applications and 50 EP patent grants for the narrower area of "plant biotechnology" reported by Graff et al. (2003).

The percentage of PCT applications in agricultural biotechnology has been on the rise. In 1985, agricultural biotechnology applications were 4.0% of the total submitted. By 1990, they were 7.5% of the total, and in 2000 had risen to 9.7% of the total. In 2000, ag-biotech patents granted in EPO were 18.5% of the total number of patents granted. Clearly further conclusions based on patent data, with respect to the commercial and public good consequences encompassing the changing geographical and institutional origins of biotechnology innovations on a global scale, and their spillovers or transfer to other countries, could well be sensitive to the scope of applications covered in the source set of documents.

8. Conclusion

Agricultural research is unique in its global nature, in the prominent role of the "South", in the rough equality of shares of Public North, Public South and Private North, and in its record of sustained high productivity. In some parts of the world, the share of private investment has increased recently, in response to newly available opportunities in biotechnology, and new, complementary technical and legal means of protecting innovations. But the increased bioscience focus of private effort is on the genetic transformation of a small set of large-volume crops including soybeans, corn, cotton and canola. Even if consumer opposition to transgenic technologies has peaked, or agricultural biotechnology finds attractive applications subject to less consumer opposition, for the foreseeable future, biotechnology research on transformation of the other major food staples (including wheat, rice, potatoes, cassava and bananas) will be largely left to the public sector.

This does not mean that intellectual property protection will be largely irrelevant for most major food staples. An array of means of protecting innovations is available in agriculture, and their scope is becoming global under TRIPS and subsequent agreements. Patenting and licensing is actively pursued by public as well as private institutions in agriculture. But in neither sector is licensing a dominant source of research funding. On the other hand, cumulative, uncertain and conflicting claims are making freedom to operate a real constraint on developing new technologies to commercialization, especially for the "orphan" crops in the North and the staples in the South that lie beyond the set of crops that attract high private-sector attention. The PIPRA and BIOS initiatives are interesting responses to this problem.

To the extent that agricultural biotechnology fulfills its great promise, most of the benefits will likely be realized in feeding and nourishing a large and growing world population at low cost, and not in producer profit margins that are in any way comparable to those earned in pharmaceuticals. Intellectual property rights do have an increasingly important role in encouraging agricultural innovation, but continued and increased funding of public and non-profit research and development, beyond a handful of high-value commercial crops, will remain a crucial source of continued technical progress in global agriculture.

References

Adusie, E.O., Norton, G.W. (1990). "The magnitude of agricultural maintenance research". Journal of Production Agriculture 3 (1), 1–6.

Alex, G. (2004). Compilation from unpublished USAID financial files. United States Agency for International Development, Washington, DC.

Alston, J.M., Dehmer, S., Pardey, P.G. (2005). "International initiatives in agricultural R&D – The changing fortunes of the CGIAR". In: Pardey, P.G., Alston, J.M., Piggott, R.R. (Eds.), Agricultural R&D in the Developing World: Too Little, Too Late? International Food Policy Research Institute, Washington, DC. In press.

Alston, J.M., Norton, G.W., Pardey, P.G. (1995). Science under Scarcity: Principles and Practice for Agricultural Research Evaluation and Priority Setting. Cornell University Press, Ithaca, NY.

Alston, J.M., Pardey, P.G. (1993). "Market distortions and technological progress in agriculture". Technological Forecasting and Social Change 43 (3/4), 301–319.

Alston, J.M., Pardey, P.G. (1999). "The economics of agricultural R&D policy". In: Alston, J.M., Pardey, P.G., Smith, V.H. (Eds.), Paying for Agricultural Productivity. Johns Hopkins University Press, Baltimore, MD. Ch. 2.

Alston, J.M., Venner, R.J. (2002). "The effects of the U.S. Plant Variety Protection Act on wheat genetic improvement". Research Policy 31, 527–542.

Alston, J.M., Marra, M.C., Pardey, P.G., Wyatt, T.J. (2000). "A meta analysis of rates of return to agricultural R&D: Ex Pede Herculem?" IFPRI Research Report No. 113. International Food Policy Research Institute, Washington, DC.

Arora, A., Fosfuri, A. (2005). "Pricing diagnostic information". Management Science 51 (7), 1092–1100.

Arrow, K.J. (1962). "Economic welfare and the allocation of resources for invention". In: The Rate and Direction of Inventive Activity: Economic and Social Factors. Arno Press, New York.

Atkinson, R.C., Beachy, R.N., Conway, G., Cordova, F.A., Fox, M.A., Holbrook, K.A., Klessig, D.F., McCormick, R.L., McPherson, P.M., Rawlings III, H.R., Rapson, R., Vanderhoef, L.N., Wiley, J.D., Young, C.E. (2003). "Public sector collaboration for agricultural IP management". Science 301, 174–175.

Barton, J. (2003). "Non-obviousness". IDEA: The Journal of Law & Technology 43 (3), 475–508.

Beintema, N.M., Stads, G.J. (2004). "Investing in Sub-Saharan Africa agricultural research: Recent trends", 2020 African Conference Brief No. 8. International Food Policy Research Institute, Washington, DC.

Benkler, Y. (2004a). "Commons-based strategies and the problems of patents". Science 305, 1110–1111.

Benkler, Y. (2004b). "Sharing nicely: On sharable goods and the emergence of sharing as a modality of economic production". Yale Law Review 114, 273–356.

Bessen, J. (2004). "Open source software: Free provision of complex public goods". Massachusetts Institute of Technology, Cambridge, MA.

Binenbaum, E., Nottenburg, C., Pardey, P.G., Wright, B.D., Zambrano, P. (2003). "South–North trade, intellectual property jurisdictions, and freedom to operate in agricultural research on staple crops". Economic Development and Cultural Change 51 (2), 309–336.

Boettiger, S., Graff, G., Pardey, P.G., van Dusen, E., Wright, B.D. (2004). "Intellectual property rights for plant biotechnology: International aspects". In: Christou, P., Klee, H. (Eds.), Handbook of Plant Biotechnology. John Wiley and Sons, Chichester. Ch. 56.

Boldrin, M., Levine, D. (2002). "The case against intellectual property". American Economic Review 92 (2), 209–212.

Boyle, J. (2003). "The second enclosure movement and the construction of the public domain". Law and Contemporary Problems 66, 33–74.

Brahmi, P., Saxena, S., Dhillon, B.S. (2004). "The protection of plant varieties and farmers' right act of India". Current Science 86 (3), 392–398.
Brush, S.B., Stabinsky, D. (Eds.) (1996). Valuing Local Knowledge: Indigenous People and Intellectual Property Rights. Island Press, Covelo, CA.
Bush, V. (1945). Science: The Endless Frontier, A Report to the President. U.S. Government Printing Office, Washington, DC.
Carrizosa, S., Brush, S.B., Wright, B.D., McGuire, P. (Eds.) (2004). Accessing Biodiversity and Sharing the Benefits: Lessons from Implementing the Convention on Biological Diversity. IUCN, Gland, Switzerland and Cambridge, UK.
CFIA (2002). "10-year review of Canada's plant breeders' rights act". Canadian Food Inspection Agency. Available at http://www.inspection.gc.ca/english/plaveg/pbrpov/10yrenglish.pdf.
CGIAR (1998). "Shaping the CGIAR's future: Summary of proceedings and decisions". Consultative Group for International Agricultural Research (CGIAR) International Centers Week, October 26–30. CGIAR Secretariat, Washington, DC.
Chang, H.F. (1995). "Patent scope, antitrust policy, and cumulative innovation". RAND Journal of Economics 26 (1), 34–57.
Chin, J.C., Grossman, G.M. (1990). "Intellectual property rights and North–South trade". In: Jones, R.W., Krueger, A.O. (Eds.), The Political Economy of International Trade: Essays in Honor of Robert E. Baldwin. Basil Blackwell, Cambridge, MA.
Coase, R.H. (1937). "The nature of the firm". Economica (New Series) 4, 386–405.
Coase, R.H. (1960). "The problem of social cost". Journal of Law and Economics 3, 1–44.
Cohen, W.M., Nelson, R.R., Walsh, J.P. (2000). "Protecting their intellectual assets: Appropriability conditions and why U.S. manufacturing firms patent (or not)", NBER Working Paper No. 7552. National Bureau of Economic Research, Washington, DC.
Commission on Intellectual Property Rights (2002). "Integrating intellectual property rights and development policy". http://www.iprcommission.org/graphic/documents/final_report.htm.
Constantine, J.H., Alston, J.M., Smith, V.H. (1994). "Economic impacts of California's one variety cotton law". Journal of Political Economy 102 (5), 951–974.
Cornelli, F., Schankerman, M. (1999). "Patent renewals and R&D incentives". RAND Journal of Economics 30 (2), 197–213.
Correa, C.M. (2000). Intellectual Property Rights, the WTO and Developing Countries: The TRIPS Agreement and Policy Options. Zed Books, London, New York, Penang, Malaysia.
Costello, C., Ward, M. (2003). "Search, bioprospecting, and biodiversity conservation: Comment", Mimeo. University of California, Santa Barbara.
CPVO (2003). Annual Report, 2002. Community Plant Variety Office, Paris.
Dalton, R. (2004). "Bioprospects less than golden". Nature 429, 598–600.
Deardorff, A.V. (1992). "Welfare effects of global patent protection". Economica 49, 35–51.
de Janvry, A., Graff, G., Sadoulet, E., Zilberman, D. (2000). "Technological change in agriculture and poverty reduction", Concept paper for WDR on Poverty and Development 2000/2001. University of California, Berkeley, pp. 6–7. Available at http://www.worldbank.org/poverty/wdrpoverty/background/dejanvry.pdf.
de Laat, E. (1996). "Patents or prizes: Monopolistic R&D and asymmetric information". International Journal of Industrial Organization 15, 369–390.
Delmer, D.P., Nottenburg, C., Graff, G.D., Bennett, A.B. (2003). "Intellectual property resources for international development in agriculture". Plant Physiology 133, 1666–1670.
Denicolo, V. (2000). "Two-stage patent races and patent policy". RAND Journal of Economics 31, 488–501.
Diaz, C.L. (2000). "Regional approaches to implementing the convention on biological diversity: The case of access to genetic resources". In: Proceedings of Concerted Action Programme on the Effectiveness of International Environmental Agreements and EU Legislation, Barcelona.
Diez, M.C.F. (2002). "The impact of plant varieties rights on research: The case of Spain". Food Policy 27, 171–183.
Diwan, I., Rodrik, D. (1991). "Patents, appropriate technology, and North–South trade". Journal of International Economics 30, 27–47.

Dutfield, G. (2002). "Trade, intellectual property and biogenetic resources: A guide to the international regulatory framework", Background paper for the Multi-Stakeholder Dialogue on Trade, Intellectual Property and Biological and Genetic Resources in Asia, April 2002. International Centre for Trade and Sustainable Development (ICTSD), Geneva.
Ehrlich, P.R. (1968). The Population Bomb. Ballantine, New York.
Eisenberg, R.S. (2003). "Patent swords and shields". Science 299, 1018–1019.
Erbisch, F.H., Maredia, K.M. (2004). Intellectual Property Rights in Agricultural Biotechnology. CAB International.
Esquinas-Alcázar, J. (1996). "Farmers' rights", Paper presented at the Symposium on the Economics of Valuation and Conservation of Genetic Resources for Agriculture, May. Università di Roma Tor Vergata, Rome, Italy.
Fan, S., Chan-Kang, C., Qian, K., Krishnaiah, K. (2003). "National and international agricultural research and rural poverty: The case of rice research in India and China", EPTD Discussion Paper No. 109. International Food Policy Research Institute, Washington, DC.
FAO (2000). "FAOSTAT, on-line version". Food and Agriculture Organization of the United Nations, Rome. Accessed January 17, http://faostat.fao.org/default.htm.
Fowler, C. (1994). Unnatural Selection: Technology, Politics, and Plant Evolution. Gordon and Breach, Yverdon, Switzerland.
Fowler, C., Mooney, P. (1990). Shattering: Food, Politics, and the Loss of Genetic Diversity. University of Arizona Press, Tucson, AZ.
Friedman, D.D., Landes, W.M., Posner, R.A. (1991). "Some economics of trade secret law". Journal of Economic Perspectives 5 (1), 61–72.
Gallini, N.T. (1992). "Patent policy and costly imitation". RAND Journal of Economics 23, 52–63.
Gilbert, R.J., Shapiro, C. (1990). "Optimal patent length and breadth". RAND Journal of Economics 21 (1), 106–112.
Godden, D. (1998). "Growing plants, evolving rights: Plant variety rights in Australia". Australasian Agribusiness Review 6, 1–54.
Gordon, H.S. (1954). "The economic theory of a common-property resource: The fishery". Journal of Political Economy 62, 124–142.
Graff, G., Zilberman, D., Bennett, A., Wright, B., Zilberman, D. (2001). "Towards an intellectual property clearinghouse for ag-biotechnology", IP Strategy Today No. 3, BioDevelopments. International Institute, Ithaca, NY.
Graff, G.D., Cullen, S.E., Bradford, K.J., Zilberman, D., Bennett, A.B. (2003). "The public–private structure of intellectual property ownership in agricultural biotechnology". Nature Biotechnology 21 (9), 989–995.
GRAIN (2004). "Bilateral agreements imposing TRIPS-plus intellectual property rights on biodiversity in developing countries". Accessed May 2005, http://www.grain.org/rights/tripsplus.cfm?id=68.
Green, J., Scotchmer, S. (1995). "On the division of profit in sequential innovation". RAND Journal of Economics 26, 20–33.
Griliches, Z. (1957). "Hybrid corn: An exploration in the economics of technological change". Econometrica 25 (4), 501–522.
Grossman, G., Lai, E.L.C. (2004). "International protection of intellectual property". American Economic Review 94 (5), 1635–1653.
Heller, M.A., Eisenberg, R.S. (1998). "Can patents deter innovation? The anticommons in biomedical research". Science 280, 698–701.
Helpman, E. (1993). "Innovation, imitation, and intellectual property rights". Econometrica 61 (6), 1247–1280.
Hopenhayn, H.A., Mitchell, M.F. (2001). "Innovation variety and patent breadth". RAND Journal of Economics 32, 152–166.
Huffman, W.E., Evenson, R.E. (1993). Science for Agriculture: A Long-Term Perspective. Iowa State University Press, Ames, IA.
Huang, J., Rozelle, S., Pray, C., Wang, Q. (2002). "Plant biotechnology in China". Science 295, 674–677.

Jaffe, A.B. (2000). "The U.S. patent system in transition: Policy innovation and the innovation process". Research Policy 29, 531–557.

Jaffe, A.B., Lerner, J. (2004). Innovation and Its Discontents: How Our Broken Patent System Is Endangering Innovation and Progress, and What to Do about It. Princeton University Press, Princeton, NJ.

Janis, M.D., Kesan, J.P. (2002). "U.S. plant variety protection: Sound and fury...?". Houston Law Review 39 (3), 727–778.

Jefferson, R. (2001). "Transcending transgenics – Are there 'babies in the bathwater,' or is that a dorsal fin?". In: Pardey, P.G. (Ed.), The Future of Food: Biotechnology Markets and Policies in an International Setting. International Food Policy Research Institute, Washington, DC. Ch. 5.

Jefferson, R., Byth, D., Correa, C., Otero, G., Qualset, C. (1999). "Genetic use restriction technologies", UNEP/CBD/SBSTTA/4/9/Rev. 1. Convention on Biological Diversity, United Nation Environmental Program (UNEP).

Josling, T. (2005). "Geographical indications: Protection for producers or consumer information?", Invited paper presented at the 2005 annual meeting of the Australian Agricultural and Resource Economics Association, Coffs Harbour, Australia.

Juma, C. (1989). The Gene Hunters: Biotechnology and the Scramble for Seeds. Princeton University Press, Princeton, NJ.

Kingwell, R. (2001). "Charging for the use of plant varieties". Australian Journal of Agricultural and Resource Economics 45 (2), 291–305.

Kingwell, R., Watson, A. (1998). "End-point royalties for plant breeding: Cornucopia or coueism?", Paper presented at the Australian Agricultural and Resource Economics Association annual meeting, Armidale.

Klemperer, P. (1990). "How broad should the scope of patent protection be?". RAND Journal of Economics 21 (1), 113–130.

Kloppenberg, J. (1988). First the Seed: The Political Economy of Plant Biotechnology. Cambridge University Press, Cambridge.

Knudson, M.K., Pray, C.E. (1991). "Plant variety protection, private funding, and public sector research priorities". American Journal of Agricultural Economics 73 (3), 882–886.

Knudson, M.K., Ruttan, V.W. (1988). "Research and development of a biological innovation: Commercial hybrid wheat". Food Research Institute Studies 21 (1), 45–68.

Koo, B. (1998). "The economics of plant genetic resources: The effects of alternative intellectual property protection systems and advances in biotechnology", Ph.D. thesis. University of California, Berkeley.

Koo, B., Nottenburg, C., Pardey, P.G. (2004). "Plants and intellectual property: An international appraisal". Science 306, 1295–1297.

Koo, B., Wright, B.D. (2005). "Dynamics of patenting research tools and tool-using innovations", Mimeo. International Food Policy Research Institute, Washington, DC.

Koo, B., Pardey, P.G., Qian, K., Zhang, Y. (2006). "An option perspective on generating and maintaining plant variety rights in China". Agricultural Economics 35 (1), 35–48.

Kremer, M. (1998). "Patent buyouts: A mechanism for encouraging innovation". Quarterly Journal of Economics 113, 1137–1167.

Kremer, M., Zwane, A.P. (2005). "Encouraging private sector research for tropical agriculture". World Development 33 (1), 87–105.

Kryder, R.D., Kowalski, S.P., Krattiger, A.F. (2000). "The intellectual and technical property components of pro-vitamin A rice (GoldenRice^{TM}): A preliminary freedom-to-operate review", ISAAA Briefs No. 20. The International Service for the Acquisition of Agri-Biotech Applications (ISAAA).

Lai, E.L.C., Qiu, L.D. (2003). "The North's intellectual property rights standard for the South?". Journal of International Economics 59, 183–209.

Landes, W.M., Posner, R.A. (2003). The Economic Structure of Intellectual Property Law. Belknap Press, Cambridge, MA.

Langinier, C., Moschini, G. (2002). "The economics of patents". In: Rothschild, M., Newman, S. (Eds.), Intellectual Property Rights in Animal Breeding and Genetics. CAB International, Wallingford.

Lanjouw, J.O. (2003). "Intellectual property and the availability of pharmaceuticals in poor countries". Innovation Policy and the Economy 3 (3), 91–129.

Lei, Z., Juneja, R., Wright, B. (2006). "Implications of intellectual property protection for academic agricultural biologists", Mimeo. University of California, Berkeley.

Lerner, J., Tirole, J. (2002). "Some simple economics of open source". Journal of Industrial Economics 50, 197–234.

Levin, R.C., Klevorick, A.K., Nelson, R.R., Winter, S.G., Glibert, R., Griliches, Z. (1987). "Appropriating the returns from industrial research and development". Bookings Papers on Economic Activity 3, 783–831.

Lindner, R.K. (2004). "Privatised provision of essential plant breeding infrastructure". Australian Journal of Agricultural and Resource Economics 48 (2), 301–321.

Machlup, F. (1958). "An economic review of the patent system", Study No. 15 of the Sub-Committee on Patents, Trademarks, and Copyrights of the Committee on the Judiciary. U.S. Senate, 85th Congress, 2nd Session, Washington, DC.

Machlup, F., Penrose, E. (1950). "The patent controversy in the nineteenth century". Journal of Economic History 10 (1), 1–29.

Margeules, D. (2002). Personal communication.

Maskus, K.E. (2000). Intellectual Property Rights in the Global Economy. Institute for International Economics, Washington, DC.

Masters, W.A. (2003). "Research prizes: A mechanism to reward agricultural innovation in low-income regions". AgBioForum 6 (1&2), 71–74.

Matutes, C., Regibeau, P., Rockett, K. (1996). "Optimal patent design and the diffusion of innovations". RAND Journal of Economics 27, 60–83.

Maurer, S.M. (2005). "The right tool(s): Designing cost-effective strategies for neglected disease research". Goldman School of Public Policy, University of California, Berkeley.

Merges, R.P., Nelson, R. (1990). "On the complex economics of patent scope". Columbia Law Review 90, 839–916.

Moschini, G., Lapan, H. (1997). "Intellectual property rights and the welfare effects of agricultural R&D". American Journal of Agricultural Economics 79, 1229–1242.

Mowery, D.C. (1998). "The changing structure of the US national innovation system: Implications for international conflict and cooperation in R&D policy". Research Policy 27, 639–654.

Mowery, D.C., Nelson, R.R., Sampat, B.N., Ziedonis, A.A. (2001). "The growth of patenting and licensing by U.S. universities: An assessment of the effects of the Bayh–Dole act of 1980". Research Policy 30 (1), 99–119.

Murray, F., Stern, S. (2004). "Do formal intellectual property rights hinder the free flow of scientific knowledge? A test of the anti-commons hypothesis", Mimeo. Massachusetts Institute of Technology, Cambridge, MA.

National Academy of Sciences (1996). "Intellectual property rights and research tools in molecular biology", Summary of a workshop held at the National Academy of Sciences. National Academy Press, Washington, DC.

Nature (2004). "Open-source biology". Nature 431, 491.

Nordhaus, W. (1969). Innovation, Growth, and Welfare. MIT Press, Cambridge, MA.

Nottenburg, C. (2004). "Schmeiser v. Monsanto". In: Navigating the Patent Maze, posted 22 July 2004; http://www.cougarlaw.com.

Nottenburg, C., Pardey, P.G., Wright, B.D. (2002). "Accessing other people's technology for non-profit research". Journal of Agricultural and Resource Economics 48 (3), 389–416.

O'Donoghue, T. (1998). "A patentability requirement for sequential innovations". RAND Journal of Economics 29, 654–679.

O'Donoghue, T., Scotchmer, S., Thisse, J. (1998). "Patent breadth, patent life, and the pace of technological progress". Journal of Economics and Management Strategy 7, 1–32.

Olmstead, A.L., Rhode, P.W. (2002). "The red queen and the hard reds: Productivity growth in American wheat, 1800–1940". Journal of Economic History 62, 929–966.

Paddock, W., Paddock, P. (1967). Famine 1975! America's Decision: Who Will Survive? Little, Brown, Boston, MA.

Pardey, P.G., Beintema, N.M. (2001). "Slow magic: Agricultural R&D a century after Mendel", IFPRI Food Policy Report. International Food Policy Research Institute, Washington, DC.
Pardey, P.G., Koo, B., Nottenburg, C. (2004). "Creating, protecting, and using crop biotechnologies worldwide in an era of intellectual property". Minnesota Journal of Law, Science & Technology 6 (1), 213–252.
Pardey, P.G., Roseboom, J., Craig, B.J. (1992). "A yardstick for international comparisons: An application to national agricultural research expenditures". Economic Development and Cultural Change 40 (2), 333–349.
Pardey, P.G., Wright, B.D., Nottenburg, C., Binenbaum, E., Zambrano, P. (2003). "Intellectual property and developing countries: Freedom to operate in agricultural biotechnology". In: Pardey, P.G., Koo, B. (Eds.), Biotechnology and Genetic Resources Policies, Briefs 1–6. International Food Policy Research Institute, Washington, DC. Brief No. 3.
Pardey, P.G., Alston, J.M., Chan-Kang, C., Castello Magalhães, E., Vosti, S.A. (2004). "Assessing and attributing the benefits from varietal improvement research in Brazil", IFPRI Research Report No. 136. International Food Policy Research Institute, Washington, DC.
Pendleton, C.N. (2004). "The peculiar case of terminator technology: Agricultural biotechnology and intellectual property protection at the crossroads of the third green revolution". Biotechnology Law Report 23 (1), 1–29.
Perrin, R.K., Kunnings, K.A., Ihnen, L.A. (1983). "Some effects of the U.S. plant variety protection act of 1970", Economics Research Report No. 46. Department of Economics and Business, North Carolina State University, Raleigh, NC.
Plant, A. (1934). "The economic theory concerning patents and inventions". Economica 1, 30–51.
Plucknett, D.L., Smith, N.J.H., Ozgediz, S. (1990). Networking in International Agricultural Research. Cornell University Press, Ithaca, NY.
Quilkey, J.J. (1970). "The merino export embargo – A comparative static analysis". Australian Journal of Agricultural Economics 14 (1), 63–71.
Rausser, G.C. (1999). "Public/private alliances". AgBioForum 2 (1), 5–10.
Rausser, G.C., Small, A.A. (2000). "Valuing research leads: Bioprospecting and the conservation of genetic resources". Journal of Political Economy 108 (1), 173–206.
Roa-Rodríguez, C. (2003). "Promoters used to regulate gene expression", Version 1. CAMBIA web site: http://www.bios.net/daisy/promoters/768.
Rockefeller Foundation (1999). "Food gains for the world's poor are being threatened by furor over genetically modified (GM) foods", Press release. Rockefeller Foundation, Washington, DC. Accessed February 2005, http://www.rockfound.org/display.asp?context=1&Collection=4&DocID=59&Preview=0&ARCurrent=1.
Roe, T.L., Pardey, P.G. (1991). "Economic policy and investment in rural public goods: A political economy perspective". In: Pardey, P.G., Roseboom, J., Anderson, J.R. (Eds.), Agricultural Research Policy: International Quantitative Perspectives. Cambridge University Press, Cambridge. Ch. 1.
Root, W., De Rochemont, R. (1976). Eating in America: A History. William Morrow and Company, New York.
Runge, C.F. (2004). "Enclosure, intellectual property and life sciences research". Journal of World Intellectual Property 7 (6), 807–827.
Ruttan, V.W. (1982). Agricultural Research Policy. University of Minnesota Press, Minneapolis, MN.
Sampat, B.N. (2004). "Genomic patenting by academic researchers: Bad for science?", Mimeo.
Samuelson, P. (2004). "Intellectual property arbitrage: How foreign rules affect domestic protections". University of Chicago Law Review 71, 223–339.
Scherer, F.M. (1972). "Nordhaus' theory of optimal patent life: A geometric reinterpretation". American Economic Review 62, 422–427.
Schultz, T.W. (1964). Transforming Traditional Agriculture. Yale University Press, New Haven, CT.
Scotchmer, S. (1996). "Protecting early innovators: Should early innovators be patentable?". RAND Journal of Economics 27, 322–331.
Scotchmer, S. (1999). "On the optimality of the patent renewal system". RAND Journal of Economics 30, 181–196.

Scotchmer, S. (2004a). "The political economy of intellectual property treaties". Journal of Law, Economics and Organization 20 (2), 415–437.
Scotchmer, S. (2004b). Innovation and Incentives. MIT Press, Cambridge, MA.
Simpson, R.D., Sedjo, R.A., Reid, J.W. (1996). "Valuing biodiversity for use in pharmaceutical research". Journal of Political Economy 104 (1), 163–185.
Shavell, S., van Ypersele, T. (2001). "Rewards versus intellectual property rights". Journal of Law and Economics 44, 525–547.
Shoemaker, R. (2001). "Economic issues in agricultural biotechnology", Agricultural Information Bulletin No. 762. Economic Research Service, United States Department of Agriculture, Washington, DC.
Smith, P. (2002). "Patent rights and trade: Analysis of biological products, medicinals and botanicals, and pharmaceuticals". American Journal of Agricultural Economics 84 (2), 495–512.
Sunding, D., Zilberman, D. (2001). "The agricultural innovation process: Research and technology adoption in a changing agricultural sector". In: Gardner, B., Rausser, G. (Eds.), Handbook of Agricultural Economics, vol. 1. Elsevier Science. Ch. 4, pp. 208–250.
Taubman, A. (2001). "The way ahead: Developing international protection for geographical indications", Paper presented to the DNPI/WIPO Symposium on the International Protection of Geographical Indications. National Directorate for Industrial Property and World Intellectual Property Organization, Montevideo and Geneva.
Taylor, M.R., Cayford, J. (2004). "American patent policy, biotechnology, and African agriculture: The case for policy change". Harvard Journal of Law and Technology 17 (2), 321–407.
Thambisetty, S. (2002). "Patents and human genome research in developing countries – Problems and proposals". Journal of World Intellectual Property 5, 5.
The Crucible II Group (2001). Seeding Solutions, vol. 2, Options for National Laws Governing Access to and Control Over Genetic Resources. International Development Research Centre, International Plant Genetic Resources Institute and Dag Hammarskjöld Foundation, Ottawa, Rome and Uppsala.
Ulrich, A., Furtan, H., Schmitz, A. (1986). "Public and private returns from joint venture research: An example from agriculture". Quarterly Journal of Economics 10, 103–129.
Ulrich, A., Furtan, H., Schmitz, A. (1987). "The cost of a licensing system regulation: An example from Canadian prairie agriculture". Journal of Political Economy 95, 160–178.
UNDP (1999). Human Development Report 1999. United Nations Development Programme and Oxford University Press, New York/Oxford.
UPOV (2003a). "Plant variety protection statistics for the period 1998–2002", C/37/7. International Union for the Protection of New Varieties of Plants, Geneva.
UPOV (2003b). Plant variety database CD 2003/02. International Union for the Protection of New Varieties of Plant, Geneva.
U.S. Census Bureau/Bureau of Economic Analysis (2001). U.S. International Trade in Goods and Services. U.S. Department of Commerce, Washington, DC.
U.S. Department of Agriculture (2004). Certificate status database. Agricultural Marketing Service, Science and Technology Program, Plant Variety Protection Office, Beltsville, MD.
U.S. Patent and Trademark Office (2004). US patent database. Office of Electronic Information Products, Alexandria, VA.
Usher, D. (1964). "The welfare economics of invention". Economica 31 (123), 279–287.
Vogel, J. (1994). Genes for Sale: Privatization as a Conservation Policy. Oxford University Press, Oxford, UK.
von Hippel, E. (2002). "Open source software projects as user innovation networks", Working paper. MIT Sloan School of Management.
Walsh, J.P., Arora, A., Cohen, W.M. (2003). "Research tool patenting and licensing of biomedical innovation". In: Merrill, S., Levin, R., Meyers, M. (Eds.), Patents in the Knowledge-Based Economy. National Academies Press, Washington, DC.
Walsh, J.P., Cho, M., Cohen, W.M. (2005). "View from the bench: Patents and material transfers". Science 309, 2002–2003.

Weitzman, M.L. (1974). "Price vs. quantities". Review of Economic Studies 41 (4), 477–491.
World Bank (1996). World Development Report 1996: From Plan to Market. World Bank, Washington, DC.
World Bank (2002). World development indicators database, CD-ROM version. World Bank, Washington, DC.
World Bank (2004). Unpublished project data files. World Bank, Washington, DC.
Wright, B.D. (1983). "The economics of invention incentives: Patents, prizes, and research contracts". American Economic Review 73 (4), 691–707.
Wright, B.D. (1984). "The resource allocation problem in R&D". In: Tolley, G., Hodge, J., Oehmke, J. (Eds.), The Economics of R&D Policy. Praeger Press, New York. Ch. 22.
Wright, B.D. (1998). "Public germplasm development at a crossroads: Biotechnology and intellectual property". California Agriculture 52 (6), 8–13.
Yang, Y. (1998). "Why do southern countries have little incentive to protect northern IPRs?". Canadian Journal of Economics 31 (4), 800–816.

Chapter 49

PRIVATE AGRICULTURAL RESEARCH

CARL E. PRAY

Rutgers University, New Brunswick, NJ

KEITH O. FUGLIE

Economic Research Service, Washington, DC

DANIEL K.N. JOHNSON

Department of Economics and Business, Colorado College, Colorado Springs, CO

Contents

Handbook of Agricultural Economics, Volume 3
Edited by Robert Evenson and Prabhu Pingali

DOI: 10.1016/S1574-0072(06)03049-0

Abstract

This chapter serves to assess the quantity and impact of private agricultural research in developing countries, alongside the policies which affect that research. To provide context, an outline of the history, size and structure of private research presents data on research expenditures by region, by agency, by crop and by industrial sector. Evidence on research inputs reflects not only the importance of private research, but also the wide national and regional differences in both inputs and implementation of programs. Research output and intellectual property rights (IPR) protection are also strikingly different across time and space, as seen in data on agricultural IPRs by region, by industrial origin, and by crop type. The diffusion of agricultural innovations among farmers is described using data on approvals of genetically modified crops, and on cultivated areas actually planted. Studies of the subsequent impacts of agricultural research show that private sector research is making an important contribution to agricultural productivity growth in some countries and could make an even larger impact if government policies were more favorable. The conclusion provides a review of economic and policy determinants affecting private incentives, and some policy prescriptions to encourage further investment.

Keywords

agricultural research, developing nation, high yield variety, genetically modified crop, hybrid, patent, plant breeders right

JEL classification: Q160, O130, O320, O330, O340

1. Introduction

The private sector has played a significant role in fostering technical innovation in agriculture. In fact, in most industrialized countries private investment in agricultural research has surpassed that of the government. The revolution in agricultural biotechnology has given even greater impetus to private investments in agricultural research. In developing countries, private investment in research is growing faster than spending on public research but still appears to play a relatively small and limited role as a source of agricultural productivity growth.

The opportunities and challenges created by scientific discoveries related to recombinant DNA have brought private agricultural research issues to the forefront. A few private companies in industrialized countries, the United States in particular, have invested aggressively in applying biotechnology and are now providing almost all new biotechnology to developed and developing country farmers. Debates surrounding biotechnology have focused attention on the gap between the amount and the effectiveness of private research in developed and developing countries, the role of multinational corporations in agricultural development, the problems with policies such as intellectual property rights, liability indemnity, food safety and environmental regulations, and the role of government research.

The major objective of this chapter is to assess the quantity and impact of private research in developing countries. We also assess policies that encourage or discourage the private sector to invest in agricultural research. To meet these objectives the chapter is organized as follows: in Section 2 the history, size and structure of private research in developing countries is described, including available data on the types of research undertaken by the private sector. Section 3 describes the outputs of this research, and Section 4 looks at the diffusion of these outputs among developing country farmers. Section 5 presents evidence on the impact of private technology on agricultural growth and farm income and also reviews how economic gains from new proprietary technology are shared between farmers and firms supplying the technology. Section 6 reviews the key economic and policy determinants affecting private incentives to invest in this research. The final section concludes with some policy prescriptions of how governments can encourage the private sector to invest more in agricultural research.

2. History, size and structure of private agricultural research

In this chapter we restrict ourselves mainly to agricultural research conducted for and by privately owned[1] profit-oriented firms, especially those supplying inputs to producers of crop and livestock commodities. In some cases the suppliers of new, proprietary technology for agriculture may also be producers of the commodities (such as large tropical

[1] These include companies owned by individuals and companies whose stock is publicly traded, but where governments hold at most minority shares in these enterprises.

plantations or integrated livestock companies), but in most cases these firms specialize in the development, manufacture and distribution of agricultural inputs, such as chemicals, machinery, pharmaceuticals, seeds and feeds, and sell these inputs to farms. With the exception of the seed industry, which is part of the agricultural sector, these inputs are produced by the industrial sector and used in the agricultural sector.

The beginnings of private research in developing countries were in the agricultural sector itself. In Asia the first private sector research started in the colonial period when large private plantations started to have difficulties with their production practices. An early example was the tea industry in eastern India, where tea leaf production was reduced by an insect pest. Tea planters organized and taxed themselves so that they could hire an entomologist to develop solutions to the problem. They also hired staff to help identify new varieties and cultural practices to increase their productivity and profits. Similar pest, variety selection, and agronomic problems prompted rubber plantations, oil palm plantations, and coffee plantations to start research programs, usually as part of an association of private plantations. Often these associations sought assistance from colonial governments to force all plantations to contribute funds and prevent free riding. Consequently, many of the research programs developed into quasi-government agencies such as the Rubber Research Institute of Malaya.

Some early research in Asia was also carried out by agricultural processing industries like the tobacco industry. In the British colonies the British American Tobacco Company developed a network of small experiment stations for identifying good cigarette and cigar tobacco varieties and developed the cultural practices that would provide the company with the quality of tobacco it wanted.

In the southern cone of Latin American agricultural research was initiated by some of the *haciendas*, or large estates and ranches. These private enterprises were large enough that they could afford to have their own crop and animal breeders who could develop improved varieties and breeds for them, as well as to sell seed and breeding stock to neighboring farms. Private wheat and maize breeding developed in this way: companies that developed in this manner in the 1920s and flourished in the 1930s continue today in well-known names such as Buck and Morgan.

Plantations in Asia and Africa and the haciendas of Latin America were able to profit from research because the innovations that they developed were used on their own farms and reduced costs or increased demand enough to pay for the research. While technology might spillover to neighboring farms, the original innovators could stay a step ahead and might go into the business of selling the technology to other farmers. For example, Buck and Morgan moved quickly from supplying wheat and maize varieties on their own estates to becoming important seed companies in Argentina.

Around the Caribbean the majority of private research was conducted by banana multinational corporations such as United Fruit and United Brands. They perfected the Cavendish variety which still dominates the industry.

After World War II, private research in the agricultural input industries grew in importance. In Asia, rapid population and per capita income growth led to increased demand for food and agricultural commodities. Because land was limited, Asian farmers had to

Table 1
Growth of public and private agricultural research in Asia (million 1995 US dollars)

	Public			Private			Private share of total in 1995 (%)
	1985	1995	Annual growth rate (%)	1985	1995	Annual growth rate (%)	
India	206	348	5.2	26	56	7.7	13.9
China	403	485	1.9	0	16	–	3.2
Indonesia	62	81	2.7	3	6	6.9	6.9
Malaysia	44	64	3.7	14	17	1.9	21.0
Philippines	17	38	8.0	6	11	6.1	22.4
Thailand	67	127	6.4	11	17	4.4	11.8
Pakistan	n.a.	n.a.	n.a.	2	6	11.0	n.a.
Total	800	1142	3.6	62	128	7.2	10.1

Source: Pray and Fuglie (2002).

increase yields through the use of modern plant varieties, fertilizer, pesticides, and irrigation. In Latin America local population and income growth, plus increased demand from Asia and Europe, also induced farmers to increase production. Latin American farmers – particularly the large ones – often had plenty of land but not enough labor so they greatly increased their demand for tractors and other types of machinery. The induced increase in demand for industrial inputs was initially met in developing countries mainly by importing technology. Later, larger developing nations started producing their own technology (inputs) often in state-owned corporations or through policies which encouraged or required foreign firms to invest and transfer technology locally. It was easier for countries that already had large manufacturing sectors to develop capacity to supply agricultural inputs.

Quantitative estimates of the amount of private sector research are not available until well after World War II. The most detailed evidence available on private agricultural research in developing countries comes from a survey of seven Asian countries conducted in the mid-1990s by Pray and Fuglie (2002). For all of these countries, there is a clear pattern of growth but also significant differences among them (Table 1). In the three largest countries – China, India, and Indonesia – private investment grew much more rapidly than public investment between the mid-1980s and the mid-1990s, but still accounted for only a small fraction of total agricultural research in these countries. In the smaller Southeast Asian economies – Malaysia, the Philippines and Thailand – public research grew more rapidly than private research. By the mid-1990s, the private sector accounted for about 10% of total agricultural research in these Asian countries.

The Pray and Fuglie (2002) survey also found that growth in private research was uneven across subsectors. The agricultural chemical industry conducted the most private research, followed by the processing and plantation industries (Table 2). Research by the agricultural chemical industry – primarily plant protection chemicals but also some on fertilizer and biotechnology – experienced the most rapid growth, tripling in real

Table 2
Growth of private agricultural research by sector in seven Asian developing countries (million 1995 US dollars)

	1985	1995	Annual growth rate (%)
Agricultural machinery	3.9	7.5	6.5
Agricultural chemicals	14.5	47.0	11.8
Livestock/animal health	5.4	15.9	10.8
Plant breeding	8.2	16.4	6.9
Plantation & processing	21.2	40.8	6.5
Total	53.2	127.5	8.7

Source: Pray and Fuglie (2002).

Table 3
Spending on agricultural research by domestic and foreign firms in Asian developing countries in 1995 (million 1995 US dollars)

	Private agricultural research expenditures			
	By local firms	By foreign firms	Total	Foreign firms' share (%)
India	38.7	16.8	55.5	30
China	0	16.0	16.0	100
Indonesia	2.6	3.5	6.1	58
Malaysia	15.0	1.6	16.6	10
Philippines	3.2	7.3	10.5	69
Thailand	6.4	11.0	17.4	63
Pakistan	3.9	1.8	5.7	31
Total	69.8	58.0	127.8	45

Source: Pray and Fuglie (2002).

terms between the mid-1980s and mid-1990s. Private livestock research grew almost as rapidly. Private spending for research on plantation crops and agricultural machinery grew substantially slower. Private research was narrowly concentrated on a few agricultural commodities, especially oil palm, rubber, maize, poultry, pigs, and some vegetables.

Pray and Fuglie (2002) found that both local and multinational firms played important roles in agricultural technology development in these countries. On average, foreign firms represented about 45% of all private agricultural research spending in the seven Asian countries, ranging from 100% in China to 10% in Malaysia (Table 3). Foreign firms were concentrated in the subsectors which had the fastest growth – chemicals,

Table 4
Shares of agricultural research expenditure by government institutes, universities, farmer associations and private companies in selected countries in 1995 (percentage of total expenditure)

Countries	Government institutes	Universities	Farmer associations	Private companies
Argentina	89	5	0	6
Brazil[a]	63	29	0	8
Chile	75	20	1	4
Colombia[b]	61	2	29	8
Ecuador	52	5	7	36
Mexico	50	17	5	28
Peru	65	20	10	5
Venezuela	80	10	1	9

Source: Echeverría, Trigo and Byerlee (1996).
[a]1991.
[b]1993.

livestock, and plant breeding, while multinational companies played a smaller role in private plantation and agricultural machinery research.

Outside of Asia, there is much less detailed evidence on the trends and scope of private agricultural research. Case studies from Latin America suggest no clear trend in private agricultural research for the region as a whole. At least through the early 1990s, foreign multinational seed and chemical corporations dominated agricultural research by the private sector in Latin America. Falconi and Elliot (1995) estimated that total spending by private companies represented less than 15% of total agricultural research in the region. Table 4, constructed by Echeverría, Trigo and Byerlee (1996), suggests that this could be less than 10% for Latin America given that in three of the largest research systems – Argentina, Brazil, and Colombia – the share of agricultural research by the private sector is less than 10%. In Brazil, private agricultural research investment appears to have been roughly constant in real terms in the 1980s and 1990s and remained less than 10% of total agricultural research in the country. In Chile, private agricultural research spending after the 1980s was constant or declined slightly as the economy was liberalized and trade restrictions reduced. Instead of conducting their own research, Chilean firms found they could import technology directly from northern hemisphere countries that had similar agricultural growing conditions. Nevertheless, for Latin America as a whole, the importance of agricultural research financed and conducted by private companies and producer groups rose during the 1990s due to a decline in publicly-funded research [Echeverría, Trigo and Byerlee (1996)].

The evidence from Africa is sparse but suggests that private investment in agricultural research is very limited. The only African countries with significant amounts of private agricultural research appear to be South Africa, Egypt and Kenya. In South Africa, private agricultural research spending as a share of total agricultural research declined

Table 5
Public and private expenditures for maize breeding in the late 1990s (millions of US dollars/year)

	Private sector	Public sector
Latin America		
Domestic	7.095	11.079
Multinationals	19.531	
Total	26.626	11.079
Eastern and Southern Africa		
Domestic	0.814	5.878
Multinationals	1.562	
Total	2.376	5.878
East, South, and Southeast Asia		
Domestic	1.858	4.307
Multinationals	7.830	
Total	9.688	4.307
All regions		
Domestic	9.767	21.263
Multinationals	28.923	7–13[a]
Total	38.690	28–34

Source: Morris (2002, Figure 1, Table 3, and Table 5).
[a]CIMMYT maize research expenditure estimates based on research personnel costs.

in the 1990s and was less than 4% of the total by the end of the decade [Liebenberg and Kirsten (2006)]. In Kenya, private agricultural research reportedly grew in the 1990s, mainly on export commodities like tea and coffee [Ndii and Byerlee (2005)].

One survey of maize breeding by the International Center for Maize and Wheat Improvement (CIMMYT) used a common methodology across all developing countries in the late 1990s [Morris (2002)], gathering information on the number of scientists and expenditures on maize breeding in both the public and the private sectors. It found that the regional distribution of maize research is similar to that of private sector research generally – most private research is in Latin America and Asia with very little in Africa (Table 5). Latin America, where maize is an important food and industrial crop and is grown by both subsistence and large commercial farms, has by far the most private maize research with more that $26 million per year (compared with $11 million annually by the public sector). In Asia, the private sector invested $9.7 million per year in maize research, about twice the amount of public institutions. Within the private sector, multinationals are dominant in all three regions, spending almost three times as much on research as local companies, making maize unique among crops grown in developing countries in that the private sector significantly outspends the public sector in plant breeding. Even including maize research by CIMMYT together with research by na-

Table 6
Number of field trials of genetically modified crops through mid-2003, by region

	Latin America	Africa	Asia	Transitional economies	Total LDC countries	Total field trials – all countries	LDC share of total
Corn	603	263	22	98	986	5,564	0.18
Canola	22	8	3	12	45	1,358	0.03
Potato	36	25	16	28	105	1,169	0.09
Soybean	192	61	3	7	263	954	0.28
Cotton	147	116	56	0	319	985	0.32
Tomato	38	6	23	0	67	686	0.10
Sugar beet	7	1	0	29	37	397	0.09
Tobacco	16	1	36	11	64	371	0.17
Wheat	19	5	1	6	31	367	0.08
Rice	14	0	42	0	56	252	0.22
Other	141	39	41	10	231	1,989	0.12
Total	1,235	525	243	201	2,204	14,127	0.16

Source: Compiled by authors from national government statistics.

tional programs, private research spending still exceeds that of the public sector. The only exception is Africa, where private research once again trails the public sector.

Since the mid-1980s, in some developing countries a major change in private sector research has been the growth of private biotechnology research, as crop applications. Almost all of this research is very applied, such as the backcrossing of genetically modified varieties with local varieties or the conducting of field trials of genetically modified varieties that were developed in private laboratories in developed countries. In a few cases private companies have conducted more basic biotechnology research in developing countries. For example, Monsanto has a major laboratory in Bangalore, India, which conducts strategic research in support of their worldwide research program.

Although the size of private investment in agricultural biotechnology in developing countries is not known, one way we can measure private sector activity is through the number of field trials conducted with genetically modified plant varieties. Testing of genetically modified varieties in developing countries started in the late 1980s in China and a few Latin American countries, and plant breeding to cross genetically modified varieties with local varieties started about the same time. With the exception of China most of these field trials were conducted by private firms. Table 6 shows the cumulative number of field trials with genetically modified crops through mid-2003 by crop and by region. Latin American had far more field trials than other regions, but somewhat surprisingly, Africa comes next, with Asia having the fewest number of field trials. The strong showing of Africa is primarily due to South Africa where 90% of the African trials were conducted. Asia (with the exception of China) has been much slower than Latin America or South Africa at approving field trials and at commercializing the use of genetically modified crops.

Table 7
Estimates of global agricultural research spending in the mid-1990s (billion 1993 US dollars per year)

	Public	Private – Pardey and Beintema	Private – James	Total
Industrialized countries	10.2	10.8	11	21.0–21.2
Developing countries	11.5	0.7	2	12.2–13.5
Total	21.7	11.5	13	33.2–34.7

Source: Public agricultural research expenditures from Pardey and Beintema (2001). Private agricultural research expenditures from Pardey and Beintema (2001) and James (1997).

Based on these surveys of private agricultural research in developing countries, some general trends can be identified. Overall, private research expenditures are small relative to the size of the agricultural sector and relative to research spending by the public sector. Pardey and Beintema (2001) estimated that the private sector's share of total agricultural research spending in developing countries was about 5%. In Asia and Latin America, it was estimated to be somewhat higher than 5% while in Africa it was considerably less. Between 1985 and 1995 agricultural research expenditures by the private sector in Asia grew twice as fast as publicly-financed agricultural research but in Africa and Latin America growth in private sector research was very limited. Since the mid-1990s biotechnology has stimulated more private research in plant breeding, but it is not clear whether biotechnology has been able to induce more research in related industries such as agricultural chemicals. However, it is clear from the trends in both private and public research expenditures that growth was not nearly sufficient to close the gap in research spending between developed and developing countries. Spending on public and private agricultural research together accounted for less than 1% of agricultural GDP in developing countries, compared with 5.5% of agricultural GDP in industrialized countries [Pardey and Beintema (2001)].

Total global investment in agricultural research was estimated at about $33–35 billion annually in the mid-1990s (Table 7). The private sector is thought to have accounted for slightly over one-third of this total, or $11.5–13 billion [Pardey and Beintema (2001); James (1997)]. While public agricultural research spending was fairly evenly divided between industrialized and developing countries ($10.2 billion and $11.5 billion, respectively), private agricultural research was heavily concentrated in the industrialized nations. Pardey and Beintema (2001) estimated that aggregate spending on private agricultural research in developing countries was $0.7 billion, while James (1997) put this figure at $2 billion, compared with about $11 billion in private research in industrialized countries. The main difference in these estimates of private research in developing countries is that Pardey and Beintema (2001) based their figure on a compilation of R&D surveys of firms in developing countries, while James (1997) formed his estimate by assuming a certain share of agricultural input sales by private companies was allocated to research. One way of interpreting these differences is that Pardey and Beintema

(2001) estimated the amount of private research *in* developing countries, while James (1997) captured private research spending *for* markets in developing countries (much of which may take place in private laboratories in industrialized countries).

In addition to the aggregate size and the geographic distribution of private agricultural research it is important to recognize that foreign firms account for at least half of the private research in developing countries. The Asian research expenditures data collected by Pray and Fuglie (2002) makes this clear for that region, while CIMMYT maize research data indicates the same is true for Africa and Latin America. Patent and Plant Variety Protection data below reinforce the importance of the role of foreign firms in developing and transferring agricultural technology to developing nations.

3. Private agricultural research output

One might assume that private agricultural research outputs – innovations that can be used by farmers – are closely related to research expenditures. However, this may not be the case if some research programs are more productive than others, or if much of the technology was developed by multinational firms that conduct part of their research in their home countries. Therefore, in this section of the chapter we report various measures of private agricultural research output.

Research output comes in several forms. Some of it adds to the body of knowledge in a discipline through academic journal articles, but this type of research output is quite unusual for private firms in developing countries. Most private research output is in the form of new technology such as plant or animal varieties, improved machinery, new chemicals, improved animal feed and pharmaceuticals, or new production practices. It is often possible to obtain government or industry statistics on the total sales of many of these inputs, but only a small part of these input sales are actually the sales of new technology. Thus, to identify the actual outputs of research we have to find other measures of innovation. Our primary measures of output of private research are the number of patents, plant breeders' rights (PBR), new varieties of hybrid maize, and regulatory approvals for commercialization of genetically modified (GM) crops. While patents cover a broad range of technologies including mechanical, chemical, and biological products or processes, in contrast PBRs, hybrid varieties, and GM crop approvals refer specifically to biological research outputs embodied in seed.

As a means of protecting and promoting intellectual achievements, patents have a history stretching back 500 years in the United Kingdom, and are enshrined in the United States Constitution. However, until recently, patent law applied only partially to agriculture. New and useful agricultural machinery patents date back to the early days of patent systems, and patent protection was given for agricultural chemicals and animal pharmaceuticals when these industries emerged in the mid-20th century. In contrast, the extension of patent protection to biological inventions has come about more recently. New varieties of plants were first given patent rights in the United States with the Plant Patent Act of 1930, which established a special category of patents and only extended coverage to asexually propagated varieties not occurring in nature. Finally, in 1980 a landmark decision by the U.S. Supreme Court (Diamond vs Chakrabarty) held that an invention could not be treated unpatentable simply because it was comprised of living matter [Fuglie et al. (1996)]. There are now more than 13,000 plant-related patents listed in the U.S. alone, of which well over half are still under patent protection.

Most nations have established patent systems to encourage innovation (183 countries are currently members of the World Intellectual Property Organization, or WIPO), and the members of the World Trade Organization (WTO) are committed to extending intellectual property protection to agricultural (including biological) inventions under the Trade Related Aspects of Intellectual Property Rights (TRIPS) agreement. While national definitions vary slightly, the typical requirements for a patent to be granted are that they describe technology that is new, useful and nonobvious to an expert in the field. Some countries provide a less stringent type of patent, called a petty patent, in which the nonobvious criterion is weakened, and this distinction has often been used in developing nations to foster local innovation. By permitting requirements that are substantially lower than for utility patents, there is a recognition that petty patents are adaptive innovations, building on existing technology to make it more suitable for local conditions. Petty patents have been used extensively for agricultural innovations in, for example, Brazil and Korea. China and Hungary also have widely used petty patent systems, but they have been used more for industry than for agriculture [Johnson and Evenson (1999)].

Plant breeders' rights are another form of intellectual property protection available for new varieties of crops. PBR are generally less stringent than patents, typically permitting farmers' privilege (farmers retain the right to use harvested material for the next crop cycle) and breeders' exemption (breeders are permitted to use a protected variety as a source for further research and breeding). PBR also have an international agreement, the International Union for the Protection of New Varieties of Plants (which is known by its French acronym as UPOV). UPOV was first formed by European signatory states in 1961, and has since been joined by a total of 62 countries, including 18 transitional economies and 16 developing countries.

As an indicator of research output, patent data have the advantage of being able to reveal contributions to agriculture performed in other sectors. As described in the previous section, there is actually a fairly small amount of private research done by agricultural firms themselves, while a substantial amount is developed by other sectors such as the chemical and machinery industries. The challenge in measuring agricultural-use patents, however, is that patents are not registered by their sector of use, but by a product classification system useful primarily to lawyers and patent examiners.

To register and classify patents, most countries have adopted the standard International Patent Classification (IPC) system, which originated in 1976. Unfortunately, the IPC is a unique product definition system that does not correspond to other industry classification systems. For example, the IPC grouping B05 includes all goods or processes for "spraying or atomizing in general; applying liquids or other fluent materials to surfaces, in general", and so will include products and processes from a variety of different industries, from cosmetics atomizers to agricultural pesticide sprayers. However, between 1972 and 1995 the patent examiners at the Canadian Intellectual Property Office simultaneously assigned each of the 300,000 patents granted by their office an IPC code, along with an industry of manufacture (IOM) code and sector of use (SOU) code based on their professional expertise and consultation with the applicant. Evenson and a series of collaborators [see the Johnson and Evenson (1997) volume] made use of this dataset to create a set of concordance tools starting with the Yale Technology Concordance for IPC to Standard Industrial Class, and culminating most recently with the OECD Technology Concordance (OTC) for IPC to International Standard Industrial Class [Johnson (2002)].

The fact that the Canadian data contain many (actually, a majority) of foreign patents means that although the OTC is based on Canadian data, its use does not superimpose the industrial structure of Canadian inventions when applied to patent data from other nations. The probabilities of the OTC indicate a technical relationship between IPC (a product/process definition) and IOM or SOU (an industry definition), but permit flexibility for the data to display the industrial composition of patenting in any nation. Tests of the procedure's validity across nations, across technology types, and across time periods have shown it to be very robust [see papers in the Johnson and Evenson (1997) volume; Johnson and Evenson (1999); Johnson and Santaniello (2000); Johnson (2002)], creating results virtually identical to slower manual classifications. Thus, with the probabilities generated from the Canadian patent data, it is possible to derive estimates of the number of patents in each industry of origin and industry of use simply by applying the OTC probabilities to the number of patents assigned to each IPC.

Using the OTC methodology, Table 8 presents estimates of the number of patents for use in agriculture, across several developed, transitional and developing nations. In parallel, the table also reports plant breeders' rights (PBR) issued for new crop varieties in the same countries. The table distinguishes between domestic and foreign applicants, and traces each nation over time to point out several key patterns. The United States and Korea show domestic and foreign inventions for agriculture growing significantly over time with a dramatic rise in foreign patenting. Among developing countries, China and Brazil show a similar pattern of growth in private agricultural research output. However, a noticeable difference between China and Brazil is that domestic inventors were awarded the majority of patents in China (79% of the number shown in Table 8) whereas in Brazil foreign inventors received 82% of patents. India, Hungary and Bulgaria showed increasing domestic patents but declining foreign patenting while the Philippines had declining domestic but rising foreign patenting. Both domestic and foreign patenting declined in Argentina and Egypt. Thus the data may imply that there may be considerably more private sector innovation in Latin American than is indicated by research expenditures, as reported in the previous section of the chapter.[2] Brazil in particular registered a large number of patents with applications to agriculture although the

[2] However, this observation may equally be due to a difference in patent office standards, a factor fairly hard to measure.

Table 8
Number of agricultural-use patents and plant breeders' rights issued in selected countries

		Domestic owners		Foreign owners	
		Ag patents	PBR	Ag patents	PBR
Argentina	1974	71	No	37	No
Argentina	1990	8	na	19	na
Argentina	2000	na	58	na	58
Brazil	1974	83	No	159	No
Brazil	1995	79	No	225	No
Brazil	2000	133	70	621	5
Bulgaria	1974	1	No	13	No
Bulgaria	1993	6	No	11	No
Bulgaria	2000	na	74	na	0
Hungary	1974	30	14	85	18
Hungary	1995	42	20	26	37
Hungary	2000	na	18	na	31
Egypt	1977	8	No	63	No
Egypt	1995	5	No	22	No
Egypt	2000	3	No	13	No
Kenya	1975	0	No	6	No
Kenya	1989	0	No	6	No
Kenya	2000	na	24	na	45
South Africa	1974	52	12	128	0
South Africa	1995	28	76	168	164
South Africa	2000	30	60	72	110
China	1990	244	No	120	No
China	1995	334	No	222	No
China	2000	1115	57	299	5
India	1976	15	No	30	No
India	1992	16	No	17	No
India	2000	44	No	10	No
Korea	1979	41	No	1	No
Korea	1995	101	2	112	2
Korea	2000	na	233	na	0
Philippines	1976	22	No	49	No
Philippines	1990	5	No	53	No
USA	1975	1166	75	242	1
USA	1990	1471	275	460	174
USA	2000	1819	446	1076	550

Source: Agricultural-use patents from author calculations using method of Johnson (2002). Plant breeders' rights are the number of plant variety protection certificates issued for new crop varieties between 1998 and 2002 reported by the International Union for the Protection of New Varieties of Plants (UPOV).

estimate of private research expenditure in Brazil is relatively low. The fact that most of the agricultural patents in Brazil were of foreign origin suggests that the private research was being done elsewhere and then transferred to Brazil.

One general result from patent and PBR data in Table 8 is that agricultural technology of foreign origin is very important for most countries, including the United States. In a study of citations of biotechnology patents granted in the United States, Johnson and Mareva (2002) found that knowledge flows between nations diminished with physical distance but that the importance of distance has faded over time. They also document increased networking between inventors, emphasizing the interconnectedness of key inventors, key firms and key locations. The implication for developing nations is that integration into international research circles has become increasingly central, and more feasible now as physical distance is dimming in relevance.

It is difficult to draw many conclusions on private sector research output from the data on the number of plant breeders rights issued. Only a few developing countries, namely Argentina, Hungary, and South Africa – have had PBR legislation for a long period of time. Instead, for most nations PBR legislation has only recently been established. Further, several countries (Egypt, India, and the Philippines) do not yet have a plant variety protection system in place.

The primary sectors responsible for generating domestic patents applied to agriculture (the industries of origin for agricultural-use patents) are shown in Table 9. It is clear that most innovations used in agriculture do not originate in the agricultural sector (which is included in the "other" category). An average of just over 1% originate in agricultural sectors, approximately comparable with the contribution of the electronics/electrical sectors. In fact, more than 80% of agricultural patents are supplied by other sectors, primarily the chemical/drug and machinery sectors. There is an interesting dichotomy among countries regarding which sector is the more important source of agricultural inventions. For countries with relatively large endowments of agricultural land per worker, the machinery sector is a major source of technology. Brazil, Argentina and South Africa fit this pattern. For countries with limited land resources but large agricultural labor forces, the chemical sector is a more important source of technology. Fertilizers, pesticides, and veterinary pharmaceuticals are the kinds of inputs supplied by the chemical/drug sector that allow more intensive use of scarce land. China, India, Egypt, and the Philippines are examples of countries following this pattern.

Another indicator of private sector agricultural research output is the number of maize varieties released by private companies in developing countries. As mentioned above, data on PBRs are an incomplete measure of private outputs in plant breeding because for most developing countries plant breeders rights have only recently been established, if at all. However, hybrid maize has been the major focus of private plant breeding research around the world. Hybrid varieties provide appropriability without patents or PBRs because hybrid seed cannot be reproduced easily by farmers or other seed companies. Table 10 shows the number of proprietary maize varieties that were being sold by the private sector in developing countries in the late 1990s. These include almost 500 varieties in Latin America, 330 in Asia, and a mere 25 in Africa. 98% of these varieties

Table 9
Industrial origin (IOM) of domestic patents for agricultural use as share of all domestic patents for agricultural use

		Electronic/ electrical	Chemicals/ drugs	Machinery	Other
Argentina	1974	0.01	0.02	0.83	0.14
Argentina	1990	0.01	0.11	0.71	0.16
Brazil	1974	0.01	0.11	0.74	0.14
Brazil	1995	0.02	0.11	0.66	0.21
Brazil	2000	0.02	0.25	0.53	0.20
Bulgaria	1993	0.01	0.28	0.53	0.18
China	1990	0.02	0.52	0.28	0.18
China	1995	0.02	0.37	0.43	0.18
China	2000	0.02	0.56	0.26	0.16
Egypt	1977	0.00	0.79	0.17	0.04
Egypt	1995	0.00	0.59	0.34	0.07
Egypt	2000	0.01	0.53	0.34	0.11
Hungary	1974	0.03	0.37	0.41	0.18
Hungary	1995	0.16	0.37	0.25	0.22
India	1976	0.02	0.39	0.49	0.10
India	1992	0.01	0.55	0.36	0.09
India	2000	0.01	0.66	0.20	0.12
Korea	1979	0.01	0.72	0.19	0.08
Korea	1995	0.02	0.34	0.41	0.23
Philippines	1976	0.02	0.59	0.32	0.07
Philippines	1990	0.01	0.80	0.14	0.05
South Africa	1974	0.01	0.18	0.67	0.14
South Africa	1995	0.02	0.29	0.50	0.20
South Africa	2000	0.02	0.31	0.47	0.20
USA	1975	0.01	0.35	0.48	0.16
USA	1990	0.02	0.44	0.40	0.14
USA	2000	0.04	0.39	0.38	0.19

Source: Calculations by authors using methodology from Johnson (2002).

are hybrids and only 2% are open-pollinated varieties (which can be saved by farmers for use as seed). The varieties have been adapted to all of the major ecosystems except highlands. They have also been bred for all of the important maturity groups, grain colors and grain qualities. Interestingly, 58% of all of the private varieties and 70% of the private varieties adapted to tropical regions contained CIMMYT germplasm in their

Table 10
Characteristics of maize varieties developed by private seed companies[a,b]

	Latin America	Eastern and Southern Africa	South, East, and Southeast Asia	All regions
Total varieties (number)	498	25	330	853
Type of material				
Open-pollinated (%)	0.03	0.08	0.00	0.02
Hybrid (%)	0.97	0.92	1.00	0.98
Ecological adaptation				
Lowland tropical (%)	0.47	0.04	0.91	0.59
Subtropical/mid-altitude (%)	0.25	0.78	0.07	0.21
Highland (%)	0.01	0.13	0.00	0.01
Temperate (%)	0.27	0.04	0.02	0.19
Maturity range				
Extra early/early (%)	0.19	0.17	0.79	0.42
Intermediate (%)	0.36	0.30	0.15	0.28
Late/extra late (%)	0.46	0.52	0.06	0.31
Grain color				
White grain (%)	0.38	0.88	0.10	0.29
Yellow/other color grain (%)	0.62	0.12	0.90	0.71
Grain texture				
Flint/semi-flint (%)	0.59	0.21	0.74	0.63
Dent/semi-dent (%)	0.41	0.79	0.26	0.37
Other (%)	0.00	0.00	0.00	0.00
Containing CIMMYT germplasm				
All materials (%)	0.73	0.21	0.19	0.58
Non-temperate materials (%)	0.89	0.15	0.18	0.70

Source: Morris (2002).
[a]Includes all proprietary varieties being sold during the late 1990s.
[b]Varieties sold in more than one country counted only once each (no duplicates).

parentage. Public-sector germplasm enhancement is clearly important for private sector breeding efforts in developing countries.

Another measure of private-sector research output for agriculture is the number of approvals of genetically modified plants and animals that have been granted by government authorities. Agricultural biotechnology has a unique measure of research output – a “genetic modification (GM) event” (the insertion of a gene or group of genes into a specific plant or animal) – which can be counted because commercial use of GM events must be approved by government regulators. By the end of 2004, at least 54 agricultural GM events had been approved for commercial use in transitional and developing countries (Table 11a for commercial use, Table 11b for environmental release). These included use as food, feed, or in production. The main GM events approved are herbicide tolerance (HT) in maize, cotton and soybeans (10 events each) and insect resistance in maize (8 events) and cotton (8 or 10 events). All of the insect resistance events are

Table 11a
Agricultural "GM events" approved for commercial use in transitional and developing countries (includes approvals for environmental release, food, feed, and marketing)

	Maize			Cotton		Soybean	Other	Total
Trait:	IR	HT	IR + HT	IR	HT	HT		
Argentina	2	3	2	1	1	1		10
Brazil						1		1
Bulgaria								0
China	1	1		1 (3)	1	1	1 (3)	6 (10)
Colombia								0
Czech Rep.						1		1
Egypt								0
Hungary								0
India				1				1
Kenya								0
Korea Rep.	1	1				1		3
Mexico				3		1	1	5
Philippines	3	4	3	1	2	1	4	18
Russia						1		1
South Africa	1	1	2	1	1	1		7
Uruguay						1		1
Total	8	10	7	8 (10)	5	10	6 (8)	54 (58)

Source: Agbios GM Database, http://www.agbios.com/dbase (accessed December 17, 2004).
All of the GM events were commercialized by the private sector except (4) events in China. A "GM event" occurs when a gene or group of genes have been transferred to a plant or animal through biotechnology. IR = insect resistance (all based on insertion of a Bt gene into crop varieties). HT = herbicide tolerance.

based on the use of a gene from the bacterium Bacillus thurigiensis (Bt). Monsanto is by far the dominant player in the commercial application of agricultural biotechnology in these countries, responsible for 39 of the GM events. Other companies with GM events include Sygenta, DeKalb, Bayer, Mycogen, Dow, and Calgene. China is the only country in which GM events developed by government institutions have been commercialized. They include two GM events for insect-resistant cotton and two GM events for vegetables.

4. Diffusion of private agricultural technology

New technology may not necessarily be adopted. In this section we review evidence on the diffusion of research outputs by the private sector. Data that separates the spread of technology developed by private and public sectors is limited, however. Information on adoption that we present in this section is primarily of inputs in which private sector

Table 11b
Agricultural "GM events" approved for environmental release in transition and developing countries

	Maize			Cotton		Soybean	Other	Total
Trait:	IR	HT	IR + HT	IR	HT	HT		
Argentina	2	3	2	1	1	1		10
Brazil						1		1
Bulgaria								0
China				1				1
Colombia								0
Czech Rep.								0
Egypt								0
Hungary								0
India				1				1
Kenya								0
Korea Rep.								0
Mexico				1		1	1	3
Philippines				1				1
Russia								0
South Africa	1	1	1	1	1	1		6
Uruguay						1		1
Total	3	4	3	6	2	5	1	24

Source: Agbios GM Database, http://www.agbios.com/dbase (accessed December 17, 2004).
All of the GM events were commercialized by the private sector. A "GM event" occurs when a gene or group of genes have been transferred to a plant or animal through biotechnology. IR = insect resistance (all based on insertion of a Bt gene into crop varieties). HT = herbicide tolerance.

innovations are dominant – agricultural chemicals, agricultural machinery, hybrid maize and poultry varieties, and genetically engineered crops.

Two of these technologies – fertilizer and tractors – were first adopted in developing countries in the early part of the 20th century but were given a major impetus by the Green Revolution. High yielding varieties of rice, wheat and other crops responded favorably to high doses of fertilizers and pesticides [Hayami and Ruttan (1985)]. Time series statistics on fertilizer and tractor use are available from the Food and Agriculture Organization (FAO) by region and are summarized in Figures 1 and 2, respectively. The adoption of fertilizer and tractors show wide differences across Asia, Africa and Latin America: growth in input use across all three regions accelerated in the mid-1970s and growth continued Asia in the 1980s while leveling off in Latin American and Africa. In the 1990s, fertilizer and tractor use in Asia continued to grow, and in Latin America fertilizer use started to grow again while tractor use remained stagnant. In Africa the use of tractors and fertilizers actually declined after 1980 – the number of tractors in use in Africa in 2001 was the same as in 1969. Fertilizer use in Africa peaked in 1981 and then leveled off or declined slightly.

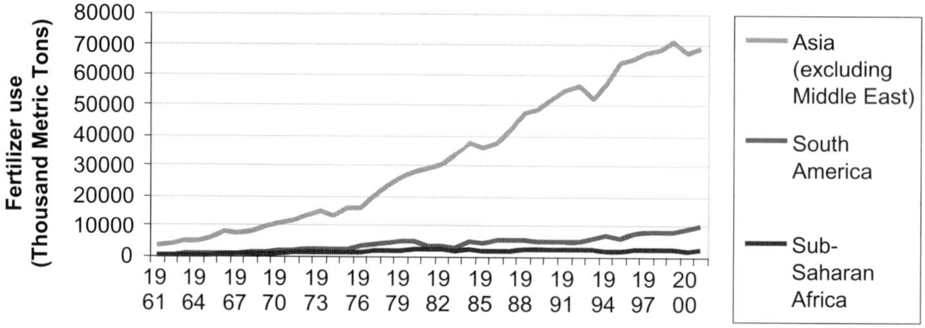

Figure 1. Fertilizer use by region. *Source*: FAOSTAT.

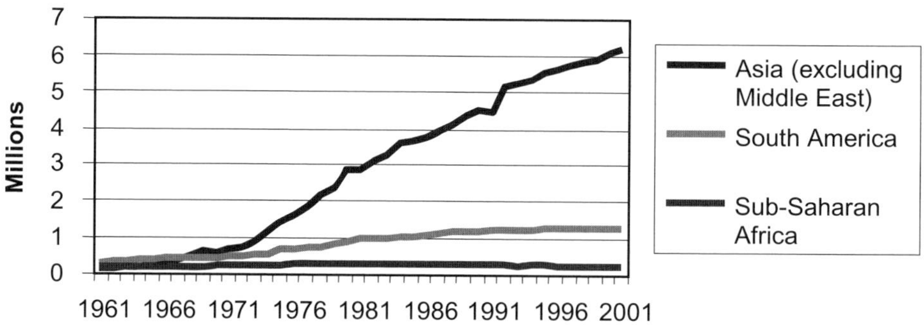

Figure 2. Number of tractors in use by LDC region. *Source*: FAOSTAT.

Another agricultural technology primarily developed and diffused by the private sector is hybrid maize. With the exception of China (where 84% of the maize area is currently planted to hybrid varieties developed by the public sector), the great majority of area in hybrid maize worldwide is planted with private varieties. The spread of hybrid maize in developing countries started in the 1950s in temperate countries such as Argentina, Chile, South Africa, Zimbabwe. Later, hybrid maize gradually moved to the highlands of tropical countries and then into tropical lowland as disease and pest constraints were overcome. The hybrids were primarily developed by private or quasi-private organizations. However, the parent material for the private hybrids relied heavily on basic breeding done by public institutions. In particular, a number of major breakthroughs in pest and disease resistance and other traits that allowed hybrid maize to spread to tropical regions were made by public sector.

Table 12 shows the percentage of total maize area planted with hybrid varieties for major regions of the world. In the United States and Europe, about 98% of the maize area is planted to hybrid varieties. In developing countries, the spread of hybrid varieties has been most significant in temperate areas: 94% of the maize area in China, 55% of the maize area in the southern cone of Latin America, and 45% of maize area in eastern

Table 12
Maize area planted to traditional and modern varieties in the late 1990s[a]

Region[b]	Maize area (000 ha)	Planted to farm-saved seed[c] (%)	Planted to modern varieties		
			Open-pollinated[d] (%)	Hybrids (%)	Total (%)
U.S., Canada and Europe	43,414	1.0	1.0	98.0	99.0
Southern Cone Countries	14,862	38.5	6.3	55.2	61.5
Andean Zone	2,203	58.4	8.7	32.9	41.6
Mexico, Central America and the Caribbean	10,055	79.0	2.2	18.8	21.0
Western and Central Africa	9,047	64.0	32.3	3.7	36.0
Eastern and Southern Africa	14,910	47.5	6.9	45.7	52.6
China	25,939	1.0	5.2	93.8	99.0
South Asia	8,207	50.0	24.6	25.4	50.0
Southeast Asia	8,144	37.8	24.8	37.4	62.2
All developing countries	94,182	37.6	11.5	51.0	62.4
All non-temperate regions[e]	65,731	52.8	14.8	32.4	47.2

Source: Morris (2002).

[a]Includes data for 48 countries covered by surveys conducted by CIMMYT and IITA, plus northern China, Côte d'Ivoire, Pakistan, and Egypt. Countries with 100,000 ha or more planted to maize that were not included are Turkey (545,000 ha), DPR Korea (496,000 ha), Morocco (341,000 ha), Myanmar (203,000 ha), Afghanistan (200,000 ha), Somalia (200,000 ha), Madagascar (190,000 ha), Iran (133,000 ha), and Burundi (115,000 ha).

[b]Year of coverage: Latin America = 1996; Eastern and Southern Africa = 1997; Western and Central Africa = 1998; East, South, and Southeast Asia = 1999.

[c]Includes landraces and very old open-pollinated varieties and hybrids grown from advanced-generation recycled seed.

[d]Includes area planted to recycled seed from open-pollinated varieties.

[e]Excludes Southern Cone countries, South Africa, and northern China.

and southern Africa were planted to hybrid varieties by the late 1990s. Adoption of hybrid maize in tropical lowland regions (South Asia, Southeast Asia, Mexico, Central America and the Caribbean) is considerably lower, at less than 30% of the total maize area.

All of the genetically modified (GM) crops that had entered commercial use by 2004 were owned and marketed by the private sector, with the exception of about half of the GM cotton area in China (public-sector varieties accounted for about half of the GM cotton area and private varieties owned by Monsanto and Delta & Pineland accounted for the other half). In fact, the first GM crops to be grown commercially in developing countries were developed by public-sector breeders in China in the early 1990s. The first GM crops introduced by the private sector were herbicide-tolerant soybeans in Argentina and insect-resistant cotton in Mexico, both introduced in 1996. Since that

Table 13
Area (million ha) planted to transgenic crops in transitional and developing countries

Country	1996	1997	1998	1999	2000	2001	2002	2003
Argentina	0.1	1.4	4.3	6.7	10.0	11.8	13.5	13.9
Brazil	–	–	–	–	–	–	–	3.0
China	–	0	<0.1	0.3	0.5	1.5	2.1	2.8
S. Africa	–	–	<0.1	0.1	0.2	0.2	0.3	0.4
India	–	–	–	–	–	–	<0.1	0.1
Romania	–	–	–	<0.1	<0.1	<0.1	<0.1	<0.1
Uruguay	–	–	–	–	<0.1	<0.1	<0.1	<0.1
Mexico	<0.1	<0.1	0.1	<0.1	<0.1	<0.1	<0.1	<0.1
Bulgaria	–	–	–	–	<0.1	<0.1	<0.1	<0.1
Indonesia	–	–	–	–	–	<0.1	<0.1	<0.1
Colombia	–	–	–	–	–	–	<0.1	<0.1
Honduras	–	–	–	–	–	–	<0.1	<0.1
Ukraine	–	–	–	<0.1	–	–	–	–
Philippines	–	–	–	–	–	–	–	<0.1
Total	0.1	1.5	4.4	7.1	10.7	13.5	16.0	20.4

Source: James (2003).

time GM crops have been approved for commercial production in at least ten developing and transitional economies. While the first generation of private-sector GM varieties were primarily varieties developed and first grown in the United States, they are now being replaced with new varieties that are better adapted to local conditions. Most of the adoption of GM crops in developing countries has occurred in temperate regions, especially Argentina, southern Brazil, northern China, and South Africa (Table 13). So far there has been very little adoption of GM crops in tropical regions.

The pattern of global diffusion of genetically modified crops stands in sharp contrast to the spread of hybrid maize. Hybrid maize was first grown in the United States and Western Europe in the 1920s but took several decades to reach temperate developing countries and then several more decades to be adapted to tropical environments. Genetically modified crops, on the other hand, were grown in temperate developing countries at virtually the same time they became available in developed countries. Further, GM crops have not been adopted in Europe due in part to concerns about food and environmental safety. However, GM crops have yet to penetrate into tropical regions due to lack of appropriate regulatory systems, poor adaptation, and concerns that adoption of GM crops may restrict agricultural exports to Europe.

Another indicator of the adoption of private agricultural technology in developing countries is the spread of commercial hybrid poultry breeds. Modern breeds of poultry are based on the same hybrid technology that dominates much of the world's maize production: crosses from inbred parent stock create chicks with hybrid vigor but which

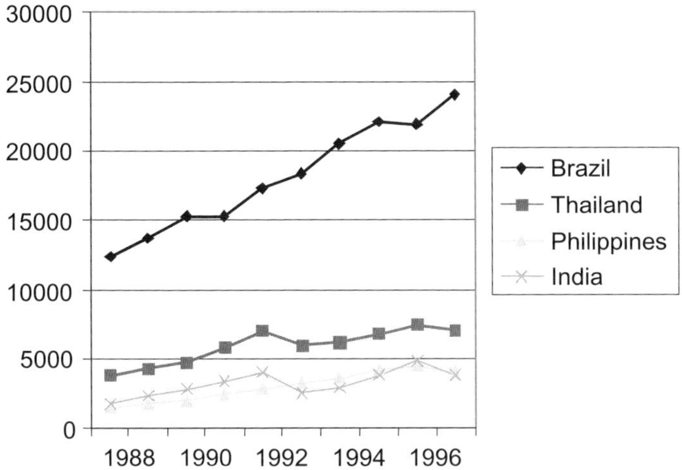

Figure 3. Markets for broiler chicks (thousands). *Source*: Narrod and Pray (2002).

cannot be used to produce off-spring with similar high yield. Therefore, poultry farmers need to repurchase chicks from breeding companies each season.

Private companies based in the United States and Europe began to introduced hybrid poultry breeds to developing countries in the early 1960s. By 1988, 99% of Brazilian poultry, 81% of Thailand's poultry, 70% of Philippine's poultry and 62% of India's poultry were from breeds developed by seven private companies based in the United States or Europe [Narrod and Pray (2002)]. Most of the rest of India's poultry breeds were developed through the joint venture using some parent breeding material from the US and Europe. Even China has based its poultry industry on breeds from a few foreign companies (unlike for hybrid maize, where China developed its own public-sector varieties). Thailand, along with India, has developed local private breeding capacity through joint ventures with foreign firms. Despite the high levels of penetration of modern poultry breeds by 1988, Figure 3 shows that the numbers of breeders' stock in these countries continued to expand. Because it is much less location-specific, hybrid poultry spread easily from temperate to tropical countries unlike crop technologies.

5. Impact of private agricultural research

The spread of modern agricultural inputs such as improved seed, fertilizer, and machinery have significantly raised agricultural production and productivity in developing countries. While there is a large literature on the impact of public agricultural research [see Alston et al. (2000)], there have been relatively few studies examining the impact of private research on agriculture. The few studies that exist focus primarily on the impact of private research within the United States [Evenson and Huffman (1993); Chavas,

Aliber and Cox (1997)]. In this section we review empirical evidence on the economic impact of private agricultural research in developing countries.

Methods for measuring the economic impact of agricultural research include econometric approaches in which productivity differences over time or across countries are modeled as a function of research expenditures and other variables, and case studies that select a particular innovation and use microeconomic (farm level) data to examine its effect on farm productivity, costs and returns. One issue of interest is the relative importance of the private sector in transferring agricultural technology across national borders, especially from developed to developing countries. Studies of developed nations have found that international spillovers of technology are a major determinant of agricultural productivity [Schimmelpfennig and Thirtle (1999)]. Another issue of particular interest involving private research is the extent to which private firms capture the economic benefits of new technology. Failure to profit from sales of technology reduces the incentive for the private sector to invest in research, but on the other hand if few benefits spill out to producers or consumers of agricultural commodities, then the research may have more limited social value.

5.1. Econometric estimates of impact of private research and technology transfer

It is possible to derive some rough estimates of the aggregate impact of the spread of modern manufacture inputs (chemicals and machinery) on agricultural productivity in developing countries from the results of agricultural growth accounting studies [see Evenson and Avila (forthcoming)]. These studies decompose changes in output into components attributable to growth in use of conventional inputs (primarily land, labor, livestock, fertilizers, and machinery) and a residual, known as total factor productivity or TFP, that is attributed to changes in technology, education, and other 'non-conventional' inputs. Contributions to agriculture from private research are partly captured by the increased use of modern inputs like fertilizer and machinery. The contribution to agricultural growth due to increased use of modern inputs is given by the growth rate in input use times its factor share. Based on the regional average growth rates and factor shares for chemical and machinery inputs reported by Evenson and Avila (forthcoming) it appears that increases in manufactured agricultural inputs added about 0.5% to average annual agricultural growth rates in Latin America and Asia during 1980–2001, but added very little (less than 0.05%) to agricultural growth in Africa. As noted above, the use of modern inputs in agriculture has been very slow in Africa.

Private research also increases agricultural output by improving total factor productivity. TFP grows when the quality of conventional inputs improves, such as when a superior chemical or machine is developed. Pioneering work on the impact of private research is a study by Evenson, Pray and Rosegrant (1999), who examined sources of growth in total factor productivity (TFP) in Indian agriculture. First, they estimated changes in TFP in crop production for each district in 13 states of India from 1956 to 1987. They then regressed TFP against public and private research, extension, and other variables. The private sector research variable included both local and imported mechanical and chemical technology. To capture technology spill-ins from the international agricultural research centers, they included as an explanatory variable the share of cropland planted with high-yielding varieties (HYV) of rice and wheat. They found that while public research and extension contributed most to productivity growth, private research accounted for about 11% of productivity growth over the whole period. The diffusion of HYV was also an important contributor to TFP but less important than private research. Since the measure of private research includes both local and imported technology, the authors were unable to assess the role of international technology transfer by the private sector.

A study by Johnson and Evenson (1999) found that domestic public and private research both had significant and positive impacts on agricultural TFP. Domestic public research appeared to be a substitute for foreign public research while domestic private research was a complement to foreign private research (in other words, private research done within the country appeared to help to absorb private research done outside the country). They documented some spillover of private and public research flows from developed to developing nations, but virtually none between less developed nations themselves.

In another study that used data on patents with potential agricultural use as a measure of private research, Johnson and Evenson (2000) examined factors affecting agricultural TFP across ninety developing nations over a period of 33 years. They found clear evidence of technology spillovers from developed to developing countries, highest for developing regions with similar production and climate patterns, large markets (high GDP per capita and high GDP), and high educational attainment. They found no evidence to suggest that domestic and foreign research differed in impact, although domestic research was more effective among nations with low levels of technological infrastructure. They estimate average growth impacts for agriculture due to research which ranged from 2.43% per year in South East Asia to 1.43% annually in Sub-Saharan Africa. Foreign research spillovers accounted for over 80% of those total growth estimates in all cases, leaving less than 20% attributable to domestic research.

Other studies using econometric methods with country-level data have examined the impact of private research on productivity changes in particular commodities, especially hybrid maize and hybrid poultry, two commodities where the private sector is a dominant source of new technology. Pray and Echeverría (1988) used data from 50 developed and developing countries from 1960 to 1985 to examine the impact of pubic and private research and hybrid seed imports on maize yields. They found a significant effect on maize yield from research conducted by multinational seed companies and from seed imports. Using a more robust model and additional variables, Echeverría (1991) confirmed the result of the earlier study. He also showed that hybrid seed technology could be transferred directly among temperate countries through seed imports (seed imports had a significant impact on maize yields while the location of research by multinational companies was not significant) while additional adaptive research was required to move technology from temperate to tropical regions (seed imports did not contribute to maize yield but the location of private research did). In a study of hybrid varieties of maize, sorghum and pearl millet in the semi-arid tropics of India, Pray and Ramaswami (2001) found a strong positive impact from private research by Indian seed companies on crop yields and farm profits. The study also found that the seed companies made significant use of enhanced germplasm developed by the International Crop Research Institute for the Semi-Arid Tropics (ICRISAT) as parent material for their hybrid varieties.

In a study of the impact of private research on productivity in the poultry industry, Narrod and Pray (2002) examined factors affecting differences in poultry productivity across developed and developing countries and over time. Using various measures for private poultry research (research expenditures, patents, and franchising), they found consistent and positive effects of private research on poultry productivity. Their findings imply that poultry technology is relatively easy to transfer across countries, even from temperate to tropical regions. They found that public sector research did not have a significant effect on poultry productivity, although public provision of veterinary services did.

5.2. Micro-level studies of private research impact and the distribution of benefits between private firms and farms

A number of studies have examined the impact of private R&D using farm production or household survey data to assess impact of technology adoption on farm yield, cost of production and profits. The CIMMYT impact study of modern maize varieties estimated that by 2000, maize breeding by international, national, and private sector researchers collectively had added about 1 t/ha, on average, to the 58.8 million hectares in developing countries where modern maize varieties had been adopted [Morris (2002)]. Of this, 48 million hectares were planted to hybrid varieties, many of which were developed by private companies. Gross economic benefits from adoption of modern maize varieties was estimated to be between \$3.7 billion and \$11.1 billion annually. While the study was not able to separate the contributions of CIMMYT, national programs, and the private sector (many of the private varieties had germplasm from CIMMYT and national programs in their pedigrees), this estimate gives an indication of the potential scale of impact of private maize breeding on developing nations.

Another study on the impact of hybrid crop varieties, Pray et al. (1991), examined the impact of private breeding of pearl millet and sorghum in the semi-arid tropics of India. Using farm survey data, the authors found that private pearl millet breeding made a particularly important contribution to farmers' income and welfare by increasing yields with hybrids that were high yielding and resistant to disease that affected all of the public hybrids. The authors also assessed the distribution of economic benefits from the improved varieties among farmers and seed companies. For hybrid sorghum, seed companies captured at most 18.5% of the benefits through higher seed prices, while 81.5% of benefits went to farmers. For hybrid pearl millet, seed firms captured only about 6% of benefits. More than 90% of the benefits from private pearl millet research went to farmers.

The controversies surrounding biotechnology and the role of multinationals in supplying that technology have led to a number of studies of the impact of adoption of genetically modified crops. All of the genetically modified crops that have been commercialized so far, except in China, were developed and marketed by private firms. Most research on this topic has focused on impacts in developed countries, especially on the impact of GM cotton, maize, and soybeans in United States. In a review of these studies, Marra (2001) concluded that:

- Growing insect resistant (Bt) cotton is likely to reduce pesticide use in most years in most U.S. states, and is more likely than not to be profitable in most of the U.S. Cotton Belt.
- Insect resistant (Bt) maize provides a small but significant yield increase in most years across the U.S. Maize Belt, and in some years and some places the increase is substantial.
- Although growing herbicide tolerant soybeans may result in a small yield loss, in most years and locations the savings in pesticide and tillage costs more than offset the lost revenue from the loss in crop yield.

In developing countries, the same three innovations (insect resistant (IR) cotton and maize, and herbicide tolerant (HT) soybeans) account for nearly all of the area under GM crops. The impact of farm adoption of these varieties has been the subject of a number of detailed studies. The results of these studies are summarized in Table 14. HT soybeans did not increase yields in Argentina but did reduce pesticide use and increased farm profits. Adoption of IR cotton led to increased yields, lower pesticide use, and higher farm profits in all four countries examined.. The only difference was that in countries that already had low levels of pesticide application with nontransgenic cotton (Argentina, Mexico and South Africa), adoption of IR cotton increases in total production cost per hectare. In China, where pesticide use is relatively high, adoption of IR cotton reduced production cost.

In addition to the financial benefits for farmers, several of these studies report positive impacts from adoption of private GM varieties on farmers' health and the environment. The adoption of HT soybeans in Argentina led to a change in the type of herbicide used from one that is highly toxic (toxicity class II) to a relatively benign chemical (toxicity class IV). In addition, adoption of HT soybeans in Argentina is associated with a large increase in use of low-tillage cultivation, which has significantly reduced soil erosion [Qaim and de Janvry (2002)]. In China the adoption of IR cotton led to a dramatic fall in insecticide use which significantly reduced the number of reported pesticide poisonings [Pray et al. (2002)].

The spread of private GM varieties, which are often protected by patents, has raised concerns about the monopoly power of multinational companies to extract excessive profits from farmers adopting the technology. Recent studies of the impact of private GM varieties have assessed the share of total benefits from the GM varieties that go to farmers and the share retained by private companies through premiums charged for the seed and restrictions on its use. In all of the studies shown in Table 14, except the IR cotton case in Argentina, the share of benefits accruing to farmers (where total ben-

Table 14
Impact of adoption of genetically modified crops on farm yields, costs and profits in developing countries

Country	Crop	Year of survey	Changes in			
			Yield (t/ha)	Pesticide use	Cost ($/ha)	Profit ($/ha)
Argentina	HT soybeans		No change	−11 kg/ha	+21	+23
Argentina	IR cotton	1999/00	+0.49 (seed)	−16 kg/ha	+97	+35
		2000/01	+0.52	−19	+96	+4
Mexico	IR cotton	1997	+0.04 (lint)	−117 $/ha	+	+8
		1998	+0.29	−94	+	+582
S. Africa	IR cotton	1998/99	+0.08	−14 R/ha	+	+79 R
		1999/00	+0.15	−44	+	+277
China	IR cotton	1999	+0.19 (seed)	−49 kg/ha	−47	+60
		2000	+1.04	−27	+7	+66
		2001	+0.34	−53	−31	+43

Source: HT soybean in Argentina [Qaim and Traxler (2002)]. IR cotton in Argentina [Qaim and de Janvry (2002). IR cotton in Mexico [Traxler et al. (2001)]. IR cotton in South Africa [Ismael, Bennett and Morse (2001)]. IR cotton in China [Pray et al. (2002)].
HT soybean = herbicide tolerant soybeans. IR cotton = insect resistant cotton (all based on the Bt gene).

efits are measured as the change in farm profits plus the additional profits of the GM seed supplier) was far greater than the benefits to the suppliers of the technology. In the Argentina IR cotton case, the seed/biotechnology company was able to prevent other seed companies and farmers from saving or selling seed, and charged the same price it received in the United States for its GM cotton seed. The result was a low profit margin for farmers and very slow spread of IR cotton in Argentina [Qaim and de Janvry (2002)]. The next highest share of benefits going to the seed/biotechnology company was in South Africa, where the company captured about one-third of total benefits from IR cotton and farmers received two-thirds of total benefits [Pray and Schimmelpfennig (2001)]. In the other cases, IR cotton in China and HT soybeans in Argentina, farmers capture between 80 and 90% of the total benefits [Qaim and Traxler (2002); Pray et al. (2001)]. The ability of private developers of new technology to extract profits from input sales is affected by a number of considerations. Although one company has been the major supplier of GM varieties in developing countries, so far its profits from GM seed do not appear to be excessive. First, the option farmers have to stay with existing technology provides a limit on what a company can charge for new technology. Moreover, offering the technology at a lower price (and therefore higher profits to adopters), will increase the rate of its diffusion. This is of particular concern for a new and controversial technology like GM crops: companies may be willing to forgo some short-term profits in order to increase public acceptance of GM technology. Finally, in some cases, other sources of GM technology quickly entered the market, such as in China where public research institutions developed their own lines of IR cotton soon after its initial introduction by the private multinational company.

6. Incentives for private agricultural research and the role of public policy

The evidence on private agricultural research in developing countries suggests that it has historically been less than public research but that its overall importance to agricultural development has been increasing over time. But private sector investment in agricultural research is highly uneven – it is concentrated on certain types of technologies and inputs, a few commodities, and in certain countries. Among various classes of agricultural inputs, the private sector is the major developer of new agricultural chemicals, farm machinery, and veterinary pharmaceuticals. Among commodities, private research in developing countries is concentrated on maize, poultry, genetically modified varieties, and a few plantation crops. Within developing countries, most private research is done in large, middle income countries, with very little in low income countries.

The driving force behind private sector investment in agricultural research is the calculation by firms of potential profits versus the costs of research. The potential profits from private research improve in the presence of sizable expected demand for the products of research, the availability of exclusion mechanisms to appropriate part the benefits from the new product or process, and a favorable business environment that provides legal certainties and permits efficient operations. Potential demand for inputs and consumer products developed through research, and thus market size, varies among countries depending on the size of the farm sector, the purchasing power of prospective buyers, local agro-climatic conditions, and sectoral and macroeconomic policies that influence input and output prices and market transactions costs. Local agro-climatic conditions set the bounds on the type of technologies that could be adopted and thus shape the nature of local demand.

The costs of developing and bringing a technology to market in a country are strongly influenced by the stock of existing scientific knowledge, local availability of skilled labor and specialized technical capacities, and by industrial and regulatory policies. Industrial policies influence the size of markets open to the private sector, the degree of market competition, and the prices of research inputs and outputs. Regulations on new product testing for efficacy, safety and seed certification procedures can greatly affect the time and cost of commercializing research outputs.

6.1. Appropriating benefits of agricultural research

The invention of hybrid corn in the 1920s, and the subsequent application of hybrid multiplication to other crop and animal species, changed the incentive structure for inventors of agricultural technology. In hybridization, two 'inbred' parent lines are crossed to produce seed that possesses hybrid 'vigor' – this seed produces a crop that is superior in yield to both of its parents. But if the harvest is kept to grow a subsequent crop, the yield advantage is lost as reproduction dissipates the genetic construct of the original hybrid seed. Thus, a breeder who invested years of effort to select a superior hybrid among thousands of combinations of parent lines could protect that investment by restricting access to information about the cross and the parent material. Farmers (or other seed companies) have to repurchase new seed from the original breeder each season in order to reap the higher yield. With this system, developers of biological technology have a means of maintaining proprietary control over their technology products and extracting some of the economic benefits of their private research endeavors.

However, hybridization is only economical in the production of a few crops and animal species. Most crops are self-pollinating (e.g., rice, wheat, soybean) or are grown from clonal planting material (potatoes, cassava, and most tree crops). Means of commercially exploiting hybrid vigor with these commodities have been extended to rice, wheat and some vegetables, but with only limited success. To provide an incentive for private improvement of a wider range of agricultural commodities, most countries have extended legal protection for intellectual property in new plant varieties. Beyond the two forms of intellectual property rights (IPRs) discussed above as applicable to agriculture, trade secrets are also occasionally a viable alternative, although it is probably the least socially desirable alternative, because it involves no public disclosure, which limits knowledge diffusion and follow-on innovation. Furthermore, it is often not practical for agricultural innovations when the research output (seed) also provides a means for reproducing the technology. Patents require full disclosure but grant the patent owner exclusive rights to use the technology for a limited period of time, usually 20 years. As indicated earlier, PBRs are considerable weaker than patents, usually allowing farmers the right to retain seed for their own use as well as allowing other breeders to use the variety as parent material in breeding other varieties. Enforcement of IPRs has traditionally been a legal question left to national authorities or litigation, but a novel mechanism has recently been developed that places enforcement more clearly in the hands of inventors themselves. Inventors may now use biological means, or genetic use restriction techniques (GURT) to prevent unauthorized copying through the insertion of self-limiting characteristics into varieties. By limiting the viability of saved seed, these uses of "terminator genes" effectively dictate the sphere of plant breeders' rights, regardless of national law [Goeschl and Swanson (2000)]. This enhanced level of control is expected to initially encourage further private research, but could impede subsequent research.

Most studies of the economic effects of agricultural IPR have focused on whether there is a link between the establishment of IPR and increased private investment in innovation. Studies of agricultural IPR in developing countries have found some positive effects on private innovative activity so long as other requisites are in place. The establishment of PBRs in Argentina did not stimulate private research until the plant breeding industry effectively organized itself to provide enforcement [Van Wijk (1996)]. In Chile, PBR law did not increase local research but encouraged the introduction of new varieties bred elsewhere, especially from similar ecological conditions such as the United States

[Van Wijk (1996)]. Garcia (1998) found PBRs did stimulate private plant breeding research in Brazil, and Aquino (1998) obtained similar results for Mexico. In their survey of private research in seven Asian countries, Pray and Fuglie (2002) concluded that agricultural IPRs may increase private incentives to invest in agricultural research as long as countries first allow the private sector to compete in agricultural input markets, establish a productive public research system, and maintain good legal institutions. One of the few econometric studies to compare international experience with agricultural IPR is Pray, Courtmanche and Govindasamy (2001). Using time series data from 37 countries, they found that PBRs and the ability to patent plants were positively associated with the spread of applied research in agricultural biotechnology.

6.2. *The role of public research*

An important issue concerning government research policy is whether public spending on agricultural research discourages private research by providing competing technologies or encourages private research through expanding technology opportunities available for commercialization. Evidence from developing countries points toward complementarities between public and private research. By establishing strong national agricultural research programs and universities, governments effectively expand the stock of scientific knowledge and personnel, reducing the cost of research inputs to private firms. One of the strongest pieces of evidence on public-private research complementarities is the degree to which crop germplasm developed by national and international agricultural research programs is used by private breeders. Pray et al. (1991) found that improved inbred lines of sorghum and pearl millet developed by the International Crops Research Institute for the Semi-Arid Tropics (ICRISAT) were the major source of parent material for high-yielding hybrid varieties developed by private seed companies in India. Improved maize germplasm from CIMMYT found its way into private-sector hybrids used in Latin America, Asia and Africa. Of all private-sector maize hybrids being sold in developing countries in the late 1990s, 58% contained CIMMYT germplasm [Morris (2002)]. Sometimes public research has served to overcome an important local constraint, considerably improving the profitability of private research investments. In the 1970s, the Rockefeller Foundation and Kasetsart University, Thailand, developed maize germplasm resistant to downy mildew, one of the most important diseases of maize in the lowland tropics of Asia. This helped induce several foreign and local seed companies to begin breeding programs in hybrid maize in Southeast Asia using the Kasetsart germplasm as parent material [Pray (1987)]. Private breeding of hybrid rice also borrowed extensively from elite rice germplasm developed by national programs and the International Rice Research Institute (IRRI). By making many of the initial investments to develop sources of major traits for crop improvement, these public breeding efforts significant reduced the private sector's marginal cost of developing new commercial varieties.

Whether public research encourages or competes against private research depends critically on research policy. In a recent study on the "crowding out" hypothesis in U.S.

agriculture, Fuglie and Walker (2001) used a simultaneous equations model to examine the allocation of public and private plant breeding scientist years across commodities. They found that more public research on basic breeding induced more private research on that crop, but more public research on applied varietal development reduced private breeding for that commodity. They also found that the public sector reduced its research on a commodity as private research on that crop increased. In many countries, however, there is increasing pressure on public research programs to raise their own revenues through sales of agricultural technology. This could place pressure on national programs to compete in commercial markets in direct competition with private research firms. The evidence though points toward the greater importance of public research in expanding the general pool of scientific knowledge and resources for commercial exploitation.

6.3. *Industrial and regulatory policies*

Governments have often placed explicit restrictions on private-sector activity in certain agricultural markets, such as through establishing public monopolies to supply inputs and market products. Opening up agricultural markets to the private sector in China and India in the 1980s provided a boost to private research, but restrictions on foreign participation in seed markets still reduce incentives for multinational companies to transfer technology, especially in China [Pray and Fuglie (2002)]. However, for many developing countries that in the 1980s and 1990s eliminated public monopolies on seed marketing, this alone did not provide sufficient incentives for the private sector to significantly increase its investment in plant breeding, with the important exception of hybrid maize [Byerlee and Echeverría (2002)].

Recent controversies about the potential safety and environmental impacts of biotechnology have emphasized both the need for regulation and the costs that those regulations can impose. Establishment of regulatory systems for testing genetically modified crops has constrained the transfer of biotechnology to developing countries, but there has been little empirical research on the effects of regulations on incentives for private agricultural research. In one of the few studies on this issue, Gisselquist, Pray and Nash (2002) assessed the impact of regulations on private research and technology transfer in the seed, poultry and irrigation machinery industries in Turkey, India, Bangladesh and Zimbabwe. Through case studies tracing the effects of regulatory liberalization in these countries, they were able to document that reducing regulations stimulated private research and technology transfer which led to improvements in farm productivity.

7. Conclusions

Private research is growing rapidly in developing countries of Asia and Latin America, but overall it is dwarfed by the amount of money that is being spent on public research. There are a few areas such as maize breeding, however, where there is more private research than public research. Private research in developing countries is split fairly evenly between national research companies and foreign-owned companies. Private research can also be broken down into research conducted by firms in the agricultural sector such as seed companies and plantations, and research conducted by firms in the industrial sectors such as chemicals and machinery. Research expenditure data from Asia suggests that both parts play a large role in developing improved technology for agriculture while more of the patented technology used in agriculture comes from industrial sectors.

Growth of technology from the private sector is determined not only by research within the country but also by flows of technology that come from the outside. Most agricultural chemical technologies, tractor technologies, and new biotechnologies were developed in the U.S., Europe and Japan and, if adopted, have been adapted to local conditions through local applied research.

The few studies that are available on the impact of private research on agricultural productivity show that both local private research and imported private technology have had an important positive impact on productivity growth. The only study that has directly estimated the impact of private research on total factor productivity in a developing country is the Evenson, Pray and Rosegrant (1999) study for India. The results showed that even in the period 1956 to 1987, before the recent rapidly growth of private research, private research and imported private technology accounted for significant share of agricultural productivity growth in India.

The key policy questions focus on what governments can do to induce more private research by national and foreign companies and what can it do to induce more technology transfer by the private sector.

First, it is clear that there must be a large market for new agricultural inputs. For governments this means privatizing public sector supply of inputs and liberalizing controls on prices. Pray and Fuglie (2002) found this to be an important component of growth in private research in Asia. In addition, governments must ensure against monopolies and oligopolies by either government or private companies. Agriculture-related industries must be able to get access to foreign capital and the technology that it brings with it.

Second, it is clear that firms must be able to capture a reasonable share of the benefits from their research investments. Intellectual property rights such as patents and plant breeders' rights, although far from perfect, are ways in which the government can strengthen firms' ability to capture benefits.

Third, a strong public research system is critical for stimulating growth in private research and technology transfer investments. Public sector research in developing countries provides the basic inputs for private sector research such as elite breeding lines with key genetic traits and a pool of skilled scientists and technicians needed to staff private sector research programs. The public research systems also provides the scientists who help develop a country's science and technology policies and provide the expertise for patent and regulatory systems.

Fourth, consumers and environmentalists are demanding assurance that food is safe and does not harm the environment. Governments must provide this assurance to the extent that they can, while balancing the costs that cut off the supply of safe, new technology which may feed people and form the basis of sustained economic growth.

The countries where agricultural production and productivity have grown rapidly in recent years are the countries that have most of these systems and policies in place. Brazil, Argentina, India, and China all have large agricultural markets. They have gradually liberalized their agriculture and the agricultural input industries over the last two decades to allow local and foreign firms to participate in these markets. Of these four countries, China has been the most restrictive on foreign investment in the agricultural sector although it has compensated somewhat by spending heavily on the public research system to import and copy foreign technology. Intellectual property rights are still weak but have been strengthened in all of these countries. Finally, all four countries have made major investments in public sector research and have the natural advantage of having large temperate regions which makes borrowing agricultural technology from the U.S. and Europe relatively easy (compared to the tropics).

Another set of countries with mid-sized agricultural economies has also prospered with the help of the private sector. Two notable examples are South Africa and Thailand. South Africa has the benefit of a temperate climate, industrial strength, and a large commercial agricultural sector. Its policies provide an attractive target for foreign investment in agricultural research and technology transfer, in part because it has strong intellectual property rights, no agricultural input parastatals, and a functioning regulatory regime that is not excessively burdensome or expensive. It has had a strong public agricultural research system, although that has eroded somewhat in recent years. Thailand, because it has an export driven economy, has also been open to importing technology and encouraging foreign investment in agriculture. It was one of the early importers of modern poultry production systems. Likewise, Thailand encouraged private maize breeding companies to set up research facilities in Thailand. Thailand's public research system played a supportive role by investing in basic breeding to overcome key pest and disease constraints facing maize production in the lowland tropics. These examples suggest that there remains substantial scope through policy measures to increase the role and contribution of the private sector to agricultural development in many developing countries.

References

Alston, J.M., Marra, M.C., Pardey, P.G., Wyatt, T.J. (2000). "Research returns redux: A meta-analysis of the returns to agricultural R&D". Australian Journal of Agricultural and Resource Economics 44, 185–215.

Aquino, P. (1998). "Brazil". In: Morris, M. (Ed.), Maize Seed Industries in Developing Countries. Lynne Rienner, Boulder, CO, pp. 213–230.

Byerlee, D., Echeverría, R. (Eds.) (2002). Agricultural Research Policy in an Era of Privatization. CABI International, Cambridge, UK.

Chavas, J.-P., Aliber, M., Cox, T.L. (1997). "An analysis of the source and nature of technical change: The case of U.S. agriculture". Review of Economics and Statistics, 482–492.

Echeverría, R.G. (1991). "Impact of research and seed trade on maize productivity". In: Pardey, P., Roseboom, J., Anderson, J.R. (Eds.), Agricultural Research Policy: International Quantitative Perspectives. Cambridge University Press, Cambridge.

Echeverría, R.G., Trigo, E.J., Byerlee, D. (1996). "Institutional change and effective financing of agricultural research in Latin America", Technical Paper No. 330. World Bank, Washington, DC.

Evenson, R.E., Avila, A.F. (forthcoming). "FAO data-based TFP measures". In: Evenson, R.E., Pingali, P. (Eds.), Handbook of Agricultural Economics, vol. 4. Elsevier, Amsterdam. In press.

Evenson, R.E., Pray, C.E., Rosegrant, M. (1999). Agricultural Research and Productivity Growth in India. International Food Policy Research Institute, Washington, DC.

Falconi, C., Elliot, H. (1995). "Public and private R&D in Latin American and the Caribbean". In: Peters, G.H., Hedley, D.D. (Eds.), Agricultural Competitiveness: Market Forces and Policy Change. International Association of Agricultural Economists, Dartmouth, UK.

Fuglie, K.O., Walker, T.S. (2001). "Economic incentives and resource allocation in U.S. public and private plant breeding". Journal of Agricultural and Applied Economics 33 (3), 459–473.

Fuglie, K.O., Ballenger, N., Day, K., Klotz, C., Ollinger, M., Reilly, J., Vasavada, U., Yee, J. (1996). "Agricultural research and development: Public and private investments under alternative markets and institutions", Agricultural Economics Report No. 735. Economic Research Service, U.S. Department of Agriculture, Washington, DC.

Garcia, P. (1998). "Mexico". In: Morris, M. (Ed.), Maize Seed Industries in Developing Countries. Lynne Rienner, Boulder, CO, pp. 231–250.

Gisselquist, D., Pray, C., Nash, J. (2002). "Deregulating technology transfer in agriculture: Impact on technical change, productivity, and incomes". World Bank Research Observer 17 (2), 237–265.

Goeschl, T., Swanson, T. (2000). "Genetic use restriction technologies and the diffusion of yield gains to developing countries". Journal of International Development 12, 1159–1178.

Hayami, Y., Ruttan, V.W. (1985). Agricultural Development: An International Perspective, revised ed. Johns Hopkins University Press, Baltimore, MD.

Huffman, W.E., Evenson, R.E. (1993). Science for Agriculture. Iowa State University Press, Ames, IA.

Ismael, Y., Bennett, R., Morse, S. (2001). "Farm level impact of Bt cotton in South Africa". Biotechnology and Development Monitor 48, 15–19.

James, C. (1997). "Progress in public–private sector partnerships in international agricultural research and development", ISAAA Briefs No. 4-1997. International Service for the Acquisition of Agri-Biotech Applications, Ithaca, NY.

James, C. (2003). "Global status: Distribution of transgenic crops, by country", ISAAA Brief. International Service for the Acquisition of Agri-Biotech Applications, Ithaca, NY.

Johnson, D.K.N. (2002). "The OECD Technology Concordance (OTC): Patents by industry of manufacture and sector of use", Working Paper 2002/5 for Organization for Economic Cooperation and Development, Directorate for Science, Technology and Industry.

Johnson, D.K.N., Evenson, R.E. (1997). "Innovation and invention in Canada". Economic Systems Research 9, 177–192.

Johnson, D.K.N., Evenson, R.E. (1999). "R&D spillovers in agriculture: Measurement and application". Contemporary Economic Policy 17, 432–456.

Johnson, D.K.N., Evenson, R.E. (2000). "How far away is Africa? Technological spillovers to agriculture and productivity". American Journal of Agricultural Economics 82, 743–749.

Johnson, D.K.N., Mareva, M. (2002). "It's a small(er) world: The role of geography and networks in biotechnology innovation", Working Paper No. 2002-01. Department of Economics, Wellesley College.

Johnson, D.K.N., Santaniello, V. (2000). "Biotechnology inventions: What can we learn from patents?". In: Santaniello, V., et al. (Eds.), Agriculture and Intellectual Property Rights: Economic, Institutional and Implementation Issues in Biotechnology. CABI Publishing.

Liebenberg, F., Kirsten, J. (2006). "South Africa: Coping with structural changes". In: Pardey, P.G., Alston, J.M., Piggott, R.R. (Eds.), Agricultural R&D in the Developing World: Too Little, Too Late? International Food Policy Research Institute, Washington, DC.

Marra, M.C. (2001). "The farm level impacts of transgenic crops: A critical review of the evidence". In: Pardey, P.G. (Ed.), The Future of Food: Agricultural Biotechnology: Markets and Policies in an International Setting. Johns Hopkins University Press and International Food Policy Research Institute, Baltimore, MD, pp. 155–184.
Morris, M.L. (2002). Impact of International Maize Breeding in Developing Countries, 1966–98. CIMMYT, Mexico, DF.
Narrod, C.A., Pray, C.E. (2002). "Technology, policies, and the role of the private sector in the global poultry revolution", Unpublished working paper. Department of Agricultural, Food, and Resource Economics, Rutgers University.
Ndii, D., Byerlee, D. (2005). "Realizing the potential for private-sector participation in agricultural research in Kenya". In: Ndiritu, C.G., Lynam, J.K., Mbabu, A.N. (Eds.), Transformation of Agricultural Research Systems in Africa: Lessons from Kenya. Michigan State University Press, East Lansing, MI.
Pardey, P.G., Beintema, N.M. (2001). Slow Magic: Agricultural R&D a Century after Mendel. International Food Policy Research Institute, Washington, DC.
Pray, C.E. (1987). "Private sector agricultural research and technology transfer in Thailand: A preliminary survey". New Jersey Agricultural Experiment Station P-02001-3-87, Rutgers University, New Brunswick, NJ.
Pray, C.E., Courtmanche, A., Govindasamy, R. (2001). "The importance of intellectual property rights in the international spread of private sector agricultural biotechnology", Report to the World Intellectual Property Organization.
Pray, C.E., Echeverría, R.G. (1988). "Transferring hybrid maize technology: The role of the private sector". Food Policy (November), 366–374.
Pray, C.E., Fuglie, K.O. (2002). Private Investment in Agricultural Research and International Technology Transfer in Asia. International Potato Center, Rutgers University, and Economic Research Service, Bogor, Indonesia.
Pray, C.E., Ramaswami, B. (2001). "Technology, IPR, and reform options: A case study of the seed industry with implications for other input industries". The International Food and Agricultural Marketing Review, Special Issue 2 (3/4), 407–420.
Pray, C.E., Schimmelpfennig, D. (2001). "South Africa trip report – Role of biotech in S. African agriculture", Unpublished report. Department of Agricultural, Food and Resource Economics, Rutgers University.
Pray, C.E., Ribeiro, S., Mueller, R.A.E., Parthasarathy Rao, P. (1991). "Private research and public benefit: The private seed industry for sorghum and pearl millet in India". Research Policy 20, 315–324.
Pray, C.E., Huang, J., Ma, D., Qiao, F. (2001). "Impact of Bt cotton in China". World Development 29, 813–825.
Pray, C.E., Huang, J., Hu, R., Rozelle, S. (2002). "Five years of Bt cotton in China – the benefits continue". The Plant Journal 31 (4), 423–430.
Qaim, M., de Janvry, A. (2002). "Bt cotton in Argentina: Analyzing adoption and farmers willingness to pay", Presented at the AAEA Meeting, Long Beach, CA, July 2002.
Qaim, M., Traxler, G. (2002). "Roundup ready soybeans in Argentina: Farm level, environmental, and welfare effects", Presented at the 6th ICABR Conference, Ravello, Italy, July 2002.
Schimmelpfennig, D., Thirtle, C. (1999). "The internationalization of agricultural technology: Patents, R&D spillovers, and their effects on productivity in the European Union and United States". Contemporary Economic Policy 17, 457–469.
Traxler, S., Godoy-Avila, G., Falck-Zepeda, J., Espinosa-Arellan, J. (2001). Transgenic Cotton in Mexico: Economic and Environmental Impacts. Auburn University, Auburn, AL.
Van Wijk, J. (1996). "The impact of plant breeders' rights in developing countries: Results of a study in five Latin American countries". In: van Wijk, J., Jaffe, W. (Eds.), Intellectual Property Rights and Agriculture in Developing Countries. University of Amsterdam, the Netherlands.

Chapter 50

PLANT BIOTECHNOLOGY FOR DEVELOPING COUNTRIES

ROBERT HERDT

Cornell University, Ithaca, NY

GARY TOENNIESSEN

Rockefeller Foundation, New York, NY

JOHN O'TOOLE

Tulsa, OK

Contents

Handbook of Agricultural Economics, Volume 3
Edited by Robert Evenson and Prabhu Pingali

DOI: 10.1016/S1574-0072(06)03050-7

Abstract

This paper reviews the tools applied in plant biotechnology and explores the prospects for biotechnology to generate benefits for developing countries. Possible near-term applications are identified. Needed capability in biological research, intellectual property management and biosafety are outlined. The experience of the Rockefeller Foundation in helping to build capacity to use the tools in developing countries is described.

Plant biotechnology includes four primary sets of techniques that enhance the capacity of scientists to modify the genetic composition of plants – plant tissue culture, marker assisted breeding, genomics and genetic engineering. These complement other techniques that have long been used by plant breeders and before them farmers to develop crop varieties. Genetic engineering has attracted critical attention because it enables the transfer and functioning of DNA from one species to another, even from bacteria or animals to plants; and although most biological scientists hold there is no significance to the origin of DNA, this possibility has embroiled biotechnology in controversy. The concentration of variety development, seed production and seed sales in less than half a dozen multinational companies, another development that critics find troubling, is an important consequence of extending patenting to plants.

Keywords

technological change, poverty, hunger, plant science

JEL classification: O13, O32, O34, Q16

1. Introduction

The first genetically engineered crop varieties were commercially grown in the United States in 1996 and by 2005 such crops covered over 49 million hectares there and another 34 million hectares in China, South Africa, Argentina, India, Brazil and other developing countries [James (2005)]. At the beginning of 2005 they were planted on very little land in Western Europe and Japan, where opposition to so-called genetically-modified organisms or "GMOs" has been strong, although farmers in Spain and Romania planted more than 100,000 hectares in each country in 2004. Cotton and soybeans are the main genetically engineered crops grown in developing countries, but maize is also grown in South Africa and the Philippines.

While food crops genetically engineered to grow well and give major benefits under most developing country conditions have yet to be created, there is little doubt that such crops could be created if appropriate resources were devoted to the task. Given that 80% of the world's population lives in developing countries, that over 800 million people in those countries are chronically hungry and that their population will increase by roughly 3 billion over the next thirty years, it seems there may well be a role for genetically engineered crops to increase food production [United Nations (2002)].

Over the past 50 years, there have been substantial increases in food production and reductions in poverty in the developing world. Despite these favorable trends, however, malnutrition remains one of the biggest health problems in developing countries. About 800 million people consume less than 2000 calories a day and are chronically undernourished [FAO (2003)]. A recent analysis indicates that 127 million pre-school children suffer from vitamin A deficiencies, leading to blindness and early death [West (2002)]. Iron deficiency is common, with about 400 million women of childbearing age afflicted by anemia. Those women give birth to underweight children and are more likely to die in childbirth. Roughly 24,000 people die each day from hunger and hunger-related causes, three-quarters of them children. Genetically engineered crops could help overcome these problems.

While many correctly argue that the root cause of such hunger is poverty, economists who study how poor agrarian societies overcome hunger and poverty agree that because poor people are concentrated in the rural sector and their livelihoods are based in agriculture, raising the productivity of small-scale farming is a key requirement to overcome poverty [Johnston (1970); Lewis (1954); Mellor (1966); Pinstrup-Andersen (2002); Thirtle, Lin and Piesse (2003); Timmer (1997)]. Growing productivity in agriculture provides the base for growing rural non-farm jobs and increased income. China and India each have over 500 million people living on small-scale farms. Sub-Saharan Africa has over 400 million and this number is increasing rapidly, despite rapid urbanization. In some of the poorest countries in Sub-Saharan Africa, over 90% of the population depends on small-scale farming for their livelihoods. In Asia and Africa, over one billion people in extreme poverty earning less than a dollar a day live in rural areas and are dependent on agriculture for their meager incomes. They are hindered by traditional farming methods, increasingly depleted soils, shrinking plots of land, scarce and unreliable water, inequitable land distribution patterns, and inefficient or unfair markets. Yet they have few, if any, good non-agriculture-dependent livelihood options. Plant biotechnology has the potential to increase the productivity of options offered by agriculture.

2. Biotechnology tools

Agricultural biotechnology is used to modify the genetic code of crops, which, together with the inputs of nutrients, water, sunshine, carbon, pests and pathogens determine crop production. These factors vary, sometimes gradually over generations and sometimes violently from day to day. Farmers manage all that variation in the course of production. But production is only the beginning of farming, because translating crop production into income depends on roads, credit, and access to fertilizer, input and output markets, control over land, institutions and policies. Hence, genetics are but one of the important factors determining crop production, and it would be hubris to claim that simply changing crop genetics through biotechnology would solve the world's problems of hunger and poverty. Rather, biotechnology is a set of powerful tools that can facilitate the production, multiplication and distribution of agricultural crops with the genetic ability to be more productive under the conditions in which they are grown. Experience has shown that such crops are powerful contributors to increasing food production and incomes [Barker and Herdt (1985); Byerlee and Moya (1993); Evenson and Gollin (2003)].

Four principal forms of plant biotechnology have the potential to improve the lives of poor farmers:

(1) Tissue culture, the production of whole plants from plant cells;
(2) Marker-aided selection, identifying the presence and following the inheritance of previously identified DNA fragments in living plants;
(3) Genetic engineering, inserting fragments of DNA into plant cells and then producing a whole plant with a different genetic composition from the original cells;
(4) Genomics, using genetic information to understand what fragments of DNA confer what traits in whole organisms.

The first three are being applied to crops of agricultural importance in the developing world while genomics and related methods in bio-informatics are currently generating vast quantities of data in developed country labs but are still at an early stage of application.

2.1. Tissue culture

Protocols for regenerating whole plants from single plant cells or clumps of cells were first developed over three decades ago. These are especially valuable for multiplying planting stock of tree crops that take many years to produce seeds or which under normal production conditions do not produce seeds, like banana. Conventional vegetative propagation has long been used with such plants. Techniques developed over the past few decades are widely used today to produce tiny plants (micro-propagation) of orchids, pineapple, banana and a wide range of temperate zone trees and shrubs. Used properly under sterile conditions, these techniques have the added advantage of eliminating nearly all diseases from the regenerated plants. Tissue culture can greatly speed the dissemination of improved varieties of crops such as cassava, sweet potato, coffee and banana that have low multiplication ratios under traditional vegetative propagation. Profitable new industries based on such micro-propagation have been established in developed countries as well as in Asia and Latin America and are beginning to be established in Africa.

Anther culture is a special form of tissue culture that can speed crop breeding for seed-propagated plants. Plant breeders have long known that crossing plants from related but different species will give progeny that combine the traits of both, as in the cross between wheat and rye that produced triticale (http://www.worldbank.org/html/cgiar/newsletter/april97/8tritic.html). Often termed "wide crosses" such interbreeding of different species occurs infrequently in nature. If forced by breeders the result is generally a plant that has very few fertile seeds that in turn have progeny with low fertility and low yields, due to poor chromosome pairing. Although triticale and some rice varieties were produced by tedious inter-specific crosses without using anther culture, the technique can overcome low fertility problems much more quickly and give regenerated plants with perfectly paired chromosomes that are fertile and contain DNA (and genetic traits) derived from both parents of the original cross. The seeds behave just like the seeds of ordinary varieties so, for self pollinating crops like rice, farmers can save a portion of their harvest as seed for subsequent plantings.

2.2. *Marker-aided selection*

Genetic traits that may be quite economically important may also be difficult or time-consuming to detect. For example, grain quality cannot be assessed directly until a plant is fully mature and produces grain; root length cannot be measured without digging up plants or growing them in special apparatus; disease resistance cannot be gauged without the presence of the pathogen and conditions that encourage disease development. Plant breeders have long used visible plant traits or "markers" that are closely associated with such hard-to-observe traits to ease the task of determining which of thousands of progeny that *may* carry a trait actually *do* carry it.

Molecular biologists have identified a number of different ways to use DNA sequences as markers, based on the ability to detect sequences at specific locations on the chromosomes of an organism. For plant breeding purposes a useful DNA marker is one that is easily detectable, is genetically linked to one or more useful traits, and generates some reproducibly different signal (such as bands on a gel) for each of the two parent plants used in a cross. Using such markers, breeders can determine the inheritance of linked traits in progeny plants at the seed or seedling stage even if the trait is expressed only in the mature plant.

An on-going goal of plant breeders is to incorporate several different genes for resistance to a single pest into a single variety to increase the durability of the resistance, the logic being that even if a pest evolves the ability to overcome one resistance gene, it will take longer to overcome several. Marker aided selection is a powerful tool to speed this process by intercrossing plants having different genes for disease resistance and testing whether each of the desired genes is present in individual progeny. Without markers to identify their presence it is impossible to know whether the additional resistance genes are present in an individual plant because the plant will look the same (resistant), whether from one, two or more genes. With marker-aided crossing a desired trait can be moved to a superior variety in four to six generations rather than ten or more, as required by more conventional methods. In January 2002, the government of Indonesia released two new rice varieties, 'Angke' and 'Conde', which were derived by disease resistance breeding augmented with MAS to combine a number of different bacterial blight resistance genes into farmer grown and well-adapted varieties [Bustamam et al. (2002)].

2.3. *Genetic engineering*

Not all genetic traits can be transferred through wide crossing or anther culture; in such cases genetic engineering may be used. It is a set of techniques that enable scientists to identify and move genes from one organism to another, either from the same or different species and has the capability of introducing a new trait that provides the plant with a completely new capability. A good example is the introduction of a gene that produces a *Bt* toxin that kills specific lepidopteran insect pests. Varieties transformed with this trait are the most widely grown among genetically engineered crops. It has been introduced into corn to control the larvae of various pests that feed on leaves, stems or roots and into cotton to control insects that chew on the cotton boll (flower/fruit).

Genetic engineering is the most controversial of the agricultural biotechnologies, in part because it is new and in part because the possibility of moving genes across species, especially from animals to plants, is viewed by some as unnatural. Although scientists assure the public that all DNA from all organisms is composed of the same four chemically identical nucleotides, many non-specialists have misgivings. As with all new technologies, there is no way to know the long-term impacts. Even railroads engendered vigorous opposition to their introduction on the basis that there was no way to prove there were no long-term ill effects on the human body from traveling at sustained speeds over 30 miles per hour [Fogel (2000)]. Some well-publicized demonstrations of potential negative effects on natural ecosystems [evaluated in NAS (2001)], questions about the use of antibiotic selection procedures in some transgenics, and concern about potential food allergies especially after the StarLink episode [Fox (2001)] all contribute. Although there have been a number of examinations of these concerns [Kendall et al. (1997); National Academy of Sciences et al. (2000); Nuffield Council on Bioethics (1999)], to date they have not overcome the misgivings that have kept genetically engineered crops from being produced in Europe.

There has, nonetheless, been a steady increase in the area of genetically engineered crops in the United States and elsewhere. First planted in 1996, the worldwide area reached 90 million hectares (222 million acres) in 21 countries in 2005. Farmers in the United States plant the largest area, some 50 million hectares, but in Argentina, Canada, Brazil, Paraguay, India and China farmers grew more than a million hectares in 2005 and in 7 other countries grew more than one million hectares. Soybean and corn are the most important, comprising 74% of the global area, with cotton and canola each exceeding one million hectares [James (2005)].

Public research institutions in countries such as China, India and Brazil, which have both excellent scientific capacity and favorable intellectual property regimes, are likely to become the primary employers of plant biotechnology to deliver useful new varieties of tropical food crops to farmers with limited purchasing power. The private sector is increasingly concentrating on a handful of major crops and profitable markets. And, owing to proprietary property and regulatory constraints, public sector institutions in industrialized countries find it increasingly difficult to commercialize products of plant biotechnology without corporate sponsors. This would be a change from the system that generated most crop varieties currently being grown in developing countries.

2.4. Plant genomics

Genomics is the most recent of the four biotechnology tools, dealing with the complete DNA sequences of species. Although the precise route by which it will make its practical contribution to agriculture is still uncertain, the scientific community is convinced it has immense potential and is providing strong support for research. Most sources use words like "ultimately", "promise" and "daunting challenge" when talking about genomics. Presently it is an area of advanced science defined by reference to its techniques and hoped-for results rather than its practical products. According to the National Academy of Sciences, "Genomics is the science and technology associated with large-scale DNA sequencing of the complete set of chromosomes of a species – its genome – and the interpretation of that sequence information. The genome is the blueprint from which an organism is built. The power derived from determining whole genome sequences is ultimately the power to understand how an organism works" [NAS Board on Life Sciences (2002)].

The genome of the model plant *Arabidopsis thaliana* has been sequenced, but plant scientists are now faced with the daunting challenge of understanding what each of the ~25,000 genes of this model plant does and how these genes relate to those of other plant species, including crop plants. The direct action of any single gene, the protein produced, is identical but the way one gene's action is combined with others to generate a plant trait is much more complex and the most valuable plant traits – yield, drought tolerance, durable resistance – are the results of many genes acting together. Hence, plant genomics holds the promise of describing the entire genetic repertoire of plants. "Ultimately, plant genomics may lead to the genetic modification of plants for optimal performance in different biological, ecological, and cultural environments for the benefit of humans and the environment" [Cornell (2005b)].

3. Crop variety development in developing countries

During the twentieth century the public sector played an important role in agricultural research and, in particular, in production of crop varieties in the United States. But for the first half of the century little was done for agriculture in developing countries. Most farmers planted food crop varieties they had inherited and selected the best performing plants for their next season's seed. The agricultural researchers in the colonial services of Britain and France, located in colonies in the tropics and sub-tropics, largely concentrated on export crop research.

Beginning in 1943, the Rockefeller Foundation assisted Mexico to develop its own improved agricultural technology. After World War II the United States government began efforts to help developing country agriculture by transferring crops and machines from the U.S. However, "In the 1950s most foreign aid programs had a simplistic view of agricultural development – build extension systems and community structures to absorb American technology, particularly new crop varieties and practices . . . It soon became apparent that this technology was not transferable without some adaptive research. By the 1960s, the importance of technological change in agriculture was beginning to be recognized" [Mellor (1998)].

The Rockefeller and Ford Foundations conceived of the International Rice Research Institute (IRRI) to carry out variety development and other research on rice in the tropics to raise yield and incomes in Asia. IRRI was established in the Philippines in 1960 and by 1965 had created and released IR8, the first semi-dwarf tropical rice, from a cross of an Indonesian with a Chinese variety. By the end of the decade the two foundations were supporting three other international agricultural research centers, in Mexico, Nigeria and Colombia. In the 1950s Dr. Orville Vogel, the USDA wheat breeder at Washington State University, crossed a dwarf variety of wheat he had obtained from Japan with North American wheat to produce the first semi-dwarf winter-habitat varieties with higher yield potential. Dr. Vogel shared a few of his early generation seeds with Dr. Norman Borlaug in Mexico. There, through much breeding effort, the semi-dwarf trait was transferred to the local spring-habitat wheat varieties [Toenniessen (2003)].

The first Mexican semi-dwarfs wheats were released in 1962. Shortly thereafter they were shared with India and Pakistan, where they performed surprisingly well, and the Green Revolution in Asia was under way [Hanson, Borlaug and Anderson (1982)]. By the early 1970s the semi-dwarf varieties of rice and wheat had been developed, released and spread to 30 million of hectares [Dalrymple (1975)], with such unprecedented speed that the phenomena was dubbed the Green Revolution.

In 1971 bilateral donors and the World Bank had joined with the foundations to establish the Consultative Group on International Agricultural Research (CGIAR) and by 1976 over $62 million was being made available annually by 26 donor organizations

[Baum (1986)]. In 2004 the system consists of 15 centers supported at a cost of about $350 million annually, producing knowledge, plant breeding lines and other global public goods freely available anywhere in the world.

3.1. The continuing Green Revolution

Today, the vast majority of improved varieties of staple food crops grown in developing countries are the product of public sector international agricultural research collaborations. Evenson and Gollin have summarized a series of impact studies of this international network and examined the development and adoption in developing countries of modern varieties of eleven crops over the period 1960–2000. As in the case of rice and wheat, many of these varieties employed dwarfing genes that gave them shorter, stiffer stems, channeled more of the products of photosynthesis into the grain, and made the plants more responsive to fertilizer and hence varieties with higher yield potential.

International plant breeding programs took on an increasing complexity of objectives over time. Initial efforts were aimed at increasing yield potential. Once that was achieved, it became evident that if the first generation of varieties were to remain successful, they would need genetic resistance to insects and diseases – so-called biotic stresses. Even as breeders were achieving success in developing this second generation of varieties it became evident that closer adaptation to local conditions would be helpful so that objective motivated a third generation of locally-adapted, resistant, high-yielding varieties.

From 1960 to 2000, over 400 public breeding programs in over 100 countries released over 8000 modern varieties of the eleven crops [Evenson and Gollin (2003)]. More than 35% of these varieties were based on crosses made at the CGIAR international centers. Most of the hybrid maize, sorghum and millet marketed by local seed companies in developing countries were based on "platform" varieties generated by these international public sector breeding programs. Table 1 provides a summary of estimated rates of adoption of the first generation of high yield potential varieties by 2000. Breeding and

Table 1
Estimated adoption of first generation modern varieties by 2000 (% of crop area)

	Asia	Latin America	Mid East N. Africa	Sub-Saharan Africa
Wheat	90–95	90–95	90–95	85–90
Rice	75–80	80–85	40–50	25–30
Maize	50–60	65–75	40–50	25–30
Other cereals	40–50	50–60	40–50	30–40
Protein crops	50–60	50–60	40–50	20–30
Root crops	50–60	70–80	40–50	30–40

Source: Evenson and Gollin (2003).

release of second and third generation modern varieties followed the first generation by 10 to 20 years and their adoption follows proportionately. The table shows that adoption has proceeded most completely in Asia and Latin America and much less in Africa. Wheat, the one commodity where new varieties have been widely adopted in Africa is a relatively minor crop grown on about 8 million hectares, about the same area as rice. Adoption of new varieties of maize, which covers 26 million hectares, roots and tubers with 20 million and protein crops with 17 million hectares is all much lower in Africa. But aside from Africa, varieties produced by modern plant breeding have penetrated developing countries very widely.

Since 1962, when intensive international plant breeding efforts first were initiated, average yields of rice, maize and wheat have more than doubled in much of the developing world (Table 2). Given that the three crops together provide more than half of the food energy consumed in the developing countries, this was a major contribution to food security in many countries. With increased production food prices dropped, and average caloric intake rose. In Asia the proportion of the population suffering from chronic hunger *dropped* from 40% to 20% while the total population more than doubled [United Nations (2002)]. Life expectancy in the developing world was 40 years in 1960 and by 2000 had increased to 64, in part because of better nutrition, although in the most recent years HIV–AIDS has brought it sharply down in Sub-Saharan Africa.

As the table makes clear, adoption of modern crop varieties and the benefits from them were not evenly distributed. They performed best with an adequate supply of water and fertilization. In Asia and Latin America, poor urban consumers, who spend a large proportion of their income on food, clearly benefited from the lower prices. Farmers who adopted the varieties and whose productivity rose more than prices declined also gained additional income. Farmers who produce most of their own food benefited directly from increased productivity and farmers who buy most of their food and sell cash crops benefited from lower food prices. Many people and large regions of Asia experienced economic growth [Barker and Herdt (1985)]. But farmers who primarily grow and sell staple food crops but did not adopt the new technology or adopted but had limited productivity gains, probably suffered economic losses as prices declined.

Table 2
Developing world staple food crop yields, kg/ha, 1961–1965 and 1999–2003

	Wheat		Rice		Maize		Sorghum		Potato	
	1961–65	1999–03	1961–65	1999–03	1961–65	1999–03	1961–65	1999–03	1961–65	1999–03
Asia	850	2750	1800	3950	1250	3650	650	975	8750	1450
L. America	1500	2550	1750	3550	1225	2850	1500	2850	7200	15250
MidEastNA	950	2050	3400	5750	1850	5620	1025	850	11400	21950
SSAfrica	790	1500	1250	1650	850	1250	750	825	6250	7550

Source: FAOSTAT.

In Sub-Saharan Africa there was little contribution from the Green Revolution and only minimal increases in yields. Significant increases in production were achieved by extending the area under cultivation and mining the soil of plant nutrients through shorter fallow periods. But production has not kept pace with population growth in Africa and a decade-long drop in per capita food production continues. Today, Africa faces a food crisis and an environmental crisis, both resulting from low input, low yield agriculture.

Small-scale farmers in Africa who got little from past innovations need what Gordon Conway has called a "Doubly Green Revolution" [Conway (1999)]. He conceives of this as long-term sustainable advances in productivity and profitability per unit of land, labor, and capital generated through the application of science to create production systems well adapted to local agro-ecologies. Conway calls for such new agricultural technologies to focus on foods consumed by the poor, be scale neutral, minimize external inputs, preferentially use inputs internal to the farm, focus on traits important to poor farmers, benefit mixed cropping systems, and enhance human nutrition. Achieving the Doubly Green Revolution will require an approach that employs farmer participatory methods, draws on the best of agro-ecological research combined with judicious use of fertilizer to help restore soil fertility, and combines it with crop genetic improvement achieved through both conventional plant breeding and biotechnology [DeVries and Toenniessen (2001)].

3.2. Near-term crop biotechnology possibilities for developing countries

Genetic engineering has great potential for revolutionizing developing world crop production but is just beginning to make an impact. The developing world plants over twice as much land to crops as the developed world but has about one-third of the world area in genetically engineered crops as shown in Table 3.

Genetically engineered cotton is the most widely grown genetically engineered crop in developing countries. Studies in China, India and South Africa [Huang et al. (2003); Qaim (2003); Thirtle et al. (2003)] show that farmers growing cotton engineered with the *Bt* gene use less insecticide and get higher profits than obtained with conventional cotton. Nonetheless, controversy about its value continues, especially in India

Table 3
Global area of transgenic crops in industrial and developing countries

	2004 million ha	% of world	2005 million ha	% of world	Change million ha	% Change
Industrial countries	53.4	66	56.1	62	2.7	5
Developing countries	27.6	34	33.9	38	6.3	23

Source: James (2005).

[Jayaraman (2003)]. United States farmers who grow soybeans engineered with a herbicide resistance gene use less herbicide and get higher profits [Gianessi et al. (2002)]. Potential benefits are large in developing countries as yet untouched by genetically engineered crops. For example, it is estimated that if farmers in five countries of West Africa would grow *Bt* cotton, their incomes would be between \$21 and \$200 million higher [Cabanilla, Aboulaye and Sanders (2003)].

However, without deliberate public intervention as in the green revolution, it is unlikely that developing countries will get large benefits from biotechnology on food crops because relatively little effort is directed toward that goal. Most research using biotechnology to create crop varieties is being carried out by private companies and they have little incentive to create food crop varieties for poor countries because potential profits from them are likely to be small. The development aid agencies are doing relatively little to help developing countries build their capacity to use the biological tools of biotechnology; CGIAR centers are each conducting a limited amount of biotechnology research. Together the public research efforts on biotechnology are small and have generated a limited number of promising technologies.

In the East African highlands, where banana is a staple crop, micro-propagation of improved and disease-free banana seedlings is commercialized on a small-scale. The farmers who are purchasing and growing the seedlings have increased their production, increased their income and increased their employment of rural laborers [Wambugu and Kioime (2001)].

By using anther culture to produce thousands of lines with different combinations of traits derived from Asian rice and African rice, WARDA has been able to identify many highly promising lines, which farmers then evaluate in a participatory varietal selection process. By 2002, the first of these "New Rices for Africa" (NERICA) were being grown by over 20,000 upland farmers in Guinea where they are more than doubling yields [Obilana and Okumu (2005)]. Farmers in other countries of West Africa are also adopting them and it looks like these varieties might be the beginning of a Doubly Green Revolution for Africa. They are promising enough that in 2003 the African Development Bank provided a \$30 million loan to disseminate the NERICAs (http://www.afdb.org/knowledge/pressreleases).

Current anther culture research on cassava is aimed at generating in-bred lines for crossing to produce advanced hybrid varieties. This could lead to yield increases from hybrid cassava similar to the major advances that occurred with hybrid maize. In the case of cassava, such hybrid varieties would most likely be disseminated to national programs as "clean" true seed and then disseminated to farmers as cuttings.

Marker-aided selection (MAS) also holds great promise in breeding for complex quantitative (multi-gene) traits like drought tolerance. To achieve a desired quantitative trait, the genes controlling the trait, or quantitative trait loci (QTLs), must be present in their most favorable format. By mapping these loci and using their markers to track their occurrence in large numbers of genotypes, it is possible to identify the markers associated with those plants having the most favorable genetic make up. The right combination of QTLs can then be duplicated in a breeding program using the markers.

Currently many research groups worldwide are attempting to demonstrate the success of MAS in breeding for drought tolerance in various cereal crops. A key challenge faced by these groups is determination of QTLs that enhance performance across varying combinations of water-stress conditions, plant growth stages and environments, with some efforts focusing on the genetic control of the drought tolerance that has been successfully introduced into maize varieties in southern Africa [Ribaut et al. (2002)].

Recent approaches for improving drought tolerance in pearl millet have focused on the development of QTL molecular markers for drought tolerance during the vulnerable flowering and grain-filling stages [Yadav et al. (2002)]. One QTL, which explained 23% of yield under water deficits, was common across environments and has been integrated into pearl millet breeding programs using markers. In sorghum, drought that occurs after flowering is particularly detrimental to yields and the 'stay green' trait (i.e., lack of green leaf senescence) has been associated with greater drought tolerance. Four QTLs are consistently associated with the 'stay green' trait in field experiments and are reported to explain 53% of the phenotypic variation in drought tolerance [Sanchez et al. (2002)]. Transgenic cotton varieties containing *Bt* genes are now being grown commercially in China, South Africa, Mexico, Argentina, Indonesia, and India. Pray and colleagues have followed the adoption of *Bt* cotton in China, which began in 1977 [Huang et al. (2003); Pray, Hu and Rozell (2002)]. By 2001, 3.5 million Chinese farmers planted 1.5 million hectares of *Bt* cotton, about 30% of China's cotton area. The rapid spread of *Bt* cotton was driven by the farmers' demand for a technology that increases yield, reduces insecticide use and costs, reduces insecticide poisonings and requires less labor. The use of insecticide fell by 20,000 tons in 1999 and 78,000 tons in 2001, the latter amounting to about a quarter of all of the insecticide sprayed in China before the adoption of *Bt* cotton. Initial yield increases were in the 5–10% range and modest increases continued over time, suggesting that farmers are learning to manage *Bt* varieties better. As yet no indications of insect pests becoming resistant to *Bt* cotton have emerged, although it would be unusual if resistance did not eventually emerge and strategies to slow resistance are an important recommendation in production. Cost savings for farmers are now beginning to push down the price of cotton, so consumers will also benefit.

In China the Beijing-based Biotechnology Research Institute of the Chinese Academy of Agricultural Sciences originally developed many of the *Bt* cotton varieties [Fang et al. (2001)]. Public sector institutions in China have produced and field-tested transgenic varieties of over fifteen different crops including many minor crops [FAO (2004)].

A powerful use of genetic engineering is to introduce a trait for which naturally occurring variation is absent. An example is Golden Rice, which has been engineered to synthesize provitamin A (β carotene) in the grain. Vitamin A deficiency affects some 150 million children in Asia. Half million are permanently blinded every year from xerophthalmia and 1 to 2 million children die unnecessarily from vitamin A deficiency [Sommer (2001)]. Golden rice could help overcome this tragedy. The first Golden Rice had a number of inadequacies, but there have been substantial advances since then [Beyer et al. (2002)]. Beta-carotene synthesis is now achieved by adding two genes, daffodil phytoene synthase (*psy*) and bacterial phytoene desaturase (*crt*I), together with rice endosperm-specific promoters. Mannose is now used as a selective agent so that the new lines contain no antibiotic resistance. These new 'clean' lines are being crossed to local varieties in Asia that are well adapted to grow in the regions where vitamin A deficiency is prevalent.

3.3. Capacity-building for biotechnology

If developing countries are to take advantage of the potentials offered by plant biotechnology and to participate in creating a doubly green revolution, they will have to have their own scientists who can work at the cutting edge of technology. This requires scientists with excellent training, laboratories, with reliable water and electric power, access to the latest scientific information and interchange with other cutting edge scientists.

Not much attention has been paid to how to accelerate the growth of this kind of capacity in developing countries. Carl Eicher charges that "after fifty years of experience, most donors remain confused about how to package, coordinate and deliver aid to accelerate agricultural and rural development in Africa ... The present donor approach to strengthening agricultural research and extension while ignoring investments in African faculties and Universities of Agriculture is a conceptually flawed capacity building model" [Eicher (2003)]. While not directed at building biotechnology capacity, Eicher might have easily been talking about it. Each donor has different ideas about what capacity is needed and how it can best be strengthened.

Developing countries, in particular those in Sub-Saharan Africa, are falling behind in the capacity to develop, import, regulate, and use genetically engineered crops [Berg et al. (2003); Mugabe (2000)]. This is partly because some are inhospitable markets for biotechnology products but that is, in turn, partly because they have few scientists able to inform their decision makers about the potential benefits and disadvantages of biotechnology. Agricultural development assistance from all donors increased steadily from the 1960s until the late 1980s when it reached a peak of $12.5 billion (in 2002$), since then it has been declining with falling proportions devoted to biological agricultural research [Herdt (2005)]. There has been little donor funding for biotechnology and a high fraction of that has been for disseminating information about biotechnology rather than training or research. One long-standing European information effort, the Biotechnology and Development Monitor, is published from the Netherlands. USAID has financed the Agricultural Biotechnology Support Project (ABSP). The Global Knowledge Center on Crop Biotechnology (http://www.isaaa.org/kc) is an electronic source for agricultural biotechnology information.

ABSP was launched by USAID in 1991 with the aim of assisting developing countries in the development and management of the tools and products of agricultural biotechnology. It mobilizes the talents of a number of US universities and other institutions to work with scientists and authorities in developing countries. It provides short-term training and seminars in biotechnology policy, intellectual property rights, biosafety, and regulatory systems. During the first ten years it supported research designed to apply biotechnology to cucurbits, potato, sweet potato and tomato; and in 2003 the project was refunded and refocused [Cornell (2005a, 2005b)]. A companion effort directed at increasing the capacity of countries to deal with institutional issues, the Program for Biosafety Systems (PBS) was created by USAID in 2003. Its mission complements ABSP by assisting developing countries to enhance biosafety policy, research, and capacity [IFPRI (2005)]. The PBS program will be initially implemented in Bangladesh, India, Indonesia, Philippines, and East and West Africa, the same geographic areas in which ABSP is working.

The Rockefeller Foundation's experience in building plant biotechnology capacity in Asia may have some lessons for contemporary challenges. About 1982, in the very early years of plant biotechnology, the Foundation decided to undertake a comprehensive program directed at building a developing country knowledge base for rice biotechnology. Through a series of strategically placed grants some of the world's premier laboratories were invited to participate in the program. The early output from these labs and others spurred on by Rockefeller attention and the promise of rice biotechnology was impressive: rice was the first cereal to be regenerated from a protoplast in 1986; a comprehensive molecular (RFLP) map was produced by 1988 and experimental genetic transformation was accomplished between 1988 and 1990 [Toenniessen (2003)]. These rapid developments were heartening and encouraging enough to expand the goals of the program from the discovery of scientific fundamentals to their transfer to CGIAR-supported centers and rice researchers in rice consuming countries. By 1990, four primary objectives were driving the program:

1. To assure that the scientific tools of biotechnology were developed for tropical rice;
2. To create sufficient biotechnology capacity in rice-dependent countries to meet current and future challenges to rice production;
3. To better understand the consequences of agricultural technological change in Asia, in part to help set priorities for biotechnological applications; and
4. To apply this knowledge and capacity to the production of improved rice varieties that will enable farmers to produce more abundant supplies of nutritious food while causing less environmental damage.

The program was research driven but capacity-building suffused all its activities. All participants became associated with the program through research, and capacity was built by deliberately involving developing country scientists in the research. Developing country scientists were selected for training opportunities in high income countries labs by proactively matching candidates with host scientists based on common research interests and the needs of their home institutions. Grants were given to developing country institutions to facilitate the return of those trained. Every grantee had to report to the entire community of program participants at periodic international meetings where they were brought face-to-face with the world's leading plant biotechnologists, and those whose performance fell short of program standards did not get subsequent support. The program meetings increased the probability of face-to-face discussions of scientists from different countries and hence the evolution of collaborative research proposals.

The Foundation fostered this outcome by providing grants that funded research and training together. The benefits of participating apparently were attractive to both scientists in the developed and developing countries. During the life of the program, over 700 scientists from 77 different developing country institutions in 16 different developing countries participated.

The program provided information to participants including distribution of theses, reprints, books, patents and the Rice Biotechnology Quarterly, a newsletter serving program participants worldwide. At international meetings the participants could hear about the latest rice biotechnology science before its publications as well as the latest thinking about research priorities established from social scientists in the network. Thus the international collaborative research and training mechanism supported by the program resulted in synergies and benefits far greater than might have been expected given the financial support available from the one donor.

The program made grants of almost $105 million between 1984 and 2000, an average of $6.2 million per year. During 1984–1989 the highest proportion of funds, about 40%, went to research centers in the industrialized countries while labs in the developing countries received about 10%. In the early years about 15% and from 1990 onward over 25% of the total was devoted to the education and training of scientists from developing countries. As the scientists trained in the early years returned, funding of developing country labs picked up, making up about 40% of funding during the last four years of the program. One feature was the application of a deliberate priority setting process that balanced the potential offered by a particular trait or target of biological research with the cost and time estimated necessary to achieve success [Herdt (1991)]. The priorities were discussed at the network meetings and provided highly visible targets for those seeking applied research grants. Due to the highly integrated program implementation, the other funds were used to create bridging opportunities like meetings and workshops, which along with the priority setting research, led to international collaborations among industrialized country, developing country and the international agricultural research centers.

The rice biotechnology program helped create the capacity for rice-dependent nations, largely in Asia, not only to use biotechnology on rice but to give local scientists background needed to understand and provide a basis for dealing with the international issues of biosafety and intellectual property rights. In 1998, when it became evident that rice biotechnology research had enough stakeholders on a global basis to ensure its continuity, the Foundation decided it would conclude the rice program in an orderly way and redirect its resources to needs in Africa [Normile (1999)].

The Foundation conducted an examination of how the accomplishments of the rice program might be used in Africa and an examination of the opportunities and needs for plant genetic improvement in Africa [DeVries and Toenniessen (2001)]. An internal review found that research outputs for "drought tolerance in rice", one of the highest priority traits, had remaining unmet potential for rice and also promise for application in maize, a crop of major importance in Africa. After a feasibility study involving international experts and a series of workshops [Ito, O'Toole and Hardy (1999)] the "Resilient

Crops for Water-Limited Environments" program was launched in 2000 with a focus on rice in Asia and maize in Africa. An acute lack of suitable field drought screening methods and knowledge of soil and plant-water relations among rice and maize breeders were identified as the primary obstacles for successful use of biotechnology to genetically improve cereal drought tolerance [O'Toole (2004)]. The new program provided specialized drought screen facilities and training in the use of molecular markers and the resulting work has established the efficacy of the drought screening techniques for both maize and rice [Chandrababu et al. (2003); Lanceras et al. (2004); Ribaut et al. (2002)].

The resilient crops work absorbed an additional Rockefeller Foundation investment of $20 million in cereal biotechnology, much of it valuable for application in Africa. In addition, the Foundation established a program to train 50 African plant breeders over a 5–8 year period on African plant breeding challenges [Rockefeller Foundation (2003)]. This will provide a core of trained Africans with technical understanding of the biological challenges.

4. Institutional capacity

Developing countries face several challenges in addition to creating the scientific capacity to work with plant biotechnology. Until the 1980s, seeds of the plants that breeders use as the starting point for their work had been obtained from friends and colleagues around the world without charge and with minimal government or other regulation. But the ability to precisely identify ancestors through DNA together with the extension of property rights to plants have combined to make the production of improved varieties the business of profit making companies [Herdt (1999)]. Developing country plant breeding systems must be able to deal with issues of proprietary property and to navigate the international agreements on property rights and biosafety that now exist.

4.1. Proprietary property

The genetic improvement of plants is a process in which each enhancement is based directly on preceding generations and requires the physical use of the plant material itself. Much of the value in today's seeds was developed over the centuries, as farmers selected their best plants as a source of seed for the next crop. Traditionally, these land races and the indigenous farmer knowledge associated with them were free of charge to collectors and hence the world community. In exchange, public sector research and breeding programs, like those of Drs. Vogel and Borlaug, added valuable traits and scientific knowledge and provided others with improved breeding lines. In the absence of any foolproof way to trace the origin of seeds they were public goods, available to farmers and plant breeders alike. Now, however, the rules of the game are changing.

DNA-based tools allow the identification of the ancestry of seeds with legal certainty and United States courts have approved the patenting of seeds created by plant breeding.

In industrial countries, over the past decade applied crop-biotechnology research and the production of improved varieties have increasingly become functions of the "for-profit" private sector [Barton and Berger (2001)]. This has led to a significant increase in plant science and crop improvement research, but the results of such research are generally protected by intellectual property rights (IPR) of various forms including patents, material transfer agreements, plant breeders' rights, and trade secrets. Increasingly, this is true of results from public sector research as well.

Industrial countries have made IPR an important component of international trade negotiations, using them to exploit their competitive advantage in research and development. Countries joining the World Trade Organization, for example, must have IPR systems that include protection of crop varieties, according to the Trade Related Aspects of Intellectual Property Rights (TRIPS) provisions. Most developing countries had until January 1, 2006, to implement such IPR systems.

Because poor farmers typically do not purchase new seed for each planting and an abrupt change in practices is unlikely, it is important that developing-country IPR laws are modeled on plant variety protection systems that include provisions allowing farmers to save and replant seed and plant breeders to use varieties for further breeding. This is in contrast to the utility patent system that extends protection to the seed and progeny of patented plants. With that system breeders must negotiate licenses before legally using patented varieties as breeding material.

Ironically, the major IPR change affecting the operations of the international agricultural research system comes from a public, not a private sector initiative. To promote technology transfer and product development in the United States, the 1980 Bayh–Dole Act gave universities and other publicly funded research institutions the right to obtain patents on and commercialize inventions made under government research grants. Similar arrangements have emerged in Europe, Japan, Australia and most other industrialized countries. The result is that while many biotechnology discoveries (e.g., pathogen-derived plant resistance to virus infection) and enabling technologies (e.g., *Agrobacterium* and biolistic transformation methods) were generated with public funding in public universities, these discoveries are no longer in the public domain. Rather, they are patented by the researcher or research institution and then licensed, often exclusively, to the for-profit sector. Discoveries and technology (i.e., applied biological knowledge) flows from the public sector to the for-profit sector. When and if it flows back, it usually come under material transfer agreements (MTAs) that significantly restrict its use, usually for research purposes only, and often include reach-through provisions to capture results of future research.

Since crop genetic improvement is a derivative process, each incremental improvement made through biotechnology now comes with a number of IP constraints, with new IP added with each transfer or further improvement. To deal with this predicament, the seed industry is becoming greatly centralized through mergers and acquisitions in a global oligopoly dominated by just a few seed and biotechnology firms, some of which are also the major producers of agricultural pesticides. These mergers were made in part to accumulate the IP portfolios necessary to produce biotechnology-derived finished crop varieties with "freedom to operate" and in part to gain control over a new technology that is threatening their pesticide markets.

The publicly funded agricultural research community for the most part lacks freedom to turn discoveries into products for farmers because of these IP restrictions, despite their leading inventive role. After analysis of agricultural biotechnology patents of the US, Europe and Japan Graff states that: "Collectively, the universities and government institutions of the public sector have created a set of IP that is larger in number and estimated value than even the largest of the individual corporate portfolios. However, these public-sector holdings remain highly fragmented across many organizations, and no single institution owns a package of technologies sufficient to enable the development of a novel transgenic plant variety" [Graff et al. (2003)]. Leading academic researchers are primarily interested in research accomplishments. They readily sign agreements to gain access to the latest tools but most MTAs contain provisions restricting further transfer of research products. Many universities have "technology transfer offices" where maximizing licensing and royalty income is just as important as technology transfer, and often achieved by granting exclusive licenses. The net result has been that improved plant materials produced by academics are highly IP encumbered and commercially useful only to companies having an IP portfolio covering most of the technologies used. For example, some forty patents and six material transfer agreements were potential constraints to the dissemination of the first generation of Golden Rice [Kryder, Kowalski and Krattiger (2000)].

International agricultural research centers have concentrated on building their own biotechnology capacity and as yet do not have significant IP portfolios. One consequence is that the traditional flow of materials through the international system is breaking down. Previously, new technologies and improved plant materials had flowed from public sector researchers in developed countries to international centers and developing country crop improvement programs. Africa, in particular, is being short changed of the benefits of biotechnology because, unlike Asia and Latin America, its public sector has little capacity to use biotechnology for the benefit of poor farmers, even in countries where the IP is not protected. Africa is much more dependent on partnering with others but publicly funded researchers in industrial countries are no longer partners who can freely share their most important discoveries and products. But some new initiatives hold out the promise of offsetting these limitations.

The Center for the Application of Molecular Biology to International Agriculture or CAMBIA, is an Australian non-profit organization working to assist developing country scientists to achieving lasting solutions to food security, agricultural and environmental problems. CAMBIA "envision a situation in which the broadest community of researchers and farmers are empowered with dramatic new technologies to become innovators in developing their own solutions to the challenges they face – solutions for which they feel ownership ... providing education and technical training is an important aspect of its mission. CAMBIA aims to create technologies that meet the requirements of the most difficult environments and the least equipped laboratories. Hosting students, Postdoctoral Fellows and visiting scientists from international laboratories ensures that CAMBIA has a real understanding of these conditions" (http://www.cambia.org). CAMBIA hosts one of the world's largest and most comprehensive searchable patent databases, including full text of all PCT, European and US patents relevant to biotechnology and provides a no-cost patent search service through the Internet (http://www.cambiaip.org/Home/welcome.htm).

In mid-2003, a dozen leading universities that have generated much of the intellectual property in crop biotechnology, but have in the past entered into exclusive IP licensing agreements with the private sector, came together and declared their determination to create a new organization to steward such innovations in the public interest [Atkinson et al. (2003)]. The new organization, PIPRA (Public Intellectual Property Resource for Agriculture), is currently developing its program (http://www.pipra.org) envisioned as a system whereby universities will license innovations to private companies for some specific purposes ("field of use") and retain rights to use the innovations for other purposes, especially for applications in developing countries and to specialty crops for U.S. farmers – purposes consistent with their missions. This could enable them to commercialize new transgenic varieties of crops now being tested in university green houses. If commercialized, current experimental strawberries, apples, lettuce and other crops could be grown without pesticides, benefiting both local farmers and the environment. Their creation has been paid for with taxpayer dollars, but they are not being brought to market because the universities have not retained the IP rights to commercialize them, and the companies that hold the rights are only interested in major crops like corn, soybean and cotton.

The African Agricultural Technology Foundation (AATF) is a "not-for-profit foundation facilitating and promoting partnerships with public and private sector entities designed to remove many of the barriers that have prevented smallholder farmers in Sub-Saharan Africa from gaining access to existing agricultural technologies that could help improve food security and reduce poverty" (http://www.aftechfound.org). AATF was designed to promote public–private partnerships that benefit African agriculture operations. It is based in Africa and is led, managed and directed by Africans. Not designed to produce or distribute finished products, rather it will be a focal point where Africans can access new materials and information on which technologies can be built. It is a way of giving poor nations the tools to determine what new technologies exist in the public and private sectors, which ones are most relevant to their needs, how to obtain and manage them, and how to develop nationally appropriate regulatory and safety regimes within which to introduce these technologies.

4.2. Regulatory systems

Poor management of IPR is only one of the ways the public sector has been handing over control of agricultural biotechnology to the big multinational corporations. Increasingly onerous and expensive biosafety regulations are another cause. In the USA, the cost of obtaining regulatory approval of a new crop variety with a transgenic event can exceed $5 million [Lichtenberg (2000); Redenbaugh and McHughen (2004)]. Even the big companies are abandoning research programs where the size of the market does not warrant this level of investment. Small seed and biotechnology companies are essentially priced out of the market unless they partner with the multinationals, and the public sector may be left out as well. Ironically, it is often environmental and consumer groups that warn against corporate control of agriculture who also work to establish regulations so costly that only big multinational corporations can afford to obtain regulatory approvals.

If developing countries establish as complex and costly biosafety regulations as the wealthy countries, they too are likely to find themselves highly dependent on multinational corporations as their primary source of advanced new crop varieties. Here, 'Golden Rice' again serves as a good example. If the developing countries require 'Golden Rice' to be approved first in the rich countries where it was invented or have biosafety regulations that are as costly as those in rich countries, there is little prospect that the public research institutions developing it will be able to afford the cost of that kind of regulatory approval. Government regulations must strike a balance between protecting consumers and the environment and allowing commercialization of products that may substitute for chemical pesticides and improve incomes and health.

Regulatory uncertainties and constraints have affected commercialization of transgenic crops produced by national researchers in developing countries. In Thailand scientists working for the National Center for Genetic Engineering and Biotechnology have produced transgenic local varieties of papaya, highly resistant to the strains of papaya ring spot virus prevalent in Thailand. These varieties have undergone three years of field tests and performed very well but approvals to commercialize have repeatedly been delayed, just one example of the challenges facing developing countries [Damrongchai (2004)].

As with intellectual property rights, the public sector needs to find better and less expensive ways of addressing legitimate regulatory concerns if it is to continue to play an important role in producing new crop varieties for the hundreds of millions of small-scale farmers who will not be served by the big companies. If not, the public sector in agriculture may find itself in the same situation as the public sector in health – generating exciting research results but seeing them used only by the private sector to develop products that can generate profits.

4.3. Public acceptance and farmer adoption

Opposition to food from "genetically modified organisms" (GMOs) is a major constraint to the adoption of plant biotechnology, particularly in Europe, but there have been significant concerns raised in the United States as well [Rissler and Mellon (1996)]. In part the opposition arises because none of the GM products currently on the market provide benefits to consumers or even to food processors or retailers. Current GM crops primarily benefit seed suppliers, farmers, and the rural environment through reduction in insecticide use. Appeals to the world's need for more plentiful and cheaper food evoke little sympathy in the overfed wealthy nations. Orchestrated campaigns against

GM foods have found a receptive audience among developed country urban consumers concerned about food and environmental safety and control of the food supply by a few large multinational firms.

Many possible fears have been expressed about GM foods, reviewed usefully in [Uzogara (2000)]. Scientific examinations of reasons to fear possible ill effects on human health have found little evidence to support such concerns. A recent review conducted with European Commission funding carried out by researchers from 8 different countries in Europe concluded as follows: "DNA, once it has been introduced into the recipient organism, is indistinguishable from the host DNA in its physical and chemical properties and behaves identically; in order to assess the impact of the transfer of GMO-DNA on food safety, it is necessary to understand the gene transfer processes occurring in nature and the mechanisms behind them, including their occurrence at different stages along the food chain; horizontal gene transfer (HGT) is at the origin of the variety of life itself and there is very little reason to assume that consumption of transgenic food or feed adds any particular generalized risk; uptake of ingested DNA by mammalian somatic cells has been demonstrated, there is so far no evidence that such DNA may end up in germ line cells as a consequence of the consumption of food" [Eede et al. (2004)].

In some developing countries there has been reluctance to import GM grain or accept GM American corn as food aid for fear that some might be planted, cross pollinate with local crops and affect potential future food exports to Europe. Given the opposition of most of Europe to GM there is some basis for such fear, even in places that have no current exports to Europe [Nuffield Council on Bioethics (2004)]. While impossible to predict, it may be that any country seeking to export to some markets in the future may have to have certifiable segregation of GM and non-GM foods, adding yet another challenge to those faced by developing countries seeking to build export earnings.

Concerns that genes from GM plants will spread to non-engineered varieties or weedy relatives are a source of special concern in developing countries that are the centers of origin, for example Mexico for corn, China and Southeast Asia for rice [Nuffield Council on Bioethics (2004); Pew Initiative on Food and Biotechnology (2004)]. There is little doubt that genes will flow; at issue is the distance they will move, the time it will take and the consequences of such gene flow. Experiments can illuminate the issues of rate and distance but the long-term consequences can only be fully known as the event plays out, thereby providing the rationale for taking a precautionary approach.

The interests of the majority of people in developing countries may well be different from those in more wealthy countries [Paarlberg (2001)]. In many, a majority of the population are farmers as well as consumers. They would recognize the benefits and risks of GM crops as farmers and the benefits and risks of GM foods as consumers. As such, their assessments will likely differ from most people in Europe and North America and will include the economics of production as well as food and environmental safety. They need to be given the opportunity to make those judgments.

The potential nutritional advantage of Golden Rice has made it the focus of both corporate promoters seeking a positive public image for biotechnology and people concerned with malnutrition in developing countries. However, even assuming that all the issues of nutrition, intellectual property, biosafety and seed production are resolved, researchers recognize that golden rice varieties will only be adopted by farmers if they provide advantages to the farmers – higher yield or higher price than varieties without the trait. The persistence of "old" crop varieties for decades after their release and first adoption attest to this. In India over 80% of the rice area is planted to "new" varieties, but, farmers continued to plant traditional varieties on the rest of the area [Janaiah and Hossain (2002)]. Newer is not necessarily better, whether genetically engineered or conventionally produced. If Golden Rice is to be adopted governments are likely to have to provide incentives for its production.

5. Conclusions

The days when plant breeders routinely shared early generation breeding lines and when fledgling breeders from throughout the world could go to the USA or Mexico for training and take home the newest semi-dwarf varieties to test in their own countries are gone and unlikely to ever return. New training modes and new institutions for the development and exchange of germplasm have replaced the old.

Plant tissue culture, marker assisted breeding and genetic engineering complement techniques that have long been used by plant breeders to generate genetic modification, a key step in producing improved crop varieties. Genetic engineering enables the transfer of DNA from one species to another, making it the object of both fear and hope. In 2005 some 49 million hectares were planted to genetically engineered crops in the US and another 39 million hectares in Canada and four large developing countries. Despite the fears expressed by many, negative health and environmental consequences have not materialized in the decade since their widespread planting. Adaptation by insects and weeds to host resistance incorporated through genetic engineering is likely to emerge the longer crops with a particular resistance gene are planted; but this is not a challenge peculiar to GM. On the business side, variety development and seed sales have become concentrated in a less than half a dozen multinational firms. Small developing countries with few scientists face significant obstacles to using crop biotechnology, but at the same time must make investments to understand and regulate them and participate in international treaties, including the World Trade Organization.

Profit incentives and the private sector no doubt generate and deliver useful products. Reasonable regulations of new technologies and education of farmers in their application can enhance and prolong the usefulness of new products but they impose very high costs. In today's global markets, some charge that property rights, regulations and liability concerns have gone too far and make access by the poor to new agricultural technologies too difficult. Half a dozen big multinational companies have ever increasing control over the seeds of the world's main food crops. Whether this will lead to an escalation in the cost of food in wealthy countries like similar concentration in pharmaceuticals has led to an escalation in the price of drugs is yet to be seen, but such a scenario is feared by many.

Getting good farm technology to over two billion poor small-scale farmers in developing countries in a way that is responsible and sustainable is likely to remain a public sector responsibility. It will require that governments, public research institutions, non-governmental organizations and corporations devise new ways of doing business through partnerships that include the interests of the majority of the world's people from developing countries as well as the concerns of the technology providers, users who can pay, and consumers from rich countries.

References

Atkinson, R.C., Beachy, R.N., Conway, G., Cordova, F.A., Fox, M.A., Holbrook, K.A., Klessig, D.F., McCormick, R.L., McPherson, P.M., Rawlings III, H.R., Rapson, R., Vanderhoef, L.N., Wiley, J.D., Young, C.E. (2003). "Public sector collaboration for agricultural IP management". Science 301, 174–175.

Barker, R., Herdt, R.W. (1985). The Rice Economy of Asia. Resources for the Future, Washington, DC.

Barton, J., Berger, P. (2001). "Patenting agriculture". Issues in Science and Technology 17, 43–50.

Baum, W.C. (1986). Partners Against Hunger: Consultative Group on International Agricultural Research. The World Bank, Washington, DC.

Berg, T., Syed, M., Ferstad, H.G.O., Tsegaye, B., Waktola, A. (2003). "Why are developing countries falling behind in biotechnology?". Center for International Environment and Development Studies, Nairobi, Kenya.

Beyer, P., Al-Babili, S., Ye, X., Lucca, P., Schayb, P., Welsch, R., Potrykus, I. (2002). "Golden rice: Introducing the *beta*-carotene biosynthesis pathway into rice endosperm by genetic engineering to defeat vitamin a deficiency". Journal of Nutrition 132, 506S–510S.

Bustamam, M., Tabien, R., Suwarno, A., Abalos, M., Kadir, T., Ona, O., Bernardo, M., VeraCruz, C., Leung, H. (2002). "Asian rice biotechnology network: Improving popular cultivars through marker-assisted backcrossing by the NARES". Beijing.

Byerlee, D., Moya, P. (1993). "Impacts of international wheat breeding research in the developing world 1966–1990". CIMMYT, Mexico, DF, p. 87.

Cabanilla, L.S., Aboulaye, T., Sanders, J.H. (2003). "Economic cost of non-adoption of Bt cotton in West Africa: With special reference to Mali". Ravello, Italy.

Chandrababu, R., Nguyen, B.D., Chamarerk, V., Shanmugasundaram, P., Chezhian, P., Jeyaprakash, P., Ganesh, S.K., Palchamy, A., Sadasivam, S., Sarkarung, S., Wade, L.J., Nguyen, H.T. (2003). "Genetic analysis of drought resistance in rice by molecular markers: Association between secondary traits and field performance". Crop Science 43, 1457–1469.

Conway, G. (1999). The Doubly Green Revolution. Cornell University Press, Ithaca, NY.

Cornell (2005a). Agricultural Biotechnology Support Project. Cornell University, Ithaca, NY.

Cornell (2005b). "Cornell genomics initiative: Focus area plant genomics". Cornell Genomics Initiative, http://www.genomics.cornell.edu/focus_areas/plant/index.cfm.

Dalrymple, D.G. (1975). "Measuring the Green Revolution: The impact of research on wheat and rice production". Foreign Development Division, Economic Research Service, U.S. Department of Agriculture, Washington, DC, p. 40.

Damrongchai, N. (2004). "Implementing biosafety regulations in Thailand: The role of Biotec". Thailand BIOTEC, http://www.biotec.or.th/~nares/home/isnar_paper.htm.

DeVries, J., Toenniessen, G. (2001). Securing the Harvest: Biotechnology, Breeding and Seed Systems for African Crops. CAB International, Wallingford, UK.

Eede, G.V.D., Aartsb, H., Buhkc, H.-J., Corthierd, G., Flinte, H.J., Hammesf, W., Jacobseng, B., Midtvedth, T., Vosseni, J.V.D., Wrightj, A.V., Wackernagelk, W., Wilcksl, A. (2004). "The relevance of gene transfer to the safety of food and feed derived from genetically modified (GM) plants". Food and Chemical Toxicology 42, 1127–1156.
Eicher, C.K. (2003). "Flashback: Fifty years of donor aid to African agriculture". InWent, IFPRI, NEPAD and CTA, Pretoria, South Africa.
Evenson, R.E., Gollin, D. (2003). Crop Variety Improvement and Its Effect on Productivity. CABI Publishing, Wallingford, UK, p. 522.
Fang, X., Cheng, D., Xu, J., Fan, T. (2001). "Commercial implementation of intellectual property rights of Chinese transgenic insect-resistant cotton with Bt gene and Bt + Cpt genes". Journal of Agricultural Biotechnology 9, 103–106.
FAO (2003). The State of Food Insecurity in the World 2003. Food and Agriculture Organization of the UN, Rome.
FAO (2004). Biotechnologies in Developing Countries: BioDeC Database. Food and Agriculture Organization of the UN, Rome.
Fogel, R.W. (2000). The Fourth Great Awakening & the Future of Egalitarianism. University of Chicago Press, Chicago.
Fox, J.L. (2001). "EPA re-evaluates StarLink license". Nature Biotechnology 19, 11.
Gianessi, L.P., Silvers, C.S., Sankula, S., Carpenter, J.E. (2002). "Plant biotechnology: Current and potential impact for improving pest management in U.S. agriculture an analysis of 40 case studies". National Center for Food and Agricultural Policy, Washington, DC.
Graff, G.D., Cullen, S.E., Bradford, K.J., Zilberman, D., Bennett, A.B. (2003). "The public-private structure of intellectual property ownership in agricultural biotechnology". Nature Biotechnology 21, 989–995.
Hanson, H., Borlaug, N., Anderson, R. (1982). Wheat in the Third World. Westview Press, Boulder, CO.
Herdt, R. (1991). "Research priorities for rice biotechnology". In: Khush, G., Toenniessen, G. (Eds.), Rice Biotechnology. CAB International, Wallingford, UK.
Herdt, R. (1999). Enclosing the Global Plant Genetic Commons. The Rockefeller Foundation, New York.
Herdt, R.W. (2005). "Priorities for plant disease epidemiology and developing country food security". European Journal of Plant Pathology 115, 1.
Huang, J., Hu, R., Pray, C., Qiao, F., Rozell, S. (2003). "Biotechnology as an alternative to chemical pesticides: A case study of Bt cotton in China". Agricultural Economics 29, 55–68.
IFPRI (2005). "Program for biosafety systems". International Food Policy Research Institute.
Ito, O., O'Toole, J.C., Hardy, B. (1999). Genetic Improvement of Rice for Water-Limited Environments. International Rice Research Institute, Los Banos (Philippines).
James, C. (2005). "Preview: Global status of commercialized biotech/GM crops: 2005", ISSA Briefs No. 34. International Service for the Acquisition of Agri-Biotech Applications (ISAAA), Ithaca, NY, p. 7.
Janaiah, A., Hossain, M. (2002). "The productivity impact of the Green Revolution: The Indian experience". Foundation for Advanced Studies on International Development, Tokyo, Japan.
Jayaraman, N. (2003). "Biotech cotton failure". Multinational Monitor 24, 6–8.
Johnston, B.E. (1970). "Agriculture and structural transformation in developing countries: A survey of research". Journal of Economic Literature 3, 369–404.
Kendall, H.W., Beachy, R., Eisner, T., Gould, F., Herdt, R.W., Raven, P.H., Schell, J.S., Swaminathan, M.S. (1997). Bioengineering of Crops, Report of the World Bank Panel on Transgenic Crops. The World Bank, Washington, DC.
Kryder, D., Kowalski, S., Krattiger, A. (2000). "The intellectual and technical property components of pro-vitamin A rice (Golden Rice)". International Service for the Acquisition of Agri-Biotech Applications (ISAAA), Ithaca, NY.
Lanceras, J., Pantuwan, G., Jongdee, B., Toojinda, T. (2004). "Quantitative trait loci associated with drought tolerance at the reproductive stage in rice". Plant Physiology 135, 384–399.
Lewis, W.A. (1954). "Economic development with unlimited supplies of labor". Manchester School of Economic and Social Studies 22, 139–191.

Lichtenberg, E. (2000). "Costs of regulating transgenic pest-protected plants". In: Genetically Modified Pest-Protected Plants: Science and Regulation. National Academy of Sciences, Washington, DC.
Mellor, J.W. (1966). The Economics of Agricultural Development. Cornell University Press, Ithaca, NY.
Mellor, J.W. (1998). "Foreign aid and agriculture-led development". In: Eicher, C.K., Staatz, J.M. (Eds.), International Agricultural Development. Johns Hopkins University Press, Baltimore, MD, pp. 55–66.
Mugabe, J. (2000). "Biotechnology in developing countries and countries with economies in transition: Strategic capacity building considerations". African Centre for Technology Studies, Nairobi, Kenya.
NAS (2001). "Studies evaluating the impact of Bt corn pollen on monarch and black swallowtail large populations". In: Proceedings of the National Academy of Sciences of the United States, p. 98.
NAS Board on Life Sciences (2002). The National Plant Genome Initiative: Objectives for 2003–2008.
National Academy of Sciences, Royal Society of London, Brazilian Academy of Sciences, Chinese Academy of Sciences, Indian National Science Academy, Mexican Academy of Sciences, Third World Academy of Sciences (2000). Transgenic Plants and World Agriculture. National Academy Press, Washington, DC.
Normile, D. (1999). "Rockefeller to end network after 15 years of success". Science 286, 1468–1469.
Nuffield Council on Bioethics (1999). "Genetically modified crops: The ethical and social issues". Nuffield Council on Bioethics, London.
Nuffield Council on Bioethics (2004). "The use of genetically modified crops in developing countries". Nuffield Council on Bioethics, London, UK, p. 144.
Obilana, A.B., Okumu, B.N. (2005). Interspecific Hybridization between African and Asian Rice Species: Evaluation Study Report. WARDA and UNDP, Cotonou, Benin.
O'Toole, J.C. (2004). Rice and Water: The Final Frontier. Rockefeller Foundation, New York.
Paarlberg, R.L. (2001). The Politics of Precaution: Genetically Modified Crops in Developing Countries. Johns Hopkins University Press/International Food Policy Research Institute, Washington, DC.
Pew Initiative on Food and Biotechnology (2004). "Gene flow: What does it mean for biodiversity and centers of origin". Pew Initiative on Food and Biotechnology, Mexico.
Pinstrup-Andersen, P. (2002). "Food and agricultural policy for a globalizing world: Preparing for the future". American Journal of Agricultural Economics 84, 1201–1214.
Pray, C., Hu, R., Rozell, S. (2002). "Five years of Bt cotton in China". Plant Journal 31, 423–430.
Qaim, M. (2003). "Bt cotton in India: Field trial results and economic projections". World Development 31, 2115–2128.
Redenbaugh, K., McHughen, A. (2004). "Regulatory challenges reduce opportunities for horticultural biotechnology". California Agriculture 58, 106–115.
Ribaut, J., Banzinger, M., Betran, J., Jinang, C., Edmeades, G., Dreher, K., Hoisington, D. (2002). "Use of molecular markers in plant breeding: Drought tolerance improvement in tropical maize". In: Kang, M. (Ed.), Quantitative Genetics, Genomics and Plant Breeding. CABI International, Wallingford, UK.
Rissler, J., Mellon, M. (1996). The Ecological Risks of Engineered Crops. MIT Press, Cambridge, MA and London.
Rockefeller Foundation (2003). Annual Report 2003. The Rockefeller Foundation, New York.
Sanchez, A., Subudhi, P., Rosenow, D., Nguyen, H. (2002). "Mapping QTLs associated with drought tolerance in sorghum". Plant Molecular Biology 48, 713–726.
Sommer, A. (2001). Vitamin A Deficiency Disorders: Origins of the Problem and Approaches to Its Control. Johns Hopkins University Bloomberg School of Public Health, Baltimore, MD.
Thirtle, C., Lin, L., Piesse, J. (2003). "The impact of research-led agricultural productivity growth on poverty reduction in Africa, Asia and Latin America". World Development 31, 1959–1975.
Thirtle, C., Beyers, L., Ismael, Y., Piesse, J. (2003). "Can GM-technologies help the poor? The impact of Bt cotton in Makhathini Flats, Kwazulu-Natal". World Development 31, 717–733.
Timmer, C.P. (1997). "How well do the poor connect to the growth process?". Harvard Institute for International Development, Cambridge, MA.
Toenniessen, G. (2003). "Opportunities for and challenges to plant breeding adoption in developing countries". National Agricultural Biotechnology Conference, Pullman, WA.
United Nations (2002). World Population Prospects: The 2002 Revision. United Nations Population Division, New York.

Uzogara, S.G. (2000). "The impact of genetic modification of human foods in the 21st century". Biotechnology Advances 18, 179–206.

Wambugu, F., Kioime, R. (2001). "The benefits of biotechnology for small-scale banana farmers in Kenya". ISAAA, Ithaca, NY.

West, K.P. (2002). "Extent of vitamin A deficiency among preschool children and women of reproductive age". Journal of Nutrition 132, 2857S–2866S.

Yadav, R., Hash, C., Bidinger, F., Cavan, G., Howarth, C. (2002). "Quantitative trait loci associated with traits determining grain and stover yield in pearl millet under terminal drought-stress conditions". Theoretical and Applied Genetics 104, 67–83.

PART 4

MARKETS, INSTITUTIONS AND TRANSACTION COSTS

Chapter 51

EFFICIENCY AND EQUITY EFFECTS OF LAND MARKETS

KEIJIRO OTSUKA

Foundation for Advanced Studies on International Development, Tokyo, Japan

Contents

Handbook of Agricultural Economics, Volume 3
Edited by Robert Evenson and Prabhu Pingali

DOI: 10.1016/S1574-0072(06)03051-9

Abstract

It is widely believed that land markets, including both land sales and tenancy markets, are neither efficient nor conducive to social equity. It is often argued that tenants, particularly share tenants, do not have proper incentives to work and invest, partly because of the disincentive effects of output sharing and partly because of the tenure insecurity. It is also widely accepted that land sales transactions tend to exacerbate the social equity and rural poverty by facilitating the concentration of land ownership by hands of a few wealthy landlords. Based on these presumptions, land reform programs have been implemented in a number of developing countries.

This article critically reviews these presumptions both theoretically and empirically. Firstly, we identify why land tenancy transaction is more common than land sales transactions and why share tenancy is more common than leasehold tenancy. Secondly, we critically review the theories of share tenancy, tenure security, and adjustment costs of farm size. Thirdly, we empirically review the efficiency and equity effects of land markets as well as the impacts of conventional land reform programs.

It has become clear from the literature review that land reform polices have generally failed to improve land use efficiency and social equity. It is also found that tenancy contracts, including share tenancy, are generally efficient and conducive to social justice. In conclusion, we propose to encourage tenancy transactions, in general, and share tenancy, in particular.

Keywords

land tenancy, share tenancy, land use efficiency, land reform

JEL classification: D23, D82, O15, O17

1. Introduction

For a long time it was widely believed among policymakers and development economists in Asia, as well as in other parts of the world, that land markets, including both land sales and tenancy markets, are neither efficient in the use and allocation of land nor conducive to social equity. It was often argued that tenants, particularly share tenants, do not have proper incentives to work and invest, partly because of the disincentive effects of output sharing and partly because of the tenure insecurity.[1] It was also widely accepted that land sales transactions tend to exacerbate the social equity and rural poverty by facilitating the concentration of land ownership by hands of a few wealthy landlords.

Land tenancy has been particularly common in Asian countries, where land reform was attempted for the twin objectives of reducing rural poverty by transferring land from the landed class to landless tenants and marginal farmers, and of achieving higher production efficiency by converting share tenancy to owner cultivation or leasehold tenancy with tenure security. Tenancy reform usually ruled out the practice of share tenancy, regulates the leasehold rent at a low level, and prohibited the eviction of tenants. The land redistribution policy set the ceiling on land holding and transferred the ownership right of land in excess of the ceiling to actual cultivators. In order to "protect" or preserve the status of land reform beneficiaries, new leasing and subleasing contracts were restricted by law. Another common feature is that area under owner cultivation was exempted from land reform, which permitted large owner-cultivation with employment of hired labor. These features of the land reform programs are widely shared by land reforms laws in a number of Asian countries [Warriner (1969); Ladejinsky (1977); Herring (1983); Prosterman and Riedinger (1987); Hayami, Quisumbing and Adriano (1990); Riedinger (1995)].

There are three presumptions justifying these programs. First, the tenants belong to the poorest segment of rural society and therefore the transfer of land property rights from landlords to tenants serves to the goal of greater social equity. Second, large "owner-cultivated" farms, even with employment of hired labor, are not inefficient. Third, share tenancy is less efficient than leasehold tenancy and owner cultivation.

In general, however, these presumptions are empirically incorrect. First, the poorest of the rural poor are landless laborers [David and Otsuka (1994)]. Second, share tenancy is generally not very inefficient compared with leasehold tenancy and owner-cultivation, unless tenancy contracts are distorted by land reform laws [Hayami and Otsuka (1993); Sadoulet, de Janvry and Fukui (1994, 1997)]. Third, by suppressing the option of tenancy contracts, land reform tends to induce large owner cultivation with employment of hired labor, which is inefficient [Binswanger and Rosenzweig (1986); Hayami, Quisumbing and Adriano (1990); Binswanger, Deininger and Feder (1995)].

[1] Deininger and Binswanger (2001) point out that "Land Reform Policy Paper" issued by the World Bank in 1975 disregarded the role of land tenancy because of the presumption that land tenancy suffers from both static and dynamic inefficiency.

This is likely to explain, at least partly, why the inverse correlation often exists between farm size and productivity, in which family-labor dependent small farms are more productive than hired-labor dependent large farms. Fourth, land reform tends to block the agricultural ladder for landless laborers to ascend by suppressing tenancy transactions, thereby perpetuating the rural poverty [Hayami and Otsuka (1993); Otsuka (1993)].

Thus, conventional land reform is unlikely to be conducive to production efficiency and social equity. Moreover, the experience of land reform programs is replete with the repeated failures in terms of the implementation of intended programs. As will be discussed in this paper, their partial implementation or the threat thereof has also had serious adverse consequences on the rural poverty.

Recently a new consensus has emerged on the desired principle of land reform among groups of economists who have been working on land and agrarian issues for many years [Hayami, Quisumbing and Adriano (1990); Hayami and Otsuka (1993); Binswanger and Deininger (1993); Deininger and Binswanger (2001); de Janvry et al. (2001); Sadoulet, Murgai and de Janvry (2001)]. That is, land tenancy transactions, which are often suppressed or even prohibited by land reform law, should be promoted for both the efficiency and equity. A major purpose of this article is to provide theoretical and empirical bases for such an argument by conducting a comprehensive and updated literature review.

Although we pay special attention to Asia, land market issue are becoming increasingly important in Africa due to increased scarcity of land and in Latin America where access to land for the landless workers and minifundists is becoming the major issue in the current phase of land reform [Carter and Salgado (2001); de Janvry, Sadoulet and Wolford (2001)]. Yet the majority of existing studies on land markets and land reform are concerned with Asia. It is hoped that the past experience in Asia provides valuable lessons for other regions of the Third World.

The organization of this article is as follows. Before undertaking reviews of theoretical and empirical literature, we provide an overview of the agrarian economies in Section 2. In particular, using the agricultural census from selected countries, we explore the inequality of operational landholdings and the significance of tenancy transactions in the 1990s in comparison with the 1970s. In Section 3, we theoretically examine the functions of land and labor markets in rural economies with a view to identifying the reasons why land tenancy transactions, particularly share tenancy contracts, are important in practice compared with land sales transactions and labor contracts in allocation of resources among rural households. In this section we also consider theoretically how efficiency of land market transactions, including share tenancy contracts, can be tested empirically. In Section 4, we examine empirically the efficiency of share tenancy, efficiency of land allocation among agricultural households, and equity effects of land tenancy transactions by conducting a review of the empirical literature. Finally, based on major findings of this study, we consider the desirable land policy options for achieving greater production efficiency, social equity, and prospects for long-term economic development.

Table 1
Distribution of farms and farmland by land-tenure status in major regions of the Third World

	Asia		Africa		Latin America	
	1970	1990	1970	1990	1970	1990
Number of countries enumerated[a]	6	6	4	4	8	8
Number of farms (million)	100.9	150.0	10.0	8.4	8.9	13.7
Average operational farm size (ha)	2.3	1.6	1.6	0.8	60.5	48.3
Distribution of farmland						
Owner-cultivation	83.2	87.3	9.2	42.3	61.4	75.6
Pure tenancy	5.6	3.2	3.1	7.7	17.2	7.8
Owner-cum-tenancy	16.7	8.7	29.1	9.9	5.9	2.4
Other[b]	0.1	0.6	58.7	40.1	15.9	18.7
Percentage of share tenancy in tenanted land	67.1	51.6	0.	1.3	14.0	n.a.[c]

Sources: (1) Bangladesh, *Report on the Agricultural Census of Bangladesh, 1977*; *1978 Land Occupancy Survey of Bangladesh*; *Census of Agriculture 1996*. (2) India, *National Sample Survey, No. 215, 26th Round, 1971–72*; *All India Report on Agricultural Census 1980/71*; *Agricultural Census 1990–91*. (3) Indonesia, *1973 Agricultural Census*; *1993 Agricultural Census*. (4) Philippines, *1971 Census of Agriculture*; *1991 Census of Agriculture*. (5) Thailand, *1978 Agricultural Census Report*; *1993 Agricultural Census*. UN-FAO, *1970 World Census of Agriculture, Analysis and International Comparison of the Results*, Rome, 1971; UN-FAO, *1990 Results of National Censuses of Agriculture*, http://www.fao.org/ES/ESS/census/webhome.htm.

[a]Countries enumerated are: *Asia*: Bangladesh, India, Indonesia, Pakistan, Philippines, and Thailand; *Africa*: 1970, Cameroon, Reunion, Swaziland, and Zaire; 1990, Dem. Rep. of Congo, Egypt, Guinea, and Reunion; *Latin America*: Brazil, Colombia, Honduras, Mexico, Panama, Peru, Uruguay, and Venezuela.

[b]Farms operated by squatters, under tribal on communal tenure forms, and under other single forms of tenure.

[c]"n.a." means not available.

2. An overview of agrarian economies

In this section, we provide an overview of the agrarian structure in the Third World in terms of average farm size, inequality of operational landholdings, and the importance of tenant farms and share tenancy, using the agricultural census data around 1970 and 1990. While the consistent census data are available for major countries in Asia and Latin America, the availability of data is extremely limited for Africa in which we could collect the data only from four and different countries in 1970 and 1990. Therefore, we need to exercise proper caution to interpret the African data shown in Table 1.

Broadly, owner-cultivation is the most common form of land tenure in both Asia and Latin America. In Asia pure owner-cultivation comprises well more than 80%, and if we can add owner-cultivated portion of owner-cum-tenancy land, total owner-cultivation areas will be close to 90%. Furthermore, its importance has increased over time not only in Asia but also in Latin America. The average farm size is very small in Asia, which reflects the fact that Asian agriculture is dominated by peasant farming based on family labor. This peasant mode is augmented by the use of tenancy contracts that facilitates land transfers from relatively land abundant households to households with little land so as to make the ratio of farmland to family labor more or less equal across farms. Labor contracts that facilitate the development of large farms through labor transfers from land scarce and landless households to large landowners are less common as a means to equalize the ratio of land to labor. Although owner cultivation also dominates in Latin America, the average farm size in this region is far greater than in Asia, even though the dominance of latifundia has been reduced by redistributive land reform [de Janvry, Sadoulet and Wolford (2001)]. It is also noteworthy that squatters included in the other category represent a significant share of the land-tenure distribution in Latin America. Africa is different from Asia and Latin America in that communal tenure accounts for large areas. Average farm size in Africa is small, partly because the farm size data do not include large tracts of fallow land. Since many farmers under customary tenure institutions do not possess rights to rent out and sell land, land markets are generally inactive.

The incidence of tenancy has declined in both Asia and Latin America. In addition, the importance of share tenancy has significantly declined in Asia.[2] In some countries, such as India, the decreases in tenancy is likely due to the replacement of formal tenancy by informal or concealed tenancy to evade land reform regulations [e.g., Radhakrishnan (1990); Ray (1996); Thorat (1997); Thimmaiah (2001)]. If the declining incidence of land tenancy is real, we must inquire the causes of such changes. There are, however, few empirical studies analyzing the determinants of land tenure choice statistically. In Africa, there is some indication that the incidence of tenancy is on the increase.

The size distribution of farms in Asia is totally different from that of Latin America (compare Tables 2 and 3). In Asia many farms are below one hectare and land organized in middle size farms above 10 hectares are generally less than 5%. In contrast, in Latin America, small farms below 5 hectares constitute well less than 10% of farmland, while more than 70% of land is held by large farms above 50 hectare. Correspondingly, the Gini coefficient measuring inequality in the distribution of farmland is generally lower than 0.5 for Asian countries as compared with greater than 0.8 for Latin American countries. These data reflect the fact that most agricultural production in Asia involve large number of peasants organized around family farms, whereas in Latin America most agricultural production takes place on large plantation or haciendas using hired labor.

These structural differences have given rise to different land issues among the three major regions of the Third World. In Africa, the chief concern is how to develop individualized property rights in land consistent with proper investment and natural resource-conservation incentives [Otsuka and Place (2001)]. In Latin America it is related to improvement of access to land for the landless workers employed by large estates, as well as for minifundists, and the enhancement of competitiveness of land reform beneficiaries [de Janvry, Sadoulet and Wolford (2001)]. The traditional agrarian problem

[2] The figures shown in Table 1 are weighted averages, so that low incidence of share tenancy in such large country as Thailand affects the average figure significantly.

Table 2
Distribution of farms and farmland by operational farm size and the extent of tenancy in selected countries in Asia

Country	Year of survey	Average operational farm size (ha)	Percentage of farms and farmland[a]				Gini coefficient of land concentration	Percentage of tenanted land to total farmland		Percentage of share tenancy to total tenanted land
			Below 1 ha		Above 10 ha					
			Farms	Area	Farms	Area		Pure tenancy	Total[b]	
Bangladesh	1976/77	1.4	49.7	28.8	n.a.[c] (9.4)[d]	n.a.[c] (1.6)[d]	0.42	0.4	44.1	91.0
	1996	0.6	80.8	41.1	0.1 (32.4)[d]	1.4 (16.5)[d]	0.57	0.6	42.3	73.0
India	1970/71	2.3	50.6	9.0	3.9	30.9	0.64	2.4	8.5	68.6
	1990/91	1.6	59.4	15.0	1.7	17.3	0.58	n.a.[c]	(5.1)[e]	66.5
Indonesia	1973	1.0	70.4	30.0	5.9	10.3	0.55	1.7	23.2	60.0
	1993	0.9	70.8	29.8	0.2	3.4	0.56	4.5	18.0	n.a.[c]
Philippines	1971	3.6	13.6	1.9	4.9	33.9	0.51	20.6	31.6	79.3
	1991	2.1	36.6	7.3	2.4	23.3	0.54	19.0	51.3.	67.8
Thailand	1978	3.7	16.4	2.5	6.0	23.6	0.46	6.0	15.5	32.3
	1993	3.4	21.5	3.6	4.5	23.2	0.50	(7.2)[e]	(7.2)[e]	19.2

Sources: (1) Bangladesh, *Report on the Agricultural Census of Bangladesh, 1977*; *1978 Land Occupancy Survey of Bangladesh*; *Census of Agriculture 1996*. (2) India, *National Sample Survey, No. 215, 26th Round, 1971–72*; *All India Report on Agricultural Census 1980/71*; *Agricultural Census 1990–91*. (3) Indonesia, *1973 Agricultural Census*; *1993 Agricultural Census*. (4) Philippines, *1971 Census of Agriculture*; *1991 Census of Agriculture*. (5) Thailand, *1978 Agricultural Census Report*; *1993 Agricultural Census*.

[a]Since farm size classes differ form country to country, interpolations were made.

[b]Area in "pure" tenant farms plus area in owner-cum-tenant farms.

[c]"n.a." means not available.

[d]Farm size above 3 ha.

[e]Total area under tenancy.

Table 3
Distribution of farms and farmland by operational farm size and the extent of tenancy in selected countries in Latin America

Country	Year of survey	Average operational farm size (ha)	Percentage of farms and farmland[a]				Gini coefficient of land concentration	Percentage of tenanted land to total farmland		Percentage of share tenancy to total tenanted land
			Below 5 ha		Above 50 ha					
			Farms	Area	Farms	Area		Pure tenancy	Total[b]	
Brazil	1970	59.7	36.8	1.3	16.3	84.6	0.84	6.1	10.2	n.a.[c]
	1996	72.8	36.6	1.0	18.9	87.8	0.85	3.3	(3.3)[d]	n.a.[c]
Colombia	1970/71	26.3	59.6	3.7	8.4	77.7	0.86	5.3	11.5	49.4
	1988	23.3	51.6	3.6	10.7	70.6	0.79	2.6	6.2	n.a.[c]
Peru	1971/72	16.9	78.0	8.9	1.9	79.1	0.91	4.5	13.6	0[e]
	1994	20.1	68.1	6.3	2.9	77.6	0.87	0.9	3.5	n.a.[c]
Uruguay	1970	214.1	14.3	0.2	37.6	95.8	0.82	19.1	46.3	4.7
	1990	287.9	8.1	0.1	50.2	97.4	0.77	12.2	40.8	n.a.[c]
Venezuela	1971	91.9	43.8	0.9	13.6	92.5	0.91	4.5	2.4	n.a.[c]
	1997	60.0	48.4	1.6	14.5	89.3	0.88	1.7	2.8	n.a.[c]

Sources: UN-FAO, *1970 World Census of Agriculture, Analysis and International Comparison of the Results*, Rome, 1971; UN-FAO, *1990 Results of National Censuses of Agriculture*, http://www.fao.org/ES/ESS/census/webhome.htm.

[a]Since farm size classes differ form country to country, interpolations were made.

[b]Area in "pure" tenant farms plus area in owner-cum-tenant farms.

[c]"n.a." means not available.

[d]Area in owner-cum-tenant farms is not available.

[e]Less than 0.05%.

in Asia has been concerned with the landlord–tenant relationship [Ladejinsky (1977)]. Of course, these are overly simplified view. For example in some parts of Africa and Latin America, the landlord–tenant relationship is also a serious problem. Yet by far the majority of existing studies on land market issues relate to the Asian context.

In Asia, the average operational farm size has been declining, particularly in Bangladesh, India, and the Philippines due to increasing population pressure. In contrast, average farm size did not decrease much in Latin America. Thus, how to allocate increasingly scarce land resources efficiently through land markets is a truly serious issue in Asia. It can be also noticed that the proportion of pure tenant households was comparatively low in Asia except in the Philippines, implying that the majority of tenants were part owners holding own land rather than the landless. It is also indicated in the literature that the landless households do not have much access to land through land tenancy, particularly in South Asia [e.g., Sharma and Dreze (1996); Sarap (1998)]. Why this is the case, however, is not clear from the existing studies. As may be expected, share tenancy is more common than leasehold tenancy in Bangladesh, India, Indonesia, and the Philippines. The prevalence of share tenancy can be explained by the preference of smaller farmers to reduce income risk. In fact, share tenants are typically small and marginal farmers [Thorat (1997); Sadoulet, Murgai and de Janvry (2001)]. Share tenancy is uncommon in Thailand and some areas in other countries, where absentee landlordism is pervasive, because absentee landlords cannot effectively monitor tenants' conducts and check actual harvests.

Thus, although this paper is concerned with the efficiency and equity effects of land market transactions in the Third World, it has focus on Asia in view of the utmost importance of land market issues.

3. Theoretical framework

In order to understand the allocative role of land market transactions in rural economies, we begin with the presentation of a simple model of land and labor transactions under the assumption of perfectly competitive markets and absence of production risk. Needless to say, however, markets are not perfectly competitive in practice. Following an examination of the simple model, we argue that tenancy transactions are common, precisely because land sale and labor markets are more seriously imperfect. Second, we demonstrate that once we consider uncertainty and the lack of efficient risk markets, share tenancy tends to be preferred over fixed-rent tenancy. Third, we point out that tenancy contracts may fail to provide proper investment incentives, if tenure is insecure. Fourth, we contend that to the extent that tenancy transactions are inefficient and hence costly, allocative inefficiency of land among farming households arises. In order to avoid confusion, we use the terms land sales and land tenancy markets distinctively and when we use the term land markets, we refer to land market transactions inclusive of both land sale and tenancy transactions.

3.1. A simple model of land and labor transactions

Let us assume that there are only two farmers (or two groups of many identical farmers), whose production functions are identical and characterized by constant returns to scale.[3] Endowments of land and labor, however, are different between the two farmers. For simplicity, we consider only two inputs, i.e., labor (L) and land (A).[4] Production function can be written as

$$Q_i = F(L_i, A_i) \quad (i = \mathrm{I}, \mathrm{II}),$$

where Q is output and F exhibits production function with positive first and negative second derivatives ($F_1, F_2 > 0$; $F_{11}, F_{22} < 0$). Factor markets are assumed to be competitive and farmers maximize their net revenue (Π) given fixed endowment of owned land ($\mathbf{A}$) and family labor ($\mathbf{L}$), which is expressed as

$$\Pi_i = PF(L_i, A_i) - r(A_i - \mathbf{A}_i) - w(L_i - \mathbf{L}_i),$$

where P, r, and w refer to product price, rental price of land, and wage rate, respectively. If desired cultivation area of land exceeds the endowed land area ($A_i > \mathbf{A}_i$), the farmer rents in land. Otherwise, the farmer rents out part of the owned land. The similar relations hold for the employment of labor.

The first-order net revenue maximization conditions lead to the following familiar relations:

$$PF_1(L_{\mathrm{I}}, A_{\mathrm{I}}) = PF_1(L_{\mathrm{II}}, A_{\mathrm{II}}) = w,$$
$$PF_2(L_{\mathrm{I}}, A_{\mathrm{I}}) = PF_2(L_{\mathrm{II}}, A_{\mathrm{II}}) = r.$$

The equality of the values of marginal products between two farmers shown above ensures the Pareto optimality.

Under the assumption of constant-returns-to-scale production function, marginal products depend only on factor proportions. Thus, the efficient resource allocation can be achieved either by land tenancy transaction or by exchange of labor through competitive markets.[5] That is, the same efficient outcome can be achieved, if the land-abundant farmer leases out land to or hires labor from other farmer. It is also obvious that if the land-abundant farmer sells a portion of his/her land to the land-deficient farmer at the

[3] The assumption of constant-returns-to-scale production function is not unrealistic in the absence of large mechanization, as argued by Hayami and Otsuka (1993) and Deininger and Binswanger (2001). Hayami (2001) also argues that even large-scale plantations do not have scale advantage over appropriate contract farming by small farms.

[4] For more general models, which consider imperfection of labor, land tenancy, credit, and insurance markets, see Kevane (1996), as well as Sadoulet, Murgai and de Janvry (2001).

[5] More generally, under the assumption of constant-returns-to-scale production function, imperfection in one factor market does not lead to allocative inefficiency [Kevane (1996)].

price reflecting the present value of future returns to land, the equality of the marginal products of land can be achieved. In terms of the present value, the income distribution between the two farms is the same as in the case of renting. Thus, perfect markets in land tenancy, land sale, and labor employment can lead to identical and equally efficient outcome.

3.2. On the dominance of land tenancy transactions

The land tenancy transaction is the most common way of adjusting different factor endowments among farming households, particularly in Asia. Skoufias (1995) aptly points out that the dominance of land tenancy transactions can be attributed to relatively less efficient functioning of land sales and labor market transactions than that of land tenancy market.[6]

In the real world, it is unlikely that labor market transaction leads to the efficient resource allocation, because it is generally costly to supervise and enforce hired labor in certain critical tasks in agricultural production. According to the theory formulated by Feder (1985) and Eswaran and Kotwal (1986), large farmers employ hired labor because of the limited endowment of family labor relative to owned land. Hired wage laborers, however, do not have strong work incentives, as they receive the same wage regardless of how hard they work. Thus, it is not possible to enforce their work effort without explicit supervision. Furthermore, it is likely that the supervision cost of hired labor contract increases more than proportionally with farm size. Therefore, the high enforcement cost of hired labor will lead to lower production efficiency on large farms, even though those farms would have the advantage of better access to the credit market owing to the ownership of land that can be used as collateral.[7]

High enforcement cost of hired labor does not imply that casual labor markets are inactive. Since it is easy to observe work effort or inspect the outcome of work in such simple tasks as weeding, transplanting, and harvesting, daily-wage workers, as well as piece-rate workers, are widely used for these activities. It is, however, costly to employ hired labor for the tasks that require care and judgment, such as land preparation, fertilizer application, supervision of a group of hired laborers, and water and pest control in spatially dispersed agricultural environments. Imperfect supervision and labor enforcement in these activities lead to shirking of hired wage labor, which leads to inefficiency of farm operation dependent on hired labor employment. These tasks

[6] The similar point is made later by Sadoulet, Murgai and de Janvry (2001).

[7] Feder (1985) and Eswaran and Kotwal (1986) assume that tenancy does not exist, because the landless do not have sufficient access to credit market to pay for family consumption and purchased inputs and, hence, cannot undertake tenant cultivation. But landlords can and often do provide credit to their tenants, particularly under share tenancy. Therefore, the imperfection of credit market alone cannot justify the choice of labor contract over tenancy contract.

therefore are carried out usually by family labor on small farms [Hayami and Otsuka (1993)].

Large farms are often managed by the use of permanent laborers in South Asia, who generally perform the same tasks as family labor on small farms.[8] It is under such conditions that we observe the so-called inverse relationship between productivity and farm size in South Asia. Transfer of labor from small to large farms cannot eradicate the inefficiency, so long as labor shirking prevents the achievement of truly efficient resource allocation. In other words, labor market generally fails in achieving efficient resource allocation in agriculture.

Even if labor market fails to function, efficiency outcome can be achieved, if land sales market functions well. If productivity of land is lower on larger farms, there must be agreeable land prices, at which the sellers (i.e., large landowners) and the buyers (small cultivators) can gain through market transactions, particularly in the absence of credit market imperfections. It is well known, however, that land market is inefficient in the real world. This is evident from inactive or even non-existent land markets in many places.

Since credit markets are actually imperfect and small farmers and the landless workers tend to be rationed out from credit transactions due to the paucity of collateral, the present value of land perceived by the poor, land-scarce households may be exceeded by the value perceived by the rich, land-abundant households. Binswanger and Rosenzweig (1986) and Binswanger and Elgin (1988) also argue that since land can be used as collateral for obtaining credit, the price of land exceeds the present value of future agricultural profits accrued to land by the amount of benefit accrued from the use of land as collateral. Thus, buyer of land cannot cover the cost of land purchase solely from the future agricultural profits. In order for the land transaction to take place, buyer must have own fund or additional saving to purchase land. If potential buyers are poor small farmers, they would not possess such extra funds. In consequence, they may not be able to purchase land from large landowners, even if they are more efficient than large farmers. More often than not, we observe purchase of land by wealthy merchants and landowners from poor small farmers, even though their farms are relatively efficiently managed. Thus, it is often the difference in wealth but not relative factor endowments that governs land sales transactions.

Land has also additional values other than those derived from its use in agriculture and collateral, such as store of wealth against inflation and source of stable employment [de Janvry et al. (2001)]. Such additional values will further increase the gap between the land price and the present value of profits in agricultural production. This, in turn, will hinder the market transactions of land, as they will create the problem of financing the fund to purchase land. The transactions of land may be also costly, because of

[8] Eswaran and Kotwal (1985) attempt to justify the choice of permanent labor contracts. Their model, however, precludes the option of tenancy contracts. If such option is incorporated, it can be shown that permanent labor contract will never be chosen [Otsuka, Chuma and Hayami (1993)].

the imperfect information about the quality of land and inaccurate land registration system. Added to these is covariate risk in agriculture; many farmers in the same locality commonly want to sell their lands in bad crop years and they want to buy in good crop years, so that land transactions seldom take place [Binswanger and Rosenzweig (1986)]. When the transaction takes place, it is often distress sale from the poor households to wealthier households at the time of extreme adversity.

Thus, resource allocation in rural economies will be neither efficient nor equitable, unless land tenancy market functions effectively [Sadoulet, Murgai and de Janvry (2001)].

3.3. Advantages of share tenancy

In our simple model, we assumed that land service is transacted at rental price of r. Such contract is called fixed-rent leasehold tenancy, in which rent is fixed per unit of rented land. In practice, share tenancy is generally a more prevalent form of tenancy not only in contemporary Asia, as we have seen, but also historically in Europe and the U.S. [Hayami and Otsuka (1993)]. It appears that share tenancy is also becoming common in Africa, as population pressure increases scarcity value of land.[9] The prevalence of share tenancy can be explained by the fact that this contract substitutes for missing or imperfect markets in rural areas.

First of all, share tenancy serves as a role of insurance by sharing output, which shares revenue risk between landowner and tenant. This is beneficiary for the contracting parties, because insurance market is practically absent in rural areas despite the fact that agricultural production is subject to vagary of weather. If fixed-rent contract is chosen, the tenant has to bear all the revenue risk. Share tenancy contract automatically shares the production and price risk and, hence, it has an obvious advantage over fixed-rent leasehold tenancy in reducing income risk for risk averse tenant cultivators.

Secondly, share tenancy contract is often interlinked with production credit contract explicitly as well as implicitly through cost-sharing arrangements [Bardhan (1980, 1984)]. Usual practice of cost-sharing is for landowner to provide fertilizer and other purchased inputs at the time of their application and to deduct the amount of output equivalent to the cost of inputs before sharing output at the time of harvesting. Thus, the landlord provides de facto production credit to his tenant through cost sharing. This arrangement is efficient and advantageous for the contracting parties to the extent that the landowner has better access to cheaper sources of credit. The repayment is easily assured, if the landlord has the relevant production knowledge and borrower's information. Occasionally landlord also provides consumption credit, if the tenant, who is reliable, is temporarily in need of cash.

Third, landlord also provides non-tradable inputs such as management know-how. Since landlord receives a share of output, he has an interest in increasing output. Thus,

[9] See, for example, Gavian and Ehui (1999) for Ethiopia and Otsuka and Quisumbing (2001) for Ghana.

he will provide his management knowledge to his tenant, particularly if tenant is inexperienced in farming. This is helpful for a young landless farm worker, who has had little farming experience, as the market of such knowledge is non-existent in rural areas.

The fact that the share tenancy contract provides insurance service, production credit, and management knowledge implies that this contract is friendly to the rural poor [de Janvry et al. (2001); Sadoulet, Murgai and de Janvry (2001)]. In fact, it is universally observed that the share tenancy contracts typically transfer land from land-rich to land-scarce households. There is therefore no wonder that share tenancy is generally prevalent in rural areas.

3.4. *Competing models of share tenancy*

Share tenancy is often prohibited by land reform law. The theoretical justification for the prohibition of share tenancy lies in the Marshallian argument that share tenancy is inefficient due to disincentive effect of output sharing. We now review the essence of this argument and its theoretical shortcoming based on literature surveys of land and labor contracts by Otsuka and Hayami (1988), Otsuka, Chuma and Hayami (1992), and Hayami and Otsuka (1993).

The Marshallian view of share tenancy asserts that since the marginal return to tenant's labor is proportionally less than the value of marginal product of labor under share contract, the work incentives to the tenant are thwarted. Figure 1 depicts this argument, where VMPL refers to the value of marginal product of tenant's labor and α(VMPL) corresponds to the marginal return curve for the tenant, where α is tenant's output share. Share tenant prefers to apply L^{**} of labor, where α(VMPL) is equal to his opportunity wage rate (w), rather than the socially optimum amount of labor, which is L^*.[10]

The situation is analogous to the effect of 50% income tax on wage-workers, if the output is shared at commonly observed rate of 50:50. In contrast, it is commonly accepted that owner farming and the fixed-rent leasehold contract, in which the tenant pays a fixed sum to the landlord and claims the residual, do not distort work incentives.

An implicit assumption of the inefficiency view of share tenancy is that it is prohibitively costly for the landlord to observe and enforce tenant's work effort, so that tenant determines his effort to maximize his own utility without fear of punishment on labor shirking. In practice, like owner farmers, tenants engage not only in simple tasks but more importantly in care-intensive activities, such as pest and water control. In other words, the tenant's effort represents his conscientious effort to apply his labor even for allocative and managerial decision-making. Given a spatially dispersed and ecologically diverse production environment in agriculture, it is not an easy task for landowners to supervise a tenant's work effort. Therefore, it is not surprising that the inefficiency view of share tenancy had been widely accepted since the day of Adam Smith to the present.

[10] Since a major role of share tenancy is risk sharing, the model should take into account the risk-averse behavior of tenant [Hayami and Otsuka (1993)]. The essence of the argument, however, is unchanged, even if we disregard the risk and uncertainty.

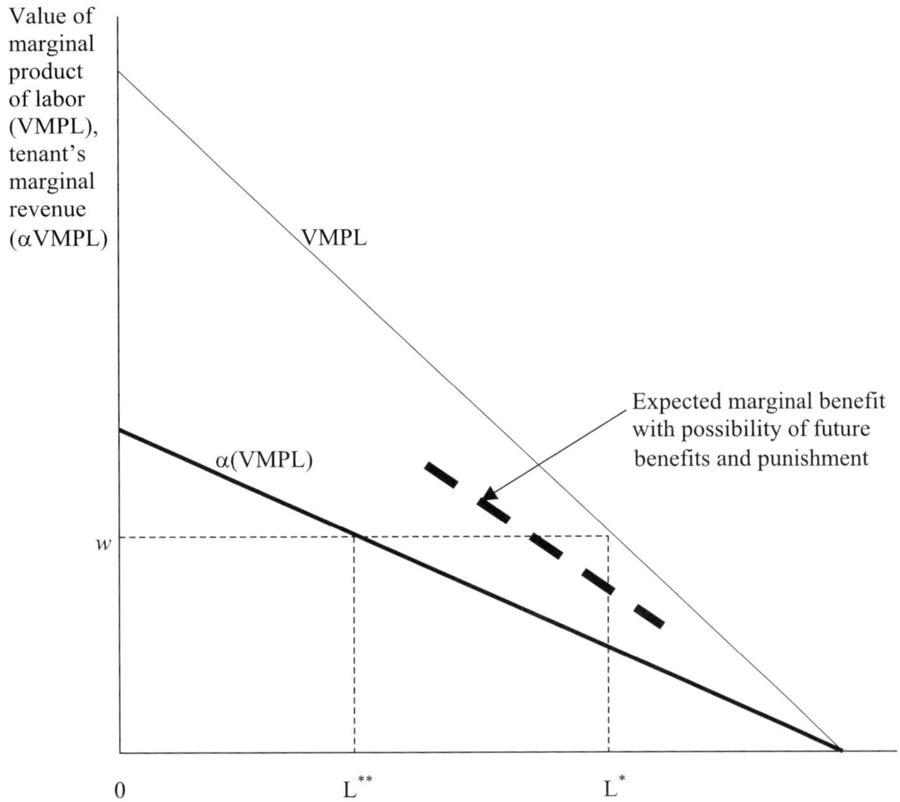

Figure 1. Illustration of efficiency of share tenancy contracts.

The source of incentive problem lies in the difficulty the landlord has in observing and enforcing the tenant's work effort. If the work effort is observable without cost, as assumed by Cheung's (1969) model of share tenancy, the landlord can specify and enforce the tenant's work effort at the desired level, which is L^*. In equilibrium, no inefficiency is shown to arise under share tenancy in this situation.

While the assumption of costless enforcement is unrealistic, the assumption of prohibitively costly enforcement seems too restrictive. A model of Otsuka, Chuma and Hayami (1993) introduces the notion of monitoring function of a tenant's work effort, which posits that the probability of detecting a tenant's shirking is an increasing function of the degree of his shirking and landlord's supervision time. The monitoring efficiency will depend on landlord's ability, associated with his farming experience, size of landholding, and residential proximity to his tenant farms, and the technological characteristics of production. As long as the monitoring of a tenant's work effort is feasible but costly, as certainly is, some extent of shirking cannot be avoided in this extended model.

Whether the equilibrium level of shirking is closer to the level envisaged by the Marshallian view or to zero formulated by Cheung depends not only on the probability of detecting a tenant's shirking, but also on the amount of punishment. If there is no punishment, the share tenant will shirk as much as he wants, even if the landlord specifies the desired effort level. Thus, the threat of penalty must exist to enforce the terms of the contract effectively. It is, however, difficult to impose explicit payment of heavy penalty on a tenant's shirking in practice, in part because of the possibility of tenant's desertion and in part because of the absence of an effective legal enforcement mechanism of contact (e.g., court).

The existing studies on agrarian organizations by Hayami and Kikuchi (1982), Binswanger and Rosenzweig (1986), and Hayami (1997) argue that the loss of reputation may play an important role of punishment in agrarian communities, where people are long settled and know one another quite well. If tenant's dishonest behavior is detected in such a community, not only will his current contract be terminated but also he will lose the opportunity to contract with someone else in the future. To the extent that the landlord assures the tenant utility higher than his opportunity utility obtainable elsewhere, the fear of losing the present contractual opportunity may represent a potentially strong penalty on a tenant's shirking. Conceptually, marginal return to tenant's effort, which takes into account the future benefits arising from the increased probability of renewal of contracts and reduced probability of penalty, may be depicted by dotted curve located above the traditionally conceived marginal return curve, i.e., α(VMPL).

While the majority of the existing models of share tenancy consider one-season contracts, the above argument suggests that it is more appropriate to consider long-term contract or contracts with the possibility of contract renewal. Within this generalized framework, the case envisaged by the Marshallian view of share tenancy corresponds to the special case where the duration of the contract is restricted to be one season regardless of a tenant's conduct in the current season. Since punishment on shirking is absent in this case, share tenant will shirk. The efficiency view of share tenancy may well have relevance, if the contract is long term and if the expected cost of losing reputation is extremely high.

In my view, the empirical issue is not whether share tenancy is socially inefficient or not, but whether it is *significantly* inefficient compared with leasehold tenancy and owner farming. Another implication of our arguments is that empirical analysis of the efficiency of share tenancy is incomplete unless the enforcement mechanism of the contract is analyzed in addition to the comparative analysis of resource allocation and productivity.[11]

[11] Analogous to the Marshallian argument, it is also argued that share tenancy discourages long-term investment and the adoption of new profitable technology because of the disincentive effects of output sharing. Our theoretical arguments developed in the text can be applied to these inefficiency issues. These hypotheses, however, are not empirically supported [Hayami and Otsuka (1993)].

3.5. Land tenure security and investment

Land tenure insecurity refers to the probability that land-owning or land-using household loses land rights in future. If share tenant does not have tenure security, so that his contract is likely to be terminated at the end of one cropping season, his work incentive will be low. Also if land tenure is insecure, there is no guarantee that the person who has invested in land improvement will receive benefits from investment in future [Feder and Noronha (1987); Feder and Feeny (1993)]. The weak land rights, including rights to rent out, bequeath, and sell, are shown to weaken incentives to invest in long-term land improvements [Besley (1995)].[12] Similarly, if share tenancy contract is insecure, tenant may not have strong incentives to invest to improve farm productivity [Banerjee, Gertler and Ghatak (2002)].

In China even after the demise of people's communes, individual land rights are still weak. If a piece of land is leased out, it may be reallocated to other village members in future by the community leaders [Liu, Carter and Yao (1998); Kung (1995, 2000); Brandt et al. (2002)]. A major question is whether such land tenure insecurity in China deters investment in land improvement significantly, which may reduce growth prospect of the Chinese agriculture. Therefore, a number of empirical studies have been conducted on this subject, as will be reviewed in the next section.

Similarly, individual land rights are weak under customary land tenure institutions, in which village chief or leaders of extended family have ultimate rights to allocate land use rights to community or family members. Thus, whether investments in land improvement are discouraged significantly under such land tenure systems is an important policy question [Otsuka and Place (2001)].

Tenure insecurity is expected to affect not only long-term farm investments but also other activities in rural economies. An important example is the insecurity of land ownership rights in Thailand due to the lack of official land titles, which reduces farmers' access to formal-sector credit because land cannot be used as collateral [Feder et al. (1988)]. When land lease rights are restricted, as in the case of China, land tenancy contracts will be short-term and, hence, they are likely to be fixed-rent rather than share contracts, because the possibility of renewing contracts plays an important role in reducing moral hazard under share contracts, as discussed before. Furthermore, restrictions on land transfer rights in China will reduce incentives to migrate from villages to distant towns, because it is likely that migrants will lose their land rights in the next round of land reallocation, unless land is continued to be cultivated by their relatives or close friends. Thus, by increasing the cost of land tenancy transactions, the land tenure insecurity can hinder efficient allocation of not only land but also human resources in rural areas.

[12] Besley's (1995) study is concerned with the effect of weak individual land rights on commercial tree planting in customary land areas of Ghana, where cocoa agroforestry is a dominant farming system. Follow-up studies were conducted in Ghana by Quisumbing et al. (2001a, 2001b) and Otsuka et al. (2003). Their new results do not support the earlier findings of Besley.

3.6. Adjustment costs and allocative efficiency

As we have seen earlier in this section, land transaction through tenancy arrangement from land-abundant households to land-scarce households can potentially increase allocative efficiency of land and labor resources among farming households. At the same time, land transaction helps reduce rural poverty, if indeed the land-scarce households, which generally suffer from poverty, gain access to land. Land tenancy transaction, however, may not be able to achieve the high degree of allocative efficiency among farming households because of the high transaction costs.

First, since share tenant has an incentive to shirk, landowner must supervise them, which is costly, though less costly than wage laborer employed on the time-rate contract. Second, although it is not widely recognized, it is critically important for landowners to inspect the amount of harvested output accurately to avoid tenant's cheating. Thus, landowners or their relatives usually go to the fields to check outputs on harvesting days. Third, landowners must inspect the use of land and other fixed assets to avoid the abuse by tenants. Labor shirking, output under-reporting, and abuse of land will be particularly serious, if landowner's land rights are insecure, so that he can offer only short-term contracts for the fear of losing land rights under longer-term contracts. In such cases, tenant's misconducts cannot be penalized by termination of the contract.

Problems of shirking and under-reporting of output can be avoided by offering fixed-rent contracts, but at the cost of forgoing risk-sharing opportunity and the risk of greater mining of land and fixed assets by tenants. In practice, the absentee landowners usually offer fixed-rent tenancy contracts to avoid the problem of supervising tenants under sharing arrangements. Because of the transaction costs associated with land tenancy contracts, landowners may be willing to rent out much smaller land area than socially optimum. The question is how important such insufficient renting is.

As is specified by Skoufias (1995),[13] one way of incorporating the cost of renting in a model of land renting decision is to assume that area rented in (R) is a fraction of the difference between the desired cultivation area in the absence of transaction costs (A^*) and owned area of land ($\mathbf{A}$):

$$R = \theta(A^* - \mathbf{A}),$$

where θ is adjustment coefficient, which is equal to or less than unity; A^* is supposed to be a function of family labor endowment and farm management ability of a cultivator, among other things; and $\mathbf{A}$ is assumed to be predetermined. If adjustment cost is nil, θ will be close to unity. Otherwise, θ will be significantly smaller than unity. According to Skoufias (1995), the adjustment coefficient can be identified from the estimated coefficient of $\mathbf{A}$, if the rent area function is estimated after specifying the functional form of A^* properly. If θ is estimated to be significantly less than unity, the function of land tenancy market is identified to be significantly inefficient.

[13] See Kevane (1996) and Sadoulet, Murgai and de Janvry (2001) for more general models of market imperfections in agrarian economies.

The second way of assessing the efficiency of land tenancy market (as well as labor market) is to compare the input use intensity, yield, and profit across large and small farms. If the land tenancy market dose not function well, then it is likely that labor intensity and crop yields are larger on smaller farms, which use primarily family labor, than on larger farms, which tend to rely on hired labor.[14] If such differences in resource allocations are not observed statistically, the null hypothesis of well-functioning land tenancy market cannot be rejected.

Supplementary to the second method of testing is to examine the significance of Marshallian inefficiency by comparing input use intensity and yield under share tenancy with those under owner-farming or fixed rent tenancy. If significant inefficiency of share tenancy is found, it is likely that land allocation among farming households is also significantly inefficient.

4. A review of empirical studies

In this section, we review the empirical literature to determine the efficiency of share tenancy and land tenancy markets, their effects on equity, and the effects of tenure security on long-term investments.

4.1. Efficiency of share tenancy

In order to test the hypothesis of inefficient resource allocation under share tenancy, a large number of case studies have been conducted, particularly in South and Southeast Asia. They generally compared average physical output per unit of land between share tenancy and owner farming and between share tenancy and leasehold tenancy. Production data were often classified by the presence of irrigation and the size of cultivation area to control for the quality differences in land and the effect of scale of farm operation. While most empirical studies compared the physical yields of the same crop, some compared the total value of different crops per unit of cropped area [e.g., Laffont and Matoussi (1995)]. The latter comparison is inaccurate as the test of inefficiency arising from the undersupply of tenant's work effort, as the crop choice affects the marginal product curves of the work effort. Hayami and Otsuka (1993) conclude, based on an exhaustive survey of empirical literature, that there is no evidence to support the hypothesis that yields under share tenancy are lower than under owner farming or fixed-rent leasehold tenancy. In fact, the distributions of percentage differences in yields between share tenancy and owner farming or leasehold tenancy, which are created by using the descriptive data shown in published journal articles,

[14] The intensity of purchased input use may be higher on larger farms due to greater access to credit by larger farms [Feder (1985); Eswaran and Kotwal (1986); Kevane (1996)]. In this case, yield could be higher on larger farms.

are largely symmetrical around the mean of zero, suggesting that the yield under share tenancy tends to be equalized with that under owner farming and leasehold tenancy.[15] Note, however, that the past studies often did not adequately control for the effects of relevant factors for the comparison of yields, such as land quality and ability of cultivators. Therefore, caution must be taken to interpret the empirical evidence. However, there is no reason to assume that the omitted factors affect the yield comparison routinely in one direction. To be conservative, the accumulated evidence may be taken to imply that share tenancy is not as inefficient as the inefficiency theory assumes.

Hayami and Otsuka (1993), as well as their earlier studies [Otsuka and Hayami (1988); Otsuka, Chuma and Hayami (1992)], argue that significant shirking of tenants tends to be prevented by long-term contractual and personal relationships between landlords and tenants and by the landlord's ability to enforce terms and conditions of contracts. In fact, tenants and landlords are often relatives and long-term neighbors and friends, and landlords are generally experienced farmers residing in the villages or nearby towns. In other words, self-selection of share tenancy contracts by those landlords who are adept at monitoring tenants' work leads to relatively efficient outcome under share tenancy. If landlords are absentee or inexperienced in farming, fixed-rent contracts tend to be chosen [e.g., Jodha (1984); Sharma and Dreze (1996); Sadoulet, Murgai and de Janvry (2001)]. Furthermore, in China where individual land rights are still insecure, tenancy is relatively uncommon and, when rented, contracts are very short-term fixed-rent leasehold contracts among relatives and friends [Brandt et al. (2002); Benjamin and Brandt (2000); Deininger and Sougqing (2001a); Kung (1995, 2000); Liu, Carter and Yao (1998)]. Such findings are consistent with our theoretical predictions.

More recently, case studies in Thailand by Sadoulet, de Janvry and Fukui (1994, 1997); in India by Sharma and Dreze (1996); in Ethiopia by Gavian and Ehui (1999), Pender and Fafchamps (2001), and Holden and Yohannes (2002); and in Ghana by Quisumbing et al. (2001a, 2001b) and Otsuka et al. (2003) provide added support for the empirical generalization that share tenancy is not significantly inefficient.

Yet, there are two well-known studies, which report the significant production inefficiency of share tenancy based on comparison of yields of owner-operated and share-tenanted fields of the same operators in India [Bell (1977); Shaban (1987)]. In India, the land reform program, known as the land-to-the-tiller program, was implemented to transfer the land from large landlords to tenants who actually cultivate the land [Dantwala and Shah (1971); Ladejinsky (1977); Herring (1983); Appu (1975)]. The important assumption of this program is that there is only a single tenant on each plot of tenanted land. In many areas in India, however, landlords shifted tenants every year

[15] It must be pointed out that the yield data taken from past case studies are average yields of the sample, rather than the original sample data, so that the distribution referred to in the text may be conceptually closer to the sampling distribution than the distribution of samples.

or every season to prevent them from claiming that they are actual tillers of any particular plot of land [Ladejinsky (1977)]. Villages surveyed by the International Center for Research in Semi-Arid Tropics or ICRISAT are located in such areas [Cain (1981); Jodha (1984); Walker and Ryan (1990)]. As the landlords attempted to circumvent the implementation of the land-to-the-tiller program, the share tenancy contracts became insecure. It is precisely under these conditions that Bell and Shaban found the significant inefficiency of share tenancy.[16]

Under the one-period contract with no possibility of contract renewal, future penalty on tenant's shirking is bound to be limited. As we discussed earlier, this is exactly an implicit assumption of the inefficiency theory of share tenancy. Thus, Bell's and Shaban's findings should not be interpreted as evidence of the general inefficiency of share tenancy, but more legitimately as evidence of the inefficiency of distorted short-term share tenancy contracts.

Significant inefficiency of share tenancy is likely to be more of a consequence of land reform regulations than of the inherent difficulty of enforcing contractual terms under a share contract. In general, significant inefficiency of share tenancy is not found in Southeast Asia where land reform programs have seldom been implemented except in the Philippines. In this country land reform was vigorously implemented in major rice growing areas [Otsuka (1991)], and significantly lower residual profit per ha is found under share tenancy [Estudillo and Otsuka (1999)].

Banerjee, Gertler and Ghatak (2002) recently report that rice yield increased by as much as 51 to 62% after tenancy reform in West Bengal from 1979 to 1991, which strengthened the tenure security and improved output sharing rate for tenants, typically from 50 to 75%. They attribute such large yield gains to the enhanced incentives to work and invest in long-term land improvement because of the tenancy reform, implying that share tenancy used to be significantly inefficient. However, Saha and Saha (2001) report that the government has been active in facilitating access to credit for the beneficiaries of tenancy reform, and Rogaly, Harris-White and Bose (1999), Gazdar and Sengupta (1999), and Webster (1999) commonly point out that the provision of credit facilitated the purchase of tubewells by share tenants, which led to the adoption of recent high-yielding varieties and shift to higher-yield *boro* (or dry) season cropping. Thus, although we cannot deny the possibility of significant yield-enhancing effects of the tenancy reform, it seems misleading to attribute the large yield gains primarily to such effects.

4.2. *Efficiency of land allocation and use*

To my knowledge, there are only a few studies, which estimate the land rental area adjustment coefficient (θ), discussed in the previous section. Using the ICRISAT data, earlier Pant (1983) and later Skoufias (1995) find that the adjustment coefficient is significantly smaller than unity. This is consistent with the finding of Shaban (1987) that

[16] Shaban (1987) uses the ICRISAT data.

share tenancy is significantly inefficient in the ICRISAT villages. To the extent that the efficiency of share tenancy is impaired by land reform program, finding by Pant and Skoufias should not be interpreted as evidence that tenancy markets do not generally achieve efficient land allocation among farm households. On the other hand, Kevane (1996) obtains the adjustment coefficient, which is not significantly different from unity in his study of farm households in Western Sudan, where tenancy transactions are active.[17] Obviously, the same type of statistical analysis must be implemented in many other areas, before we reach a clear conclusion on the allocative efficiency of land tenancy markets. Although the adjustment coefficient is not estimated, Sharma and Dreze (1996) find that tenancy leads to remarkable convergence of land–labor ratio in different land tenure groups in an Indian village, indicating the ability of tenancy markets to equalize resource use intensity among households.

A large number of empirical studies have been conducted in Asia to analyze the relationship between farm size and yield or value added per unit of area or input use intensity. While the significant inverse relation is not generally found in Southeast Asia [David and Otsuka (1994)], it is found in South Asia, especially in India [Bhagwati and Chakravarty (1969); Sen (1975); Berry and Cline (1979); Dyer (1996/97); Heltberg (1998)]. The observed inverse correlation is largely explained by differences in land quality and crop mix; large farmers tend to cultivate less fertile land and grow crops of lower output value [Verma and Bromley (1987); Bhalla and Roy (1988); Newell, Pandya and Symons (1997)].[18] Yet, a significant inverse correlation remains even after controlling for land quality and other differences associated with farm size [Carter (1984); Heltberg (1998)]. It is often pointed out that the inverse correlation disappeared in India after the Green Revolution because larger farmers apply larger amount of purchased inputs.[19] According to Newell, Pandya and Symons (1997), the inverse relation between farm size and value added per hectare disappears, but the inverse relation between farm size and labor input per hectare remains significant even within a village in India. Ramasamy, Paramasivam and Kandaswamy (1994) also obtain the similar results from their village study in Tamil Nadu in India. These findings strongly indicate the larger use of family labor and lower use of purchased inputs by smaller farmers, reflecting the advantage of relatively abundant family labor endowment and disadvantage of unfavorable access to credit markets. Heltber's (1998) careful analysis of household panel data in Pakistan clearly supports our interpretation.

If the inverse relation exists, the transfer of land from larger farmers to smaller farmers will result in higher production efficiency as well as more equitable distribution of

[17] Although Deininger and Sougqing (2001a) also estimate the similar land renting area function in China using Tobit estimation procedure, they do not test the significant difference of the estimated θ coefficient from unity.

[18] See also Benjamin (1995) for the case of Java.

[19] Empirical evidence, however, is not necessarily strong. See, for example, recent reviews by Dyer (1996/97) and Heltberg (1998).

income. The question is why inefficient large farms do not lease out their land to smaller farmers and the landless.

In India, the land reform program applied to tenant cultivated land with the exemption of owner cultivated land using hired labor [Khusro (1973); Dantwala and Shah (1971); Appu (1975); Ladejinsky (1977); Herring (1983)]. Since regulated land rent was set at a level significantly lower than the market rate, landlords were motivated to evict tenants in order to undertake owner cultivation. According to Bhalla (1976), Dantwala and Shah (1971), Ladejinsky (1977), and Bardhan (1989), many landlords actually evicted tenants and converted them to permanent laborers. At the all India level the percentage of farm area under tenancy declined from about 20% in the pre-reform period of the mid-1950s to about 12% in the mid-1960, at least partly because of the implementation of the land reform program [Narain and Joshi (1969)].

Although the available empirical evidence from India may not be sufficiently conclusive, it seems that land reform programs in India induced the tenant eviction and suppressed the opportunity to rent out the land of large farms. If so, the land reform not only perpetuated the status of the landless as agricultural laborers, thereby aggravating the rural poverty, but also distorted resource allocations reflected in the observed inverse relationship.[20]

In the Philippines, more concrete evidence is available. According to Otsuka (1991), 20 to 50% of tenants were evicted at the time of land reform implementation in selected villages in Central Luzon and Panay Island. At the same time, a large number of share tenants have been converted to leaseholders and amortizing owners and these land reform beneficiaries received significantly higher income than the remaining share tenants [Otsuka, Cordova and David (1992)]. Because of the prohibition of new tenancy and sub-tenancy, however, land reform beneficiaries, who cultivate large areas relative to the endowment of family labor, began to employ permanent labor [Hayami and Otsuka (1993)]. Cultivation of large farms by permanent labor, however, is revealed to be significantly inefficient [Otsuka, Chuma and Hayami (1993)]. In this way, the inverse correlation was newly created by land reform implementation in rice growing areas of the Philippines.

In China, where tenancy transaction is still suppressed, Benjamin and Brandt (2000) find significant inverse relation between farm size and labor intensity. As in Carter's (1984) study in India, their finding suggests the emergence of inefficiency in resource allocation among farm households.[21] This may pose serious problem in Chinese agriculture in future, because farmers are actively seeking non-farm jobs in China [Yao (2000)], which requires the efficient reallocation of land among farm households to maintain the productivity of the farm sector. In Japan, malfunction of land rental market

[20] According to Besley and Burgess (2000) who use state level data on the incidence of poverty, tenancy reform, but not land redistribution, contributed to the reduction in rural poverty significantly.

[21] Benjamin and Brandt (2000), however, do not observe the inverse relation between farm size and yield. This may well be due to greater access of larger farmers to cheap credit markets, as in the case of India [Newell, Pandya and Symons (1997)].

due to rent controls and other government interventions after the post-World War II land reform is one of the major causes underlying the preservation of small-scale inefficient farms until today [Otsuka (1992)]. Using household data, Dong (2000) finds from the estimation of Cobb–Douglas production function that agricultural production is subject to significant diseconomies of scale in China, which is consistent with the emergence of the inverse correlation between farm size and productivity. In this connection, it is noteworthy that land tenancy is most active in Punjab and Haryana where the economy has undergone the most dynamic changes in India [Thorat (1997)].

In sum, although there is no concrete evidence to support the efficient function of land tenancy markets in allocation of land resources among farming households, there is fairly strong evidence that suppression of land tenancy transactions lead to the significant inefficiency of land allocation among farming households.

4.3. Land tenure security and investment

Despite the fact that land ownership rights of large landowners are insecure in India, surprisingly no solid empirical study has been conducted as to the effect of tenure security on long-term investment in this country. In contrast, there is serious concern among researchers on Chinese agriculture with the possible adverse effects of land tenure insecurity on investment, due to periodical reallocation of lands among households. In China, however, major agricultural investments, such as irrigation investment, are carried out collectively even now. A series of case studies using household survey data by Li, Rozelle and Brandt (1998), Li, Rozelle and Huang (2001), and Jacoby, Li and Rozelle (2002) find that the tenure insecurity reduces organic manure application, whose effects on yield increases are expected to last for more than one cropping season. Although whether manure application is a major investment can be questioned, Deininger and Sougqing (2001b) find the same tendency in their estimation of investment function, including various investment expenses. Therefore, it is likely that long-term investments in land improvement are discouraged in China because of the insecurity of tenure.

There are a large number of case studies, which analyze the impact of tenure security on long-term investments in communal or customary land tenure areas primarily in Africa and a few in Asia. Contrary to the finding of Besley (1995), the majority of studies find that tenure insecurity does not affect tree planting and other long-term investments under customary land tenure systems [Place and Hazell (1993); Braselle, Gaspart and Platteau (1998); Otsuka and Place (2001); Gray and Kevane (2001); Quisumbing et al. (2001a, 2001b); Otsuka et al. (2003); Suyanto, Tomich and Otsuka (2001); Holden and Yohannes (2002)].[22] Moreover, Barrows and Roth (1990) and Place and

[22] There are a few exceptions [Gavian and Fafchamps (1996); Hayes, Roth and Zepeda (1997); Place and Otsuka (2001)]. Place and Otsuka argue that if investment is socially minor, e.g., plating a few frit trees in the middle of fields, social institutions, which grant stronger land rights, will not emerge.

Migot-Adholla (1998) find that land titling or registration did not have significant effects on long-term investment and resource allocations in Sub-Saharan Africa. This is because indigenous land rights have been strengthened with increasing scarcity of land resources in African communities. More specifically, Otsuka and Place (2001), as well as Sjaastad and Broley (1997), argue that the effort to invest in profitable land improvement is commonly rewarded by enhanced individualized land rights *after* investment, according to communal rules. This expectation provides proper investment incentives even when land rights are weak *before* investment. The strengthening of individual land rights can occur, if the outcome of investment is easily observable by community members (e.g., established trees). Thus, when we analyze the effect of tenure security on long-term investment, we must consider not only the prevailing level of tenure security but also the expected level after investments are made.

4.4. Equity effects of tenancy transactions

The case studies of land rental markets find through regression analyses that small farmers rent in land from large farmers in Bangladesh [Hossain (1978)], in India [Pant (1983); Skoufias (1995)], in China [Deininger and Sougqing (2001b)], in Sudan [Kevane (1996)], and in Paraguay [Carter and Salgado (2001)], even though the extent of land reallocation may fall short of the social optimum because of the significant adjustment costs. There are, however, studies reporting that tenancy contracts tend to be made between farmers belonging to the similar land ownership classes in India [e.g., Swamy (1988); Sharma and Dreze (1996); Sarap (1998)]. Also in areas where mechanization takes place, which leads to expansion of the optimal scale of production, the "reverse" tenancy is observed in which tenants are large landowners renting in land from landlords who are small landowners, as observed in postwar Japan [Hayami and Kawagoe (1989)] and in the most advanced rural area of India [Singh (1989)]. On the whole, however, the empirical evidence supports our argument that tenancy transactions contribute to the equalization of operational landholdings relative to family labor endowment across farming households in developing countries [Sadoulet, Murgai and de Janvry (2001)].

As would be expected, there are a small number of studies reporting the active land sales market. One exception is a study of West Bengal where it is found that large landowners prefer to sell land to former tenants as the tenancy reform strengthened tenure security and reduced crop share for landowners [Rawal (2001)]. A recent study in Sumatra by Suyanto, Tomich and Otsuka (2001) finds that purchased land area accounts for 23% of commercial tree areas, whereas tenanted area accounts for 21%. The study sites are subject to matrilocal tradition, in which husband moves to wife's village. In recent years, however, sons inherit land contrary to the matrilineal tradition [Quisumbing and Otsuka (2001)]. Thus, young men who are about to marry with wives in distant villages and towns sell their land. An important finding of Suyanto, Tomich and Otsuka (2001) is that the Gini coefficients of operational landholdings are 0.47

for purchased land, 0.17 for rented land, and 0.38 for the overall landholdings.[23] This contrast seems to imply that land sales transactions contribute to in-equalization of landholdings, whereas land tenancy transactions contribute to equalization. Such tendencies are confirmed also in Paraguay, Nicaragua, and Chile by Carter and Salgado (2001). The equitable role played by tenancy transactions is also observed in Western Ghana [Quisumbing et al. (2001b)].

It should not be an exaggeration to say that land tenancy transactions contribute to equitable distribution of landholdings by transferring land from large to small farmers. Thus, land tenancy should be encouraged as a means to promote equity in rural villages. It must be emphasized, however, that the landless laborers cannot participate in land tenancy transactions in some parts of India [e.g., Sharma and Dreze (1996); Sarap (1998)]. This suggest that removal of government restrictions on tenancy transactions alone is likely to be insufficient for achieving equitable operational land distribution, given traditional institutions, such as the cast system.

5. Concluding remarks

Our literature review demonstrated that huge amount of theoretical and empirical knowledge has been accumulated on the efficiency of share tenancy, allocative efficiency of tenancy transactions, the existence of inverse correlation between farm size and productivity, and the effects of tenure security on long-term investments. It is clear, however, that these issues have been addressed independently in the majority of the existing studies, even though they are closely related. For example, if share tenancy is inefficient, it is likely that land allocation among farming households will be inefficient because of the high transaction costs. In consequence, the inverse correlation is likely to arise. Furthermore, if the cause for the inefficiency of share tenancy is the insecurity of land ownership, investment decisions will be also affected. Thus, in order to gain deeper understanding of the role of land markets in achieving allocative efficiency and equitable land distribution, an integrated approach, which considers not only the efficiency effects of land tenancy contracts, land transactions, and land tenure security, but also their effects on land distribution between land-abundant and land-scarce households, is called for.

It is clear from our literature review that so far land reform polices have generally failed to improve land use efficiency and social equity. In fact, the implementation of the conventional redistributive land reform is often ineffective and, where effective, creates adverse consequences, such as the eviction of tenants, excessive employment of hired labor, and the insecurity of tenancy contracts. In contrast, land tenancy, including share tenancy, is found to be efficient in transferring land from land-abundant to land-scarce

[23] All these Gini coefficients, which are called pseudo-Gini coefficients. are computed using the raking of households in terms of total landholdings, but not landholdings of specifically acquired land category.

rural households. It is therefore highly advisable to promote tenancy transactions for the sake of both efficiency and equity.[24]

We expect that promotion of tenancy transactions will reduce inefficiency of production on large farms and the inefficiency of share tenancy caused by inappropriate tenancy regulations. Such policies alone, however, may not contribute significantly to the enhancement of production efficiency, because tenants are likely to have lesser access to cheap credits. Therefore, the failure in credit markets must be removed, in order to make liberalization of tenancy truly effective and equitable.[25]

Tenancy issues are important not only in Asia but also in Sub-Saharan Africa, where individual land rights including rights to rent out and sell have been strengthened with increased population pressure, and in Latin America, where land titling to the beneficiaries of earlier redistributive land reforms assumes greater importance [Carter and Salgado (2001); de Janvry, Sadoulet and Wolford (2001)]. Proper understanding of the role and the limitation of tenancy transactions will be necessary for implementing appropriate land policies in African and Latin American countries in years to come.

While the activation of land tenancy market is essential for both the efficiency and equity, it cannot rectify the fundamental inequity arising from the unequal distribution of landownership. A large number of economists recommend newly designed land reform programs, which attempt to induce land market transactions from large landowners to land-poor households by deliberate policy interventions [Hayami, Quisumbing and Adriano (1990); Binswanger and Deininger (1993); Deininger (1999); Deininger and Binswanger (2001); Sadoulet, Murgai and de Janvry (2001)]. The World Bank experiments market-assisted land reform, which provides grant or subsidy to the poor to purchase land [Deininger (1999)]. Hayami, Quisumbing and Adriano (1990) propose the progressive land tax, which intends to induce land sales by large landowners. Recently, Gaiha (1993) and Ray (1996) join Hayami, Quisumbing and Adriano (1990) in supporting such land tax scheme. Although it is critically important to assess relative advantages of subsidy and tax programs, it is beyond the scope of the present study. In conclusion, I want to emphasize that, judging from the poor performance of redistributive land reforms, it is worth considering tax or subsidy schemes as new means of achieving socially desirable land distribution without sacrificing production efficiency significantly.

[24] Interestingly, a group of economists in India support the idea to promote tenancy transactions for both the greater efficiency and equity in rural areas, in view of the inefficient use of large tracts of land by large landowners under the current system [e.g., Radhakrishnan (1990); Rao and Gulati (1994); Thimmaiah (2001); Thorat (1997); Vyas (1994)].

[25] Considering that the poverty associated with growing landlessness is becoming increasingly a serious concern in Asia [e.g., Hossain, Sen and Rahman (2000); Hossain, Gascon and Marciano (2000); Isvilanonda, Ahmad and Hossain (2000); Janaiah, Bose and Agarwal (2000)], it is imperative to assist the landless to become tenants by targeted programs.

References

Appu, P.S. (1975). "Tenancy reform in India". Economic and Political Weekly 10 (33-5), 1339–1375.

Banerjee, A.V., Gertler, P.J., Ghatak, M. (2002). "Empowerment and efficiency: Tenancy reform in West Bengal". Journal of Political Economy 110 (2), 239–280.

Bardhan, P.K. (1980). "Interlocking factor markets and agrarian development". Oxford Economic Papers 32 (1), 82–98.

Bardhan, P.K. (1984). Land, Labor, and Rural Poverty: Essays in Development Economics. Columbia University Press, New York.

Bardhan, P.K. (Ed.) (1989). The Economic Theory of Agrarian Institutions. Clarendon Press, Oxford.

Barrows, R., Roth, M. (1990). "Land tenure and investment in African agriculture: Theory and evidence". Journal of Modern African Studies 28 (2), 265–297.

Bell, C. (1977). "Alternative theories of sharecropping: Some tests using evidence from Northeast India". Journal of Development Studies 13 (4), 317–346.

Benjamin, D. (1995). "Can unobserved land quality explain the inverse productivity relationship?". Journal of Development Economics 46 (1), 51–84.

Benjamin, D., Brandt, L. (2000). "Property rights, labor markets, and efficiency in a transition economy: The case of rural China", Mimeo. University of Toronto.

Berry, R.A., Cline, W.R. (1979). Agrarian Structure and Productivity in Developing Countries. Johns Hopkins University Press, Baltimore, MD.

Besley, T. (1995). "Property rights and investment incentives". Journal of Political Economy 103 (5), 913–937.

Besley, T., Burgess, R. (2000). "Land reform, poverty reduction, and growth: Evidence from India". Quarterly Journal of Economics 115 (2), 389–430.

Bhagwati, J.N., Chakravarty, S. (1969). "Contributions to Indian economic analysis: A survey". American Economic Review 59 (Suppl.), 1–73.

Bhalla, S. (1976). "New relations of production in Haryana agriculture". Economic and Political Weekly 11 (13), 23–30.

Bhalla, S.S., Roy, P. (1988). "Mis-specification in farm productivity analysis: The role of land quality". Oxford Economic Papers 40 (1), 55–73.

Binswanger, H.P., Deininger, K. (1993). "South African land policy: Legacy of history and current options". World Development 21 (9), 1451–1475.

Binswanger, H.P., Deininger, K., Feder, G. (1995). "Power, distortions, revolt, and reform in agricultural land relations". In: Behrman, J., Srinivasan, T.N. (Eds.), Handbook of Development Economics, vol. III. Elsevier, Amsterdam.

Binswanger, H.P., Elgin, M. (1988). "What are the prospects for land reform?". In: Mazumder, A., Valdes, A. (Eds.), Agriculture and Government in an Interdependent World. Dartmouth Publishing Co.

Binswanger, H.P., Rosenzweig, M.R. (1986). "Behavioral and material determinants of production relations in agriculture". Journal of Development Studies 22 (3), 503–539.

Brandt, L., Li, G., Huang, J., Rozelle, S. (2002). "Land rights in China: Facts, fictions, and issues". China Journal 47 (1), 67–97.

Braselle, A.-S., Gaspart, F., Platteau, J.-P. (1998). "Land tenure securities and investment incentives: Further puzzling evidence from Burkina Faso", Working paper. Faculty of Economics, Social and Management Sciences, University of Namur, Belgium.

Cain, M. (1981). "Risk and insurance: Perspectives on fertility and agrarian change in India and Bangladesh". Population and Development Review 7 (3), 435–474.

Carter, M.R. (1984). "Identification of the inverse relationship between farm size and productivity". Oxford Economic Papers 36 (1), 131–145.

Carter, M., Salgado, R. (2001). "Land market liberalization and the agrarian question in Latin America". In: de Janvry, A., Gordillo, G., Platteau, J.-P., Sadoulet, E. (Eds.), Access to Land, Rural Poverty, and Public Action. Oxford University Press.

Cheung, S.N.S. (1969). The Theory of Share Tenancy. University of Chicago Press, Chicago.
Dantwala, M.L., Shah, C.H. (1971). Evaluation of Land Reforms. University of Bombay Press, Bombay.
David, C.C., Otsuka, K. (1994). Modern Rice Technology and Income Distribution in Asia. Lynne Rienner, Boulder, CO.
Deininger, K. (1999). "Making negotiated land reform work: Initial experience from Colombia, Brazil, and South Africa". World Development 27 (2), 651–672.
Deininger, K., Binswanger, H.P. (2001). "The evolution of the World Bank's land policy". In: de Janvry, A., Gordillo, G., Platteau, J.-P., Sadoulet, E. (Eds.), Access to Land, Rural Poverty, and Public Action. Oxford University Press.
Deininger, K., Sougqing, J. (2001a). "Land rental markets as an alternative to reallocation? Equity and efficiency consideration in the Chinese land tenure system", Mimeo. World Bank.
Deininger, K., Sougqing, J. (2001b). "The impact of property rights on investment, risk coping, and policy preferences: Evidence from China", Mimeo. World Bank.
de Janvry, A., Sadoulet, E., Wolford, W. (2001). "The changing role of the state in Latin American land reform". In: de Janvry, A., Gordillo, G., Platteau, J.-P., Sadoulet, E. (Eds.), Access to Land, Rural Poverty, and Public Action. Oxford University Press.
de Janvry, A., Platteau, J.-P., Gordillo, G., Sadoulet, E. (2001). "Access to land and land policy reforms". In: de Janvry, A., Gordillo, G., Platteau, J.-P., Sadoulet, E. (Eds.), Access to Land, Rural Poverty, and Public Action. Oxford University Press.
Dong, X.Y. (2000). "Public investment, social services, and productivity of Chinese household farms". Journal of Development Studies 36 (3), 100–122.
Dyer, G. (1996/97). "Output per acre and size of holding: The logic of peasant agriculture under semi-feudalism". Journal of Peasant Studies 24 (1/2), 102–131.
Estudillo, J.P., Otsuka, K. (1999). "Green Revolution, human capital, and off-farm employment: Changing sources of income among farm households in Central Luzon, 1966–94". Economic Development and Cultural Change 47 (3), 497–523.
Eswaran, M., Kotwal, A. (1985). "A theory of two-tier labor markets in agrarian economies". American Economic Review 75 (1), 162–177.
Eswaran, M., Kotwal, A. (1986). "Access to capital and agrarian production organization". Economic Journal 96 (382), 482–498.
Feder, G. (1985). "The relationship between farm size and farm productivity: The role of family labor, supervision, and credit constraints". Journal of Development Economics 18 (2/3), 297–313.
Feder, G., Feeny, D. (1993). "The theory of land tenure and property rights". In: Hoff, K., Braverman, A., Stiglitz, J.E. (Eds.), The Economics of Rural Organization: Theory, Practice, and Policy. Oxford University Press, Oxford, UK.
Feder, G., Noronha, R. (1987). "Land rights systems and agricultural development in Sub-Saharan Africa". World Bank Research Observer 2 (2), 143–169.
Feder, G., Onchan, T., Chalamwong, Y., Hongladarom, C. (1988). Land Policies and Farm Productivity in Thailand. Johns Hopkins University Press, Baltimore, MD.
Gaiha, R. (1993). "Design of poverty alleviation strategy in rural areas", FAO Economic and Social Development Paper No. 115. United Nations Food and Agriculture Organization.
Gavian, S., Ehui, S. (1999). "Measuring the production efficiency of alternative land tenure contract in a mixed crop–livestock system in Ethiopia". Agricultural Economics 20 (1), 37–49.
Gavian, S., Fafchamps, M. (1996). "Land tenure and allocative efficiency in Niger". American Journal of Agricultural Economics 78 (2), 460–471.
Gazdar, H., Sengupta, S. (1999). "Agricultural growth and recent trends in well-being in rural West Bengal". In: Rogaly, B., Harris-White, B., Bose, S. (Eds.), Sonar Bangla: Agricultural Growth and Agrarian Changes in West Bengal and Bangladesh. Sage Publication, New Delhi.
Gray, L.C., Kevane, M. (2001). "Evolving tenure rights and agricultural intensification in southwestern Burkina Faso". World Development 29 (4), 573–587.
Hayami, Y. (1997). Development Economics: From the Poverty to the Wealth of Nations. Oxford University Press, Oxford, UK.

Hayami, Y. (2001). "Ecology, history, and development: A perspective from rural Southeast Asia". World Bank Research Observer 16 (2), 169–198.
Hayami, Y., Kawagoe, T. (1989). "Farm mechanization, scale economies, and polarization: The Japanese experience". Journal of Development Economics 31 (2), 221–239.
Hayami, Y., Kikuchi, M. (1982). Asian Village Economy at the Crossroads. Johns Hopkins University Press, Baltimore, MD.
Hayami, Y., Otsuka, K. (1993). The Economics of Contract Choice: An Agrarian Perspective. Clarendon Press, Oxford, UK.
Hayami, Y., Quisumbing, A.R., Adriano, L.S. (1990). Toward an Alternative Land Reform Paradigm: A Philippine Perspective. Ateneo de Manila University Press, Makati, Philippines.
Hayes, J., Roth, M., Zepeda, L. (1997). "Tenure security, investment and productivity in Gambian agriculture: A generalized probit analysis". American Journal of Agricultural Economics 79 (2), 369–382.
Heltberg, R. (1998). "Rural market imperfections and the farm size–productivity relationships: Evidence from Pakistan". World Development 26 (10), 1807–1826.
Herring, R.J. (1983). Land to the Tiller, The Political Economy of Agrarian Reforms in South Asia. Yale University Press, New Haven, CT.
Holden, S., Yohannes, H. (2002). "Land redistribution, tenure insecurity, and intensity of production: A study of farm households in southern Ethiopia". Land Economics 78 (4), 573–590.
Hossain, M. (1978). "Factors affecting tenancy: The case of Bangladesh agriculture". Bangladesh Development Studies 6 (2), 139–162.
Hossain, M., Gascon, F., Marciano, E.B. (2000). "Income distribution and poverty in rural Philippines: Insights from repeat village study". Economic and Political Weekly 35, 62–80.
Hossain, M., Sen, B., Rahman, H.Z. (2000). "Growth and distribution of rural income in Bangladesh". Economic and Political Weekly 35, 1–24.
Isvilanonda, S., Ahmad, A., Hossain, M. (2000). "Recent changes in Thailand's rural economy". Economic and Political Weekly 35, 43–61.
Jacoby, H.G., Li, G., Rozelle, S. (2002). "Hazards of expropriation: Tenure insecurity and investment in rural China". American Economic Review 92 (5), 1420–1447.
Janaiah, A., Bose, M.L., Agarwal, A.G. (2000). "Poverty and income distribution in rainfed and irrigated ecosystems: Village studies in Chhattisgarh". Economic and Political Weekly 35, 103–118.
Jodha, N.S. (1984). "Agricultural tenancy in semiarid tropical India". In: Binswanger, H.P., Rosenzweig, M.R. (Eds.), Contractual Arrangements, Employment, and Wages in Rural Labor Markets in Asia. Yale University Press, New Haven, CT.
Kevane, M. (1996). "Agrarian structure and agricultural practice: Typology and application to western Sudan". American Journal of Agricultural Economics 78 (1), 236–245.
Khusro, A.M. (1973). "Farm size and land tenure in India". Indian Economic Review 4 (2), 123–145.
Kung, J.K.-S. (1995). "Equal entitlement versus tenure security under a regime of collective property rights: Peasants' preference for institutions in post-reform Chinese agriculture". Journal of Comparative Economics 21 (1), 82–111.
Kung, J.K.-S. (2000). "Common property rights and land redistribution in rural China: Evidence from a village survey". World Development 28 (4), 701–719.
Ladejinsky, W. (1977). In: Walinsky, L.J. (Ed.), Agrarian Reform as Unfinished Business: Selected Papers of Wolf Ladejinsky. Oxford University Press, Oxford, UK.
Laffont, J.-J., Matoussi, M.S. (1995). "Moral hazard, financial constraints, and sharecropping in El Oulja". Review of Economic Studies 62 (3), 381–399.
Li, G., Rozelle, S., Brandt, L. (1998). "Tenure, land rights, and farmer investment incentives in China". Agricultural Economics 19 (1), 63–71.
Li, G., Rozelle, S., Huang, J. (2001). "Land rights, farmer investment incentives, and agricultural production in China", Mimeo.
Liu, S., Carter, M., Yao, Y. (1998). "Dimensions and diversity of property rights in rural China: Dilemmas on the road to further reform". World Development 26 (10), 1789–1806.

Narain, D., Joshi, P.C. (1969). "Magnitude of agricultural tenancy". Economic and Political Weekly 4 (39), A139–A142.
Newell, A., Pandya, K., Symons, J. (1997). "Farm size and the intensity of land use in Gujurat". Oxford Economic Papers 49 (2), 307–315.
Otsuka, K. (1991). "Determinants and consequences of land reform implementation in the Philippines". Journal of Development Economics 35 (2), 339–355.
Otsuka, K. (1992). "Agricultural reforms and development in Asia: A comparative perspective". In: Garnaut, R., Guoguang, L. (Eds.), The Economic Reform and Internationalization: China and the Pacific Region. Allen and Unwin, Sydney and London.
Otsuka, K. (1993). "Land tenure and rural poverty in Asia". In: Quibria, M.G. (Ed.), Rural Poverty in Asia: Priority Issues and Policy Options. Oxford University Press, Hong Kong.
Otsuka, K., Chuma, H., Hayami, Y. (1992). "Land and labor contracts in agrarian economies: Theories and facts". Journal of Economic Literature 30 (4), 1965–2018.
Otsuka, K., Chuma, H., Hayami, Y. (1993). "Permanent labor and land tenancy contracts in agrarian economies: An integrated analysis". Economica 60 (237), 57–77.
Otsuka, K., Cordova, V.G., David, C.C. (1992). "Green revolution, land reform, and household income distribution in the Philippines". Economic Development and Cultural Change 40 (4), 719–741.
Otsuka, K., Hayami, Y. (1988). "Theories of share tenancy: A critical survey". Economic Development and Cultural Change 37 (1), 31–38.
Otsuka, K., Place, F. (2001). Land Tenure and Natural Resource Management: A Comparative Study of Agrarian Communities in Asia and Africa. Johns Hopkins University Press, Baltimore, MD.
Otsuka, K., Quisumbing, A.R. (2001). "Land rights and natural resource management in the transition to individual ownership: Case studies from Ghana and Indonesia". In: de Janvry, A., Gordillo, G., Platteau, J.-P., Sadoulet, E. (Eds.), Access to Land, Rural Poverty, and Public Action. Oxford University Press, Oxford, UK.
Otsuka, K., Quisunbing, A., Payongayong, E., Aidoo, J.B. (2003). "Land tenure and the management of land and trees: The case of customary land areas of Ghana". Environment and Development Economics 8 (1), 77–104.
Pant, C. (1983). "Tenancy and family resources: A model and some empirical evidence". Journal of Development Economics 12 (1/2), 27–39.
Pender, J., Fafchamps, M. (2001). "Land lease markets and agricultural efficiency: Theory and evidence from Ethiopia", Mimeo. International Food Policy Research Institute.
Place, F., Hazell, P. (1993). "Productivity effects of indigenous land tenure systems in Sub-Saharan Africa". American Journal of Agricultural Economics 75 (1), 10–19.
Place, F., Migot-Adholla, S.E. (1998). "The economic effects of land registration on smallholder farms in Kenya: Evidence from Nyeri and Kakamega districts". Land Economics 74 (3), 360–373.
Place, F., Otsuka, K. (2001). "Tenure, agricultural investment, and productivity in customary tenure sector of Malawi". Economic Development and Cultural Change 50 (1), 77–100.
Prosterman, R.L., Riedinger, J.M. (1987). Land Reform and Democratic Development. Johns Hopkins University Press, Baltimore, MD.
Quisumbing, A., Otsuka, K. (2001). "Land inheritance and schooling in matrilineal societies: Evidence from Sumatra". World Development 29 (12), 2093–2110.
Quisumbing, A., Payongayong, E., Aidoo, J.B., Otsuka, K. (2001a). "Women's land rights in transition to individualized ownership: Implications for tree resource management in Ghana". Economic Development and Cultural Change 50 (1), 157–182.
Quisumbing, A., Aidoo, J.B., Payongayong, E., Otsuka, K. (2001b). "Agroforestry management in Ghana". In: Otsuka, K., Place, F. (Eds.), Land Tenure and Natural Resource Management: A Comparative Study of Agrarian Communities in Asia and Africa. Johns Hopkins University Press, Baltimore, MD.
Radhakrishnan, P. (1990). "Land reforms: Rhetoric and reality". Economic and Political Weekly 25 (47), 2617–2621.

Ramasamy, C., Paramasivam, P., Kandaswamy, A. (1994). "Irrigation quality, modern variety adoption, and income distribution: The case of Tamil Nadu in India". In: David, C.C., Otsuka, K. (Eds.), Modern Rice Technology and Income Distribution in Asia. Lynne Rienner, Boulder, CO.

Rao, C.H.H., Gulati, A. (1994). "Indian agriculture: Emerging perspectives and policy issues". Economic and Political Weekly 29 (53), A158–A169.

Rawal, V. (2001). "Agrarian reform and land markets: A study of land transactions in two villages of West Bengal, 1977–1995". Economic Development and Cultural Change 49 (3), 611–629.

Ray, S.K. (1996). "Land system and its reforms in India". Indian Journal of Agricultural Economics 51 (1), 220–237.

Riedinger, J.M. (1995). Agrarian Reform in the Philippines: Democratic Transition and Redistributive Reform. Stanford University Press, Palo Alto, CA.

Rogaly, B., Harris-White, B., Bose, S. (1999). "Introduction: Agricultural growth and agrarian change in West Bengal and Bangladesh". In: Rogaly, B., Harris-White, B., Bose, S. (Eds.), Sonar Bangla: Agricultural Growth and Agrarian Changes in West Bengal and Bangladesh. Sage Publication, New Delhi.

Sadoulet, E., de Janvry, A., Fukui, S. (1994). "Efficient share tenancy contracts under risk: The case of three rice-growing villages in Thailand". Journal of Development Economics 45 (2), 225–243.

Sadoulet, E., de Janvry, A., Fukui, S. (1997). "The meaning of kinship in sharecropping contracts". American Journal of Agricultural Economics 79 (2), 394–406.

Sadoulet, E., Murgai, R., de Janvry, A. (2001). "Access to land via land rental markets". In: de Janvry, A., Gordillo, G., Platteau, J.-P., Sadoulet, E. (Eds.), Access to Land, Rural Poverty, and Public Action. Oxford University Press, Oxford, UK.

Saha, U.S., Saha, M. (2001). "Case study. Regulating the sharecropping system: Operation barga". In: de Janvry, A., Gordillo, G., Platteau, J.-P., Sadoulet, E. (Eds.), Access to Land, Rural Poverty, and Public Action. Oxford University Press, Oxford, UK.

Sarap, K. (1998). "On the operation of the land market in backward agriculture: Evidence form a village in Orissa, eastern India". Journal of Peasant Studies 25 (2), 102–130.

Sen, A.K. (1975). Employment, Technology and Development. Oxford University Press, Oxford, UK.

Shaban, R.A. (1987). "Testing between competing models of sharecropping". Journal of Political Economy 95 (5), 893–920.

Sharma, N., Dreze, J. (1996). "Sharecropping in a North Indian village". Journal of Development Studies 33 (1), 1–39.

Singh, I. (1989). "Reverse tenancy in Punjab agriculture: Impact of technological change". Economic and Political Weekly (June 24), A86–A92.

Sjaastad, E., Broley, D.W. (1997). "Indigenous land rights in Sub-Saharan Africa: Appropriation, security, and investment demand". World Development 25 (4), 549–562.

Skoufias, E. (1995). "Household resources, transaction costs, and adjustment through land tenancy". Land Economics 71 (1), 42–56.

Suyanto, S., Tomich, T., Otsuka, K. (2001). "Agroforestry management in Sumatra". In: Otsuka, K., Place, F. (Eds.), Land Tenure and Natural Resource Management: A Comparative Study of Agrarian Communities in Asia and Africa. Johns Hopkins University Press, Baltimore, MD.

Swamy, D.S. (1988). "Agricultural tenancy in the 1970s". Indian Journal of Agricultural Economics 43 (4), 555–568.

Thimmaiah, G. (2001). "New perspectives on land reform in India". Journal of Social and Economic Development 3 (2), 179–198.

Thorat, S. (1997). "Trends in land ownership, tenancy, and land reform". In: Desai, B.M. (Ed.), Agricultural Development Paradigm for the Ninth Plan under New Economic Environment. Oxford & IBH Publishing, New Delhi.

Verma, B.N., Bromley, D.W. (1987). "The political economy of farm size in India: The elusive quest". Economic Development and Cultural Change 35 (4), 791–808.

Vyas, V.S. (1994). "Agricultural policies for the nineties: Issues and approaches". Economic and Political Weekly 29 (26), A54–A63.

Walker, T.S., Ryan, J.G. (1990). Village and Household Economies in India's Semi-Arid Tropics. Johns Hopkins University Press, Baltimore, MD.
Warriner, D. (1969). Land Reform in Principle and Practice. Oxford University Press, Oxford, UK.
Webster, N. (1999). "Institutions, actors and strategies in West Bengal's rural development: A study on irrigation". In: Rogaly, B., Harris-White, B., Bose, S. (Eds.), Sonar Bangla: Agricultural Growth and Agrarian Changes in West Bengal and Bangladesh. Sage Publication, New Delhi.
Yao, Y. (2000). "The development of land lease market in rural China". Land Economics 76 (2), 252–266.

Chapter 52

LABOR: DECISIONS, CONTRACTS AND ORGANIZATION*

JAMES ROUMASSET and SANG-HYOP LEE

Department of Economics, University of Hawaii at Manoa, Honolulu, HI

Contents

* We are grateful to Julie Anderson Schaffner, Robert Evenson, and Wallace Huffman for extensive suggestions and to Kimberly Burnett and Sean D'Evelyn for research assistance.

Handbook of Agricultural Economics, Volume 3
Edited by Robert Evenson and Prabhu Pingali

DOI: 10.1016/S1574-0072(06)03052-0

Abstract

We assess the development economics of on-farm employment with an eye toward policy implications. What do we know and what additional research is needed? The older tradition of labor market dualism and some of the more modern research are seen to share a characteristic of misplaced exogeneity, and calls for asset redistribution and institutional regulation may need to be tempered by more fundamental explanations. Understanding labor contracts as a facilitator of specialization on the farm and in the larger economy is key. Integrating the wedge model of farm behavior with agency-cost explanations of organization will provide a powerful analytical tool. Ultimately, a general equilibrium view with endogenous institutions will deepen our understanding of why total costs of coordination increase even as turnover costs per worker decline and how public policy can facilitate that cooperation.

Keywords

labor contracts, transaction costs, farm size, specialization, extent of the market

JEL classification: J43, Q12, J41, O13

1. Introduction

The purpose of this chapter is to synthesize economic wisdom about the nature and causes of on-farm employment, labor contracts, and the policy implications thereof. We attempt to do justice to Adam Smith by recognizing that the division of labor is central to the nature of the agricultural firm, the evolution of rural institutions, and agricultural development. Questions addressed include the following. What are the key restrictive assumptions in the conceptual framework that have shaped most theoretical and empirical work to date? How might some of these assumptions be relaxed in order to enrich empirical inquiry and deepen our understanding of the farm employment relationship? What are the implications of on-farm labor relations and productivity for regulations of contracting, land reform and other government interventions?

On-farm employment issues are important for rural development and poverty alleviation strategies. Many rural residents are landless laborers, who tend to be poor, and whose entire way of life – income, the way they spend their day, the uncertainty they face each day about what comes next – is shaped by what happens in rural labor markets. Many other rural residents are small farmers, who also tend to be poor, for whom the ability to work in agricultural labor markets may help them maintain consumption when shocks hit and for whom the ability to hire in agricultural labor markets may be important in making productive use of assets they already own, and in shaping the attractiveness of possible investments. We think that development in general involves a reallocation of labor from traditional to commercial agriculture and from agriculture to non-agriculture (whether rural or urban non-agricultural), and labor markets are important in bringing this reallocation about. Economists have taken a particular interest in agricultural labor contracts. The complexity and diversity of labor contracts makes it difficult to assess the level and direction of change in both the cost of labor to employers and the level of living attained by agricultural laborers. Moreover, specific features of labor contracts have led observers to be concerned about both inefficiency and exploitation in these markets.

In Section 2, we review the conventional wisdom about labor supply and demand in the agricultural household and the nature and consequences of contracts for hired labor. Studies can be roughly classified into two schools of thought. One is Development Microeconomics [e.g., Bardhan and Udry (1999)] which rests on modern theory and recent empirical evidence, but which often employs blackboard economics to discover preconceived allocative inefficiencies. The other is the efficiency school [e.g., Yang (2003)], which aims to provide fundamental explanations of the nature and causes of household decision-making and the organization of production, albeit with models that presume efficiency. This leads naturally to the discussion in Section 3 of promising new directions of research and remaining challenges. Section 4 provides some thoughts on policy and directions for future research.

2. Theoretical and empirical issues

2.1. The wedge-model farm-household decisions

The essence of the Coase Theorem is that without transaction costs economic organization is indeterminate. Large farms could hire labor or workers could equivalently rent land. Moreover, there would be no need to consider the farm-household as a decision-making unit. Farms would maximize profits and the household would (separately) spend household income, be it earned from the farm or elsewhere. This would leave us unable to explain the stylized facts of farmer behavior and agricultural organization, however, and unable to meaningfully assess the consequences of proposed policy reforms.

In the presence of transaction costs, farm and household decisions are interdependent and the farm-household model becomes an appropriate centerpiece for decision-making regarding family and hired labor. Consider a single-period farm-household model:

$$\max u(a, c, l; z) \quad \text{w.r.t. } a, c, l, L_h, L_f$$

subject to budget, time, and production constraints

$$p_s s + w_s L_s + y = c + p_b b + w_h L_h,$$
$$\bar{L} = L_f + L_s + l,$$
$$f(L_h, L_f; A) + b = a + s,$$

where

a: food consumption,
c: non-food consumption,
l: leisure,
z: farm characteristics,
L_h: labor hired to work on farm at wage w_h,
L_f: labor supplied by the household to work on farm,
L_s: labor supplied by the household to work off farm at wage w_s,
y: other exogenous income,
p_s: price of food sold by the household,
s: quantity of food sold by the household,
p_b: price of food bought by the household,
b: quantity of food bought by the household,
$\bar{L}$: total labor endowment of the household,
$f(L_h, L_f; A)$: food production function,
A: farm characteristics.

To highlight the role of labor, we assume that non-labor inputs are fixed.

Rearranging budget and time constraints yields

$$c + wl = w\bar{L} + \pi + y,$$

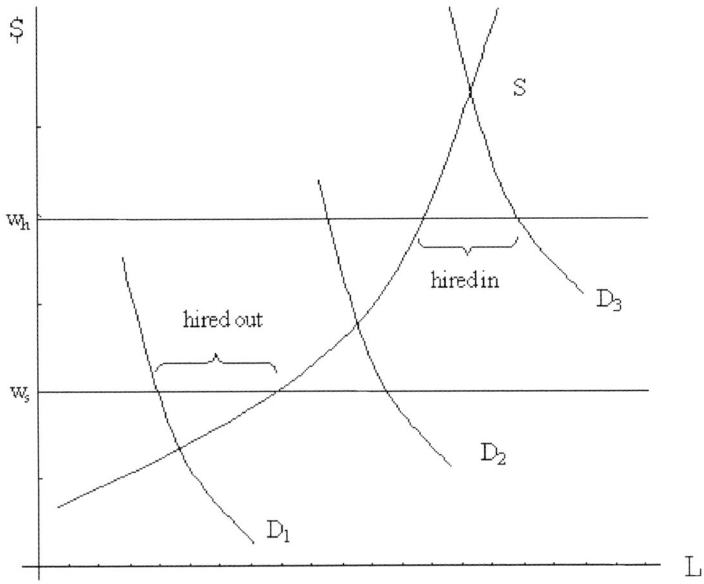

Figure 1. Segmentation of farm labor demand and household supply.

where w is the shadow wage with $w_s \leqslant w \leqslant w_h$, and $\pi = pf(L_f, L_h; A) - w(L_f + L_h)$ is shadow profit with shadow price $p_s \leqslant p \leqslant p_b$. The utility-maximizing farm-household can be said to be equivalently maximizing shadow profits. This is illustrated below.

Figure 1 depicts three possible cases, which shows the household labor supply schedule of a representative farm household and three labor-demand schedules, depending on (quality-adjusted) farm size.

For D_1, the family exports its excess labor ($L_h = 0$), and the relevant shadow price of labor is w_s, the selling wage after deducting journey to work and other necessary expenses from the nominal wage. For D_3, the farm-household imports hired labor ($L_s = 0$), and the shadow wage is w_h, the hiring wage after including the employer's agency cost, recruiting and supervision costs and the residual costs of labor shirking (see Section 2.2). In both cases, the costs embodied in low alternative wages force family to work on other farms or to hire workers beyond the point when marginal product of labor equals nominal wage rate. If labor demand intersects household supply in the intermediate range between w_h and w_s, the shadow wage rate is given by the household's marginal opportunity cost of labor. Accordingly, the rational farm household can be said to be maximizing shadow profits based on the shadow-wage schedule:

$$w = \begin{cases} w_s, & L < L_1, \\ w_h, & L > L_2, \\ S, & L_1 < L < L_2. \end{cases}$$

In this version of the model, both family and hired labor are implicitly in effective labor units, i.e., after deducting any costs of labor shirking.

This shadow-profit-maximization form of the model can be used to estimate farm demand for labor and other inputs without having to fully specify household demands and supplies. By assuming that farm-households do not switch categories in response to small changes in price, the "wedge model" can also be used to determine the composition between hired and family labor, as well as labor demand. Using simulations based on African parameters, de Janvry, Fafchamps and Sadoulet (1991) estimate the demand elasticity of leisure with respect to output price to be 0.27 for market-oriented farm households but only 0.06 for self-sufficient farmers. The elasticity of demand for hired labor was 0.61 for market-oriented farm-households and zero (by definition) for the self-sufficient group. In situations where a substantial proportion of farm households are self sufficient, these results help to reconcile observations of non-responsiveness with the assumption of (shadow) profit maximization.

Several factors contribute to the size of unit transaction costs (e.g., per worker/day). Employers pay labor turnover costs, including search, negotiation, and training. They pay implicit or explicit supervision costs and costs of the residual shirking that remains. Workers pay for the journey to work and some of the costs of dedicated tools and equipment. As with commodity taxation, it is not the nominal incidence that matters but the fact that a wedge is driven between price paid and price received. From the perspective of The New Institutional Economics, the organization of farm production will be largely determined by an effort to economize on transaction costs, e.g., the institutions of long-term contracts and the occasional provision of on-farm worker residences. However, transaction cost economics has not advanced to the point of measuring transaction costs and incorporating those measures into a fully specified model (see below for a discussion of pioneering efforts).

In the remainder of Section 2, we review some of the issues addressed in the labor and development economics literature regarding agricultural employment. These include the efficiency of hired labor, the separability and substitutability of hired and family labor, the form of labor contracts, efficiency wages and the choice of permanent vs casual labor. Many of these studies are motivated by policy questions. If patterns of agricultural employment reveal substantial inefficiency, this is taken as evidence that policy reforms are warranted. We shall see, however, that the studies do not always make appropriate use of transaction cost economics. A preliminary extension to non-substitutable labor is discussed in Section 3.

2.2. *Farm size, transaction cost, and efficiency*

We begin with the question of whether hired labor is inefficient relative to family labor. A number of studies have suggested that it is.[1] Utilizing family labor economizes on

[1] See, e.g., Binswanger and Rozenweig (1986), Binswanger, Deininger and Feder (1995).

recruiting and supervision costs, the latter because family labor stands to lose from both quality and effort shirking. These labor market imperfections result in the productive superiority of family farms [Deininger (2003, p. 84)] and to the characterization of hired labor as inefficient [Otsuka (2007, this volume)]. In contrast, Benjamin (1992) finds that hired labor is neither significantly more nor less productive than family labor.

The empirical case for inefficiency rests largely on the notorious inverse relationship between size and productivity [Berry and Cline (1979)]. Recent evidence is mixed, however. Some studies confirm the inverse relationship [e.g., Burgess (2001); Udry (1997)]. Others fail to reject constant returns to scale [e.g., Dow and Putterman (2000); Wan and Cheung (2001)].

But the inverse relationship is also consistent with the efficient allocation and employment of labor. First-best efficiency predicts that landlords will equate the marginal product of labor across diverse land qualities by adjusting the size of family farms thus leading to the observation of higher per hectare yields on smaller farms [e.g., Roumasset and James (1979)]. Indeed, Benjamin (1995) shows that the inverse relationship is at least partly due to the bias induced by omitting land quality from the regressions.[2] Deininger asserts, however, that the inverse relationship persists even after controlling for land quality with proxies such as land value. But land value is not an accurate indicator of land's potential agricultural productivity, nor is distance-to-market and other proxies. Lacking a perfect measure, one cannot confidently reject the hypothesis that the inverse relationship is due to land quality nor conclude that the relationship implies higher productivity of small-farm labor.

A second-best efficiency explanation for the inverse relationship is that, from Figure 1, the shadow price of labor for farm households that hire labor at the margin is higher than that for households who supply all of the farm labor, especially so for households who supply labor to other farms as well as their own [Sah (1986)]. To the extent that the inverse relationship is sourced in this cause, no inefficiency is indicated. In the second-best equilibrium, shadow prices vary over space, time, and economic agents. Using a first-best standard of efficiency risks drawing policy implications that have efficiency-decreasing consequences.

Future documentation of the inverse relationship should distinguish between family and commercial farms. Feder (1985), Eswaran and Kotwal (1986), Carter and Wiebe (1990), and Deininger (2003) discuss the possibility that the inverse relationship could reverse for larger farms, noting that their disadvantages in the labor market could be outweighed by their advantages in credit and other markets. Indeed, Uy (1979) finds an inverse relationship on family farms but a positive relationship between productivity and farm size on commercial farms.

Hopefully, future studies of agricultural employment patterns will shift from the indirect uses of transaction costs discussed above to empirical studies that directly estimate transaction cost models of both farm behavior and organization. There is a paucity of

[2] See also Roumasset (1976, Ch. 4).

studies that do so. For example, Frisvold (1994) includes hours of family supervision as a measure of supervision cost and controls for other sources of productivity differences, concluding that family labor is indeed more productive than hired labor, even before deducting the costs of supervision.[3] A related question concerns the possible tradeoff between the quantity of supervision and the wage rate. A handful of results in both developed and developing countries are inconclusive and give contradictory results, however [Rebitzer (1995); Ewing and Payne (1999); Neal (1993); Kruse (1992)].

The apparent inconsistency of results may be rooted in measurement problems. For example, Evenson, Kimhi and DeSilva (2000) and DeSilva, Evenson and Kimhi (2002), following the urban-industrial hypothesis of Schultz (1951), estimate a measure of supervision intensity (number of direct supervision hours divided by hours of hired work) on Philippine farms as a function of distance to market and other proxies for unit transaction costs. Relatedly, Vakis, Sadoulet and de Janvry (2003, Peru) consider four types of transaction costs in output markets transportation, information and search, bargaining, and monitoring and enforcement costs and then use proxy variables for each. For example, the information and search cost is proxied by the selling time.

While these studies make a start at the important business of measuring transaction costs, there is an apparent need for conceptual clarity regarding the theoretical construct and an appropriately corresponding metric of transaction costs. An interesting question, for example, is whether recruiting and supervision costs are substitutes. To the extent that market institutions are well developed, it is possible to recruit workers who are less prone to shirking, so that the cost of recruiting a worker with a given propensity to shirk is less. To test this hypothesis, one might estimate the hours of family supervision as a function of recruiting cost and control for other sources of farm-specific productivity differences.

In order for the empirical literature on hired labor to progress, two improvements are needed. First, the different types of transaction costs must be distinguished. Transaction costs have been defined by Nobel Laureate Kenneth Arrow as costs of running the economic system and are the economic equivalent of friction in physical systems [Williamson (1985)]. The primary category of transaction costs is contracting costs, including the costs of participant-selection, negotiation, and enforcement. Lower costs of transportation, communication and institutional innovations that lower enforcement costs facilitate falling *unit transaction costs* per worker. But as intensification and specialization increase, for example, as the number of workers per hectare rises, *transaction expenditures* increase, even as unit transaction costs fall. Making this distinction is essential for future empirical work.

The second needed improvement is to recognize that the choice of hired vs family labor is endogenous and that the two kinds of labor will naturally differ in both tasks and skills. In the simple version of the wedge model portrayed in Figure 1, household

[3] Although his study uses a direct measure of supervision cost, supervision time, it would be useful in future studies to include task-specific information about levels and modes of supervision.

and hired labor are assumed to be perfect substitutes, and labor is hired because of the rising opportunity cost of household labor. An additional reason for hiring labor is that it facilitates specialization such as teams of workers that move from farm-to-farm doing the same task.[4] On the prototypical farm in which both family and hired labor are employed, economics implies that there will be a non-random division of tasks between family and hired labor according to the comparative advantages of each.

The perspicacious reader may notice a family resemblance between the enterprise of testing whether share tenancy is significantly less efficient than wage and/or rent contracts and that of testing the relative efficiency of family and hired labor. Both are reminiscent of the old structure, conduct, performance paradigm in which structure was implicitly assumed to be exogenous so that one could compare the relative conduct and performance of alternative organizational forms. This paradigm was replaced by contestable market theory and other innovations, including The New Institutional Economics, wherein organizational structure is treated as endogenous. The prospects of meaningful empirical work on labor productivity in developing agriculture likewise await the development of an appropriate structural model wherein farm organization, specialization between family and hired labor, and choice of contracts across tasks and economic environment are understood as parts of an endogenous whole. We return to this theme in Section 3.

2.3. *Separability and substitutability*

In the presence of positive transactions costs, household labor supply and household-farm labor demand are not fully separable. There is also no reason to suppose that household and farm labor are equally productive in any and all tasks. But because these simplifying assumptions facilitate tractability, there is a substantial literature on testing for these conditions. For example, Lopez (1986) categorically rejects the "hypothesis" that production and consumption decisions are independent and finds that "important gains in explanatory power result from estimating the consumption and production sectors jointly", including the estimation of labor-supply elasticity.

In contrast, Benjamin (1992, India) tests whether farm employment is uncorrelated with household structure. Under the null hypothesis of separation, farmers choose labor demand by equating the value of labor's marginal product to the market wage. The alternative hypothesis is that labor is supplied by the household and demanded by the farm according to a shadow wage, which is a function of household composition. Benjamin could not reject the null hypothesis.[5] In contrast, Shapiro (1990, Zaire) and Gavian and Fafchamps (1996, Nigeria) reject separability. Shapiro (1990) tests whether land cultivated per household worker is affected by household size and composition, while Gavian and Fafchamps (1996) test whether yield is affected by household manpower.

[4] See, e.g., Roumasset and Uy (1980).

[5] The effectiveness of Benjamin's test is limited, however. 40% of the households in the SUSENAS data set are landless and are thereby subject to a constraint, which may or may not be binding, not to hire labor.

Singh, Squire and Strauss (1986) assume separability between output and food consumption and review estimates of labor-supply elasticity with respect to the price of the farm output. They find significantly negative labor-supply elasticities for Asia but roughly zero for Africa. This is consistent with the assumption that labor supply and demand decisions are interdependent for Sub-Saharan African households but quasi-separable for Asia in the sense that labor demand only affects labor supply via the income effect on leisure.

Even the Asian results do not imply that the "wedge" model can be rejected, however. It may simply be the case that within the range of agricultural price variation considered, most farm-households did not switch from one category to another (hiring-in to self-sufficient or vice versa). That is, since shadow wages remained unchanged over the range of output prices considered, labor supply and demand were largely independent for that particular exercise.[6]

Thus, the appropriate interpretation of Benjamin's (1992) and Singh, Squire and Strauss's (1986) non-rejection of separability is that separability is not necessarily a bad approximation. However, the choice of whether or not to abstract from transaction costs depends largely on the question at hand, i.e. the extent to which the assumption of full separability, while false in the sense of Friedman (1953), is nonetheless illuminating. For example, Evenson and Roumasset (1986) find that farm-households that hire workers have a higher demand for children *ceteris paribus* relative to households who supply labor to other farms. The significant difference in fertility across the two groups provides indirect support for the assumption of non-separability.

The studies just mentioned implicitly assume away differences in inherent productivity in order to focus on separability. Pitt and Rosenzweig (1986, Indonesia) test the more heroic assumption that hired and household labor are perfect substitutes, i.e. whether there is separability *and* if there are no inherent differences in productivities across tasks. Their test is based on the effect of illness on farm profit. If labor time in efficiency units can be hired at constant cost per unit to perfectly substitute for changes in the farmer's labor supply, then farm profits will be independent of the farmer's health status although household income decreases. The results suggest that, although the illness of either spouse significantly decreases household labor supply, there is little effect on farm profits after accounting for opportunity costs. They conclude that hired and family labor *are* perfect substitutes.[7] On the other hand, Deolalikar and Vijverberg (1983, 1987) reject the perfect substitutability hypothesis. Their result is somewhat clouded by the confounding difficulty posed by seasonal variations in labor productivity. Marginal productivity of labor might be higher for peak-season, where hired labor dominates, than

[6] Carter and Yao (2002) also argue that Benjamin's test of "global" separability is only appropriate if one of factor market is completely absent and all households are constrained by their absence. However, if transaction costs differentially constrain some, but not all households, then his test is inappropriate. Local separability tests are preferred in such cases.

[7] This result is contingent on a reliable estimate of opportunity cost, however. If, as other studies suggest, there are substantial transaction costs associated with hired labor and if adult household labor is inherently more productive, opportunity cost is extremely difficult to measure.

for slack season tasks that are more often performed by family members. As Deolalikar (1988) suggests, labor productivity differentials may also be nutrition-based. If farm household members are better fed than hired laborers, this may render them more productive. More importantly for Section 3 below, household members may specialize in different tasks and have correspondingly different skills.

In any case, non-rejection of perfect substitutability only establishes that perfect substitutability is not necessarily a bad approximation for all applications and problems. The researcher is left to judge whether this abstraction is suitable for the problem at hand. Substitutability would seem to be unsuitable, however, for illuminating the division of labor, the centerpiece of economic development from Adam Smith to Xiaokai Yang.[8] In the early stages of development, hired labor tends to specializes in tasks that are routine, standardized, and arduous.[9] Farm operators specialize in tasks where labor and management are jointly provided [e.g., supervising hired labor, applying fertilizer, diagnosing and treating pest problems; Eswaran and Kotwal (1985a), Hayami and Kikuchi (1999)]. Labor is hired, after all, precisely because its cost is less than the lowest shadow price for a household member performing the task in question with equal effectiveness.

As development proceeds and the division of labor increases, so does the accumulation of human capital. The marginal product of human capital accumulation in routine tasks diminishes more rapidly than that of discretionary managerial tasks, however, leading to the hypothesis that the differential between hired wages and implicit operator wages increases with modernization.[10] The difference is less to the extent that hired labor also becomes increasingly skilled with modernization, especially where it is used to operate machinery and other farm equipment. Indeed, the wedge model suggests that quality differences between household and hired labor will vary with cross-sectional differences such as land quality and that the quality differential will change over time, depending on what is changing the composition between family and hired labor. For example, an increasing opportunity cost of family labor may have one implication and institutional innovations in the organization of hired labor may have another. We revisit this theme briefly in Section 3.

Indeed, Jacoby (1993, Peru), Skoufias (1994, India), Sonoda and Maruyama (1999, Japan), and Abdulai and Regmi (2000) find evidence of non-substitutability. They estimate farm-household supply and demand for labor and derive the shadow price of labor from its marginal product. The result is that shadow price of household labor is roughly twice that of hired labor. Presumably, this is due to both skill differences *and* transaction cost considerations.[11] Neither explanation justifies the conclusion that hired labor is less efficient than household labor.

[8] In particular, see Yang (2003).

[9] See, e.g., Roumasset and Smith (1981).

[10] Schaffner (2001).

[11] For a simple version of the wedge model with skill differences, see Roumasset and Smith (1981).

2.4. *Share, piece-rate and wage contracts*

In the incomplete-contract theory of the firm, capital typically hires labor, instead of the other way around, because of the high agency costs of renting capital equipment [Putterman (1984); Hart (1988)]. In agriculture, "land hires labor" for similar reasons. More accurately, the owner of land and its capital improvements hires some combination of management and labor (or neither one). Presumably because of difficulties in modeling and measuring management, the literature has focused on hired labor. What are the alternative contractual forms for hiring labor and what are their implications for efficiency? Contract dimensions include the average levels of wages and their seasonal variations, basis of payment (piece rate, share, time rate), means of payment (cash, food, crop land, housing, protection, etc.), implicit length (casual/day, season, year, longer-term/lifetime), tasks, and the nature of supervision. There has been particular emphasis on the basis of payment. As we shall see, however, this has resulted in a confounding of hired labor contracts in isolation and the more complete view of the nature of farm organization.

The economics of agricultural labor contracts has evolved from the theory of share tenancy, which has itself been (misleadingly) portrayed as a labor contract. Both empirical evidence and theory are often used in support of the view that share tenancy is inefficient. For example, Shaban (1987) finds that owner-managed plots are more productive than sharecropped plots on the same farm.[12] Stiglitz (1994, 2002) declares that share tenancy is as distortionary as a 50% tax on income and would not exist were it not for extreme asset inequality.

The standard theoretical model assumes worker's income to be a linear function of the worker's share of output, i.e. $Y = \alpha + \beta Q$. Three particular contracts are typically the focus of interest. A standard wage contract, wherein $\alpha = wL$ and $\beta = 0$, has the disadvantage of labor shirking but the advantage that risk is borne by the landlord. If $\alpha < 0$ and $\beta = 0$, payment is representative of a fixed-lease contract. This has the advantage of avoiding labor-shirking, but, assuming that the tenant is more risk-averse than the landlord, misallocates risk-bearing. If $\alpha = 0$ and $0 < \beta < 1$, say $1/2$, the payment represents a share tenancy contract. If both labor shirking and risk-bearing are serious problems, share tenancy can emerge as the optimal compromise – the cost of risk-bearing under share contracting is less than half of that under wage contracting, and the agency costs of labor shirking are less than half of what they would be under the fixed-lease arrangement [Stiglitz (1974); Hayami and Otsuka (1993)].

There are several problems with this theory, however. First, the model does not imply, as claimed by Stiglitz (1974), that the optimal share, β, varies positively with the tenant's degree of risk aversion. Stiglitz implicitly uses a measure of risk aversion at the

[12] Efficiency-based explanations of this finding have apparently not been considered. First, there are substantial differences in moisture stress and other elements of land quality within Asian farms [Roumasset (1976)]. Second, in the presence of transaction costs, the average product of labor may be higher on owner-managed plots even though the net marginal products are the same on owner-managed and sharecropped plots.

margin instead of an index of the degree to which the tenant's utility function displays risk aversion. Moreover, the canonical theory does not allow risk aversion to induce an increase in the effort expended on risk management activities.[13] Even when these extensions are allowed, the labor-shirking versus risk-bearing model is incapable of explaining the empirical distributions of tenant shares, which cluster around 50%, with a smaller cluster around 2/3.[14] But the larger problem is that the theory fails to recognize the nature of share tenancy, a typically long-term contractual arrangement for bringing management together with land and that facilitates the tenant's learning-by-doing about production decisions [Reid (1976); Murrel (1983); Eswaran and Kotwal (1985a); Roumasset (1995)]. Share tenants themselves hire substantial amounts of labor, especially for the more arduous and routine tasks. Modeling share tenancy as an alternative to wages thereby abstracts from its essence.

To the extent that the choice of labor contracts has been mischaracterized, a productive research area involves the more substantive documentation of their nature. What are the alternative contractual arrangements whereby the farm manager, be he owner or tenant, hires labor? Workers may be paid according to time, according to work accomplished (piece rates) or as a share of the harvest. Piece rates are commonly used in situations where the product of labor is easily observable, for example, sizing and sharpening the cane stalks prior to planting, and the planting of stalks at uniform spacing. These tasks are tantamount to intermediate products delivered to the farm operator, who pays according to quantity. This institution economizes on minimum agency cost, i.e. the minimum sum of supervision cost and minimum shirking cost. For tasks that are not amenable to ex post inspection, supervision is used to concurrently monitor the labor activity in question and workers are paid according to the time spent on an activity, not its result. Figure 2 illustrates how agency theory can be used to explain the distribution of contracts according to the nature of the task.

Whereas the wedge model takes unit transaction costs as given and examines the implications for shadow prices and farm-household decisions regarding inputs and outputs, agency theory leaves these costs endogenous in order to focus on organizational form. This allows comparisons of the ability of alternative organizational forms to minimize agency costs in particular situations. Panels one and two of Figure 2 show that piece rates may dominate wages for tasks wherein labor produces an observable product such as transplanted rice. The lower two panels show the opposite tendency when the immediate results of the task are not readily observable, for example, application of fertilizer. Together, the four panels illustrate the comparative-statics proposition that if tasks are sufficiently easy to monitor through *ex post* inspection then the corresponding

[13] See, e.g., Deweaver and Roumasset (2002) for further discussion.

[14] Deweaver and Roumasset (2002) show that, for parameters representative of the Philippine case, the model predicts that optimal tenant's share declines from one to 80% as the tenant goes from risk neutrality to moderate risk aversion and increases back to one as risk aversion increases further.

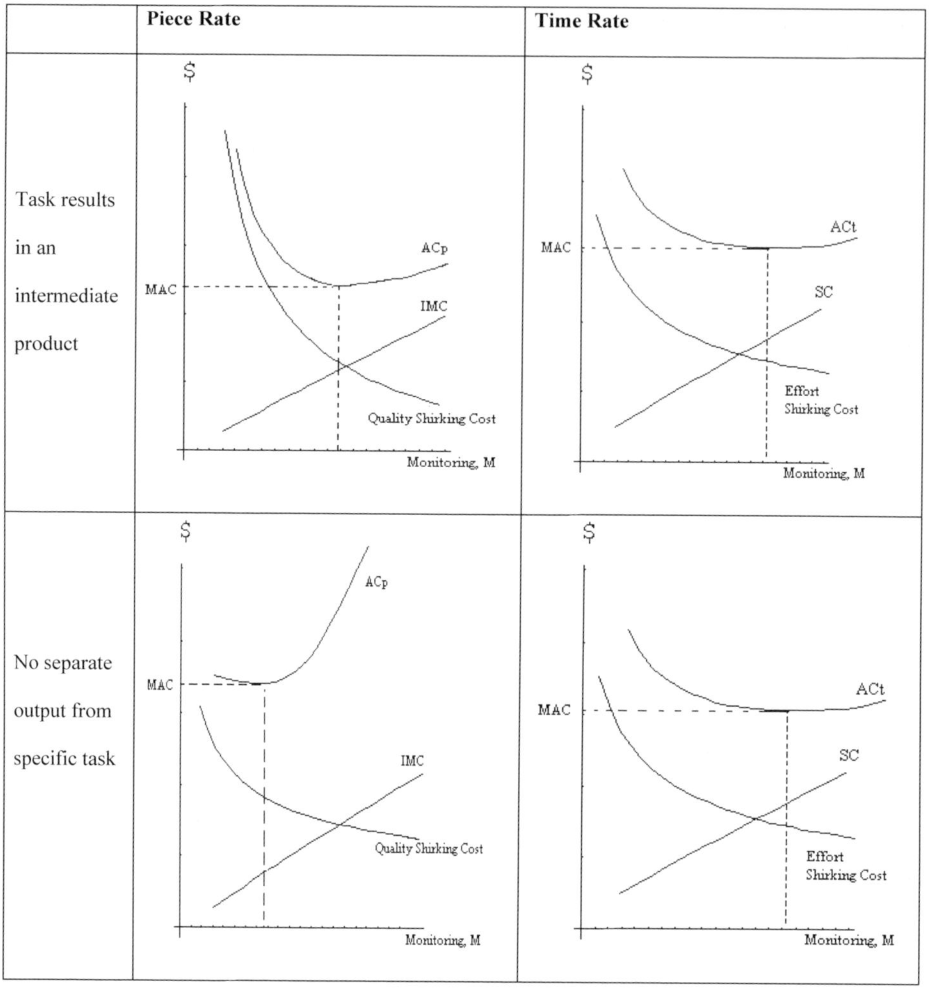

Figure 2. Specialization of contracts by task. IMC: information and monitoring cost (change SC to IMC); MAC: minimum agency cost.

agency cost at optimal monitoring will be lower than the agency cost of wage contracts. The opposite is true for tasks that are hard to monitor.[15]

It is often assumed that another advantage of piece rates is that it obviates the need to estimate a worker's productivity and adjust compensation accordingly. This is not the case, however, if productivity per unit of output also varies. For example, in Chilean

[15] Statistical analysis of sugarcane contracts in the Philippines confirms this tendency [Roumasset and Uy (1980)]. See also Pencavel (1977).

grape harvesting, more productive workers are paid higher piece rates [Newman and Jarvis (2000)].

Agency theory is also a useful tool for understanding share contracting, but first we must distinguish the alternative forms thereof. We first consider *task share-contracting* wherein the workers are paid a share of the harvest for performing a set of tasks. This is called *ceblokan* in Indonesia and *gama* in the Philippines [Hayami and Kikuchi (1982)]. Originally workers were contracted to do only harvesting. But as real wages fell, labor tasks were added, instead of the share declining. Ceblokan workers eventually did harvesting and threshing, but weeding and some of the land preparation as well. Shares remained constant, typically at 1/6. Land preparation and weeding are both somewhat observable as intermediate products, albeit imperfectly so, due to possibilities of quality shirking, and limited in-task monitoring may be used to supplement *ex post* inspection. Task share-contracting is, accordingly, an intermediate institution between piece rates and time rates.[16]

Sharecropping in the postbellum American South may be thought of as itself intermediate between *share tenancy*, wherein the tenant does most of the management as well as much of the labor, and task share-contracting. Cotton farming one year was largely a repetition of cotton farming the year before and the little management that needed to be done was largely provided by the landlord [e.g., Reid (1973)]. The institution of sharecropping thus blurs the distinction between hired labor and tenancy and illustrates the difficulty of separating labor contracts from the organization of farm production more generally.[17] The essence of incomplete contract theory [Hart (1988); Hart and Moore (1990)] is that the internal organization of the firm can be viewed as a process of incentivizing some agents with a program of evaluation and rewards/penalties and others with a share of the residual (or some combination). It is therefore somewhat misleading to compare alternative schemes for paying one type of agent in isolation from others, inasmuch as paying one agent a residual detracts from the amount that can be paid to another.

Incomplete contract theory can be usefully applied to the organization of labor within the agricultural firm. There are two kinds of farm workers. Casual workers are hired primarily on a task-by-task basis. Because the product of their work is verifiable, they are often paid piece rates. In effect, they supply intermediate inputs (e.g., by converting sugar cane stalks into sharpened stalks ready for planting), and are viewed accordingly as being outside of the firm. Permanent workers work on multiple tasks throughout the growing season. They perform non-verifiable tasks and are indirectly incentivized by either a share of the residual or by wage premia and repeat contracts. They, and the worker-manager tenant, are part of the agricultural firm. Share workers who perform

[16] The more standard view that share contracting is a kind of piece rate [e.g., Otsuka (2007, this volume)] does not distinguish between paying according to the intermediate or final product, nor between share tenants and workers.

[17] Ergo the titles "The nature of the agricultural firm" [Roumasset (1995)] and *The Nature of the Farm* [Allen and Lueck (2002)].

multiple tasks represent somewhat of an intermediate case, but by the criteria just given can be considered inside the firm.

The division of labor in the agricultural firm is warranted by the extent of intensification, which in turn is warranted by factors such as land quality. Where specialization is not warranted, agency costs may be minimized by paying the worker (sharecropper) a share of the residual. In the case of share tenancy, the residual is going for management (supervision and decision-making) as well as labor in management intensive tasks (e.g., the application of chemicals). Where even more specialization is warranted, the owner may take the entire residual, sometimes hiring both management and labor [Roumasset (1995)]. We return to the intensification-specialization theme in Section 3.2.

From the perspective of agency theory, it is therefore meaningless to estimate the relative efficiency of labor contracts or forms of tenure, although some studies attempt to do so.[18] Unless government policy renders some contracts illegal, piece rate, time rate, and share contracts will tend to be chosen for those tasks in which they are most efficient.

2.5. *Efficiency wages*

In most of the agricultural contracts literature, the participation constraint is given by the spot market wage and assumed to be binding [e.g., Newbery and Stiglitz (1979); Braverman and Stiglitz (1982); Mitra (1983)]. This seemingly harmless assumption renders labor contracts such as share tenancy twice damned. Not only are these contracts inefficient,[19] they are inequitable as well. The entire economic surplus is appropriated by the landlord, and the tenant is left no better off than landless workers.

The literature on efficiency wages provides reasons for participation constraints to be non-binding, however. According to this theory, employers operate more efficiently if wages are above the equilibrium level, even in the presence of surplus labor. Higher wages attract higher quality workers and lower worker turnover and shirking. In the context of low-income agriculture, efficiency wage theory is also used to focus on the link between wages and worker health. Better-paid workers eat a more nutritious diet and are consequently healthier and more productive. Thus, an employer may find it more profitable to pay workers higher than reservation wages in order for them to increase their productivity. Giving workers a meal, however, lowers and may eliminate the need for a worker's valuation of the meal combined with his monetary wage to exceed the reservation utility level. Because employers may provide meals with much lower cost than market price, employers have an additional incentive to provide meals if they can shift the cost to wages.

[18] See, e.g., Bell (1977), Shaban (1987), and Binswanger, Deininger and Feder (1995).

[19] According to Stiglitz (1994), share, wage, and rent contracts are all "constrained Pareto inefficient", in the sense that the inefficiency lost from trading-off labor-shirking against risk-bearing could be avoided by land-to-the-tiller reform [Greenwald and Stiglitz (1986)].

To investigate the relationship between wages, nutrition, and productivity, assume that working hours are determined by household utility maximization, but that effort is a function of individual nutrient intake. The production function is given by $X = F(Nh\lambda(c))$, where λ represents effort, N is the number of workers, h is hours worked, and c is calorie intake. The first derivatives of λ and F are both positive, and the first order condition of profit maximization is $\lambda/w = \lambda'$. Mirrless (1975) and Stiglitz (1976) assume that all calories are proportional to wages such that wages can be substituted for c in the formulation above. Now it may be worthwhile for the employer to pay more than the worker's reservation wage to enhance workers performance. The employer can also increase the proportion of wages that are spent on the workers' food. The employer's problem is to select a point on the worker's reservation indifference curve (between money and food) to maximize profits. In this case of undifferentiated workers, the participation constraint is binding, and unemployment is not a consequence of, say, making a daily meal part of the compensation package. Indeed household calorie availability has a strong impact on output and wages [Strauss (1986); Sahn and Alderman (1988); Deolalikar (1988); Bhargava (1997); Croppenstedt and Muller (2000); Huffman and Orazem (2007, this volume)]; labor force participation is significantly lower among those in poor health [e.g., Lavy, Palumbo and Stern (1995); Schonenbaum (1995); Dow et al. (1997)]; and people in better health are more likely to undertake strenuous tasks [Pitt, Rosenzweig and Hassan (1990); Bhargava (1997); Thomas and Strauss (1997)].

On the other hand, Dasgupta and Ray (1986) show that, if worker productivity is dependent upon nutrition, a separating equilibrium may be obtained wherein only wealthier workers are employed, i.e. that unemployment is not inconsistent with competitive markets. These findings have been used to support the argument, similar to that of Stiglitz, *supra* note 15, that land reform would not only improve income distribution, but would enhance productivity. This is "blackboard economics" [Coase (1988)], however, since the model was neither motivated by nor has been used to explain actual patterns of employment.[20] The policy inferences concerning land reform are premature without a fuller understanding of the determinants of rural organization. Moreover, assuming that government can engage in wealth transfers without injury to efficiency or justice ignores both economics[21] and philosophy,[22] and commits Nirvana Fallacy [Demsetz (1969)] as well.

2.6. *Casual vs permanent workers*

Another application of efficiency wages concerns the distinction between tied and casual workers. In Eswaran and Kotwal (1985b), the growing season is modeled as

[20] "An inspired theoretician might do as well without empirical work, but my own feeling is that inspiration is more likely to come from the patterns . . . of data" [Coase (1994, p. 13)].

[21] Redistributive lump-sum transfers are not feasible [Samuelson (1947)].

[22] Confiscation of property, even with partial compensation, violates rectificatory justice [Rawls (1971)].

consisting of two parts. In the first stage, labor tasks require discretion and are inherently difficult to monitor (e.g., crop choice, planting date, water management, and fertilization). Work in the second period is easier to monitor (e.g., weeding, harvesting, and threshing) and is accordingly contracted out to casual workers. The employer can infer the performance of first period worker by observing the output at the end of period 2. The employer incentivizes the first period worker by paying something above the participation constraint. This provides a meaningful and credible threat that an underperforming worker will be terminated.[23] This model begs many questions, including the accuracy of the performance estimate, how the difficulties of contracting around a potentially subjective estimate can be surmounted, and bargaining problems that may arise as a result of learning-by-doing on the part of the worker. Also, the model seems to imply an absentee landowner. Otherwise, the landowner or the farm operator/tenant would have the option of performing many of the discretionary tasks himself.

While Eswaran and Kotwal (1985b) focus on labor shirking and incentive problems, Bardhan (1983) considers transaction costs and risk allocation between employers and employees. The employer hires some tied laborers for both seasons, paying them above their marginal product in the lean season in exchange for their commitment of labor supply during the peak season. If the peak season labor demand is higher than what is supplied by tied laborers, the employer then enters the casual market, hiring the additional labor at the spot wage rate revealed at that time. Employers conserve on transaction costs and pay less during the peak season than the expected spot wage. Risk-averse workers avoid a low or zero income in the slack season. Anderson (1993) and Schaffner (1995) develop similar models showing that "attached" labor in Brazil and even seemingly serfdom in feudal systems are consistent with voluntary choice. These competitive models show that as labor markets tighten during the process of economic development, the variation between expected slack and peak wages may increase thus leading to an increased incidence of permanent labor without the contrivance of feudal exploitation. Employers have an interest in reducing discounted labor costs by providing consumption smoothing across seasons to their employees, and under certain assumptions, their incentive to do so will increase with the level of wages. It would be natural to expect that as economic development continues that the inter-seasonal variation between wage rates declines again and that other institutions for consumption-smoothing would become available. Both of these would lead to a Kuznets curve phenomenon for permanent workers.

Summarizing Section 2, the conventional wisdom is that labor is linked to land by inefficient mechanisms. Hired labor is thought to be less efficient than household labor resulting in small family farms being more efficient than commercial farms. Even in the modern economics of imperfect information, labor contracts are thought to be only "pairwise efficient", but socially inefficient and exploitative. A commonly inferred policy recommendation is that land reform is needed such that a much higher percentage of

[23] See the discussion in Bardhan and Udry (1999, Ch. 4).

agricultural labor is provided by households that own land. Thus the predominant view of the 1970s remains largely popular in current times. The earlier view puts excessive reliance on the second welfare theorem. More recent literature shows that land reform is far from being a lump-sum redistribution and is likely to engender severe moral hazard and rent-seeking. This has led to suggestions for more "market-friendly reform", although these remain somewhat amorphous. At the same time, the theory supporting the conclusion of social inefficiency fails to embed the theory of labor contracts in the internal organization of the firm and fails to allow for governance mechanisms to mitigate against labor shirking. As a consequence, empirical estimates are based on incomplete structural foundations.

As Yoram Ben-Porath once observed, economic knowledge proceeds from a dialectic feedback between theory and evidence. Theory often begins as an explanation of stylized facts. The resulting theoretical structure disciplines empirical investigation, leading to new patterns lead, in turn, to advances in the theory. And so it goes. Looking forward, then, new theories will hopefully overcome the limitations reviewed above and to explain the role of agricultural organization in facilitating the division of labor and the corresponding role of human capital in skill acquisition and differentiation. They will allow for specialization between different participants in the agricultural firm, will not require a sharp distinction between discretionary and non-discretionary tasks, and will allows for a richer tapestry of contracting. Some suggestions along these lines are offered in Section 3.

3. Extensions: Toward a co-evolutionary view of agricultural organization

We have seen that a policy-informative understanding of on-farm labor allocation and contracting needs to be based in a more general understanding of the agricultural firm. In this section, we advance some preliminaries regarding the co-evolution of on-farm specialization, production intensification, and labor contracts. The central hypothesis is that the intensification of production warranted by relative price changes and increased productivity in turn warrants greater specialization of agricultural production. Labor contracts and other organizational arrangements can be understood as the institutions that facilitate the warranted specialization.

3.1. Interdependency of markets, contracts, and farm-household organization

At the outset, it may be useful to acknowledge desirable attributes of a model of labor markets and agricultural contracts. First, labor contracts and the division of labor must be made consistent with rational behavior of the household-farm unit, for example, by extending the wedge model to allow for specialization. The decisions of how much labor to utilize in farm production, the composition between household and hired labor, and the contracts to be used in various tasks are interdependent (the composition depends on the size of the wedge which depends on the contract).

Second, we need a theory of those labor market conditions that the farm-household takes as exogenous. These include not only the market wage suitable for the time and place in question, but the labor institutions that are available. This can be accomplished by constructing the aggregate demand/supply profile for the relevant location, say a village. Just as the shadow wage of the individual household is determined by whether the household is a net exporter or importer of labor, so is the market wage of the village. For example if the village is a net exporter of labor to the town, then its market wage is given by the wage in the town minus the journey to work and analogous transaction costs. This market position may vary by season and type of labor (even in a competitive setting, a village may be a net importer of workers of one type and be an exporter of another).

3.2. The co-evolution of contracts, markets, and specialization

As noted in Section 2, one cannot meaningfully assess the efficiency of hired labor in absence of a theory of why labor is hired, the nature of specialization among hired workers and between household and hired workers, and the formation of human capital. We turn first to the classic question regarding the division of labor and the extent of the market.

Farm-household labor supply and demand decisions are conditioned by the nature of labor markets. The neoclassical theory of labor markets is of little help, focusing as it does on the polar extreme of what happens to the set when contracting costs are zero and the number of agents becomes large. Idiosyncrasy in contracting disappears in the model. Formally, the core shrinks to the competitive equilibrium as the number of agents goes to infinity. But the essence of economic development has to do with how the economy gets from the other polar extreme – self-sufficient farm-households – to a market economy. What are the intermediate forms? What drives institutional change from one form to another?

Notable exceptions that focus on the nature of economic organization as specialization proceeds have been collected in Buchanan and Yoon (1994); see especially the excerpt from Smith's *Wealth of Nations*, and the papers by Young (1928), Stigler (1951), Romer (1987), and Borland and Yang (1992). These studies extend and formalize Adam Smith's theory of specialization and the extent of the market. Not only is the division of labor within the firm warranted by higher firm output, but economic growth warrants both more final and intermediate products which in turn imply greater specialization. In Yang's formalization of the classical theory [see also Yang and Ng (1993); Yang (2003)], the division of labor evolves according to product specialization and through expansion in the number of both final and intermediate products. For example, as population grows, the size of the potential market lowers the average fixed cost of a firm specializing in a new intermediate product, eventually overcoming that barrier to specialization. Similarly, growth in the potential size of the market lowers the average fixed cost of specializing in one product and (later) increasing the number of products. Even with a fixed number of final products and only one intermediate product, the number

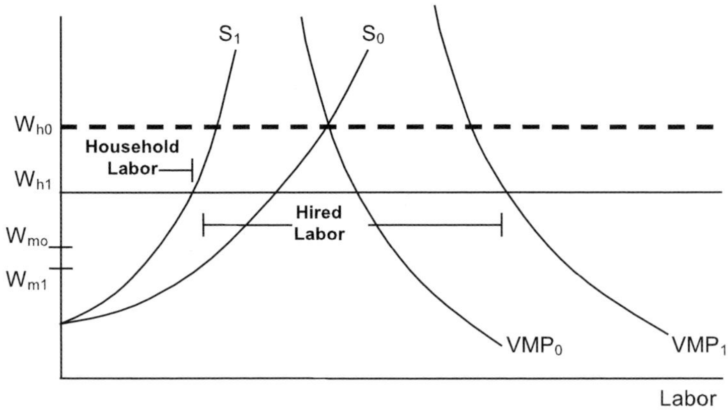

Figure 3. HYVs and the advent of labor markets.

of transactions increases from zero to the square of the number of final products as the economy proceeds from autarky to complete specialization [Yang (2003, p. 436)]. Adding intermediate products and more final products implies that transactions increase explosively with the size of the economy.[24] This provides a theoretical explanation for the observation that the transaction sector grows faster than the economy [North (1986); North and Wallis (1982)].[25]

One application of endogenous specialization theory concerns the emergence of hired labor. It has often been claimed that new technology such as new varieties of rice and wheat have resulted in the polarization of the peasantry and the emergence of a rural proletariat.[26] The new technology is said to have favored large commercial farms, to disenfranchise peasant agriculturalists, and to increase rents at the expense of the wage bill. This is reminiscent of the proposition that the modernization triad – commercialization, technological change, and population pressure – enrich capitalists and create an army of unemployed. By failing to recognize the endogeneity of commercialization and technological change, the underlying reasoning commits the fallacy of *post hoc ergo propter hoc*. The fact that wages apparently fell after, or in conjunction with, the adoption of new technology and the emergence of rural labor markets does not mean that falling wages were the result of those technological and institutional changes.

Figure 3 provides an alternative perspective, illustrating how the emergence of rural labor markets can be induced by land scarcity and the associated intensification of production. The graph shows how the dramatic increases in hired labor observed after the

[24] One theoretical limitation of the Yang framework is that it takes falling unit transaction costs as the engine of growth but leaves transaction costs as exogenously determined. Nonetheless, it may be a fruitful conceptual framework for explaining the division of labor in agriculture.

[25] See also the discussion in Yang (2003, p. 438) and Roumasset et al. (1995).

[26] See, e.g., the discussion in Hayami and Kikuchi (1982).

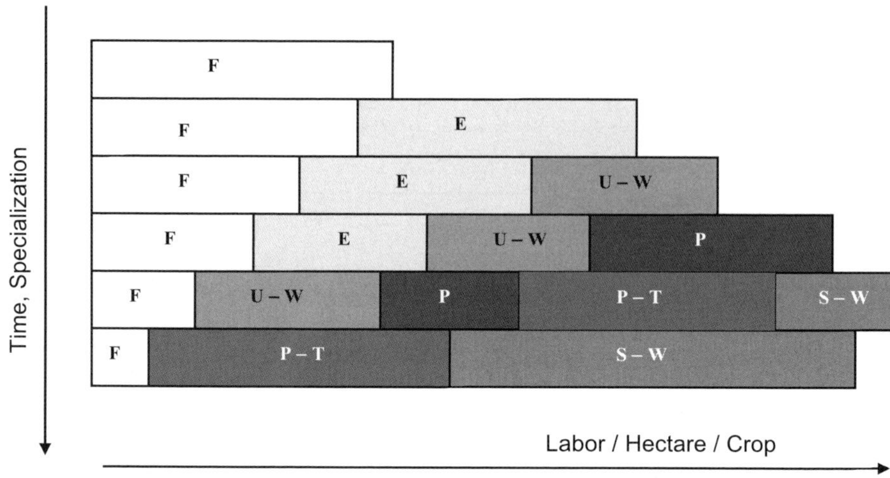

Figure 4. Evolution of labor contracts. *Notation*: F: family labor; E: exchange labor; U – W: undifferentiated wage labor; P: piece rate labor; P – T: piece rate with team labor; S – W: specialized wage labor.

introduction of high yielding varieties of rice and wheat were consistent with efficiently induced innovation in contrast to the Marxian proposition that modernization led to the immiserization of the peasantry. First, the intensification of production, ultimately caused by increasing land scarcity and accommodated by the new technology, increased the demand for labor per hectare. This is illustrated by the shift in the demand curve to the right. Secondly, increased farmer incomes resulted in increased schooling of farm children. This combined with the increased specialization among farm workers lowered the supply of farm-household labor per hectare. Thirdly, the market wage went down from W_{m0} to W_{m1} as population growth, including in-migration (e.g., into the irrigated rice areas of the Philippines and Indonesia) increased by more than enough to supply the increased labor demand. Fourthly, the transaction cost wedge between the market wage and the gross hiring wage decreased due to the advent of labor contractors and other new institutions of labor contracting that reduced agency costs per worker (see below). These third and fourth factors are illustrated by a downward shift in the gross hiring wage (from W_{h0} to W_{h1}). In the Philippine case, the combination of these four factors all reinforcing one another led to a tripling of hired labor in a mere 10 years, beginning in the late 1960s.[27]

The endogenous fall in the transaction cost wedge can be understood by studying the evolution of labor contracts that facilitate increasing specialization. Figure 4 provides a stylized evolutionary-pattern of labor contracts that facilitate an increase in both vertical

[27] For further details see Hayami and Kikuchi (1982), Roumasset and Uy (1980), and Roumasset and Smith (1981).

and horizontal specialization. During stage I, labor is provided by the farm household and exchange arrangements with residents in the same village. During stage II, the next three rows of Figure 4, hired labor emerges. At first labor is hired on a wage basis and workers are not differentiated with respect to task. As horizontal specialization increases, piece rate workers are hired for selected tasks (those which are relatively easy to monitor) and undifferentiated wage labor declines. The third phase of stage II involves a further decrease in undifferentiated wage labor, a decline in individually hired piece workers, and the advent of two new contracts. In *piece-rate with team* labor, the farm operator contracts with a labor contractor who also serves as team leader and supervisor. The other new form is for skilled laborers who specialize in particular tasks and were paid in wages. These new forms come to dominate the other forms of hired labor in stage III. Piece rate with teams continued to replace individual piece-rate contracts, and specialized wage labor replaced undifferentiated wage labor and most of household labor.[28] The last row of the labor-contract pyramid illustrates capital-for-labor substitution overcoming overall input intensification such that hired labor per hectare declines.

The explanation of the above dynamic pattern of labor contracts is similar to the agency theory explanation of the spectrum of agricultural firms ranked according to percentage of hired labor and other indicators of specialization.[29] In both cases, the objective is to explain a spectrum of contracts ranked according to specialization. In the cross-section case, the same preconditions for production intensity (e.g., land quality) also predispose a more specialized organizational form. As the profit maximizing level of inputs increases, more production management is warranted, indicating an organizational form wherein the manager is rewarded with a larger share of the residual. That is, the agency costs associated with shirking of non-labor inputs increase, moving toward better quality land, and those costs are best economized by supervising labor and incentivizing managers.

Figure 5 illustrates a generalized agency cost explanation of the labor-contract pyramid.[30] Agency costs are the sum of transaction expenditures that improve the quality of information and enforcement [e.g., monitoring and bonding ala Jensen and Mekling (1976)] and the residual shirking costs that still remain. In order to generalize the concept of agency costs, which are internal to the firm, to the entire economy, we may define transaction costs as the sum of agency costs and foregone gains from specialization, i.e. the difference between real income associated with a particular economic organization and that which is possible in the first-best solution with costless specialization.[31] Piece

[28] For statistical documentation and further discussion, see Setboonsarng (1991) and Roumasset et al. (1995).

[29] The spectrum proceeds from pure owner operator households to lease worker to various forms of share tenancy to lease-manager to owner-manager and finally to corporate agriculture with separation between ownership and management [Roumasset (1995)].

[30] See DeSilva, Evenson and Kimhi (2002) and Roumasset et al. (1995).

[31] This inclusive definition of transaction costs is consistent with the proposition that economic organization evolves so as to minimize transaction costs [see, e.g., Williamson (1985) and North (1986)]. It is also consistent with Arrow and Williamson definitions ("the costs of running the economic system" and "the equivalent of friction in physical systems").

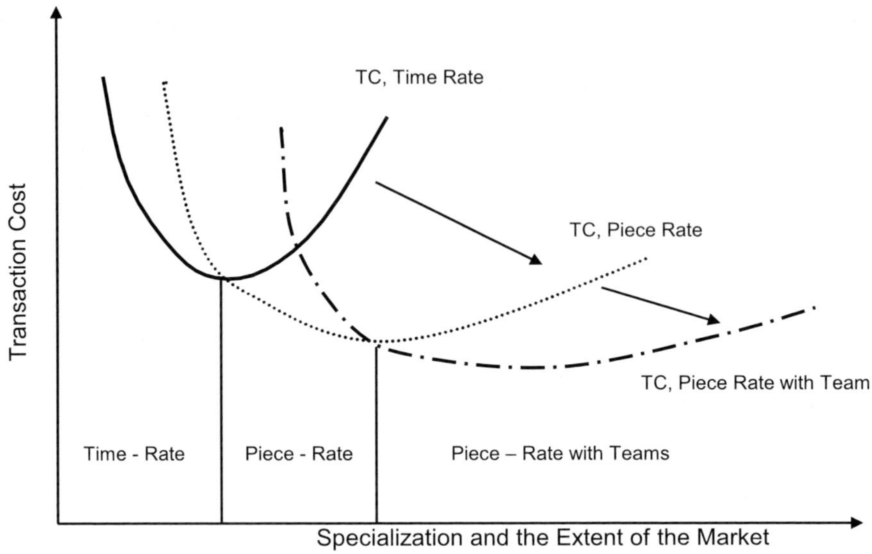

Figure 5. Efficiency explanation of labor contract evolution.

rate with teams facilitates the most horizontal specialization. The teams move from farm to farm and village to village doing only one task. Vertical specialization increases as well, with the team leader occupying a new level in the organizational hierarchy. However, this institution requires the greatest set-up costs with the team leader selecting, training and incentivizing team members. Undifferentiated wage labor facilitates the least specialization, with the worker "specializing" in all tasks not performed by household labor, primarily the farm operator. But the set-up costs are correspondingly low. The piece-rates arrangement without a separate team leader is intermediate in both regards.

When each of the agency cost curves is added to the same decreasing cost of foregone specialization curve, we get the transaction cost curves as shown and the corresponding lower envelope. As illustrated in the figure, wage labor is dominant at low levels of specialization. It leaves unexploited opportunities for specialization but economizes on set up costs. It is possible to hire specialized wage workers for different tasks but the training and other set up costs would be prohibitive unless the labor market were sufficiently extensive and integrated to spread the fixed costs among many employers. Accordingly, the transaction costs of wage labor are shown as steeply rising when used to facilitate specialization without deep markets. Even without full market development, piece rate with teams facilitates extensive horizontal specialization. Set-up costs are substantial but are specific to the operation in question and can be spread across many farms. Accordingly this contract is dominated at low levels of specialization but dominates other forms when the market is sufficiently extensive. Since the more specialized forms of labor organization require higher set up costs, there is the potential for induced institutional change to generate growth [see, e.g., Borland and Yang (1992)].[32]

As agriculture intensifies and labor markets become increasingly integrated, labor contracts become increasingly specialized, eventually with labor contracts made on a task-by-task basis. Thus intensification, the extent of the market and specialization are co-evolutionary. The diagram also helps to resolve the fundamental paradox that the transaction sector grows as economic development proceeds [North and Wallis (1982)]. Lower unit costs of transportation and communication (*unit transaction costs*), are falling and improved institutions lower the *agency costs* (supervision plus residual shirking costs) per unit of labor hired. But because more labor is hired and because specialization increases the number of contracts (even normalized by yield per hectare), *transaction expenditures* increase.

We have sketched the rudiments of a theory of how the division of labor and markets co-evolve. Two aspects of the theory remain to be developed. The first concerns an extended version of the agricultural household model wherein specialization between family and hired workers is recognized and where contracts facilitate said specialization. As hired labor becomes increasingly important (stage II of Figure 4), it specializes in tasks that are easy to monitor, e.g., through *ex post* inspection, and/or in tasks wherein learning-by-doing is task-specific, e.g., straight-row planting in rice. Family labor specializes in tasks wherein the agency costs of hired labor are high and/or in tasks wherein learning-by-doing is farm-specific, e.g., applying fertilizer according to the propensities toward moisture stress on different parts of the farm.[33] Another aspect concerns the duration of contracts. As labor markets tighten and the rule of law matures, the employment relationship becomes longer lived, thus facilitating investment in farm-specific human capital [Schaffner (2001)]. As economies enter the stage of modern economic growth [Kuznets (1966)], the falling relative price of capital induces capital-using and labor-saving technology. Hired labor may remain important but is of a different nature. Labor skills are increasingly farm-specific and involve working with capital equipment. Labor management becomes increasingly a problem of avoiding asset-abuse. Efficiency wages become more pervasive in order to increase the duration of the employment relation and incentivize firm-specific investment. Relatedly, the spread between wages of the highest paid employees and those of casual workers at the bottom of the wage ladder increases.[34]

Specialization may also evolve in different ways in different economies. For example in the U.S., the average farm of 161 acres may be managed with only family labor and a single permanent worker due to the greater extent of mechanization. In this case, there

[32] In the conventional framework [Binswanger and Ruttan (1978)], institutions induced by changing factor prices suffice only to render factor substitution more elastic in the long run. The non-convexity introduced through institutional set-up costs allows for growth by shifting the meta-production function outward.

[33] See Acemoglu, Antras and Helpman (2005) for a promising analytical approach, albeit in the context of industrial workers.

[34] Again, see Schaffner's (2001) comparison of workers in the U.S. and Columbia.

is a high degree of product specialization but learning-by-doing may have more to do with handling farm equipment and idiosyncratic farm characteristics. And large farms may be managed more like corporations with permanent workers specializing on tasks within the organization instead of selling their skills to multiple farms.

An expanded model of the household farm that incorporates both the transaction cost and specialization considerations above could be applied to the stylized facts of labor supply and demand at the micro level, including the role of human capital. For example, Huffman (1980)[35] shows that U.S. farmers with more education switched to off-farm work at a more rapid rate in response to a rising differential between skilled and unskilled wages in the non-farm sector. Yang (1997) finds a similar pattern for China. However, Rosenzweig (1980, India) finds that, for landholding households, schooling has a *negative* and significant impact on off-farm work of both males and females. The contrast presumably reflects the limited opportunities and high transaction costs associated with integrating into the modern sector of India at the time of the study and to the domination of the rural hired labor market by unskilled and casual employment. Where rural markets are more developed, the return to human capital is substantial, even among those who remain in the rural market [Rubin and Perloff (1993); Newman and Jarvis (2000)]. Human capital may be most important in increasing the rate of learning-by-doing (i.e., it is complementary with experience) and found in those situations where efficiency wages are used to motivate both skill acquisition and the length of the employment relationship.

3.3. Unresolved questions and conceptual challenges

Horizontal and vertical specialization are mutually reinforcing. Intel producing microprocessors for Dell computers exemplifies vertical specialization as modeled by Stigler (1951). In the early stages of such vertical specialization, the end user adjusts production to accommodate a standardized intermediate product. As the market expands, intermediate producers are able to tailor their products (horizontal specialization) to the individual niches of varieties of the final product and still capture sufficient economies of scale.[36] We have seen that the division of agricultural labor sometimes exhibits a similar pattern. Vertical specialization allowing transplanting and tasks to be produced as distinct intermediate "products" simultaneously facilitates horizontal specialization with different workers performing different tasks.

As mentioned above, the evolution of specialization may take different forms in different economies. The labor-contract pyramid has evolved in land-scarce and labor-abundant economies. In labor-scarce and capital abundant economies such as the U.S. or in land-abundant or in the capital scarce economies of Africa, specialization may evolve differently (although the communal form of agriculture observed in Africa corresponds to the stage one of our pyramid). A natural agenda for labor market development

[35] Huffman (2001) endogenizes human capital formation for the farm household in a multi-period model.

[36] Van Assche (2004).

involves documenting and explaining these various patterns. Similarly, how does the bimodal distribution of land holdings in Latin America [Otsuka, Chuma and Hayami (1992); Tomich, Kilby and Johnston (1995)] affect the evolution of specialization? Once documented, these patterns will provide promising opportunities to apply and extend the new economics of specialization [Yang (2003); Matsui and Postlewaite (2000)].

These different patterns of specialization will have implications for the rates and types of human capital formation. Another area for future research regards specialization by gender. There is a widespread tendency across developing countries for women to specialize in home-based activities such as milking, collecting fodder, and meal preparation. This is more so in Muslim cultures such as Pakistan [see, e.g., Fafchamps and Quisumbing (2003)] and less so in other countries, where women take often take responsibility for production and marketing activities. For example, in the Peruvian Sierra women specialize in livestock production [Jacoby (1991)], in animal husbandry in Trinidad [Harry and Thorpe (1990)], in raising food crops, pigs and chicken in Ecuadorian Amazon [Pichon (1993)], and in vegetable marketing [Eder (1991)]. In Laos, women pound most of the animal feed, feed the pigs and poultry, and sell pigs, chickens, ducks and eggs in the market [Schenk-Sandbergen and Choulamany-Khamphoui (1995)]. Specialization by age is another potential research area, e.g., explaining why younger and older workers are more likely to choose effort-intensive piece-rate jobs and middle-aged workers to choose time-rates [Rubin and Perloff (1993)].

Ultimately, the micro-analytic theory of agricultural labor will be married to the general equilibrium theory of agricultural development such that the economic environment that is exogenous to the theory discussed above can be endogenized. In particular, one seeks a relationship between the extent and forms of hired labor to the underlying relative factor scarcities along the lines of induced institutional change [Binswanger and Ruttan (1978)]. How does economic organization evolve with the stages of changing factor scarcities? The Kuznets stylized facts (e.g., rising relative wage and capital–labor ratio) characterize the third such stage. Several developing countries have experienced an earlier stage of Malthusian involution, stage II, and some have as of yet not escaped. At an even earlier stage of development, when land is still abundant relative to labor, the internal and external economies of scale permitted the average product of labor and per capita incomes to rise, as shown in stage I of Figure 6. To the extent that internal economies associated with increasing division of labor were responsible for the rising average product of labor in recently settled regions, marginal product pricing was infeasible [Day (1967)]. This provides another reason why hired labor was not observed.

As the "land frontier" is reached at the end of stage I, cultivated hectarage ceases to grow, and aggregate yield increases come from Ricardian intensification [Crisostomo (1971); Hayami and Ruttan (1985)]. This renders marginal product pricing feasible, affording a *fundamental transformation* to decentralized decision-making and allowing economic organization by markets. As population pressure continues to build up against the land constraint, wages relative to implicit rents fall, inducing land-saving innovations including more intensive land preparation, increased cropping intensities, organic and chemical fertilization, and the technological and institutional changes dis-

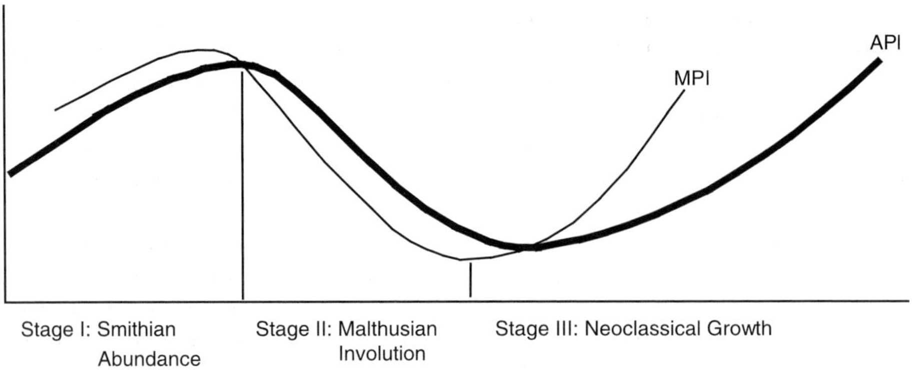

Figure 6. Stages of institutional evolution.

cussed in Section 3.2. To the extent that these changes are induced, they can mitigate the downward pressure on wages, but not reverse it.

How then does an economy get "from Malthus to Solow" [Hansen and Prescott (1998)]? The answer lies in Smithian external economies of specialization [Yang (2003)] and "Boserup Effects" [Boserup (1965, 1981, 1987)] whereby growth induces an increase in cropping intensity, in turn fostering changes in economic organization and the non-agricultural economy that may have the combined effect of increasing labor productivity.[37]

Another element needed is the theory of general equilibrium with transaction cost wedges. A prototype is provided by the theory of competitive equilibrium with exogenous unit transportation costs [e.g., Foley (1970)]. It is relatively straightforward to provide such a theory assuming internal equilibria. But a theory should allow many potential transactions not to be made. This is *economics as if zero's really mattered*.[38] Shadow prices in such a system are idiosyncratic but related. A complete solution requires first determining which parts of the economy are integrated and which are not. The prices of each integrated segment of the economy are systematically related to a normalized price. Which parts of the economy are integrated depends on all the parameters of the system, including policy.

Unit transaction costs (the size of the wedge) can be made endogenous using the concepts of minimum agency costs (Figure 2) and other contracting costs. From the perspective of the Coasean firm, economic organization evolves so as to minimize the sum of agency costs within the firm and contracting costs with suppliers of labor and intermediate products outside of the firm [Jensen (2000)]. Together, these are the costs of specialization. Efficient economic organization is that which minimizes the sum of

[37] See Krautkramer (1994) for a formal model of how population growth can induce labor-productivity-enhancing increases in cropping intensity.

[38] Yang (2003) and Yang and Ng (1993) call this "infra-marginal economics".

these and the forgone cost of specialization. Defining this sum as transactions costs reconciles Arrow's (1969) definition of transaction costs as "costs of running the economic system" with the proposition that institutions evolve to minimize transaction costs [Williamson (1985)].

One of the unsolved mysteries about agricultural contracts is why they are so simple. The linear payment schedule in Section 2.4 already represents a simplification. One might be inclined to generalize in order to explain more complicated arrangements that cannot be linearly characterized. But the puzzle turns out to be in the other direction. Share contracts for both tenants and workers are remarkably simple. First, the distribution of β's clusters tightly around two modes. Most tenancy contracts have tenant shares equal to $1/2$. Most of the remaining contracts are clustered tightly around $2/3$ [Hayami and Otsuka (1993); Deweaver and Roumasset (2002)] with an even smaller cluster at $1/3$.[39] In both cases, α is zero. Even in contracts for share workers, β is usually one-sixth, and α is zero.[40] This striking degree of simplification remains a mystery that begs explanation. Why are these particular fractions so prevalent and why are hybrid contracts, i.e. with both α and β being nonzero, so rare?[41]

4. Policy considerations and directions for further research

In the not-so-distant past, land reform was justified by two stylized facts. The first was the inverse relationship between yield per hectare and farm size, said to be caused by dualism in agricultural labor markets. The second was the mere existence of share tenancy, thought to be inefficient and exploitative. These claims are now recognized as founded on *ad hoc* theorizing, and more fundamental explanations of the stylized facts have been recognized [e.g., Sah (1986)]. In the "new dualism" a more market-friendly, albeit still interventionist, land reform is justified by the claim that commercial farms are inefficient, due to the inefficiency of hired labor [Deininger (2003)] along with a belief that asset redistribution is an effective instrument of poverty reduction. The tendency to leap to policy implications from a single explanation of a stylized fact perseveres. Not only do explanations need to be more complete in the sense described, but multiple explanations, with potentially different implications, should be entertained.

Politicians, and many academics, have the incurable disease of top-downism. As recognized by Adam Smith, they are forever designing rules, regulations, and insti-

[39] When wages are extremely low (e.g., as they were in Bangladesh in the late 1970s), or when the output elasticity of capital improvements is high (e.g., for perennials such as coconuts), the tenant's share is typically $1/3$ [Roumasset (1995)].

[40] Remarkably, the first written account of share contracts, in the *Constitution of Athens*, circa 800 B.C., records that share workers were known as *hectomori*, or "sixth-parters" [Roumasset (1995)].

[41] See Eggleston, Posner and Zeckhauser (2000) for a tentative explanation of simplicity in non-agricultural contracts.

tutions to be coercively imposed on the economy.[42] For example, despite decades of failed land reform legislation that have resulted in untold waste and injustice, land reform efforts continue to this day. The palliative for top-down tinkering with institutional design is an understanding of institutional choice and evolution. More specifically, we need a theory of how agricultural organization evolves from a self-sufficient peasant economy to a more specialized and intensive market economy. As also envisioned by Adam Smith, the division of labor affords a window into market development generally. Specialization is limited by the extent (size) of the potential market, and the size of the market is limited by population, incomes, and transaction costs.

On the other hand, a healthy respect for the role of efficiency in institutional change should not lead to one to ignore the conventional role of government in the provision/internalization of public goods/externalities and the less conventional role of facilitating economic cooperation more generally. In particular, investing in agricultural research and legal as well as physical infrastructure will stimulate the co-evolution of the division of labor and the corresponding institutional change. As specialization proceeds, more and more complex patterns of coordination are facilitated. In Reardon, Berdequé and Timmer's (2005) supermarket metaphor, for example, farmers are increasingly linked to specific retailers by means of complex chains that transform farm products over space, time, and form; thereby replacing the cumbersome and costly method of indirect coordination via inventories. The transaction sector that produces such transformation actually grows, even as the per-unit costs of coordination fall [North and Wallis (1982)]. The agricultural development that ensues from this approach is likely to have a high growth elasticity of poverty reduction [e.g., Lipton and Ravallion (1995)]. Not only does the facilitation strategy generate the traditional pro-poor linkages associated with lower food prices and higher demand for labor, but it aids workers whose wages are net of lower unit transaction costs as well as small farmers who benefit from falling transaction costs being subtracted from their sales and added to their purchases. The alternative of central design may actually fragment economic connectivity and stagnate efficiency-enhancing evolution.

References

Acemoglu, D., Antras, P., Helpman, E. (2005). "Contracts and the division of labor", MIT Working paper.

[42] "The man of system, on the contrary, is apt to be very wise in his own conceit; and is often so enamored with the supposed beauty of his own ideal plan of government, that he cannot suffer the smallest deviation from any part of it. He goes on to establish it completely and in all its parts, without any regard either to the great interests, or to the strong prejudices which may oppose it. He seems to imagine that he can arrange the different pieces of a great society with as much ease as the hand arranges the different pieces on a chessboard. He does not consider that the pieces upon the chess-board have no other principle of motion besides that which the hand impresses upon them; but that, in the great chess-board of human society, every single piece has a principle of motion of its own, altogether different from that which the legislature might chuse to impress upon it" [Smith (1976, VI.ii.2.17)].

Abdulai, A., Regmi, P.P. (2000). "Estimating labor supply of farm households under nonseparability: Empirical evidence from Nepal". Agricultural Economics 22, 309–320.
Allen, D.W., Lueck, D. (2002). The Nature of the Farm: Contracts, Risk, and Organization in Agriculture. MIT Press, Cambridge, MA.
Anderson, J. (1993). "Rural labor legislation and permanent agricultural employment in northeastern Brazil". World Development 21, 705–719.
Arrow, K. (1969). "The organisation of economic activity: Issues pertinent to the choice of market versus non-market allocation". The Analysis of and Evaluation of Public Expenditures. The PPB System. Joint Economic Committee, 91st Cong., pp. 59–73.
Bardhan, P. (1983). "Labor-tying in a poor agrarian economy: A theoretical and empirical analysis". Quarterly Journal of Economics 98, 501–514.
Bardhan, P., Udry, C. (1999). Development Microeconomics. Oxford University Press, Oxford.
Bell, C. (1977). "Alternative theories of sharecropping: Some tests using evidence from northeast India". Journal of Development Studies 13, 317–346.
Bhargava, A. (1997). "Nutritional status and the allocation of time in Rwandese households". Journal of Econometrics 77 (1), 277–295.
Benjamin, D. (1992). "Household composition, labor markets, and labor demand: Testing for separation in agricultural household models". Econometrica 60, 287–322.
Benjamin, D. (1995). "Can observed land quality explain the inverse productivity relationship?". Journal of Development Economics 46 (1), 51–84.
Berry, R.A., Cline, W.R. (1979). Agrarian Structure and Productivity in Developing Countries. Johns Hopkins University Press, Baltimore, MD.
Binswanger, H.P., Deininger, K., Feder, G. (1995). "Power, distortions, revolt, and reform in agricultural land relations". In: Behrman, J., Srinivasan, T.N. (Eds.), Handbook of Development Economics, vol. II. Elsevier Science, Amsterdam.
Binswanger, H.P., Rozenweig, M.R. (1986). "Behavioral and material determinants of production relations in agriculture". Journal of Development Studies 22, 503–539.
Binswanger, H.P., Ruttan, V.W. (1978). Induced Innovation: Technology Institutions and Development. Johns Hopkins University Press, Baltimore, MD.
Borland, J., Yang, X. (1992). "Specialization and a new approach to economic organization and growth". American Economic Review, Papers and Proceedings of American Economic Association 82 (2), 386–391.
Boserup, E. (1965). The Conditions of Agricultural Growth. The Economics of Agrarian Change under Population Pressure. Earthscan Publications, London.
Boserup, E. (1981). Population and Technological Change: A Study of Long-Term Trends. University of Chicago Press, Chicago, IL.
Boserup, E. (1987). "Agricultural growth and population change". In: Economic and Demographic Relationships in Development. Essays Selected and Introduced by T.P. Schultz (1990). Johns Hopkins University Press, Baltimore, MD.
Braverman, A., Stiglitz, J.E. (1982). "Sharecropping and the interlinking of agrarian markets". American Economic Review 72, 695–715.
Buchanan, J.M., Yoon, Y.J. (Eds.) (1994). The Return to Increasing Returns. University of Michigan Press, Ann Arbor, MI.
Burgess, R. (2001). "Land and welfare: Theory and evidence from China", Working Paper. London School of Economics.
Carter, M.R., Wiebe, K. (1990). "Access to capital and its impact on agrarian structure and productivity in Kenya". American Journal of Agricultural Economics 72, 1146–1150.
Carter, M.R., Yao, Y. (2002). "Local versus global separability in agricultural household models: The factor price equalization effect of land transfer rights". American Journal of Agricultural Economics 84, 702–715.
Coase, R. (1988). The Firm the Market and the Law. University of Chicago Press, Chicago, IL.

Coase, R. (1994). Essays on Economics and Economists. University of Chicago Press, Chicago, IL.

Crisostomo, C.M. (1971). "New rice technology and labour absorption in Philippine agriculture". Malayan Economic Review 16 (2), 117–158.

Croppenstedt, A., Muller, C. (2000). "The impact of farmers' health and nutritional status on their productivity and efficiency: Evidence from Ethiopia". Economic Development and Cultural Change 48 (3), 475–502.

Dasgupta, P., Ray, D. (1986). "Inequality as a determinant of malnutrition and unemployment: Theory". Economic Journal 96, 1011–1034.

Day, R.H. (1967). "The economies of technological change and the demise of sharecropping". American Economic Review 75, 427–449.

Deininger, K. (2003). Land Policies for Growth and Poverty Reduction: A World Bank Policy Research Report. World Bank and Oxford University Press.

de Janvry, A., Fafchamps, M., Sadoulet, E. (1991). "Peasant household behavior with missing markets: Some paradoxes explained". Economic Journal 101, 1400–1417.

Deolalikar, A.B. (1988). "Nutrition and labor productivity in agriculture: Estimates from rural south India". Review of Economics and Statistics 70 (3), 406–413.

Deolalikar, A.B., Vijverberg, W.P. (1983). "The heterogeneity of family and hired labor in agricultural production: A test using direct-level data from India". Journal of Economic Development 8 (2), 45–69.

Deolalikar, A.B., Vijverberg, W.P. (1987). "A test of heterogeneity of family and hired labour in Asian agriculture". Oxford Bulletin of Economics and Statistics 49 (3), 291–305.

Demsetz, H. (1969). "Information and efficiency: Another viewpoint". Journal of Law and Economics 12, 1–22.

DeSilva, S., Evenson, R.E., Kimhi, A. (2002). "Labor supervision and transaction costs: Evidence from Bicol rice farms", Mimeo.

Deweaver, M., Roumasset, J. (2002). "Risk aversion as effort incentive: A correction and primafacie test of the canonical theory of share tenancy". Economics Bulletin 15 (4), 1–16.

Dow, G., Putterman, L. (2000). "Why capital suppliers (usually) hire workers: What we know, and what we need to know". Journal of Economic Behavior and Organization 43, 319–336.

Dow, W., Gertler, P., Schoeni, R.F., Strauss, J., Thomas, D. (1997). "Health care prices, health and labor outcomes: Experimental evidence?", Labor and Population Program Working Paper No. 97-01. RAND, Santa Monica.

Eder, J.F. (1991). "Agricultural intensification and labor productivity in a Philippine vegetable growing community: A longitudinal study". Human Organization 50, 245–255.

Eggleston, K., Posner, E., Zeckhauser, R. (2000). "The design and interpretation of contracts: Why complexity matters". Northwestern University Law Review 95 (1), 91–132.

Eswaran, M., Kotwal, A. (1985a). "A theory of contractual structure in agriculture". American Economic Review 75 (3), 352–377.

Eswaran, M., Kotwal, A. (1985b). "A theory of two-tier labor markets in agrarian economics". American Economic Review 75 (1), 162–177.

Eswaran, M., Kotwal, A. (1986). "Access to capital and agrarian production organization". Economic Journal 96, 482–498.

Ewing, B.T., Payne, J.E. (1999). "The trade-off between supervision and wage: Evidence of efficiency wages from the NLSY". Southern Economic Journal 66, 424–432.

Evenson, R.E., Kimhi, A., DeSilva, S. (2000). "Supervision and transaction costs: Evidence from rice farms in Bicol, the Philippines", Center Discussion Paper No. 814. Economic Growth Center, Yale University.

Evenson, R.E., Roumasset, J. (1986). "Markets, institutions and family size in rural Philippine households". Journal of Philippine Development 23, 141–162.

Fafchamps, M., Quisumbing, A. (2003). "Social roles, human capital, and the intrahousehold division of labor: Evidence from Pakistan". Oxford Economic Papers 55 (1), 36–80.

Feder, G. (1985). "The relation between farm size and farm productivity: The role of family labor, supervision, and credit constraints". Journal of Development Economics 18, 1361–1372.

Foley, D. (1970). "Economic equilibrium with costly marketing". Journal of Economic Theory 2 (3), 276–291.

Friedman, M. (1953). "On the methodology of positive economics". In: Essays in Positive Economics. University of Chicago Press, Chicago, IL.
Frisvold, G.B. (1994). "Does supervision matter? Some hypothesis tests using Indian farm-level data". Journal of Development Economics 43 (2), 217–238.
Gavian, S., Fafchamps, M. (1996). "Land tenure and allocative efficiency". American Journal of Agricultural Economics 78, 460–471.
Greenwald, B.C., Stiglitz, J.E. (1986). "Externalities in economies with imperfect information and incomplete markets". Quarterly Journal of Economics 101 (2), 229–264.
Hansen, G., Prescott, E. (1998). "Malthus to Solow", NBER Working Paper No. 6858.
Harry, I., Thorpe, T. (1990). "Agriculture in Trinidad: A comparison of male and female participants". Social and Economic Studies 39 (2), 79–103.
Hart, O.D. (1988). "Incomplete contracts and the theory of the firm". Journal of Law, Economics, and Organization 4 (1), 119–140.
Hart, O.D., Moore, J. (1990). "Property rights and the nature of the firm". Journal of Political Economy 98, 1119–1158.
Hayami, Y., Kikuchi, M. (1982). Asian Village at the Crossroads: An Economic Approach to Institutional Change. Johns Hopkins University Press, Baltimore, MD.
Hayami, Y., Kikuchi, M. (1999). A Rice Village Saga: Three Decades of Green Revolution in the Philippines. Barnes and Noble, New York.
Hayami, Y., Otsuka, K. (1993). The Economics of Contract Choice: An Agrarian Perspective. Clarendon Press, Oxford.
Hayami, Y., Ruttan, V.W. (1985). Agricultural Development: An International Perspective. Johns Hopkins University Press, Baltimore, MD.
Huffman, W.E. (1980). "Farm and off-farm work decisions: The role of human capital". Review of Economics and Statistics 62 (1), 14–23.
Huffman, W.E. (2001). "Human capital: Education and agriculture". In: Gardner, B.L., Rausser, G.C. (Eds.), Handbook of Agricultural Economics, vol. 1A. Elsevier Science, Amsterdam.
Huffman, W.E., Orazem, P.F. (2007). "Agriculture and human capital in economic growth: Farmers, schooling and nutrition". In: Evenson, R., Pingali, P. (Eds.), Handbook of Agricultural Economics, vol. 3, Agricultural Development: Farmers, Farm Production, and Farm Markets. Elsevier Science, Amsterdam. This volume.
Jacoby, H. (1991). "Productivity of men and women and the sexual division of labor in peasant agriculture of the Peruvian Sierra". World Development 37 (1–2), 265–287.
Jacoby, H. (1993). "Shadow wages and peasant labor supply: An econometric application to the Peruvian Sierra". Review of Economic Studies 60, 903–921.
Jensen, M.C. (2000). A Theory of the Firm: Governance, Residual Claims and Organizational Forms. Harvard University Press, Cambridge, MA.
Jensen, M.C., Mekling, W. (1976). "Theory of the firm: Managerial behavior, agency costs and ownership structure". Journal of Financial Economics 3 (4), 305–360.
Krautkramer, J. (1994). "Population growth, soil fertility, and agricultural intensification". Journal of Development Economics 44, 403–428.
Kruse, D. (1992). "Supervision, working conditions, and the employer size-effect". Industrial Relations 31, 229–249.
Kuznets, S. (1966). Modern Economic Growth: Rate, Structure, and Spread. Yale University Press, New Haven, CT.
Lavy, V., Palumbo, M., Stern, S. (1995). "Health care in Jamaica: Quality, outcomes, and labor supply", Living Standard Measurement Study Working Paper No. 116. World Bank, Washington, DC.
Lipton, M., Ravallion, M. (1995). "Poverty and policy". In: Behrman, J., Srinivasan, T.N. (Eds.), Handbook of Development Economics, vol. 3B. North-Holland, Amsterdam.
Lopez, A. (1986). "Structural models of the farm household that allow for interdependent utility and profit-maximization decisions". In: Singh, I., Squire, L., Strauss, J. (Eds.), Agricultural Household Models: Extensions. Applications, and Policy. Johns Hopkins University Press, Baltimore, MD.

Matsui, A., Postlewaite, A. (2000). "Specialization and the distribution of income". Games and Behavior 33, 72–89.

Mirrless, J. (1975). "A pure theory of underdeveloped economies". In: Reynolds, L. (Ed.), Agriculture in Development Theory. Yale University Press, New Haven, CT, pp. 84–108.

Mitra, P.K. (1983). "A theory of interlinked rural transactions". Journal of Public Economics 20, 167–191.

Murrel, P. (1983). "The economics of sharing: A transaction cost analysis of contractual choice in farming". Bell Journal of Economics 14, 283–293.

Neal, D. (1993). "Supervision and wages across industries". Review of Economics and Statistics 75, 409–417.

Newbery, D.M., Stiglitz, J.E. (1979). "Share cropping, risk sharing and the importance of imperfect information". In: Roumasset, J., et al. (Eds.), Risk, Uncertainty and Agricultural Development. Agricultural Development Council, New York.

Newman, C., Jarvis, L. (2000). "Worker and firm determinants of piece rate variation in an agricultural labor market", Agricultural and Resource Economics Working Paper No. 00-028. University of California, Davis.

North, D. (1986). "Measuring the transaction sector in the American economy, 1870–1970". In: Engerman, S.L., Gallman, R.E. (Eds.), Long Term Factors in American Economic Growth. University of Chicago Press, Chicago, IL, pp. 95–161.

North, D., Wallis, J. (1982). "American government expenditures: A historical perspective". American Economic Review 72 (2), 336–340.

Otsuka, K. (2007). "Efficiency and equity effects of land markets". In: Evenson, R., Pingali, P. (Eds.), Handbook of Agricultural Economics, vol. 3, Agricultural Development: Farmers, Farm Production, and Farm Markets. Elsevier Science, Amsterdam. This volume.

Otsuka, K., Chuma, H., Hayami, Y. (1992). "Land and labor contracts in agrarian economies: Theories and facts". Journal of Economic Literature 30 (4), 1965–2018.

Pencavel, J. (1977). "Work effort, on the job screening, and alternative methods of remuneration". In: Ehrenberg, R. (Ed.), Research in Labour Economics. JAI Press, Greenwich, CT.

Pichon, F.J. (1993). "Agricultural settlement, land use, and deforestation in the Ecuadorian Amazon frontier: A microlevel analysis of colonists' land-allocation behavior", Ph.D. dissertation. University of North Carolina at Chapel Hill, Chapel Hill, NC.

Pitt, M.M., Rosenzweig, M.R. (1986). "Agricultural prices, food consumption, and the health and productivity of farmers". In: Singh, I., Squire, L., Strauss, J. (Eds.), Agricultural Household Models: Extensions, Applications, and Policy. Johns Hopkins University Press, Baltimore, MD.

Pitt, M.M., Rosenzweig, M.R., Hassan, M.M. (1990). "Productivity, health, and inequality in thee intrahousehold distribution of food in low-income countries". American Economic Review 80 (5), 1139–1156.

Putterman, L. (1984). "On some recent explanations of why capital hires labor". Economic Inquiry 22, 171–187.

Rawls, J. (1971). A Theory of Justice. Harvard University Press, Cambridge, MA.

Reardon, T., Berdegué, J.A., Timmer, C.P. (2005). "Supermarketization of the 'Emerging Markets' of the Pacific Rim: Development and trade implications". Journal of Food Distribution Research 36 (1).

Rebitzer, J.B. (1995). "Is there a trade-off between supervision and wages? An empirical test of efficiency wage theory". Journal of Economic Behavior and Organization 28, 107–129.

Reid, J.D. (1973). "Sharecropping as an understandable market response: The post-bellum South". The Journal of Economic History 33 (1), 106–130.

Reid, J.D. (1976). "Sharecropping and agricultural uncertainty". Economic Development and Cultural Change 24, 549–576.

Romer, P. (1987). "Growth based on increasing returns due to specialization". American Economic Review 77, 875–890.

Rosenzwig, M.R. (1980). "Neoclassical theory and the optimizing peasant: An econometric analysis of market family labor in developing economy". Quarterly Journal of Economics 94, 31–56.

Roumasset, J. (1976). Rice and Risk: Decision-Making among Low-Income Farmers in Theory and Practice. North-Holland, Amsterdam.

Roumasset, J. (1995). "The nature of the agricultural firm". Journal of Economic Behavior and Organization 26, 171–177.
Roumasset, J., James, W. (1979). "Explaining variations in share contracts: Land quality, population pressure and technological change". Australian Journal of Agricultural Economics 23 (2), 116–127.
Roumasset, J., Smith, J. (1981). "Population, technological change and the evolution of labor markets". Population and Development Review 7, 401–419.
Roumasset, J., Uy, M. (1980). "Piece rates, time rates and teams: Explaining patterns in the employment relation". Journal of Economic Behavior and Organization 1, 343–360.
Roumasset, J., Setboonsarng, S., Wickramasinghe, U., Estudillo, J., Evenson, R. (1995). "Specialization and the co-evolution of agricultural markets", Institutional Reform and the Informal Sector Working Paper. University of Maryland.
Rubin, D.K., Perloff, J.M. (1993). "Who works for piece rate and why". American Journal of Agricultural Economics 75, 1036–1043.
Sah, R.K. (1986). "Size, supervision and patterns of labor transactions". Journal of Philippines Development 13 (1–2).
Sahn, D.E., Alderman, H. (1988). "The effects of human capital on wages and the determinants of labor supply in a developing country". Journal of Development Economics 29 (2), 157–184.
Samuelson, P.A. (1947). Foundations of Economic Analysis. Harvard University Press, Cambridge, MA.
Schaffner, J.A. (1995). "Attached farm labor, limited horizons and servility". Journal of Development Economics 47, 241–270.
Schaffner, J.A. (2001). "Job stability in developing and developed countries: Evidence from Colombia and the United States". Economic Development and Cultural Change 49, 511–535.
Schenk-Sandbergen, L., Choulamany-Khamphoui, O. (1995). Women in Rice Fields and Offices: Irrigation in Laos. Jirangrachata Co., Bangkok.
Schonenbaum, M. (1995). "The health status and labor force behavior of the elderly in Taiwan", Comparative Study of the Elderly in Asia, Research Report No. 95-32. Population Studies Center, University of Michigan.
Schultz, T.W. (1951). "A framework for land economics – The long view". Journal of Farm Economics 33, 204–215.
Setboonsarng, S. (1991). "The evolution of rural labor institutions in a rice-based economy: A simultaneous equation model with selectivity bias", Ph.D. Dissertation. University of Hawaii.
Shaban, R. (1987). "Testing between competing theories of sharecropping". Journal of Political Economy 95, 893–920.
Shapiro, D. (1990). "Farm size, household size and composition, and women's contribution to agricultural production: Evidence from Zaire". Journal of Development Studies 27, 1–21.
Singh, I., Squire, L., Strauss, J. (Eds.) (1986). Agricultural Household Models: Extensions, Applications, and Policy. Johns Hopkins University Press, Baltimore, MD.
Skoufias, E. (1994). "Using shadow wages to estimate labor supply of agricultural household". American Journal of Agricultural Economics 76, 215–227.
Smith, A. (1976). The Theory of Moral Sentiments. Liberty Classics, Indianapolis, IN.
Sonoda, T., Maruyama, Y. (1999). "Effects of the internal wage on output supply: A structural estimation for Japanese rice farmers". American Journal of Agricultural Economics 81, 131–143.
Stigler, G. (1951). "The division of labor is limited by the extent of market". Journal of Political Economy 59, 185–193.
Stiglitz, J.E. (1974). "Incentives and risk sharing in sharecropping". Review of Economic Studies 41, 219–255.
Stiglitz, J.E. (1976). "The efficiency wage hypothesis, surplus labor and the distribution of income in LDCs". Oxford Economic Papers 28, 185–207.
Stiglitz, J.E. (1994). Whither Socialism? The MIT Press, Cambridge, MA.
Stiglitz, J.E. (2002). Globalization and It's Discontents. Norton, New York.
Strauss, J. (1986). "Does better nutrition raise farm productivity?". Journal of Political Economy 94 (2), 297–320.

Thomas, D., Strauss, J. (1997). "Health and wages: Evidence on men and women in urban Brazil". Journal of Econometrics 77 (1), 159–186.

Tomich, T., Kilby, P., Johnston, B. (1995). Transforming Agrarian Economies: Opportunities Seized, Opportunities Missed. Cornell University Press, Ithaca, NY.

Udry, C. (1997). "Recent advances in empirical micro research in poor countries: An annotated bibliography". Journal of Economic Education 28 (1), 58–75.

Uy, M. (1979). "Contractual choice and internal organization in Philippine sugarcane production", M.A. thesis. University of the Philippines.

Vakis, R., Sadoulet, E., de Janvry, A. (2003). "Measuring transactions costs from observed behavior: Market choices in Peru", CUDARE Working Paper No. 962. Department of Agricultural & Resource Economics, UCB.

Van Assche, A. (2004). "A theory of modular production networks", Mimeo.

Wan, G.H., Cheung, E. (2001). "Effects of land fragmentation and returns to scale in Chinese farming sector". Applied Economics 33 (2), 183–194.

Williamson, O. (1985). The Economic Institutions of Capitalism: Firms, Markets, and Relational Contracting. Free Press, New York.

Yang, D.T. (1997). "Education and off-farm work". Economic Development and Cultural Change 45, 613–632.

Yang, X. (2003). Economic Development and the Division of Labor. Blackwell, New York.

Yang, X., Ng, Y.-K. (1993). Specialization and Economic Organization: A New Classical Microeconomic Framework. Elsevier Science, New York.

Young, A.A. (1928). "Increasing returns and economic progress". Economic Journal 38, 527–542.

Chapter 53

FERTILIZERS AND OTHER FARM CHEMICALS*

PAUL W. HEISEY

Economic Research Service, U.S. Department of Agriculture, Washington, DC

GEORGE W. NORTON

Department of Agricultural and Applied Economics, Virginia Polytechnic Institute and State University, Blacksburg, VA

Contents

* The views expressed in this chapter are those of the authors and are not to be attributed to the U.S. Department of Agriculture.

Handbook of Agricultural Economics, Volume 3
Edited by Robert Evenson and Prabhu Pingali

DOI: 10.1016/S1574-0072(06)03053-2

Abstract

Demand for fertilizer in developing countries has expanded at a rapid rate over the past forty years. The relative scarcity of agricultural land has been a major underlying cause of this expansion in demand. More proximate causes include the development of complementary Green Revolution technologies – high yielding, fertilizer responsive seed varieties and expansion of irrigation or better water control within irrigated systems. At the same time, real fertilizer prices have declined over time, driven by technical change in fertilizer production. Expansion of fertilizer consumption has been particularly high in many Asian countries, and particularly low in much of Sub-Saharan Africa, where infrastructural and institutional constraints have restricted use.

Pesticide use has also expanded in developing countries, albeit in more localized circumstances. Relative scarcity of agricultural labor has been one cause of increased herbicide demand. Disease pressure and the availability of disease-resistant cultivars have influenced insecticide and fungicide demand. Integrated pest management (IPM), over the past 20 years, and genetically modified crops, over the past five to ten years, are new technologies that have the potential to curb the growth in pesticide use.

Price policies, environmental policies, and related investments in agricultural research and development, infrastructural expansion, or education all influence the markets for fertilizer and other farm chemicals in developing countries. One major policy issue is how to reduce or eliminate fertilizer subsidies at the same time that measures are taken to increase demand in areas such as Sub-Saharan Africa where fertilizer use is below the social optimum. At the same time, in intensive agricultural systems where agricultural chemical use is high, resource degradation and human health risks from pesticide use compromise productivity growth. In areas of both high and low use of chemical inputs, meeting the production and environmental challenges of the future will require increasing reliance on knowledge-intensive technology.

Keywords

fertilizers, pesticides, agricultural development, market liberalization, environmental policy

JEL classification: O130, Q120, Q160, Q550, Q560

1. Introduction

In 1961, developing countries consumed only 12% of the world total consumption of manufactured fertilizer nutrients.[1] Industrialized countries used over 70% of the world total, with the countries now sometimes referred to as the transitional economies of Eastern Europe and the former Soviet Union making up the balance. The United States was the world's largest single country user, accounting for about a quarter of all nutrient consumption. Application rates (kg of nitrogen, phosphorus, and potassium nutrients per crop hectare) were over ten times as high in industrialized countries as in developing countries.

By 1977, total fertilizer nutrient consumption in developing countries overtook consumption in the current transitional economies, and by 1984, it surpassed consumption in industrialized countries for the first time. About the same time, China overtook the United States to become the world's leading consumer of fertilizer nutrients. By 2001, developing countries consumed 64% of total nutrients and industrialized countries 31%. Consumption in transitional economies, which had reached a peak share of 27% in the mid-1980s, had collapsed after 1990 and only constituted a little more than 5% of the total. China alone accounted for about a quarter of all nutrient consumption, and India for about one-eighth. India stood just behind the U.S., with its 14% share, as the third largest user of fertilizer. Application rates in industrialized countries were still higher than in developing countries, but only by about 20 to 25% on average.

In contrast to data on fertilizers, data on pesticide use (in particular herbicides, insecticides, and fungicides) over time by region and country are limited, especially for pesticide usage prior to 1990. Some data do exist, however, and it is clear that pesticide use has grown over time in developing countries as a proportion of total use [FAOSTAT (2004)]. As of the mid-1990s, developing countries consumed approximately 25% of all pesticides, 85% of it for agricultural purposes [World Resources Institute (1999)].

This chapter surveys the current state and historical development of markets for fertilizers and pesticides in developing countries. The chapter begins with a discussion of fertilizer consumption by region and by crop, as well as a brief summary of pesticide consumption. The next section reviews determinants of consumption on both the demand and supply sides. Very broadly, fertilizers can be seen as substitutes for land, and all other things equal, one would expect high fertilizer consumption to develop in

[1] Unless otherwise indicated, in this chapter total fertilizer consumption will be measured in terms of nutrients: N for nitrogen, P_2O_5 for phosphorus, and K_2O for potassium. Application rates will generally be measured as nutrients per hectare. This latter measure confounds application rates per fertilized hectare with percentage of all cropland that receives some fertilizer, but it is used for two reasons: data availability and the simplicity of a single measure. Basic data sources include FAO's FAOSTAT data base (http://faostat.fao.org/faostat/collections?subset=agriculture) and Fertilizer Use by Crop (2002), a joint publication of the International Fertilizer Industry Association (IFA), the International Fertilizer Development Center (IFDC), the International Potash Institute (IPI), the Phosphate and Potash Institute (PPI), and the Food and Agricultural Organization of the United Nations (FAO).

land-short, labor-abundant agricultural economies. In contrast, herbicides can be seen as substitutes for labor-intensive hand weeding, and thus they might be most widely used in land-abundant, labor-short agriculture. Broad input substitution effects appear less determinative for consumption of insecticides and fungicides; instead, one major cause for widespread use of these latter two pesticides may be whether strong host-plant resistance to the relevant pests is widely available in cultivars planted by farmers. However, other interacting factors also influence the use of farm chemicals, and the determinants of consumption may shift over time. Factor substitution and other causes for farm chemical use will be analyzed in greater detail below.

Following the discussion of farm chemical consumption, the next substantive section considers two issues in market development. Price and regulatory policy, in particular subsidies for chemical inputs such as fertilizer, have been widely debated in the literature, and these debates will be summarized. Environmental issues in the use of fertilizers and pesticides have been prominently recognized for agriculture in developed countries. The nature of negative externalities from chemical use in developing countries, as well as possible policy responses, will also be addressed. The chapter concludes with a brief summary.

2. Consumption of fertilizers and pesticides

2.1. Fertilizer consumption by region

As noted, the growth of fertilizer consumption in developing countries has been extremely rapid. This subsection summarizes trends in developing country fertilizer use. The following section, on the determinants of fertilizer use, will outline some of the reasons for the observed trends and reasons for differences between those trends in different parts of the world.

Table 1 presents summary data for 1999–2001. The two most populous countries in the world, China and India, account for a significant percentage of world fertilizer use. Fertilizer use in the rest of Asia (excluding the Middle East) and Latin America is also notable. On the other hand, fertilizer use in Sub-Saharan Africa is very low as compared to use in the remainder of the developing world. For most of this chapter, data for South Africa are included with data for Sub-Saharan Africa. The differences between Sub-Saharan Africa and the rest of the world would be even more striking were South Africa to be excluded.[2]

[2] In many data sets, in particular in FAO's agricultural data base, data for South Africa are excluded from summary statistics for Sub-Saharan Africa. It is true that South Africa's history, agricultural structure, and pattern of input use make it markedly different from much of the rest of Sub-Saharan Africa. On the other hand, it seems very strange indeed to include such countries as Argentina and Brazil as "developing countries" and not to include South Africa when certain aspects of their general and agricultural economies appear quite similar. At times when considering the rest of Sub-Saharan Africa separately from South Africa illustrates a particular point, the effects of including or excluding South Africa will be noted.

Table 1
Total world fertilizer consumption, 1999–2001

Country/region	Million metric tons of nutrients, 1999–2001 average
China	35.3
India	17.4
Other Asia	14.1
Middle East/North Africa	6.5
Sub-Saharan Africa	2.0
Latin America/Caribbean	12.1
All developing	*87.4*
Transitional	7.3
Industrialized	43.0
World	137.7

Source: Calculated from FAOSTAT data.

Rapid changes in overall application rates drove many of the changes in world fertilizer consumption between 1961 and 2001.[3] At the world level, nutrient application rates in developing countries increased extremely rapidly over most of this period. On the other hand, application rates in industrialized countries grew quickly until the late 1970s, then stabilized or even fell slightly. Changes in transitional economies were even more pronounced. A remarkable increase in the rate of application until the late 1980s was followed by an even more remarkable collapse in this rate by the mid-1990s. Today, the application rate in the transitional economies is far lower than in other world regions. Together, these trends constitute changes in the mean rate of application at the world level, which rose until the late 1980s, fell with the collapse of use in the transitional economies, and then recovered somewhat with continued growth in developing country application rates (Figure 1).

In 1961, application rates in all major aggregations of developing countries were very low, under 10 $kg\,ha^{-1}$ in all instances.[4] The recorded application rate in India was the lowest, even slightly under that in Sub-Saharan Africa. Since that time, there have been four distinct patterns in the trends in application rates (Figure 2). The rate of nutrient application in China grew extremely rapidly, surpassing the average rate in industrialized countries by about 1980. Since the mid-1990s, application of nutrients in China has fallen somewhat, but it still remains very high at between 200 and 250 $kg\,ha^{-1}$.

[3] These changes included both an increase in the rate of application to fertilized cropland as well as an increase in the proportion of cropland including fertilizer. Data are insufficient, however, to permit a detailed separate exploration of both trends.

[4] In this chapter, developing countries are often disaggregated as follows. The large countries of China and India are separated out. Middle Eastern countries are grouped with North Africa into "Middle East/North Africa". The remaining countries of South and Southeast Asia are referred to as "Other Asia". Sub-Saharan Africa and Latin America are the other developing regions in the disaggregation.

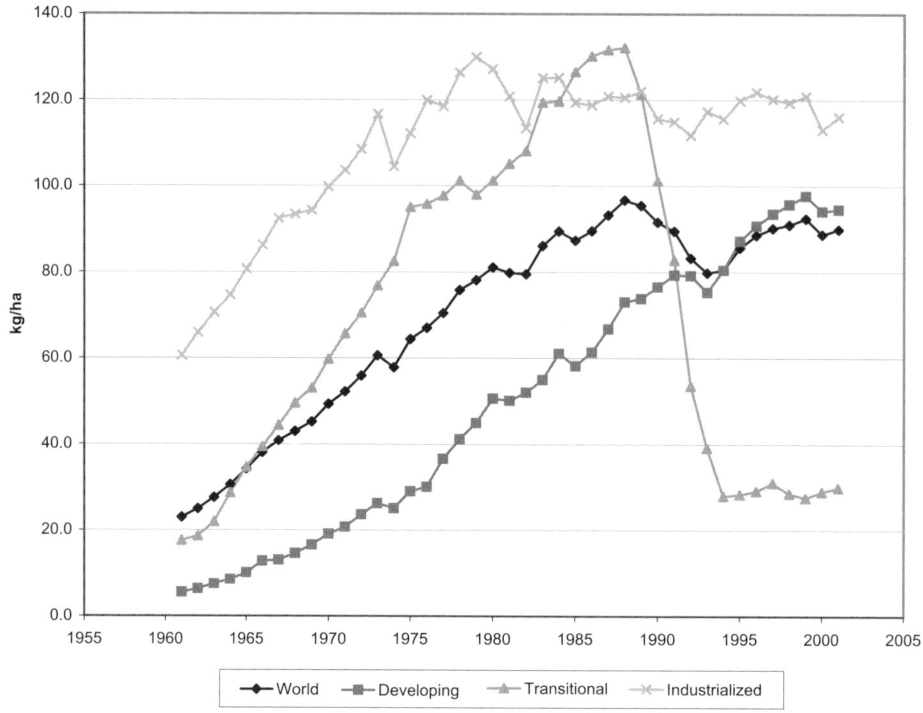

Figure 1. World fertilizer application rates, 1961–2001. *Source*: Calculated from FAOSTAT data.

Application rates in India and Other Asia grew fairly steadily for forty years. The nutrient application rate in India started out very low, and was less than that in Other Asia for many years, but by the late 1990s the mean rates in the two entities were very similar, at just around 100 kg ha^{-1}.

Application rates in Latin America and in the Middle East/North Africa region also showed similar upward trends until the early to mid-1980s, when the rate of growth slackened somewhat. The Latin American rate of application had been higher than the developing country average up until this point, but at that time it fell below the average for all developing countries. The application rate in Latin America picked up again in the late 1990s, while the trend in the Middle East/North Africa remained slower. Today, the average rate of nutrient application is about 70 kg ha^{-1} in Latin America and between 50 and 60 kg ha^{-1} in the Middle East/North Africa region.

The fertilizer application rate in Sub-Saharan Africa has lagged well behind application rates in the rest of the world. It grew slowly until the early 1980s, then declined slightly after that period. This result was driven first by the application rate in South Africa, which fell after about 1980. The application rate in the rest of Sub-Saharan Africa continued to grow until about 1990, when it started to fall slightly. Today, the average application rate in Sub-Saharan Africa is about 11 nutrient kg ha^{-1}. With South

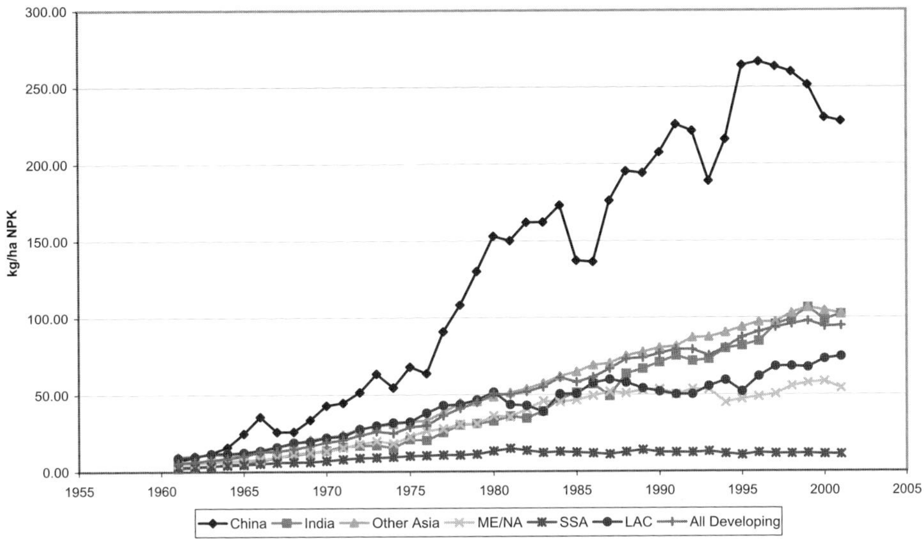

Figure 2. Fertilizer application rates, developing countries. *Source*: Calculated from FAOSTAT data.

Africa (average application rate just under 50 kg ha^{-1}) excluded, the average application rate in Sub-Saharan Africa is about 8 kg ha^{-1}.

2.2. *Consumption by crop*

The International Fertilizer Development Center (IFDC), International Fertilizer Association (IFA), and the FAO have worked together for about 15 years to publish, on five separate occasions, estimates of fertilizer use by crop for a large number of countries.[5] The latest report was made in 2002, with estimates for the late 1990s. The methodology used is described by Harris (1998). These estimates are usually in rough correspondence with FAO estimates for total fertilizer consumption for individual countries, but the two estimates are not completely congruent.[6] In this chapter, the primary source is the latest estimates [IFA et al. (2002)]. The most important omissions in that report are the three large transitional economies of the former Soviet Union, the Russian Federation, Ukraine, and Kazakhstan. This probably biases estimates for the transitional economies, but since transitional economies now account for such a small share of world fertilizer use, it does not affect world estimates significantly.

[5] In the latest edition, the International Potash Institute and Phosphate and Potash Institute have also collaborated in producing these estimates.

[6] In a few cases, for example the U.S., these estimates can be compared with detailed estimates produced by the country's economic and statistical reporting services [Daberkow, Taylor and Huang (2003)].

Table 2
Fertilizer use by crop (percentage of total fertilizer use)

Crop	Developing countries	Transitional economies[a]	Industrialized countries	World
Wheat	16.7	24.1	20.2	18.1
Rice	22.2		1.6	14.4
Maize	10.6	5.8	25.9	15.8
Other cereals	2.5	28.0	9.7	5.7
Total cereals	*52.0*	*57.9*	*57.4*	*54.1*
Soybeans	3.6	0.2	5.0	4.0
Pulses	2.0	0.4	0.6	1.5
Cotton	3.9		2.0	3.2
Cassava	0.4			0.3
Potatoes	1.4	5.5	2.4	1.8
Other crops	36.6	36.0	32.6	35.2
Total	100.0	100.0	100.0	100.0

Source: Calculated from data in IFA et al. (2002).
[a]No data available for large countries of the former Soviet Union – Russia, Ukraine, Kazakhstan, etc.

Worldwide, and in all major regional aggregations, cereals account for over half of all fertilizer consumption (Table 2). Wheat, the crop most widely grown across different world regions, is the largest single crop user of fertilizer.[7] Maize, which leads fertilizer demand in industrialized countries with around one-quarter of the total, is second most widely fertilized at the world level, and rice, which leads in the developing nations with over one-fifth of the total fertilizer applied, is third at the world level. Among the crops not reported in the table, fodder crops probably rank fourth in total fertilizer consumption worldwide [FAO (2000)]. Other important crops that contribute to world fertilizer demand in both developing countries and at the world level include cotton and soybeans (Table 2). Vegetables and sugarcane, not listed in Table 2, might also have some notable impact on total fertilizer demand [FAO (2000)].

A crop's importance to total fertilizer demand is determined both by the area planted to that crop as well as the rate at which it is fertilized. Table 3 indicates fertilization rates for various crops, and Figures 3–5 give more details for wheat, rice, and maize. On average for the developing countries, wheat receives fertilizer at a higher amount per hectare than any of the other cereals. Fertilizer application rates for wheat are at roughly the same level in developing, transitional, and industrialized countries (Table 3). However further disaggregation shows that application rates on wheat in China are very high, at over 200 kg ha^{-1}, very similar to application rates in Western Europe. Wheat in India and other wheat growing countries in Asia is fertilized at about the world average of 120 to 130 kg ha^{-1}. In Latin America, the overall average application rate for wheat

[7] The wheat share estimate for the transitional economies would probably be even larger if Russia, Ukraine, and Kazakhstan were included.

Table 3
Fertilizer use by crop (kilograms total nutrients per total hectares planted)

Crop	Developing countries	Transitional economies[a]	Industrialized countries	World
Wheat	135	106	118	127
Rice	116	–	224	120
Maize	96	98	270	152
Sorghum	26	–	115	38
Millet	16	–	IC[b]	16
Barley	69	91	132	110
Soybeans	86	105	72	79
Pulses	49	63	111	54
Cotton	130	IC[b]	177	137
Cassava	36	–	–	36
Potatoes	218	107	314	230

Source: Calculated from data in IFA et al. (2002).
[a]No data available for large countries of the former Soviet Union – Russia, Ukraine, Kazakhstan, etc.
[b]IC: Insufficient country coverage.

is roughly at the overall average for Canada and the U.S., around 90 kg ha^{-1}.[8] The average application rates for wheat in the Middle East/North Africa and Sub-Saharan Africa are the lowest (Figure 3).

Most of the world's rice is grown in developing countries, so the developing country average of just under 120 kg ha^{-1} is close to the world average (Table 3). Once again, China's application rate of around 220 kg ha^{-1} is one of the world's highest, not far from Japan's rate of about 240 kg ha^{-1}. In the case of rice, however, other Asian countries, and particularly India, apply fertilizer at somewhat under the world average. The Latin American application rate is slightly higher than the application rate for the rest of Asia, but the rate in Sub-Saharan Africa is particularly low (Figure 4).

Maize receives a substantial amount of fertilizer in developing countries, although the average application rate, under 100 kg ha^{-1}, is considerably below the average rate in industrialized countries (270 kg ha^{-1}) or the world average (150 kg ha^{-1}) (Table 3). In contrast with rice and wheat, maize production in China is only fertilized at a little over 160 kg ha^{-1}, just slightly above the world average. The relatively high rate of application in the Middle East/North Africa is influenced particularly by intensive application in irrigated maize production in Egypt (almost 270 kg ha^{-1}) and in Turkey (just under 170 kg ha^{-1}), where some maize is irrigated. Total maize area in this region represents only a small fraction of the developing world's maize production. In other developing regions where maize area is higher, average application rates are particularly low in

[8] Application rates in Australia, another major industrialized wheat producer, are also considerably lower than in western Europe, but IFC et al. (2002) data do not disaggregate fertilizer use by cereal crop for Australia.

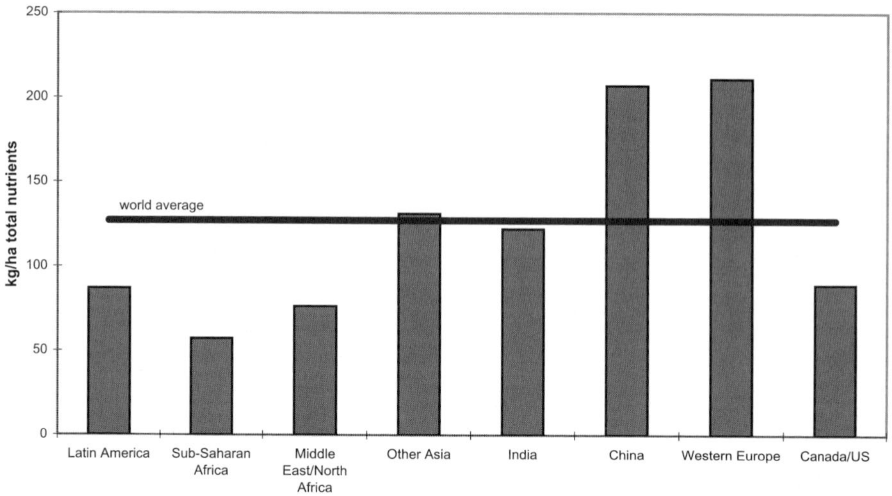

Figure 3. Fertilizer use per hectare, wheat. *Source*: Calculated from data in IFA et al. (2002).

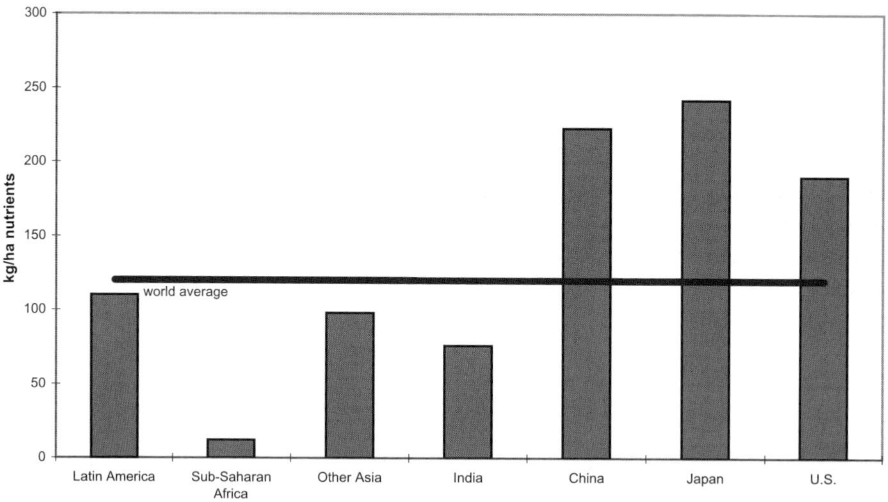

Figure 4. Fertilizer use per hectare, rice. *Source*: Calculated from data in IFC et al. (2002).

Sub-Saharan Africa (38 $\mathrm{kg\,ha^{-1}}$) and India (27 $\mathrm{kg\,ha^{-1}}$) (Figure 5). If South Africa were omitted, the application rate in Sub-Saharan Africa would fall to 26 $\mathrm{kg\,ha^{-1}}$, and if both South Africa and Zimbabwe were left out, the rate would fall to 15 $\mathrm{kg\,ha^{-1}}$.[9]

[9] An earlier and more thorough study focusing specifically on fertilizer on maize in Sub-Saharan Africa [Heisey and Mwangi (1997)] found a higher mean application rate (33 $\mathrm{kg\,ha^{-1}}$) even though South Africa

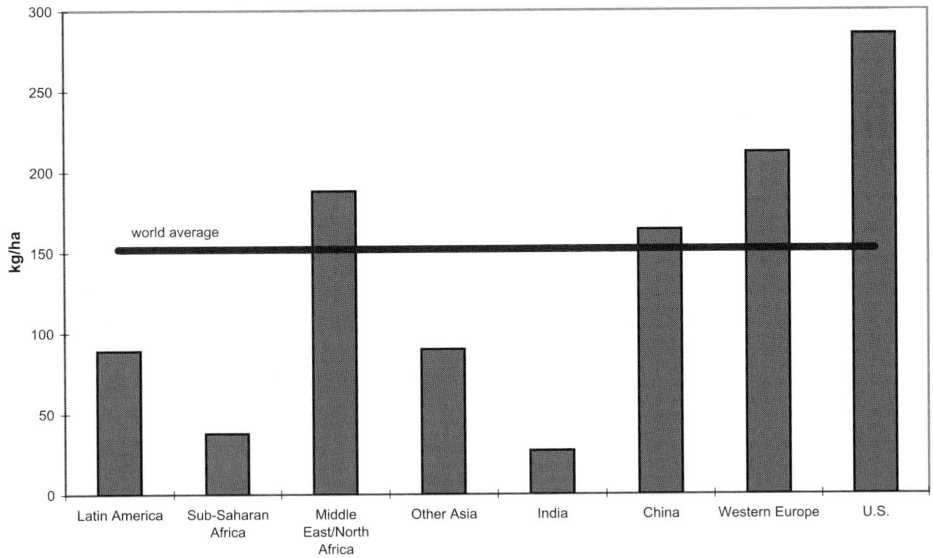

Figure 5. Fertilizer use per hectare, maize.

Among other crops, cereals like sorghum and millet and root crops like cassava are fertilized at relatively low rates in developing countries. On the other hand, potatoes and cotton are fertilized at relatively high rates in both developing and industrialized countries. In both developing and industrialized countries, soybeans receive somewhat less fertilizer per hectare than the major cereals, because of soybeans' ability to fix nitrogen; but they do receive more on average than the relatively unfertilized crops (Table 3).

2.3. Pesticide consumption

Data on use of pesticides – in particular herbicides, insecticides, and fungicides – is far less available for developing countries than is data for fertilizer use. The FAO has attempted to collect data on pesticides for over 30 years, often with unsatisfactory results. The FAOSTAT database now records worldwide pesticide consumption data for 1990–2001, but even in these years country coverage tends to be sporadic. In this chapter, averages of active ingredient use are presented for 1994–1997, because these are the years for which the largest number of important agricultural countries in most world regions have records. The data do not include China, which is well known to be a large consumer of pesticides, particularly insecticides [U.S. Embassy, Beijing (1996)].

was not included in the study. The decline for Sub-Saharan Africa as a whole over a roughly ten year period appears to be related particularly to a policy-related collapse in the fertilizer application rate to maize in Nigeria.

Table 4
Total world pesticide consumption, 1994–1997

Country/region	Thousand metric tons of active ingredients, 1994–1997 average		
	Herbicides	Insecticides	Fungicides
China	NA	NA	NA
India	6.8	37.2	9.4
Other Asia	24.4	41.0	19.2
Middle East/North Africa	9.7	19.5	14.1
Sub-Saharan Africa	11.7	9.7	9.0
Latin America/Caribbean	85.8	39.8	31.8
All developing	*138.4*	*147.2*	*83.6*
Transitional	35.6	7.9	23.2
Industrialized	337.8	163.4	190.4
World	511.8	318.4	297.2

Source: Calculated from FAOSTAT data.

The data do not permit very thorough analysis of trends in pesticide use, either. Fragmentary information suggests the pattern of levels and trends over time for pesticides has been similar to that for fertilizers: (1) considerably higher use in industrialized countries than in developing countries (Table 4); (2) rapid increase in use in industrialized countries, followed by considerable leveling off of growth in the 1980s [see Osteen et al. (2003) for details for the U.S.]; and (3) continued increase in consumption in developing countries [World Resources Institute (1999)].

In both industrialized and developing countries, pesticide use tends to be crop specific, considerably more so than for fertilizers. Partly for this reason, the data in Table 4 are presented in terms of total use of active ingredients, rather than application rates per hectare over an entire region.

Herbicide use tends to be relatively less crop-specific than the other major pesticide categories. In the mid-1990s, herbicide use in industrialized countries was about two and a half times the level in developing countries. Latin America was the developing region with the highest use of herbicides. Herbicide application in India, Middle East/North Africa, and Sub-Saharan Africa was particularly low (Table 4). If South Africa were excluded, Sub-Saharan African consumption would be even lower; South Africa accounts for 69% of African consumption of herbicides.

Insecticides are the pesticides most widely used in developing countries, and their total consumption is only slightly lower than in industrialized countries. Insecticide use tends to be particularly crop specific, notably being applied to rice in Asia, cotton, potato production in the Andes, and to a lesser extent to maize. Once again, Sub-Saharan Africa's consumption is lower than in the rest of the world, although in this case other African countries provide a greater share of consumption than South Africa, which makes up 42% of the total (Table 4).

In terms of total active ingredients, fungicides are the least widely used major class of pesticides in developing countries. Fungicide application in developing regions is highest in Latin America and lowest in India and Sub-Saharan Africa (Table 4). In the case of fungicides, South Africa constitutes 58% of the African total.

3. Determinants of consumption

This section discusses economic and other factors that can affect the consumption of fertilizer and other farm chemicals. The majority of the discussion will focus on fertilizer, but some of the factors that specifically influence the use of pesticides will be considered in the ultimate sub-section.

3.1. Level of analysis

Economic analysis of input demand, in particular fertilizer demand in the development of the agricultural sector, is usually conducted on one of three different but related levels. These are market development and institutional studies, agricultural sector modeling, and farm household adoption and use. There is no strict segregation between these approaches; features from one methodology can be used to inform another one.

The first approach is both an economic and institutional analysis of the development or obstacles to development of the market for fertilizers. Examples of this kind of analysis can be found in the work of Desai and colleagues [e.g., Desai and Stone (1987); Desai (1988); Desai and Gandhi (1988); Desai (1991)] and more recently in the work of Kelly and colleagues, who focus specifically on Sub-Saharan Africa, where fertilizer market development has lagged market development in other regions [e.g., Crawford et al. (2003); Kelly, Adesina and Gordon (2003)]. Desai emphasizes four processes that determine changes in fertilizer consumption: (1) those that influence the agronomic potential for fertilizer use; (2) those that convert the potential into farmers' effective demand for fertilizer; (3) those that determine the growth of aggregate fertilizer supply; and (4) those that develop the fertilizer distribution system [Desai and Stone (1987)]. Kelly, Adesina and Gordon (2003) ask how input use in Africa can be expanded, and whether this can be done in a way that is consistent with growth in private sector input markets. They posit a series of constraints acting on both farmers and potential retail-level input suppliers, and suggest different policies are in order depending on which constraints are most binding.

Standard economic analysis at the sectoral level is the second major approach to analyzing the economic factors determining fertilizer consumption. Demand for fertilizer is a derived demand in agricultural production. The analyst will usually set up and estimate an econometric model based on the production function or one of its duals, for example the profit function. This model may be constructed for the agricultural sector, the crop sector, or a particular crop. Although such models take fertilizer price as given, some analysts have looked at the economics of agricultural supply. Both fertilizer supply and demand will be considered in more detail in the next two sub-sections.

Finally, adoption studies focus on the farm household level. The literature on technology adoption in developing countries has been reviewed by Feder, Just and Zilberman (1985) and Feder and Umali (1993), and a summary of the economics of adoption can be found in Sunding and Zilberman (2001). Within this context, fertilizer use has often been analyzed in conjunction with the entire Green Revolution technological package based on high yielding crop varieties (HYVs) and sometimes featuring an analysis of water use as well as HYVs and fertilizer. This chapter will not review this literature exhaustively, but will refer to a few farm household based production studies in its discussion of fertilizer demand.

3.2. Determinants of fertilizer supply

Real fertilizer prices have declined nearly continuously over the past 100 years, despite occasional brief sharp upward movements such as the one that occurred in the early 1970s. Prices will be considered in more detail below, after both supply and demand factors have been presented. Nonetheless, technological change in fertilizer production has been a key factor in driving the real price of fertilizer lower. Major drivers of this change include a sharp decrease in the price of electricity, a major cost of fertilizer production until the mid-1960s; advances in transport and packaging; and a shift to higher analysis fertilizer, in other words fertilizer containing a higher proportion of plant nutrients. The most important savings stemmed from discoveries in applied chemistry and mechanical engineering that could be utilized in the production of ammonia – a key compound in the production on nitrogenous fertilizers – and phosphoric acid and superphosphates, which are used in the production of phosphorous fertilizers [Tomich, Kilby and Johnston (1995)].

The economics of fertilizer production investment depend heavily on potential domestic demand, the availability of local feedstocks, the cost of capital, and ex-factory pricing policy. Fertilizer production is characterized by considerable economies of scale. By the late 1980s, for example, an ammonia/urea plant needed to produce about 500,000 metric tons of urea per year to operate at maximum efficiency. Plants took three or four years to come on-line and another two years to reach full capacity [Segura, Shetty and Nishimizu (1986); Vlek (1990)]. Most fertilizer producing countries plan output to meet domestic demand, with exports occurring as a residual. As a result, world fertilizer prices have tended to be more volatile than those for other commodities, making the decision about whether to rely on the world market or to initiate domestic production a difficult one, even for countries where domestic demand potential is large [Ahmed, Falcon and Timmer (1989)].

Nonetheless, Tomich, Kilby and Johnston (1995) argued that from 1950 through 1990 most developing countries with large agricultural sectors over-invested in nitrogen production capacity. From 1950 through 1970 this was a result of technological obsolescence and comparative disadvantage. From 1970 through 1990 the gap between fertilizer production technologies in developing and industrialized countries was not so great, but comparative disadvantage still prevailed. For twenty-seven of the thirty years up to about 1990 Tomich, Kilby, and Johnston contended that "the maximum point of [world] price variation for nitrogen fertilizer fell below the [economic] cost of domestic manufacture" for most developing countries. Even the 1974–1975 price spike for fertilizers, initiated by the dramatic OPEC price increase as well as the sharp increase in world cereals' prices, was driven more by speculation than by any underlying changes in real long run supply and demand parameters [Sheldrick (1990)].

For countries that import fertilizer, the cost of fertilizer is dependent on transport costs, among other things. Landlocked countries are at a particular disadvantage for importing such a bulky input. Ideally a larger, unified market composed of more than one country could allow greater competition in fertilizer importation, economies of scale, and research cooperation [Gisselquist and Van Der Meer (2000)]. For individual markets, size of the market determines optimal strategies for assuring fertilizer supply through importation – which range downward from granulation based on imported and local raw materials, importation of bulk products and raw materials with local blending and bagging, importation of bulk products with local bagging, through importation of bagged products [Vlek (1990)].

Further down the marketing chain, the seasonality of demand for fertilizer and the bulkiness of the product lead to relatively slow stock turnover, considerable storage requirements, accompanying high finance charges, and resultant risk for distributors and dealers. At intermediate stages in the distribution channel, distributors have their own credit requirements as well as the need to work closely with credit agencies or to offer credit to end users [Shepherd (1989)]. All told, in smaller fertilizer markets with substantial internal marketing costs, the farm gate price of fertilizer might be two or more times greater than the price at the port of entry [Shepherd and Coster (1987); Bumb (1988); Jayne et al. (2003)].

3.3. Demand side factors

In a simple profit maximizing model, both the decision to adopt fertilizer and the application rate are determined by the interaction between agronomic response and the nutrient–grain price ratio, suggesting both technical and price factors are important in explaining total demand. In mature, well functioning fertilizer markets, the principal factors influencing aggregate demand are the total area of planted cropland and the mix of crops planted, fertilizer prices, and commodity prices and policies [Denbaly and Vroomen (1993)]. Other determinants of farmers' fertilizer use at the field level include "soil characteristics, climate and weather, crop rotations, application technology, and nutrient management practices" – in other words, primarily technical factors determining agronomic response [Daberkow, Taylor and Huang (2003)].

In developing fertilizer markets, a variety of explanatory factors can be important influences on the basic price and technology determinants of fertilizer use. The Green Revolution is often thought of as consisting of three complementary inputs – high yielding, fertilizer responsive seed varieties (HYVs); fertilizer; and expansion of irrigation or better water control within irrigated systems [Hayami and Ruttan (1985); Byerlee (1996)]. Certainly all other things equal – which they rarely are in reality – fertilizer use is considerably higher for HYVs than it is for so-called "farmers'" varieties, and higher on irrigated than on rainfed land.[10] In a review of international agricultural research for nine crops, Evenson and Gollin (2003) found strong complementarities between the use of modern varieties and fertilizer in Asia and Latin America, but not in Africa. The complexity of agricultural development patterns across regions and crops, and the presence of research spillovers from more favored to more marginal environments [Maredia and Byerlee (2000); Renkow (2000)] have meant, however, that though modern variety, fertilizer use, and more reliable moisture tend to move together, the correlations are by no means equal to one.

Other inputs, notably labor, may be related to fertilizer use, especially in a developing market for fertilizer. A number of studies [Parikh (1990); Chaudhary and Mufti (1999); Bhattacharyya and Kumbhakar (2003); Lamb (2003)] have investigated the relationships between labor and fertilizer markets in South Asia. These studies differ in whether they find fertilizer and labor as substitutes or complements. Though differences in findings may stem partly from methodological differences, some may be related to the differences between on-farm and off-farm labor use. Parikh (1990), for example, found for Bangladesh fertilizer use and family labor to be complementary, while fertilizer and hired labor are substitutes. Lamb (2003) agreed that generally own-farm labor and fertilizer use are viewed as complementary, but he discovered in the riskier environment of semi-arid India that improved opportunities for off-farm labor boosted fertilizer demand on-farm. He argued that in this risky environment, off-farm labor markets were an important tool for income smoothing, and this promoted fertilizer use.

This finding suggests that particularly in the earlier stages of fertilizer adoption and use, a widely cited constraint in technology adoption, risk, might be important. Several observers have concluded, however, that after adoption of fertilizer risk aversion reduces fertilizer application by no more than 20% of the "optimal" rates [Binswanger and Sillers (1983); Shalit and Binswanger (1985); Roumasset et al. (1989)]. Furthermore, the real constraints might be those on cash or credit availability, which can cause farmer behavior that resembles risk aversion [Masson (1972); Binswanger and Sillers (1983)]. However, studies looking at the effects of risk have generally not considered production risk in more marginal areas, where it might have a larger effect on fertilizer use, nor have they considered price risk in a general equilibrium context [Ahmed, Falcon and Timmer (1989)]. Finally, in the early stages of fertilizer use, farmers may lack knowledge of profitable inputs, or they may not have knowledge and skills necessary to use these inputs efficiently [Kelly, Adesina and Gordon (2003)].[11]

[10] Within rainfed land categories, fertilizer use is higher in areas with higher, more reliable rainfall than in more marginal areas.

[11] In the early years of the Green Revolution in Pakistan, farmers could not always identify different fertilizers when asked to do so by researchers, who carried small plasticine packets with fertilizer samples. By the mid-

Regional differences in fertilizer use also reflect differences in land versus labor scarcity, with fertilizer (together with improved seeds and water) increasingly substituting for relatively scarce land in densely populated Asian countries, as compared to say less densely populated countries in Africa and Latin America (Figure 2). Induced innovation driven by relative resource price changes has led, over time, to somewhat predictable regional differences in fertilizer consumption [Hayami and Ruttan (1985)].

In summary, while the underlying economic models for fertilizer demand are the same whether at the adoption, market development, or market maturity phases, the variables influencing fertilizer response, price ratios, and factor substitutability may differ. In terms of policy modeling, it seems likely that in many cases, in developing countries fertilizer demand is more responsive to price changes or policy shifts [Parikh (1990); Bhattacharyya and Kumbhakar (2003)] than it is in industrialized countries [Choi and Helmberger (1993); Abler and Shortle (1995); Rougoor et al. (2001)].

3.4. Prices

As noted earlier, technical change in fertilizer production has driven real fertilizer prices lower over much of the last century [Tomich, Kilby and Johnston (1995)]. The past thirty years of F.O.B. prices for N and P_2O_5 nutrients from urea and triple superphosphate, respectively, are shown in Figure 6. Even abstracting from the speculative price spike of 1974–1975, the real world price of fertilizer nutrients has continued downward, with smaller transitory price increases around 1980 and in the mid to late 1990s.

As is well known, however, real cereals prices have also fallen over much of the last century [Evenson and Avila (forthcoming)]. Figure 7 shows the ratio of the price of nitrogen from urea to the world crop price for rice, wheat, and maize. World rice prices are higher than wheat prices, which in turn are higher than maize prices. These differences are reflected in the relative position of the three series. Note as well that the price spikes of 1974–1975 adversely affected the nitrogen–crop price ratios for wheat and maize, but did not do so for rice – the relatively thin world rice market suffered a huge but temporary run-up in price sufficient to offset the increase in the urea price. Trends for all three ratios were analyzed for 1977 to 2003, to avoid giving undue significance to the events of 1974–1975. The nitrogen–wheat price ratio did fall at a significant rate of 1.3% per year over the period from 1977 to 2003, and the nitrogen–maize price ratio fell at a rate of 0.9% per year, which was also significant. However, the nitrogen–rice price ratio was essentially unchanged over this period.

World market prices are not transmitted directly to farmers, however. In any particular country at any particular point in time, there may be a substantial difference between the world market price for a given nutrient and the price paid at the farm gate. Factors

1980s, farmers in several irrigated areas of Pakistan could fairly accurately compare the nutrient contents of different nitrogen fertilizers, but were considerably less accurate in comparing the nutrient contents of different phosphatic fertilizers [unpublished data from a study of post-Green Revolution technical change focusing on varietal turnover; the information on varieties was published in Heisey (1990)].

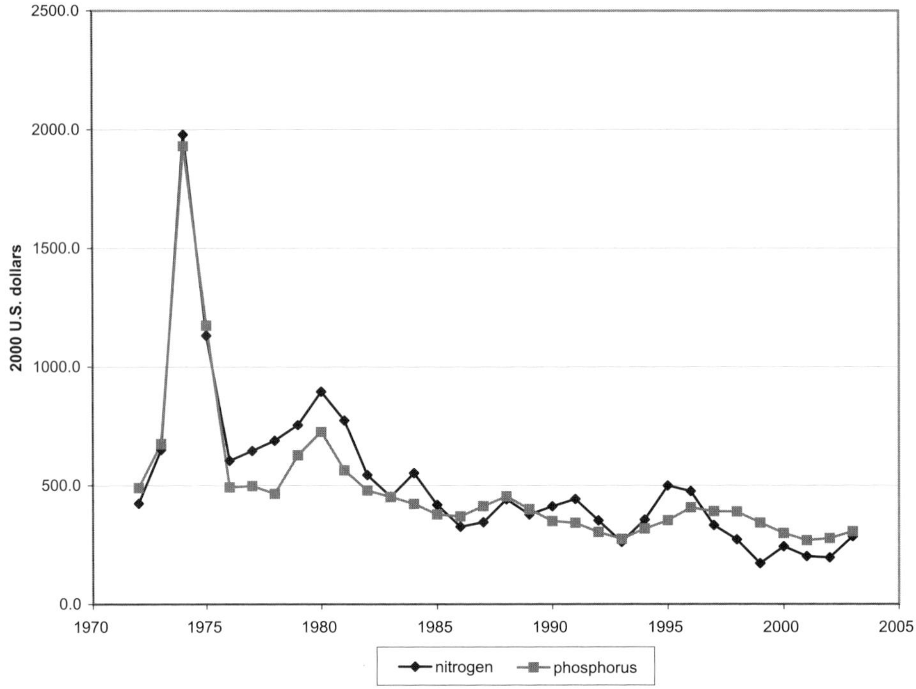

Figure 6. World fertilizer prices, 1972–2003, real terms. *Sources*: World Bank "Pink Sheets" and CIMMYT Economics Program database, urea and TSP prices; U.S. Department of Commerce, U.S. GDP deflator.

that raise the price of fertilizer to farmers include distribution costs, which include transportation costs [Bumb (1988)] and considerable credit requirements throughout the fertilizer marketing channel [Shepherd (1989)]. Particularly in Sub-Saharan Africa, poor physical infrastructure is a major factor in raising transportation costs [Vlek (1990); Spencer (1994); Mwangi (1997)]. In fertilizer-importing countries, small volume also can raise prices [Bumb (1988)]. Donor-required or government policy-created limitations on origin, transporters, or fertilizer types can lead to large marketing costs and margins [Gerner and Harris (1993); Marfo and Tripp (1997)], as can lack of competition in fertilizer distribution systems [Pinstrup-Andersen (1993)].

Policy can also drive a wedge, positive or negative, between world market and farm gate prices, and between financial and economic prices. Many countries have subsidized fertilizer prices, credit, or both. Much of the policy literature on developing-country fertilizer consumption has focused on direct subsidies, perhaps because their effects on government agricultural budgets are relatively easy to observe [Dalrymple (1975); Barker and Hayami (1976); Parish and MacLaren (1982); Chambers (1985); Sidhu and Sidhu (1985); Shalit and Binswanger (1985); Miller and Tolley (1989); Desai (1991); Ndayisenga and Schuh (1995); Nwosu (1995); Omamo and Mose (2001);

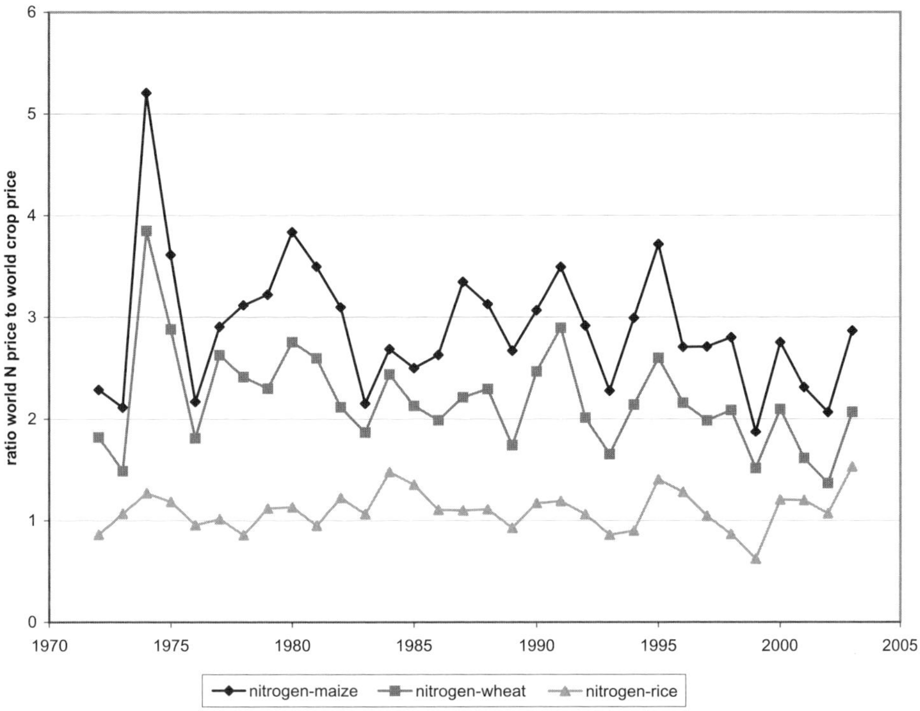

Figure 7. Fertilizer–crop price ratios, 1972–2003. *Sources*: World Bank "Pink Sheets" and CIMMYT Economics Program Database.

Kelly, Adesina and Gordon (2003)]. However, indirect policy effects such as cross-subsidization among different government trading operations (e.g., subsidization of input trading through output marketing profit margins) and pan-territorial pricing can also cause farm gate prices to differ from economic ones [Shepherd (1989); Bumb et al. (1994)]. Finally, exchange rate overvaluation can have complex effects because it can affect prices for both fertilizer and crops, and because the degree to which these goods are tradable differs from country to country.

A comprehensive historical analysis of the effects of these factors on farm gate fertilizer prices is beyond the scope of this review. The common consensus has been that small import volumes and high transportation costs have made fertilizer prices particularly high for African farmers [Shepherd and Coster (1987)]. While available data tend to confirm this conclusion, it also appears that fertilizer prices in many African countries were subsidized up until the mid-1980s or mid-1990s, when policy changes reduced or ended the subsidies [Heisey and Mwangi (1997)]. Recently, however, soil nutrient depletion in much of Sub-Saharan Africa has led some analysts to reconsider the case for fertilizer subsidies in the region [Sachs et al. (2004)].

FAOSTAT does provide data on farm gate fertilizer prices in various countries. Coverage is not complete across time or across countries. For this review, real farm gate prices for nitrogen from urea for a number of countries around the world were calculated using the FAO database.[12] Prices were converted to 2000 U.S. dollars using the "first deflate, then convert" rubric [Craig, Pardey and Roseboom (1991)]. In this approach, country-specific GDP deflators, usually obtained from IMF data, were used to deflate or inflate prices to 2000 units in local currency, then local currency units were converted to U.S. dollars. Although many of the time series were broken or otherwise incomplete, this analysis could in some cases cover as much as a thirty year period from 1972 through 2001.

For purposes of comparison, the U.S. farm gate nitrogen price from the FAOSTAT database averaged roughly 70% above the world nitrogen price from 1972 to 2001. In Asia, India, Pakistan, and Indonesia all had nitrogen prices that averaged *under* the world price over this period, and the Philippines and Thailand had prices that averaged 20 to 40% above the world price. In the most recent years, the gap between farm gate and world prices in the Philippines and Thailand did appear to be somewhat greater than this. In the Middle East and North Africa, all countries had nitrogen prices that averaged from below to roughly equal with the world price.

In Sub-Saharan Africa, four of the five countries – all except South Africa – had nitrogen prices that were below the world price for much of the earlier part of the available time series, and above the world price for most of the latter portion. Nitrogen prices in South Africa have been higher than the world price since the FAO series for South Africa began in 1985, and the margin has averaged about 85%. In the other African countries, the switch from apparent subsidization to perhaps more market-determined prices took place earliest in Kenya, latest in Nigeria. The series from Côte d'Ivoire ended in 1996, but for three of the other countries – Ghana, Kenya, and Nigeria – farm gate nitrogen prices now do tend to average over twice the F.O.B. world market price, as might have been predicted by students of fertilizer marketing margins in Africa. Farm gate fertilizer prices are still quite variable, however.

In Latin America, the time series for Argentina, Brazil, and Peru in the FAOSTAT data base are too short to draw definitive conclusions, although it appears that particularly in Brazil, hyperinflation made it difficult to calculate the appropriate farm gate price. The margin over the world price has averaged about 90% for Argentina, as it has in Peru. Farm gate nitrogen prices in Mexico and Colombia, in comparison, have averaged only about 20 to 50% above the world price.

In summary, three features have characterized fertilizer prices over time. First, the general trend in world prices for fertilizer nutrients has been downward. In some cases, but not all, this decline has been somewhat steeper than the decline in commodities

[12] These countries were China, India, Indonesia, Pakistan, Philippines, and Thailand in Asia; Egypt, Iran, Morocco, and Turkey in the Middle East/North Africa; Côte d'Ivoire, Ghana, Kenya, Nigeria, and South Africa in Sub-Saharan Africa; and Argentina, Brazil, Colombia, Mexico, and Peru in Latin America. The same database was used to measure farm gate prices paid in the U.S., for comparative purposes.

prices. Second, a plethora of factors cause farm gate fertilizer prices to diverge from world market prices. In many cases, farm gate prices differ significantly from economic prices. Third, the relationship between pricing and the adoption and use of fertilizer is a complex one. Earlier periods in which fertilizer was subsidized in some African countries did not inevitably lead to sustained higher consumption in Sub-Saharan Africa, although ending subsidies has undoubtedly cut back consumption in parts of Africa.[13]

3.5. *Determinants of pesticide consumption*

Analyzing the economics of pesticide demand is somewhat more complicated than analyzing the demand for fertilizer. At the sectoral level, pesticides have a marginal productivity that can be related to input and output prices, so pesticides might look like other inputs. Pesticide prices certainly influence producer decisions to apply pesticides as opposed to using other methods of pest control, and the marginal productivity of pesticides on agricultural production can be measured [Tjornhom, Norton and Gapud (1998); Teague and Brorsen (1995)]. At the farm level, the use of pest management practices, including pesticides, should be influenced by pest infestations, yield and quality losses caused by those infestations, as well as by crop prices and the costs of pesticides and alternative control methods [Osteen et al. (2003)]. There are three model specification issues in the estimation of pesticide demand, however. These are interaction of direct production inputs such as labor with damage control inputs (e.g., pesticides) in damage abatement; justification of *a priori* exclusion of production inputs from the abatement function; and the rationale for and consequences of alternative stochastic specifications. "Misspecification of the stochastic element in the production function can overestimate the marginal physical productivity of pesticides and grossly underestimate the responsiveness of demand to increases in pesticide prices" [Saha, Shumway and Havenner (1997)].

In developing countries, the relatively low cost of some pesticides as a means of pest control has led to increased use. Pesticides may also be seen as a relatively cheap form of insurance, and higher than optimal use may result [Gandhi (1997)]. In the initial stages of the use of pesticides, farmers may be quite unfamiliar with them. This can lead to health and environmental risks, which will be addressed at greater length below. But another information problem that might influence demand directly is the potential for adulteration.

Several specific features particularly influence pesticide demand. As noted, pesticide use is in general more crop specific than fertilizer use. Pesticide use also tends to be specific to the nature and incidence of the pest. Herbicide use is perhaps less crop specific than the use of other pesticides, but in developing countries the use of herbicides seems to be most common among commercialized, larger scale producers of major field crops.

[13] Nigeria is probably the most prominent exemplar of large policy-related swings in fertilizer use, with subsidies promoting use [Smith et al. (1994)] but large decreases in consumption following subsidy removal.

Its use is heavily influenced by labor costs as herbicides substitute for hand-weeding. Insecticide use is most common in rice and cotton, and it is also used in crops such as maize, potatoes, or fruits and vegetables. Fungicide use in developing countries is the lowest both in comparison with the other major pesticide groups and in comparison with fungicide use in industrialized countries. Host-plant resistance has been part of the breeding strategy in the development of many new varieties in different crops. In developing countries, fungicide use features particularly in Andean potato production, and in many instances in potato, vegetable, or fruit production in other developing regions. The growth in cut flower production in developing countries has also resulted in significant application of both insecticides and fungicides.

In theory, increased pesticide use should not be closely tied to the use of HYVs, because of the breeding focus on host-plant resistance. In at least one major case, however, Asian rice production, pesticide use was strongly promoted in many early Green Revolution technology packages [Widawsky et al. (1998)]. In some cases, donor agencies have also supplied pesticides to developing countries as part of their aid packages [Tobin (1996)].

Pesticide consumption is influenced by availability of substitute approaches to pest management. Its use has several potential negative health and environmental externalities as discussed below. Partly in response to these externalities, as well as to problems with increased pest resistance to pesticides, alternative pest management approaches have been developed to reduce dependence on pesticides. The longest standing and continually evolving approach has been integrated pest management (IPM). Integrated pest management is an eco-system-based approach that attempts to foster long-term prevention of pests or their damage through combinations of practices such as biological control, habitat management, altered cultural practices, and breeding in pest resistance. Pesticides are applied only after monitoring indicates they are needed based on established guidelines.

Use of IPM has expanded in developed countries over the past 30 plus years and in developing countries over the past 20. IPM has been fostered in developing countries through research and education programs in national agricultural research systems, international agricultural research centers, non-governmental organizations, and other programs supported by international donor and technical support agencies such as the Food and Agricultural Organization of the United Nations (FAO) and U.S. Agency for International Development.

Despite its spread, effects of IPM on total pesticide use are difficult to estimate, even for developed countries. In the United States, where IPM use has grown steadily since the 1970s, pesticide use increased rapidly from about 200 million pounds of active ingredient in the early 1960s to about 550 million pounds in 1982, and has stabilized since then in a range from 475 to 575 million pounds [USDA/ERS (2004)]. It currently stands at around 500 million pounds. Without IPM, usage might have continued on more of an upward trend, although such a conclusion is speculative. A review of 61 evaluations of IPM programs for cotton, soybeans, vegetables, fruits, peanuts, tobacco, corn, and alfalfa did find that pesticide use decreased, on average, for seven out of eight of those commodities (the exception being corn) [Norton and Mullen (1994)]. However, careful econometric evaluation of the aggregate impact of IPM on pesticide use in agriculture has been hindered by a variety of confounding factors, not the least of which is the difficulty of defining a single consistent IPM variable.

In developing countries, pesticide use has continued to trend upward in recent years, with little apparent impact of IPM in the aggregate. IPM research programs in these countries were limited to a few isolated cases before the mid-1980s, and the primary IPM tactic was varieties resistant to specific pests in a small subset of crops. Today, many IPM programs in developing countries are associated with high-input crops such as cotton, fruits, and vegetables. Rice in Asia, potatoes in the Andes, and soybeans in Brazil are other crops with notable IPM programs.

While IPM has been spreading gradually throughout the developing world, particularly as a result of international donor support, relatively few small-scale farmers have been reached to date. FAO began supporting IPM programs in the 1980s for small farmers, first in Indonesia and later elsewhere in the developing world. While realizing some success, progress has been slow in reaching the masses, in part because many of the programs have involved intensive season-long training which has proven expensive and has failed to yield much spread of IPM information beyond the initial trainees [Feder, Murgai and Quizon (2004)]. Other promising approaches to developing and communicating IPM information are currently being applied in Asia, Africa, Latin America, and Eastern Europe, but are still functioning on a somewhat limited scale [Norton et al. (2005); IRRI (2003)]. Rice has been the target of many IPM programs, including the FAO program mentioned above. One program that has been locally effective in reducing pesticide use relies on simple IPM messages delivered to Vietnamese rice farmers through mass media techniques such as radio and village dramas [IRRI (2003)].

IPM faces strong competition from pesticides that are directly or indirectly subsidized in many developing countries. Even without subsidies, as labor costs continue to gradually rise with income levels, herbicide use in particular expands as a means of reducing those costs. This transition from hand weeding to herbicides has already occurred in many parts of Latin America and is gradually moving across Asia. IPM may be an increasingly important means of slowing the spread of herbicides in the future.

Biotechnology applications have also just begun to have effects on pesticide use in developing countries. Genetically modified cotton in China, Mexico, India, South Africa, and Indonesia has led to reductions in insecticide use [Traxler (2004)]. Transgenic cotton containing a gene from the bacterium *Bacillus thuringiensis* (*Bt*) that is resistant to certain insect pests was first introduced commercially in Mexico and the United States in 1996, and the global area planted to Bt cotton increased from less than 1 million ha in 1996 to 4.6 million ha in 2002 [FAO (2004)]. Pray and Huang (2003) estimated that pesticide use on cotton in China was reduced by 67% as a result of adopting *Bt* cotton. Such a reduction represents approximately 25% of all pesticides applied in China. Commercial use of *Bt* cotton in India only began in 2003, but Qaim and Zilberman (2003) found significant reductions in pesticide use in field trial data.

Table 5
Genetically engineered crops in developing countries

Country	Area planted to genetically modified crops, 2003 (million ha)	Crops planted
Argentina	13.9	Herbicide tolerant soybeans, *Bt* maize, *Bt* cotton
Brazil	3.0	Herbicide tolerant soybeans
China	2.8	*Bt* cotton
South Africa	0.4	*Bt* cotton, *Bt* maize, herbicide tolerant soybeans
India	0.1	*Bt* cotton
Uruguay	>0.05	Herbicide tolerant soybeans, *Bt* maize
Mexico	<0.05	*Bt* cotton, herbicide tolerant soybeans
Philippines	<0.05	*Bt* maize
Colombia	<0.05	*Bt* cotton
Honduras	<0.05	*Bt* maize
Indonesia	<0.05	*Bt* cotton

Source: James (2003).

Transgenic soybeans with herbicide tolerance were introduced first in the United States and Argentina in 1996. They currently cover 99% of the soybean area in Argentina and are also widely used in a few other developing countries such as Brazil [Traxler (2004); Qaim and Traxler (2005)]. These RoundupReady™ (RR) soybeans contain a gene from the soil bacterium *Agrobacterium tumefaciens*, which make them tolerant to glyphosate, a broad spectrum herbicide produced by Monsanto. Unlike in the United States, in Argentina, RR soybeans have not resulted in a reduced quantity of herbicides applied, but they have caused costs to drop significantly, and most importantly, have resulted in a switch from highly toxic class I herbicides to less toxic class IV herbicides, a significant environmental benefit. Glyphosate has virtually no residual activity and rapidly decomposes in the soil.

These examples may portend significant reductions in pesticide use (or at least in the rate of growth) in the future in developing countries, and a movement away from the most toxic chemistries as additional adoption of transgenic crops occurs. To date, the overall impact has been small, mirroring the limited adoption of GM crops in these countries (Table 5).

4. Issues in market development

Policy discussions concerning the use of fertilizer and other farm chemicals in developing countries usually coalesce around one of two broad sets of issues. The literature relating to price and regulatory policy tends to apply particularly to fertilizer. Environmental issues, on the other hand, tend to be particularly important in pesticide use, although fertilizer use also has environmental dimensions.

4.1. Price and regulatory policy for fertilizer and other farm chemicals in developing countries

Since Schuh's (1974) paper, policy analysts have recognized the importance of macroeconomic factors in agriculture as well as traditional price policy features such as taxes and subsidies [Timmer (1986)]. The consensus view of agricultural policy is that the net effect of policy is to tax agriculture in developing countries, where there are many farmers, and subsidize it in industrialized countries, where there are few [de Gorter and Swinnen (2002)]. Subsidies and taxes are often justified by market failure arguments, but Bates (1981) introduced the concept of "government failure" into the agricultural policy discussion for developing economies. These general agricultural policy debates concerning the relative importance of market failure and government failure give context to the literature on fertilizer policy in developing countries [Dorward et al. (2004)].[14]

Fertilizer market development in developing countries has been subsidized in two ways. First, governments in some countries with relatively large potential fertilizer markets have subsidized local fertilizer production [Ahmed, Falcon and Timmer (1989); Tomich, Kilby and Johnston (1995)]. Second, fertilizer prices at the farm gate have often been subsidized for a number of reasons. In a world of market and information failures, where policy makers often choose non-efficiency objectives, a subsidy on inputs might be justified [Shalit and Binswanger (1985)]. If the government's goal is to achieve food self-sufficiency, in many cases a subsidy on fertilizer is relatively more efficient than a subsidy on output [Barker and Hayami (1976); Parish and MacLaren (1982); Chambers (1985); Sidhu and Sidhu (1985)]. Another major argument for subsidies is to encourage adoption in cases where learning costs or other constraints within the fertilizer market tend to slow or halt movement toward a socially optimal level of fertilizer use [Shalit and Binswanger (1985); Miller and Tolley (1989)]. Using parameters derived primarily from Asian, and not Latin American or African experience, Miller and Tolley show that the social benefits from such an optimal subsidy policy are expected to be relatively small.

Subsidies did contribute to the development of fertilizer production capacity in many countries [Desai (1991); Tomich, Kilby and Johnston (1995)], but most observers conclude that fertilizer subsidies have not been a particularly efficient means of encouraging fertilizer adoption by farmers [Dalrymple (1975); Ndayisenga and Schuh (1995)]. For at least one case, India, Desai (1988, 1991) argued that the observable effects of subsidy policies were considerably greater in the development of domestic production capacity than they were on fertilizer adoption. In Sub-Saharan Africa, the period of heavy subsidies that lasted to about the mid-1980s certainly did not promote rapid growth in fertilizer consumption [Heisey and Mwangi (1997)], although in individual cases such as Nigeria subsidies did play a role in the expansion of seed-fertilizer technology [Smith et al. (1994)].

[14] This part of the discussion will focus on fertilizer, as the literature on policies for pesticides concentrates on environmental issues. These environmental issues will be discussed in the next sub-section.

In developing countries, liberalization of fertilizer markets is sometimes thought to begin with the removal of subsidies to fertilizer production or consumption. Experience to date, though, suggests that subsidy removals are easier to institute, often with donor pressure, in small fertilizer markets such as in many countries of Sub-Saharan Africa, than they are in large, relatively more mature markets such as those in Asia.

Market liberalization also goes beyond exchange rate liberalization, liberalization of external tariffs, and reduction or elimination of internal taxes and subsidies. Governments have in many instances provided fertilizer distribution services themselves, or regulated the fertilizer market in many ways. These can include restricting the number of companies that can import fertilizer, requiring private sector importers to obtain permits from the Ministry of Agriculture, controlling foreign exchange allocation, restricting participation in internal markets through registration requirements or other means, or restricting the types and compositions of fertilizers that are imported or distributed [Gerner and Harris (1993); Pinstrup-Andersen (1993); Marfo and Tripp (1997); Gisselquist and Van Der Meer (2000)]. Liberalization of some or all of these restrictions on fertilizer trade can be an important part of changing fertilizer policy.

4.2. *Experiences with market reform*

Available literature on market reform for agricultural inputs illustrates two general points. First, changes in policy for an input such as fertilizer often take place within a larger context of general market liberalization. On the other hand, even within the fertilizer market itself, market reform often stops short of complete liberalization.

China's large fertilizer market, for example, was marked for many years by fertilizer manufacture in heavily subsidized state-owned enterprises. Distribution also occurred through state-operated sales networks. In the 1980s, fertilizer imports were allowed for the first time, but there was still strong state control of imports. Only the state-owned Sinochem imported fertilizer, with quantities determined explicitly by the State Council. Tariffs on fertilizer imports remained high. Nonetheless, increased import volumes contributed to reduced fertilizer prices in China. In the early 1990s, the elimination of factory subsidies and the promotion of ownership reforms also increased domestic supply and this also helped lower fertilizer prices. Wholesalers and retailers were commercialized, and private trade in fertilizers was allowed. Fertilizer markets became more integrated, with prices in one region somewhat more likely to move in the same direction as prices in other regions. With China's accession to the WTO, further liberalization of the fertilizer market is to be expected. Tariff rate quotas for fertilizer imports are to become open, and allocation of these quotas to non-state enterprises will be allowed [Qiao et al. (2003)].

In countries as diverse as Egypt and Mexico, fertilizer market liberalization was part of general price liberalization in the agricultural sector. Analyses of these liberalizations tend to focus on price changes *per se*, and accompanying institutional reforms are only given brief mention. In Egypt, for example, fertilizer prices had been constant in nominal terms and declining in real terms for many years before the beginning of structural adjustment reforms in 1986. At the same time, producer prices for most major crops were well below world market prices. Although fertilizer prices were allowed to rise with structural adjustment, liberalization of prices and relaxation of controls on area allotments and marketing were expected to lead to increases in producer welfare for all major crops except cotton [Baffes and Gautam (1996)].

The Mexican case also illustrates the importance of simultaneous output market liberalization. Two quite similar analyses in terms of agricultural sector modeling reached opposite conclusions about policy-induced changes in fertilizer consumption in Mexico. Baffes (1998) focused on price liberalization and structural reforms instituted by the Mexican Government from the mid-1980s and concluded that fertilizer consumption in Mexican agriculture could fall anywhere from 14 to 36%. Williams and Shumway (2000) concentrated on trade liberalization under the North American Free Trade Agreement (NAFTA) and predicted a substantial (62% from 1991 to 2005) increase in fertilizer use.[15] In both cases the results were driven by assumed changes in output prices. Mexico's historical fertilizer consumption data as recorded in FAOSTAT shows a decline in the late 1980s and early 1990s, including a sharp, probably weather-induced drop in 1996. By the late 1990s and early 2000s, however, Mexican fertilizer consumption returned to mid-1980s levels. This suggests, although it does not prove, that trade liberalization may have begun to offset the effects of internal market liberalization by the late 1990s.

Sub-Saharan Africa is the locus of some of the most intricate debates concerning fertilizer policy. This is probably because fertilizer consumption in Sub-Saharan Africa, on average, is much lower than consumption in any other major world region, and many smallholder farmers may not use manufactured fertilizer at all. At the same time, much of Sub-Saharan Africa can no longer be regarded as land abundant, especially when land quality is taken into account [Binswanger and Pingali (1988)], and soil nutrient depletion is a common consequence of most African agriculture as fallow periods shorten or disappear [Smaling (1993); Stoorvogel, Smaling and Janssen (1993)]. Integrated soil fertility management combining organic nutrients with inorganic fertilizers might partially address this issue [Place et al. (2003)]. Integrated soil fertility management is knowledge-intensive, however, and in any case a considerable increase in inorganic fertilizer consumption would still be an important part of such a strategy if depleted soils are to be restored and food production is to keep pace with population growth [Janssen (1993)].

Although fertilizer subsidies had a very limited effect on increasing fertilizer use in Sub-Saharan Africa, the combined effect of such policy changes as subsidy removal and exchange rate liberalization was often to reduce fertilizer consumption [Heisey and Mwangi (1997)]. Privatization in the face of a rapidly shrinking market could not always

[15] Williams and Shumway's analysis is also interesting because unlike most other sector-level policy studies, it considered the effects of trade liberalization on pesticide use. In this case, Williams and Shumway predicted that pesticide use in Mexico would fall.

counteract this reduction [Bumb et al. (1994)]. On the other hand, if a market was better developed and the public sector provided adequate support, in some cases the private sector could import and deliver fertilizer at a lower cost than parastatals [Truong and Walker (1990); Omamo and Mose (2001); Jayne et al. (2003)].

The Kenyan case is particularly instructive. In Kenya, where fertilizer prices already tended to be higher than in other African countries, fertilizer import quotas and licenses were abolished and foreign exchange allocation was liberalized in the early 1990s. Internal prices were decontrolled. There was a substantial market entry at both the importation and distribution levels, and market innovations resulted, for example the introduction of trading in smaller packet sizes [Omamo and Mose (2001); Jayne et al. (2003)]. Marketing margins were reduced, although internal marketing costs were still high [Jayne et al. (2003)]. Production potential, credit availability, and the regional maize price were important factors in explaining dealer sales. It appeared that agricultural production was shifting in the direction of greater regional comparative advantage, although the net effect of various policy changes on fertilizer consumption was initially hard to determine [Omamo and Mose (2001)]. However, Freeman and Omiti (2003) and Jayne et al. (2003) argued that in the aggregate, the policy changes stimulated fertilizer use.

Both Zambia and Ethiopia illustrate that "liberalization" or "privatization" can be undercut if the government continues to participate in large-scale fertilizer distribution under other guises after parastatals have been abolished or curtailed. In Zambia, fertilizer consumption declined, largely because of the unprofitability of fertilizer use in maize after the withdrawal of maize transport subsidies. In Ethiopia, fertilizer consumption increased, both because of highly concessional loans for fertilizer imports and profitable technical packages in at least some crops and regions [Jayne et al. (2003)].

Several common threads run throughout these otherwise quite disparate cases. First, "market liberalization" may consist of varying combinations of specific policy changes, and in any particular country not all components of a market may be liberalized. In some cases, liberalization may even be undercut by other government actions. Thus, judging the success or failure of any given shift in policy is likely to be case specific and to require detailed analysis [Jayne et al. (2003)]. Second, the profitability of technology requiring fertilizer, driven both by output price and by the technological package, is extremely important in determining whether market liberalization will lead to increased or decreased fertilizer consumption in the long run. Analyses that focus only on fertilizer markets, but ignore price and policy changes in output markets, are likely to miss many of the important factors explaining changes in fertilizer use.

Finally, some policy analysts have begun to question a strict dichotomy between state solutions and market solutions to low use of inputs such as fertilizer. Particularly if poverty reduction is one goal of policy, government may need "to intervene in financial, input and output markets, not necessarily to participate directly in these markets itself, but to reduce the transaction risks and costs facing private agents". It is important, however, "that governments quickly withdraw from expensive and distortionary interventions" [Dorward et al. (2004)]. This chapter has argued that rapid subsidy withdrawal is not often easily accomplished. Policy might aim instead at expanding input use in such a way that it supports the growth of private sector input markets that are more sustainable than government marketing in the long run. If so, investment in roads, irrigation, basic education, market information systems, and research and development, as well as in improved institutions such as contract law and enforcement, and systems of grades and standards, might be more important than direct provision of inputs or credit to farmers [Kelly, Adesina and Gordon (2003); Crawford et al. (2003)].

4.3. Fertilizer, pesticides, and negative externalities in developing countries

In industrialized countries, nutrient runoff into surface water or leaching into groundwater can pose significant negative externalities from fertilizer use. Policies that can affect farmers' land use and production decisions can include voluntary measures; regulation; direct controls on outputs, inputs, or technology; and incentives such as taxes, subsidies, or marketable discharge permits. Incentive measures are particularly difficult to implement because the sources of agricultural pollutants are diffuse, or "non-point" [Ribaudo and Johansson (2006); Scheierling (1995)]. Growing awareness of these problems has probably been one factor in more efficient nutrient management and the leveling of fertilizer consumption in developing countries.

Pollution from nutrient runoff has not attracted much attention from economists studying environmental issues in developing countries, although the potential for such problems in intensive agricultural systems does exist [Ongley (1996); Agrawal (1999)]. Rather, the focus on fertilizer in intensive agriculture has been on whether or not declining responses to fertilizer use indicate degradation of the resource base and declines in agricultural productivity. Obviously, if fertilizer response has the standard concavity and fertilizer application increases over time, average fertilizer productivity will decrease even if the quality of the resource base remains constant. A more relevant indicator of potential resource degradation would be the marginal response to fertilizer when measured at the same level of fertilizer application and the same level of other inputs. Degradation of the agricultural resource base could cause this measure to fall over time [Byerlee (1992)]. The empirical evidence is mixed, with some suggestions that this marginal rate over time has fallen in intensive, multi-crop rice systems, but has not necessarily fallen in wheat [Cassman and Pingali (1995); Chaudhary and Harrington (1993)].[16]

In less intensively farmed systems of the developing world, where fertilizer use is still low or non-existent, it is very likely that increased fertilizer use will have a net positive impact on the environment. Nutrient mining and increased use of marginal lands are likely to have far more serious negative environmental consequences than those foreseen from increased fertilizer use [Matlon and Spencer (1984); Dudal and Byrnes (1993)].

[16] There is little doubt, however, that in general intensively cropped systems in developing countries are suffering some degradation of the resource base [Cassman and Pingali (1995); Huang and Rozelle (1995); Ali and Byerlee (2001)].

Negative externalities from pesticide use in developing countries include water and surface pollution, as well as human health risks from pesticide residues in food, as they do in industrialized countries. However, by far the greatest current problem in developing country pesticide use is human health risks to the farm population, caused by excessive exposure to pesticides. Poisonings and deaths can occur because of inadequate pesticide handling practices, unsafe equipment, overuse of pesticides, and the use of pesticides more toxic than those applied in industrialized countries [World Resources Institute (1999)]. Economists and other scientists have documented the human health risks from pesticide use in Asian rice systems [Rola and Pingali (1993); Pingali and Roger (1995)], Pakistan's cotton production [Khan et al. (2002)]; and Andean potato cultivation [Crissman, Antle and Capalbo (1998)]. Furthermore, economic analysis has demonstrated that the negative impacts of pesticides on farmers' health in turn have negative impacts on farm productivity, so that the net results of reduction of pesticide use would be increased farm productivity [Antle and Pingali (1994)], especially when the direct productivity of pesticides is relatively low [Antle, Cole and Crissman (1998)].

Reductions in pesticide use could result from a variety of policy options, including the incentive policy of a tax on pesticides. Regulatory policies might include restricting pesticide application to trained personnel, curtailing or ending the use of highly toxic pesticides, or including use regulations on product labels. Over the past 15 years, many developing countries have begun to develop and enforce regulatory policies on pesticide use, but in several countries these policies are still rudimentary. Many pesticides sold in developing countries are products that are no longer used in industrialized nations because of health and environmental concerns. As policies tighten in these countries, use of the most toxic chemicals is projected to decline.

Technology and education policy options include public support for integrated pest management, greater use of host-plant resistance, and programs to promote the safe handling of pesticides and the use of protective clothing [Antle and Pingali (1994); Antle, Cole and Crissman (1998); Widawsky et al. (1998); Tjornhom, Norton and Gapud (1998)]. Exchange rate policy has also been found effectively to subsidize pesticide use [Lee and Espinosa (1998); Tjornhom, Norton and Gapud (1998)], so reforms of exchange rate policy can also have positive effects where pesticides have been overused.

5. Summary

The market for fertilizer in developing countries, measured by nutrient consumption, has expanded at a striking rate over the past 40 years. Fertilizer use has been stimulated in particular both by demand side factors such as the introduction of fertilizer responsive HYVs, and by supply side factors such as the reduction in fertilizer price as the result of technological advance in fertilizer production. The expansion in fertilizer consumption has been uneven, however, as it has been particularly high in many Asian countries. At the same time, fertilizer use in Sub-Saharan Africa has lagged far behind that in most other developing regions.

Data for developing country pesticide use is far more fragmentary, and such use is more specialized than fertilizer use, but it is likely that growth rates in pesticide use have also been high. Over the past 20 years, gradual expansion of integrated pest management (IPM) in developing countries has conceivably reduced that rate of growth from what it might have been otherwise, although there has been little apparent impact on the aggregate upward trend in pesticide use. Within the past five to ten years, genetically modified varieties in a few crops (cotton, soybeans, maize) have also begun to diffuse in a few developing countries. These countries, however, have tended to be those with large agricultural economies. The genetically modified varieties grown in countries like Argentina, Brazil, China, South Africa, and India have the potential to reduce pesticide use, or to shift herbicide use to less toxic herbicides.

At the broadest level, fertilizer demand has been driven by the relative scarcity of agricultural land. This explains some, but not all of the differences in consumption between many Asian countries on the one hand, and many Sub-Saharan African countries on the other. Governments, too, have stimulated or influenced fertilizer consumption through a wide variety of policies, including subsidization of both local fertilizer-production capacity and fertilizer use by farmers. While such policies undoubtedly increased fertilizer use in many countries with large agricultural sectors, considerable debate surrounds the question of whether this increase in use could have been obtained at less social cost. Market liberalization, both in terms of fertilizer-specific measures and more economy-wide changes, has become the common policy rubric over the past 20 years. "Liberalization", however, means different things in different contexts and needs to be defined carefully in policy studies.

The low use of fertilizer in Sub-Saharan Africa, where greater use would be expected to have both positive productivity and environmental effects, has been a particular policy concern. Once viewed as land abundant, Sub-Saharan Africa has been marked by reduction or elimination of soil-regenerating fallow periods and increased nutrient mining. When land measures are adjusted for production potential, a number of African countries surpass many Asian countries in intensity of land use. For a variety of reasons, including infrastructural and other institutional constraints, fertilizer use in these countries has remained low, however. Many analysts agree that sustainable increases in fertilizer use in this region are more likely to accompany broad policy changes such as investments in physical and market infrastructure, basic education, research and development, and improved institutions rather than direct provision of inputs or credit to farmers.

At the opposite side of the developing country continuum, environmental issues have become increasingly important over time in intensive agricultural systems where the use of agricultural chemicals has had a longer history and where application rates are high. In some cases, marginal fertilizer response measured at constant application rates might be declining over time because of resource degradation. In another example, pesticide use has been found in some instances not only to harm human health but to have significant negative effects on agricultural productivity as well.

Despite the disparate physical, technological, and economic circumstances in developing country markets for fertilizers and other farm chemicals, several common threads emerge. Meeting the production and environmental challenges of the future will require increasing reliance on knowledge-intensive technology. Although the balance of investments is likely to differ with the stages of market development, continued or increasing investments in education, agricultural research and development, infrastructure, and institutional development will be necessary to meet these challenges in intensive as well as in more marginal agriculture.

References

Abler, D.G., Shortle, J.S. (1995). "Technology as an agricultural pollution control policy". American Journal of Agricultural Economics 77 (1), 20–32.

Agrawal, G.D. (1999). "Diffuse agricultural water pollution in India". Water Science and Technology 39 (3), 33–47.

Ahmed, J.K., Falcon, W.P., Timmer, C.P. (1989). "Fertilizer policy for the 1990s", Development Discussion Paper No. 293 AFP. Harvard Institute for International Development, Cambridge, MA.

Ali, M., Byerlee, D. (2001). "Productivity growth and resource degradation in Pakistan's Punjab". In: Bridges, E.M., Hannam, I.D., Oldeman, L.R., Penning de Vries, F.W.T., Scherr, S.J., Sombatpanit, S. (Eds.), Response to Land Degradation. Science Publishers, Enfield, NH, pp. 186–199.

Antle, J.M., Cole, D.C., Crissman, C.C. (1998). "Further evidence on pesticides, productivity and farmer health: Potato production in Ecuador". Agricultural Economics 18, 199–207.

Antle, J.M., Pingali, P. (1994). "Pesticides, productivity, and farmer health: A Philippine case study". American Journal of Agricultural Economics 76, 418–430.

Baffes, J. (1998). "Structural reforms and price liberalization in Mexican agriculture". Journal of International Development 10, 575–587.

Baffes, J., Gautam, M. (1996). "Price responsiveness, efficiency, and the impact of structural adjustment on Egyptian crop producers". World Development 24 (4), 765–771.

Barker, R., Hayami, Y. (1976). "Price support vs. input subsidy for food self-sufficiency in developing countries". American Journal of Agricultural Economics 58 (4), 617–628.

Bates, R.H. (1981). Markets and States in Tropical Africa. University of California Press, Berkeley, CA.

Bhattacharyya, A., Kumbhakar, S.C. (2003). "Measuring price responsiveness under a profit maximization framework". Journal of Quantitative Economics, New Series 1 (1), 66–81.

Binswanger, H.P., Pingali, P. (1988). "Technological priorities for farming in Sub-Saharan Africa". World Bank Research Observer 3, 81–98.

Binswanger, H.P., Sillers, D.A. (1983). "Risk aversion and credit constraints in farmers' decision making: A reinterpretation". Journal of Development Studies 20, 5–21.

Bumb, B.L. (1988). "Fertilizer supply in Sub-Saharan Africa: An analysis". In: Tshibaka, T.B., Baanante, C.A. (Eds.), Fertilizer Policy in Tropical Africa. International Fertilizer Development Center and International Food Policy Research Institute, Lomé, Togo.

Bumb, B.L., Teboh, J.F., Atta, J.K., Asenso-Okyere, W.K. (1994). "Ghana policy environment and fertilizer sector development", IFDC Technical Bulletin No. T-41. International Fertilizer Development Center and Ghana Institute of Statistical, Social, and Economic Research, Muscle Shoals, AL.

Byerlee, D. (1992). "Technical change, productivity, and sustainability in irrigated cropping systems of South Asia: Emerging issues in the post-Green Revolution era". Journal of International Development 4 (5), 477–496.

Byerlee, D. (1996). "Modern varieties, productivity, and sustainability: Recent experiences and emerging challenges". World Development 24 (4), 697–718.

Cassman, K.G., Pingali, P.L. (1995). "Extrapolating trends from long-term experiments to farmers fields: The case of irrigated rice systems in Asia". In: Barnett, V., Payne, R., Steiner, R. (Eds.), Agricultural Sustainability in Economic, Environmental and Statistical Terms. John Wiley and Sons, London.
Chambers, R. (1985). "Least cost subsidization alternatives". American Journal of Agricultural Economics 67 (2), 251–256.
Chaudhary, M.A., Mufti, S.S. (1999). "Translog cost function estimation of farmer production and employment relationships". Pakistan Economic and Social Review 37 (1), 39–60.
Chaudhary, M.K., Harrington, L.W. (1993). "The rice–wheat system in Haryana: Input–output trends and sources of future productivity growth". CCS Haryana Agricultural University Regional Research Station (Karnal) and CIMMYT, Mexico, DF.
Choi, J.-S., Helmberger, P.G. (1993). "How sensitive are crop yields to price changes and farm programs?". Journal of Agricultural and Applied Economics 25 (1), 237–244.
Craig, B.J., Pardey, P.G., Roseboom, J. (1991). "Internationally comparable growth, development and research measures". In: Pardey, P.G., Roseboom, J., Anderson, J.R. (Eds.), Agricultural Research Policy: International Quantitative Perspectives. Cambridge University Press, Cambridge, UK, pp. 131–172.
Crawford, E., Kelly, V., Jayne, T.S., Howard, J. (2003). "Input use and market development in Sub-Saharan Africa: An overview". Food Policy 28, 277–292.
Crissman, C.C., Antle, J.M., Capalbo, S.M. (Eds.) (1998). Economic. Environmental, and Health Tradeoffs in Agriculture: Pesticides and the Sustainability of Andean Potato Production. Kluwer Academic Publishers, Boston, MA.
Daberkow, S., Taylor, H., Huang, W.-Y. (2003). "Nutrient use and management practices". In: Agricultural Resources and Environmental Indicators. Resource Economics Division, Economic Research Service, U.S. Department of Agriculture. Ch. 4.4, http://www.ers.usda.gov/publications/arei/ah722/arei4_4/AREI4_4nutrientmgt.pdf.
Dalrymple, D.G. (1975). "Evaluating fertilizer subsidies in developing countries", AID Discussion Paper No. 30. USAID, Washington, DC.
de Gorter, H., Swinnen, J. (2002). "Political economy of agricultural policy". In: Gardner, B.L., Rausser, G.C. (Eds.), Handbook of Agricultural Economics, vol. 2B, Agricultural and Food Policy. North-Holland, Amsterdam.
Denbaly, M., Vroomen, H. (1993). "Dynamic fertilizer nutrient demands for corn: A cointegrated and error-correcting system". American Journal of Agricultural Economics 75 (1), 203–209.
Desai, G.M. (1988). "Policy for rapid growth in the use of modern agricultural inputs". In: Mellor, J., Ahmed, R. (Eds.), Agricultural Price Policy for Developing Countries. Johns Hopkins University Press, Baltimore, MD, pp. 204–218.
Desai, G.M. (1991). "Fertilizer policies: A perspective on price and subsidy issues", Paper presented at the National Consultation on Fertilizer Pricing: An Aspect of Public Policy, Jaipur, India, 30 September–2 October 1991.
Desai, G.M., Gandhi, V. (1988). "Fertilizer consumption in Sub-Saharan Africa: An analysis of growth and profile of use". In: Tshibaka, T.B., Baanante, C.A. (Eds.), Fertilizer Policy in Tropical Africa. International Fertilizer Development Center and International Food Policy Research Institute, Lomé, Togo.
Desai, G.M., Stone, B. (1987). "Fertilizer market development and national policy in China and India: A comparative perspective", Paper presented at the IFA-FADINAP Southeast Asia and Pacific Regional Fertilizer Conference, Kuala Lumpur, Malaysia, 22–25 July.
Dorward, A., Kydd, J., Morrison, J., Urey, I. (2004). "A policy agenda for pro-poor agricultural growth". World Development 32 (1), 73–89.
Dudal, R., Byrnes, B.H. (1993). "The effects of fertilizer use on the environment". In: van Reuler, H., Prins, W.H. (Eds.), The Role of Plant Nutrients for Sustainable Food Crop Production in Sub-Saharan Africa. VKP (Dutch Association of Fertilizer Producers), Leidschendam, The Netherlands.
Evenson, R.E., Avila, A.F. (forthcoming). "FAO data-based TFP measures". In: Evenson, R.E., Pingali, P. (Eds.), Handbook of Agricultural Economics, vol. 4. Elsevier, Amsterdam.
Evenson, R.E., Gollin, D. (2003). "Assessing the impact of the Green Revolution, 1960 to 2000". Science 300, 758–762.

FAO (2000). Fertilizer Requirements in 2015 and 2030. Food and Agricultural Organization of the United Nations, Rome.

FAO (2004). The State of Food and Agriculture, 2003–2004. Food and Agricultural Organization of the United Nations, Rome.

Feder, G., Just, R.E., Zilberman, D. (1985). "Adoption of agricultural innovations in developing countries: A survey". Economic Development and Cultural Change 33 (2), 255–298.

Feder, G., Murgai, R., Quizon, J.B. (2004). "Sending farmers back to school: The impact of farmer field schools in Indonesia". Review of Agricultural Economics 26 (1), 45–62.

Feder, G., Umali, D.L. (1993). "The adoption of agricultural innovations". Technological Forecasting and Social Change 43, 215–239.

Food and Agricultural Organization of the United Nations (2004). FAOSTAT database. http://apps.fao.org/faostat/collections?version=ext&hasbulk=1&subset=agriculture.

Freeman, A., Omiti, J. (2003). "Fertilizer use in semi-arid areas of Kenya: Analysis of smallholders' adoption behavior under liberalized markets". Nutrient Cycling in Agroecosystems 66, 23–31.

Gandhi, V.P. (1997). "Growth and economics of pesticide use in India: Overview – Analysis of the environment, patterns and market potential", Working Paper No. 1372. Indian Institute of Management, Ahmedabad, India.

Gerner, H., Harris, G. (1993). "The use and supply of fertilizers in Sub-Saharan Africa". In: van Reuler, H., Prins, W.H. (Eds.), The Role of Plant Nutrients for Sustainable Food Crop Production in Sub-Saharan Africa. VKP (Dutch Association of Fertilizer Producers), Leidschendam, The Netherlands.

Gisselquist, D., Van Der Meer, C. (2000). Regulations for Seed and Fertilizer Markets: A Good Practice Guide for Policymakers. The World Bank, Washington, DC.

Harris, G. (1998). "An analysis of global fertilizer application rates for major crops". In: Proceedings of the FAO/IFA/IFDC Informal Meeting on Fertilizer Use by Crop, Rome, Italy, 2–3 March.

Hayami, Y., Ruttan, V.W. (1985). Agricultural Development: An International Perspective, revised and expanded ed. Johns Hopkins University Press, Baltimore, MD.

Heisey, P.W. (Ed.) (1990). Accelerating the Transfer of Wheat Breeding Gains to Farmers: A Study of the Dynamics of Varietal Replacement in Pakistan, CIMMYT Research Report No. 1. CIMMYT, Mexico, DF.

Heisey, P.W., Mwangi, W. (1997). "Fertilizer use and maize production in Sub-Saharan Africa". In: Byerlee, D., Eicher, C.K. (Eds.), Africa's Emerging Maize Revolution. Lynne Rienner Publishers, Boulder, CO.

Huang, J., Rozelle, S. (1995). "Environmental stress and grain yields in China". American Journal of Agricultural Economics 77 (4), 853–864.

International Fertilizer Industry Association (IFA), International Fertilizer Development Center (IFDC), International Potash Institute (IPI), Phosphate and Potash Institute (PPI), and Food and Agricultural Organization of the United Nations (FAO) (2002). Fertilizer Use by Crop. FAO, Rome.

International Rice Research Institute (IRRI) (2003). http://www.peopleandplanet.net/doc.php.

James, C. (2003). "Preview. Global status of commercialized transgenic crops: 2003", International Service for the Acquisition of Agri-Biotech Applications (ISAAA) Brief No. 30. ISAAA, Ithaca, NY.

Janssen, B.H. (1993). "Integrated nutrient management: The use of organic and mineral fertilizers". In: van Reuler, H., Prins, W.H. (Eds.), The Role of Plant Nutrients for Sustainable Food Crop Production in Sub-Saharan Africa. VKP (Dutch Association of Fertilizer Producers), Leidschendam, The Netherlands.

Jayne, T.S., Govereh, J., Wanzala, M., Demeke, M. (2003). "Fertilizer market development: A comparative analysis of Ethiopia, Kenya, and Zambia". Food Policy 28, 293–316.

Kelly, V., Adesina, A.A., Gordon, A. (2003). "Expanding access to agricultural inputs in Africa: A review of recent market development experience". Food Policy 28, 379–404.

Khan, M.A., Iqbal, M., Ahmad, I., Soomro, M.H. (2002). "Economic evaluation of pesticide use externalities in the cotton zones of Punjab, Pakistan". Pakistan Development Review 41 (4), 683–697.

Lamb, R.L. (2003). "Fertilizer use, risk, and off-farm labor markets in the semi-arid tropics of India". American Journal of Agricultural Economics 85 (2), 359–371.

Lee, D.R., Espinosa, P. (1998). "Economic reforms and changing pesticide policies in Ecuador and Colombia". In: Crissman, C.C., Antle, J.M., Capalbo, S.M. (Eds.), Economic, Environmental, and Health Trade-offs in Agriculture: Pesticides and the Sustainability of Andean Potato Production. Kluwer Academic Publishers, Boston, MA.
Maredia, M.K., Byerlee, D. (2000). "Efficiency of research investments in the presence of international spillovers: Wheat research in developing countries". Agricultural Economics 22 (1), 1–16.
Marfo, K., Tripp, R. (1997). "Maize technology development and Ghana's economy". In: Byerlee, D., Eicher, C.K. (Eds.), Africa's Emerging Maize Revolution. Lynne Rienner Publishers, Boulder, CO.
Masson, R.T. (1972). "The creation of risk aversion by imperfect capital markets". American Economic Review 62, 77–86.
Matlon, P.J., Spencer, D.S.C. (1984). "Increasing food production in Sub-Saharan Africa: Environmental problems and inadequate technological solutions". American Journal of Agricultural Economics 66 (5), 671–676.
Miller, T., Tolley, G. (1989). "Technology adoption and agricultural price policy". American Journal of Agricultural Economics 71 (4), 847–857.
Mwangi, W. (1997). "Low use of fertilizers and low productivity in Sub-Saharan Africa". Nutrient Cycling in Agroecosystems 47 (2), 135–147.
Ndayisenga, F., Schuh, G.E. (1995). "Fertilizer policy in Sub-Saharan Africa: Recurring issues and recommendations", Mimeo. Sasakawa-Global 2000 Draft Paper. Minneapolis, MN.
Norton, G.W., Mullen, J. (1994). "Economic evaluation of integrated pest management: A literature review", Virginia Cooperative Extension Publication No. 448-120. Virginia Polytechnic Institute and State University, Blacksburg, VA.
Norton, G.W., Heinrichs, E.A., Luther, G.C., Irwin, M.E. (2005). Globalizing Integrated Pest Management: A Participatory Research Process. Blackwell Press, Ames, IA.
Nwosu, A.C. (1995). "Fertilizer supply and distribution policy in Nigeria". In: Ikpi, A.E., Olayemi, J.K. (Eds.), Sustainable Agriculture and Economic Development in Nigeria. Winrock, Morrilton, AR.
Omamo, S.W., Mose, L.O. (2001). "Fertilizer trade under market liberalization: Preliminary evidence from Kenya". Food Policy 26, 1–10.
Ongley, E.D. (1996). "Control of water pollution from agriculture", FAO Irrigation and Drainage Paper No. 55. FAO, Rome.
Osteen, C., Fernandez-Cornejo, J., Greene, C., Jans, S., Padgitt, M. (2003). "Pest management practices". In: Agricultural Resources and Environmental Indicators. Resource Economics Division, Economic Research Service, U.S. Department of Agriculture. Ch. 4.3, http://www.ers.usda.gov/publications/arei/ah722/arei4_3/AREI4_3pestmgt.pdf.
Parikh, A. (1990). "Fertilizer consumption in Bangladesh: Tobit analysis of farm level data". Indian Economic Review 25 (1), 1–15.
Parish, R., McLaren, K. (1982). "Relative cost-effectiveness of input and output subsidies". Australian Journal of Agricultural Economics 26 (1), 1–13.
Pingali, P.L., Roger, P.A. (1995). Impact of Pesticides on Farmer Health and the Rice Environment. Kluwer Academic Publishers, Boston, MA.
Pinstrup-Andersen, P. (1993). "Fertilizer subsidies: Balancing short-term responses with long-term imperatives". In: Russell, N.C., Dowswell, C.R. (Eds.), Policy Options for Agricultural Development in Sub-Saharan Africa. CASIN/SAA/Global 2000, Mexico, DF.
Place, F., Barrett, C.B., Freeman, H.A., Ramisch, J.J., Vanlauwe, B. (2003). "Prospects for integrated soil fertility management using organic and inorganic inputs: Evidence from smallholder African agricultural systems". Food Policy 28, 365–378.
Pray, C.E., Huang, J. (2003). "The impact of Bt cotton in China". In: Kalaitzandonakes, N. (Ed.), The Economic and Environmental Impacts of Agbiotech: A Global Perspective. Kluwer–Plenum Academic Publishers, New York.
Qaim, M., Traxler, G. (2005). "Roundup ready soybeans in Argentina: Farm level and aggregate welfare effects". Agricultural Economics 32 (1), 73–86.

Qaim, M., Zilberman, D. (2003). "Yield effects of genetically modified crops in developing countries". Science 299, 900–902.

Qiao, F., Lohmar, B., Huang, J., Rozelle, S., Zhang, L. (2003). "Producer benefits from input market and trade liberalization: The case of fertilizer in China". American Journal of Agricultural Economics 85 (5), 1223–1227.

Renkow, M. (2000). "Poverty, productivity, and production environment: A review of the evidence". Food Policy 25 (2), 463–478.

Ribaudo, M., Johansson, R. (2006). "Water quality: Impacts of agriculture". In: Agricultural Resources and Environmental Indicators. Resource and Rural Economics Division, Economic Research Service, U.S. Department of Agriculture. Ch. 2.2, http://www.ers.usda.gov/publications/arei/eib16/eib16_2-2.pdf.

Rola, A.C., Pingali, P.L. (1993). Pesticides, Rice Productivity, and Farmers' Health. International Rice Research Institute, Los Baños, Philippines.

Rougoor, C.W., Van Zeijts, H., Hofreither, M.F., Bäckman, S. (2001). "Experiences with fertilizer taxes in Europe". Journal of Environmental Planning and Management 44 (6), 877–887.

Roumasset, J.A., Rosegrant, M.W., Chakravorty, U.N., Anderson, J.R. (1989). "Fertilizer and crop yield variability: A review". In: Anderson, J.R., Hazell, P.B.R. (Eds.), Variability in Grain Yields: Implications for Agricultural Research and Policy in Developing Countries. Johns Hopkins University Press, Baltimore, MD.

Sachs, J.D., McArthur, J.W., Schmidt-Traub, G., Kruk, M., Bahadur, C., Faye, M., McCord, G. (2004). "Ending Africa's poverty trap". Brookings Papers on Economic Activity 1, 117–216.

Saha, A., Shumway, C.R., Havenner, A. (1997). "The economics and econometrics of damage control". American Journal of Agricultural Economics 79 (3), 773–785.

Scheirling, S.M. (1995). "Overcoming agricultural pollution of water: The challenge of integrating agricultural and environmental policies in the European Union", World Bank Technical Paper No. 269. The World Bank, Washington, DC.

Schuh, G.E. (1974). "The exchange rate and U.S. agriculture". American Journal of Agricultural Economics 56 (1), 1–13.

Segura, E.L., Shetty, Y.T., Nishimizu, M. (Eds.) (1986). Fertilizer Producer Pricing in Developing Countries: Issues and Approaches. Industry and Finance Series, vol. 11. The World Bank, Washington, DC.

Shalit, H., Binswanger, H.P. (1985). "Is there a theoretical case for fertilizer subsidies?", Discussion Paper No. ARU 27. Research Unit, Agricultural and Rural Development Department, The World Bank, Washington, DC.

Sheldrick, W.F. (1990). "Structure of the international urea market and the outlook for urea prices and price fluctuations". FAO Commission on Fertilizers, Rome.

Shepherd, A. (1989). "Approaches to the privatization of fertilizer marketing in Africa". Food Policy 14 (2), 143–154.

Shepherd, A., Coster, R. (1987). "Fertilizer marketing costs and margins in developing countries", Mimeo. FAO/FIAC (Fertilizer Industry Advisory Council), Rome.

Sidhu, J.S., Sidhu, D.S. (1985). "Price-support versus fertiliser subsidy: An evaluation". Economic and Political Weekly 20 (13), A17–A22.

Smaling, E.M.A. (1993). "Soil nutrient depletion in Sub-Saharan Africa". In: van Reuler, H., Prins, W.H. (Eds.), The Role of Plant Nutrients for Sustainable Food Crop Production in Sub-Saharan Africa. VKP (Dutch Association of Fertilizer Producers), Leidschendam, The Netherlands.

Smith, J., Barau, A.D., Goldman, A., Mareck, J.H. (1994). "The role of technology in agricultural intensification: The evolution of maize production in the northern Guinea savanna of Nigeria". Economic Development and Cultural Change 42 (3), 537–554.

Spencer, D.S.C. (1994). "Infrastructure and technology constraints to agricultural development in the humid and sub-humid tropics of Africa", EPTD Discussion Paper No. 3. IFPRI, Washington, DC.

Stoorvogel, J.J., Smaling, E.M.A., Janssen, B.H. (1993). "Calculating soil nutrient balances in Africa at different scales. I. Supranational scale". Fertilizer Research 35, 227–235.

Sunding, D., Zilberman, D. (2001). "The agricultural innovation process: Research and technology adoption in a changing agricultural sector". In: Gardner, B.L., Rausser, G.C. (Eds.), Handbook of Agricultural Economics, vol. 1A, Agricultural Production. North-Holland, Amsterdam.
Teague, M.L., Brorsen, B.W. (1995). "Pesticide productivity: What are the trends?". Journal of Agricultural and Applied Economics 27 (1), 276–282.
Timmer, C.P. (1986). Getting Prices Right: The Scope and Limits of Agricultural Price Policy. Cornell University Press, Ithaca, NY.
Tjornhom, J.D., Norton, G.W., Gapud, V. (1998). "Impacts of price and exchange rate policies on pesticide use in the Philippines". Agricultural Economics 18 (2), 167–175.
Tobin, R.J. (1996). "Pest management, the environment, and Japanese foreign assistance". Food Policy 21 (2), 211–228.
Tomich, T.P., Kilby, P., Johnston, B.F. (1995). Transforming Agrarian Economies: Opportunities Seized, Opportunities Missed. Cornell University Press, Ithaca, NY.
Traxler, G. (2004). "The economic impacts of biotechnology-based technological innovations", FAO-ESA Working Paper 04-08. FAO, Rome.
Truong, T.V., Walker, S.T. (1990). "Policy reform as institutional change: Privatizing the fertilizer subsector in Cameroon". In: Brinkerhoff, D., Goldsmith, A. (Eds.), Institutional Sustainability in Agriculture and Rural Development: A Global Perspective. Praeger, New York.
U.S. Department of Agriculture/Economic Research Service (USDA/ERS) (2004). http://www.ers.usda.gov/briefing/agchenicals/questions/pmqa4.htm.
U.S. Embassy, Beijing (1996). "Agricultural pesticide use in China". U.S. Embassy, Beijing, Environment, Science and Technology Section. http://www.usembassy-china.org.cn/sandt/mu5pest.htm.
Vlek, P.L.G. (1990). "The role of fertilizers in sustaining agriculture in Sub-Saharan Africa". Fertilizer Research 26, 327–339.
Widawsky, D., Rozelle, S., Jin, S., Huang, J. (1998). "Pesticide productivity, host-plant resistance and productivity in China". Agricultural Economics 19, 203–217.
Williams, S.P., Shumway, C.R. (2000). "Trade liberalization and agricultural chemical use: United States and Mexico". American Journal of Agricultural Economics 82 (1), 183–199.
World Resources Institute (1999). "Intensification of agriculture: Chemical inputs". http://www.wri.org/wr-98-99/agrichem.htm.

Chapter 54

AGRICULTURAL MECHANIZATION: ADOPTION PATTERNS AND ECONOMIC IMPACT*

PRABHU PINGALI

Agricultural and Development Economics Division, FAO, United Nations, Rome, Italy

Contents

* The assistance of Hideyuki Nakagawa and Annelies Deuss is gratefully acknowledged.

Handbook of Agricultural Economics, Volume 3
Edited by Robert Evenson and Prabhu Pingali

DOI: 10.1016/S1574-0072(06)03054-4

Abstract

Over the past half a century developing regions, with the exception of Sub-Saharan Africa, have seen labor-saving technologies adopted at unprecedented levels. Intensification of production systems created power bottlenecks around the land preparation, harvesting and threshing operations. Alleviating the power bottlenecks with the adoption of mechanical technologies helped enhance agricultural productivity and lowered the unit cost of crop production even in the densely populated countries of Asia. Economic growth and the commercialization of agricultural systems is leading to further mechanization of agricultural systems in Asia and Latin America. Sub-Saharan Africa continues to have very low levels of mechanization and available data indicate declining rather than increasing levels of adoption, even among the countries that were the early trendsetters, such as Kenya and Zimbabwe. This chapter documents the trends and sequential patterns in the adoption of mechanical technology, assesses the evidence on the productivity and equity impact of mechanization, and discusses the implication for mechanization policy.

Keywords

agricultural mechanization, adoption, labor productivity, impact

JEL classification: O13, O14, O31, O32, O38

1. Introduction

Over the past half a century developing regions, with the exception of Sub-Saharan Africa, have seen labor-saving technologies adopted at unprecedented levels. Intensification of production systems created power bottlenecks around the land preparation, harvesting and threshing operations. Alleviating the power bottlenecks with the adoption of mechanical technologies helped enhance agricultural productivity and lowered the unit cost of crop production even in the densely populated countries of Asia. Mechanization of agricultural operations was very selective and sequential; power-intensive operations such as land preparation, threshing and milling were readily mechanized. While operations that require more human judgment, such as weeding, continued to be done by hand under low wage conditions.

Economic growth and the commercialization of agricultural systems is leading to further mechanization of agricultural systems in Asia and Latin America. The advanced countries of East Asia have a completely mechanized rice production system, while the rapidly growing countries of Southeast Asia are moving in that direction [Pingali (1998)]. Middle income countries of Latin America, such as Brazil, Chile and Mexico, are observing a similar rapid shift to labor-saving technologies, both mechanical as well as chemical. Conservation tillage in association with herbicide use has resulted in significant cost savings in cereal production systems in Brazil and Argentina [Ekboir (2000)]. Also in Latin America, vertical integration of the post-harvest processing industry is leading to the replacement of small village-based post-harvest facilities with large-scale processing plants [Balsevich et al. (2003)].

Sub-Saharan Africa continues to have very low levels of mechanization and available data indicate declining rather than increasing levels of adoption, even among the countries that were the early trendsetters, such as Kenya and Zimbabwe. The persistent low levels of mechanization in relatively land abundant Sub-Saharan Africa has been a longstanding puzzle in the literature on agricultural mechanization [Pingali, Bigot and Binswanger (1987)]. The explanation is in the driving forces of agricultural intensification and the incentives for increasing productivity growth. Agricultural areas facing relatively inelastic demand conditions, due to low population densities and/or poor market infrastructure, tend to persist in low intensity, low yield subsistence production systems. The move to mechanical technologies for land preparation is not cost-effective in such societies. Attempts to expand the area under cultivation and to modernize agriculture by bringing tractors into such areas have consistently failed. Tractors by themselves are not an effective tool for inducing the process of agricultural intensification and productivity growth.

The critics of mechanization have argued that the widespread use of labor-saving technologies has had serious equity consequences in terms of the displacement of labor and tenant farmers. Existing evidence indicates, however, that the equity consequences have not been as severe or as widespread as they are presumed to be. The mechanization of power-intensive operations have had minimal equity effects even in the labor surplus economies of Asia. The switch from manual labor to mechanical or chemical technologies for control-intensive operations, such as weeding, has had adverse equity effects in low-wage countries. However, where markets have been allowed to function with minimal government intervention, control-intensive operations continue to be performed by human labor until wages rise due to increased labor withdrawal from the agricultural sector. Serious equity consequences are invariably associated with policies that inappropriately promote mechanization, such as subsidized credit for tractor purchase.

This chapter documents the trends and sequential patterns in the adoption of mechanical technology (Section 2), assesses the evidence on the productivity and equity impact of mechanization (Section 3), and discusses the implication for mechanization policy (Section 4).

2. Trends and patterns in agricultural mechanization

Agricultural operations can be grouped according to the relative intensity with which they require power, or energy, in relation to the control functions of the human mind, or judgment. Operations such as land preparation, transport, pumping, milling, grinding, and threshing are power intensive, while weeding, sifting, winnowing, and fruit harvesting, for example, are control-intensive operations. The shift of the source of power from human to animal to mechanical power is dependent on the level of power intensity and control intensity of the operation [Binswanger (1984)]. Table 1 offers a comparison of operations according to their power intensity and control intensity and the sequence of their transfer to the new power source.

In both land scarce and land-abundant economies the power intensive operations are the ones to be mechanized first, while the mechanization of control intensive operations occurs much later and is closely associated with the wage rate. The sequence, in which power intensive operations are mechanized, however, differs according to land

Table 1
Comparison of agricultural operations according to their power and control intensity

Nature of operation and source of power	Low control intensity, high power intensity	Intermediate intensity	High control intensity, low power intensity
Stationary operations	Grinding, milling, crushing Water lifting Threshing, wood cutting		Sifting, winnowing
Mobile operations	Transport Primary tillage	Harvesting root crops Harvesting grain crops Secondary tillage and interculture	Weeding and harvesting tea, coffee, and apples Seeding

Source: Pingali and Binswanger (1987).

endowments. In land-scarce economies the pumping operation is generally the first to be mechanized, using diesel and electric pumps. Pumps provide a supply of water for double or triple cropping and allow the expansion of cultivation onto marginal lands. Mechanical mills, tillage, and transport equipment follow the adoption of pumps. In land-abundant economies, the use of pumps is delayed until conditions of land scarcity emerge; milling and transport operations are the first to be mechanized, followed by tillage where its mechanization is feasible. The threshing operation, although power intensive, is generally not mechanized where wages are low and harvested volumes are small. Even when land is abundant, therefore, if agricultural production is mainly for subsistence, threshing is the last of the power intensive operations to be mechanized.

The mechanization of power-intensive operations has taken place rapidly even in countries with high population densities and low wages, such as India, Bangladesh, and the Philippines [Herdt (1983); Pingali and Binswanger (1987)]. Mechanization reduced the costs of power-intensive operations significantly as well as ensured their timely completion. Mechanization of control intensive operations is more closely associated with the wage rate. In land-scarce economies in which nonagricultural demand for labor is low, operations such as weeding, interculture, and harvesting continue to be done by human and animal power. Cultivators became prevalent in Japan during the late 1950s, when agricultural wages rose in response to rapid post-war industrialization. It was only in the 1970s that rice transplanters and harvest combines began to be used in Japan [Ohkawa, Shinohara and Umemura (1965)]. In India, and the Philippines, where tractors do tillage and transport, interculture continues to be done by hand and animal-drawn equipment, while harvesting is done only by hand. In land-abundant economies where market opportunities are good and the wage rates are high, many of the control-intensive operations are transferred to mechanical power. This selective nature of the adoption of a new source of power – was also observed in Europe and the United States [Hurt (1982)]. Numerous examples of the sequential adoption of mechanical technologies, for both the developed as well as the developing world can be found in Binswanger (1984), Pingali, Bigot and Binswanger (1987), and Pingali (1998).

2.1. Power tillers/tractors

The levels of adoption of mechanical equipment for land preparation vary significantly by continent, by crop and by farming system. Figure 1 provides a comparison across the three continents of the trends in tractor adoption since 1960 using data from FAO. Intensively cultivated lands of Asia and Latin America have experienced significant levels of mechanization, while adoption rates in Africa, especially Sub-Saharan Africa are extremely low. In fact, one observes a reversal in tractor adoption among some African countries that were thought to be early adopters. In 2002 average tractor use in Sub-Saharan Africa was around 1.3 per 1000 hectares of cultivated land, while in South Asia it was around 9.1 and in Latin America it was around 10.4 tractors for the same time period. Tractor use in Sub-Saharan Africa peaked at 1.9 per 1000 hectares in 1986 and has gradually declined since then. Permanent cultivation systems under grain crops,

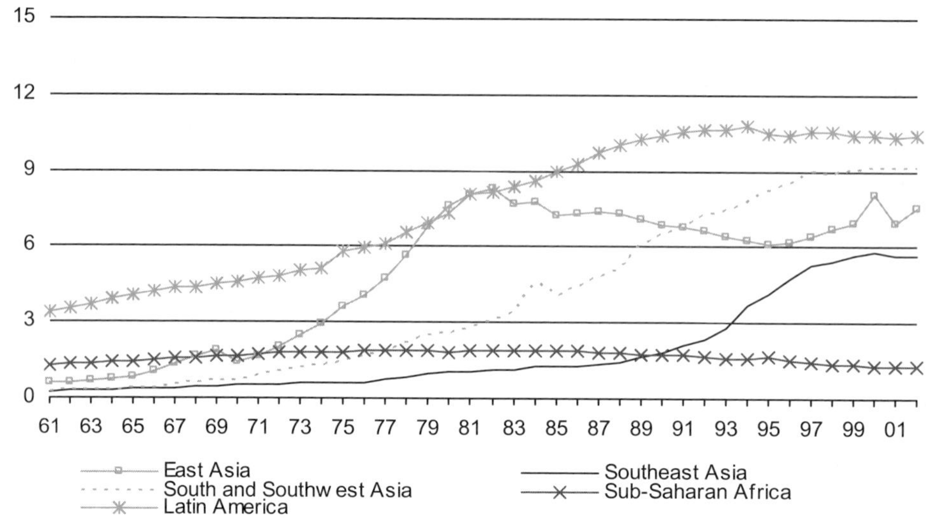

Figure 1. Number tractors per 1000 hectare of crop land – by region. *Source*: FAO (2005).

such as rice, wheat and maize, tend to have higher levels of mechanization relative to less intensively cultivated systems under root and tuber crops.

2.1.1. Asia

In the case of Asia one needs to make a distinction between power tiller and tractor use. Intensive wetland rice production systems in East and Southeast Asia have witnessed a switch from animal drawn plows to power tillers. While four wheel tractors are commonly used for non-rice crops and for dryland environments. South Asia on the other hand has relied much more extensively on four wheel tractors, although power tiller numbers are rising in Bangladesh [Salokhe and Ramalingam (1998); Hossain, Bose and Mustafi (2002)].

Japan and Korea led the rest of Asia in the speed and extent of mechanization and set a pattern that other countries followed. In Japan by 1960, the mechanization of pumping and threshing had already been completed, and the use of power tillers had just started to take off. The number of power tillers on Japanese farms grew from 750,000 units in 1960 to 2.5 million units in 1965 [Kisu (1983)]. By 1965 there was one power tiller for every 2 ha of crop land in Japan [Herdt (1983)]; by 1989, Japan had more than one power tiller per hectare of riceland [Mizuno (1991)]. The Korean experience was similar: number of power tillers rose from a little more than 1000 in 1965 to around 290,000 by 1980 [Cho (1983)]. By 1970, Korea had approximately one power tiller for every 10 ha of riceland [Herdt (1983)], and by 1989, one power tiller for 2 ha of riceland [APO (1991)]. The process of decollectivization in China has led to rapid mechanization of farm operations using power tillers and other small machines. By 1992 China had around 220 power tillers per thousand hectares of riceland [CSO (1993)]. In 1994, 71% of the farm machinery belonged to individuals, 10% to the state, and 19% to collectives [Salokhe and Ramalingam (1998)].

In Southeast Asia, mechanization of ricelands took off in the early 1980s. Power tiller use rose from 26 and 14 per thousand ha of riceland in Thailand and the Philippines, respectively, in 1980 [Herdt (1983)] to approximately 56 and 20 tillers per thousand ha by 1990 [APO (1991)]. Indonesia, Myanmar, and Vietnam have been slower to switch to power tillers for preparing ricelands, reporting less than one power tiller per thousand hectares of rice lands in 1990. Although the situation has changed since then in Vietnam, liberalization of the agriculture sector in the early 1990s has lead to a rapid increase in the adoption of modern rice technologies, including power tillers [Pingali, Hossain and Gerpacio (1997)]. Vietnam today, has the highest number of tractors per 1000 hectares of cultivated land relative to other countries in Southeast Asia. For Thailand and the Philippines, national average figure do not reflect the substantial variation in power tiller adoption by rice environment. The irrigated rice bowl provinces of the two countries tend to be highly mechanized, having more than one power tiller per 10 ha of land, whereas the less favorable rice environments continue to rely on animal power. South Asia is very diverse in terms of the mechanization of land preparation, both across countries and within countries.

Although a superficial look at national average figures would seem to indicate that South Asian countries continue to rely on animal draft power, the intensively cultivated "Green Revolution" provinces, such as the Indian Punjab, tend to be on the same mechanization pathway as similar rice-growing regions in Southeast Asia [Pingali (1998)]. Mechanized land preparation is most advanced in Sri Lanka and least in Nepal. As mentioned earlier, unlike Southeast Asian countries, in South Asia the mechanization of land preparation has emphasized four-wheel tractors rather than power tillers. The larger tractors are more conductive to the rice – non-rice crop rotations that are common in South Asia, more suitable for operating rental markets over a larger geographic area, and more amenable to use as transport vehicles. Tractor numbers in India rose from 0.19 per 1000 hectares in 1961 to 9 per 1000 hectares by 2000, a level that is now at par to that of other developing countries in East and Southeast Asia (Figure 1). Liberalized agricultural equipment import policies in Bangladesh have lead to a dramatic increase in small pump and power tiller use. The 1983–1984 Agricultural Census reported nonexistence of farm machinery except irrigation equipment. The 1996 Agricultural Census, on the other hand, reported ownership of 150,000 power tillers or tractors by 140,000 rural households who constitute 1.2% farm households in the country [Hossain, Bose and Mustafi (2002)]. Mandal (2002) estimates that since then around 15,000 power tillers have been imported annually.

2.1.2. Africa

The first tractors appeared in Sub-Saharan Africa during the period between the wars. They were used initially on settler farms and government-run farms. After 1945, however, tractors began to be used by African farmers also. Most of the postwar imports of tractors were financed by the fund for farm machinery allocated through the Marshall Plan. Since then bilateral and multilateral aid has been supporting the import of tractors for agricultural purposes. Between 1945 and 1981 there were three significant spurts in the number of tractors in use, with intermediate periods of slow growth [Pingali, Bigot and Binswanger (1987)].

The first spurt came around 1945 under colonial influence, and the countries that started to promote the use of tractors during the period 1945–1955 – Zimbabwe, Kenya, Zambia, and Malawi – can be called the first generation of tractor users. The use of tractors in these countries spread from colonial farms to private farms owned by native Africans. The second spurt of tractorization came between 1958 and 1970 and it can be characterized as state sponsored mechanization in some newly independent countries such as – Tanzania, Ethiopia, Ghana, and Cote d'Ivoire. In many of these countries tractors were provided through cooperative farms, state farms, or tractor-hire services.

The third spurt in tractor numbers took place between 1970–1980 when oil and other resource exporting countries, such as Nigeria, Cameroon and the Democratic Republic of Congo tried to re-distribute the gains from resource exports to rural areas. Tractor were purchased and provided to farmers either through subsidized credit schemes or through state sponsored hire-schemes [Pingali, Bigot and Binswanger (1987)].

In a number of countries throughout Sub-Saharan Africa, including Burkina Faso, Niger, Rwanda, Burundi, and Liberia, there was never an organized effort to increase the use of tractors, and the number of tractors is very small. As mentioned earlier, Sub-Saharan Africa has, in fact, fewer tractors per thousand inhabitants than either Asia or Latin America (Figure 1) and the numbers are declining even in countries that experienced the early boost in tractor numbers. The use of tractors was and continues to be restricted to a small commercial farm sector.

Pingali, Bigot and Binswanger (1987) summarize the experiences of tractor projects in Sub-Saharan Africa from 1945 until the early 1980s (Table 2). They identified thirty tractor projects, of which seven were smallholder projects, thirteen were tractor-hire schemes, and ten were block cultivation schemes. Tractor-hire schemes are government-sponsored rental programs for multifarm use of equipment. Block cultivation schemes are group of farms being managed and operated as a single unit, often with mechanization and other modern inputs. The following conclusions can be drawn from Table 2: In many tractor project areas no tractors can be found today. Where any tractors are still being used, their use is inevitably associated with rice cultivation. But even these surviving tractors today are privately owned. The transition from the hand hoe to animal-draft power, where its use is appropriate, continued to be made despite the emphasis on tractors. Of the seventeen attempts to bypass animal traction for tractorization only three succeeded, all of them associated with low-land rice cultivation. None of the block cultivation schemes has ever been successful. For an evaluation of the performance of tractors in Sub-Saharan Africa, see Labrousse (1971), Bonnefond (1967), and Kline et al. (1969).

Given that the rapid spread of mechanical equipment has historically been associated with an abundance of land, why has the spread of mechanization in Sub-Saharan

Table 2
Experiences with tractor projects in Sub-Saharan Africa

Number of projects or areas	Individual farm enterprise	Tractor-hire service	Block cultivation schemes
Initial number of projects	7	13	10
Number of areas in which tractors are still used in the original organization form in the 1980s	None	5	None
Number of areas in which tractors are now used under private operation	None	4	2
Number of areas that had animal traction originally	3	7	2
Number of areas in which animal traction is in use in the 1980s	6	9	3
Number of areas in which continued use of the tractor is associated with rice cultivation	1[a]	9	None

Source: Pingali and Binswanger (1987).
[a]Tractor-hire scheme.

Africa been slower than in countries such as India and China, where labor is abundant and wages are low? Pingali, Bigot and Binswanger (1987) explain the above puzzle in terms of the lack of farm level incentives for the intensification of agricultural systems and hence a low demand for substituting out of existing power sources which in Africa is mainly human. Intensification of agricultural systems is associated with rising demand for agricultural products, triggered by growing populations and/or improved access to domestic or international markets. Farming communities facing an inelastic demand for their products, tend to persist in farming practices that are of low intensity and low productivity. Power requirements in such systems are low and can easily be met by human labor. As farming intensities increase, the number of tasks that need to be performed increases as does the energy requirement for each of the tasks, hence the adoption mechanical technology. Pingali, Bigot and Binswanger (1987) observe a positive correlation between the evolution of farming systems and mechanization based on the extensive field research in Sub-Saharan Africa. The numerous failed experiences across Africa indicate clearly that tractors cannot be used as an instrument for driving the process of intensification. Where the demand side factors are in place, agricultural intensification and the adoption of mechanical power occurs in Africa in a similar pattern to Asia and Latin America.

2.2. *Milling and other post-harvest operations*

Postharvest processing operations are extremely labor-intensive and tedious to perform. Miracle (1967) reported, for instance, that to grind a week's supply of maize meal –

thirty pounds – by hand would take from eight to fourteen hours. The same operation would take half an hour with a hand mill and perhaps not more than ten minutes with a motorized mill. The same is true of dehusking rice, crushing sugarcane, grinding groundnuts, and ginning cotton. In most parts of the world these operations have been transferred to stationary machines powered by water, wind, steam, and – more recently – internal combustion engines or electricity.

Water was first used for milling, pounding, and grinding in the first century B.C. in China, and its use for these purposes was fairly widespread between the second and fourth centuries A.D. Water-powered milling had also been adopted in all corners of Europe by the twelfth century A.D. Wind power was used concurrently for milling and lifting water in Europe. With the advent of steam power, mills were transferred to this source of power in the nineteenth and early twentieth centuries in both Europe and the United States. By the outbreak of the U.S. civil war, steam power had almost completely replaced horses and oxen for powering sugar and rice mills and to turn cotton gins. With steam power, three men and a cotton gin could remove the seeds from 1000 to 4500 pounds of cotton a day, which was about a hundred times the amount they could gin without steam power [Hurt (1982, p. 101)].

In South Asia, animals have long driven Persian wheels, sugarcane crushers, and oil crushers, but the animals used in these operations are gradually being replaced by diesel and electric engines. In India in 1973 the number of stationary engines for power-intensive operations was about twenty times the number of tractors (India 1975 and 1976). In all of Asia mechanical milling of large traded quantities of rice had already been introduced in the late nineteenth century, usually by steam engines, later by internal combustion engines. Smaller rice mills have swept across Asia since the 1950s, and it is hard to find villages where rice is still pounded by hand.

Mechanical mills were introduced in Africa after World War I and spread rapidly through the continent. There is documentation of the earlier existence of water mills in Angola and Kenya [Manners (1962); Jones (1959)]. Jones (1959) and Miracle (1967) have reported widespread use of mechanical mills, both hand and motorized, across Sub-Saharan Africa. Pingali, Bigot and Binswanger (1987) provide evidence from Sub-Saharan Africa indicating that low intensity of farming is not a constraint to the adoption of mechanized mills. This is mainly because the labor input required for milling is independent of the intensity of farming, and mills are rarely owned by the individual households who use them. The service is provided on a charge-per-unit basis by private entrepreneurs or village cooperatives. Since mills do not face any of the timeliness problems associated with plows, efficient rental markets are easily established and the cost of the equipment is spread over many users.

2.3. *Harvesting and threshing operations*

The mechanization of harvesting and threshing tends to follow a two phase pattern. In the first phase, harvesting continues to be done by hand, but threshing is increasing conduct by mechanical means. The second phase is characterized by the adoption of harvester combines that provide for the complete mechanization of both operations. The first phase starts even under low wage conditions, but where peak season labor scarcity emerges around the harvesting period due to growing off-farm employment opportunities. Peak season labor scarcity problems are aggravated in areas where two or more crops are grown on the same field each year. In this case, the labor peak for harvesting blends into the labor peak for land preparation and seeding (transplanting) the next crop. The switch to harvester-combines occurs as economies grow and rising real agricultural wages are observed.

In this connection it is useful to examine technological change in threshing operation in early U.S. agriculture as documented by Hurt (1982). Until 1850 colonial farmers, particularly those in New England, used the hand-held flail to thresh grain from the heads. The flail consisted of a short wooden club attached to a long handle by means of a piece of leather. In the midwestern states, where harvested quantities were much larger, farmers used oxen or horses to tread the grain from the heads.

The first horse-powered threshing machine was patented in the United States in 1791, but it was only between 1820 and 1830 that a number of small, simple, inexpensive, and locally made hand- and horse-powered threshing machines began to appear on the market. Most farmers, however, found threshing with the flail to be cheaper than investing in a threshing machine, because the work could be done during the winter, when there was an abundance of cheap farm labor. It was only with the advent of contract threshing operations that mechanical threshing became profitable. Threshing machines were owned by an entrepreneur, who sent a thresher with an itinerant crew from farm to farm. Although contract threshing imposed an immediate cash expense on the farmer, it did free him/her from the capital investment necessary to purchase a threshing machine and enabled him/her to get the crops to the market before prices fell.

Steam-powered threshing machines preceded steam-powered tractors by more than ten years. In less than a decade steam had almost entirely replaced horses for power. By 1880 the Bureau of the Census estimated that 80% of the grain in the principal wheat-producing states was threshed by steam-powered machines. Steam-powered threshing machines were followed in the 1930s by tractor-powered harvester-combines. The adoption of threshers in Europe followed the same pattern. In 1950 agriculture in Japan was in the early stages of mechanization, with many small pedal threshers and power tillers. This happened in response to rising wages caused by a rapidly growing industrial sector. By 1960 there were one thresher per 2.5 ha and one power tiller per 12 ha in Japan [Herdt (1983)].

The use of mechanical threshers did not emerge in South Asia and parts of Southeast Asia until the late 1960s. this is not surprising, since wages were low, capital costs high, and harvested volumes small. But where the green revolution raised wages and increased harvested volumes, small threshers were rapidly adopted in Indian Punjab, the Philippines, and Central Thailand as soon as efficient designs were available. By the early 1980s the new threshers were penetrating into other South Asian regions [Walker and Kshirsagar (1981)]. As in the United States in the nineteenth century, these threshers are owned by private entrepreneurs who thresh on a contract-hire basis.

Mechanical threshers are still rare in African agriculture. Threshers are not yet profitable in Africa because the harvested quantities per person are small. Therefore there are two conditions under which threshers would be profitable in Africa: improvements in access to markets or seed varieties leading to increases in harvested output and rapidly rising wages, which would increase incentives for adopting labor-saving technology.

Harvester-combines have been in use in East Asia and in Malaysia for several decades now and are emerging in Thailand, but their spread has been limited in the rest of Asia by low harvest wages and small plot sizes. Although harvest wages are relatively low compared with those in developed countries of East Asia, they are rising, and the prospects are that over the next decade mechanization of the harvest operation will be demanded in Southeast Asia.

Small harvesters are seen as an intermediate step in the transition to harvester-combines in much of Southeast Asia. In the absence of land consolidation and the re-design of riceland to form large contiguous fields, the prospects for large-scale adoption of the harvester-combines are limited. Small harvesting machines have been used in Malaysia since the mid-1980s; they began emerging in Thailand in the early 1990s and are expected to be gradually available commercially in other Southeast Asian countries [Pingali (1998)].

2.4. Labor substitution for "control-intensive" operations

As discussed earlier, the adoption of labor saving technologies for control intensive operations, such as planting, weeding, etc., is only profitable as wages rise. In the 19th and the first half of the 20th century mechanical alternatives were the only means available for labor savings for these operations. However, since the later half of the last century chemicals, such as herbicides, changes in crop establishment practices, such as direct seeding, and knowledge intensive crop management practices, such as integrated pest management, have emerged as alternatives to mechanical equipment for control intensive operations. The following is a discussion of the attempts to seek labor savings for control intensive operations in rice systems in Asia.

Rice transplanting and weeding operations are both control intensive as well as require high levels of labor, especially female labor, during a narrow window of time in the crop cycle. Several attempts have been made to mechanize control-intensive operations in rice production systems since the Green Revolution [IRRI (1986)]. Mechanical transplanters, weeders, and fertilizer applicators were some of the technologies released by the rice research systems in Asia. With the exception of mechanical transplanters (machine-driven) in East Asia, most attempts at mechanizing crop establishment and crop care activities have failed. The failures can be attributed to the higher cost of using the mechanical technology relative to the alternatives available, as discussed below.

Since 1970 the use of mechanical rice transplanters spread rapidly across Japan, Korea, and China (Taiwan), as wages rose. The use of transplanters in Japanese rice production took off in 1970, and by 1989 there were 2.2 million transplanters in use nationwide [Mizuno (1991)], approximately one transplanter for every 2 ha of arable land. In Korea by 1989, approximately 66% of the area planted to rice was mechanically transplanted [Chang (1991)]. Taiwan has seen similar rapid growth in transplanter use since the late 1970s [Peng (1983)].

The alternative to mechanical transplanting, as a means of saving labor in crop establishment, is to broadcast pregerminated rice seedlings into the field. Direct seeding is not generally possible in the temperate environments, however, because the cold spring temperatures are not conducive to seedling establishment and growth in the field. Rice seedlings are grown under controlled temperatures conditions and transplanted into the field when they are old enough to resist cold stress; by this time, a few weeks in the season, temperatures are warmer.

Mechanical transplanting has not taken off in the humid tropics of Southeast Asia, even under high-wage conditions, such as Malaysia, because direct seeding is possible. Direct seeding eliminates the transplanting operation because germinated seed is broadcast onto prepared (puddled) paddy soils. Most of the irrigated rice in Malaysia is wet seeded. Wet and dry seeding methods are becoming increasingly popular in other rice-producing countries of southeast Asia. Even though chemical weed control costs are higher for direct-seeded rice, the savings in labor cost have been more than the additional cost of herbicide. Empirical studies in the Philippines and Vietnam indicate that farmers who use wet seeding relied more heavily on herbicides than farmers who transplanted. Similarly, in Peninsular Malaysia, 95% of farmers who direct-seeded their rice applied herbicides, at an average herbicide cost of \$30/ha. On the other hand, only 46% of farmers who transplanted used herbicides, with average expenditure on herbicides being \$4/ha [Moody (1994); Pingali (1998)]. In low wage South Asian countries manual transplanting of rice, using mainly female labor, persists.

Mechanical weeders were introduced into rice systems in the mid-1970s, but their record of adoption was very poor, primarily because herbicides – a substantially cheaper source of weed control – were available. While high-wage countries have used herbicides for several decades, recent trends indicate increasing use in the so-called low-wage countries of Asia (FAOSTAT). The Philippines, Thailand, Malaysia and India more than doubled their herbicide imports in the 1980s [Pingali (1998)]. In these countries, herbicide has been adopted in association with rice direct seeding, which in turn was adopted in response to rising wages to replace the traditional transplanting system.

Attempts to enhance the efficiency of chemical fertilizer use through mechanical technologies allowing deep placement of fertilizer have also generally not succeeded. The long-term decline in global fertilizer prices has reduced farmers' interest in improving fertilizer use efficiency and investing in equipment to do so. The poor performance of fertilizer placement technology has been observed even in the high-income countries of East Asia. Given the proliferation of nonmechanical options for improved crop management, the future for the adoption of machines for control-intensive operations seems rather bleak.

3. Impacts of agricultural mechanization

The productivity impact of the switch to mechanical technologies for agricultural operations is measured in terms of changes in yields, labor savings, area expansion (in terms of increases in cropping intensities), and quality of enhancement of the marketed output. The equity impact, on the other hand, is measure in terms of labor displacement and income distribution effects, particularly for the landless labor households and for women. The productivity and equity impacts of mechanization vary depending on the power intensity of the operation that is being mechanized, the region's land and labor endowments, and the country's level of economic development. Since the mechanization of power-intensive operations has been well under way throughout Asia, several studies have documented the impact, and evidence from these studies is presented below. Few studies of control-intensive operations and quality enhancement technologies are available because the introduction and adoption of these technologies has been sparse. The power-intensive operations considered here are land preparation, threshing, harvesting (small harvesters), and milling (small mechanical mills). The evidence presented below indicates that, for power-intensive operations, the productivity benefits of mechanization consist mainly of labor savings, and the equity implications are minimal, even in labor-abundant, low-wage economies.

3.1. Land preparation

The movement from using animal-drawn plows to tractors or power tillers is considered efficient if yield per hectare increases and/or if labor hours required for land preparation per hectare are reduced. Yield increases are possible only when mechanization improves tilling quality. However, the available evidence indicates that generally no significant yield difference exists between animal draft and tractor tillage. Herdt (1983) found that yield differences between animal draft and tractor farms were negligible after differences in fertilizer use were considered (Table 3). This is consistent with results from South Asia, where more than 50% of farms using tractors had significantly higher yields, but in almost all cases these higher yields were associated with greater fertilizer use [Binswanger (1978)]. If we find no yield differences between animal draft and tractor farms, we must conclude that the transition to tractor-drawn plows is rarely motivated by improvement in tillage quality. Area expansion and/or labor saving must be the driving forces for such a transition.

Pingali, Bigot and Binswanger (1987) reviewed 24 studies on labor use by operation on farms relying on animal draft power and farms relying on tractors in Asia. They found changes in labor use by operation, in total labor use, and shifts in the levels of labor use between operations. Twenty-two of the 24 studies reviewed reported lower total labor use per hectare of crop production for tractor farms compared to animal draft farms. Twelve studies reported reductions in labor use of 50% or more.

The greatest reduction in labor use was for land preparation, with all studies reporting reduction in labor input exceeding 75%. It is instructive to consider cases in which the

Table 3
Summary of studies comparing rice yields on farms that used animal or hand for land preparation with farms that used machinery

Author	Area	Comparison	Reported yield (t/ha)	Fertilizer (urea) (kg/ha)	Adjusted yield (t/ha)
Pudasaini	Nepal (without pumps)	Bullock vs tractors	1.7 2.1	16 164	1.7 1.4
Pudasaini	Nepal (with pumps)	Bullock vs tractors	2.1 2.3	214 264	2.1 2.1
Sinaga	West Java, Indonesia (1979–80 wet season)	Animal vs tractors	4.9 4.9	323 323	4.9 4.9
Sinaga	West Java, Indonesia (3 seasons, 1978–80)	Manual vs tractors	3.8 3.9	285 308	3.8 3.8
Tan and Wicks	Nueva Ecija, Philippines (1979 wet season)	Water buffalo vs tractors	2.6 4	89 129	2.6 3.8
Anuwat	Central Thailand (irrigated–transplanted)	Bullock vs tractors	2.6 2.8	32 48	2.6 2.6
Anuwat	Central Thailand (rainfed broad-cast)	Bullock vs tractors	0.2 0.2	3 2	0.2 0.2
Alam	Bangladesh	Bullock vs power tiller	1.5	n.a.	1.5
Deomampo and Torres	Central Luzon, Philippines	Before vs after tractor and tillers	2.2 2.6	57 79	2.2 2.1
Antiporta and Deomampo	Philippine provinces	Animals vs tractors tillers	2.6 2.8	86 117	2.6 2.5

Source: Herdt (1983).

labor used for land preparation was reduced by 50% or more and trace the effect on other operations. Consider weeding first. Of the 14 cases with 50% or higher reduction in labor for land preparation, only two reported a reduction in weeding labor greater than 25%. Ten reported reduction in weeding labor smaller than 25%, and two reported increases in weeding labor relative to farms relying on animal traction. The situation for harvesting is quite similar. Of the 14 cases with 50% or higher reduction in labor for land preparation, only one resulted in an equal reduction in harvesting labor, nine reported labor reduction less than 25%, and one found increased labor requirements.

These results indicate that labor savings resulting from the transition to tractors are confined mainly to land preparation, where one observes a substantial reduction in labor peak. However, where aggregate area expansion effects exist (including an increase in cropping patterns), one could expect a rightward shift in the demand curve for weeding and harvesting labor, despite the lower per hectare requirements. Rice cropping intensities have increased significantly from the joint mechanization of land preparation and threshing, especially in humid tropical Southeast Asia [Juarez and Pathnopas (1983)]. With the introduction of modern rice varieties and irrigation infrastructure in Southeast Asia, two to three crops of rice can be grown on the same plot of land per year. Because the first crop is usually harvested during the rainy months, the danger that grain would spoil was very high if threshing was not done soon after harvest. Also, fields had to be cleared of the previous crop before irrigation water could be released. Without the use of power tillers and threshers, high intensity cropping could not have been sustained in much of Southeast Asia.

What are the equity consequences of a shift to tractors/power tillers from animal-drawn plows for land preparation? The answer to this question depends on the answers to four related questions. First, did tractor owners expand the size of their operation by displacing tenant farmers? Second, was there a power bottleneck around land preparation before tractors arrived? Third, has total labor use on tractor owners' farms increased or decreased since tractors arrived? Finally, has there been a net transfer of income from tractor owners to agricultural labor? Each of these questions will be examined in turn.

3.1.1. Aggregate area expansion

Aggregate area expansion occurs only when private farms extend into fallow or reclaimed land or when cropping intensity is increased on a given plot of land. Where private area expansion occurs at the cost of other farmers (mainly through the displacement of tenant farmers), there are no benefits from aggregate area expansion; rather, adverse equity consequences are observed. Any inference on the aggregate expansion in agricultural output as a result of an increase in cultivated land area brought about by tractor use would require additional information on where the additional area comes from.

Lockwood et al. (1983) found in Faisalabad, Pakistan, that 70% of the area expansion on tractor farms came from tenant displacement. The remaining 30% came from increasing area rented in, leading to further displacement of tenants. Eighty-eight of the original 105 tenants lost their land when the landowners bought their first tractor. The average size of tenant farm declined from 4.4 to 3.4 ha. This study confirms the earlier findings in the same area of McInerney and Donaldson (1975). Jabbar, Bhuiyan and Bari (1983) reported that 81% of power tiller owners in Mymensingh, Bangladesh increased their cultivated area by expanding into previously rented-out land.

Where uncultivated or fallow land was not available, the promotion of tractors for land preparation led to high levels of tenant displacement. The exception as noted earlier is the growth in rice cropping intensity owing to mechanization, which did not lead to tenant displacement and resulted in positive employment benefits [David and Otsuka (1994)].

3.1.2. Power bottleneck

Mechanization of the land preparation can lead to significant productivity gains with minimal equity effects in areas with a power bottleneck during land preparation. Such bottleneck could occur because of shortage of either labor or draft animals. Labor shortages are most common in sparsely populated areas and in areas with good nonfarm employment opportunities. Peak season labor shortages have also occurred with an increase in rice cropping intensities, as discussed previously, around the harvest operation for one crop and land preparation for the next. Shortage of draft power may occur in sparsely or densely populated areas (in the latter case, shortages occur because maintenance of draft animal is relatively expensive because of the high cost of fodder). Significant productivity benefits accrue to the alleviation of labor and power bottlenecks, through mechanization. The adoption of tractors (power tillers) in areas with a power bottleneck during land preparation could lead to general improvement in income levels.

3.1.3. Total labor use

It was observed earlier in this section that tractor use reduces labor use per hectare for land preparation, but leads to an increase in area cultivated by tractor farms. Where uncultivated or fallow land is available, or where cropping intensities increase, the use of tractors (or power tillers) for land preparation could lead to an increase in labor employment.

3.1.4. Income transfer

Income distributional effects of mechanized land preparation depend on the nature of final demand and the extent of output growth. Consider first areas with an inelastic demand for the food that is produced. These are typically small, closed economies, where neither exports nor imports of food (in this case, rice) are allowed. In such economies the total demand for food is determined only by domestic demand, and any increase in output will lead to a decline in price.

Where final demand is inelastic, aggregate area expansion caused by mechanized land preparation results in a transfer of income from tractor-owning households to landless-labor households. This happens for two reasons: (i) expansion into previously uncultivated areas (or an increase in cropping intensities) results in increased employment opportunities, and (ii) real price of food output declines as a result of output expansion, and since the landless labor are net purchasers of food, the effect of this price decline is analogous to increasing their income. The direction of income transfer is reversed where the opportunities for aggregate area expansion are limited. Here, tractor farms expand into land previously cultivated by tenants and the result is net labor displacement rather than an increase in aggregate output. Where final demand is elastic (open economies, or economies with large domestic demand), aggregate area expansion does not lead to a decline in prices, but a limited amount of income transfer to labor is observed as a result of expanded employment opportunities. If opportunities for aggregate area expansion are not available, the effects are the same as discussed in the preceding paragraph.

3.2. *Mechanization of post-harvest operations*

Next to land preparation, harvesting, threshing and milling are the most arduous operations in rice production. Consequently, where mechanical technologies for these operations exist and can be profitably used, they tend to be adopted fairly rapidly. Small mechanical mills, for instance, spread spontaneously across the world without any government program promoting them. What have been the productivity and equity consequences of mechanizing post-harvest operations?

3.2.1. *Threshers*

The late 1970s and 1980s have seen the rapid spread of mechanical threshers in parts of Southeast Asia and parts of India [Duff (1986); Walker and Kshirsagar (1981)]. In any given area, the private profitability (efficiency) of using a mechanical thresher over hand beating and animal or tractor treading is determined by yield benefits of mechanized threshings, marketable surplus generated on the farm, and by labor wages and availability during the harvesting–threshing period.

Proponents of the thresher technology usually argue that the mechanical thresher presents a significant increase in realized yields due to: (i) a more complete threshing of grain than manual or treading techniques; (ii) a reduction in losses caused by repeated handling of both threshed and unthreshed materials; and (iii) an increase in cropping intensity resulting from a lower turnaround time with mechanical thresher use. On-farm experiments comparing manual and mechanical threshing have shown that mechanical threshers reduce grain loss by 0.7 to 6% of total yield [Toquero and Duff (1985)]. However, there have been no studies of actual farmer thresher use to see if such savings are observed in practice. Cropping intensity effect has been discussed above, and to the extent that threshers contribute to sustaining high cropping intensities, thereby allowing more timely completion of operations they have a positive effect on yields.

The existing evidence indicates that the primary motivation for mechanical thresher use is the labor saving benefits. The adoption of a portable axial-flow thresher in the Philippines resulted in a decline in threshing labor requirements from 7.69 labor days per ton of grain for foot treading to 0.81 labor day per ton. In Thailand, the adoption of a large axial-flow thresher resulted in labor savings of 3.50 labor days per ton of grain relative to the traditional threshing by buffalo treading. Consequently, large gains in labor productivity were observed in both cases [Duff (1986)].

Whether the above labor savings actually lead to efficiency gains or not depends on the real wage of threshing labor, the timely availability of labor and other sources of power for threshing, and the nature of final demand. The following generalization is possible: Mechanical threshing is economically efficient where the farmer faces an elastic demand curve for his output and a power bottleneck exists during the harvesting–threshing period either due to high land–labor ratios or due to opportunities for non-farm rural and urban employment. It is important to remember that a combination of elastic demand and a threshing power bottleneck is required for mechanical threshers to be profitable. Equity considerations in thresher adoption center around the source of the labor that is displaced. Surveys in Nueva Ecija, Philippines, showed that postharvest labor on mechanized farms was 25% lower than on farms in which rice was manually threshed. Disaggregated into family and hired labor, the data reveal that much of the labor savings came at the cost of landless households whose labor services declined by 31% [Duff (1986)]. A similar decline in the use of hired labor was observed in Laguna and Iloilo, Philippines [Juarez and Pathnopas (1983)]. Ahammed and Herdt (1983) used a general equilibrium model to estimate the nationwide employment implications of thresher use (among other production methods) in the Philippines. They found that a 1% increase in rice production would generate employment effect of 16,000 labor years if manual threshing is used. The employment effect would be 22% lower (12,400 labor years) if portable threshers are used. Taking all sectors of the economy into account, substituting mechanized threshing for manual threshing would reduce the employment-generating potential of increased rice production by 7%. This overall adjustment depends on urban employment opportunities, the ability of labor to migrate between regions and between sectors, and industrial policies.

A similar analysis in Thailand found that all the labor savings came from family labor and not for hired labor [Sukharomana (1983)]. This is because Thailand has a very favorable land-to-labor ratio and therefore, a very small landless labor class. Traditionally, the threshing operation is done by family labor with buffalo or tractor treading. The use of large axial-flow threshers on a contract basis has resulted in the release of family labor for other activities.

For both family and hired labor, the thresher had a differential impact on men, women, and children. Mechanical threshers considerably reduced the arduousness of post-production tasks. The lighter nature of the work made it possible for women and children to substitute for men in the threshing operation [Ebron (1984)]. Where off-farm employment opportunities exist, thresher use can result in increased incomes of labor households since male workers may be released for other income generation while women and children provide threshing labor.

3.2.2. *Milling*

The use of small mechanical mills for dehusking paddy or for pounding grain into flour is perhaps the least controversial of all forms of mechanization. Handpounding of grains, the traditional alternative to mechanical milling, is an extremely labor-intensive task usually performed by women. The efficiency gain in the shift to mechanical mills comes from the resulting labor savings. Because much of the rural milling is done piece-meal for home consumption only and because traditional handpounding is done by female household members, switching to small mechanical mills results in substantial gains in leisure time for women. Consequently, one observes the widespread use of small rural mills on a contractual basis.

With one significant exception, small mechanical mills have increased efficiency of food production without adverse equity effects. The exception is Bangladesh where significant displacement of hired labor has occurred as a consequence of mechanical mills [IBRD (1987)]. Traditional rice milling in Bangladesh is done by a foot-operated mortar and pestle known as a 'dheki', which is usually owned by large landowners and operated on a contractual basis by women from landless and low-income households. The ratio of milling costs with the 'dheki' to those with the mill is about 12:1, not counting the transport costs of bringing the rice to the mill. Thus, the mill owners can charge much lower rates for milling than 'dheki' operators.

The rapid spread of mechanized milling has benefited large landowner households, subsistence-farming households, and urban consumers including the urban poor, whose rice prices have been reduced. Female members of large surplus farms have more leisure time, because they no longer have to supervise hired 'dheki' operators. Female members of subsistence farms who previously operated the 'dheki' for their home consumption are relieved from time-consuming and physically-demanding labor. Nonetheless, women from landless families who work for wages have suffered as a result of the mills because of absent alternative remunerative employment. The policy challenge in Bangladesh has been to find alternative employment opportunities for sustaining the incomes of the women of landless and low-income households, without slowing down the growth in small mechanical mills.

3.2.3. *Herbicides*

Throughout Asia, the ratio between price of herbicide and wage rate has been declining steadily overtime, making herbicide use economically attractive. In Iloilo province, Philippines, the cost of weed control by herbicide in wet seeded rice is less than one-fifth of the cost of a single hand weeding [Moody (1994)]. Similarly, in West Java, Indonesia, and Mekong Delta, Vietnam, the cost of hand weeding is 3 to 5 times the cost of herbicides. Economic analysis of weed control practices in the Philippines indicate that the marginal benefit cost ratio associated with herbicide usage is as high as 16 [Naylor (1996)].

Although economic and technological factors are likely to lead to increased substitution of labor by herbicides, such a substitution may result in short term adverse social consequences. Due to the need to complete weeding within a short time window, farmers traditionally relied on hired labor for weeding. As small and marginal farmers are the main source of hired labor, their income and employment will be adversely affected unless they can find an alternative use of their time. Increase in cropping intensity made possible by improved technology has in the past been able to absorb most of the displaced labor [David and Otsuka (1994)]. The extent to which substitution of labor by capital in agriculture will occur without imposing welfare costs on certain groups depends mainly on government policies on exchange rate, pricing of inputs and outputs and macro-economic policies. To the extent that herbicides are made cheaper relative to labor due to distortionary price policies, substitution of herbicides for labor is socially undesirable. The social costs are of course lower in societies experiencing a rapid withdrawal of labor from the agricultural sector.

4. Implications for mechanization policy

4.1. Tractors are a poor instrument for stimulating agricultural growth

Over the last century there have been several dozen attempts at introducing tractors into areas that are sparsely populated, particularly in Africa, in an attempt to rapidly expand the area under cultivation and increase production. These attempts have consistently failed because the market infrastructure and economic incentives that induce a production response were not present. The end result was a "boom and bust cycle", a rapid expansion in output that was invariably followed by a collapse in local prices and a subsequent abandonment of the newly opened land and the tractors. Tractors ought to be seen for what they are: a potential tool in the production process and not a driver of economic change.

4.2. Agricultural mechanization policy ought to be seen within the context of an overall agricultural growth strategy

A broad based strategy for agricultural growth provides incentives and infrastructure that enables farmers to enhance productivity growth. Macroeconomic policies that tend to discriminate against or tax the agricultural sector need to given as much attention as agriculture sector specific policies. Policies that enhance rural infrastructure are necessary for sustainable productivity growth and overall rural development. Farmers respond to improvements in incentives and market conditions by changing production practices and investing in new technologies, including mechanical technologies. Broadbased rural infrastructure development also reduces the costs to private entrepreneurs that supply technology and inputs as well as market output.

4.3. *The demand for motorizing power intensive operations, such as tillage and threshing, is closely associated with the intensification of farming systems, while the mechanization of control-intensive operations, such as weeding, is driven by rising real wages*

The need for increased energy requirements with the intensification of agricultural system has been extensively documented in the literature [Boserup (1965); Pingali and Binswanger (1987)]. The movement from land-extensive cultivation systems, such as shifting cultivation systems, to land-intensive permanent cultivation systems increases both the number of tasks performed and the intensity with which they are performed. For operations that require high levels of power, such as tillage, human labor is gradually replaced by animal and then tractor power. In intensively cultivated systems mechanization of power intensive operations is profitable, even under low wage conditions. In such systems, human labor continues to be used for seeding, weeding, crop care, and harvesting. The co-existence of mechanical and human labor disappears as wages rise due to economic growth and the increasing availability of off-farm employment opportunities. Mechanical seeders, herbicides, and harvester-combines substitute for human labor as economies grow. The sequential adoption of mechanization, first for power intensive and then for control intensive operations, is not a historical artefact, it is a farmer response induced by the changing relative prices of factor inputs.

4.4. *Promotion of small stationary machines for power-intensive operations such as milling and pumping can have significant benefits for the poor*

Small mills have spontaneously and successfully penetrated even the most remote villages around the developing world. Mechanization has released labor, invariably family labor especially women from the arduous task of de-husking, pounding and milling grain, often on a daily basis. Poor households benefit the most, since the released labor can be reallocated for other income earning activities or for leisure. Governments can play a catalytic role in the further spread of mills, first in promoting research and development on mills that are more energy efficient and improve on quality; second in providing credit and other support for rural entrepreneurs to acquire and operate rural mills.

The adoption of small pumps is less spontaneous yet equally crucial for the livelihood of poor rural households. Pumps help stabilize food supplies in drought prone lands, and where enabling conditions exist, the commercialization of smallholder agriculture. The adoption of pumps resulted in the intensification of cropping in the Indo-Gangetic Plains that extend through Northern India and Bangladesh. Small holders were able to grow a dry season crop of rice or vegetables exclusively for the market, hence stimulating overall growth in the rural economy. Governments can play a similar catalytic role, as with mills, in helping small farmers acquire pumps.

4.5. Clearly established property rights could minimize the risk of displacement of small farmers from their land

In both land scarce as well as land abundant economies, tractor ownership is associated with an expansion in area cultivated. In the absence of clearly established rights to land, tractor-induced farm size expansion could come at a cost to the poor. In land scarce economies, tractor adoption has resulted in the displacement of tenant farmers, while in land abundant economies, traditional access rights to land have been impinged upon as tractor farms expand into uncultivated or fallow lands. Inappropriate promotion of mechanization, through subsidized tractor prices or cheap credit programs, tends to aggravate the negative equity impacts without commiserate productivity benefits. Property rights give small and tenant farmers the bargaining power to prevent encroachment or to seek compensation. Formal land titles empower small farmers further by providing them the collateral necessary for acquiring credit for the purchase of machinery and other agricultural inputs.

4.6. Adoption of labor saving technology does not always imply labor displacement

Mechanization has often been seen as having a negative impact on agricultural employment and therefore detrimental for densely populated "labor surplus" countries. In reality the picture is not as straightforward, whether labor displacement occurs or not depends on: the operation, the labor market, and the policy environment. As discussed in this chapter, mechanization of power intensive operations, water lifting, tillage, milling, etc., have minimal labor displacement effects. The adoption of water pumps tend to increase cropping intensity and hence labor requirements. Mechanized land preparation shifts the demand for labor from land preparation to weeding and harvesting operations. While mechanical mills, release female family labor from the arduous task of hand pounding grain. The mechanization of control intensive operations such as weeding and harvesting could have negative employment effects if promoted under low wage conditions. However, if the adoption of labor saving technologies for these operations occurs in response to rising wages, due to growth in the non-farm sector, then the labor displacement consequences are small. Labor displacement problems are most pronounced when mechanization policy is inappropriate and machines are promoted where they are not required, such as low intensity farming systems, or where wage rates and the opportunity cost of labor are low.

4.7. Public sector run tractor promotion projects, including tractor-hire operations, have neither been successful nor equitable

Public sector record as a promoter of tractor use and as a supplier or tractor services has been poor both conceptually as well as operationally. The pervasive misconception that tractors are a shortcut to agricultural modernization has resulted in the inappropriate promotion of tractors in environments where private farmer decisions would not lead to intensification or tractor use. Public sector tractor projects tend to displace the private sector (or prevent its growth) as a supplier of equipment, spare parts and maintenance services. Being donor driven, public sector projects do not build a self-sustaining system for the long term supply of tractors and associated services. Hence the service collapses at the end of the project. Public sector run tractor hire services are a particular case of operational inefficiency and poor longevity. Where economic conditions are right, the private sector has been an efficient provider of equipment and mechanization services. The public sector can play an important catalytic role in promoting private sector supply as well as private initiatives in equipment research and development. The latter in the particular context of unfavorable production environments and communities facing special needs, such as AIDS affected populations.

4.8. Alleviating supply side constraints to mechanization is important, but only where the demand conditions are right and the enabling environment is in place

Lack of or low level of adoption of mechanization is often attributed to supply side constraints, such as the lack of equipment and spare parts suppliers and skilled mechanics that can provide maintenance services. However, its demand side factors, such as unfavorable relative factor prices and market access conditions that are a more plausible explanation of the poor spread of mechanical technologies, especially in Sub-Saharan Africa. The private sector, where it has been allowed to operate freely and where the enabling conditions are in place, has successfully met the demand for equipment and spare parts, as well as for repair and maintenance. Governments ought to play a facilitating role in reducing the transactions costs involved, for farmers as well as for small entrepreneurs, in the acquisition and maintenance of mechanical technologies.

4.9. Conservation agriculture is not a panacea for farming systems that are not mechanized today

Conservation agriculture, which is generally taken to imply the systematic application of planting without tillage, cover crops, and crop rotation, is seen by some as an opportunity for bypassing the need for mechanical power for land preparation. However, it is completely false to link conservation agriculture with mechanization strategy. In mechanized conservation agriculture access to a tractor and to a no-till drill is required and thus brings with it the same set of issues as for conventional cropping systems. In shifting cultivation systems, the question of adopting conservation agriculture is moot since the practices followed by these farmers, such as the incorporation of fallow vegetation minimal land preparation and the use of a dibble stick for seeding are practices that are consistent with principles of conservation agriculture. However, benefits have been shown in hand-till and draft animal systems. No-till technology for manual and draft animal planting systems have been developed that show a reduction in labor requirements coupled with yields that are less impacted by soil moisture deficits. Which is not to say that conservation agriculture has been proven successful in all agro-climatic zones and associated farming systems. In some cases there is competition for plant residues and potential conflicts between the needs of conservation agriculture and pastoral livestock that are yet to be resolved. Farmers in remote locations who face poor market access conditions are unlikely to find the package of practices that conservation agriculture recommends more remunerative. However, this is a problem for mechanization as a whole, not just for conservation agriculture. For a farmer to increase expenditure on inputs requires a market for increased production, with a return that justifies the increased input expense.

4.10. Global integration of food and input markets can have positive as well as negative consequences for small farm mechanization

Global integration of input markets implies cheaper access to mechanical and other technologies. Technologies developed elsewhere can be more easily transferred and adapted to country specific agro-ecological and farming system conditions. On the other hand, the global integration of food markets exemplified by the global spread of super markets could have more ambiguous effects for the small farmer. Modern food systems impose high standards for quality and safety that the post-harvest equipment and handling practices on small farms may not be able to meet. Scale economies may become increasingly important in meeting the quantity and quality requirements of super markets and hence lead to the displacement of smallholders from the emerging commercial food systems. Whether innovations in post-harvest technologies aimed at smallholders can reverse this situation is an open question.

References

Ahammed, C.S., Herdt, R.W. (1983). "Impacts of farm mechanization in a semi-closed input–output model of the Philippine economy". American Journal of Agricultural Economics 35 (3), 516–525.

APO (1991). Utilization of Farm Machinery in Asia, Report of APO Multi-Country Study Mission, 19–29 June. Asian Productivity Organization, Tokyo, Japan.

Balsevich, F., Berdegue, J.A., Flores, L., Mainville, D., Reardon, T. (2003). "Supermarkets and produce quality and safety standards in Latin America". American Journal of Agricultural Economics 85 (5), 1147–1154.

Binswanger, H.P. (1978). The Economics of Tractors in South Asia. Agricultural Development Council and the International Crops Research Institute for the Semi-Arid Tropics, Hyderabad, India.

Binswanger, H.P. (1984). "Agricultural mechanization: A comparative historical perspective", Staff Working Paper No. 673. World Bank, Washington, DC.

Bonnefond, P.H. (1967). "Problems of profitability of an agricultural development action of the SATEC operation in Mossi". Ministere de Plan: Office de la Recherche Scientifique et Technique Outre-Mer (ORSTROM), Bondy, France (in French).

Boserup, E. (1965). Conditions of Agricultural Growth. Aldine Publishing Company, Chicago, IL.

Chang, J.Y. (1991). "Country papers: Republic of Korea". In: Utilization of Farm Machinery in Asia, Report of APO Multi-Country Study Mission, 19–29 June. Asian Productivity Organization, Tokyo, Japan.

Cho, K.H. (1983). "Other country report on the status of farm mechanization – Republic of Korea". In: Farm Mechanization in Asia. Asian Productivity Organization, Tokyo, Japan.

CSO (1993). China Statistical Yearbook. China Statistical Office, Beijing.

David, C., Otsuka, K. (1994). Modern Rice Technology and Income Distribution in Asia. International Rice Research Institute, Los Baños, Philippines.

Duff, B. (1986). "Changes in small-farm rice threshing technology in Thailand and the Philippines", Working Paper No. 120. International Rice Research Institute, Los Baños, Laguna, Philippines.

Ebron, L.Z. (1984). "Changes in harvesting–threshing labor arrangements in Nueva Ecija", Master thesis. College of Economics and Management, University of the Philippines at Los Baños, College, Laguna, Philippines.

Ekboir, J. (2000). "Innovation systems and technology policy: Zero tillage in MERCOSUR", Economic Program Paper. CIMMYT, Mexico City, Mexico.

FAO (2005). FAOSTAT database. Available at http://faostat.fao.org/faostat/default.jsp?version=int&hasbulk=1; accessed July 2005.

Herdt, R.W. (1983). "Mechanization of rice production in developing Asian countries: Perspective, evidence and issues". In: Consequences of Small Farm Mechanization. International Rice Research Institute, Los Baños, Laguna, Philippines.

Hossain, M., Bose, M.L., Mustafi, B.A.A. (2002). "Adoption and productivity impact of modern rice varieties in Bangladesh", Paper presented at the Workshop on Green Revolution in Asia and Its Applicability to Africa, FASID, Tokyo, Japan, December 8–10, 2002.

Hurt, R.D. (1982). American Farm Tools: From Hand Power to Steam Power. Sunflower University Press, Manhattan, KS.

International Bank for Reconstruction and Development (IBRD) (1987). Agricultural Mechanization: Issues and Options. The World Bank, Washington, DC.

IRRI (1986). Annual Report. International Rice Research Institute, Los Baños, Laguna, Philippines.

Jabbar, M.A., Bhuiyan, M.S.R., Bari, A.K.M. (1983). "Causes and consequences of power tiller utilization in two areas of Bangladesh". In: Consequences of Small Farm Mechanization. International Rice Research Institute, Los Baños, Laguna, Philippines.

Jones, W.O. (1959). Manioc in Africa. Stanford University Press, Stanford, CA.

Juarez, F., Pathnopas, R. (1983). "Comparative analysis of thresher adoption and use in Thailand and the Philippines". In: Consequences of Small Farm Mechanization. International Rice Research Institute, Los Baños, Laguna, Philippines.

Kisu, M. (1983). "Mechanization of rice farming". In: Farm Mechanization in Asia. Asian Productivity Organization, Tokyo, Japan.

Kline, C.K., Green, D.A.G., Donahue, R.L., Stout, B.A. (1969). "Agricultural mechanization in equatorial Africa", College of Agriculture and Natural Resources Research Report No. 6. Michigan State University, East Lansing, MI.

Labrousse, G. (1971). "Historic–economic aspects of the development of mechanized cultivation of food crops in tropical countries". Agricultural Research Center, Bambey, Senegal.

Lockwood, B., Munir, M., Hussain, K.A., Gardezi, J. (1983). "Farm mechanization in Pakistan: Policy and practice". In: Consequences of Small Farm Mechanization. International Rice Research Institute, Los Baños, Laguna, Philippines.

Mandal, M.A.S. (2002). "Agricultural machinery manufacturing and farm mechanization: A case of rural non-farm economic development in Bangladesh". In: Proceedings of the International Workshop on Fostering Rural Economic Development through Agriculture Based Enterprises and Services, GTZ, Berlin, Germany, 20–22 November 2002.

Manners, R.A. (1962). "Land use, trade and the growth of market economy in Kipsigis country". In: Bohannan, P., Dalton, G. (Eds.), Markets in West Africa. Northwestern University Press, Evanston, IL.

McInerney, D.P., Donaldson, G.F. (1975). "The consequences of farm tractors in Pakistan", World Bank Staff Working Paper No. 210. The World Bank, Washington, DC.

Miracle, M.P. (1967). Agricultural Growth in the Congo Basin. University of Wisconsin Press, Madison, WI.

Mizuno, T. (1991). "Agricultural mechanization trends in Japan". In: Utilization of Farm Machinery in Asia, Report of APO Multi-Country Study Mission, 19–29 June. Asian Productivity Organization, Tokyo, Japan.

Moody, K. (1994). "Postplanting weed control in direct seeded rice". In: Ali, A.H. (Ed.), Direct Seeding Practices and Productivity. Malaysian Agricultural Research and Development Institute, Penang, Malaysia.

Naylor, R. (Ed.) (1996). Herbicides in Asian Rice: Transitions in Weed Management. Institute of International Studies, Stanford University, Palo Alto, CA.

Ohkawa, K., Shinohara, M., Umemura, M. (Eds.) (1965). Estimates of Long Term Economic Statistics of Japan since 1868. Agriculture and Forestry, vol. 9. Toyo Keizai Shinposha, Tokyo.

Peng, T.S. (1983). "Status of farm mechanization: Republic of China". In: Farm Mechanization in Asia. Asian Productivity Organization, Tokyo, Japan.

Pingali, P.L. (1998). "Dealing with growing labor scarcity: The adoption and impact of mechanical and chemical technologies in the Asian rice economies". In: Pingali, P.L., Hossain, M. (Eds.), Impact of Rice Research. International Rice Research Institute, Los Baños, Philippines.

Pingali, P.L., Bigot, Y., Binswanger, H. (1987). Agricultural Mechanization and the Evolution of Faring Systems in Sub-Saharan Africa. The World Bank, Washington, DC.

Pingali, P.L., Binswanger, H.P. (1987). "Population density and agricultural intensification: A study of the evolution of technologies in tropical agriculture". In: Johnson, G., Lee, R. (Eds.), Population Growth and Economic Development. National Research Council, Washington, DC.

Pingali, P.L., Hossain, M., Gerpacio, R.V. (1997). Asian Rice Bowls: The Returning Crisis? CAB International, Wallingford, UK.

Salokhe, V.M., Ramalingam, N. (1998). "Agricultural mechanization in the South and Southeast Asia", Paper presented at the International Conference of the Philippine Society of Agricultural Engineers, Los Baños, Philippines, 21–24 April 1998.

Sukharomana, S. (1983). "The impact of farm power strategies in Thailand". In: Farrington, J., Abeyratne, F., Gill, G. (Eds.), Farm Power and Employment in Asia. Agricultural Development Council, Bangkok, Thailand, pp. 139–150.

Toquero, Z.F., Duff, B. (1985). "Physical losses and quality deterioration in rice post-production systems", IRRI Research Paper No. 107. International Rice Research Institute, Los Baños, Laguna, Philippines.

Walker, T.S., Kshirsagar, K.G. (1981). "The village level impact of machine threshing and implications for technology development in semi-arid tropical India", ICRISAT Economics Program Progress Report No. 27. ICRISAT, Hyderabad, India.

Chapter 55

TRANSFORMATION OF MARKETS FOR AGRICULTURAL OUTPUT IN DEVELOPING COUNTRIES SINCE 1950: HOW HAS THINKING CHANGED?*

THOMAS REARDON

Department of Agricultural Economics, Michigan State University, East Lansing, MI

C. PETER TIMMER

Center for Global Development, Washington, DC

Contents

* No senior authorship assigned. We are grateful to Robert Evenson, Jean Kinsey, and Julio Berdegué for useful comments on earlier versions.

Handbook of Agricultural Economics, Volume 3
Edited by Robert Evenson and Prabhu Pingali

DOI: 10.1016/S1574-0072(06)03055-6

Abstract

This chapter traces the evolution of the agricultural economics literature on agrifood output markets over the past 50 years, emphasizing research approaches and policy issues. The analysis of agrifood systems encompasses the demand and supply side of output markets. The analysis in this chapter is set within the conceptual framework of how the agrifood economy develops during the agricultural and structural transformation. The related paths of development of policy and policy issues, and research themes and methods, are analyzed in parallel. The chapter examines the shift from broad and atomistic "commodity" markets to differentiated and more concentrated "product" markets over the half century. Spurred by massive retail sector foreign direct investment (FDI) to which was added competitive investments from domestic capital, a profound retail transformation has occurred in the past decade – the "supermarket revolution". This revolution has been the leading edge of globalization of domestic agrifood systems, not, as the literature currently emphasizes, opening to international trade. The chapter ends with a focus on the challenge for researchers in the next several decades, especially the need for new research methodologies that are suitable for understanding the role and influence of a small number of large-scale, multi-national firms, and for analyzing the impacts of the consolidation of the downstream segments of the agrifood system – the food industry – on upstream segments of the domestic agrifood systems, on rural development, and on trade.

Keywords

food and agricultural markets, agrifood systems, development, agricultural and food policy, structural adjustment, supermarkets, food industry, small farmers

JEL classification: O100, Q130, L100, L200, N010

1. Introduction

Agricultural output markets in developing countries have changed fundamentally and rapidly since 1950 in the organization, institutions, policies, and technologies conditioning and embodied in these markets. This chapter analyzes these changes from the perspective of how the profession was and is organizing its thinking about the issues, i.e. what were and are the prevailing conceptual paradigms? We concentrate on markets for food commodities, as they have received the most analytical and policy attention, but non-food commodities enter the discussion as well. To be inclusive, we refer to agricultural output markets and view them in the framework of the agrifood system, wherein the upstream segments (e.g., farmers and first-stage processors) are the supply side of the output market and the downstream segments (wholesalers, second-stage processors, retailers, consumers) are the demand side of the output market.

A basic bifurcation emerged over the five-decade period, between "commodity" and "product" markets.[1] This divide reflects a shift from a traditional to a modern phase. The early stage was characterized by:

- informal and traditional domestic food markets with many small producers in the production segment,
- direct sale or local brokers for the rural market or traditional wholesale to the urban market in the wholesale segment, and
- small shops and wetmarkets or central markets as the retail segment of the output market. At this stage, most goods traded beyond the farm are undifferentiated, unprocessed, commodities.

At the later stage, the "modernized" domestic food markets still incorporate many small producers, but the market agents are on average larger and differentiated by capital intensity. There is usually a continued role for traditional brokers amid the rise of modern wholesale markets and non-traditional wholesalers. The latter are specialized wholesalers focused on modern food industry clientele, and are often directly dedicated to a single type of client such as supermarket chains. At the end of the period we see consolidation downstream in the agrifood system (in the processing and retail segments), including the rapid rise of large-scale processors, supermarkets, and food service chains.

Naturally, this transformation from the traditional to the modern phase does not occur at a uniform rate across regions, or countries, or zones, or between rural and urban areas. The transformation can be observed as a diffusion process, of new organizations, institutions, and technologies in the food system – which occurs first and fastest in urban areas of the richer developing countries. These then ripple out into their own rural

[1] We use "commodity" to mean "standardized agricultural products that have had little or no processing and often raw materials for further processing... unbranded...." [Schaffner, Schroder and Earle (1998, p. 6)]; a given commodity, say wheat, is minimally differentiated and buyers incur little cost in switching among suppliers, controlling for transport costs, e.g., bulk wheat. By contrast, "products" are subsets of a given commodity, differentiated with some attribute, such as organic or not, processed or not, branded or not, variety A versus variety B.

towns and into other developing countries. We characterize this throughout the chapter as transformation in three waves over space and time – with the first wave in much of South America, East Asia outside China, and northern–central Europe, the second wave in much of Central America and Mexico, Southeast Asia, southern–central Europe, and South Africa, and the third wave in South Asia, China, Eastern Europe, and parts of Africa. These waves are correlated with socioeconomic variables and policy factors such as the rate of liberalization of foreign direct investment (FDI).

The result is a shift in the *distribution* of the characteristics of the food system – with what can be called the degree of modernization (ranging from traditional to modern) of the technologies, organization, and institutions of the agrifood system on the vertical axis, and times and space on the horizontal axis. This distribution was firmly anchored in an average of "traditional" in 1950, with a relatively narrow distribution around that mean. It is now approaching a mean of near-modern or modern phase characteristics, but with a much wider distribution (over degree of modernization) over the geographical areas represented by the three waves. But this shift to modern phase characteristics had occurred in much of the developing world by 2004, the year in which the world's population became more than half urban.

We pay special attention to the frontrunners here – the first and second wave areas – in particular the urban areas and countries with faster food market transformation, because these are the nodes of change, the pacesetters. Focusing on these "nodes of change" allows us to "see around the corner" what output market changes are likely to occur next in the slower changing areas. This foresight permits policy and program steps to be taken to help food system actors cope with and undertake anticipatory policies and strategies to deal with that fast-approaching change.

We use the structural transformation of the economy as our analytical lens, and extend it from impact on the primary production sector to understand transformation in the markets faced by producers [Timmer (1988)]. These markets include wholesale, processing, and retail segments downstream from producers in the food system. This transformation of the marketing system is analyzed as a set of structural, organizational, and institutional changes.

The output market transformation comprises five sets of changes:

(1) shift from *commodity to product markets* as the "center of gravity"[2] of the food system;
(2) rapid *organizational change* in the markets including consolidation in the downstream segments of the agrifood system – i.e. the processing and retail segments – with the often spectacular rise of supermarkets, a retail revolution in food markets in developing countries that is fundamentally altering their nature and structure;

[2] The shift noted is the shift of the "central tendency" of the food system; while the opportunities for differentiation into products are greater for the food industry than for farmers, the former influences the latter. Of course, a shift of a central tendency of the food system implies that there are many left out of the shift, many poor commodity farmers in all regions that merely produce for the commodity market and cannot yet differentiate into products.

(3) rapid *institutional change* in the markets, with the rise of contracts, and private grades and standards for food quality and safety;
(4) induced rapid *technological and managerial change* by and among suppliers, wholesalers, and retailers in the context of the above organizational and institutional changes; and
(5) the concomitant distributional, technological, and welfare impacts of changes in the "downstream" segments of the food system on the "upstream" segments of the food system, in particular on farmers.

The chapter proceeds as follows. Section 2 presents our analytical framework, briefly, as it draws on extensive work by both authors. Section 3 provides an equally brief review of the main themes in the output market literature from the 1950s to the 1980s.[3] We divide this phase that focuses mostly on commodity markets into three strains of literature. These trace the main themes of policy debate from growth, to equity, to market-sustainability. Accompanying the policy debates, and often preceding them, were methodological innovations that gave analysts more rigorous tools for addressing each of these issues.

The bulk of the chapter, and its major contribution, is in Sections 4 and 5. Here we focus on what is being called the "supermarket revolution" which started in developing countries in the 1990s, and which shifts our analytical focus from commodity to product markets and to the technological, organizational, and institutional changes in the food system induced by this shift. Section 6 concludes with implications for research methodologies and policy.

2. Analytical framework

We use two theoretical perspectives to understand the transformation of markets in developing countries over the past half century: (1) the structural transformation as a regular economic process, with both demand and technological drivers; and (2) induced technological, organizational, and institutional changes, which respond to the impact of those drivers. We use the "new institutional economics" meanings of these terms, where "organization" is the structure of relations among producers, and "institutions" are the rules of the game governing economic relations [North (1990); Williamson (1975, 1996)].

As economic growth proceeds and per capita incomes rise, there is a gradual but persistent change in the product composition, or intersectoral structure, of the economy.

First, there is a decrease in the share of food output in total output. This process is driven by changing demand, via Engels Law (showing that the share of food in total

[3] Years and periods are given as rough approximations in this chapter because the literature and debate differed somewhat in terms of timing over regions and countries.

household or national expenditure declines as income per capita increases) and technological change that drives, and is driven by, changing factor prices [via Rybczynski's Theorem (1955); Hayami and Ruttan (1971)].

Second, accompanying the smaller relative role of the food sector in both production and consumption in all countries (despite the opportunity for international trade to permit specialization) is the process of urbanization. In most countries, this has involved the active migration of rural workers to urban jobs, and the difficulties of this process have been the focus of most policy attention to the structural transformation. Continuing efforts by rich countries to protect their agricultural sectors in the face of this unrelenting pressure show how deeply rooted are the problems of equilibrating incomes between rural and urban areas [Gardner (2002)].

Third, there is an increase in specialization and gains from efficiency that translate into rising incomes. Specialization and reduced transaction costs that come from infrastructure investments and better communication technologies bring increases in market size and economies of scale. In turn, there is consolidation and increases in firm scale over time.

Finally, the *agricultural transformation* mirrors and is driven by the overall structural transformation, and simultaneously drives it. On the one hand, there is a change in *intra*sectoral composition, with a decrease over time in the share of staples in total agricultural output and a concomitant increase in processed staples and consumption of non-staples, such as fruit and vegetables, meat and fish, dairy, and oils and fats. This increase is demand driven, reflecting Bennett's Law,[4] and technology driven, reflecting lower costs of storage, transportation, and processing. On the other hand, there is an increase in specialization in production and marketing. With this specialization comes an increase in market size and a shift from autarchy and subsistence consumption to commercialized agriculture, which is needed to feed both rural areas and rapidly growing cities [Timmer (1997); Pingali and Rosegrant (1995)].

Changes in relative factor scarcities during agricultural transformation induce change in the factor bias of new technologies [Binswanger and Ruttan (1978)]. As population increases and land available per capita falls, and as industry develops, there is at first a shift from extensive technology systems (using land, and saving on labor and capital) toward either extensive systems that use capital and land and save labor (e.g., Argentina) or toward intensive systems that save land, use labor, and save capital (initial stages of intensification) followed (where capital costs decline) by production systems that save land, use labor, and use capital (e.g., Green Revolution technologies in Asia).

The literature [such as Timmer (1988) and Hayami and Ruttan (1971)] has focused on the structural transformation and its impact on the agricultural sector (the upstream segment of the agrifood system). In this chapter we apply that basic framework to transformation of markets for agricultural output (adding the downstream segments of the

[4] Bennett's Law states that as income per capita rises, the share of the food budget spent on starchy staples declines and the share to non-staples such as meat, fish, dairy, fruits, vegetables, and oils increases [Timmer, Falcon and Pearson (1983)].

agrifood system). This literature on the structural transformation yields several hypotheses that we support below.

First, following the paths of specialization and product diversification, we expect a shift over time from commodity markets toward product markets for specific segments of consumers and uses.[5] Product differentiation becomes important and the marketing system both supports and creates it. Following the path of specialization and increase in scale, we expect a shift from atomistic merchants (petty traders, open-air spot markets) toward segmentation into traditional wholesalers and small-scale retailers, and eventually toward further segmentation and consolidation into specialized wholesalers and "modern retailers" including supermarkets and convenience store chains. Scale effects initiate this expectation but technological change, especially in retail and distribution technology (such as information technology and other logistics technology change) to lower coordination costs, with concomitant organizational and institutional changes in the supply chain, will be the ultimate drivers.

Reduction in the relative cost of capital changes the factor bias of marketing technology. For example, the adoption of modern logistics platforms by supermarket chains and large-scale food processing companies allows large-volume procurement. Computer systems used by leading supermarket chains in most developing countries allow "efficient consumer response"[6] techniques and thus inventory holdings by supermarkets can be lowered. Telecommunications allow long-range commerce, and changes in shipping and storage technologies in the mid–late 1980s allowed fresh apples, strawberries and asparagus, for example, to be shipped from the southern hemisphere to northern markets [Codron (1992)]. This new marketing technology reduces transaction costs and increases market integration. Of course, the diffusion of these technologies is correlated with the "waves" noted above, hence fastest in the regions of the first and second waves.

[5] This shift can be in any or all segments of the agrifood system; for example, the farmer might sell an undifferentiated commodity to the first-stage processor who then differentiates it into various products, or the farmer might differentiate products.

[6] "Efficient consumer response" (ECR) is a voluntary, industry-wide movement that started circa 1992 in the US and diffused to other industrial countries and now into emerging markets. It focuses on continued improvement in the product supply chain to supermarkets. The three tenets are (1) providing consumer value, (2) removing costs that do not add value, and (3) maximizing value and minimizing inefficiency throughout the supply chain. Key actions are the reduction of inventories and paper transactions, and streamlining of product flow. Order cycles are shortened through the use of electronic data interchange (EDI, a computer-to-computer exchange of business data) and UPC bar codes to automatically identify products. This automated the order-creation and -entry phases of the order cycle, taking days out of the total cycle time. Order cycles were further reduced through what were called "strategic partnerships" where retailers and manufacturers would work together as a team to set up ways of achieving performance goals that exceeded existing industry practices. There is extensive use of point-of-sale data and other consumer-oriented data, giving an accurate read on consumer demand. These data are passed by way of EDI to the manufacturer so products can be made in quantities based on actual consumer demand, and then distributed to the end consumer in the most efficient manner – hence Efficient Consumer Response. (http://www.fmi.org, excerpted and paraphrased from the website, accessed May 11, 2005.)

The nature of this transformation of markets is further influenced, beyond the demand and technology drivers just noted, by organizational and institutional changes. Those changes fall into the following categories, drawing on and applying various strands of the literature.

Increased competition is driven by policies of market privatization and economic liberalization, and liberalization of foreign direct investment (FDI). These in turn lead to an influx of foreign competitors and rising sectoral concentration. When coupled with the technological change in the retailing and distribution sectors noted above, changes in "mechanisms of governance" [Williamson (1975, 1996)] in agrifood marketing systems are induced, in two important dimensions:

(1) organizational change including shift from spot markets (including, for example, traditional wholesale markets) and traditional brokers to use of specialized/dedicated wholesalers and various degrees of partial vertical integration, including preferred supplier lists; and
(2) institutional change in the adoption of contracts and other "vertical restrictions" [Carlton and Perloff (2000)] and the formulation and imposition of private standards (of quality and safety of agrifood products) as instruments of strategic market differentiation and coordination mechanisms [Reardon et al. (2001); Farina and Reardon (2000); Mainville, Zylbersztajn and Reardon (2005)].

These institutional and organizational changes increase coordination and lower transaction costs in the chain. The pressure to lower transaction costs and increase system flexibility to extend the scope of markets and permit a shift from commodity to product markets induces organizational change in output markets.

3. Transformation of markets: Commodity market development, 1950s to 1980s

While one can debate whether economic theories and research drive policy debate or vice versa, it is useful to categorize the phases of the "commodity market development" period by the policy debates that dominated in different sub-periods, and organize research literature by those periods. Obviously, some strains of research cut across these periods, but most have their centers of gravity in particular phases linked to specific debates about how to accomplish general development objectives. Moreover, we find that most development debates (whether about agricultural technology, markets, etc.) have three phases: growth and efficiency, then distribution and equity, and then sustainability (in environmental, economic and political terms). We find that the debates over development of commodity markets were no exception.

To summarize the story, analysis from the 1950s to the 1980s tended to:

(1) focus mainly on basic grains and on commodity markets, domestically and internationally;
(2) focus on wholesalers (where the marketing agent was specified at all) with little attention to the retail sector, which was assumed to be small-scale, traditional, and essentially a passive element of the marketing system. Where there was a

specific discussion of the marketing agent, the analytical lens was through oligopsonistic or monopsonistic competition;

(3) focus on public policies (standards, marketing boards, buffer stocks and stabilization schemes, administered prices, subsidies, and so on) and little on private sector or firms and their strategies; and

(4) focus on spot markets and atomized suppliers, rather than markets vertically coordinated by institutions such as contracts or actors organized into joint ventures and associations.

3.1. Phase 1 – 1950s–1960s: Growth as the policy objective for grain commodity markets

In this first phase, most developing countries were newly independent ex-colonies, and a Lewis-inspired vision of agriculture was common among governments, where peasant farming was not considered capable of generating large commodity surpluses [Lewis (1954)]. Mirroring the Ricardian debate more than a century earlier, there was a general tendency to promote the growth of international and national commodity markets in order to move food to the growing cities and feed urban workers. Overall economic transformation was intended, and market policy was used to provide cheap food. This was done by several methods:

- Infrastructure (roads, ports, trains) was built to move grain, and to carry out grain transactions (public markets and storage infrastructure).
- Mirroring the logic of dismantling the Corn Laws (tariffs on imported grain) in England in the first half of the 19th century to reduce the cost of imported grain (thought to be a spur to industrialization), exchange rates were allowed to become overvalued as a mechanism to cheapen grain imports.
- Commodity exports (especially tropical products such as coffee, cocoa, rubber and rice) were taxed in a manner reminiscent of the Physiocrats (France) or the Meiji Restoration (Japan), approaches that taxed land to provide government revenues and permit infrastructure development.
- Public grades and standards, mainly for bulk grains, were established or extended and refined to facilitate long-distance trade.

The body of research that informed the above policy debate, with its center of gravity in this phase, had several main strands.

First, there was a strand of research on the determinants of, and the effectiveness of policy instruments to spur, the *expansion of international and domestic commodity markets*. Research included analysis of public commodity grades and standards, for example, Lowell Hill's work in the 1960s–1990s [Hill (1990)] and improvements in infrastructure on output market development [see Ahmed and Donovan (1992) for a review of this literature in developing regions].

Second, there were, starting in the mid-1950s, the beginnings of concern for, and analysis of, the impacts of output market structure (in particular regarding hypotheses of oligopsony and monopsony) on efficiency of markets and equity with respect to distributional effects on farmers. J.C. Abbott, in his chapter "The development of marketing institutions" [Abbott (1967, p. 364)] pointed out that:

> "EXPLICIT analysis of the role of marketing in the development of the less advanced countries began only a few years ago. Previously it had been the subject of considerable administrative intervention – often without a clear economic basis, as Bauer[7] in the mid-fifties made clear.... The first to point out specifically that distribution was of greater importance in the economic development of backward areas than its neglect in the literature would suggest was Holden, following his work with Galbraith in Puerto Rico [Holden (1953)]."

Abbott pointed out that "market organization came of age" in 1959 in a symposium in the *Journal of Farm Economics* by four economists who had worked in developing countries, such as Cook (1959) in the Philippines and Martin (1959) in Pakistan and India – and in a special issue of The State of Food and Agriculture [FAO (1959)]. Abbott noted that these two forays into the literature emphasized that it was crucial to understand the downstream segments of the food system – the market structure and incentives facing farmers – rather than just the supply, or production side. That call for the importance of markets – still echoed in today's debates nearly a half-century later – spurred a series of field studies at the micro and meso level to understand the empirics of food markets in the 1960s and 1970s.

The research started in this phase and while it continued into the 1970s/1980s, we discuss its neoclassical and institutional economics strands here, and return to its new-institutional-economics manifestations in the 1990s below. The root of this research mainly consisted of "mapping" market structures and identifying the actors and the market conditions (prices, margins, volumes) along the supply chain or subsector. The work had its roots in three general categories of economics literature developing before and during the period: (a) the analyses of economic and technical relations in the inter-industry supply chain and linkages literature [such as Rasmussen (1957) and Hirschman (1958)]; (b) the economics of the structure-conduct-performance of subsectors [Bain (1959)]; (c) the economics of transaction costs, coordination, and institutions in firms, industries, and subsectors [Coase (1937); Arrow (1969); Williamson (1975); North (1981)].

Drawing from the above three strains of general economics and development literature, and inspired by the "markets" forays in the literature in the late 1950s, were a series of innovative empirical studies on agrifood output markets, supply chains, and subsectors in Africa, Asia, and Latin America in the 1960s and 1970s. A useful survey of this work is in Martin (1992) and a brief methodological survey is in Timmer, Falcon and Pearson (1983). Haggblade (2005) provides a detailed review of this work since the 1960s in Africa, Asia, and Latin America, classing this body of literature as studies of "vertical networks" (a general term applied to the networks variously, and often in

[7] See Bauer (1957).

parallel lines of work, as "subsectors" or "filières" or "supply chains" or "commodity chains" depending on whether the research originates in Latin or Anglo-Saxon institutions, or is done by agricultural economists or business schools). Examples of work focused on the structure (degree of market concentration) and pricing efficiency of traditional markets included Jones (1968) in Nigeria, Cummings (1967) and Lele (1971) in India, Timmer (1974) in Indonesia, and Shepherd (1967) and Scott (1985) in Peru.

There was also an early introduction of social capital, trust, and reputation as a determinant of traditional market performance that one sees in Abbott (1962), Schmid and Shaffer (1964), and Van der Laan (1975). This latter work, combined with the institutions and transaction costs perspective noted above, was manifested in a re-emergence and quantitative analysis of these factors seen in work in the 1990s and on, such as the work on contracts, transaction costs, and social capital in African markets of Fafchamps (2004). This latter extension into the present of the line of thought concerning institutions, merged with and was given impetus by the "new institutional economics" in the 1990s and later, discussed below.

Much of the applied agrifood systems fieldwork of the 1960s and 1970s focused on the interface between domestic commodity markets and peasant producers. This involved tracing the structure of marketing systems, testing the "market efficiency" hypothesis in search of exploitation, and exploring market power and marketing margins of traditional market intermediaries in developing countries.

The lines of debate were drawn early. One strand of the literature [for example, Bauer (1957)] posited that traditional output market intermediaries in developing countries were operating efficiently without evidence of exploitative monopsony. This school contended that wholesalers in peasant food output markets were not earning super-profits, and thus were not exploitative; their high margins reflected what in the general literature were starting to be termed "transaction costs". This essentially mirrored at the markets level of Schultz's (1964) finding of peasants as "poor but efficient" with apparent inefficiency simply reflecting constraints and costs. This is the precursor of later work on transaction costs in marketing channels in developing countries in the 1990s such as Fafchamps (2004), Kopicki et al. (2004), and Escobal (2005). An example of this "pole" is the work of Shepherd (1967) and Scott (1985) in potato markets in Peru.

Another strand of literature [for example, Abbott (1962)] posited that various structural and institutional features of intermediation in the market produced a tendency toward monopsony or oligopsony, and these generated disincentives for farmers. This approach reflected the "working hypothesis" of developing country researchers and policymakers in those decades, and contended that there was evidence of monopsonistic (often interpreted in the literature as "exploitative") traders and thus inefficient and inequitable output markets. An example of this pole is the work of Wharton (1962) in Malaya, Mears (1957) in Indonesia, or of Harriss (1979) in India.

Abbott (1967) himself argued that while the evidence was not fully conclusive, the existence of some information indicating constraints faced by farmers in markets due to local monopsony or oligopsony of wholesalers implied that there was a need for investment in market institutions [a call echoed in similar terms in World Bank (2002)], calling for "public provision of market information and advice, credit institutions, and local warehousing facilities, or by reducing barriers to the entry of new trading enterprises and fostering the growth of alternative marketing channels, as through cooperatives" [Abbot (1967, p. 370)].

In fact, the evidence marshaled by and the arguments of the market economists finding evidence of market inefficiency in general, and in extreme cases of monopsony, was used as a key justification for support among developing country governments and multilateral organizations such as the World Bank of the need for food marketing parastatals to obtain funds for sales promotion, research and extension, to raise the bargaining power of farmers, to improve marketing organization, to regulate standards, to raise the scale of operation [Abbott (1967)], a key policy element in the latter part of this period and the next phase. Note that in the later, structural adjustment phase of the 1980s, the first strand of thought (hypothesizing non-exploitativeness of traders) would be resurrected by the World Bank and other donors and governments to justify the dismantling of the parastatals.

3.2. *Phase 2 – 1970s to early 1980s: The policy objective becomes improving income distribution from the transactions in grain commodity markets*

In this second phase, much of Asia and parts of Latin America were well-advanced in overall economic transformation, and were well into the middle stages of agricultural transformation, accelerated by the Green Revolution. Africa and parts of South Asia, the "third wave" areas, remained at earlier stages of transformation. The prior policies continued, aimed at inducing expansion in the commodity output markets, but now new policies were added to induce specific distributional outcomes from the growth in commodity markets. Input markets entered the debate as they provided key support for the agricultural transformation using Green Revolution technology. The distributional objectives for output markets and the growth goals for input markets were pursued by several methods beyond those used in the previous phase:

(1) Cheap international credit became available in the 1970s, just as global commodity markets boomed, generating funds to pay for active interventionist measures.
(2) Input subsidies were used to induce growth of capital, fertilizer, seed, and machinery markets.
(3) Consumer food subsidies on the demand side and producer price supports on the supply side were used to cushion the "food price dilemma".
(4) Oil price shocks in the mid-1970s and early 1980s and famines in South Asia in the 1970s and Africa in the 1980s focused attention on the need to stabilize markets and buffer the poor from food supply shocks. That led to debates about stabilization schemes and buffer stocks from the mid-1970s on [Newbery and Stiglitz (1981)]. Domestic development of commodity markets emphasized price stabilization. Similar concerns were generated by instability in non-grain commodity markets, with attendant attention to commodity market stabilization schemes. Note that in both of these dimensions, "stabilization" had a different

connotation than its use in the mid- to late 1980s and 1990s in structural adjustment.

(5) There was induced organizational change with the establishment of parastatals to administer the above systems, although many of these dated to the 1960s in Asia. Mirroring the distribution-related debates of the period over, for example, the diffusion of Green Revolution technologies (with much of the debate over non-adoption and lagging adoption), such as in Lipton (1977), as well as land tenure, there was policy and research attention to the distribution-related issues of market outcomes, especially the role of exploitative and oligopsonist merchants. This debate over distribution was often stimulated by the efforts by national governments, by the World Bank, and other multilateral organizations to support the formation of parastatals as a way to "get around" the consolidated, inefficient, and traditional marketing system. Policymakers often thought the traditional marketing system was too intransigent to change through normal economic and policy forces.

The research that informed the above debates was focused on three questions:

(1) What are the determinants of *efficiency and stability* in commodity markets?
(2) What are the *equity impacts* on consumers and producers of commodity market structure and organization? And
(3) What are the *tradeoffs between the efficiency and equity impacts* of food policy changes that are mediated in commodity markets?

Specific strains of research in this phase were as follows.

First, as noted above, there was a continuation of empirical research on the efficiency and the equity impacts of traditional wholesale market systems, including analysis of marketing margins and market power of traditional market intermediaries in developing countries. This research was in parallel with, and related to, the farming systems research of the time.

Second, spurred by policymakers' concern over the instability of commodity markets and the effects that this has on fiscal and external accounts as well as urban and rural welfare, a major body of research emerged on how changes in price and consumption variability are conditioned by the degree of inter-spatial and inter-temporal market integration, in turn determined by infrastructure and market liberalization. This work tested for relationships between market integration and price transmission. The policy debates that stimulated this research included the impact of public investments in infrastructure, regional integration policies, and market information systems on market integration and price stability.

There were two sub-strands of this literature, macro and micro. The macroeconomic work focused on the link between the emerging food security debate and grain buffer stock schemes and food aid impacts of the 1970s and 1980s. This work constituted a major blossoming of research on commodity price movements, testing transmission effects, lags, integration of markets, and the economics of buffer stocks. See, for example, Valdes (1981), Newbery and Stiglitz (1981), and Berck and Bigman (1993).

The micro- and meso-analytic work focused on econometric analysis of inter-spatial and inter-temporal price integration. The work tested for efficiency of transmission of price shocks. General (not specific to developing regions) research in this line is reviewed in Fackler and Goodwin (2001). The work on these themes started in developing regions in earnest in the 1970s [such as Timmer (1974)], and blossomed thereafter with improvements in econometric techniques – and also because policy issues such as drought shocks, food aid, and market liberalization kept the issue of market integration and price shock transmission on the "front burner" into the 1980s and 1990s. An important contribution to the latter was made in Ravallion's analysis of the 1974 famine in Bangladesh, as this developed rigorous new econometric tools for measuring market integration [Ravallion (1986)]. Examples of this work include Alderman (1993), Badiane and Shively (1998), and Abdulai (2000) in Ghana, Timmer (1996) in Indonesia. Work along this line, such as Barrett (1997) in Madagascar, Fafchamps (2004), and Platteau (2000) suggest the presence of constraints to market integration due to transaction costs, highly personalized exchange, and market entry barriers.

The combination of market liberalization and integration has shown mixed results on price and consumption variability. Some work shows increases in price variability, such as Barrett's work in Madagascar, and some showing decreased consumption and price variability, such as McIntire's work in the Sahel, and cereals market integration work in East Asia (Timmer) and South Asia (Ravallion).

Third, toward the end of the period, Timmer and colleagues launched the "food policy" approach that linked farmer and consumer decision-making models, through the marketing system, to macro economic and trade outcomes at one end of the food system and to poverty outcomes at the other [Timmer, Falcon and Pearson (1983)]. This literature was to the marketing sector what the Schultz (1978) volume *Distortions in Agricultural Incentives* was to the production sector, the first hard look at the efficiency and equity impacts of the market and farm policies that had been put in place earlier under very different rationales. These two transitional literatures helped usher in the next phase of policy and research discussed below, relating the "food policy" perspective to the debate around the effects of structural adjustment policies.

3.3. Phase 3 – early–mid 1980s to the early–mid 1990s: Structural adjustment of markets and "getting prices right"

Toward the end of the 1970s and into the early 1980s, there was a momentous change in the center of gravity of policy regarding commodity markets. This change centered on a first wave of market reforms focused on the abandonment of the "Cheap Food Policies" that had been used to support early industrialization in developing countries in the 1950s until the late 1970s. Accompanying these policies were attendant input and output market subsidies, price controls, and parastatals.

Once subsidies and parastatals were unaffordable they were reduced or eliminated by "Structural Adjustment Programs" in the 1980s and 1990s, bringing in the era of "Liberalized Food Price Policies". "Getting Prices Right" in output and factor markets became the development mantra for well over a decade [Timmer (1986)]. The rate of policy change was (very) roughly correlated with the waves of transformation as noted above, with the latest reform occurring in Africa.

These programs were aimed at encouraging and sustaining a new spurt in growth from the grain commodity markets and beginning the transition from commodity to product markets, including fostering the emergence of competitive markets for non-staples. The key factors were as follows:

(1) Cheap international credit was no longer available on a large scale after the mid-1980s, when recycling of petro-dollars ended.

(2) Commodity and input market liberalization began, first in Latin America and Asia, then in Africa, and later in the socialist countries of Central and Eastern Europe and then East Asia.[8] The latest of these processes began in the 1990s when liberalization began in China's grain markets [Rozelle et al. (1997)], a process that continued during the recent process of policy changes as conditions of Chinese accession to WTO.

(3) Input subsidies were phased out along with dismantlement of input parastatals. This was intended to induce investments by the private sector into input market distribution facilities. In theory, this private capacity would lower input costs through greater competition and efficiency.

(4) Consumer food subsidies on the demand side and producer price supports on the supply side were phased out because they could not be financed by public budgets, and donors were no longer willing to finance them.

(5) International markets were deregulated via GATT and later WTO. There were major trade initiatives conceived in this phase and then executed in the next (NAFTA, MERCOSUR). It was assumed that international market liberalization would substitute for the functions of administered stabilization schemes of the 1980s. Commodity grain trade into developing countries rose with lower barriers, both from policy interventions and improved transportation technology. Even faster, non-commodity trade grew from and to developing regions, facilitated by lower trade barriers, and by major technological improvements in storage, processing, and shipping that greatly increased the efficiency and distance of trade and domestic marketing. The forerunner of this trend was apple exports from Chile to the northern hemisphere [Codron (1992)].

(6) An extremely important trend, little noticed at the time, was the deregulation of foreign direct investment (FDI) in the marketing sector, and this began in earnest in the early to mid-1990s. This trend was part of the broader liberalization of capital markets in most developing countries, which in turn was part of structural adjustment programs. This deregulation soon induced the wave of FDI in the agrifood sector that initiated the globalization period discussed below.

[8] It should be noted that the policy reform speed and completeness varied considerable over countries (within a given "wave") and sectors [Kherallah et al. (2002)].

The lines of research on markets in this period were as follows.

First, there was a development of micro analysis linked to meso- and macro-level changes – relating household and market analyses – such as farm household models to test the income and supply responses to subsidy removal [Singh, Squire and Strauss (1986)], and food consumption analysis and other partial equilibrium analyses related to food subsidy removal in output markets, on producers and consumers. An example of this work is Pinstrup-Anderson (1985).

Second, there was an application of the food policy analysis introduced in the prior phase in a multi-market agricultural sector framework [such as Braverman (1986), work on Senegal] and a general equilibrium framework [such as De Melo and Robinson (1982), de Janvry and Sadoulet (1987), and Taylor (1990)].

Third, there was an extension of methods with increased complexity to test market integration (for example, the GARCH models) and applied to markets that were recently liberalized to test for the efficiency and transmission effects of the liberalization and other shocks [Barrett (2001)].

3.4. Phase 4 – 1990s: Second generation of reforms, "getting institutions right" and the rise of new institutional economics research on developing country markets

Toward the end of the period and into the 1990s, a second wave of reform was initiated with a new mantra: "getting institutions right" [Williamson (1996)]. There were several sources of inspiration for and justification of institutional reforms focused on reducing transaction costs and increasing property rights.

First, there was emerging evidence that "market failures" persisted after "getting prices right" in the previous phase [Stiglitz (1989)]. To correct those failures, governments could no longer turn to the sort of price policies they had used before price policy reform. Instead, they sought other non-price-policy measures, such as support to NGOs to provide market assistance to those left out of the benefits of the earlier reforms, and institutional and regulation changes to reduce transaction costs. A decade of emphasis on the need for these reforms was crowned by the World Bank's *World Development Report 2002: Building Institutions for Markets.*

Second, the reforms were inspired on the theoretical side by work of North (1981, 1990) and Williamson (1975, 1996). They focused on the need to develop public and private institutions (such as property rights and contracts) as essential to development both historically in currently developed regions, and currently in developing regions. There was also a resurrection of the theme of social capital and market development, treated in the institutional economics and economic anthropology literature of the 1960s, as noted above.

The combination of the above two factors inspired (and was inspired by) the rapid development over the 1990s and to the present of the New Institutional Economics (NIE) approach to market development [see Williamson (1996) for an inspired exposition on the confluence of market development economics and the NIE approach]. Examples of this work include Dorward, Kydd and Poulton (1998), Harriss, Hunter and Lewis

(1998), and Fafchamps (2004). In general, this work tends to be micro analytic, and to employ similar models (for example, farm household models) to those used in the earlier phases, but with an emphasis on the introduction of variables reflecting transaction costs, social capital, and so on. Moreover, NIE was fused with the technical and institutional-economics agrifood system analysis inherited from the 1960s to the 1980s in vertical and horizontal network analyses with careful consideration of governance mechanisms, such as commodity chain analysis [Kaplinsky and Morris (2001)] and the "netchain" work of Lazzarini, Chaddad and Cook (2001). There has also been an extension of earlier work on interlinked and interlocked agrarian markets by Bardhan (1980) and Braverman and Stiglitz (1982), such as recent work by Giné and Klonner (2002) analysis of product and credit market linkages for technology upgrading by small-scale fishermen in the traditional fishing sector of Tamil Nadu.

4. Phase 5 – mid-1990s to the present: Globalization and product market development in developing countries

The first half of the 1990s is a watershed because, by then, structural adjustment and stabilization programs had substantially reduced, even eliminated, controls on and state interventions in commodity output markets in many developing countries. With subsequent liberalization of capital markets, the combined effect was to bring very deep changes in output markets over the following decade, inducing a further shift in the center of gravity from being in a traditional, commodity system, to the effective initiation of a modernized food system and the emergence of differentiated product markets. This fundamental transformation was highlighted in the introduction. The analysis of changes in output markets has to be part of an analysis of changes in the entire agrifood system, otherwise the drivers of change cannot be identified. From this perspective, the main change in output markets in the 1990s emerged from the *demand* side of the market, in particular the "downstream segments" of the food system, including the components of what is commonly termed the "food industry" – the retailers, processors, and wholesalers.

The key point of this section is that trade liberalization was *not* the only, and perhaps even not the most important, factor by which "globalization" changed agrifood systems and thus output markets in developing countries during the 1990s and beyond. Yet trade liberalization is at the center of the debate on how globalization is affecting developing countries' food systems and thus farmers [Cline (2004)]. In fact, two factors are having much greater impact on domestic agrifood systems:

(1) the fundamental restructuring of *domestic* food markets, in particular changes in the processing, retail, and food service segments. Typically, domestic food markets provide at least 85% of food consumed domestically; and
(2) the impact of foreign direct investment (FDI) on this restructuring has been basic and fundamental, with an immense flood of FDI, after it was liberalized in most countries in the 1990s, in these downstream segments. These investments

altered domestic markets in developing countries far more powerfully than has the change in international trade regimes for domestic products. This story is told below.

4.1. The impact of globalization on output markets in developing countries: Trade liberalization is just the "tip of the iceberg" – FDI liberalization was crucial

The key message of this subsection is that the primary impact of globalization on domestic food markets came not through the trade effect, but through direct changes wrought on domestic food markets by FDI liberalization. To make that point, we begin by noting globalization's effects on trade, and then comparing trade with the domestic market economy.

4.1.1. The effects of globalization on international trade

In most developing countries, the earliest measures toward trade liberalization were taken in the latter half of the 1980s and the first half of the 1990s. The primary impact on commodity markets was to reduce tariffs and export taxes, consumer subsidies, and producer price controls. These measures were embedded in individual countries' structural adjustment programs as well as in various multilateral agreements to which countries subscribed, such as GATT, then WTO, MERCOSUR, NAFTA, etc. Globally, trade liberalization – combined with higher incomes and improvements in supply, transport, and storage – had several effects, with important qualifications made below.

During the last two decades of the 20th century, food trade doubled in volume and value. Between 1980 and 2002, the value of exports of all agrifood products increased from $200 billion to $400 billion [FAOSTAT (2004)]. Population increased by 50% during the same period. Only 30% of that trade increase was from increased exports by Africa, Asia, and Latin America, even though their population share is roughly 80%. Most of the effect of globalization on agrifood trade was within OECD countries.

There was a concomitant, and fundamental, shift in the overall product composition of trade, from mainly commodity staples in 1980 to mainly product non-staples in 2000. The share of bulk grains in world agricultural trade dropped from 45% in 1980 to 30% in 2000. An exception was the rapid increase in trade in soybeans (a key input to livestock production), with the entry of Brazil as a major exporter, and large increases in demand from Asia, and especially China. Reductions in trade barriers and economies of scale through increases in trade reinforced the downward trend in basic grain prices on the world market that had been set in train decades before by technological changes in production and transport. From peak in the early 1950s to trough at the turn of the Millennium, the real prices of rice and wheat dropped nearly tenfold.

Part of the overall trade growth was a rapid increase in trade in fruits and vegetables, and meat and fish. This was relatively concentrated, on the exports side, in a total of just 8 to 10 countries (mostly the US and several countries in Europe and a few countries in Latin America and Asia). There was a similarly rapid (mirroring) increase in imports [Regmi and Gehlhar (2005a)]. These increases were driven on the demand side by income effects via Bennett's Law, and on the supply side by the stimulus from this demand, but facilitated by improved transportation and storage (to permit trade in off-season fruit) and major investments and government support (investment and tax credits, infrastructure, etc.) in non-traditional exports of produce and seafood/fish and meat from developing countries. Chile and Guatemala were prime examples of successful exporters [see Barham et al. (1992)], as was Kenya [Jaffee and Morton (1995)]. In general, Latin America and East/Southeast Asia were the primary winners from this increase in non-cereal trade, and again, only a few countries in each of them, with Africa and South Asia less so.

There was an explosion of trade in processed foods such as juice, baked goods, snacks, meat, and beverages. These were 18% of global food trade in 1980 and rose to 34% by 2002 [Regmi and Gehlhar (2005a)]. A crucial point, however, is that the growth in trade in processed products is eclipsed by growth in sales by foreign subsidiaries of developed country firms to consumers in developing countries; for example, local sales by foreign subsidiaries of US processed food firms are five times the exports of processed food from the US to the rest of the world [Regmi and Gehlhar (2005b)]. The major mechanism for this rapid growth in sales has been foreign direct investment (FDI), discussed further below.

4.1.2. *International trade compared to the domestic market*

Despite the increase in global trade volumes, the share of trade in the overall global food economy (that is, the *shares* of imports and exports in consumption and output, respectively) did not change much between 1980 and 2001. Moreover, global food trade is dwarfed by the size of the domestic food markets in developing regions. According to the estimates in FAOSTAT's food balance sheets (that report tons of raw product equivalents), trade remained throughout the two decades as a very minor part of overall food consumption. Focusing just on cereals, meat, and produce, in developing countries overall tonnage of food trade was about 15% of consumption in 1980 and was still about 15% in 2001. For cereals, 19% of grain consumed in developing countries was imported in 1980 compared with 20% in 2001. Exports of grain were 15% in 1980 compared with just 6% in 2001. For vegetables, just 1% of consumption in developing countries was imported in 1980 and the share was the same in 2001. Exports of vegetables were 2% of output in 1980, and this share increased to 3% in 2001. The trend is similar for fruits, with 5% of consumption imported in 1980 compared with the same share in 2001. Fruit exports were somewhat more important: 11% of fruit output was exported in 1980 compared with 12% in 2001. For meat, in both 1980 and 2001, meat exports and imports were roughly 4% of output and consumption, respectively. Several caveats are in order: (1) for individual products the share of trade can be much higher; (2) usually the share of trade in total marketings is greater than the share in consumption; and (3) in general one can consider these sorts of data as merely approximate estimates, even just orders of magnitude.

Over the same period, by contrast, changes in domestic patterns of consumption have been dramatic. From 1980 to 2001, as the population of developing countries rose from 3.25 billion to 4.8 billion, the share of cereals in a subtotal of the diet which includes cereal, vegetables, fruit, and meat, fell from 60 to 45%. During this time, per capita consumption of cereals rose only slightly, from 157 to 162 kilograms (kg). By contrast, intake per capita of vegetables, fruit and meat doubled, increasing from 51 to 111 kg, 38 to 55 kg, and 14 to 28 kg, respectively. These consumption changes are an obvious manifestation of Bennett's Law, but show the power of demand changes driven by rising income. Caloric intake per capita rose only 13% over the two decades, while protein intake rose 25% and fats/lipids rose 33%.

Thus, what we find overall is that international trade basically did not change very much in its incidence in food production or consumption and in volume terms had a minor effect on local food markets. Of course, trade has other kinds of effects on domestic markets for particular items, such as transmission effects of increases in quality or safety standards, on varieties, and so on. At the same time, there was a huge change in diets mediated by an increase in non-staple market volumes, an increase in processed food markets, and a shift from staple grain trade to processed grain trade, especially in the form of processed snacks. These changes were accomplished in domestic food markets, in local processing and retail firms (whether owned domestically or by foreigners), and by local agricultural producers.

4.1.3. *The importance of the domestic food market implies that the main way that globalization affects the food systems of developing countries is via FDI*

Research and debate about "globalization" effects on domestic food markets and producers have focused on international product market liberalization, i.e. trade [Cline (2004)].[9] But a strong case can be made that this is actually a minor element of the impact of globalization on *domestic* food and agricultural markets, on farmers, and on consumers in developing countries, compared with the liberalization of capital markets that induced an avalanche of FDI from developed into developing countries.

In 1980, total FDI was roughly $1 billion a year into Asia, and the same in Latin America; agrifood investments followed this general pattern for total FDI. By 1990, the amount was roughly $10 billion a year into each of these two regions that were frontrunners in globalization. By 2000, the figure was $80–90 billion [UNCTAD (2001)]. The growth in FDI far, far outstripped growth in product trade. Global FDI rose an average of 13% per year during those decades, versus "just" 7% yearly growth in exports of goods and nonfactor services. In 1998, FDI was $440 billion, and the total stock of investments by multinationals was 3.4 trillion dollars. The developing countries received 26% of that FDI in 1980, and that rose to 37% in 1997 [Mallampally and Sauvant

[9] It is indicative of current thinking that the Cline volume mentions foreign direct investment on only a single page, and then as an ancillary to trade liberalization [see Cline (2004, p. 101)].

(1999)]. This investment was highly heterogeneous across regions, explaining in large part the differences in the speed of change in output market structure in the 1990s. FDI to East Asia rose from \$10 to \$65 billion from 1990 to 1998, and during the same period it went from \$7.5 to \$70 billion to Latin America. The increase was only from \$0.5 to \$5 billion each to South Asia and Africa, but FDI increased from \$1 to \$25 billion to Eastern Europe [Business Week (2000)].

Another important aspect is that before the liberalization of FDI throughout the 1990s in retail, processing and distribution, most agrifood FDI was "upstream" in operations such as Chiquita banana plantations.[10] But with globalization and relatively free entry for FDI, the key difference now is that the great bulk of investment is downstream, in processing, retail, and food services. These investments consequently have had far more influence on the structure of the domestic food system than had the enclave-type upstream investments. This is a central point of this chapter. The returns to FDI in the food industry are usually much higher than those in primary production, as reflected in Mexican experience, explaining why most FDI in the food sector in Mexico, as elsewhere, is downstream, in the food industry [Shwedel (2003)].

It is thus through the vector of FDI in processing and retail that globalization is really being felt by domestic food markets in the developing countries, and so we focus on these vectors below in understanding the transformation in food output markets in the past decade. We focus on the rapid rise of supermarkets as the key factor explaining the major changes in food markets that occurred with globalization and concomitant FDI. We focus on the retail segment both because there are parallel changes in the food processing sector, and because the retail segment change in many cases led processing sector change or was symbiotically intermeshed with it.

4.2. *The rise of supermarkets: Downstream changes in food systems in the 1990s drive agrifood market transformation*

Supermarkets[11] are traditionally viewed by development economists, policymakers, and practitioners as the rich world's place to shop. But supermarkets are no longer just niche players for rich consumers in the capital cities of developing countries. The rapid rise of supermarkets in these countries in the past 5–10 years has transformed agrifood markets – at different rates and depths across regions and countries.

In this section we describe this transformation of agrifood systems in Africa, Asia (excluding Japan), and Latin America. We focus on the determinants of and patterns in

[10] This topic is discussed below in the context of liberalization of retail FDI in China, 1992, and periodically up to full liberalization in 2004 [Hu et al. (2004)]; Argentina and Mexico, 1994; Indonesia, 1998; and, partially, in India, 2000. Manalili (2005), for example, notes the very important effect on supermarket sector development of liberalization of retail FDI in the Philippines in 2000.

[11] This is a term we use as shorthand for large-format modern retail stores, such as supermarkets, hypermarkets, and discount stores. Our discussion focuses on large-format because convenience stores tend to have only a small share (circa 5–10%) of modern retail sector sales.

the diffusion of supermarkets, and then we discuss the evolution of procurement systems of those supermarkets. It is these procurement systems that are having the most visible consequences for agrifood systems, and hence on output markets.

4.2.1. Determinants of supermarket diffusion in developing countries

The determinants of the diffusion of supermarkets in developing regions can be conceptualized as a system of demand by consumers for supermarket services, and supply of supermarket services – hence investments by supermarket entrepreneurs. Both functions have as arguments incentives and capacity variables.

On the demand side, several forces drive the observed increase in demand for supermarket services (and are similar to those observed in Europe and the United States in the twentieth century). On the "demand incentives" side are: (1) urbanization, with the entry of women into the workforce outside the home, increased the opportunity cost of women's time and their incentive to seek shopping convenience and processed foods to save home preparation time; and (2) supermarkets, often in combination with large-scale food manufacturers, reduced the prices of processed products.

On the "demand capacity" side, several variables were key: (1) real per capita income growth in many countries of the regions during the 1990s, along with the rapid rise of the middle class, increased demand for processed foods, which is the entry point for supermarkets, as they could offer greater variety and lower cost of these products than traditional retailers due to economies of scale in procurement; and (2) rapid growth in the 1990s in ownership of refrigerators meant an increased ability to shift from daily shopping in traditional retail shops to weekly or monthly shopping. Growing access to cars and public transport reinforced this trend.

The supply of supermarket services was driven by several forces, *only a subset of which overlap with the drivers of initial supermarket diffusion in Europe and the United States*. On the "supply incentives" side, as discussed below, the development of supermarkets was very slow before (roughly) the early–mid 1990s, as only domestic/local capital was involved. In the 1990s and after, foreign direct investment (FDI) was crucial to the take-off of supermarkets. The incentive to undertake FDI by European, U.S., and Japanese chains, and chains in richer countries in the regions under study (such as chains in Hong Kong, South Africa, and Costa Rica) was due to saturation and intense competition in home markets and much higher margins to be made by investing in developing markets. For example, Carrefour earned three times higher margins on average in its Argentine compared to its French operations in the 1990s [Gutman (2002)]. Moreover, initial competition in the receiving regions was weak, generally with little fight put up by traditional retailers and domestic-capital supermarkets, and there are distinct advantages to early entry, hence occupation of key retail locations.

On the "supply capacity" side: (1) there was a deluge of FDI in the retail sector that was induced by the policy of full or partial liberalization of retail sector FDI undertaken in many countries in the three regions in the 1990s and after.[12] Overall FDI grew rapidly in the 1990s in these regions; and (2) retail procurement logistics technology and inventory management were revolutionized in the 1990s, such as with the introduction of Efficient Consumer Response. This was led by global chains and is diffusing now in developing regions through knowledge transfer and imitation and innovation by domestic supermarket chains.

These changes were in turn key to the ability to centralize procurement and consolidate distribution (these, and other improvements in organization and institutions of procurement, are discussed further below) in order to "drive costs out of the system", a phrase used widely in the retail industry. Substantial savings were thus possible through efficiency gains, economies of scale, and coordination cost reductions. These efficiency gains fuel profits for investment in new stores, and, through intense competition, reduce prices to consumers of essential food products.

4.2.2. *Patterns of supermarket diffusion in developing countries*

The incentive and capacity determinants of demand for and supply of supermarket services vary markedly over the three regions, within individual countries, and within zones and between rural and urban areas at the country level. Several broad patterns are observed.

First, the overall image of the *spatial* pattern of diffusion of supermarkets is of waves of diffusion rolling along that mirror the waves of output market transformation that we laid out in the introduction. While there is significant variation in trends over countries in a given area such as South America (contrasting, for example, Brazil and Bolivia), and within individual countries over zones and between rural and urban areas, several broad patterns are clearly observed. From earliest to latest adopter of supermarkets over emerging market areas, there have been three waves of diffusion.

(1) Experiencing supermarket-sector "takeoff" in the early to mid-1990s, the first-wave countries include much of South America and East Asia outside China (and Japan), Northern–Central Europe, and South Africa – a set of areas where the average share of supermarkets in food retail went from roughly only 10–20% circa 1990 to 50–60% on average by the early 2000s [Reardon and Berdegué (2002); Reardon et al. (2003a)]. Compare that to the 70–80% share that supermarkets have in food retail in 2005 in the US, UK, or France, and one sees a process of convergence. Examples include frontrunners where the supermarket takeoff started in the early 1990s, like Argentina with a 60% supermarket share in food retail in 2002 [Gutman (2002)], Brazil with 75% [Farina (2002)], Taiwan with 55% in 2003 [Chang (2005)], and Czech Republic with 55% [Dries, Reardon and

[12] For example, partial liberalization of retail trade occurred in China in 1992, with full liberalization of the sector by the end of 2004; Brazil, Mexico, Argentina in 1994; various African countries via South African investment after apartheid ended in the mid-1990; Indonesia in 1998; and partial liberalization in India in 2000.

Swinnen (2004)]. While a small number of supermarkets existed in most countries during and before the 1980s, they were primarily local firms using domestic capital,[13] and tended to exist in major cities and wealthier neighborhoods. That is, they were essentially a niche retail market serving 5–10% of national food retail sales in 1990 (for at-home consumption, that is, not bought at restaurant-retail for consumption away from home). However, by 2000, supermarkets had risen to occupy 50–60% of national food retail among these frontrunners, almost approaching the 70–80% share for the United States or France. South America and parts of developing East Asia and transition Europe had thus seen in a single decade the same development of supermarkets that the United States experienced in five decades.

There is a second set of countries perched at the tail end of the first wave and near the start of the second wave, that we class with the first wave, with their supermarket "takeoff" in the mid-1990s, such as Costa Rica and Chile now with circa 50% [Reardon and Berdegué (2002)], South Korea with 50% in 2003 [Lee and Reardon (2005)], Philippines and Thailand with approximately 50% each [Manalili (2005); Thailand Development Research Institute (2002)], and South Africa with 55% [Weatherspoon and Reardon (2003)].

(2) The second-wave countries include parts of Southeast Asia, Central America and Mexico, and Southern–Central Europe where the share went from circa 5–10% in 1990 to 30–50% by the early 2000s, with the take-off occurring in the mid- to late 1990s; examples include Mexico [40% share of supermarkets in total food retail, Reardon et al. (2005)], Colombia [47% share, de Hernandez (2004)], Guatemala [36% in 2002, Orellana and Vasquez (2004)], Indonesia with 30% [Rangkuti (2003)], and Bulgaria with 25% [Dries, Reardon and Swinnen (2004)].

(3) The third-wave countries include countries where the supermarket revolution take-off started only in the late 1990s or early 2000s, reaching about 10–20% of national food retail by circa 2003; they include some of Africa (see below) and some countries in Central and South America [such as Nicaragua, see Balsevich (2005), Peru, and Bolivia], Southeast Asia [such as Vietnam, see Tam (2004)], China, India, and Russia. The latter three are the foremost destinations for retail FDI in the world [Burt (2004)] and each a fascinating third-wave case that will influence the development of agrifood markets in developing regions on into the 21st century. (1) China had no supermarkets in 1989, and food retail was nearly completely controlled by the government; the sector began in 1990, and by 2003 had climbed meteorically to a 13% share in national food retail, 71 billion dollars of sales, 30% of urban food retail, and growing the fastest in the world, at 30–40% per year! [Hu et al. (2004)]. Many of the driving forces for supermarketization were in place (rising incomes, urbanization) and it merely took a progressive

[13] The existence of these early supermarket chains serving a tiny niche market in some developing areas, for example in Puerto Rico, was noted as early as 1953 by Holden (1953) in the Holden–Galbraith study.

privatization of the retail market and even more importantly, a progressive liberalization of retail FDI that started in 1992 and culminated in 2004, that drove immense competition, even a full-out race, in investment among foreign chains and between foreign chains and domestic chains that is not only in full swing but has greatly accelerated in 2005 with the full liberalization of FDI that occurred as a condition to accession to WTO by China. (2) Russia is a similar case, with a late start due to policy factors holding back the takeoff despite propitious socioeconomic conditions, and then a very rapid takeoff spurred by an immense competition in investments underway now [Dries and Reardon (2005)]. (3) India is an interesting case, with its substantial middle class as a "springboard" for the spread of supermarkets; it is already considered to be among the top three retail FDI candidates in the world and is poised at the edge of a supermarketization take-off, although the share in food retail is still only 5%. At present, FDI is far from fully liberalized, and regulations concerning joint ventures in retail still dam what observers think is an imminent flood of foreign investment. Wal-Mart announced in May, 2005, that it intends to enter the retail market in India [CIES (2005)]; the Indian government immediately responded that it is highly likely that retail FDI will be liberalized by mid-2005 or soon thereafter, which will facilitate the entry of Wal-Mart [M&M Planet Retail (2005)]. South Africa's Shoprite and Hong Kong's Dairy Farm chains are also entering in 2004 and 2005, as is Germany's Metro. It seems likely that with a policy change regarding retail FDI, there will be a similar competitive race in investment in the sector from chains from around the world and around India.

Sub-Saharan Africa presents a very diverse picture, with only one country (South Africa) firmly in the first wave of supermarket penetration, but the rest either in the early phase of the "third wave" take-off of diffusion – or in what may be a pending – but not yet started – take-off of supermarket diffusion.

Kenya, Zambia, and Zimbabwe are in the early phase of the "third wave", and have substantial numbers of supermarkets, initiated by both domestic investment and FDI from South Africa. This investment was attracted by a middle class base and high urbanization rates, but supermarket penetration is still approximately where South America was in the early 1980s. The share of supermarkets in urban food retail is about 10–20% in the large/medium cities, and the share of produce hovers around 5% [see Neven and Reardon (2004), for example, for Kenya]. Even with mainly domestic investment and some South African retail capital and technology, there is still considerable uncertainty about the rate at which the supermarket sector in these countries will grow.

The great majority of Africa, however, can be classified as not yet entering a substantial "takeoff" of supermarket diffusion. At the upper end of this group are a score or so of supermarkets in countries like Mozambique and Tanzania, Uganda, and Angola, places where South African retail FDI is just starting [see Weatherspoon and Reardon (2003) for evidence on investments by the South African chain Shoprite] and may a decade or two from now be recognizable as a "fourth wave". Supermarkets in these countries show signs of early growth and are surrounded by a more general trend of the growth of self-service in large semi-traditional stores in urban areas. At the lower end of this set are the very poor countries of Africa, such as Ethiopia, Sudan, Burkina Faso, and Mali. It is unlikely that the lower end of this set of countries we will see supermarket growth for several decades. Even then, it will be dependent on higher urbanization rates, better investment climates, lower transaction costs, improved infrastructure, much more rapid income growth and political instability. It will take significant improvements in most of these areas to stimulate FDI by global supermarket chains. We have shown that supermarkets, even in places like South Africa and Kenya, spread beyond the middle class into the food markets of the urban working poor. But the supermarket sector usually requires a critical mass of middle class urban consumers to build the initial base before expanding into the rest of the urban market.

Note that the growth rates of supermarket food sales as well as retail foreign direct investment (FDI) are inversely correlated with the waves, so that the fastest growth is occurring in the supermarket sector in China (with 30–40% a year!) versus only 5–10% in the more mature, relatively saturated supermarket sectors such as those in Brazil and Taiwan.

In general, the "waves" above are correlated with socioeconomic characteristics of the areas that are related to consumers' demand for supermarket services and product diversity and quality: income and urbanization, in turn correlated with the opportunity cost of time, in particular that of women, and reduction in transaction costs through improvements in roads and transport and ownership of refrigerators. These demand-side factors are necessary, but not sufficient, to explain the very rapid spread of supermarkets in the 1990s and 2000s in these countries, most of which had at least a very small supermarket sector before 1990. That is, supply-side factors were also of extreme importance, especially the influx of retail FDI as countries liberalized FDI, and improvements in procurement systems, discussed below.

As is predictable from the diffusion model outlined above, the inter-spatial and inter-socioeconomic group patterns of diffusion have differed over large and small cities and towns, and over richer, middle, and poor consumer segments. In general, there has been a trend from supermarkets occupying only a small niche in capital cities serving only the rich and middle class – to spreading well beyond the middle class in order to penetrate deeply into the food markets of the poor. Supermarkets have also spread from big cities to intermediate towns, and in some countries, already to small towns in rural areas. About 40% of Chile's smaller towns now have supermarkets, as do many small-to-medium sized towns even in low-income countries like Kenya. And supermarkets are now spreading rapidly beyond the top-60 cities of China in the coastal area and are moving to smaller cities and to the poorer and more remote northwest and southwest and interior. As large-format stores have spread, there has also been a format diversification to meet market segments, with a trend from supermarket format to the addition of larger formats such as hypermarkets and discount stores.

In addition, within a given country the diffusion started first among the upper income consumer segments, and then moved to medium and small cities and finally to rural towns, and to the middle class and then the working urban poor. Thus, in most of the first-wave regions, supermarkets have by now penetrated into the lower-middle and lower income consumer markets and into small towns – and the second wave countries are fast approaching the same situation. This is in sharp contrast with the conventional and now outmoded vision of the supermarkets as being a luxury niche in output markets. The ability of supermarket chains to spread to smaller towns and to the food markets of the urban poor was driven by competition in the higher-income market segments, and facilitated by the cost (and thus price) reductions made possible by rapid transformation of supermarket procurement systems, discussed further below.

Second, there is substantial variation in the speed of retail market change over *product categories*. The take-over of food retailing in these regions has occurred much more rapidly in processed, dry, and packaged foods such as noodles, milk products, and grains, for which supermarkets have an advantage over mom and pop stores due to economies of scale. The supermarkets' progress in gaining control of fresh food markets has been slower, and there is greater variation across countries because of local habits and responses by wetmarkets and local shops. Usually the first fresh food categories for the supermarkets to gain a majority share include "commodities" such as potatoes, and sectors experiencing consolidation in first-stage processing and production: often chicken, beef and pork, and fish.

A rough rule of thumb emerging from empirical studies is that the share of the supermarkets in fresh produce retail is lower than its share in overall food retail, and that this gap closes as the latter rises. For instance, the share of supermarkets in fresh produce retail in Guatemala is about 10% where their share in overall food retail is about 35%, hence the market penetration rate for produce is one-third of overall food market penetration; by contrast, the shares are 50% versus 75% (produce market penetration and overall food market penetration rate by supermarkets), or two-thirds, in Brazil. The latter is the same as in France. Hence, at earlier stages of supermarket development, the freshness, convenience (near consumer residences), and lower cost of small produce shops and wetmarkets easily dominate the retail produce sector. As the supermarket sector expands and gains market share, the competition between supermarkets and wetmarkets becomes increasingly stiff, and is based on shopping experience, price, quality, freshness, and variety. In the big cities of Mexico or China the differences in prices between supermarkets and wetmarkets for commodity produce items is narrowing, often equal for key items. In a recent study by AC Nielsen of 15,000 consumers in the Asia–Pacific region, they found that supermarkets are eroding the share of the wetmarket in retail by attempting to replicate the experience of the traditional wetmarket while reducing prices to compete directly [M&M Planet Retail (2004)]. Supermarkets in the emerging market regions have been making significant inroads into these categories only in the past five years or so, and usually only after making the kinds of cost-cutting and quality-increasing procurement system changes discussed below.

Despite the slower growth in the supermarkets share of the domestic fresh produce market, it is very revealing to calculate the absolute market that supermarkets now represent, even in produce, and thus how much more in other products where supermarkets have penetrated faster and deeper. For example, Reardon and Berdegue (2002) calculate that supermarkets in Latin America buy 2.5 times more fruits and vegetables from local producers than all the exports of produce from Latin America to the rest of the world! That ratio is already 2:1 in China [Hu et al. (2004)]. *The dominance of domestic procurement should be contrasted with the nearly-exclusive focus on produce exports in government and donor programs to spur growth in agricultural diversification and access to dynamic markets.*

Third, the supermarket sector in these regions is increasingly and in most cases overwhelmingly multi-nationalized (foreign-owned) and consolidated. The multinationalization of the sector is illustrated in Latin America where global multinationals constitute roughly 70–80% of the top five chains in most countries. This element of "FDI-driven" differentiates supermarket diffusion in these regions from that in the U.S. and Europe. The tidal wave of FDI in retail was mainly due to the global retail multinationals, Ahold, Carrefour, and Wal-Mart, smaller global chains such as Casino, Metro, Makro, and regional multinationals such as Dairy Farm International (Hong Kong) and Shoprite (South Africa). In some larger countries domestic chains, sometimes in joint ventures with global multinationals, have taken the fore. For example, the top chain in Brazil is *Companhia Brasileira de Distribuição* (CBD) (in partial joint venture Casino, of France, since 1999, and half-half joint venture by 2005), and the top chain in China is the giant national chain Lianhua (based in Shanghai), with some 3500 stores by 2004, in partial joint venture with Carrefour. The extent of multi-nationalization is correlated in general with the wave stage (with the least multi-nationalization of the supermarket sector in the third wave countries), but with a tendency toward convergence.

The rapid consolidation of the sector in those regions mirrors what is occurring in the U.S. and Europe. For example, in Latin America the top five chains per country have 65% of the supermarket sector (versus 50% in the US [Kinsey (2004)] and 72% in France). The consolidation takes place mainly via foreign acquisition of local chains and secondarily by larger domestic chains absorbing smaller chains and independents. This is done via large amounts of FDI: for example, in the first eight months of 2002, five global retailers (British Tesco, French Carrefour and Casino, Dutch Ahold and Makro, and Belgian Food Lion) spent 6 billion bhat, or $120 million in Thailand [Jitpleechep (2002)]. In 2002, Wal-Mart spent $660 million in Mexico to build new stores. As above, supermarket-sector consolidation is correlated with the wave stage, again with a tendency toward convergence.

These trends of multi-nationalization and consolidation fit the supply function of our supermarket diffusion model. Global and retail multinationals have access to investment funds from their own liquidity and to international credit that is much cheaper than is the credit accessible by their domestic rivals. The multinationals also have access to best practices in retail and logistics management, some of which they developed as proprietary innovations. Global retailers adopt retailing and procurement technology generated by their own firms or, increasingly, via joint ventures with global logistics multinationals – such as Carrefour (France) does with Penske Logistics (U.S.) in Brazil. Where domestic firms have competed, they have had to make similar investments; these firms either had to enter joint ventures with global multinationals, or had to get low cost loans from their governments (e.g., the Shanghai-based national chain), or national bank loans.

4.2.3. Evolution of supermarket procurement systems: A crucial vector of change in agrifood systems

Technology change in the procurement systems of supermarkets in developing regions is a key determinant of change in the markets facing farmers. Technology – defined broadly as physical production practices as well as management techniques – diffusion in the procurement systems of supermarket chains in developing countries can also be conceptualized as a system of demand and supply for new technology. Here we focus on technology for retail product–procurement systems as these choices most affect suppliers.

Demand for technology change in food retailer procurement practices is, in general, driven by the overall competitive strategy of the supermarket chain. However, specific choices are usually taken by procurement officers, e.g., in the produce procurement division. Hence it is crucial to understand the objective function of these officers in supermarkets in developing countries. We present a tentative hypothesis based on numerous interviews with these individuals.

The decisions related to purchasing products for retail shelves rest with the procurement officers in supermarket chains. Whether in the United States, Europe, Nicaragua, Chile, or China, they are under several common "pressures" from supermarket managers, operating under intense competition and low average profit margins. They are caught between the low-cost informal traditional retailers selling fresh local products on one side, and efficient global chain competitors like Wal-Mart on the other side. The procurement officers strive to meet this pressure by reducing purchase and transaction costs and raising product quality. Reflecting the varied demand of consumers, procurement officers seek to maintain diversity, year-round availability, and products with assured quality and safety levels.

Based on those objectives, we outline a rough model for demand (by procurement officers) and supply (by the supermarket chain to those procurement divisions) of change in procurement systems (technology, organization, institutions). The demand function incentives and capacity variables are discussed first. Incentives include:

(1) the ability of the traditional wholesale system to meet procurement officer objectives without the chain having to resort to costly investments in an alternative system. Usually procurement officers find this ability low, as Boselie (2002) shows in the case of Ahold for fresh produce in Thailand. Compared with the North American or the European market, produce marketing in these regions is characterized by poor institutional and public physical infrastructure support. Pri-

vate infrastructure, such as packing houses, cold chains, and shipping equipment among suppliers and distributors is usually inadequate. Risks and uncertainties, both in output and in suppliers' responsiveness to incentives, are high. The risks may arise due to various output and input market failures, such as inadequacies in credit, third-party certification, and market information;

(2) a second incentive is the need to reduce costs of procurement by saving on inputs, in this case purchased product costs and transaction costs with suppliers; and

(3) the incentive to increase procurement of products that can be sold at higher margins, hence diversify the product line into "products" rather than mere commodities (bulk items).

Capacity to demand includes:

(1) the consumer segment served by the chain. This is crucial because higher-value products cannot be marketed to poorer consumers where only cost considerations are paramount; and

(2) the resources of the procurement office. These include the number of staff to manage procurement and thus ability to make organizational and institutional changes in procurement systems such as operating a large distribution center. A variable which reflects both incentive and capacity is the size of the chain and thus product throughput in the procurement system. Usually retailers have a "step level" or threshold throughput where they go from per-store to centralized procurement as economies of scale permit and require.

The supply of procurement technology by the chain as an overarching enterprise, to the specific product category procurement office or offices, such as the fresh foods categories, is an investment and is a function of several variables. The incentive variables include:

(1) the importance of the product category to the chain's profits and marketing strategy. For example, we observed a small chain in an intermediate city in China that invested in building a distribution center (DC) for processed/packaged foods but continues to buy fresh foods from the spot market (traditional wholesalers), while a national chain invested in a large DC for packaged/processed foods and has recently built a large DC for fresh foods as throughput has attained a critical mass, and these products have attained a threshold importance in profits and chain marketing strategy;

(2) the need for assurance of various product attributes in order to meet customers' demands, such as expansion of product choice, attribute consistency over transactions, year-around availability, quality, and safety; and

(3) the costs of the technology, such as costs of transport, construction, logistics services, etc.

The capacity variables include:

(1) the size of the chain and/or access to financial capital to make the investments; and

(2) the capacity of the chain to *manage* complex and centralized procurement systems.

The incentive and capacity determinants of demand for and supply of changes in procurement system technology vary markedly over the three regions and countries, and within countries, over chains and zones. There is substantial variation within the supermarket sector in a given country, with the 4–5 leading chains (with the majority of the market) tending to be the "early adopters" of the procurement system technology change. *These can be characterized as the "change agents" in the retail sector and by extension the output markets in the country*. The second- and third-tier chains usually tend to be late adopters, but adopt they must, in order to compete on costs with the leaders. That they often lag substantially in the adoption of the technologies leads to their lower competitiveness and hence the consolidation processes observed.

It is thus crucial to understand the broad patterns observed in the procurement technologies that result. These patterns can be described as the "four pillars" of change in the organization and institutions of the procurement systems [Reardon et al. (2003a); Berdegué et al. (2005)].

The first pillar is a trend toward centralization of procurement (per chain). As the number of stores in a given supermarket chain grows, there is a tendency to shift from a per-store procurement system, to a distribution center serving several stores in a given zone, district, country, or a given region (which may cover several countries). This is accompanied by fewer procurement officers and increased use of centralized warehouses. Additionally, increased levels of centralization may also occur in the procurement decision making process, and in the physical produce distribution processes. Centralization increases efficiency of procurement by reducing coordination and other transaction costs, although it may increase transport costs by extra movement of the actual products. China Resources Enterprise (2002), for example, notes that it is saving 40% in distribution costs by combining modern logistics with centralized distribution in its two large new distribution centers in southern China. There are similar figures from (the few available) studies elsewhere, for example, in Costa Rica and Brazil.

The top several global retailers have made or are making shifts toward more centralized procurement system in all the regions in which they operate. Wal-Mart uses a centralized procurement system in most of its operating areas. Having centralized its procurement in France, Carrefour has been moving quickly to centralize its procurement system in other countries. In 2003 and 2004 Tesco and Ahold have established large distribution centers in Poland, Hungary, and the Czech Republic. In 2001 Carrefour established a distribution center in São Paulo to serve three Brazilian states (with 50 million consumers) with 50 hypermarkets (equivalent to about 500 supermarkets) in the Southeast Region. Similarly, Carrefour is building a national distribution system with several distribution center nodes in China, while Ahold centralized its procurement systems in Thailand [Boselie (2002)].

The second pillar is the adoption of organizational innovations comprising a shift from reliance on spot markets (in particular, traditional wholesale markets and brokers) toward growing use of specialized/dedicated wholesalers. They are specialized in a product category and dedicated to the supermarket sector as their main clients. The changes in supplier logistics have moved supermarket chains toward new intermediaries, side-stepping or transforming the traditional wholesale system. The supermarkets are increasingly working with specialized wholesalers, dedicated to and capable of meeting their specific needs. These specialized wholesalers cut transaction and search costs, and enforce private standards and contracts on behalf of the supermarkets. *The emergence and operation of the specialized wholesalers has promoted convergence, in terms of players and product standards, between the export and the domestic food markets*. Moreover, there is emerging evidence that when supermarket chains source imported produce they tend to do so mainly via specialized importers. For example, hypermarkets in China tend to work with specialized importers/wholesalers of fruit, who in turn sell nearly half of their imported products to supermarket chains [Produce Marketing Association (2002)]. Similarly, Hortifruti functions as the buying arm of most stores of the main supermarket chain in Central America, as does Freshmark for Shoprite in Africa.

A related development is the trend toward logistics improvements to accompany procurement consolidation, and a shift in supply organization to implement those improvements. To defray some of the added transport costs that arise with centralization, supermarket chains have adopted (and required that suppliers adopt) best-practice logistical technology. This requires that supermarket suppliers adopt practices and make physical investments which allow almost frictionless logistical interface with the chain's warehouses. The "Code of Good Commercial Practices" signed by supermarket chains and suppliers in Argentina illustrates the use of best-practice logistics by retail suppliers [Brom (2002)]. Similar trends are noted in Asia. For example, Ahold instituted a supply improvement program for vegetable suppliers in Thailand, specifying post-harvest and production practices to assure consistent supply and improve the efficiency of their operation [Boselie (2002)].

Retail chains increasingly outsource – sometimes to a company in the same holding company as the supermarket chain – logistics and wholesale distribution function, entering joint ventures with other firms. An example is the Carrefour distribution center in Brazil, which is the product of a joint venture of Carrefour with Cotia Trading (a major Brazilian wholesaler distributor) and Penske Logistics (a U.S. global multinational firm). Similarly, Wu-mart of China announced in March, 2002 that it will build a large distribution center to be operated jointly with Tibbett and Britten Logistics (a British global multinational firm). Ahold's distribution center for fruits and vegetables in Thailand is operated in partnership with TNT Logistics of the Netherlands [Boselie (2002)].

The third pillar is the adoption by leading supermarket chains of the institutional innovation of contracts with their suppliers – in particular via their dedicated, specialized wholesalers managing a preferred supplier system for them. Such contracts are part of what the industrial organization literature terms "vertical restrictions" that fall short of full vertical integration (generally and usually avoided by both supermarket chains and food processors), but that approximate in certain ways the outcomes from vertical merger [Carlton and Perloff (2000)].

The contract is established when the retailer (via their wholesaler or directly) "lists" a supplier. That listing is an informal (usually) but effective contract[14] – in which delisting carries some cost, tangible or intangible. We have observed such contracts in all the regions under study, but this development occurred first and fastest between retailers and processors, and only very recently between retailers and fresh-product producers. Contracts serve as incentives to the suppliers to stay with the buyer and over time make investments in assets (such as learning and equipment) specific to the retailer specifications regarding the products. The retailers are assured of on-time delivery and the delivery of products with desired quality attributes.

These contracts with retailers sometimes include direct or indirect assistance for farmers to make investments in human capital, management, input quality, and basic equipment. Evidence is emerging that for many small farms these assistance programs are the only source of such inputs and assistance – in particular where public systems have been dismantled or coverage is inadequate. Moreover, there is evidence of interlinked product and factor markets emerging. For example, an interlinkage between the output and credit market is evidenced in farmers' contracts with a supermarket chain serving as a collateral substitute. An illustration is the case of Metro supermarket chain (a German chain) in Croatia intervening with the bank, noting that the suppliers would have contracts with the supermarket, to provide a "collateral substitute" so would-be strawberry suppliers could make needed greenhouse investments [Reardon et al. (2003b)].

This constitutes in practice what Eswaran and Kotwal (1985) analyze theoretically – resolution by retailers or their wholesaler agents of idiosyncratic factor market failures facing small producers via implicit or explicit contract provisions providing credit, information, technical assistance, and so on. Increasingly, supermarket chains provide these services for their preferred suppliers in Central and Eastern Europe [Reardon and Swinnen (2004)], Central America [Berdegué et al. (2005)], Thailand [Boselie (2002)], and China [Hu et al. (2004)], sometimes directly, sometimes indirectly through their specialized/dedicated wholesaler. There is evidence of such inter-linking of output and factor markets in the processing sector also, for example in the CEE [Gow and Swinnen (2001); Dries and Swinnen (2004); Swinnen (2004)]. Some cases of this are remarkable in their extent and nature. Codron et al. (2004) note a case of a Turkish retailer MIGROS which contracts with a whole village nearby its Antalya market to grow 1000 tons of tomatoes during the summer. Hu et al. (2004) describe the case of Xincheng Foods in Shanghai, acting as a specialized wholesaler for the top two chains in China. Xincheng long-term leases (from townships) 1000 hectares of prime vegetable land, hires migrant labor, installs greenhouses and uses tractors and drip irrigation (thus changing production technology), and produces in-house large quantities of high quality vegetables for the supermarket chains and export. It also has contracts with 4500 small farmers to add

[14] "Contracts" is used in the broad sense of Hueth et al. (1999), which includes informal and implicit relationships.

to its own production. This kind of operation can be described as a major "agent of change" in the Chinese agrifood economy. While this type of contracting is quite recent for produce, it has been a practice for a half decade or more among chains sourcing from processed product suppliers analyzed, for example, in Latin America in the 1980s and 1990s [Schejtman (1998); Key and Runsten (1999)].

The fourth pillar of procurement system change is the rise of private quality and safety standards implemented by supermarket chains and large-scale processors. While food retailing in these regions previously operated in the informal market, with little use of certifications and standards, the emerging trend indicates a rapid rise in the implementation of private standards in the supermarket sector and other modern food industry sectors such as medium/large scale food manufactures and food service chains. The rise of private standards for quality and safety of food products, and the increasing importance of the enforcement of otherwise-virtually-not-enforced public standards, is a crucial aspect of the imposition of product requirements in the procurement systems. In general, these standards function as instruments of coordination of supply chains by standardizing product requirements over suppliers, who may cover many regions or countries. Standards specify and harmonize the product and delivery attributes, thereby enhancing efficiency and lowering transaction costs.

Private standards of a given chain may also be designed to ensure (at a minimum) that the public standards are met in all the markets in which the retail chain operates. In that sense, they represent the confluence of the trend toward product differentiation and the deepening/extension of the market. This may be the next phase in output market transformation, the marriage of the traditional commodity market and the emerging product market.

Often private standards may be designed as substitutes for missing or inadequate public standards [Reardon and Farina (2002)]. In this respect, private standards can function as competitive arms against the informal sector (and other competitor products) by claiming superior product quality attributes. The evolution of private standards in the supermarket sector in these regions is also driven by multinational retailers' striving toward convergence between the private standards applied by the chain in developed countries and in developing countries. Not surprisingly, many small farmers and processors are finding it impossible to meet the requirements of supermarkets, and are being dropped from their procurement lists.

This institutional change is also found in the processing sector. While the era of commodity markets had a focus on public standards for extensive, homogeneous markets, the era of product markets, coupled with the consolidation and multi-nationalization of the retail and food manufacturing sectors, leads to the rise of private standards as key institutions in the new markets. Private standards, which are often process standards such as HACCP or the Nestle Quality Assurance or Carrefour Quality standards and certification, rather than outcome standards, allow retailers and manufacturers to use private standards as competitive weapons, strategic tools to differentiate products finely, while coordinating supply chains.

These private standards have arisen partly because of "missing institutions" or inadequate and/or inappropriate institutions, such as the non-differentiated wheat product standards of the Brazilian government that were sidestepped by wheat product firms in Brazil with their own new standards in the 1990s. These are especially important in non-staples, also the most dynamic markets and the ones in which process standards (HACCP, because of perishability) and product differentiation are most important. Examples include private standards of McDonalds for their suppliers, and as fast food chains rise, these private standards become the most important output market entry barriers for producers of fresh produce. In the case of Ahold in Thailand and Nestle in Brazil, for example, public standards take a back seat to private standards, and in some cases (milk products in Brazil) public standards for safety and quality are copying the private standards.

5. Who wins and who loses in the retail-driven transformation of agrifood product markets during globalization?

The rise of supermarkets has been controversial everywhere, but their rise to dominance in rich countries over decades instead of years permitted gradual adjustments by farmers, processors, wholesalers and traditional small-scale retailers to the new ways of doing business. These adjustments meant learning cultivation techniques for new crops with product and process quality standards that were simply not relevant to commodity production. They meant finding new sources of employment as small retail shops and neighborhood markets closed. Consumers have tended to shift rapidly from traditional retailers to supermarkets, pulled by convenience, sometimes better prices, often greater diversity and quality of products; it has also meant that consumers needed to make adjustments, such as going longer distances to the market and making less frequent, but larger, purchases.

These adjustments tended to be difficult in financial and personal terms, but they were manageable when spread across generations. The adjustment pressures from the rapid emergence of supermarkets (and large-scale processors) in developing countries are far more severe because the change is so fast. The parallels to the overall structural transformation of an economy, and the pressures it puts on agriculture, are obvious. During the structural transformation entire societies undergo the wrenching changes associated with agricultural modernization, migration of labor from rural to urban areas, and the emergence of urban industrial centers. The structural transformation has taken as long as three centuries in England and the United States (and is still continuing), and as little as a century in Japan.

5.1. Drivers of the distribution of net benefits from output market transformation: The broad view

During any historical epoch, there are a set of identifiable "drivers" that push economies through the structural transformation, from poor to rich. In the current era – post-World

War II to keep the timing consistent with the focus of this retrospective volume – these drivers are globalization, urbanization, and technology change. The question is, how have these three forces influenced the rapid emergence of supermarkets? There is now widespread agreement that the supermarket revolution itself has been driven by precisely these three drivers of overall economic change, but a dilemma remains in using this as an answer to the speed of change in the food retail sector. After all, globalization, urbanization and technology were equally cited for the rapid economic advances in the 19th century. What is different now?

The answer is given by changes in the relative scarcity of important economic resources, changes that are themselves driven by the new industrial organization of the global food supply chain. Multinational corporations, including global retailers and food manufacturers, are increasingly dominant in this global food supply chain – and among the key players in the supermarket revolution in developing countries.

Not surprisingly, profits in the global food supply chain tend to accrue to the relatively scarce resource in the system under analysis, and to whoever controls those resources, because scarcity has value. In the global food retail system, there are three basic possibilities for what resource is scarcest in the food system: access to farm output; access to marketing technology; or access to consumers.

First, there have been long-standing concerns that population growth will outstrip growth in food supplies. If so, the scarce resource in this system is the food commodity itself – the rice, potato, Belgian endive, bell peppers, fresh fish, or chuck steak. Because supermarket quality and safety standards are so high and rigid, the ability to supply the raw commodities that meet these standards might command a price premium and additional profits for the farmers. Beneath commodity supply, of course, is the land and labor (and knowledge and technology) required to grow the commodities. Thus, ultimately, if commodities themselves are the scarce resource, capable of earning excess profits, these profits will accrue to land, labor, or both (or to the management function that harnesses the knowledge, technology and finance, although for small farmers this tends to be in the same hands as the land and labor).

But historical evidence does not support the view that agricultural commodities will be the ultimate source of scarcity in the food system. Modern agricultural technology is land-saving, there is abundant rural labor (again, on a global level), rural finance is readily available when there is a profit to be made in lending it, and water is becoming scarce only because it is provided free in most cases. What *might* be scarce at the farm level is the *management ability* to meet high quality standards and to deliver reliably a safe product that meets environmental requirements and is fully traceable to its point of production. There are likely to be significant economies of scale to this management ability, even if there are few scale economies in the physical production of most agricultural commodities.

A second possible scarce resource is access to, or control of (through intellectual property rights), the technology/organization/institutions that lower transactions costs throughout the entire food supply chain and improve coordination. However, the technology for managing supply chains – in the food system and elsewhere –

is changing rapidly, even in the United States [Kinsey (2004)]. These technologies/organizations/institutions discussed in this chapter, permit supermarket managers exquisite control over procurement, inventory levels, and knowledge of consumer check-out profiles.

For example, information technology is a key component of the vector of technologies used by the chains. Such technology provides a powerful competitive advantage in cost control, quality maintenance, and product tracking in case of defects or safety problems. When this technology is applied globally to the food supply chain of a transnational supermarket, transactions costs will be "pushed out of the system" all the way from the food aisle, through global marketing functions, to individual farmers. In general, these technologies drive down transactions costs throughout the supply chain. But further, by reducing the need to hold large inventories, these marketing and logistics technologies reduce capital costs and risks. Because inventory is basically a form of "dead capital. improved logistics and inventory management (such as in ECR or Efficient Consumer Response) generate real capital savings as well as lower transactions costs. And both contribute to higher productivity and faster economic growth.

The important question is whether access to this technology is sufficiently restricted that it is "scarce", i.e. can excess profits be earned by controlling it? The evidence suggests that it is easily duplicated as computer power becomes cheaper and local managers learn to imitate the market leaders. Intellectual property rights (IPR) seem not to be a serious impediment to this imitation, despite supermarket chains' efforts at proprietary control. It is the *knowledge* that such techniques are feasible and available that is important, not the specific code written for a particular supermarket's computers. The parallel to the "technological treadmill" [Cochrane (1993)] so familiar to American farmers is striking: first adopters of new technology have a temporary cost advantage and thus above-average profits, but competition leads all market players to adopt it quickly, reducing the advantage of the innovator, and stimulating the innovator to adopt yet a newer technology in order to regain cost and profit advantages. This seems to be the story for marketing technology.

The third possibility for what is scarce is access to consumers themselves, and especially to knowledge of how consumers behave – what they want, and therefore, how best to serve them. As concentration in food retailing rises, there seems to be an opportunity for the leading firms – Carrefour, Wal-Mart, Metro, Tesco, Ahold, etc. – to *control* this access and thus to earn higher marketing margins and profits. This has been a longstanding worry in the United States, at least since the 1940s.

The evidence so far, both in rich and poor countries alike, is that access to consumers has been highly competitive. Market power is used to drive down costs, and these lower costs are then passed along to consumers as lower prices. Why? Because supermarkets need to increase market share to achieve the economies of scale that permit their costs to be even lower. So far, this whole system has been highly *contestable*. Economists know that contestable markets pass nearly all the benefits of the marketplace (the sum of producer and consumer surplus) through to consumers.[15]

Basic competitive forces will lead most "monopoly" profits or rents to end up in the hands of the owners of the scarcest resource. The evidence so far is that access to affluent consumers and to powerful information technology and the ability to implement organizational and institutional changes in supply chains are scarcer than the ability to produce high-quality commodities, especially when individual producers are forced to compete on a global playing field. But this does *not* mean that multinational and regional supermarket chains are earning monopoly profits because they have access to, even control of, those scarce resources. The cost of information technology is dropping with Moore's Law[16] and access to affluent consumers has turned out to be highly contestable, and thus generating competitive results, despite the industry structure. Surprisingly, the picture so far is one of intense competition and low profit rates. Consumers are the main winners in the supermarket revolution. For example, a recent study in Chile [LatinPanel study for 2004, reported in Camara Nacional de Comercio (2005)] showed that supermarkets, by charging lower prices for food compared to traditional retailers, reduced the cost of the food consumption basket of the lower and middle income consumers in Chile.

However, the continuous competitive pressure to lower costs and raise quality and differentiate products for the consumers induces supermarket chains to require continuous investments by their suppliers in order to increase quality and reduce costs at the production level. Those investments are challenging for a subset of farmers and processing firms, creating a mixed picture of the benefits of the retail transformation for the farm sector, discussed next.

5.2. *Emerging evidence of the distributional effects of the demand-side (food industry transformation) on the supply-side (producers)*

A mix of opportunities and challenges are facing farmers in the deeply transformed agrifood markets of the 1990s and 2000s. In particular, there is evidence that small farmers are particularly challenged to meet the volume, cost, quality, and consistency requirements of the increasingly dominant supermarket chains and large-scale agroprocessors. That they are challenged does not necessarily mean that there is widespread exclusion and thus upstream consolidation in the food system driven by output market transformation. The picture is mixed, and the emerging impacts are conditioned by several factors

[15] Even at this late stage in the supermarket revolution in the United States, adoption of state-of-the-art marketing technology generates annual benefits equal to half the size of the entire farm economy! This is a staggering result, driven by the calculation that Wal-Mart alone, the leader in the marketing technology revolution, lowers the annual inflation rate by roughly 1% per year [Greenhouse (2004)].

[16] Moore observed an exponential growth in the number of transistors per integrated circuit and predicted that this trend would continue (http://www.intel.com); see Moore (1965). Moore's Law is often used generically to indicate the rapid pace of change in information technology.

noted below. The results are merely preliminary because the research is just beginning on the impacts of the recent transformation of food markets upstream among farmers.

First, under certain circumstances, small farmers are included in procurement systems of large-scale agroprocessors and supermarket chains. The grower-level impacts of large-scale processing firms are far more researched than those of the direct effects on farmers of procurement practices by supermarket chains. With respect to effects of processing firm procurement practices on farmers, the literature has been most rich in the "first wave" areas, as one expects. For example, Schejtman (1998) and Key and Runsten (1999) examine the participation of small farmers in contract farming schemes of agroprocessors in Latin America in the 1980s and 1990s. They find a mixed picture with specific conditions under which these firms use contract farming (rather than full vertical integration) and in the latter, contract with small farmers. Recent work in Central and Eastern Europe by Swinnen and Dries of University of Leuven and colleagues [for example, Dries and Swinnen (2004); Swinnen (2004); Dries and Reardon (2005)] also show a mixed picture, with substantial involvement of small milk producers in Poland, but very low participation of small producers in Russia, Slovakia, and Czech Republic. In these latter countries, the exclusion of small farmers is widespread, similar to the cases of much of South America, such as in Brazil [Farina (2002)], Argentina [Gutman (2002)], and Chile [Dirven (2001)]. Swinnen and Dries and colleagues find that large-scale processors tend to rely on small farmers in particular where there is the incentive (they must, due to lack of sufficient supply from larger firms) and the capacity (sufficiently low transaction costs are in place due to effective associations). Where the capacity is insufficient, but the incentive is high, large firms often try to resolve idiosyncratic market failures facing small growers by providing technical assistance and input credit, similar to actions undertaken by supermarkets discussed above.

The evidence concerning the grower-level impacts of supermarkets on producers is far more recent and partial, but points toward a similarly mixed picture.

First, a relatively unambiguous picture is emerging of relatively rapid exclusion of small processing and food manufacturing firms in supermarket procurement systems in developing countries. While there are very few studies on this, the forces leading to exclusion seem to aim in just one direction. For example, Hu et al. (2004) note that while supermarket chains in Beijing tend to increase diversity of processed products, there is a strong tendency toward selection of a small number of medium-to-large firms capable of delivering consistent quality product at large volumes. This assures "one-stop shopping" for the chains, that is, a given firm is able to supply a diversity of product lines in order to reduce transaction costs for the chain. The chains reap economies of scale from large volumes of processed products moving through their distribution centers, and seek to work with larger firms that can ship to their centers or have their own distribution centers that they can use to distribute to stores. This is an international trend, although seen vividly in the rapidly changing Chinese supermarket sector. Hu et al. (2004) noted an example of a Beijing chain that moved from 1000 to 300 processed food suppliers in one year once they had their distribution center in place and could consolidate suppliers. Dries and Reardon (2005) note a similar tendency in Russia for dairy products, and

Balsevich (2005) for meat products in Costa Rica and Nicaragua. Moreover, the effects on the majority of growers are felt indirectly, via the choice by supermarkets of larger processors, as processed food constitutes half to two-thirds of the food sales of a typical supermarket or hypermarket in developing regions.

Second, however, changes in supermarket procurement systems also affect growers directly in the case of fresh fruit and vegetables – and thus the processing sector does not mediate the link between farmers and supermarkets or the wholesalers that serve them, which is why most studies of supermarket effects directly on producers have been (only in the past year) in this category. There have been several recent studies of supermarkets and fresh vegetable and fruit growers, in China [Hu et al. (2004)], Indonesia [Reardon et al. (2005)], Kenya [Neven (2004) and Neven and Reardon (2004)], Nicaragua [Balsevich (2005)], Guatemala [Flores (2004) and Hernández et al. (2004)], and Weatherspoon and Reardon (2003) for Africa, Katjiuongua (2004) for Zambia, and Reardon et al. (2005) for Mexico.

The results from these new studies tend to show the following:

(1) there are substantial numbers of small farmers involved in supermarket supply chains in these countries, in Latin America, Asia and Africa;
(2) as in the processing sector, however, supermarkets tend to source as much as possible from medium and larger growers where the latter are willing to supply supermarkets (where the incentive relative to the export market encourages them to supply locally, or they have "seconds" to supply to domestic markets);
(3) moreover, just as it is not the poorest and smallest farmers that tend to produce fruits and vegetables, among growers of the latter, it tends to be the upper tier in terms of assets (not necessarily in terms of size, but in terms of physical, human and organizational capital) who supply supermarkets;
(4) the net benefit of selling to supermarkets relative to selling to traditional markets tends to be much higher in products ("non-commodities") compared to commodities; the grower effects of the supermarket revolution are thus a microcosm of the general trends in output markets we have discussed in this chapter;
(5) farmers' associations or cooperatives tend to be necessary (to reduce transaction costs) but far from sufficient; growers can use these to facilitate entry into the market, but a range of other threshold investments in management, institutions to ensure collective compliance with supermarket standards, and physical capital are needed to keep farmers in the dynamic markets [see Berdegué (2001) for the case of Chile and Jano et al. (2004) for cases in Central America].

An even less researched but equally intriguing subject is *indirect* effect of food industry transformation on producers via the traditional food industry actors who now compete with the modern food industry. Traditional food industry firms alter their procurement practices from farmers, at least to some extent, in order to cut costs and raise quality to match the actions of the modern food industry and stem the flow of consumers toward the latter. As far as we know, this has not been researched systematically in any country, but the point is that there appear to be emerging "ripple effects" of technology, organization, and institutional diffusion from the epicenter of the supermarket revolution out into other segments of the market. These ripples imply that producers may well be affected both in the most dynamic demand segments (the urban market progressively dominated by supermarkets) as well as at the traditional retail and wholesale segments that compete with supermarkets.

6. What we do not know: A research agenda

As in the 1950s/60s when commodity markets were expanding, there was a major research effort just to describe the new systems, and that is happening now. There are growing efforts in the first decade of the 2000s to describe the changes described in this chapter and to document them. Most of the new structures, and how they are changing food systems even in poor countries, are simply not on the "radar screen" of research and policy debate, however, because they are developing so quickly and have emerged so recently. This topic simply has not yet entered public debate and research agendas in most places, even where it is moving the fastest. There is a pressing need to extend and deepen the analysis of agrifood output markets, with innovative approaches tailored to the transformed agrifood systems, in the next decade. The research foci and methods will of course be differentiated as a function of the "waves" of transformation as discussed in this chapter, but the essence of a needed research agenda is as follows.

A major difference between the work of description and analysis of market trends in the 1970s/80s and now is that one could do broad surveys of suppliers and retailers and even wholesalers in the earlier periods. Now, with the tendency toward consolidated processing and retail sectors, the strong tendency is toward relatively few firms in particular in the "downstream segments" of the agrifood system, and they are big and not amenable to broad anonymous surveys. That means that there is now a needed fusion of agribusiness research techniques (case studies, management and business strategy studies) and research on the structure, conduct, and performance of markets. This became apparent about a decade ago in the United States, but is only now hitting researchers working in the developing regions where the changes from the supermarket revolution and consolidation in the processing sector are happening fastest.

Moreover, there is a resurgence of industrial organization research to analyze the organizational changes in market systems and subsectors in developing countries. This is analogous to the shift to the "new trade literature" in the 1990s where it was recognized that the consolidated trading systems functioned differently than assumed in the prior literature [Evans (1989)]. Innovative analyses [such as Farina (2002); or Gow and Swinnen (2001)] are combining industrial organization and agribusiness management, meso and micro research in an innovative and appropriate way.

The key is the interface of the changes in market organization and institutions on the one hand and technology adoption by the various actors in the food system on the other. There is some emerging work here such as described in Section 5 regarding farmers' participation in supply channels to supermarkets. There is also some emerging theoretical work on this interface, such as by [Heiman, McWilliams and Zilberman (2000)].

Techniques (such as the work on the relative power of price formation between supermarkets and processors) that explore price formation in an oligopsonistic/politic context will rise in importance at the expense of work on price formation/market integration in atomistic competition settings. There is also emerging work on determinants of diffusion of new institutions such as on the development of contracts in processing and retail sectors [Zylbersztajn and Farina (1999); Zylbersztajn (2002); or Gow and Swinnen (2001)] and the continued development of interlinked and interlocked market institutions in the traditional agrifood sector [such as Giné and Klonner (2002)] and the modernized agrifood sector [such as Swinnen (2004) and Reardon and Swinnen (2004)]. There is also emerging work on transformation of agrifood system organization (such as horizontal joint ventures in produce markets in developing countries [Neven and Reardon (2002)]), and emerging analysis of the very recent rise of "new generation cooperatives" (such as one saw a decade or two earlier in developed countries) as producer organizations in developing regions [see Berdegué (2001), for example].

Another key area for future research is the interaction of the evolution of procurement systems of the firms in the consolidating and multi-nationalizing food industry (retail, processing, and food service) and trade. Reardon et al. (2005), for example, suggest, illustrating with emerging case study evidence, that the rise of supermarkets in the Pacific Rim region is already starting to affect the level, nature, and composition of trade. Supermarket procurement system change at a regional and global level appears to be beginning to modify international markets in the ways that we have described as their modification of domestic markets. This needs to be systematically explored; again, this will blur the lines between research on retail and research on international agrifood trade, two subdisciplines of agricultural economics traditionally held at arm's length.

The distinction between the local and the global, the domestic and the export market is now blurred. Development agencies and NGOs have put an emphasis either on targeting local markets, which they presumed to be nearly hermetically sealed from the global market, or they put an emphasis on the export market as a way to break local demand constraints. The presence and even the dominance of the supermarkets in the *local* food markets mean that there is increasingly a mix of the challenges and the opportunities of both, facing nearly all suppliers. Helping small farmers gear up to that challenge is crucial. That has to happen very fast to match the pace of change. For 61,000 small dairy producers in Brazil in the last 4 years of the 1990s, the response was not quick enough, and they are all out of business.

Undertaking "market oriented" development assistance programs now means dealing with a handful of big companies. Reorienting development programs and researchers to this fundamentally different reality is urgent, but not easy because each locale has unique dimensions. The export market is a logical target for development assistance, but in many cases the supermarket-market domestically is already more important and is growing much more rapidly than the export market. Supermarkets in Latin America and China already buy, from their local producers, twice as much produce as either of those areas export to the rest of the world. So development programs need to add a major component to their marketing programs to focus on supermarkets.

To stay up with change and even get ahead of it, market researchers have to innovate. As a start, this means marrying agribusiness/food industry research focused on chain structure and management strategies, with market statistical analysis and development economics modeling. It is uncommon for the researchers that have been spending a decade understanding the management of global firms and their growth strategies – the agribusiness and retail researchers – to rub shoulders with development economists. It is not too late. These groups need to collaborate with each other, crossing cultural and methodological divides to create new approaches. The same mixing needs to occur between market research and technology adoption research. It is in this latter interface that we see the great challenges from exclusion of small farmers and firms.

Finally, it is clear from the above discussion that public policy is only half the story in understanding the transformation of markets in developing countries. Policy was indeed crucial in the initial stages of liberalization, but public policy is now forced to share power with the emerging force of private institutional change. While public attention is focused on public standards and market policies such as those represented by the WTO, there has been a rapid rise in private standards that have reshaped markets in developing countries. We need to understand that change and build it into the design of development assistance programs.

The private standards imposed by large firms in developing countries are dovetailing with global private standards imposed by powerful players in the food retail and food manufacturing sector. The private standards, developed in the context of the strategic goals of the large firms, will shape food markets in the years to come. For example, in 2003, CIES, the association of large supermarkets and food manufacturers based in Paris and comprised of the top 250 supermarket chains and 250 food manufacturers, launched their new worldwide harmonization initiative on food safety standards – private standards to be imposed on their suppliers over the following two years. The combined annual sales of these 400 companies are $2.8 trillion, a sum that dwarfs development assistance or even international trade. Market-led development is now supermarket-led development.

References

Abbott, J.C. (1962). "The role of marketing in the development of backward agricultural economies". Journal of Farm Economics 44, 349–362.

Abbott, J.C. (1967). "The development of marketing institutions". In: Southworth, H.M., Johnston, B.F. (Eds.), Agricultural Development and Economic Growth. Cornell University Press, Ithaca, NY, pp. 364–399.

Abdulai, A. (2000). "Spatial price transmission and asymmetry in the Ghanaian maize market". Journal of Development Economics 63 (2), 327–349.

Ahmed, R., Donovan, C. (1992). Issues of Infrastructural Development: A Synthesis of the Literature. IFPRI, Washington, DC.

Alderman, H. (1993). "Intercommodity price transmittal: Analysis of food markets in Ghana". Oxford Bulletin of Economics and Statistics 55 (1), 43–64.

Arrow, K. (1969). "The organization of economic activity: Issues pertinent to the choice of market versus nonmarket allocation". In: The Analysis and Evaluation of Public Expenditure: The PPB System, vol. 1 [U.S. Joint Economic Committee, 91st Cong., 1st session]. U.S. Government Printing Office, Washington, DC, pp. 59–73.

Badiane, O., Shively, G.E. (1998). "Spatial integration, transport costs and the response of local prices to policy changes in Ghana". Journal of Development Economics 56, 411–431.

Bain, J.S. (1959). Industrial Organization. John Wiley and Sons, New York.

Balsevich, F. (2005). "Essays on producers' participation, access and response to the changing nature of dynamic markets in Nicaragua and Costa Rica", Unpublished doctoral dissertation. Michigan State University, East Lansing, MI.

Bardhan, P.K. (1980). "Interlocking factor markets and agrarian development: A review of issues". Oxford Economic Papers 32 (1), 79–98.

Barham, B., Clark, M., Katz, E., Schurman, R. (1992). "Nontraditional agricultural exports in Latin America". Latin American Research Review 27 (2), 43–82.

Barrett, C.B. (1997). "Liberalization and food price distributions: ARCH-M evidence from Madagascar". Food Policy 22 (2), 155–173.

Barrett, C.B. (2001). "Measuring integration and efficiency in international agricultural markets". Review of Agricultural Economics 23 (1), 19–32.

Bauer, P.T. (1957). "Economic analysis and policy in underdeveloped countries". Commonwealth Studies Center, Duke University.

Berdegué, J.A. (2001). "Cooperating to compete. Peasant associative business firms in Chile", Doctoral dissertation. Communication and Innovation Group, Department of Social Sciences, Wageningen University and Research Centre, Wageningen, The Netherlands.

Berdegué, J.A., Balsevich, F., Flores, L., Reardon, T. (2005). "Central American supermarkets' private standards of quality and safety in procurement of fresh fruits and vegetables". Food Policy 30 (3).

Berck, P., Bigman, D. (1993). Food Security and Food Inventories in Developing Countries. Oxford University Press, New York.

Binswanger, H.P., Ruttan, V.W. (Eds.) (1978). Induced Innovation: Technology, Institutions, and Development. Johns Hopkins University Press, Baltimore, MD.

Boselie, D. (2002). "Business case description: TOPS supply chain project, Thailand", KLICT International Agri Supply Chain Development Program. Agrichain Competence Center, Den Bosch.

Braverman, A., Hammer, J. (1986). "Multimarket analysis of agricultural pricing policies in Senegal". In: Singh, I., Squire, L., Strauss, J. (Eds.), Agricultural Household Models: Extensions, Applications, and Policies. Johns Hopkins University Press, Baltimore, MD, pp. 233–254.

Braverman, A., Stiglitz, J.E. (1982). "Sharecropping and interlinking of agrarian markets". American Economic Review 72 (4), 695–715.

Brom, F. (2002). "Design of a code of best commercial practices to enhance relations among supermarkets and food suppliers in Argentina", Presented at the International Conference "Supermarkets and Agricultural Development in China: Opportunities and Challenges", Shanghai, May 24–25.

Burt, T. (2004). "Global retailers expand markets". Financial Times (June 22), 15.

Business Week (2000). "Foreign direct investment in developing countries (1990–1998)". http://www.globalpolicy.org/soecon/talbes/fdinvst.htm, November 2000.

Camara Nacional de Comercio (2005). "Cae consumo de sectores medios y bajos", Noticias CNC, Article posted May 10 on website publication of the Camara Nacional de Comercio, Servicios y Turismo de Chile, Santiago, Chile. http://www.cnc.cl, accessed May 14, 2005.

Carlton, D.W., Perloff, J.M. (Eds.) (2000). Modern Industrial Organization. third ed. Addison-Wesley, Reading, MA.

Chang, C.C. (2005). "The role of retail sector in agro-food system, Chinese Taipei", Presented at the Pacific Economic Cooperation Council's Pacific Food System Outlook 2005-6 Annual Meeting in Kun Ming, China, May 11–13.

China Resources Enterprise (2002). "Retailing strategies and execution plan, July 2002", Posted on website of China Resources Enterprise, Limited, Hong Kong.

CIES (2005). "Wal-Mart Q1 results, interest in India". In: CIES Food Business Forum, May 13.
Cline, W.R. (2004). Trade Policy and Global Poverty. Center for Global Development and Institute for International Economics, Washington, DC.
Coase, R.H. (1937). "The nature of the firm". Economica 4, 386–405.
Cochrane, W.W. (1993). The Development of American Agriculture: A Historical Analysis. University of Minnesota Press, Minneapolis, MN.
Codron, J.-M. (1992). "The southern hemisphere and the expansion of world trade in temperate fruits". Agribusiness 8 (6), 585–600.
Codron, J.-M., Bouhsina, Z., Fort, F., Coudel, E., Puech, A. (2004). "Supermarkets in low-income Mediterranean countries: Impacts on horticulture systems". Development Policy Review 22 (5), 587–602.
Cook, H.L. (1959). "Market structures and economic development in the Philippines". Journal of Farm Economics 41, 1316–1322.
Cummings Jr., R.W. (1967). Pricing Efficiency in the Indian Wheat Market. Impex, New Delhi.
de Hernandez, L. (2004). "Colombia retail food sector annual, 2004", GAIN Report No. CO4011. USDA Foreign Agricultural Service, Washington, DC.
de Janvry, A., Sadoulet, E. (1987). "Agricultural price policies in general equilibrium models: Results and comparisons". American Journal of Agricultural Economics 69 (2), 230–246.
De Melo, J., Robinson, S. (1982). "Trade adjustment policies and income distribution in three archetype developing economies". Journal of Development Economics 10, 67–92.
Dirven, M. (2001). "Dairy sector clusters in Latin America". International Food and Agribusiness Management Review 2 (3), 301–313.
Dorward, A., Kydd, J., Poulton, C. (Eds.) (1998). Smallholder Cash Crop Production under Market Liberalisation: A New Institutional Economics Perspective. CAB International, London.
Dries, L., Reardon, T. (2005). "Central and Eastern Europe: Impact of food retail investments on the food chain", EBRD Cooperation Program, Report Series No. 6. FAO Investment Center, London.
Dries, L., Reardon, T., Swinnen, J. (2004). "The rapid rise of supermarkets in Central and Eastern Europe: Implications for the agrifood sector and rural development". Development Policy Review 22 (5), 525–556.
Dries, L., Swinnen, J. (2004). "Foreign direct investment, vertical integration and local suppliers: Evidence from the Polish dairy sector". World Development 32 (9), 1525–1544.
Escobal, J. (2005). "The role of public infrastructure in market development in rural Peru", Doctoral dissertation. Development Economics Group, University of Wageningen, The Netherlands.
Eswaran, M., Kotwal, A. (1985). "A theory of contractual structure in agriculture". American Economic Review 75, 353–367.
Evans, D. (1989). "Alternative perspectives on trade and development". In: Chenery, H.B., Srinivasan, T.N. (Eds.), Handbook of Development Economics, vol. 2. North-Holland, Amsterdam, pp. 1241–1304.
Fackler, P.L., Goodwin, B.K. (2001). "Spatial price analysis". In: Rausser, G., Gardner, B. (Eds.), Handbook of Agricultural Economics, vol. 1B. Elsevier Science, Amsterdam.
Fafchamps, M. (2004). Market Institutions in Sub-Saharan Africa: Theory and Evidence. MIT Press, Cambridge, MA.
FAOSTAT (2004). http://www.fao.org.
Farina, E. (2002). "Consolidation, multinationalization, and competition in Brazil: Impacts on horticulture and dairy product systems". Development Policy Review 20 (4), 441–457.
Farina, E., Reardon, T. (2000). "Agrifood grades and standards in the extended MERCOSUR: Conditioners and effects in the agrifood system". American Journal of Agricultural Economics 82 (5), 1170–1176.
Flores, L. (2004). "Small lettuce farmers' access to dynamic markets in Guatemala", Unpublished master thesis. Michigan State University, East Lansing, MI.
Food and Agricultural Organization (1959). The State of Food and Agriculture. FAO, Rome.
Gardner, B. (2002). American Agriculture in the Twentieth Century: How It Flourished and What It Cost. Harvard University Press, Cambridge, MA.
Giné, X., Klonner, S. (2002). "Financing a new technology in small-scale fishing: The dynamics of a linked product and credit market", Working paper. The World Bank, Washington, DC.

Gow, H., Swinnen, J. (2001). "Private enforcement capital and contract enforcement in transition economies". American Journal of Agricultural Economics 83 (3), 686–690.
Greenhouse, S. (2004). "Wal-Mart, a nation unto itself". New York Times (April 17), C-1.
Gutman, G. (2002). "Impact of the rapid rise of supermarkets on dairy products systems in Argentina". Development Policy Review 20 (4), 409–427.
Haggblade, S. (2005). "Subsector supply chains: Operational diagnostics for a complex rural economy". In: Haggblade, S., Hazell, P., Reardon, T. (Eds.), Transforming the Rural Nonfarm Economy. Johns Hopkins University Press, Baltimore, MD.
Harriss, B. (1979). "There is method in my madness: Or it is vice-versa? Measuring agricultural market performance". Food Research Institute Studies 16, 197–218.
Harriss, J., Hunter, J., Lewis, C.M. (Eds.) (1998). The New Institutional Economics and Third World Development. Routledge, London.
Hayami, Y., Ruttan, V. (1971). Agricultural Development. Johns Hopkins University Press, Baltimore, MD. Reprinted in 1985.
Heiman, A., McWilliams, B., Zilberman D. (2000). "Marketing institutions, risk, and technology adoption", Working paper. University of California, Berkeley.
Hernández, R., Reardon, T., Berdegué, J.A., Balsevich, F., Jano, P. (2004). "Acceso de pequeños productores de tomate a los supermercados en Guatemala", Report to USAID, DFID, and the Common Fund for Commodities, under the project RAISE/SPS, Regoverning Markets, and PFID.
Hill, L.D. (1990). Grain Grades and Standards: Historical Issues Shaping the Future. University of Illinois Press, Urbana, IL.
Hirschman, A.O. (1958). The Strategy of Economic Development. Yale University Press, New Haven, CT.
Holden, R.H. (1953). "Marketing structure and economic development". Quarterly Journal of Economics 67, 344–361.
Hu, D., Reardon, T., Rozelle, S., Timmer, P., Wang, H. (2004). "The emergence of supermarkets with Chinese characteristics: Challenges and opportunities for China's agricultural development". Development Policy Review 22 (4), 557–586.
Hueth, B., Ligon, E., Wolf, S., Wu, S. (1999). "Incentive instruments in fruits and vegetables contracts: Input control, monitoring, measurements, and price risk". Review of Agricultural Economics 21 (2), 374–389.
Jaffee, S., Morton, J. (Eds.) (1995). Marketing Africa's High Value Foods: Comparative Experiences of an Emergent Private Sector. Kendall-Hunt Publishing, Iowa.
Jano, P., Lundy, M., Balsevich, F., Berdegue, J., Reardon, T. (2004). "Farmers' organizations and supermarkets in Central America", Report. Rimisp – Latin American Center for Rural Development, Santiago, Chile.
Jitpleechep, S. (2002). "Who is who", Bangkok Post article posted August 28, 2002. http://www.siamfuture.com, accessed January 4, 2004.
Jones, W.O. (1968). "The structure of staple food marketing in Nigeria as revealed by price analysis". Food Research Institute Studies 8 (2), 95–123.
Kaplinsky, R., Morris, M. (2001). A Handbook for Value Chain Research. International Development Research Center, Canada.
Katjiuongua, H.B. (2004). "Fresh fruit and vegetable marketing: The case of supermarket procurement from tomato producers in Zambia", Report to USAID. Michigan State University, East Lansing, MI.
Key, N., Runsten, D. (1999). "Contract farming, smallholders, and rural development in Latin America: The organization of agroprocessing firms and the scale of outgrower production". World Development 27, 381–401.
Kherallah, M., Delgado, C., Gabre-Madhin, E., Minot, N., Johnson, M. (2002). Reforming Agricultural Markets in Africa. The International Food Policy Research Institute, Washington, DC.
Kinsey, J. (2004). "Supply chain innovations in the U.S. retail sector", Presented at the International Conference "Supermarkets and Agricultural Development in China: Opportunities and Challenges", Shanghai, May 24–25.
Kopicki, R., Peng, Y., Liu, Y., Li, D. (2004). The Current State of Supply Chain Development in the Western Provinces. Report Produced for the Western Region Leadership Group by the World Bank. World Bank, Beijing.

Lazzarini, S.L., Chaddad, F.R., Cook, M.L. (2001). "Integrating supply chain and network analysis: The study of netchains". Journal on Chain and Network Science 1 (1), 7–22.
Lee, J.-H., Reardon, T. (2005). "Forward integration of an agricultural cooperative into the supermarket sector: The case of Hanaro Club in Korea", Joint Working Paper. Department of Industrial Economics, Chung-Ang University, Seoul, Korea, and Department of Agricultural Economics, Michigan State University, East Lansing, MI.
Lele, U.J. (1971). Food Grain Marketing in India: Private Performance and Public Policy. Cornell University Press, Ithaca, NY.
Lewis, W.A. (1954). "Economic development with unlimited supplies of labour". The Manchester School 22 (2), 139–191.
Lipton, N. (1977). Why Poor People Stay Poor: Urban Bias in World Development. Avebury, London.
Mainville, D.Y., Zylbersztajn, D., Reardon, T. (2005). "Determinants of retailers' decisions to use public or private grades and standards: Evidence from the fresh produce market of São Paulo, Brazil". Food Policy 30 (3).
Mallampally, P., Sauvant, K.P. (1999). "Foreign direct investment in developing countries". Finance and Development 36 (1). Accessed at http://www.imf.org.
Manalili, N.M. (2005). "The changing map of the Philippine retail food sector: The impact on trade and the structure of agriculture and the policy response", Presented at the Pacific Economic Cooperation Council's Pacific Food System Outlook 2005-6 Annual Meeting in Kun Ming, China, May 11–13.
Martin, L.R. (1959). "Some marketing problems in Pakistan and India". Journal of Farm Economics 41, 1323–1326.
Martin, L.R. (1992). A Survey of Agricultural Economics Literature, vol. 4, Agriculture in Economic Development, 1940s to 1990s. University of Minnesota Press, Minneapolis, MN.
Mears, L. (1957). Rice Marketing in the Republic of Indonesia. Institute of Economics and Social Research, Jakarta.
M&M Planet Retail (2004). "Consumers forsake wet markets in Asia". Daily News by M&M Planet Retail (7 June).
M&M Planet Retail (2005). "Wal-Mart hopes for India boosted". Daily News by M&M Planet Retail (31 May).
Moore, G.E. (1965). "Cramming more components onto an integrated circuit". Electronics 38 (8), 1–4.
Neven, D. (2004). "Three essays on the rise of supermarkets and their impacts on fresh fruits and vegetables supply chains in Kenya", Unpublished doctoral dissertation. Michigan State University, East Lansing, MI.
Neven, D., Reardon, T. (2002). "Modal choice in international alliances between producers of horticultural products", Contributed paper (selected) for the 12th Annual IAMA Meetings, Chain Conference, June 2002.
Neven, D., Reardon, T. (2004). "The rise of Kenyan supermarkets and evolution of their horticulture product procurement systems". Development Policy Review 22 (6), 669–699.
Newbery, D.M., Stiglitz, J.E. (1981). The Theory of Commodity Price Stabilization: A Study in the Economics of Risk. Clarendon Press, Oxford.
North, D.C. (1981). Structure and Change in Economic History. W.W. Norton, New York.
North, D.C. (1990). Institutions, Institutional Change, and Economic Performance: Political Economy of Institutions and Decisions. Cambridge University Press, Cambridge.
Orellana, D., Vasquez, E. (2004). "Guatemala retail food sector annual, 2004", GAIN Report No. GT4018. USDA Foreign Agricultural Service, Washington, DC.
Pingali, P., Rosegrant, M. (1995). "Agricultural commercialisation and diversification: Processes and policies". Food Policy 20 (3), 171–185.
Pinstrup-Anderson, P. (1985). "Food prices and the poor in developing countries". European Review of Agricultural Economics 12 (1/2), 69–85.
Platteau, J.-P. (2000). Institutions, Social Norms, and Economic Development. Harwood Academic Publishers, London.
Produce Marketing Association (2002). "China: Produce industry market research study", Report for Emerging Markets Program. USDA-FAS.

Rangkuti, F. (2003). "Indonesia food retail sector report 2003", USDA GAIN Report No. ID3028.
Rasmussen, P.N. (1957). Studies in Inter-Sectoral Relations. North-Holland, Amsterdam.
Ravallion, M. (1986). "Testing market integration". American Journal of Agricultural Economics 68 (2), 292–307.
Reardon, T., Berdegué, J.A. (2002). "The rapid rise of supermarkets in Latin America: Challenges and opportunities for development". Development Policy Review 20 (4), 317–334.
Reardon, T., Farina, E.M.M.Q. (2002). "The rise of private food quality and safety standards: Illustrations from Brazil". International Food and Agricultural Management Review (4), 413–421.
Reardon, T., Swinnen, J.F.M. (2004). " Agrifood sector liberalization and the rise of supermarkets in former state-controlled economies: A comparative overview". Development Policy Review 22 (5), 515–523.
Reardon, T., Codron, J.-M., Busch, L., Bingen, J., Harris, C. (2001). "Global change in agrifood grades and standards: Agribusiness strategic responses in developing countries". International Food and Agribusiness Management Review 2 (3), 195–205.
Reardon, T., Timmer, C.P., Barrett, C.B., Berdegue, J. (2003a). "The rise of supermarkets in Africa, Asia, and Latin America". American Journal of Agricultural Economics 85 (5), 1140–1146.
Reardon, T., Vrabec, G., Karakas, D., Fritsch, C. (2003b). "The rapid rise of supermarkets in Croatia, with implications for the produce distribution system", Report to USAID under the RAISE/ACE Project.
Reardon, T., Echanove, F., Cook, R., Tucker, N., Berdegué, J.A. (2005). "The rise of supermarkets and the evolution of their procurement systems in Mexico: Focus on horticulture products", Working paper. Michigan State University, East Lansing, MI.
Regmi, A., Gehlhar, M. (Eds.) (2005a). New Directions in Global Food Markets. USDA/ERS Agriculture Information Bulletin No. AIB794. http://www.ers.usda.gov/Publications/aib794/.
Regmi, A., Gehlhar, M. (2005b). "Processed food trade pressure by evolution of global supply chains". Amber Waves (Publication of the USDA Economic Research Service) 3 (1), 12–19.
Rozelle, S., Park, A., Huang, J., Jin, H. (1997). "Liberalization and rural market integration in China". American Journal of Agricultural Economics 79 (2), 635–642.
Rybczynski, T. (1955). "Factor endowments and relative commodity prices". Economica 22, 336–341.
Schaffner, D.J., Schroder, W.R., Earle, M.D. (1998). Food Marketing: An International Perspective. McGraw-Hill, Boston, MA.
Schejtman, A. (1998). "Agroindustria y pequeña agricultura: Experiencias y opciones de transformación". In: Agroindustria y Pequeña Agricultura: Vínculos, Potencialidades y Oportunidades Comerciales, Serie Libros de la CEPAL, No. 46 (LC/G.2007-P). Comisión Económica para América Latina y el Caribe (CEPAL), Santiago de Chile.
Schmid, A., Shaffer, J. (1964). "Marketing in social perspective". In: Sorenson, V. (Ed.), Agriculture Market Analysis. Bureau of Business and Economic Research, Michigan State University, East Lansing, MI.
Schultz, T.W. (1964). Transforming Traditional Agriculture. Yale University Press, New Haven, CT.
Schultz, T.W. (1978). Distortions in Agricultural Incentives. Indiana University Press, Bloomington, IN.
Scott, G.J. (1985). Markets, Myths and Middlemen: A Study of Potato Marketing in Central Peru. International Potato Center, Lima.
Shepherd, G. (1967). "Are fruit and vegetable marketing margins too high?", Report of the Iowa State University Mission to Peru.
Shwedel, K. (2003). "Sector agroalimentario en México", Presented at the Seminar Conglomerados y Empresas Transnacionales, Economic Commission for Latin America and the Caribbean (CEPAL), 26–27 November, Santiago, Chile.
Singh, I., Squire, L., Strauss, J. (1986). Agricultural Household Models, Extensions, Applications and Policy. World Bank and Johns Hopkins University Press, Washington, DC.
Stiglitz, J. (1989). "Markets, market failures, and development". American Economic Review 79 (1), 197–210.
Swinnen, J. (2004). "Vertical integration, interlinking markets, and growth in transition agriculture", Paper presented at the European Association of Agricultural Economics Congress in Ghent on "Role of Institutions in Rural Policies and Agricultural Markets".

Tam, P.T.G. (2004). "Regoverning markets: Securing small producer participation in restructured national and regional agri-food systems: The case of Vietnam", Report to the Regoverning Markets Project. Nong Lam University, Ho Chi Minh City, Vietnam.
Taylor, L. (1990). "Structuralist CGE models". In: Taylor, L. (Ed.), Socially Relevant Policy Analysis: Structuralist Computable General Equilibrium Models for the Developing World. MIT Press, Cambridge, MA.
Thailand Development Research Institute (2002). The Retail Business in Thailand: Impact of the Large Scale Multinational Corporation Retailers. Thailand Development Research Institute, Bangkok.
Timmer, C.P. (1974). "A model of rice marketing margins in Indonesia". Food Research Institute Studies XIII (2), 145–167.
Timmer, C.P. (1986). Getting Prices Right: The Scope and Limits of Agricultural Price Policy. Cornell University Press, Ithaca, NY.
Timmer, C.P. (1988). "The agricultural transformation". In: Chenery, H., Srinivasan, T.N. (Eds.), Handbook of Development Economics, vol. 1. North-Holland, Amsterdam, pp. 275–331.
Timmer, C.P. (1996). "Does BULOG stabilize rice prices in Indonesia? Should it try?". Bulletin of Indonesian Economic Studies 32 (2), 45–74.
Timmer, C.P. (1997). "Farmers and markets: The political economy of new paradigms". American Journal of Agricultural Economics 79 (2), 621–627.
Timmer, C.P., Falcon, W.P., Pearson, S.R. (1983). Food Policy Analysis. Johns Hopkins University Press for the World Bank, Baltimore, MD.
UNCTAD (2001). World Investment Report 2001: Trends and Determinants. United Nations Commission on Trade and Development, Geneva.
Valdes, A. (Ed.) (1981). Food Security for Developing Countries. Westview Press, Boulder, CO.
Van der Laan, H.L. (1975). The Lebanese Traders in Sierra Leone. Mouton and Co., The Hague.
Weatherspoon, D.D., Reardon, T. (2003). "The rise of supermarkets in Africa: Implications for agrifood systems and the rural poor". Development Policy Review 21 (3), 333–355.
Wharton Jr., C.R. (1962). "Marketing, merchandising and money-lending: A note on middleman monopsony in Malaya". Malayan Economic Review 7, 24–44.
Williamson, O.E. (1975). Markets and Hierarchies: Analysis and Antitrust Implications. Free Press, New York.
Williamson, O.E. (1996). The Mechanisms of Governance. Oxford University Press, New York.
World Bank (2002). World Development Report 2002: Building Institutions for Markets. Oxford University Press, New York.
Zylbersztajn, D. (2002). "Agricultura de contratos". Jornal Valor Econômico (São Paulo, Brazil) 2, 17–54.
Zylbersztajn, D., Farina, E.M.M.Q. (1999). "Strictly coordinated food systems: Exploring the limits of the Coasian firm". International Food and Agribusiness Management Review 2 (2), 249–265.

Chapter 56

RURAL FINANCIAL MARKETS IN DEVELOPING COUNTRIES

JONATHAN CONNING

Department of Economics, Hunter College and the Graduate Center, City University of New York, New York, NY

CHRISTOPHER UDRY

Department of Economics, Yale University, New Haven, CT

Contents

Handbook of Agricultural Economics, Volume 3
Edited by Robert Evenson and Prabhu Pingali

DOI: 10.1016/S1574-0072(06)03056-8

Abstract

This review examines portions of the vast literature on rural financial markets and household behavior in the face of risk and uncertainty and limited commitment. In addition to examining household strategies and bilateral contracting we place particular emphasis on studying the important role of financial intermediaries, competition and regulation in shaping the changing structure and organization of rural markets. Our goal is to provide a framework within which the evolution of financial intermediation in rural economies can be understood.

Keywords

rural finance, financial intermediation, agricultural credit

JEL classification: O16, Q14, O17, O12

1. Introduction

The organization of rural economic activity in general, and agricultural production in particular, is strongly conditioned by the fact that inputs are transformed into outputs with considerable time lags, and that production and sale outcomes can be highly uncertain because of the vagaries of nature or the swings of volatile commodity markets. In such environments, the ability of agricultural enterprises and rural households to make long-term investments, take calculated risks, and create stable consumption streams will be shaped by the set of available financial instruments and strategies to transform one pattern of variable and uncertain resource inflows and outflows into another. If the available set of financial services is very limited, households may have to forego valuable investment and income-generating activities and suffer the consequences of volatile consumption.

Financial transactions are implicit within, and often the reason behind, many contractual and organizational forms in the rural economy. Financial innovation therefore can have dramatic consequences on the ownership and governance structures of agricultural firms and community institutions. Financing options can affect decisions such as the physical placement and scale of agricultural operations, crop choices, and the decision to invest in risky but profitable new technologies or infrastructure. They may also affect choices about the size and composition of the rural household, and decisions such as whether to migrate, how much to invest in education, or the use of child labor. The availability of financing can also be a force that shapes political dynamics within a community, for example by affecting agent's outside opportunities and bargaining power.

Making new financial services and contract forms available can be viewed as a form of opening to trade. Agents in a financially isolated rural economy have little choice but to transform one set of variable and uncertain cash flows into another using available production and storage technologies and local financial instruments. Since risks in a local rural economy are typically subject to common external shocks and the pool of savings may be limited, local markets often cannot offer very good diversification opportunities and the cost of funds may be high. The introduction of new financial instruments allows agents to face new relative price tradeoffs across time periods and state-contingent events. The new trading opportunities this creates may then allow agents to specialize in higher value income activities while at the same time allowing households to purchase smoother consumption streams. Unfortunately, agents in the rural sectors of most developing countries remain cut-off from many of the opportunities for investing, risk-taking and risk spreading that would be available through better financial integration into larger national and global financial markets [de Soto (2000)].

An important research agenda is to understand the dynamics of financial innovation. There are both winners and losers from the introduction of new financial services and opportunities for trade. Losers may include incumbent local financial service providers who may stand to lose monopoly rents or market share in the face of increased outside competition [Platteau (1997); Rajan and Zingales (2003)], or those who might fear for the collapse of local informal insurance mechanisms [Scott (1976)]. Just as common have been the calls by organized groups of borrowers or activists for political and economic authorities to intervene to regulate allegedly exploitative or harmful activities of informal moneylenders or landlords.

Whether these fears were in reality justified or not, these constituencies, or those who would claim to speak for them, have often led political movements that have opposed financial market liberalization or have lobbied to favor government interventions in rural finance. In many cases a history of heavy-handed political interventions has led rural financial markets to become repressed and distorted [Gurley and Shaw (1960); McKinnon (1973); Adams, Graham and Von Pischke (1984); Gonzalez-Vega (1984)]. Yet even where financially repressive policies have been lifted or never materialized, the entry of new financial intermediaries to help local communities realize the gains to financial trade has often proven difficult or been delayed because serious information asymmetries and enforcement problems lead to market failures that are difficult to repair. For all these reasons observers continue to disagree on the role of government in promoting or repressing efficiency-enhancing financial intermediation.

There are already several very useful surveys and edited volumes of articles covering important aspects of the now vast literature on rural financial markets and household consumption behavior in the face of risk and uncertainty. A non-exhaustive list of key survey references might include Von Pischke, Donald and Adams (1983), Binswanger and Rosenzweig (1986), Hazell, Pomareda and Valdés (1986), Bell (1988), Gersovitz (1988), Besley (1994), Stiglitz (1994), Townsend (1995a), Morduch (1995), Meyer and Nagarajan (2000), Barry and Robison (2001), Banerjee (2003), Fafchamps (2003), Dercon (2004), Armendáriz de Aghion and Morduch (2005) and the relevant chapters in Bardhan (1989), Basu (1997), Deaton (1997), Ray (1998), and Bardhan and Udry (1999). While considerable overlap with these earlier studies is inevitable in the present work, we have attempted to set this chapter somewhat apart by focusing in more depth on relatively new developments in this fast growing field of empirical and theoretical research. We have also attempted to place more emphasis than earlier studies on studying the important role of financial intermediaries, competition and regulation in shaping the changing structure and organization of rural markets, rather than simply focusing on household strategies and bilateral contracting.

What then determines the extent and efficacy of the financial instruments available to the rural economy? All financial transactions involve in one way or another the exchange of contingent claims over future resources for claims in the present, in other words they involve the sale and purchase of contingent promises or IOUs. Like many other surveys before this, we shall highlight the many difficulties that arise in buying and selling such IOUs, particularly in rural environments where problems of asymmetric information and costly enforcement are likely to be important. What we hope distinguishes this survey is our focus on the process by which the financial structure of the rural economy is transformed by financial intermediaries.

Intermediaries play an essential role in the dynamic evolution of the real production and exchange possibilities of the economy by creating new instruments and contractual forms to bridge many of the trading gaps and missing markets that information and enforcement problems create. Financial intermediaries use their own capital and specialized information and enforcement mechanisms to help transform the illiquid claims held by producers and entrepreneurs in the economy into more liquid claims that can be more readily sold to less informed investors with funds. Successful contractual forms are soon imitated and improved by new entrants and in the process new markets are developed and extended. Further opportunities for trade and specialization may be uncovered along the way, and with them, a new set of information and enforcement problems to be solved.

Whether such a virtuous circle of uncovering and completing new markets proceeds or gets stuck will depend on the nature of the underlying information and enforcement problems and on the quality of a society's laws and institutions. These affect the incentives agents have to gather information and search for new contract forms to establish, monitor and enforce the new property claims that form the basis for emergent markets and trades.

Efficiency-enhancing private intermediation is more likely to emerge in environments where individuals are able to create new instruments and contracts that are given legal or societal recognition and are impartially enforced. By contrast, in situations where property rights are hard to enforce agents will find it difficult to define and secure the commitments that allow them to appropriate the returns from searching for and completing mutually beneficial trade. Our goal is to provide a framework within which the evolution of financial intermediation in rural economies can be understood.

We begin in Section 2 with a brief discussion of prominent features of rural financial markets that will guide our subsequent arguments. Section 3, the core of the chapter, examines recent developments in the theory of rural financial markets, and where possible links these to relevant empirical literature. Section 4 concludes.

2. Salient characteristics of rural financial markets

2.1. *Fragmented or absent markets*

Development economists have spent much effort in recent years trying to measure the extent to which households appear to be insured against idiosyncratic shocks and the structure and performance of local financial contracts such as bilateral credit and insurance arrangements with landlords, moneylenders, family or friends, or group-based mutual savings and insurance arrangements such as rotating savings and credit associations (ROSCAs). While these studies have advanced our understanding of local bilateral financial contracting and mutual insurance within poor communities, the study of financial *intermediation* has remained relatively neglected. A financial intermediary expands and transforms the set of trades that can take place both within communities and across communities by carrying out monitoring and control activities and providing asset transformation services at lower cost than what could be achieved under a system of local bilateral contracts or mutual insurance arrangements [Diamond (1996)].

Rural financial markets have often times been described as *fragmented* in the sense that different segments of borrowers are observed to be systematically sorted across different loan types and lending intermediaries according to the characteristics of the borrowers, the lenders and the activities financed, and other variables in trading environment [McKinnon (1973); Hoff, Braverman and Stiglitz (1993); Meyer, Nagarajan and Hushak (1997)]. Through a combination of limited access and choice, firms in the same market end up using financial instruments that can substantially differ as to interest rate charges, the type and quantity of collateral required on loans, resources spent on monitoring and enforcing contract terms, and whether or not credit is tied to transactions on other markets. In some markets, would-be borrowers may find themselves excluded or dissuaded from obtaining access to certain credit instruments, or rationed to smaller loans than they might have optimally chosen, by collateral requirements and other non-price terms. They may then adjust by turning to substitute, but possibly more expensive financing sources or may modify their first best allocation plans in other ways.

Banerjee (2003) provides a very useful review of some of the salient empirical literature on rural financial markets. He argues (p. 4) that there is "extreme variability in the interest rate charged by lenders for superficially similar loan transactions within the same economy". Aleem (1990) similarly shows that moneylenders in a semi-urban setting in Pakistan charged highly variable interest rates to different borrowers: the standard deviation of interest rates was 40% per annum, while the average rate was 80%. Timberg and Aiyar (1984) document that Shikarpuri lenders in India charged rates varying between 20% and 120%, depending on the market. Dasgupta (1989) reports high variation in the rates charged by moneylenders, with substantial numbers of loans made at rates higher than 60%, while many others are made at rates below 30%. Ghate (1992) cites results from a case study from Thailand which finds interest rates of 2–3% per month in the Central Plain, but 5–7% p.m. in the North and Northeast. Udry (1991) finds large variation in interest rates within 4 small villages of northern Nigeria: nominal monthly interest rates exceeded 7.5% on about 20% of loans, but the median nominal interest rate was 0%.

Banerjee (2003) also reports substantial evidence of very large spreads between borrowing and deposit rates in many financial markets in developing countries. Timberg and Aiyar (1984) report spreads of approximately 16%, while Aleem (1990) reports a spread of over 40% in Pakistan. Ngugi (2001) shows that the spread in Kenya in the late 1990s ranged between 15 and 30%. In rural southern Ghana, deposit takers charge a fee to depositors, and pay no interest, while loans are made at variable rates up to 10% per year [Aryeetey and Udry (2000)].

There is also strong and growing evidence that many enterprises, particularly rural enterprises, have very high rates of return to capital that may persist over time for some enterprises because of the highly fragmented nature of financial markets. Schündeln (2004) shows that the marginal rate of return for small firms in Ghana is about 50%, while the return for large firms is less than 10%. McKenzie and Woodruf (2004) find that the rate of return is as high as 15% per month for microenterprises in Mexico. Banerjee and Duflo show that a sample of medium sized firms that borrow from a large

Indian bank have rates of return of almost 100% per year. Goldstein and Udry (1999) estimate the rate of return to capital for farmers entering pineapple production to be over 1200% per year. These examples, of course, could be multiplied.

The key challenge of theoretical work on rural financial markets, therefore, is to provide a framework to make sense of these striking features of rural financial markets. Why are there such high rates of return to capital for at least many borrowers? How do large spreads persist between deposit and borrowing rates? Why is there such a diversity of contract forms and intermediary structures? How are different households and firms matched to each, and why do such highly variable interest rates persist in equilibrium across borrowers?

Moneylenders and financial intermediaries in the rural economy include most importantly input suppliers, rural product traders (including agro-industry and exporting firms), and banks. They often invest heavily in screening and monitoring their clients, and may also intervene to significantly shape their clients' choice of technology and other production decisions. Writing in the early 20th century, British colonial officer Sir Malcolm Darling (1925) had this to observe about the rural moneylender of Punjab:

> "He is always accessible, even at night; dispenses with troublesome formalities, asks no inconvenient questions, advances promptly, and if interest is paid, does not press for repayment of principal. He keeps in close personal touch with his clients, and in many villages shares their occasions of weal or woe. With his intimate knowledge of those around him he is able, without serious risk, to finance those who would otherwise get no loan at all."

Traders and contract farming firms typically contract to market or process a farmer's harvest in exchange for credit and often other services like technical assistance and farm input sales. An important characteristic of this form of lending is that the loan contract often involves much less collateral than would a similar bank loan, and at times, no collateral other than a crop pledge. These loans are however usually quite heavily monitored in the growing season and prior to repayment. The purpose appears aimed at limiting the farmer's scope to divert resources or effort away from the financed project and toward other activities where the lender may not be able to establish clear claims. Traders are likely candidates to become financial intermediaries because in the normal course of their activities as product buyers they acquire knowledge of the farmer and the crop technology. While a separate specialized lender and separate trader might both incur costs to monitor a farmer's compliance in meeting the terms of a loan, and in meeting quality standards on delivered produce, a combined trader–lender economizes on these costs through economies of scope in monitoring. They are also often able to better value some of the items a farmer might provide as collateral. A trader for instance will be much more willing to accept a farmer's crop as collateral than a bank.

Another important aspect of this type of lending relationship is that it is intermediated finance: the trader-intermediary usually employs a combination of her own equity together with funds leveraged from less informed outside intermediaries such as banks or other creditors such as the downstream product-buying intermediaries to whom they will themselves deliver product. This type of relationship is often underpinned by, or leads to the development of, a system of bills of exchange. At the time of contracting the trader may have the farmer sign a bill of exchange for the amount of funds to be advanced. The intermediary then uses the diversified collection bills obtained from many farmers as a form of proof of contracting or security to use in raising finance from the bank or outside lender. The outside creditor(s) may then lend the intermediary a fraction of the funds required to on-lend to the farmers, using these bills and other guarantees as security. By lending only a fraction of the total required finance, the outside lender forces the intermediary to make up the difference out of her own equity thereby acquiring a sufficient stake in the borrower's project to have the incentive to monitor the farmer in ways that safeguard the value of the overall investment.

2.2. *Government interventions*

Governments have intervened in rural financial markets since the earliest days of markets. Hammurabi's codes, a set of 282 laws set in stone by the ruler of Babylon and Mesopotamia over 37 centuries ago included many to regulate the operation of credit for farmer and merchants, including caps that limited interest rates to 33 and one third percent on loans of grain and regulations that limited what could be collected on agricultural debts in the event of drought or certain other natural disasters [Goetzmann (1996)].

Some types of government intervention can clearly serve the useful purpose of promoting financial market trade between private parties. For example, transaction costs between private parties may be reduced if the state is able to provide impartial and accessible legal mechanisms for the arbitration and enforcement of contracts. Similarly, the prudential regulation and supervision of deposit-taking financial intermediaries may promote deposit mobilization and encourage efficiency-enhancing market competition between banks. Even some forms of more direct government intervention, including direct loans and government loan guarantees, may arguably at times help 'crowd-in' private sector financial intermediation or provide efficiency enhancing financial services in contexts where private sector actors may have otherwise been reluctant to operate.

Even in advanced economies, governments sponsor or directly support elaborate farm lending systems. In the United States, the Farm Loan Act of 1916 established the first government-sponsored enterprise via the creation of a system of regional Farm Loan Banks to grant loans to farm cooperative associations that lend directly to farmers. In 2001 the Farm Credit System stood behind some 91 billion dollars of loans or about 30% of farm sector's total loans outstanding [General Accounting Office (2001)]. The system continues to enjoy implicit and explicit government guarantees and tax benefits that enhance the financial services cooperative institutions ability to raise funds.

At other times, government involvement has been much more heavy-handed. Governments have, for example, enforced strict 'anti-usury' laws that capped market interest rates or acted to prohibit or constrain the participation of certain types of intermediaries. Directed credit programs have compelled national banks to open rural branches and allocate a specified fraction of total lending to agricultural lending. For decades, state-sponsored finance was the dominant form of institutional lending in most developing countries, although access by different strata of farmers varied greatly by region. A 1975 World Bank Report on Agricultural Credit, written at about the height of the Bank's lending for agricultural lending, noted that less than 1% of farmers in certain African farmers obtained access to subsidized lending, whereas in a country like Taiwan almost all farmers had access [World Bank (1975)]. The report also noted that in Latin America and Asia it was not uncommon for 70 to 80% of small farmers to have virtually no access to such credit. Where such farmers did have access to institutional credit, it had generally been used for short-term loans to finance current inputs, such as seed, fertilizer and pesticides. Although institutional lines of credit to buy animals, tractors or equipment are sometimes provided over a period of two to five years or longer, longer-term institutional credit has typically remained scarce [Deininger (2003)].

A large literature has made clear that financial repression as a result of directed credit, interest rate caps, and excessive regulation and state involvement in banking has been a principal culprit of the relative lack of more effective intermediation in developing countries [McKinnon (1973); Adams, Graham and Von Pischke (1984)]. Although often well-intentioned, these policies often created extremely poor incentives for private financial intermediaries to enter the sector or to make sensible loans. State lending programs on the other hand were very often characterized by high arrears and political capture. Cole (2004) finds that over the 1985–1999 period, agricultural lending by public banks in India grew 5–10 percentage points faster in election years than in years after an election, that election year loans were more likely to be made to districts with more heavily contested elections, that these loans were less likely to be repaid, and that they did not measurably affect agricultural output. Warning and Sadoulet (1998) point to similar evidence of political capture and loan arrears in Senegal.

Interest rate ceilings have been a particularly common intervention in rural financial markets. The traditional analysis of interest rate ceilings is that, by limiting supply and increasing demand, such interest rate caps will create an artificial excess demand for credit, leading lenders to ration supply. While interest rate caps clearly harm savers and will lead to reduced rural deposits, the impact on farm borrowers is *a priori* ambiguous since what an individual farmer loses by receiving a smaller loan may be more than compensated for by a lower interest rate. It is clear however that in many, if not most cases, the resulting rationing has often ended up favoring larger farmers at the expense of smaller ones. Many country case studies have similarly observed that the lion's share of subsidized institutional credit goes to medium and large farms, even though many of these programs are justified politically as ostensibly being aimed at helping small farmers [Von Pischke, Donald and Adams (1983); Adams, Graham and Von Pischke (1984); Burgess and Pande (2003)]. This outcome has been found to be so common that Gonzalez-Vega (1984) has dubbed it 'the Iron Law of Interest-Rate Restrictions'.

Several explanations have been offered. One explanation is that if under market conditions higher interest rates must be charged of smaller farmers to recover fixed loan processing costs, legal interest rate caps will lead lenders to ration these farmers first. Since capped rate loans transfer rents to farmers lucky enough to obtain them, the credit allocation process is also susceptible to political capture, and larger farmers will typically have the upper hand. Bates (1981) gives historical details of the pattern of political capture of marketing boards which also often direct subsidized credit toward better off farmers. Explicit corruption in the allocation of low-interest loans was also common. State-owned financial institutions were often confronted with only a soft budget constraint, and received repeated financial bailouts, further reducing the incentive to put resources into enforcing contracts or mitigating problems of asymmetric information. In this context, a movement away from bad policies is a far more important step than any pressing need for policy innovation [McKinnon (1973); Adams, Graham and Von Pischke (1984); Adams (1992); Von Pischke (1997); Rajan and Zingales (2003)].

Directed credit programs have been similarly strongly criticized as being distorting and inefficient and usually part of a package of other financial repression measures that set back the development of many rural financial markets [Adams, Graham and Von Pischke (1984)]. Although these criticisms no doubt hit the mark in describing the impact of policies in many countries, other researchers have pointed to evidence suggesting that directed credit has had several of its intended effects in some contexts. For example, the Bank of Thailand (BOT) mandates portfolio allocation targets for commercial banks to lend to the agricultural sector. These include mandates to charge interest rates for agriculture not higher than non-agriculture lending and government efforts to expand the capitalization of the Bank for Agriculture and Agricultural Cooperatives (BAAC), which has facilitated the aggressive expansion of BAAC activities. There is substantial evidence of important positive impacts of this expansion on agricultural output and farmer welfare [Fitchett (1999)].

A recent study by Burgess and Pande (2003) reviews the literature on directed credit in India and analyzes a panel of state-level data to conclude that directed credit policies had the intended effect of expanding rural bank branching and that this lowered poverty and expanded non-agricultural rural output while leaving urban policy unaffected. As the authors note, at the time of independence less than 1% of rural household debt in India came from commercial banks and the vast bulk of rural borrowing was via informal sources, with moneylenders accounting for close to 70% of the total. By 1971 the share of debt to commercial banks had grown scarcely to 3%. A Central Bank mandate adopted between 1977 and 1990 which required banks to open up four new branches in under-served areas for every new branch opened elsewhere led to significant expansion of new rural branches, which varied by state. Largely as a result of these policies by 1991 the share of rural debt held by commercial banks had increased tenfold to 29%, while the moneylender share of rural household debt more than halved from 35 to 15.7%. After carefully controlling for other factors, the authors find that rural non-farm activities expanded, wages rose, and rural poverty fell relatively more in areas in which banking services expanded relatively quickly as a consequence of this policy of directed credit.

The debate over this issue is rather like the debate over trade policies. There is little doubt that protectionism and industrial policy can lead to inefficient distortions in relative prices that lowers welfare and suppresses trade, but the experience of a few countries has lead some observers to believe that such policies may at times be used to address market failures and/or to break local market power. Similarly, despite the overall record of costly failure, it is clear that directed credit policies may serve valuable social objectives in the right circumstances.

The apparent failure of many programs of state intervention in rural financial markets, and the wave of structural adjustment programs that moved through developing countries beginning in the 1980s did away with much of the legacy of financial repression. The policy change was dramatic. For example, annual World Bank lending for agricultural credit projects was over US$ one billion in the 1980s, but fell to under $250 million by the end of the 1990s [Zeller (2003)]. There was large-scale privatization, restructuring and closure of many state banks. This was associated in many countries with more widespread liberalization of rural financial markets. The reduction of financial repression was usually associated with a package of other reforms including trade liberalization and privatization of other state-owned enterprises. These reform packages led to new financial intermediaries in many cases, but nowhere near the supply response that the most optimistic 'financial repression' school people predicted [Carter, Cason and Zimmerman (1998)].

It is evident that the development of a robust rural financial system requires both careful state attention to the fundamental institutions that undergird financial contracts, and the freedom to transact without direct state regulation [Stiglitz (1994); Rajan and Zingales (2003)]. Section 3 of this chapter is devoted to providing a framework to understand the role of public goods in property rights and contract enforcement, information sharing, and prudential regulation.

The task of promoting, improving or even creating the rural institutions required to support rural financial transactions is one of the fundamental challenges facing governments of developing countries. The range of feasible financial contracts can be expanded in the presence of institutions for information dissemination or that facilitate the verifiability and enforceability of contracts. More specifically, governments have a crucial role to play in the creation and support of reasonably impartial courts to enforce private contracts and arbitrate or settle disputes, records offices to register and title property and increase the collateralizability of assets and the registration of liens, credit bureaus to record and share credit histories, and external audit mechanisms to solve problems of verifiability. Some of these institutions can be run as government offices, some can be supported as private enterprises, while others might be private but depending on government supervision.

Some might argue that these institutions emerge wherever they are needed. Early neoclassical institutional economics sees contract and institutional innovations emerging to economize on transaction costs and information asymmetries [Demsetz (1967); North and Thomas (1973)]. This is a generalized induced innovation hypothesis, strongly related to Coasian notions of contracting. But this view has largely been abandoned [North (1990)]. Dysfunctional institutions can sometimes persist over long periods of time in any particular society, even as other societies have managed to adapt their own institutions to overcome similar problems. The path of institutional change may be strongly influenced by historical precedent and by the workings of political processes [Bowles (2003)]. In the context of discussing credit market institutions in modern day Africa Fafchamps (2004) discusses innovation failure (institutions may simply not have been invented), authority failure (central government coercion may be weak or misdirected), coordination failure (public goods require the solution to collective action problems that may remain unsolved). Institutional change may be opposed by those who see themselves as losing out in the new distribution of wealth and political power that may emerge or they may be holding out for a better bargain.

One hypothesis is that institutional failures of this sort are the consequence of the lack of a catalyzing agent or organization to coordinate actions to spur change. This has lead to some hope that state banks and government guarantees can contribute by 'crowding-in' new forms of private financial intermediation. Alternatively, nonconvexities in the technologies associated with institutional innovation can be associated with institutional failure when there is insufficient local intermediary capital, or a too-small market. In this case, the relative absence of intermediaries may be a simple function of the low levels of income and wealth in developing countries. Some authors argue that growth and financial deepening go hand in hand: as the economy grows there are more opportunities for diversification, and this in turn induces agents to invest in riskier but higher return projects, so the economy grows faster [Acemoglu and Zilibotti (1997)].

In order to move toward a theory of the evolution of the institutions that might support a flourishing rural financial sector, we must first understand the economics of financial markets in the context of incomplete information and imperfect contract enforcement. It is to that task that we now turn. We then use this general framework to examine the potential for 'crowding-in' of new forms of intermediation, and the role of new semi-formal institutions of microfinance.

3. Models of rural financial markets

Historically, a good part of the theory of rural financial transactions developed in parallel to, and sometimes ahead of, more general results in the literature on information asymmetries, and the microeconomic theory of banking and corporate finance [Stiglitz (2002)].[1] For example, Stiglitz' (1974) famous paper on "Incentives and risk sharing in sharecropping" inspired a good deal of later literature on how moral hazard could shape the structure of labor, insurance and credit and equity contracts, and Akerlof's (1970) early analysis of adverse selection or the 'lemons problem' was, by his own account, partly motivated by his observations on the operation of informal rural moneylenders in India.

[1] Freixas and Rochet (1997) provide an excellent survey of the modern literature on the microeconomics of banking and corporate finance while Dewatripont and Tirole (1994) offer a useful synthesis of the theory of prudential government regulation and intervention in financial markets.

In attempting to survey developments in the theory of agricultural financial contracting it is helpful to make one small preliminary note on methodology on how it relates to a more general microeconomic theory of contracts. The early literature on agricultural contracts that developed in the 1970s and 1980s worked extensively with what is sometimes referred to as the *state-space* formulation [Hart and Holmstrom (1987)] and with linear contract forms. A typical model might describe farm project outcomes by $x = \theta f(e)$ where e is the agent's level of input or effort into production the function $f(e)$ and θ is a multiplicative random shock drawn by 'nature' from a known probability distribution $G(\theta)$. If an outside principal (e.g., a landlord or a lender) with a stake in the project could not directly to specify the farmer's choice of e in a contract, then the principal could not be sure if a low project outcome x was due to a bad draw of θ or to the farmer's low choice of e, leading to a potential problem of moral hazard. The analysis then turned on finding the terms of a contract or sharing rule that would give the farm agent incentives to choose a given effort level anyways. A linear contract of the form $\alpha x - F$, was typically assumed where α is the share or output kept by the farmer, and F is the value of a fixed payment made either from the farmer to the principal or vice-versa. The linear formulation seemed rich enough to span a broad range of recognizable contracts forms including fixed-wage contracts ($\alpha = 0$, $F < 0$), fixed-rent or fixed-debt contracts ($\alpha = 0$, $F > 0$), pure share-contracts ($\alpha > 0$, $F = 0$) and mixed share contracts with side lump sum transfers ($\alpha > 0$, $F \neq 0$). The main advantage of the state-space formulation was that it presented technology in a familiar way that could be built directly upon existing farm household models [Singh, Squire and Strauss (1986)].

By way of contrast, the more general microeconomic theory of contracts has tended to increasingly work with what has come to be known as the *parameterized distribution* formulation pioneered by [Mirrlees (1976)] and further explored by Holmstrom (1979) and others. In this formulation, the agent's effort choice e is thought of as a parameter in the distribution of project outcomes $\pi(x_s; e)$. Through his choice of e the agent chooses or 'induces' a probability distribution over state-contingent outcomes x_s where s indicates a state of the world. This formulation abandons linearity and generalizes the production function. Although the two formulations are equivalent in so far as they can be mapped onto one another, the latter formulation has led to more general insights and now clearly dominates the field. The parameterized distribution formulation also makes it easier to work with richer contracting environments – for example, tying the agent's reward not just to the outcome of their own project but to other events in the village – rather than just simple sharing rules that assume linearity.[2] For all these reasons we shall use the parameterized distribution approach to survey the literature, even where results were first cast using the state-space approach.

[2] As explained below, this has had some important consequences, for example interpretations that were derived under the state-space approach that had to be later abandoned or modified once the same problem was explored in a less restricted contracting environment.

3.1. The complete markets benchmark

It is useful to begin with a brief review of the operation of credit and insurance markets under the assumption of complete markets. Although this is not realistic in most contexts, more appropriate models of financial contracting can be understood as departures from this benchmark case. Consider a village of N farmers, where we use the term 'village' very broadly as a metaphor for any group of individuals who are able to engage in financial trade. It could refer to a small group of farmers tied together by geographic proximity, members of an extended family or expanded trading circle. Each time 'period' will be divided into two stages, a 'pre-harvest stage' where investments are made and state-contingent contracts are exchanged but in which no production or consumption takes place and a second 'harvest and post-harvest stage' where project outcomes are realized, contracts are executed, and consumption takes place.

Villager j has access to a farm production project that will yield stochastic harvest income x_{js} where $s = 1, \ldots, S$ indexes possible states of the world and these states are distributed according to the probability distribution π_{js} that may be affected by the input and effort choices made by each of the agents in the village, which for the moment are assumed to be costlessly observed and contractible. To illustrate, if each villager's harvest project yielded one of M possible harvest incomes then there would be $S = N^M$ possible ways harvest outcomes might be realized within the village. We could label each of these realizations as a state-contingent event or state of the world s, although properly speaking the event space should be far richer. A state of the world is a complete description of a possible outcome of uncertainty. For example, a particular pattern of harvest realizations in which a particular child in a particular farm household falls sick should properly be considered a separate state of the world from an outcome with the exact same pattern of crop realizations but where the same child does not fall sick.

Let c_{js} represent consumption of villager j in state s and suppose that each individual wants to maximize expected utility $\sum_s \pi_s u_j(c_{js})$ where u_j is a standard concave and well-behaved utility function. A Pareto-efficient allocation of risk in the village can then be found by maximizing the weighted sum of the utilities of each of the N villagers, where λ_j is the individual's weight in the Pareto program. These reflect the relative strength of the entitlement that each individual has over village resources. In a non-market setting these would be given perhaps by their social status or entitlement standing within the community [Sen (1982)]. In a competitive market setting these weights would be related to the market value of their initial property claims x_{js}. A Pareto-efficient allocation is found by choosing the c_{js} to solve

$$\sum_{c_{js}} \lambda_j \sum_s \pi_s u_j(c_{js}) \tag{1}$$

subject to village-wide resource constraints in each state

$$\sum_j c_{js} \leqslant \sum_j x_{js} \quad \forall s. \tag{2}$$

The familiar first-order conditions for an optimum yield the following condition relating the marginal utilities of any two villagers j and k in any state of the world s:

$$\frac{u'_j(c_{js})}{u'_k(c_{ks})} = \frac{\lambda_k}{\lambda_j} \quad \forall j, k, s. \tag{3}$$

Since the λ's are constant welfare weights the conditions imply that marginal utilities of all villagers must move together. From this it follows that each household's consumption will be monotonically increasing in the level of average village consumption. Idiosyncratic shocks to household income will be pooled at the village level so that, conditional on average consumption, a household's consumption will be unaffected by its own idiosyncratic income. The ability to accomplish such efficient risk sharing presupposes of course the existence of elaborate mechanisms to verify states and efficiently side-contract to redistribute resources between individuals in every state of the world. In a market setting this requires the existence and efficient operation of S separate competitive asset markets to span the entire state-contingent commodity space. This is quite a requirement.

The framework above can be readily extended to multiple time periods in the fashion of Arrow and Debreu (1954) to allow state and time contingent income and consumption levels. Efficiency conditions similar to (3) would hold across time and states of nature. This implies village institutions will pool resources to buffer individuals from all idiosyncratic shocks and allow individuals to efficiently smooth consumption over time and in response to all idiosyncratic shocks. All profitable investment projects would be financed[3] and full 'separability' [Singh, Squire and Strauss (1986)] between household consumption and production decisions could be achieved. A version of the Modigliani–Miller Theorem also holds: in a world of complete markets the financial structure of the farm and of the economy more generally becomes both indeterminate and irrelevant. Just as there are an infinite number of sets of linearly independent vectors that can span a vector space of dimension S, there are as many assets that could be traded to span a state-contingent commodity space of the same dimension. Since no prediction can be made as to which of many possible sets of assets will actually be traded, no interesting comparative institutional analysis is possible. More precise predictions about the contractual structure of the economy only emerge if one adds trading frictions.

3.2. Empirical tests of efficient risk sharing

Few economists believe that the Arrow–Debreu world of complete markets provides a very accurate description of the global economy. On the other hand, many economists implicitly endorse the idea of efficient risk-sharing at smaller units of analysis, for example whenever they treat households as unitary actors, since this presupposes efficient consumption pooling. The possibility that efficient consumption pooling might

[3] Where profitability would be measured by the village's internal efficient market rate of return.

be achievable in the somewhat larger unit of an idealized 'village' economy where community members are assumed to have good information about one another enjoys considerable intellectual appeal.

The complete markets model yields a number of hypothesis that researchers have sought to test against empirical data. In a well-cited study of Indian villages, Townsend (1994) regressed household consumption on household income, village aggregate consumption, and a number of other variables. Under the null hypothesis of full risk sharing household consumption ought to be highly correlated to aggregate village income but independent of household specific shocks. His results indicated a considerable amount of risk pooling, but the hypothesis of full consumption smoothing was clearly rejected as individual households' consumptions appear to adjust considerably to idiosyncratic shocks. Using the same data and more robust methods, Ravallion and Chaudhuri (1997) also conclude that there is evidence against complete risk sharing.

Broadly similar methods have been used to examine risk sharing in a wide range of different social groups, including families, ethnic groups, and neighbors. Examples include Jalan and Ravallion (1998) using data from China, Grimard (1997) using data from Côte d'Ivoire, Suri (2003) using data from Kenya, Kazianga and Udry (2006) using data from Burkina Faso, Dercon and Krishnan (2000) using data from Ethiopia, and Gertler and Gruber (2002) using data from Indonesia. In each case, the hypothesis of Pareto-efficient risk sharing within the relevant social group is rejected, though some evidence of partial risk sharing is usually found.

Looking more directly at the transfers between households, Udry (1994) arrived at similar conclusions for households in Northern Nigeria, and Fafchamps and Lund (2003) found evidence of only limited risk-sharing in rural Philippines. Udry and Duflo (2004), Goldstein (2004) and Dercon and Krishnan (2000) report furthermore that they can reject the hypothesis of efficient risk sharing even *within* the same households in rural Côte d'Ivoire, Ghana, and Ethiopia, respectively. All of these studies point to forms of imperfect consumption smoothing and to the existence of more effective risk-sharing within particular subgroups or networks within a village. Kinship, family, clan or religious affiliation may be important because these groups can threaten to impose larger punishments on individuals break commitments to mutual insurance arrangements.

3.3. Consequences of imperfect financial markets

Even if a small tight-knit group could accomplish the feat of efficient risk pooling, individuals would still very likely remain exposed to substantial risks because the very physical proximity and closeness that is required of agents to be able to enforce state-contingent risk-sharing arrangements will typically expose these individuals to correlated risks. For example, a shortfall of rain is likely to affect most of the agricultural households in the same small dryland farming community. Locals will want to exchange risks with individuals outside of the village.

Since realistically it would quickly become prohibitively costly and complicated for each individual to separately contract directly with each of many hundreds of other individuals spread out over large distances, it is natural to expect this to create demand for the entry of specialized financial intermediaries to help lower the transaction costs of pooling risks and in this way help society to further complete the market and reap the gains to financial trade.

The entry of efficiency-enhancing financial intermediation may, however, be delayed or complicated for several reasons. The first problem is that outside financial institutions (FIs) are just that: outsiders that may not have the kind of local information and enforcement mechanisms necessary to verify and enforce detailed state-contingent contracts within the village. As reviewed in more detail below, this may end up severely limiting the set of feasible contracts a FI may be willing to offer.

Hence, either because the members of their trading networks face correlated risks and/or because financial contracting is incomplete within villages, households and individuals in rural areas are likely to be left facing considerable residual risk. This leads households to search for and adopt other, possibly quite costly, strategies to smooth income or consumption. It also creates significant latent demand for financial trade with outsiders. Income smoothing strategies include scattering plots [McCloskey (1976); Townsend (1993)], choosing a lower return but more diversified mix of crops and non-farm production activities, migration and marriage patterns [Rosenzweig and Stark (1998)], the adjustment of intertemporal labor supply in response to shocks [Kochar (1999)], labor bonding and debt peonage [Srinivasan (1989); Genicot (2002)], and many other choices. While some of these strategies might be used alongside financial trade even in a world of complete markets, when financial markets are incomplete the neat separation between household production and consumption choices will be broken, leading individuals to make costly and inefficient adjustments to production and investment plans with obvious welfare consequences.

Research pointing to evidence of such costly strategies is vast, and we will mention only a few prominent examples. Kochar (1999) showed that over three-quarters of the correlation that Townsend (1994) found between household and village aggregate consumption could be accounted for by the households' increased supply of labor to the agricultural wage market following a shock to their farm production. In other words, when hit by an idiosyncratic production shock, households appear to have smoothed consumption by smoothing income rather than via financial transactions as many readers of Townsend's work might have assumed. Adjustments to labor supply plans can of course be highly disruptive, particularly if they disrupt human capital formation projects. Jacoby and Skoufias (1997) is just one of many studies that finds evidence that children in poor households work more and attend school less in response to idiosyncratic income shocks [Jensen (2000); Beegle, Dehejia and Gatti (2003); Duryea, Lam and Levison (2003)].

To complement the literature that documents the extreme degree of fragmentation in rural financial markets (Section 2.1 above), there is some research that attempts to measure the extent of credit rationing in these imperfect rural financial markets. Many surveys have found that farmers claim that they would borrow more if additional credit were available at a given interest rate [Zeller, Diagne and Mataya (1998);

Diagne and Zeller (2001)]. However, it is sometimes difficult to interpret these counterfactual responses. Some papers have attempted to use econometric methods to measure the extent of rationing in rural credit markets [Bell, Srinivasan and Udry (1997); Kochar (1997)], but these estimates are based on identification assumptions that must be considered to be tentative. Moving our focus beyond rural finance, Banerjee and Duflo (2004) show that an arguably exogenous increase (followed by a decrease) in the availability of credit to a set of firms who borrow from a particular Indian bank was associated with an increase (followed by a decrease) in output of those firms, providing well-identified evidence of credit constraints.

Amartya Sen (1982) points to the importance of land as an asset for smoothing income when he wrote "a small peasant and a landless laborer may both be poor, but their fortunes are not tied together". He has argued that wage labor markets often collapse rapidly at the outset of a famine and whereas households with land can often fall-back upon this or other assets for subsistence purposes to buffer the shock, wage laborers have few other assets to work with. Land may also 'entitle' the owner to a larger share of the communities' diminished resources than the landless. For example, the household with land may be able to borrow in a crisis, whereas the landless or those with low social standing may not.

Households may also try to smooth consumption by accumulating or decumulating physical buffer stocks of assets such as animals, grain, land, or jewelry. When the assets are used directly in production and there are incomplete or missing rental markets, consumption-smoothing can again come at the cost of productive efficiency. Rosenzweig and Wolpin (1993) document this cost when partial consumption smoothing is achieved by households in the ICRISAT India villages through the sale and purchase of bullocks that are used in production. Kazianga and Udry (2006) provide evidence of a similar cost for households in the Burkina Faso villages surveyed by ICRISAT.

There is a broad consensus of the large empirical literature on risk and household responses to risk in rural areas of less developed countries that most households succeed in protecting their consumption from the full consequences of their risky environment. However, they do not do so to the full degree implied by either Pareto-efficient risk pooling within specified communities or by strict versions of the permanent income hypothesis.

Imperfect financial markets also shape production organization more generally. It has been clear since the development of the standard agricultural household model [e.g., Singh, Squire and Strauss (1986); de Janvry, Fafchamps and Sadoulet (1991)] that the organization of production on the household farm depends upon the nature of the financial markets available to the household. The same literature shows that household labor supply response to price changes, for example, depends upon the household's access to financial markets. Eswaran and Kotwal (1986), for instance, show how access to capital, which in turn is related to the initial distribution of land, may shape equilibrium patterns of production organization, including whether land is worked by wage laborers or tenants, and the efficiency of production. Carter and Zimmerman (2000) analyze a dynamic version of this model.

More generally, the structure of rural economic relations itself depends upon the nature of available financial contracts, which in turn of course depend upon the structure of rural economic activity. This joint causation opens up the possibility of a wide range of potential equilibria, and an important research agenda.

A series of important papers have examined the role of financial market imperfections in generating a persistent non-degenerate income distribution [Loury (1981) is an early paper that raises this issue]. Galor and Zeira (1993) and Banerjee and Newman (1993) show in an economy characterized by non-convexities in investment, capital–market imperfections can cause initial disparities in wealth to persist across generations. Moreover, the distribution of wealth affects aggregate patterns of economic activity and growth, decisively breaking down any potential separation between "efficiency" and "equity" in the analysis of economic policy.

There has been a flowering of related theoretical work on linkages between distribution and growth when financial markets are imperfect. Mookherjee and Ray (2003) and Matsuyama (2002) are also good points of entry into this literature.

More recently, there have been some attempts to link these theoretical insights to data. Banerjee and Duflo (2004) examines an array of reduced-form implications of these theories, with special emphasis on data from India. There is a rapidly growing literature that examines the empirical implications of models of the endogenous growth of financial intermediation in the context of dynamic general equilibrium models with heterogeneous agents. The most recent important papers include Jeong and Townsend (2003), who examine the microeconomic underpinnings of growth models with imperfect capital markets using data from the repeated cross-sections provided by the Thai Socio-Economic Surveys. Felkner and Townsend (2004) undertake an exercise with a similar objective, but instead using a repeated census of villages in Thailand, and with a strong emphasis on spatial relationships that are generally ignored in the theoretical literature.

3.4. Contracting under asymmetric information and imperfect enforcement

These observations lead to several obvious questions: Why do financial markets and risk sharing arrangements often fail to achieve efficient exchange even in small village communities? What explains the structure and organization of actual financial markets? Why are diversified outside financial intermediaries such as banks and insurance companies often reluctant or slow to enter rural financial markets?

The defining characteristic of all financial contracts is that they involve the exchange of state-contingent promises or IOUs. But the fear that promises may be broken can limit the set of credible promises that a would-be issuer can commit to keeping. In a world of complete markets this problem was abstracted away by simply assuming that all potential contract breaches could be immediately detected and costlessly deterred, but most of the modern literature on financial contracting focuses on how asymmetric information and limited enforcement problems may together limit the set of feasible commitments.

This theory has proven powerful and rich at providing insights with which to interpret the shape of real world financial contracts and institutional arrangements.

While the theoretical literature on asymmetric information and imperfect enforcement is rich, there has been comparatively little empirical work that attempts to characterize the exact nature and extent of imperfect information in rural financial markets. Chiappori (2000) is a useful review of relevant literature in the developed country context. Aleem (1990) provides dramatic direct evidence of the importance of screening costs for lenders. Klonner (2004) shows that asymmetric information has dramatic consequences for bidding patterns in (high-value) ROSCA auctions in a village in Southern India. Gine and Klonner (2003) examine the role of imperfect information regarding borrower type for the structure of financial markets in a coastal village in Tamil Nadu. They show that uncertainty about (fishing) entrepreneurs' ability slows the pace of costly technological innovation for relatively poor entrepreneurs. Karlan and Zinman (2004) use a randomized intervention to identify the extent of adverse selection and moral hazard in a South African credit market. They conclude that about 40% of defaults in this market can be attributed to one of these types of asymmetric information.

Asymmetric information makes it difficult for a would-be creditor or insurer to be sure whether the expected probability distribution over state-contingent payoffs associated with a contract promise is the one being represented by the seller or not, as in the case of adverse selection (private information about the agent or the project's characteristics) or moral hazard (private information about whether a specified action or contingency has occurred or not). In practice variants of each of these problems may be the concern.

A farmer may promise to work diligently to repay a loan but when that farmer's harvest fails and he declares a default a lender may not be able to tell whether this was due just to bad luck or to the farmer's mishandling of the loan. Lenders and insurers may also not be able to very easily verify whether the farmer's reported harvest failure is genuine or mis-represented. In each of these cases the problem turns around to bite the borrower or the insurer who will have a hard time obtaining credit or insurance from any source in the first place unless they find a way of credibly signaling their commitment.

Problems of commitment can also arise however even when information is perfect and symmetric because even though actions and outcomes are observed agents may still be able to simply renege or walk away from their commitments unless they face credible and effective sanctions to dissuade such opportunistic default. Some literature refers to this last problem of opportunistic default as the problem of 'limited commitment' [e.g., Ligon, Thomas and Worall (1999); Paulson and Townsend (2003)] yet many contracting problems involve an agent's limited ability to commit to fulfilling elements of a contract, whether it be to truthfully reveal their type (adverse selection), to take a specified action (ex-ante moral hazard), to truthfully report an outcome (ex-post moral hazard), or to deliver on a promise (opportunistic default).

Each of these problems is related and are all believed to play important roles in shaping the pattern of financial contracting everywhere. A very large literature now exists that has studied these problems [textbook treatments include Salanié (1997); Macho-

Stadler and Perez-Castrillo (2001); Laffont and Martimort (2003)] and the manner in which each of these problems separately contributes to shaping the set of feasible financial contracts in exclusive bilateral exchanges is by now quite well understood. There is still much new research however left when it comes to trying to understand what shapes the equilibrium pattern of financial contracting for interactions between individuals and sub-coalitions within and across larger groups.

Since this literature is vast our focus will be of necessity selective. Although we will touch briefly on the problems of adverse selection and opportunistic default along the way, we shall organize a good deal of the discussion of some of the more complicated issues of multi-agent interaction around a set of variants of a simple model moral hazard. Since in all cases the problem is one of limited commitment, each of these other contracting problems will tend to invoke similar concerns and will often be addressed with related contractual solutions (e.g., the use of collateral, monitoring, multi-period contracting, 'interlinked contracts', etc.). Even with this simple model a rich picture emerges of the structure and operation of rural credit and insurance markets, the role of financial intermediaries and the challenges they face in operating in rural areas, as well as the role of government and public policy.

3.5. Moral hazard

Stiglitz (1974) laid out one of the earliest modern treatments of moral hazard in an important paper that sought to explain the age-old question of why in some contexts sharecrop contracts might dominate fixed-rent tenancy contracts. Key assumptions driving Stiglitz' analysis are that a tenant/worker's effort choices, which affected the distribution of project outcomes, are costly and cannot be observed by the landlord, and therefore could not be specified directly in a contract. If the worker was offered either a full insurance (fixed-wage) or partial insurance (sharecrop) contract the worker had an obvious incentive to choose a lower effort level (while claiming otherwise) since he then avoided effort disutility without having to bear the full consequence of that lowered effort on expected output. Classical economists from Adam Smith to Alfred Marshall had puzzled over why such seemingly inefficient contracts persisted in practice.

In the case of a risk-neutral agent, the well-understood contract to avoid this moral hazard or 'Marshallian inefficiency' problem was to offer the tenant a fixed rent (or fixed debt) contract of the form $c_{js} = x_{js} - R$ to make him a full residual claimant. This made the agent bear the full marginal benefit and the full marginal cost of his effort choices and hence aligned the agent's incentives with those of the creditor/landlord. What Stiglitz pointed out, however, was that if the farmer was risk-averse this solution imposed too much risk on the agent to be optimal. A tradeoff existed between providing incentives and sharing risks and Stiglitz argued that a sharecrop contract might strike the right balance between the two.

It is useful to briefly review the key elements of this well-studied model but recasting it using a parameterized distribution approach to permit more flexible sharing rules than Stiglitz' original linear contract assumption. A single agent now contracts exclusively with another villager or financial institution (FI) that is assumed to be large and diversified enough to be modeled as a risk-neutral principal. Farm projects require an investment I and the agent can take one of two possible effort choices: either 'high' effort (e^H) or 'low' effort (e^L), such that $V(e^H) > V(e^L)$ where V measures the agent's disutility of effort. To simplify further I is a fixed lump sum amount and we assume the villager must borrow this entire amount. Higher effort choice leads to a higher expected project return,[4] and we assume $E[x_{js} \mid e^H] > E[x_{js} \mid e^L]$ where $E[x_{js} \mid e] = \sum x_{js}\pi_{js}(e)$.

On a competitive financial market, FI's would compete to offer exclusive loans of size I and each villager would end up choosing their most preferred feasible contract among these offers. The optimal contract therefore maximizes expected borrower utility subject to the constraints of providing clear incentives for the agent to commit to high effort and to make expected repayments sufficient to cover the lender's opportunity cost of funds:

$$\max_{c_{js}} Eu[c_{js} \mid e^H] - V(e^H), \tag{4}$$

$$E\big[(x_{js} - c_{js}) \mid e^H\big] \geqslant I(1+r), \tag{5}$$

$$Eu[c_{js} \mid e^H] - V(e^H) \geqslant Eu[c_{js} \mid e^L] - V(e^L), \tag{6}$$

where (5) is the lender's break-even or participation constraint requiring that expected repayments at least cover the opportunity cost of funds,[5] and (6) is the borrower's incentive compatibility (IC) constraint requiring that the borrower expect to earn more under the contract from choosing high effort compared to low. The optimal sharing rule c_{js} that solves this problem will be characterized by the following well-known first order conditions, one for each state s:

$$\frac{1}{u'(c_{js})} = \lambda + \mu\left[1 - \frac{\pi_{js}(e^L)}{\pi_{js}(e^H)}\right], \tag{7}$$

where λ and μ are, respectively, the Lagrange multipliers on the lender's participation constraint and the borrowers incentive constraint. When the agent's effort choice is verifiable the IC constraint would not bind ($\mu = 0$) and the efficient contract will equalize the farmer's marginal utility of consumption across all states, $u'(c_{js}) = 1/\lambda$ which requires guaranteeing the farmer a constant level of consumption $c_i = \bar{c}$ in every state. This is just an adaptation of the earlier conditions (3) to the case of a risk neutral FI. One interpretation is that the farmer 'sells the farm' and uses the proceeds to finance current investment I and obtain income to guarantee fixed consumption in the following period. In Stiglitz' analysis the risk-neutral landlord owned the project and hired a fixed wage laborer (in that context $I > 0$ could be interpreted as a wage advance).

[4] For the moment the event space S is assumed to consist simply of the set of possible outcomes on the farmer's project at each level of e and I. This will later be relaxed.

[5] One could also interpret a contract with $I \leqslant 0$ as having the farmer making a first period payment to purchase second period insurance.

When effort is not contractible and the IC binds, full insurance will not be possible. The agent's consumption in each state must now be tied to the inverse likelihood ratio $\pi_{js}(e^L)/\pi_{js}(e^H)$ in expression (7). The optimal sharing rule attempts to reward those outcomes that have the highest likelihood ratios – i.e., those most likely to have resulted when the agent chooses the effort level the contract wants to implement rather than a lower effort level – and punish those outcomes with low likelihood ratios, all tempered by the competing objective of not imposing too much costly risk on the agent. Unfortunately, one of the few clear results to emerge from this literature is that the optimal sharing c_{js} will in general be non-linear and strongly influenced by the underlying characteristics of distribution $\pi(s, e)$ and how it responds to changes in effort [Grossman and Hart (1983)]. Deriving even the simple property of monotonicity – that the farmer's return be non-decreasing with the size of the project outcome x_{js} – requires making rather strong distributional assumptions. Specifically it requires assuming a monotone likelihood ratio property (MLRP) that the expression in brackets on the right hand side of (7) be monotonically non-decreasing in output x_{js}. Intuitively, higher output levels must provide stronger signals that the agent chose a higher effort.

The prediction that contracts should be highly state-contingent has led some observers to point out somewhat of an empirical puzzle. Theory predicts non-linear and highly state-contingent optimal sharing rules that at first glance do not seem much like the simple linear sharing rules (e.g., linear sharecrops or fixed debt contracts) often described in rural contexts [Allen and Lueck (2002)]. There have been different responses to this challenge. One approach has been to point to other constraints and trading frictions, for example problems of state verification [Townsend (1979)], limited liability [Innes (1993)] and/or contract renegotiation [Matthews (2001)] place additional restrictions on the range of feasible contracts and this might help explain these simpler contract forms. Another response has been to argue that real world contracts are in fact far more state-contingent than what first meets the eye [Townsend (2003)]. For example, it is not uncommon for a lender to allow a borrower to miss a couple payments, or even to forgive a portion of the loan if the borrower has fallen on bad luck. This idea has been also explored extensively in the literature on multi-period contracting and sovereign debt lending [Grossman and Van Huyck (1988)]. Once one takes such excusable defaults into account, contracts which on the surface appear linear start to look far more state contingent, and more like theory predicts. Udry (1994) provides empirical evidence documenting a high incidence of excusable state-contingent default in rural loans in Nigeria.

Another important property of the optimal contract that explains important features of many agricultural contracts is Holmstrom's (1979) sufficient statistic result that demonstrates that optimal sharing rules should be tied not only to the outcome of the farmer's own project but also to any other signal from the environment that helps the principal draw a sharper inference about the agent's choices. For example, a lender ought to be more willing to rollover a debt following a bad harvest outcome on a farmer's project if other farmers in the area also had low harvests, but less willing if other farmers had good harvests. The purpose is to better filter signals so as to attempt to reward or punish borrowers' only for those outcomes over which they exert some control and insure them against those over which they do not. The result will be more cost-effective incentives and better insurance. This logic of tying contract terms to other verifiable signal that leads to sharper inferences has been evident in the design and regulation of agricultural contracts for centuries as is evident for example from Hammurabi's code (circa 1795 BC) which stated that "if any one owe a debt for a loan, and a storm prostrates the grain, or the harvest fail, or the grain does not grow for lack of water; in that year he need not give his creditor any grain, he washes his debt-tablet in water and pays no rent for this year".

Relative-performance evaluation (RPE) contracts, which make one farmer's reward a non-increasing function of observed outcomes on other agent's projects, build on this insight [Mookherjee (1984)]. There is convincing evidence to suggest that RPE contracts are ubiquitous and play an important role in many types of agricultural labor and financial contracts. For example, RPE contracts that tie a farmer's returns to industry averages of yield or quantity are commonly used in livestock raising and agro-industry commodities [Knoeber and Thurman (1994)]. Hueth and Ligon (2001) argue that relative performance incentives are also built into many other types of contracts via payment mechanisms that depend on market prices.

The analysis so far has implicitly assumed that (a) the agent contracts exclusively with one principal, (b) project outcomes can be observed and costlessly verified and output-contingent commitments can be costlessly enforced. The next sections discuss the consequence of relaxing both assumptions and extends the analysis to multi-period contracting.

3.6. Multi-period and repeated contracts, limited commitment, and reputation

Lambert (1983) and others have shown how the basic one period moral hazard problem can be extended into a multi-period environment with commitment. When either or both parties can commit to a multi-period sharing rule there is scope for improvement over the one-shot contract. The optimal multi-period optimal sharing rule can be interpreted as a sort of 'reputation' updating mechanism in which the amount of state-contingent default (insurance) that a creditor is willing to provide a borrower following a bad realization in any given period is made to depend in part on that borrower's past history of realizations. Contracts will have 'memory' in the sense that agents who had good (bad) realizations in the recent past will be rewarded (punished) by raising (lowering) the return they can expect following any future realization. A good reputation is like an earned privilege that provides the agent with access to future surplus. The prospect of earning, or the fear of losing, this privilege can act as an effective incentive to economize on present period incentives. The ability of the principal to commit to delivering rewards for current or past good behavior allows for the provision of both better incentives and more insurance over the life of the contract compared to a series of one period.

The longer such an agency relationship can be expected to last the more the incentive problem can be alleviated. These findings are consistent with intuition and with empirical studies such as Sadoulet, de Janvry and Fukui (1997) who found that landlords in the Philippines who contracted with tenants with whom they shared kin relationships (which among other things could proxy for the length of the expected relationship) were more likely to offer insurance within multi-period tenancy contracts.

These results depend crucially however on the assumption that each party can commit to not renegotiate or abandon their exclusive multi-period commitments. Ex-ante efficient choices are sustained by the ability of both parties to commit to not renegotiate ex-post inefficient outcomes. Without an ability to make such commitments, finite multi-period contracts cannot improve on a series of one-period contracts [Fellingham, Newman and Suh (1985)].

Commitments of any sort are often difficult to enforce via third parties or the courts. If third party enforcement is not possible, then contract obligations need to be *self-enforcing* – they need to be sustained via incentives built directly into the contract. The simple moral hazard problem illustrates how a commitment to implement a particular effort level might be sustained via incentives fashioned out of the verifiable output-contingent commitments that the principal and agent are assumed to be able to enforce directly. But sometimes even output-contingent promises will be difficult to sustain. For example, a farmer might try to hide or under-report the true outcome on his project or, even if the farmer's project could be perfectly observed, he may simply choose to default on his repayment obligation.

A large literature has studied conditions for the emergence of self-enforcing lending and mutual insurance arrangements in the context of non-cooperative indefinitely repeated games. In the simplest setting a farmer wants to obtain a loan of fixed size I. A lender will only participate if she can expect to be repaid $I(1+r)$. The loan funds a project with certain outcome $x \geqslant I(1+r)$. Financing would be efficient except that in the absence of any exogenously enforced social sanction the farmer's dominant strategy in a one-shot interaction is to take the loan and then default. Anticipating this, the lender's dominant strategy is not to lend in the first place. If, however, the interaction is repeated over an indefinite horizon it may be possible to generate incentives for the farmer to continue to repay if the threat of loan non-renewal is credible and sufficiently punishing. Suppose the farmer has a time-separable utility function with discount factor δ, that he gets zero utility in each period that he fails to get a loan, $y = x - I(1+r)$ in the periods he repays, and x when he defaults. Cooperation (repayment in every period) can then be sustained as a sub-game perfect Nash equilibrium (SPNE) so long as the following incentive constraint is met in every period:

$$\frac{u(y)}{(1-\delta)} \geqslant u(x).$$

Cooperation is more likely to be sustained the more the borrower values future consumption and the lower the size of the expected repayment $I(1+r)$ relative to x. The lender's threat of cutting off the borrower from future access is viewed as credible because this 'grim strategy' is itself a subgame-perfect equilibrium.

Notice again the importance of the implicit assumption that the borrower and the lender have an exclusive relationship. A problem arises if the lender cannot commit himself not to renegotiate after a default has occurred. If the lender could potentially make a profit by lending again to the borrower who has defaulted, then perhaps the 'grim strategy' punishment could be renegotiated in such a way that, ex-post, both parties are better off. Of course, this very possibility could restrict the penalties that can be sustained in equilibrium leading to the collapse of financial trade. We will shortly return to the issue of renegotiation.

A similar problem arises if the borrower and lender do not have an exclusive relationship. If a new potential lender were suddenly to appear on the scene, the above relationship might be undermined by the fact default may now be a less severe punishment because a borrower who defaults on one lender may now start up a relationship with a new one. Of course, the second lender will face the same problem as the first, so ironically, the mere presence of a competitor – the inability to commit to exclusive arrangements – can lead to the complete collapse of financial trade.

Lending could be restored if lenders could share default information and agree to collectively punish a defaulter. But this presupposes that others will punish a lender who does not herself punish the defaulter. In societies that have well functioning publicly funded court systems such exclusivity arrangements can often be exogenously enforced, for example by allowing existing creditors to establish liens over a farmer's future harvest or existing property. Such institutions however do not exist or work properly in many contexts. In such cases 'community punishment' arrangements have to themselves also be self-policing. Kandori (1992) pioneered the analysis of equilibria in which defection by one agent leads to sanctions by others and in which the "social norm" to punish is itself sustained via self-interested interactions. The analysis points to the likely emergence of public institutions for information dissemination such as labels that indicate reputation, membership, or license, which are revised systematically. Greif (1993) is an early example of the usefulness of this reasoning for understanding enforcement mechanisms in trade relations. La Ferrara (2003) adapted this sort of framework to conduct an interesting theoretical and empirical analysis of community enforcement across generations within kin groups in rural Ghana.

Kimball (1988), Foster (1988), and Coate and Ravallion (1993) extended the study of self-enforcing (exclusive) mutual insurance arrangements in agricultural economies when farmers operate risky projects. These showed how closely mutual insurance 'cooperatives' of different numbers of borrowers could get to the efficient risk sharing condition (3) depended on how heavily farmers discounted the future, the nature of their risk aversion, and the variance of underlying project returns. Using simulations, Kimball demonstrated that it was easier to sustain mutual insurance arrangements with a larger number of farmers but harder to enforce full risk sharing in a larger group. Coate and Ravallion demonstrated that under reasonable assumptions the extent to which such arrangements diverged from efficient risk sharing decreased as mean incomes rose, increased as incomes became more covariate, and increased with the inequality among members. One of the results to emerge from this literature is the rather high sensitivity of the performance of these arrangements to small perturbations in parameter values, for example to small changes in risk aversion, suggesting ‘wildly divergent performances of the moral economy’ [Coate and Ravallion (1993, p. 19)].

The problem of sustaining self-enforcing financial contracts when agents cannot make binding commitments not to renegotiate terms ex-post has received a good deal of recent attention. An important contribution of Coate and Ravallion (1993) was their characterization of a renegotiation–proof mutual insurance equilibrium for two households in a risky environment. They restricted attention to stationary transfers. Kocherlakota (1996), Ligon, Thomas and Worall (2002), and Genicot and Ray (2002) among others have made important extensions of this work to dynamic contracts and to many agents. The term ‘quasi-credit’ is often used to describe the arrangements that emerge because they can be interpreted as informal loans with implicit repayment made contingent on the lender’s needs and the borrower’s ability to repay [Fafchamps (1999, 2004)].

Ligon, Thomas and Worall (2002) provide an explicit characterization of such reciprocal transfer arrangements in an economy that allowed for both idiosyncratic shocks and common shocks that could be correlated over time. We showed in Equation (3) that when the agents can commit to mutual insurance, the Pareto-efficient allocation fixes the ratio of marginal utilities between any two agents. Ligon, Thomas and Worall (2002) show that the constrained-efficient allocation in a limited commitment environment involves intervals of the ratio of marginal utilities. For any given state of nature (which describes the incomes received by any two households i and j), there is an interval of the ratio of marginal utilities that can be supported in the renegotiation–proof equilibrium. The endpoints of this interval are determined by the size of the transfer that can be supported in the equilibrium: if too large a transfer were required from i to j, i would not find it in her interest to continue participating in the mutual insurance arrangement. If last period’s ratio of marginal utilities falls within that interval, then the ratio remains constant. For those two periods, at least, the allocation looks just like the full commitment equilibrium described in (3). However, if last period’s ratio of marginal utilities falls outside the interval, than it can no longer be supported in equilibrium. The ratio of marginal utilities in this period will instead be at the endpoint of the interval of ratios that can be supported, and the party that is transferring resources to the other is just indifferent between continuing the arrangement and defaulting. Ligon, Thomas and Worall (2002) test the dynamic limited commitment model against the alternatives of full insurance [and against a static limited commitment model similar to that of Coate and Ravallion (1993)] using the same dataset that Townsend (1994) had used to test risk pooling in Indian villages. They find evidence that the dynamic limited commitment model fits the data better than either alternative.

Several of these ideas have also been explored rather extensively in the literature on sovereign debt lending [Grossman and Van Huyck (1988); Kletzer and Wright (2000)]. These sorts of frameworks have also been applied to games with repeated moral hazard, or as the literature calls them, games with imperfect monitoring [Abreu, Pearce and Stacchetti (1990)], where contract terms must now be chosen to also motivate hidden actions. Paulson and Townsend (2003) have tested between models of ‘limited commitment' and moral hazard using data from a survey of rural and semi-rural households in Thailand. They find evidence that both types of imperfection matter but that 'limited commitment' concerns are more dominant for poorer households while moral hazard becomes more important as household wealth increases.

There is also a set of less formal observations regarding patterns of mutual insurance in rural areas of developing countries that point to the likely importance of imperfect enforcement, asymmetric information and limited commitment. Prime among these is the observation that mutual insurance tends to be most effective within relatively narrowly-defined groups. Fafchamps and Lund (2003) provide evidence that mutual insurance among rice farmers in the Philippines flows through social networks of friends and family. Murgai et al. (2002) show that irrigation water insurance in Pakistan is highly localized among clusters of close friends and family. If random shocks tend to be correlated among these kinds of social groups, any restriction of mutual insurance to such groups is costly in terms of the value of the insurance thereby provided. The relative ease of observing the realizations of random shocks, or of enforcing agreements within narrowly-defined social groups is probably the most important reason for this observation. Genicot and Ray (2002) provide a useful formal treatment of the endogenous formation of mutual insurance groups.

3.7. Limited liability, collateral and its substitutes

A striking empirical fact about the operation of rural financial markets that has already been mentioned is how markedly the conditions of access can vary across households and how closely financing terms are tied to production activities. It is not uncommon to encounter situations where some farmers in a given region finance the bulk of their crop activities with commercial bank loans while smaller nearby farmers growing the same crops only finance with retained earnings or via more expensive informal moneylenders. It is also not uncommon to observe farmers who can obtain very generous financing from product traders or contract farming firms for certain crops, but almost no financing at all for other profitable crop activities grown in the same area but marketed through different channels.

If markets where complete all socially profitable investment projects would be financed regardless of the initial asset holdings of the borrower or the type of crop activity. In practice, however, the terms of loan access tend to be frequently tied to the borrower's existing asset position and production mix because agricultural lenders ask for land or chattel property mortgage pledges or other guarantees. Even when no such formal pledges are made lenders may simply prefer to deal with farmers with proven assets and/or more diversified cash flows. When this is the case the initial distribution of assets can have important effects on the structure and performance of the real economy and the number and types of financial contracts and intermediaries that can emerge. To build a theory of these issues one has to understand the role of limited liability and collateral.

Consider again the simple one period moral hazard problem but suppose now that the farmer is risk-neutral. Since there is no longer a tradeoff between insurance and incentives, the first-best solution is to make the agent a full residual claimant. A fixed debt contract (FDC) that obliges the borrower to repay $I(1+r)$ regardless of the project outcome offers such a solution, leaving $c_{js} = x_{js} - I(1+r)$ to the agent. This will not be feasible, however, if the agent is unable or unwilling to make the full required fixed repayment $I(1+r)$ in some low outcome states. This would happen, for example, if project returns x_{js} plus all of the borrower's additional (or 'collateral') cashflows and liquid property assets are simply not sufficient to cover the fixed repayment amount. To remain feasible contracts must therefore satisfy limited liability constraints of the form $R_{js} \leqslant x_{js} + A_{js}$ for all s where A_{js} is the maximum value of collateral resources a farmer can credibly pledge to transfer to the creditor in state s.

In a seminal paper Stiglitz and Weiss (1981) studied the impact of adding such constraints on the shape of feasible equilibrium contracts. By placing an upper bound on an agent's exposure to consequences when projects fail, limited liability may end up encouraging agents to choose excessively risky (i.e., low effort) projects. In such circumstances, rather than help compensate for expected losses, increasing the loan interest rate may only aggravate the problem by reducing the agent's marginal reward to choosing higher effort. Stiglitz and Weiss showed that this may (but need not always) lead to a backward bending supply curve and equilibrium credit rationing. Banerjee (2003) discusses how this kind of mechanism can result in a 'multiplier' effect: relatively small changes in the economic environment (such as the opportunity cost of capital to lenders) can induce large changes in equilibrium contract terms. As interest rates rise, the pool of borrowers becomes more and more risky, leading to a quickly rising interest rate and the potential collapse of the market.

Stiglitz and Weiss (1981) presented their results using the state-space approach and assumed piece-wise linear Standard Debt Contract (SDC) of the form $R_{js} = \min(x_{js} + A, R)$ where the creditor receives fixed loan repayment R or seizes all of the borrower's project returns $x_{js} + A$ plus available collateral, whichever is smaller. This can be interpreted as a fixed debt contract with excusable partial default for all harvest outcomes below a certain level. Innes (1990, Proposition 2) worked out the analysis using the parameterized distribution approach and demonstrated, somewhat strikingly, that when limited liability constraints bind the optimal contract for a risk neutral agent is *not* a debt contract but rather an extremely fine-tuned Live-or-Die Contract (LDC) that lumps all of the agent's reward on the one project outcome with the highest likelihood ratio and punishes them up to limited liability constraint in all other states. Innes also demonstrated, however, that the more familiar Standard Debt Contract (SDC) form becomes optimal once a few reasonable additional assumptions are imposed. These are that feasible contracts also monotonicity constraints requiring that repayment levels are non-decreasing in the size of the measured project outcome (i.e., $R_{js} \geqslant R_{js'}$ for all $x_{js} \geqslant x_{js'}$) as well as the earlier mentioned MLRP condition. Monotonicity constraints can be justified as necessary to remove incentives for lenders to opportunistically sabotage or mis-measure farmer's project outcomes and/or to prevent farmers from side-contracting with other farmers to artificially raise measured project outcomes in ways that could harm the lender's interests. Since measurement disputes between farmers and traders are frequent in practice, the assumptions are not unreasonable.

The point of dwelling on this seemingly technical detail is to point to the potential pitfalls of analyses that simply *assume* rather than derive linear contracts forms. While Stiglitz and Weiss (1981) were prescient enough to assume a piece-wise linear standard debt contract form (which is enough to lead to the optimal SDC form) other researchers have sometimes proposed linear contract solutions that on later examination were shown to be sub-optimal in a more general environment. To illustrate, some papers have argued that limited liability provides an alternative explanation for the existence and prevalence of share-tenancy arrangements even in the case of risk-neutral agents [Shetty (1988); Basu (1992)]. The argument was that tenants with insufficient collateral to be able to meet fixed rent obligations to a landlord or creditor in low outcome states would be more likely to seek and obtain share contracts rather than fixed rent/debt contracts. This they argued might provide micro-foundations for Spillman's (1919) tenancy ladder hypothesis which saw younger, less well capitalized farmers entering into sharecrops before later graduating to fixed-rentals and eventually farming their own land as they accumulated wealth and experience. Spillman's hypothesis may well be true, but a sharecrop is always dominated by a SDC-style fixed rent contract with excusable partial default for low harvest realizations for risk-neutral limited liability constrained tenants because an SDC provides more high-powered incentives in those states where limited liability does not bind [see Ray and Singh (2001) for a related discussion].

Adding risk-aversion back into the analysis makes the optimal contract again considerably more state-contingent. It turns out to be difficult to derive many general characterizations regarding how optimal contract terms vary across risk-averse agents of different wealth and risk-tolerance levels because of potential counter-posed effects. Standard risk-sharing analysis suggests that the risk premium that must be paid declines with an agent's wealth under the common assumption of decreasing absolute risk aversion (DARA). At the same time, however, agents become less sensitive to a given difference in payoffs across states as wealth increases and this last effect makes it more costly to provide incentives in a moral hazard setting [Thiele and Wambach (1999)]. The relation between agent wealth and the cost of contracting can therefore in general be non-monotonic. Hence it is possible that a landlord would prefer to lease a plot of land to a poor tenant rather than to a medium-poor one, but would prefer an even wealthier tenant to either of them. If we now further add binding limited liability constraints to the problem the same landlord may switch to instead now preferring the medium poor tenant to the poor one. Boucher and Carter (2002) and Boucher and Guirkinger (2004) discuss some of these issues and apply the theory to explain how observable patterns of production specialization and credit and insurance contract choices in a rural setting change with borrower wealth. See also Mookherjee (1997), Thiele and Wambach (1999) and Madajewicz (2004) for related work.

Partly because of such complications, most models of financial intermediation study limited liability with risk-neutral agents. To streamline the survey that follows let's simplify the model further by assuming just two project outcomes and two effort levels.

The crop harvest is now the outcome of a Bernoulli trial. If the borrower is diligent the harvest succeeds and yields x_1 with probability p or fails and yields x_0 with probability $(1-p)$. If instead the farmer is non-diligent (chooses low effort), the probability of a good outcome falls to $q < p$. If we also define $B = V(e^H) - V(e^L)$ to denote the opportunity cost of high effort, the incentive constraint (6) can now be written as

$$pc_1 + (1-p)c_0 \geqslant qc_1 + (1-q)c_0 + B,$$

or, re-arranging,

$$c_1 \geqslant c_0 + \frac{B}{(p-q)}. \tag{8}$$

A borrower's consumption following a good outcome has to be made sufficiently larger than consumption following a bad outcome to generate incentives for the borrower to want to raise the probability of success by choosing diligence. Since the limited liability constraint, $c_0 \geqslant -A$, restricts how much the borrower can be punished for a project failure, the incentive constraint (6) can only be met by setting the borrower's good outcome consumption level to at least $c_1 = c_0 + b/(p-q)$. Since limited liability curbs the size of the borrower's assets that can be seized following bad outcomes, the incentive to be diligent must now be created by offering a costly incentive 'bonus' in the good state (in credit contract terms the interest rate must be lowered in non-default states).

Hence, if the limited liability constraint is binding, $c_0 = -A$, then the incentive compatibility constraint implies that a borrower with assets A must earn a minimum expected return of at least

$$E[c_s \mid e^H] = \frac{pB}{(p-q)} - A \tag{9}$$

if incentives are to be maintained.

If we normalize the farmer's reservation payoff (what they could earn elsewhere by not accepting the contract) to zero then expression (9) can be interpreted as the size of the limited liability rent needed to keep incentives in place. It measures the necessary minimum expected payment over and above the agent's (zero) reservation return that is needed to keep the borrower diligent. Substituting this into the lender's participation constraint (5) gives

$$E[x_s \mid e^H] - \left[\frac{pB}{(p-q)} - A\right] \geqslant I(1+r), \tag{10}$$

from which it is clear that the lender will only willingly participate and finance a borrower if the limited liability rent (9) does not become too large, since otherwise too little of expected project returns $E[x \mid e^H]$ will be left over to cover the lender's cost of funds $I(1+r)$. One way to assure a lender that this will not be the case is to insist that all borrowers post collateral A in excess of a minimum *collateral requirement*

$$\underline{A} = \frac{pB}{(p-q)} - \left[E[x \mid e^H] - I(1+r)\right] \tag{11}$$

which is the level of A that is just sufficient to make (10) bind exactly. If lenders compete for a borrower's business but only participate on profitable loans, then an optimal contract will have the borrower retaining $s_f = -\underline{A}$ when the project fails and $s_s = -\underline{A} + B/(p-q)$ following success.[6] The minimum collateral requirement grows with the size of the loan or the lender's cost of funds r and with the borrower's opportunity cost of diligence B. The collateral requirement will be smaller the larger is the expected project return under diligence and the 'safer' is the project under diligence compared to non-diligence (the larger is p relative to q). This last result suggests reasons why lenders may want to steer collateral-poor borrowers toward safer, but possibly lower return projects.

The practical problem with this method to obtain a borrower's commitment to diligence is that it may exclude a large number of borrowers with good projects but insufficient collateral. A good part of the vast literature on financial contracting and financial intermediation since Stiglitz and Weiss (1981), and arguably also a significant part of real-world financial innovation, can be understood as efforts to find new mechanisms to create collateral substitutes or, in slightly more technical terms, ways to relax incentive constraints so as to reduce the size of the limited liability rents that limit the range of feasible contracting. One prominent strategy is for the lender to use intermediaries or delegated monitors who can help reduce information asymmetries or engage in 'monitoring' and 'control' activities aimed at directly lowering the agent's return from moral hazard. Other methods include 'incentive diversification' strategies aimed at expanding the range of feasible punishments and rewards that can be brought to bear to sustain commitments at lower cost. The latter include contingent-renewal strategies such as the ones already discussed in the context of repeated games, as well as interlinked contracts and group loans.

3.8. *Property rights and credit supply*

An important reason that limited liability may be such an impediment to contracting is that in many societies property rights are ill defined or contested [Deininger (2003)]. In a bestselling book, de Soto (2000) argues that hundreds of millions of poor people in developing countries have de facto possession and local community recognition of property rights over housing, land and other sorts of assets which can be valued in hundreds of billions of dollars worldwide. Yet their lack of formal title limits their ability to leverage those assets on capital markets. Poor people's assets remain 'dead capital' terms, leaving them excluded from the opportunities and benefits that closer integration into competitive capital and product markets might provide. Over the past decade or so, and partly in response to this rallying cry as well as to the falling cost of new mapping and information technologies, governments and international organizations have promoted vast new property titling programs in almost every country of Latin America

[6] Innes (1990) shows how to generalize the analysis to contracts with multiple outcomes and effort levels.

and many other countries around the world, just as they have also liberalized financial markets and lifted most of the worst forms of financial repression.

Although increased property rights security has been hypothesized to lead to significant increased investment demand and credit supply responses, the empirical evidence of such effects remains surprisingly mixed. Feder and Feeny (1991) and Siamwalla (1990) present evidence of strong credit supply effects of land titling in several provinces of Thailand. On the other hand, studies done for India [Pender and Kerr (1999)], Paraguay [Carter and Olinto (2003)], Kenya [Place and Migot-Adholla (1998); Carter, Wiebe and Blarel (1994)] and Burkina Faso [Brasselle, Gaspart and Platteau (2002)] all found muted or insignificant effects of tenure security on either investment demand or credit supply. Some of these studies found that investment and credit supply effects, that at first appear positive and significant in simple reduced form regression equations, disappear once more careful attempts to control for property rights endogeneity are implemented. The issue is that title status might be influenced by farmers' investments as well as vice-versa. In their study of property rights titling on Paraguayan frontier lands, Carter and Olinto (2003) found strong evidence of a wealth bias: measured credit supply effects were significant but only for farmers above a certain wealth threshold. Possible explanations for the surprisingly muted credit supply responses found in many studies include that (a) formal credit and land sale markets may have still been thin or ill developed in the study areas due to lack of profitable opportunities, (b) land foreclosure remained legally difficult or costly even after reforms, (c) informal property rights systems already provided a good measure of security (particularly in some African regions) so the incremental effect of legal titling was small. A recent study by Field (2004) finds more marked supply responses to titling programs in an urban setting in Peru, where titling appeared to increase loan approval rates from public (but not private) lenders and, conditional on loan approval, reduced the borrower's cost of funds by an average of 9%.

The issue of legal impediments to land mortgages is complicated by the fact that restrictions on the alienability of land have often been imposed via the political process, often with the ostensible aim of protecting farmers from losing land foreclosures to moneylenders. For example, following riots by farmers in the Deccan region of India in 1875, the British colonial authorities passed new laws to protect farmers against land foreclosures by non-farming moneylenders and, a few years later, legislation to promote agricultural credit cooperatives [Darling (1925); Kranton and Swamy (1999)].

4. Rural financial intermediaries

Rural households and farm enterprises in developing countries obtain credit and insurance from a wide array of financial service providers including product traders, banks, cooperatives and mutuals, contract farming firms, and input suppliers, and they might also borrow informally from relatives, friends, landlords, shopkeepers or moneylenders. A defining characteristic of many of these financial transactions is that they involve 'active monitoring' [Tirole (2006)]. The aim is to keep agents focused on their efforts to improve the chances that their financed projects do not fail and/or to reduce the possibility that project cash flows may be diverted to other purposes rather than meeting promised repayments. Monitoring is used both as a substitute for, and in addition to, collateral guarantees and legal enforcement strategies.

With the exception of some moneylenders and other informal sources who may lend entirely out of their own funds, each of these financial service providers is typically a financial *intermediary*, financing loan advances using both their own capital as well as funds leveraged from other outside sources. They are in this sense also acting as delegate monitors for other outside investors. For example, a trader or input supplier may tap into credit advances from their own buyers or suppliers, from bank loans and overdrafts or via the discounting of bills. Larger enterprises such as agro-industrial or exporting firms may raise funds on national or international markets by selling stocks and bonds. Agricultural banks and cooperatives make loans out of own equity, but mostly using depositors' savings, or credit from other state or private lenders. Financial intermediation therefore can involve a long chain of monitored financial relationships with an investor at any given node in the chain only willing to onlend if they can be convinced that financial intermediaries further down the chain face the right incentives to carefully select and monitor borrowers and projects in ways that will uphold the value of the original investors' stakes.

Active monitoring takes many forms. For example, product traders and contract farming firms often release credit in installments timed to match the farmer's likely needs at different tasks over the crop season. An installment may be held up or sized down in response to farmer's actions to that date as perceived by the trader or company extension agent. Traders also typically make it their business to visit the farmer's fields at the time of harvest or during important input applications. Input suppliers in virtually all industries supply most of their trade credit to borrowers via in-kind loans rather than as cash advances – seed, fertilizer, or a voucher for transport services will be delivered to the farmer rather than cash [Watts (1994); Burkart and Ellingsen (2004)] – and this too can be interpreted as a form of monitoring. These practices clearly aim at making it more difficult for borrowers to divert credit or other resources to other private uses or, more generally, to raise the borrower's expected return to being diligent rather than non-diligent in the financed activity. It is in part because of the possibilities of achieving such monitored lending relationships that contract farming schemes have often been heralded as a promising mechanism for financing small farmers [Glover and Kusterer (1990); Glover (1994); Watts (1994); Carter, Barham and Mesbah (1996)].

Consider again the simple model of moral hazard and limited liability. We can characterize the optimal contract by finding an allocation of consumption that maximizes borrower expected utility $[pc_0 + (1 - p)c_1]$ subject to incentive compatibility, limited liability and lender participation constraints. The limited liability constraint remains unchanged: $c_0 \geqslant -A$. Now assume, however, that a local lender or his delegate is in a position to monitor a borrower at a cost m in a way that lowers the borrower's opportunity cost of diligence from B to $b < B$.[7] If the borrower is monitored, the incentive compatibility constraint (8) becomes more relaxed to $c_1 \geqslant c_0 + b/(p-q)$. For the moment the monitoring agent might be anyone ranging from a microfinance loan officer to a local trader-moneylender. Whether or not they have capital of their own to put at risk in the borrower's lending project will make a difference, as analyzed below.

Since an outside investor is typically not in a position to directly observe his delegate's monitoring actions, the delegate's remuneration must be tied to outcomes on the borrower's project to create monitoring incentives. Consider first the case of a trader-moneylender or contract-farming firm that monitors a borrower at cost m and invests I^m of its own capital, leveraging the remaining $I^u = I - I^m$ from an outside bank to finance a borrower's loan of size I. Consider also a monitoring contract that promises to pay the monitor $m/(p-q) - I^m(1+r)$ if the borrower's project succeeds and $-I^m(1+r)$ if the borrower fails. That is, the monitor sinks the full opportunity cost of his funds and earns a reward only when the project succeeds. The expected *monetary* cost of using a delegated monitor is therefore

$$\frac{pm}{(p-q)} - I^m(1+r) \geqslant m. \tag{12}$$

The inequality must hold if the monitor is to be willing to participate since m is the monitor's cost of participation (assuming zero reservation utility). From (12), and since $p/(p-q) > 1$, it is evident that the size of any delegation rent (i.e., expected payments in excess of the cost of monitoring m), and hence the total cost of delegating monitoring, increases with the size of m and can be lowered by asking the delegate to put more of their own capital I^m at risk.

Consider for the moment the case where $I^m = 0$, in which case the delegate has no capital of his or her own to place at risk. This would be the case of a typical loan officer who faces her own limited liability constraint that wage payments must be non-negative. Following again the steps leading to (11), but this time also considering the delegation rents needed to maintain monitor incentives, one arrives at a new minimum collateral requirement for the borrower, now as a function of m:

$$\underline{A}(m) = \frac{p(b+m)}{(p-q)} - \big[E[x \mid e^H] - I(1+r)\big]. \tag{13}$$

As long as $(b+m) < B$, then monitoring lowers the minimum collateral requirement from $\underline{A}$ in (11) to (13), and therefore expands loan access to asset-poor borrowers over a range of monitoring. Note however that delegation costs $m/(p-q)$ increase the total cost of borrowing[8] so monitored loans will be chosen only by those borrowers who

[7] This section adapts Holmstrom and Tirole (1997) which has become a workhorse model of sorts. A closely related interpretation of monitoring by Banerjee, Besley and Guinnane (1994) is that monitoring increases the borrower's probability of success, or in terms of this simple model, that it raises p relative to q. These two interpretations are complementary and yield very similar results.

[8] The implicit interest rate for a borrower with assets A is $r + m/I(p-q)$ where m is given by $A = \underline{A}(m)$ in (13). Under the assumption of competition among lenders and intermediaries, any increase in the cost of monitoring is borne by the borrower who must cover any increase in participation costs by the intermediary.

do not have enough collateral to access to pure (non-monitored) collateral-based loans. Those who do use monitored lending will choose a contract with only as much monitoring m as is minimally required to lower available collateral assets A to the minimum collateral requirement (i.e., to set $A = \underline{A}(m)$).

Several results that explain the operation and structure of rural financial markets can be derived by working within this kind of framework [Banerjee, Besley and Guinnane (1994); Holmstrom and Tirole (1997); Conning (1999); Gine (2004); Varghese (2004)]. By directly reducing the borrower's scope for moral hazard, a delegated monitor may be able to attract relatively uninformed outside investors to help finance borrowers that these same outsiders would have otherwise found unprofitable. All else equal, delegated monitors without intermediary capital will, however, be more expensive to motivate than those with capital at risk. This is because putting capital at risk allows delegates to better commit to monitor, reducing the cost of providing monitoring incentives. This helps understand why already capitalized crop traders, shopkeepers and landlords are likely to emerge as local financial intermediaries. It also points to the possibility that rural financial intermediation may be held up not for lack of locally informed agents – these can generally be found and hired locally – but for lack of local intermediary capital (which, by definition, cannot be borrowed).

To see this more formally suppose that the monitor has assets to lend $I^m > 0$ out of own capital and offers outside investors senior claims to the borrower's project returns in the event of project failure. The outside investor now lends just $I^u = I - I^m$. It is easy to demonstrate that if the delegate puts up at least $I^m = qm/(p-q)$ then the delegation *rent* is eliminated and the cost of delegation is reduced to the cost of monitoring m and (13) becomes $\underline{A}(m) = pb/(p-q) + [EX - I(1+r)] + m$. This expands loan access and lowers the cost of funds to borrowers so monitored loan contracts will be cheaper and available to more borrowers where intermediary capital is more plentiful.

Whether or not intermediary capital is available, contracts will need to be more heavily monitored to reach poorer borrowers with fewer collateral assets, since the minimum monitoring intensity m that solves $\underline{A}(m) = A$ is higher for borrowers with less A. This in turn means that delegated monitors need to acquire a deeper financial stake (higher I^m) or, where that is not possible, that monitoring delegation costs must rise more quickly than the cost of monitoring m. It follows from this that financial institutions serving asset poor borrowers will also tend to be less highly leveraged (lower I^u/I) and generally have higher (monitoring) costs per dollar loaned compared to institutions serving borrowers who are able to offer more loan guarantees.

These last observations may also help explain the continued prevalence of informal moneylenders in rural areas who charge high interest rates even in what appear to be competitive loan markets. Moneylenders can be viewed as monitoring lenders who must lend primarily out of their own equity (i.e., high I^m/I) and charge high interest rates in large part to recover monitoring costs. There is considerable empirical and case study evidence to support this characterization [Darling (1925); Aleem (1990)]. The less collateral that is available to borrowers in a community, the more intense is the required monitoring, and the larger the fraction of total lending that must be fronted by the monitor. Very marginal borrowers who have few or no collateral assets or proven cash flows will only be able to borrow from pure moneylenders, or friend and relatives (i.e., from individuals who lend entirely out of own equity and cannot become financial intermediaries in this context). This helps to explain the slow and uneven spread of commercial rural bank branching and other forms of intermediation into poor and undeserved communities. It also helps to explain why some rural microcredit lenders that specialize in lending to the poor have continued to rely on donor and government funds rather than tap into larger financial markets, even though they appear to be profitable and maintain very high repayment rates [Conning (1999); Morduch (1999a, 1999b)]. In a nutshell, the problem is that lenders' serving collateral-poor target groups cannot easily 'sell' or leverage any significant portion of their loan portfolio without diluting their own incentives to monitor and preserve the value of their loan portfolios.

Jain (1999) explores similar issues of formal–informal sector interaction in an adverse selection setting. Formal sector lenders are assumed to be able to mobilize funds at considerably lower cost than better-informed informal moneylenders, but formal lenders look for the presence of informal lending as signal or certification that the borrowers have been screened. Hence in both Jain's and the moral hazard context described above, formal and informal lenders compete but may also compliment each other's lending activities. Several papers have tried to empirically identify the extent of formal–informal interaction, including Kochar (1997), Bell, Srinivasan and Udry (1997), Conning (1996), Hauge (1998). As in all econometric study of credit market behavior these studies face the challenge of identifying credit supply and demand in a market with rationing, with the added challenge that in this context loan demand may spillover from the formal to the informal sector. Both Key (1997) and more recently Gine (2004) provide an alternative by studying estimated structural models of credit supply with formal and informal lending. Gine's model, which is estimated using data from Thai villages, provides evidence to support the view that borrowers turn to the informal sector, not so much because they face fixed costs to formal sector borrowing but because of the limited ability of formal banks to enforce contracts in village communities.

Douglas Diamond's (1984) seminal work on financial intermediation clarified an important additional mechanism for lowering delegation rents which helps to further understand the opportunities and challenges faced by rural financial intermediaries. Although Diamond's original focus was on a model of costly state verification (ex-post moral hazard), his ideas carry over easily to the moral hazard example we are employing. Diamond's insight was to note that if a delegate were placed in charge of monitoring several loan projects rather than just one, and if returns from those different projects are imperfectly correlated, then the monitor can be made to cover the losses on one loan project out of the 'bonus' that would have otherwise been received for success on another monitored loan. The size of the delegation rent per loan can then be reduced and therefore also the size of either the minimum stake that an intermediary needs to place at risk to attract outside investors. Hence, financial intermediaries with more diversified loan portfolios can achieve much higher levels of financial leverage, expand loan access, and lower the cost of borrowing.

To see this suppose that a loan agent with no initial capital ($I^m = 0$) were made to monitor n farm borrowers each with an identical but stochastically independent project. It can be easily shown[9] that the delegation rent per borrower can be lowered from $pm/(p - q)$ in expression (12) to $p^n m/(p^n - q^n)$ which quickly converges to m as the number of borrowers n is increased. In other words, a delegated monitor with no intermediary capital of her own can in principle reduce her costs per loan to the level of a less diversified monitor who does have intermediary capital by monitoring a sufficiently diversified portfolio. This 'incentive diversification' effect is important for understanding the spread of financial intermediation and has important implications in other multi-task settings including the joint-liability settings examined in a later section below.

A local intermediary is often a good monitor because he or she knows a lot about a narrow sector or geographic area. Unfortunately, the correlation across project returns within any such sector is likely to remain high, reducing the opportunities for incentive diversification opportunities identified by Diamond. This may explain why commercial financial intermediaries and microfinance have in general been much slower to penetrate into rural areas compared to urban areas where diversification is higher [Chaves and Gonzalez-Vega (1996)], or why new microfinance ventures such as Grameen Bank of Bangladesh have been more successful at funding rural non-farm activities than normal crop-cycle lending.

4.1. *Crowding-in vs crowding-out of financial services*

As we have seen, a rural financial intermediary may help to 'crowd-in' funds from less informed outsiders who where it not for the presence of these intermediary insiders might have found it unprofitable to contract directly with farm borrowers. A very general principle is at work here that Itoh (1993), Holmstrom and Milgrom (1990), Arnott and Stiglitz (1991), Varian (1990) and others have noted and that is useful for understanding the role of intermediaries in the economy and for classifying models in the literature. Loosely stated, the principle is that if a group of agents can 'side-contract' among themselves to enforce actions or reward contingencies in ways that an outside principal would not have been able to specify in a contract, then it will generally be more efficient for the outsider to contract with the group (or the 'coalition') to try to harness the benefits of those side-contracts rather than to contract separately and independently with each agent.

The simpler financial intermediary structures of the last section – with a lender, a delegated monitor/intermediary, and an agent – can be understood as applications of this general idea. The outside investor took advantage of the local intermediary's ability to 'side-contract' or 'monitor' the agent in ways that the outsider could not, allowing

[9] See Conning (2004) or Laux (2001).

the investor to economize on the cost of providing indirect incentives.[10] A good part of the earlier literature on interlinked agrarian contracts should also be understood as applications of this very general idea [Braverman and Stiglitz (1982); Bell (1988)].

The ability of agents to side-contract with other locally informed agents is *not* always helpful however. If agents can side-contract only on the same observable outcomes or actions that an outside financial institution could have included in the contract already, then side-contracts can never improve, and may possibly harm or 'crowd-out', the agents' ability to access financial services. This is because side-contracts among agents potentially undermine incentives that a financial institution might have wanted to build into a contract. For example, an outside lender may be reluctant to provide loan finance to a borrower if existing mutual aid arrangements within a village provide too much consumption insurance in states of the world when the financed project fails because this may undermine the repayment incentives the lender would like to have built into the loan contract. It is easy to construct examples where the inability to commit to not engaging in certain types of side contracts may lead to situations where outsiders hold back from providing new forms of efficiency-enhancing finance, yet at the same time the mere possibility that a new outsider might enter the market can crowd-out existing local contracting, leaving to the possibility of a decline in welfare.

The design of all financial intermediary structures and policies to promote financial intermediation can be thought of as involving decisions over which side-contracts to allow or to internalize within the institution, and which to try to disallow or regulate. A large number of papers, employing assorted types of information asymmetry problems, have been written that illustrate how certain types of unregulated competition between potential financial providers may end up 'crowding-out' or shrinking the size of financial markets [Ray and Sengupta (1989); Arnott and Stiglitz (1991); Hoff and Stiglitz (1997); Jain (1999); Kranton and Swamy (1999); Navajas, Conning and González-Vega (2003); Wydick and Mcintosh (2005)]. The issue has also been examined empirically in several studies, for example, Attanasio and Rios-Rull (2000) provide recent evidence that the introduction of Mexico's PROGRESA public social safety net program crowded-out local insurance. Morduch (1999a) offers a review of some of the issues.

An important related implication is that in contexts where exclusive contracting is difficult or costly to enforce, competition may have a potentially destructive effect on the extent and depth of lending relationships. Petersen and Rajan (1995) argue for instance that loan sizes are larger in areas where banking is more concentrated because banks are more likely to be able to enter into long-term lending arrangements with an

[10] A difference is that in the previous section the agent and the intermediary did not *costlessly* and perfectly 'side-contract' as in the scenarios typically discussed in this earlier literature. It was worthwhile to work via a costly intermediary monitor only so long as the additional cost of delegation did not exceed the benefit it conferred. Conning (2004) provides a lengthier discussion of the different implications of models with costly versus costless 'side-contracting'.

entrepreneur when there are fewer competing banks to poach away its more successful clients. Testing this idea, they find clear empirical evidence of a correlation between average loan size and bank concentration in small-business lending in the United States.

This issue may be of potential great importance for explaining the pattern of agricultural lending in many contexts and why financial intermediaries including banks and agro-industry and contract-farming firms have not penetrated as deeply into some crops and regions as they have in others. Examining financing patterns in rural Chile, Conning (1996) points to an evident relationship between agro-industry processor concentration and lending depth. In crops such as tobacco or sugar beet where there was a single monopsony product buyer, agro-industry contracts regularly financed 85–100% of small farmers' working capital needs, while in other crops such as tomatoes for canning or rice where several agro-industry firms or mills compete in any given region, contracts rarely financed more than 25% of small farmer costs. In less perishable crops with many potential buyers such as wheat and many types of legumes, contract-farming firms are virtually non-existent, even though these are crops dominated by small farm producers. In interviews, product buyers attributed part of their reluctance to deeper financing to the fear of 'leakage' of that farmers who had pledged all or part of their harvest to a trader would surreptitiously divert part of their produce to another buyer. Jaffee (1994) underscores the importance of this perceived problem, concluding that the problem of 'leakage' had led to the termination of credit in a large Kenyan horticultural scheme. Runsten and Key (1996) document how similar contracting issues shaped the structure and success or failure of different Mexican contract farming schemes.

4.2. Group loans, cooperatives, ROSCAs, and mutuals

No account of the history of rural lending would be complete without a mention of the role of credit cooperatives, farm credit and insurance mutuals, and other group-lending and insurance mechanisms including solidarity group loans and rotating savings and credit associations. Although cooperative mutual society organizations such as these have been very important, and sometimes even dominant sources of finance to the rural economy of many countries, their relative importance has waxed and waned. In some cases early successes were marred by later spectacular failures.

The potential advantages of credit cooperatives, ROSCAs, savings mutuals and group loans can be explained as before by appealing to the idea that outside financiers may prefer to contract with a coalition in order to harness the ability of members of the group to side-contract using local information and enforcement mechanisms not available to outsiders. The group entity acts as a financial intermediary, helping to crowd-in outside finance that would not be available or as large via separate individual contracts. Stiglitz (1990), Varian (1990) and others have used this type of argument to explain why joint-liability contracts to encourage costless 'peer-monitoring' within a group may provide advantages over separate individual-liability loans.[11] Greif (2004) and Levinson (2003)

[11] Surveys of the literature that discuss these concepts in the context of microfinance lending include Ghatak and Guinnane (1999), Morduch (1999a), Ahlin and Townsend (2003), Conning (2004).

argue that forms of joint-liability contracting are in fact far more important and ubiquitous than economists normally realize. Greif argues that prior to the rise of impersonal, legally enforced exchange, most long distance trading arrangements were enforced for centuries via a communal responsibility system to harness this type of side-contracting and local enforcement mechanisms. Holmstrom (1999) argues that part of the reason for the existence of firms can be understood in similar terms.

Hansmann (1996) identifies other related reasons for the rise of savings and insurance mutuals in some contexts. He argues that the demand for insurance and savings vehicles grew as the frontier expanded and farming communities where established around the United States. Although private for-profit firms tried to offer products such as life insurance policies, farmers were reluctant to enter into such long-term relationships with private or stock-owned firms for fear that these firms might in the future act opportunistically, for example by raising insurance premiums or lowering promised payouts. With a mutual insurance company on the other hand, policyholders own the firm, so what a farmer might lose on his policy he gains back as a shareholder, so the incentive for the firm to act opportunistically is sharply reduced. The ownership and governance structure of the financial institution is therefore adapted to allow for better monitoring and incentives. As Hansmann points out, the mutual ownership form can hardly be dismissed as anachronistic or utopian since even today insurance mutuals account for nearly half of all life insurance in force in the United States, and for trillions of dollars of insurance worldwide.

Credit cooperatives have also been important in the development of rural finance in many parts of the world. One of the most studied cases is that of the German credit cooperative movement which grew rapidly from a handful of small independent cooperatives in the mid-19th century to include over 19,000 cooperatives by 1914 [Guinnane (2001)]. Responding to popular demands the government passed cooperative legislation to both regulate and enable growth of the movement. In other countries governments introduced rural cooperative legislation in an effort to create new cooperative societies. The colonial government of India introduced legislation encouraging the development of agricultural cooperative credit societies in the late 19th century partly as a response to the Deccan Riots and the perception by some that rural informal moneylenders needed to forced to face more competition [Catanach (1970)].

Although the overall record is mixed, credit cooperative systems and joint liability mechanisms serve an important role in the agricultural lending systems of many developing countries. One successful system is Thailand's Bank for Agriculture and Agricultural Cooperatives (BAAC), established in 1966. This government-sponsored system extends loans to farmers, farmer's groups and cooperatives and acts as a guarantor for loans or farm credits from other sources. For many years BAAC has used joint liability groups as a substitute for more traditional land collateral for small farm loans, managing to continue to expand rapidly while maintaining high repayment incentives [Townsend (1995b)].

Despite their prevalence and frequent success in agricultural financial markets, cooperatives and mutuals have at times failed spectacularly and are clearly not always optimal ownership or contract forms. Joint-liability (JL) clauses imply that each agent's net return in a group will be an *increasing* function of the performance of other agent's projects and loans. This can create incentives for 'peer-monitoring', 'peer-selection', and 'peer-sanction'. But in many contexts relative-performance evaluation (RPE) clauses may help to lower the cost of providing incentives. Relative performance works by making each agent's payoff a *decreasing* function of the measured performance on other agent's projects, the opposite of joint liability. A bank may for example want to extend a loan repayment grace period to a farmer who reports a bad harvest when many other farmers in the same region are reporting bad outcomes, but not otherwise, because the lender can then be more certain to be providing insurance against a common adverse shock, rather than possibly bailing out a farmer for failing to be diligent. When such considerations are important, ownership forms that imply joint liability are not likely to be optimal.

Joint or group liability forms are vulnerable to other problems as well. The most commonly discussed problem is the free-rider problem [Holmstrom (1982); Braverman and Guasch (1989); Kremer (1997)]. This occurs, for example, when agents are unable to efficiently side-contract to coordinate actions, as assumed in many of the models described above. In such a context a joint-liability structure may well encourage risk-taking or a lack of diligence, as each agent faces only a fraction of the cost of changes in their actions. The free-rider problem is more likely to matter the larger is the group. Partly because of the lack of attention to incentives and oversight, some government sponsored agricultural cooperatives have failed to maintain repayment levels, sometimes resulting in later bailouts at great cost to the public. To anticipate and avoid such problems cooperative regulation and oversight exists in many countries that limits the amount of outside capital that cooperatives are allowed to raise. The benefit of such regulation is that it avoids diluting the incentives to loan monitoring within the cooperative, but this constrains cooperatives from growing very rapidly or from quickly responding to new profit opportunities compared to other types of lenders.

4.3. Policies to promote rural financial intermediation

As the above review makes clear, there is plenty of theory and evidence to suggest that financial markets frequently fail to allocate resources in a first-best fashion. Financial contract forms and intermediary structures adapt to harness local information and enforcement mechanisms to ameliorate or overcome the problems created by information asymmetries and limited enforcement, but the solutions typically fall far short of first-best optimal.

Public policy can play an important role in affecting the provision of financial services in such environments, for better or for worse. Government can provide important basic infrastructure that is needed for the operation of markets. This includes providing effective government, and a system of laws and local courts to help facilitate the creation and enforcement of property rights and contracts. This is of course not always an easy accomplishment and, creating new forms of property may at times ironically even end up crowding-out rather than crowding-in financial trade unless other types of property rights are created at the same time. We have reviewed several examples of situations where innovations that may have spurred the entry of new financial intermediaries started out well but over time led to crowding-out, or in extreme cases, even to the collapse of some markets. The problem is that unless mechanisms exist for agents to enter into exclusive contracts, increased competition may undermine financial arrangements that had previously been self-enforcing. Things that help parties to overcome or lower the cost of such problems include efforts to make information public (such as the creation of credit bureaus) and enabling legislation and courts to notarize and register liens and collateral guarantees [Fafchamps (2004)].

The prudential regulation of banks and non-bank financial institutions can play an important role in spurring financial deepening, although here again policy is a double edged sword and the potential to do more harm than good through heavy-handed intervention has been proven to be immense [Adams, Graham and Von Pischke (1984)]. Dewatripont and Tirole (1994) work with a model very similar to the monitored lending framework described above, to illustrate how prudential regulation may 'crowd-in' new forms of finance. Creditors and depositors may be more willing to invest in financial institutions if they see that regulatory or supervisory authorities are making sure that these intermediaries have the right incentives to carefully screen and monitor their borrowers. A government loan guarantee can similarly work to crowd-in private sector investment. The danger in all these mechanisms of course is that government involvement will create rather than ameliorate moral hazard, for example by encouraging banks and investors to take excessively large risks believing that the government will bail out the sector if things turn out badly. Excessively heavy regulation may easily stifle financial innovation and/or greatly raise the cost of financial services.

5. Conclusion

We have provided a brief overview of an enormous and rapidly-growing literature on rural financial markets in developing countries. The particular configurations of financial instruments and strategies that are available to rural households are extremely variable, making broad generalizations perilous. However, there is a great deal of evidence from a wide variety of rural settings that implies that financial markets are highly fragmented and imperfect. Borrowers are systematically sorted across different types of financial contracts according to their characteristics and activities. Even within single economies, the consequence is a great deal of diversity of contract form, and contract terms such as the interest rate are extremely variable.

Historically (indeed, as long as records exist) governments have intervened in rural financial markets, sometimes in a quite heavy-handed manner. The 1950s through the 1970s saw a cluster of policies that included interest rate ceilings and directed lending by state-owned and private banks being implemented in rural areas of many developing countries. A large literature on the associated *financial repression* arose that documented many of the deleterious effects of this type of intervention. Some of the fragmentation described in Section 2.1 of this chapter may be among the consequences of this widespread financial repression.

However, the fundamental determinants of the myriad imperfections that afflict rural financial markets are the difficulties that arise in transactions of contingent promises when information is asymmetric and the enforcement of contracts is not assured. We have focused attention in the chapter on one particular form of asymmetric information that can have important consequences for shape of rural financial transactions – moral hazard. We hope that it is apparent that we have done so because it permits us to discuss a wide range of important issues in the context of a single simple model, not because moral hazard is the only (or even necessarily the most important) source of asymmetric information. The core lessons that we draw from this exercise are applicable to the related contracting problems of adverse selection or opportunistic default.

In the 1980s and 1990s, structural adjustment programs adopted in many developing countries did away with many of the policies associated with financial repression. Although some promising financial innovations have taken place particularly in the realm of microfinance, most of this innovation has been focused on urban or non-farm rural activities. The response of the private agricultural financial system to these liberalizing policy changes was much less vigorous than many reformers had hoped. In order for a robust set of intermediated financial instruments to be available to rural households, government must do more than simply get out of the way of private lenders. There is a manifest need for careful state attention to the essential institutions that support rural financial intermediation.

Intermediation is more likely to emerge in situations in which new forms of financial contracts can be enforced. The range of contracts that is feasible can expand when institutions exist to facilitate the dissemination of information regarding market fundamentals (like growing conditions) or outcomes (like credit bureaus). Such institutions often, although not always, have the character of public goods and are unlikely to emerge in the absence of active but disciplined state participation. A crucial focus of new research on rural financial markets must be the broad set of issues that surround the development of these intuitions of property rights, legal enforcement of financial contracts, and information diffusion.

References

Abreu, D., Pearce, D., Stacchetti, E. (1990). "Toward a theory of discounted repeated games with imperfect monitoring". Econometrica 58 (5), 1041–1063.

Acemoglu, D., Zilibotti, F. (1997). "Was Prometheus unbound by chance? Risk, diversification, and growth". Journal of Political Economy 105 (4), 709–751.

Adams, D.W. (1992). "Taking a fresh look at informal finance". In: Adams, D.W., Fitchett, D.A. (Eds.), Informal Finance in Low-Income Countries. Westview Press, Boulder, CO, pp. 5–23.

Adams, D.W., Graham, D.H., Von Pischke, J.D. (1984). "Undermining rural development with cheap credit". In: Adams, D.-W., Graham, D.-H., Von-Pischke, J.D. (Eds.), Undermining Rural Development with Cheap Credit. In: Westview Special Studies in Social, Political and Economic Development Series. Westview Press, Boulder, CO, pp. 1–7.

Ahlin, C., Townsend, R.M. (2003). "Using repayment data to test across theories of joint liability lending", Working paper. Vanderbilt Economics Department.

Akerlof, G.A. (1970). "The market for 'lemons': Quality uncertainty and the market mechanism". Quarterly Journal of Economics 84 (3), 488–500.

Aleem, I. (1990). "Imperfect information, screening, and the costs of informal lending: A study of a rural credit market in Pakistan". World Bank Economic Review 4 (3), 329–349.

Allen, D.W., Lueck, D. (2002). The Nature of the Farm: Contracts Risk, and Organization in Agriculture. MIT Press, Cambridge, MA.

Armendáriz de Aghion, B., Morduch, J. (2005). The Economics of Microfinance. MIT Press, Cambridge, MA.

Arnott, R., Stiglitz, J.E. (1991). "Moral hazard and nonmarket institutions: Dysfunctional crowding out or peer monitoring?". American Economic Review 81 (1), 179–190.

Arrow, K., Debreu, G. (1954). "Existence of a competitive equilibrium for a competitive economy". Econometrica 22 (3), 265–290.

Aryeetey, E., Udry, C. (2000). "Saving in Sub-Saharan Africa", CID Working Paper No. 38. Harvard University.

Attanasio, O.P., Rios-Rull, J.V. (2000). "Consumption smoothing extended families", Working paper. University College London.

Banerjee, A.V. (2003). "Contracting constraints, credit markets, and economic development". In: Dewatripont, M., Hansen, L.P., Turnovsky, S. (Eds.), Advances in Economics and Econometrics: Theory and Applications, Eighth World Congress, vol. III. Cambridge University Press, Cambridge, UK.

Banerjee, A.V., Besley, T., Guinnane, T.W. (1994). "Thy neighbor's keeper: The design of a credit cooperative with theory and a test". Quarterly Journal of Economics 109 (2), 491–515.

Banerjee, A.V., Duflo, E. (2004). "Do firms want to borrow more? Testing credit constraints using a directed lending program". Department of Economics, MIT.

Banerjee, A.V., Newman, A.F. (1993). "Occupational choice and the process of development". Journal of Political Economy 101 (2), 274–298.

Bardhan, P. (Ed.) (1989). The Economic Theory of Agrarian Institutions. Oxford University Press, Oxford.

Bardhan, P., Udry, C. (1999). Development Microeconomics. Oxford University Press, New York.

Barry, P.J., Robison, L.J. (2001). "Agricultural finance: Credit, credit constraints, and consequences". In: Handbook of Agricultural Economics, vol. 1A, Agricultural Production. Elsevier Science, Amsterdam, pp. 513–571.

Basu, K. (1992). "Limited liability and the existence of share tenancy". Journal of Development Economics 38 (1), 203–220.

Basu, K. (1997). Analytic Development Economics: The Less Developed Economy Revisited. MIT Press, Cambridge, MA.

Bates, R.H. (1981). Markets and States in Tropical Africa: The Political Basis of Agricultural Policies. University of California Press, Berkeley, CA.

Beegle, K., Dehejia, R., Gatti, R. (2003). "Child labor, crop shocks, and credit constraints", NBER Working Paper No. 10088.

Bell, C. (1988). "Credit markets, contracts, and interlinked transactions". In: Chenery, H., Srinivasan, T.N. (Eds.), Handbook of Development Economics, vol. I. Elsevier Science, Amsterdam, pp. 381–423.

Bell, C., Srinivasan, T.N., Udry, C. (1997). "Rationing, spillover, and interlinking in credit markets: The case of rural Punjab". Oxford Economic Papers 49 (4), 557–585.

Besley, T. (1994). "Savings, credit and insurance". In: Behrman, J., Srinivasan, T.N. (Eds.), Handbook of Development Economics, vol. III. Elsevier Science, Amsterdam, pp. 2123–2207.

Binswanger, H.P., Rosenzweig, M.R. (1986). "Behavioural and material determinants of production relations in agriculture". Journal of Development Studies 22 (3), 503–539.

Boucher, S., Carter, M.R. (2002). "Risk rationing in moral hazard constrained credit markets", Staff Working Paper No. 445. Agricultural and Applied Economics, University of Wisconsin–Madison.

Boucher, S.R., Guirkinger, C. (2004). "Risk, wealth and sectoral choice in rural credit markets". Department of Agricultural Economics, University of California at Davis.

Bowles, S. (2003). Microeconomics: Behavior, Institutions, and Evolution. Princeton University Press, Princeton, NJ.

Brasselle, A.-S., Gaspart, F., Platteau, J.-P. (2002). "Land tenure security and investment incentives: Puzzling evidence from Burkina Faso". Journal of Development Economics 67 (2), 373–418.

Braverman, A., Guasch, J.L. (1989). "Institutional analysis of credit co-operatives". In: The Economic Theory of Agrarian Institutions. Oxford University Press, Oxford, pp. 340–355.

Braverman, A., Stiglitz, J.E. (1982). "Sharecropping and the interlinking of agrarian markets". American Economic Review 72 (4), 695–715.

Burgess, R., Pande, R. (2003). "Do rural banks matter? Evidence from the Indian social banking experiment". Department of Economics, Yale University.

Burkart, M., Ellingsen, T. (2004). "In-kind finance". American Economic Review 94 (3).

Carter, M.R., Barham, B., Mesbah, D. (1996). "Agricultural export booms and the rural poor in Chile, Guatemala, and Paraguay". Latin American Research Review 31 (1).

Carter, M.R., Cason, J., Zimmerman, F.J. (Eds.) (1998). "Development at a crossroads: Uncertain paths to sustainability after the Neoliberal Revolution", Global Studies Research Program. University of Wisconsin–Madison.

Carter, M.R., Olinto, P. (2003). "Getting institutions 'right' for whom: Credit constraints and the impact of property rights on the quantity and composition of investment". American Journal of Agricultural Economics 85 (1), 173–187.

Carter, M.R., Wiebe, K.D., Blarel, B. (1994). "Tenure security for whom? Differential effects of land policy in Kenya". In: Bruce, J.W., Migot-Adholla, S.E. (Eds.), Searching for Land Tenure Security in Africa. Kendall/Hunt Publishing, Dubuque, IA, pp. 141–168.

Carter, M.R., Zimmerman, F.J. (2000). "The dynamic cost and persistence of asset inequality in an agrarian economy". Journal of Development Economics 63 (2), 265–302.

Catanach, I.J. (1970). Rural Credit in Western India, 1875–1930; Rural Credit and the Co-Operative Movement in the Bombay Presidency. University of California Press, Berkeley, CA.

Chaves, R.A., Gonzalez-Vega, C. (1996). "The design of successful rural financial intermediaries: Evidence from Indonesia". World Development 24 (1), 65–78.

Chiappori, P.A. (2000). "Econometric models of insurance under asymmetric information". In: Dionne, G. (Ed.), Handbook of Insurance. Springer, New York, pp. 365–394.

Coate, S., Ravallion, M. (1993). "Reciprocity without commitment: Characterization and performance of informal insurance arrangements". Journal of Development Economics 40, 1–24.

Cole, S. (2004). "Fixing market failures or fixing elections? Agricultural credit in India", Working paper. Department of Economics, MIT.

Conning, J. (2004). "Monitoring by delegates or by peers? Joint liability loans under moral hazard", Working paper. Department of Economics, Hunter College.

Conning, J.H. (1996). "Financial contracting and intermediary structures in a rural credit market in Chile: A theoretical and empirical analysis", Ph.D. thesis. Yale University.

Conning, J.H. (1999). "Outreach, sustainability and leverage in monitored and peer-monitored lending". Journal of Development Economics 60 (1), 51–77.

Darling, M. (1925). The Punjab Peasant in Prosperity and Debt. Oxford University Press, London.

Dasgupta (1989). Reports on Informal Credit Markets in India: Summary. National Institute of Public Finance and Policy, New Delhi.

Deaton, A. (1997). The Analysis of Household Surveys: A Microeconometric Approach to Development Policy. Johns Hopkins University Press for the World Bank, Baltimore, MD.

Deininger, K. (2003). Land Policies for Growth and Poverty Reduction. World Bank and Oxford University Press, Washington, DC.

de Janvry, A., Fafchamps, M., Sadoulet, E. (1991). "Peasant household behaviour with missing markets: Some paradoxes explained". Economic Journal 101 (409), 1400–1417.

Demsetz, H. (1967). "Toward a theory of property rights". American Economic Review 57 (2), 347–359.

Dercon, S. (Ed.) (2004). Insurance Against Poverty. WIDER Studies in Development Economics. Oxford University Press, Oxford.

Dercon, S., Krishnan, P. (2000). "In sickness and in health: Risk sharing within households in rural Ethiopia". Journal of Political Economy 108 (4), 688–727.
de Soto, H. (2000). The Mystery of Capital: Why Capitalism Succeeds in the West and Fails Almost Everywhere Else. Basic Books, New York.
Dewatripont, M., Tirole, J. (1994). The Prudential Regulation of Banks. MIT Press, Cambridge, MA.
Diagne, A., Zeller, M. (2001). "Access to credit its impact on welfare in Malawi".
Diamond, D.W. (1984). "Financial intermediation and delegated monitoring". Review of Economic Studies 51 (3), 393–414.
Diamond, D.W. (1996). "Financial intermediation as delegated monitoring: A simple example". Federal Reserve Bank of Richmond Economic Quarterly 82 (3), 51–66.
Duryea, S., Lam, D., Levison, D. (2003). "Effects of economic shocks on children's employment and schooling in Brazil", PSC Research Report No. 03-541. Inter-American Development Bank.
Eswaran, M., Kotwal, A. (1986). "Access to capital and agrarian production organization". Economic Journal 96, 482–498.
Fafchamps, M. (1999). "Risk sharing and quasi-credit". Journal of International Trade and Economic Development 8 (3), 257–278.
Fafchamps, M. (2003). Rural Poverty, Risk and Development. E. Elgar, Cheltenham, UK/Northampton, MA.
Fafchamps, M. (2004). Market Institutions in Sub-Saharan Africa: Theory and Evidence. MIT Press, Cambridge, MA.
Fafchamps, M., Lund, S. (2003). "Risk-sharing networks in rural Philippines". Journal of Development Economics 71 (2), 261–287.
Feder, G., Feeny, D. (1991). "Land tenure and property rights: Theory and implications for development policy". World Bank Economic Review 5 (1), 135–153.
Felkner, J.S., Townsend, R.M. (2004). "The wealth of villages: An application of GIS spatial statistics to two structural economic models", Working paper. Department of Economics, University of Chicago.
Fellingham, J.C., Newman, D.P., Suh, Y.S. (1985). "Contracts without memory in multiperiod agency models". Journal of Economic Theory 37 (2), 340–355.
Field, E., Torero, M. (2004). "Do property titles increase credit access among the urban poor? Evidence from a nationwide titling program", Working paper. Harvard University.
Fitchett, D.A. (1999). "Bank for Agriculture and Agricultural Cooperatives (BAAC), Thailand (case study)". Consultative Group to Assist the Poorest (CGAP) Working Group on Savings Mobilization, Washington, DC.
Foster, A.D. (1988). "Why things fall apart: A strategic analysis of repeated interaction in rural financial markets".
Freixas, X., Rochet, J.-C. (1997). Microeconomics of Banking. MIT Press, Cambridge, MA.
Galor, O., Zeira, J. (1993). "Income distribution and macroeconomics". Review of Economic Studies 60 (1), 35–52.
General Accounting Office (2001). "Farm loan programs: Improvements in the loan portfolio but continued monitoring needed". Testimony before the Committee on the Agriculture, Nutrition, and Forestry, U.S. Senate, Washington, DC.
Genicot, G. (2002). "Bonded labor and serfdom: A paradox of voluntary choice". Journal of Development Economics 67 (1), 101–127.
Genicot, G., Ray, D. (2002). "Endogenous group formation in risk-sharing arrangements". Review of Economic Studies 70 (1).
Gersovitz, M. (1988). "Saving and development". In: Chenery, H., Srinivasan, T.N. (Eds.), Handbook of Development Economics, vol. I. Elsevier Science, Amsterdam, pp. 381–423.
Gertler, P., Gruber, J. (2002). "Insuring consumption against illness". American Economic Review 92 (1), 51–76.
Ghatak, M., Guinnane, T.W. (1999). "The economics of lending with joint liability: A review of theory and practice". Journal of Development Economics 60 (1).
Ghate, P. (1992). Informal Finance: Some Findings from Asia. Oxford University Press for the Asian Development Bank, Oxford.

Gine, X. (2004). "Access to capital in rural Thailand: An estimated model of formal versus informal credit", World Bank Policy Research Working Paper No. 3502.
Gine, X., Klonner, S. (2003). "Financing a new technology in small-scale fishing: The dynamics of a linked product credit contract", Manuscript. The World Bank Research Group.
Glover, D. (1994). "Contract farming and commercialization of agriculture in developing countries". In: von Braun, J., Kennedy, E. (Eds.), Agricultural Commercialization, Economic Development, and Nutrition. Johns Hopkins University Press for the International Food Policy Research Institute, Baltimore and London, pp. 166–175.
Glover, D., Kusterer, K. (1990). Small Farmers Big Business: Contract Farming and Rural Development. St. Martin's Press, New York.
Goetzmann, W.N. (1996). "Financing civilization". Yale School of Management.
Goldstein, M. (2004). "Intrahousehold efficiency and individual insurance in Ghana", Working paper. London School of Economics.
Goldstein, M., Udry, C. (1999). "Agricultural innovation resource management in Ghana", Working paper. Department of Economics, Yale University.
Gonzalez-Vega, C. (1984). "Credit-rationing behavior of agricultural lenders: The iron law of interest-rate restrictions". In: Undermining Rural Development with Cheap Credit, pp. 78–95.
Greif, A. (1993). "Contract enforceability and economic institutions in early trade: The Maghribi traders' coalition". American Economic Review 83 (3), 525–548.
Greif, A. (2004). "Impersonal exchange without impartial law: The community responsibility system". Chicago Journal of International Law 5 (1), 107–136.
Grimard, F. (1997). "Household consumption smoothing through ethnic ties: Evidence from Cote d'Ivoire". Journal of Development Economics 53 (2), 391–422.
Grossman, H.I., Van Huyck, J.B. (1988). "Sovereign debt as a contingent claim: Excusable default, repudiation, and reputation". American Economic Review 78 (5), 1088–1097.
Grossman, S.J., Hart, O.D. (1983). "An analysis of the principal-agent problem". Econometrica 51 (1), 7–45.
Guinnane, T.W. (2001). "Cooperatives as information machines: German rural credit cooperatives, 1883–1914". Journal of Economic History 61 (2), 366–389.
Gurley, J.G., Shaw, E.S. (1960). Money in a Theory of Finance. Brookings Institution, Washington, DC.
Hansmann, H. (1996). The Ownership of Enterprise. Harvard University Press, Cambridge, MA.
Hart, O., Holmstrom, B. (1987). "The theory of contracts". In: Bewley, T.-F. (Ed.), Advances in Economic Theory: Fifth World Congress. In: Econometric Society Monographs Series, vol. 12. Cambridge University Press, Cambridge, pp. 71–155.
Hauge, S. (1998). "The outreach impact and sustainability of group-based credit: Empirical evidence from rural Chile", Ph.D. thesis. University of Wisconsin.
Hazell, P.B.R., Pomareda, C., Valdés, A. (1986). Crop Insurance for Agricultural Development: Issues and Experience. Johns Hopkins University Press for the International Food Policy Research Institute, Baltimore, MD.
Hoff, K., Braverman, A., Stiglitz, J.E. (Eds.) (1993). The Economics of Rural Organization: Theory, Practice, and Policy. Oxford University Press for the World Bank, Oxford.
Hoff, K., Stiglitz, J.E. (1997). "Moneylenders and bankers: Price-increasing subsidies in a monopolistically competitive market". Journal of Development Economics 52 (2), 429–462.
Holmstrom, B. (1979). "Moral hazard and observability". Bell Journal of Economics 10 (1), 74–91.
Holmstrom, B. (1982). "Moral hazard in teams". Bell Journal of Economics 13 (2), 324–340.
Holmstrom, B. (1999). "The firm as a subeconomy". Journal of Law, Economics and Organization 15, 74–102.
Holmstrom, B., Milgrom, P. (1990). "Regulating trade among agents". Journal of Institutional and Theoretical Economics 146 (1), 85–105.
Holmstrom, B., Tirole, J. (1997). "Financial intermediation, loanable funds, and the real sector". Quarterly Journal of Economics 112 (3), 663–691.
Hueth, B., Ligon, E. (2001). "Agricultural markets as relative performance evaluation". American Journal of Agricultural Economics 83 (2), 318–328.

Innes, R. (1993). "Financial contracting under risk neutrality, limited liability and ex ante asymmetric information". Economica 60 (237), 27–40.
Innes, R.D. (1990). "Limited liability and incentive contracting with ex-ante action choices". Journal of Economic Theory 52 (1), 45–67.
Itoh, H. (1993). "Coalitions, incentives, and risk sharing". Journal of Economic Theory 60 (2), 410–427.
Jacoby, H., Skoufias, E. (1997). "Risk, financial markets, and human capital in a developing country". Review of Economic Studies 64 (3).
Jaffee, S.M. (1994). "Contract farming in the shadow of competitive markets: The experience of Kenyan horticulture". In: Little, P.-D., Watts, M.-J. (Eds.), Living Under Contract: Contract Farming and Agrarian Transformation in Sub-Saharan Africa. University of Wisconsin Press, Madison, pp. 97–139.
Jain, S. (1999). "Symbiosis vs crowding-out: The interaction of formal and informal credit markets in developing countries". Journal of Development Economics 59 (2), 419–444.
Jalan, J., Ravallion, M. (1998). "Transient poverty in postreform rural China". Journal of Comparative Economics 26 (2), 338–357.
Jensen, R. (2000). "Agricultural volatility and investments in children". American Economic Review 90 (2), 399–404.
Jeong, H., Townsend, R.M. (2003). "Growth and inequality: Model evaluation based on an estimation–calibration strategy", Working paper. Department of Economics, University of Chicago.
Kandori, M. (1992). "Social norms and community enforcement". Review of Economic Studies 59, 61–80.
Karlan, D., Zinman, J. (2004). "Observing unobservables: Identifying information asymmetries with a consumer credit field experiment", Working paper. Princeton Economics Department.
Kazianga, H., Udry, C. (2006). "Consumption smoothing? Livestock, insurance and drought in rural Burkina Faso". Journal of Development Economics 79 (2), 413–446.
Key, N. (1997). "Modeling and estimating agricultural household behavior under imperfect markets", University of California, Berkeley.
Kimball, M. (1988). "Farmer cooperatives as behavior toward risk". American Economic Review 78, 224–232.
Kletzer, K., Wright, B. (2000). "Sovereign debt as intertemporal barter". American Economic Review 90, 621–639.
Klonner, S. (2004). Buying fields and marrying daughters: An empirical analysis of Rosca auctions in a South Indian village". Department of Economics, Cornell University.
Knoeber, C.R., Thurman, W.N. (1994). "Testing the theory of tournaments: An empirical analysis of broiler production". Journal of Labor Economics 12 (2), 155–179.
Kochar, A. (1997). "An empirical investigation of rationing constraints in rural credit markets in India". Journal of Development Economics 53 (2), 339–371.
Kochar, A. (1999). "Smoothing consumption by smoothing income: Hours-of-work responses to idiosyncratic agricultural shocks in rural India". Review of Economics and Statistics 81 (1).
Kocherlakota, N. (1996). "Implications of efficient risk sharing without commitment". Review of Economic Studies 63, 595–609.
Kranton, R.E., Swamy, A.V. (1999). "The hazards of piecemeal reform: British civil courts and the credit market in colonial India". Journal of Development Economics 58, 1–24.
Kremer, M. (1997). "Why are worker cooperatives so rare?", Working Paper No. 6118. National Bureau of Economic Research.
La Ferrara, E. (2003). "Kin groups and reciprocity: A model of credit transactions in Ghana". American Economic Review 93 (5), 1730–1751.
Laffont, J.-J., Martimort, D. (2003). The Principal Agent Model: The Economic Theory of Incentives. Princeton University Press, Princeton, NJ.
Lambert, R.A. (1983). "Long-term contracts and moral hazard". Bell Journal of Economics 14 (2), 441–452.
Laux, C. (2001). "Limited-liability and incentive contracting with multiple projects". RAND Journal of Economics 32 (3), 514–526.
Levinson, D. (2003). "Collective sanctions". Stanford Law Review 56 (345).

Ligon, E., Thomas, J.P., Worall, T. (1999). "Mutual insurance with limited commitment: Theory and evidence from village economies", Working paper. University of California, Berkeley.
Ligon, E., Thomas, J.P., Worall, T. (2002). "Informal insurance arrangements with limited commitment: Theory and evidence from village economies". Review of Economic Studies 69 (1), 209–244.
Loury, G.C. (1981). "Intergenerational transfers and the distribution of earnings". Econometrica 49 (4), 843–867.
Macho-Stadler, I., Perez-Castrillo, J.D. (2001). An Introduction to the Economics of Information: Incentives and Contracts. Oxford University Press, Oxford.
Madajewicz, M. (2004). "Does the credit contract matter? The impact of various lending programs on poverty in Bangladesh". Department of Economics, Columbia University.
Matsuyama, K. (2002). "Explaining diversity: Symmetry-breaking in complementarity games". American Economic Review 92, 241–246.
Matthews, S.A. (2001). "Renegotiating moral hazard contracts under limited liability and monotonicity". Journal of Economic Theory 97 (1), 1–29.
McCloskey, D. (1976). "English open fields as behavior towards risk". Research in Economic History, 124–170.
McKenzie, D., Woodruff, C. (2004). "Do entry costs provide an empirical basis for poverty traps? Evidence from Mexican microenterprises", Working paper. Department of Economics, Stanford University.
McKinnon, R. (1973). Money and Capital in Economic Development. Brookings Institution, Washington, DC.
Meyer, R.L., Nagarajan, G. (2000). Rural Financial Markets in Asia: Policies, Paradigms, and Performance. Oxford University Press, Oxford.
Meyer, R.L., Nagarajan, G., Hushak, L.J. (1997). "Segmentation in the Philippine informal credit markets: A multinomial logit analysis". In: Rose, R., Tanner, C., Bellamy, M.-A. (Eds.), Issues in Agricultural Competitiveness: Markets and Policies. International Association of Agricultural Economists Occasional Paper No. 7. Ashgate, Aldershot, UK/Brookfield, VT, pp. 258–266.
Mirrlees, J.A. (1976). "The optimal structure of incentives and authority within an organization". Bell Journal of Economics 7 (1), 105–131.
Mookherjee, D. (1984). "Optimal incentive schemes with many agents". Review of Economic Studies 51 (3), 433–446.
Mookherjee, D. (1997). "Wealth effects, incentives and productivity". Review of Development Economics 1, 116–133.
Mookherjee, D., Ray, D. (2003). "Persistent inequality". Review of Economic Studies 70 (2), 369–393.
Morduch, J. (1995). "Income smoothing and consumption smoothing". Journal of Economic Perspectives 9 (3), 103–114.
Morduch, J. (1999a). "The microfinance promise". Journal of Economic Literature 47 (4).
Morduch, J. (1999b). "The role of subsidies in microfinance: Evidence from the Grameen Bank". Journal of Development Economics 60 (1), 229–248.
Murgai, R., Winters, P., Sadoulet, E., de Janvry, A. (2002). "Localized and incomplete mutual insurance". Journal of Development Economics 67 (2), 245–274.
Navajas, S., Conning, J., González-Vega, C. (2003). "Lending technologies, competition and consolidation in the market for microfinance in Bolivia". Journal of International Development 15 (6), 747–770.
Ngugi, R.W. (2001). "An empirical analysis of interest rate spread in Kenya". African Economic Research Consortium (AERC).
North, D., Thomas, R. (1973). The Rise of the Western World: A New Economic History. Cambridge University Press, Cambridge.
North, D.C. (1990). Institutions, Institutional Change, and Economic Performance. Cambridge University Press, Cambridge.
Paulson, A.L., Townsend, R.M. (2003). "Distinguishing limited commitment from moral hazard in models of growth with inequality". Federal Reserve Bank of Chicago.
Pender, J.L., Kerr, J.M. (1999). "The effects of land sales restrictions: Evidence from South India". Agricultural Economics 21 (3), 279–294.

Petersen, M.A., Rajan, R.G. (1995). "The effect of credit market competition on lending relationships". Quarterly Journal of Economics 110 (2), 407–443.
Place, F., Migot-Adholla, S.E. (1998). "The economic effects of land registration on smallholder farms in Kenya: Evidence from Nyeri and Kakamega districts". Land Economics 74 (3), 360–373.
Platteau, J.-P. (1997). "Mutual insurance as an elusive concept in traditional rural communities". Journal of Development Studies 33 (6), 764–796.
Rajan, R., Zingales, L. (2003). Saving Capitalism from the Capitalists: Unleashing the Power of Financial Markets to Create Wealth and Spread Opportunity. Crown Business, New York.
Ravallion, M., Chaudhuri, S. (1997). "Risk and insurance in village India: Comment". Econometrica 65 (1), 171–184.
Ray, D. (1998). Development Economics. Princeton University Press, Princeton, NJ.
Ray, D., Sengupta, K. (1989). "Interlinkages and the pattern of competition". In: Bardhan, P. (Ed.), The Economic Theory of Agrarian Institutions. Oxford University Press, Oxford, pp. 243–263.
Ray, T., Singh, N. (2001). "Limited liability, contractual choice, and the tenancy ladder". Journal of Development Economics 66 (1), 289–303.
Rosenzweig, M., Wolpin, K. (1993). "Credit market constraints, consumption smoothing, and the accumulation of durable production assets in low-income countries: Investment in bullocks in India". Journal of Political Economy 101 (2), 223–244.
Rosenzweig, M.R., Stark, O. (1998). "Consumption smoothing, migration, and marriage: Evidence from rural India". In: Economic Demography, vol. 2. E. Elgar, Cheltenham, UK/Northampton, MA, pp. 592–613. Distributed by American International Distribution Corporation, Williston, VT.
Runsten, D., Key, N. (1996). "Contract farming in developing countries: Theoretical aspects and analysis of some Mexican cases", Report prepared for the United Nations Economic Commission for Latin America and the Caribbean, Santiago, Chile.
Sadoulet, E., de Janvry, A., Fukui, S. (1997). "The meaning of kinship in sharecropping contracts". American Journal of Agricultural Economics 79 (2), 394–406.
Salanié, B. (1997). The Economics of Contracts: A Primer. MIT Press, Cambridge, MA.
Schündeln, M. (2004). "Firm dynamics in the presence of financing constraints: Ghanaian manufacturing". Economics Department, Yale University.
Scott, J.C. (1976). The Moral Economy of the Peasant: Rebellion and Subsistence in Southeast Asia. Yale University Press, New Haven, CT.
Sen, A. (1982). Poverty and Famines: An Essay on Entitlement and Deprivation. Oxford University Press, Oxford.
Shetty, S. (1988). "Limited liability, wealth differences and tenancy contracts in agrarian economies". Journal of Development Economics 29 (1), 1–22.
Siamwalla, A. (1990). "The Thai rural credit system: Public subsidies, private information, and segmented markets". World Bank Economic Review 4 (3), 271–295.
Singh, I., Squire, L., Strauss, J. (1986). Agricultural Household Models: Extensions, Applications, and Policy. Johns Hopkins University Press, Baltimore, MD.
Spillman, W.J. (1919). "The agricultural ladder". American Economic Review 9, 170–179.
Srinivasan, T.N. (1989). "On choice among creditors and bonded labour contracts". In: The Economic Theory of Agrarian Institutions. Oxford University Press, Oxford, pp. vi, 408.
Stiglitz, J.E. (1974). "Incentives and risk sharing in sharecropping". Review of Economic Studies 41, 397–426.
Stiglitz, J.E. (1990). "Peer monitoring and credit markets". World Bank Economic Review 4 (3), 351–366.
Stiglitz, J.E. (1994). "The role of the state in financial markets". In: Proceedings of the World Bank Annual Conference on Development Economics, 1993: Supplement to the World Bank Economic Review and the World Bank Research Observer. World Bank, Washington, DC, pp. 19–52.
Stiglitz, J.E. (2002). "Information and the change in the paradigm in economics". American Economic Review 92 (3), 460–501.
Stiglitz, J.E., Weiss, A. (1981). "Credit rationing in markets with imperfect information". American Economic Review 71 (3), 393–410.

Suri, T. (2003). "Spillovers in village consumption: Testing the extent of partial insurance", Working paper. Department of Economics, Yale University.
Thiele, H., Wambach, A. (1999). "Wealth effects in the principal agent model". Journal of Economic Theory 89 (2), 247–260.
Timberg, T.A., Aiyar, C.V. (1984). "Informal credit markets in India". Economic Development and Cultural Change 33 (1), 43–59.
Tirole, J. (2006). The Theory of Corporate Finance. Princeton University Press, Princeton, NJ.
Townsend, R.M. (1979). "Optimal contracts and competitive markets with costly state verification". Journal of Economic Theory 21 (2), 265–293.
Townsend, R.M. (1993). The Medieval Village Economy: A Study of the Pareto Mapping in General Equilibrium Models. Princeton University Press, Princeton, NJ.
Townsend, R.M. (1994). "Risk and insurance in village India". Econometrica 62 (3), 539–591.
Townsend, R.M. (1995a). "Consumption insurance: An evaluation of risk-bearing systems in low-income economies". Journal of Economic Perspectives 9 (3), 83–102.
Townsend, R.M. (1995b). "Financial systems in northern Thai villages". Quarterly Journal of Economics 110 (4), 1011–1046.
Townsend, R.M. (2003). "Microcredit and mechanism design". Journal of the European Economic Association 1 (2–3), 468–477.
Udry, C. (1991). "Rural credit in northern Nigeria". Department of Economics, Yale University.
Udry, C. (1994). "Risk and insurance in a rural credit market: An empirical investigation in northern Nigeria". Review of Economic Studies 61 (3), 495–526.
Udry, C., Duflo, E. (2004). "Intrahousehold resource allocation in Côte d'Ivoire: Social norms, separate accounts and consumption choices", Working paper. Department of Economics, Yale University.
Varghese, A. (2004). "Bank–moneylender linkages as an alternative to competition". Oxford Economic Papers.
Varian, H.R. (1990). "Monitoring agents with other agents". Journal of Institutional and Theoretical Economics 146 (1), 153–174.
Von Pischke, J.D. (1997). "The financial systems approach to development finance and reflections on its implementation", Working paper.
Von Pischke, J.D., Donald, G., Adams, D.W. (1983). Rural Financial Markets in Developing Countries: Their Use & Abuse. Johns Hopkins University Press, Baltimore, MD.
Warning, M., Sadoulet, E. (1998). "The performance of village intermediaries in rural credit delivery under changing penalty regimes: Evidence from Senegal". Journal of Development Studies 35 (1), 115–138.
Watts, M.J. (1994). "Life under contract: Contract farming, agrarian restructuring, and flexible accumulation". In: Little, P.-D., Watts, M.-J. (Eds.), Living Under Contract: Contract Farming and Agrarian Transformation in Sub-Saharan Africa. University of Wisconsin Press, Madison, pp. 21–77.
World Bank (1975). "Agricultural credit: Sector policy paper". World Bank Agricultural Policies Division, Washington, DC.
Wydick, B., Mcintosh, C. (2005). "Competition and microfinance". Journal of Development Economics 78 (2), 271–298.
Zeller, M. (2003). "Models of rural financial institutions". In: Paving the Way Forward for Rural Finance: An International Conference on Best Practices. USAID Basis, Washington, DC.
Zeller, M., Diagne, A., Mataya, C. (1998). "Market access by smallholder farmers in Malawi: Implications for technology adoption, agricultural productivity and crop income". Agricultural Economics 19 (1–2), 219–229.

PART 5

NATURAL RESOURCES

Chapter 57

SOIL QUALITY AND AGRICULTURAL DEVELOPMENT

PIERRE CROSSON†

Contents

† The late Pierre Crosson was Senior Fellow at Resources for the Future, Washington, DC. He recorded his thanks to Soren Anderson, without whose help he said he could not have written this chapter, and doubtless would have thanked Jock R. Anderson, his long-term collaborator, for helping to bring the work to the present level of completion. Jock completed this in the spirit of Crosson [Crosson, P. (1986). "Soil erosion and policy issues". In: Phipps, T., Crosson, P., Price, K. (Eds.), Agriculture and the Environment. Resources for the Future, Washington, DC, pp. 35–73] and sincerely hopes he correctly rebuilt all the citations.

Handbook of Agricultural Economics, Volume 3
Edited by Robert Evenson and Prabhu Pingali

DOI: 10.1016/S1574-0072(06)03057-X

Abstract

This chapter evaluates soil quality and agricultural development. Estimates of soil erosion, in terms of losses to farm income range from a high of $25 billion per year to a low of $100 million per year. Crosson estimates that the $100 million figure is the most reliable. Crosson notes that this estimate is quite low (roughly a quarter of a percent per year) and argues that the Soil Conservation Service in the US invested heavily in programs to prevent soil erosion.

Crosson also notes that soil erosion rates in Europe are even lower than soil erosion rates in the U.S. On a global scale, Crosson uses estimates made by Oldeman and associates at Wageningen University and concludes that global estimates of soil erosion are low (the average overall rate of erosion-based damage is 0.1% per year). A comparison of China and Indonesia, confirmed these estimates. In Sub-Saharan Africa where cropland area is still expanding rapidly, soil erosion rates are higher because of insecure property rights in land.

Keywords

soils, soil quality, soil erosion

JEL classification: C93, O93, Q29

1. Introduction

That soil quality and agricultural development are related is self-evident. Other things the same, the better the quality of the soil, the better the prospects for a robust development of agriculture. But the quality of any soil in agricultural production is variable, in part because of different natural conditions, e.g., chemical composition, depth, water-holding capacity, and in part because of different farming practices employed on the soil. Practices that control erosion and add organic matter will, over time, improve the quality of the soil. Those that promote erosion and other forms of land degradation will, over time, reduce soil quality.

In this chapter the focus is on practices that degrade soil quality and the consequent effects in reducing soil productivity. The reason for this focus is that the literature on the relationship of soil quality to agricultural development is mostly concerned with losses of quality because of practices that degrade the land. However, as will become clear, farmers all around the world are aware of the threat of soil degradation and in many places have taken steps to keep it under control.

The chapter addresses two questions, first, how severe is the effect of land degradation on soil quality, as measured by losses of soil productivity, in the U.S., Europe, on a global scale, and in a number of countries in Asia and Sub-Saharan Africa? With respect to the U.S., Europe and global-scale effects the answer, as it turns out, is that in all of these regions degradation-induced losses of soil productivity have not so far seriously impeded agricultural development. Data for Asia and Sub-Saharan Africa are not sufficient for a comprehensive judgment of soil degradation effects, but the available evidence suggests that in those areas also effects have exaggerated in much of the literature.

This is not to say that within each of the regions studied there are no cases of severe negative degradation-induced impacts on productivity [e.g., Hudson (1981)]; but the available evidence suggests that these cases are exceptions, not the rule.

The second part of the paper asks the question, why, contrary to a widely held view (detailed below) are the productivity effects of soil degradation in these several regions so low? The answer is proposed as a hypothesis to be explored. It goes as follows: for most farmers everywhere the land is one of their most important productive assets. The evidence is overwhelming that farmers around the world seek to manage their assets so that they will earn a "good" rate of return over time. It follows that farmers everywhere *should have* a strong incentive to protect the productivity of their land against the effects of degradation. But, runs the hypothesis, they *will have* such an incentive only if they have a property right in the land such that they can expect to gain the benefits of any investments they may make to protect the land against degradation. So, after presenting the evidence demonstrating, to my satisfaction, that in fact the productivity effects of soil degradation have been small in the U.S., Europe and the other regions discussed, I then investigate the state of farmers' property rights in the land in each of those regions. Let it be noted that the investigation of the property rights issue at the global scale and in countries in Asia and Sub-Saharan Africa is limited by the absence of the data needed for a completely satisfactory analysis, a situation that has long constrained debate in development circles [e.g., Anderson and Thampapillai (1990)]. But enough such data are available to make the investigation worthwhile [e.g., Blaikie (1985); Anderson and Thampapillai (1990)].

2. Impacts of land degradation on productivity: The U.S.

Legislation passed in the 1930s established the Soil Conservation Service (SCS) in the U.S. Department of Agriculture (USDA). The legislation was a response to the widely held view at the time that soil erosion was a major threat to the then current and future productivity of American agriculture. The view had been propagated primarily by Hugh Hammond Bennett, a soil scientist in the USDA and properly known as the father of the soil conservation movement in the U.S. He also was the first Administrator of the SCS. Bennett wrote [extensively in the decades leading up to his opus Bennett (1939)] and spoke in vivid, even apocalyptic, language about the dangers of soil erosion for American agriculture. In a 1928 publication [Bennett and Chapline (1928, p. 18)], for example, it was asserted that "Everyone who knows anything about it admits the problem is a serious one, but few realize how very devastating is the wholesale operation of erosion. There is necessity for a tremendous national awakening for the need for action in bettering our agricultural practices in this connection, and the need is immediate."

Despite Bennett's strong language, the fact is that at the time the SCS was established no one knew how much erosion was then occurring on the nation's agricultural land, let alone what its productivity impacts might have been [Held and Clawson (1965)]. Bennett's evidence about the seriousness of the erosion threat was strictly anecdotal, mostly based on his visual observations of the farm landscape in the area from the Great Plains to the east coast. Photographs of impressively deep gullies in much of that part of the country – particularly in the south – were used to drive the point home that the threat was indeed serious.

Crosson and Stout (1983) considered the performance of agricultural productivity measured by crop yields (crop output per unit area) in areas of the Midwest and south that according to Bennett were most seriously affected by soil erosion. They found that yields in those areas had increased in step with the impressive increases in yields in the country as a whole between the 1930s and the 1970s. Crosson and Stout concluded that Bennett had over-estimated the yield impacts of soil erosion, or their permanence, or both.

The fact is that it was not until 1977 that reasonably reliable estimates of annual cropland erosion became available. In that year the SCS conducted the first National Resources Inventory (NRI). (Comparable NRIs were subsequently conducted in 1982, 1987, 1992 and 1997.) The NRIs provided county-level estimates of annual (for the years of the surveys) cropland erosion on non-federal land for the country as a whole. By 1984 soil scientists at the University of Minnesota had developed the Productivity Index (PI) model and used it to estimate the yield effects of 100 years of soil erosion at 1977 rates on about 40 million hectares of cropland in the Midwest [Pierce et al. (1984)]. The result showed that the *cumulative* loss of yield over that period would be 4%. (Technology and other management practices were assumed to remain at the 1977 levels.) If technology-driven yield growth over that 100-year period were to grow at even half the rate of the previous 50 years, the 4% erosion-induced yield loss would be lost in the statistical chaff.

In 1989 the USDA, using the Erosion Productivity Impact Calculator (EPIC) model, published an estimate of the cumulative erosion-induced loss of crop yields over the next 100 years in the country as a whole, given 1982 cropland erosion rates and technologies and other management practices [USDA (1989)]. EPIC was developed by a USDA team at the agency's facility in Temple, Texas). The result showed that the cumulative yield loss at the end of the 100 years would be 2.3%.

Crosson (1995a) used the USDA (1989) estimate, with some adjustments, to estimate that the annual cost of erosion-induced losses of net farm income in the U.S. is about $100 million per year. If Crosson had used the PI estimate the result would not have been much different. Compared to the roughly $40 billion in farm income, net of government payments, that U.S. farmers have received over the past some years, the $100 million annual erosion-induced loss is small (0.0025%).

Pimentel et al. (1995), on the other hand, estimated that the economic costs of erosion-induced losses of farm income in the U.S. run at some $25 billion per year. Crosson (1995a, 1995b), after a careful analysis of the Pimentel et al. estimate, concluded that it could not pass even a moderately rigorous scientific scrutiny.

3. Why are the productivity impacts of erosion so low in the U.S.?

The answer, according to the hypothesis investigated here, is that American farmers have sufficiently secure property rights in their land to give them strong incentives to protect the land against erosion damage. I doubt that the assertion about the strength of property rights will be questioned. Nonetheless, there are two aspects to the issue of land management by American farmers that need some discussion.

The first arises from the fact that since the mid-1930s the USDA has invested tens of billions of dollars (prices of 2002) to subsidize the cost of erosion control measures on American farms. Although a 1977 study by the General Accounting Office [GAO (1977)] showed that a significant amount of the USDA's erosion control investment had in fact gone on land without high erosion potential, nonetheless much of that investment did go on such land. Consequently, it must be that investment is part of the explanation of why the productivity effects of soil erosion have been low. Whether in the absence of the subsidy farmers would have invested their own resources to achieve a comparable amount of erosion control is unknown.

The second aspect arises from the fact that some 40% of American farmland is worked under some kind of lease agreement [Soule, Tegene and Wiebe (2000)]. According to these authors the percentage in 1997 was 41. It has long been recognized that farmers operating leased land may have less incentive to protect the land against erosion damage than owner-operators.

Soule, Tegene and Wiebe (2000) considered this issue. They studied the behavior of 941 American maize farmers with respect to investments in erosion control. They took their data from a USDA survey, the 1996 Agricultural Resource Management Study. They classified the 941 farmers as owner-operators, share-renters, and cash-renters. The definition of owner-operators is obvious. Share-renters share some proportion of both costs and revenues with the owners they rent from, and cash-renters pay owners a fixed amount. The authors considered investments in two kinds of conservation practices: (a) conservation tillage, a practice that offers substantial control of erosion on erosive land by planting directly into the residues left from the previous year's crop. The benefits and costs of conservation tillage are essentially short-run. (b) The use of grassed waterways, strip-cropping and contour plowing as erosion control measures. The payoff to these practices is generally medium-term. (Soule, Tegene and Wiebe lacked the data needed to investigate tenure differences with respect to investment in long-term measures, e.g., terracing.)

The authors estimated differences among the three tenure classes in the "predicted probability" of adoption of the two kinds of erosion-control practices. (They did this for both highly erodible land – HEL – and non-HEL. The focus here is on their findings with respect to HEL.) The predicted probability of adoption of conservation tillage on this land by the owner-operators was 36%. For cash renters it was higher, 47%, and for share-renters it was higher still, 58%. For the medium-term practices – grassed waterways, etc. – the relationships were reversed. The highest probability of adoption, 85%, was for owner-operators; the second highest, 72%, was for cash-renters, and the percentage for share-renters was 66. With respect to the finding with respect to this set of relationships, Soule, Tegene and Wiebe (2000, p. 1003) suggest "...that the timing of benefits from conservation practices affects decisions about their adoption, and that landowners are not fully successful in requiring tenants to adopt practices with delayed benefits".

For the farmers they studied Soule, Tegene and Wiebe seem to have established with reasonable confidence that decisions to invest in erosion control practices do, in fact, vary with land tenure, and that the variation is in the direction that one would expect, that is, renters are less inclined to invest in practices with more than a short-term payoff than owner-operators. It is noteworthy, however, that even so, a substantial percentage of the cash-renters and share-renters did in fact invest in medium-term payoff practices (72% and 66%, respectively).

Considering that roughly 60% of American farmers are owner-operators and that the 40% who are renters nonetheless seem to have reasonably strong incentive to invest in erosion control practices with medium-term payoff, I conclude that the evidence supports the hypothesis that erosion damages to soil productivity are low in the U.S. primarily because farmers have incentive to keep them that way.

4. Impact of land degradation on productivity: Europe

There is little information about the scale of soil erosion in Europe and of its impact on agricultural productivity. In fact, the best source of such information is in a short article by Boardman (1998). Boardman's article is of interest not only as a source of the kind of information of interest here, but also because it tells a cautionary tale of how a certain estimate of annual erosion per hectare in Europe got embedded in the literature on the subject. Boardman demonstrates that the estimate is totally specious.

The initial work that led to the specious estimate was done by Bollinne (1982). Bollinne set out to estimate per hectare erosion rates on experimental plots in central Belgium, using four different methods. In the end, Bollinne decided not to publish average estimates of the amounts of erosion that he found because he considered the estimates not sufficiently definitive.

The next step that led up to the specious estimate was taken by Richter (1983), who interpreted Bollinne's work to mean that average erosion on agricultural land in Central Belgium was 10–25 tons/hectare/year. Based on Richter's work, the World Resources Institute [WRI (1986)] also stated that average annual erosion in Central Belgium was 10–25 tons/hectare/year. Lal, Hall and Miller (1989) took the figure from WRI but applied it to all of Belgium. Barrow (1991), based on Lal, Hall and Miller (1989), did the same thing. Finally, Pimentel et al. (1995) took Barrow's estimate, averaged it to 17 tons/hectare/year and then applied it to all of Europe. Boardman notes that the Pimentel article in which the 17 ton figure is used has been widely referenced (perhaps because it appeared in the prestigious journal *Science* – my observation, not Boardman's), and now appears to be part of the conventional wisdom about the amount of erosion in Europe.

Boardman cites another publication by Pimentel (1993) and quotes him as saying [Boardman (1998, p. 47)] that "soil loss rates in Europe range between 10 and 20 t/ha/yr". Pimentel provides no source for this estimate. Boardman notes that Pimentel (1993) includes two chapters on erosion rates in two European countries, the U.K. and Poland. After presenting the erosion estimates for those two countries in those two chapters, Boardman (1998, p. 47) states that the estimates suggest "... that Pimentel's average erosion rate is about 15 times too high if applied to those two countries". Boardman (p. 47) goes on to say that independent estimates of erosion rates from field monitoring schemes in Europe suggest that "... if we have to use average figures then 1 and 5 t/ha/yr is a realistic estimate, but that there is considerable variability in space and time ...". These numbers suggested by Boardman are substantially less than the erosion rates estimated in the various NRIs for the U.S. Since we found that the U.S. rates do not pose a serious threat to soil productivity in the U.S., it would seem to follow that the threat in Europe is even less.

I have no detailed knowledge of the structure of property rights in agricultural land in Europe, but my impression is that they are reasonably strong. If this is correct, it likely is a sufficient explanation for the finding that rates of erosion and consequent yield impacts in Europe are low.

5. Impacts of land degradation on productivity: Global scale

Brown and Wolf (1984) estimated that on a global scale, *excessive* soil erosion that is, the amount above that consistent with maintaining the productivity of the soil over the long-term, was some 26 billion metric tons per year. They did not estimate the consequent losses of soil productivity, but their estimate of the amount of excessive erosion suggests that the losses would be high.

Crosson (1995c) subjected the Brown and Wolf estimate to careful analysis and found that it was based essentially on reasonably good U.S. estimates of erosion that were then extrapolated using a string of "iffy" assumptions. Crosson did not conclude that the Brown and Wolf estimate was necessarily wrong, but that its analytical underpinnings were so weak that it could not be taken seriously.

The fact is that until the early 1990s very little was known about global-scale erosion rates, much less about their productivity impacts. In a comprehensive review of the literature available up to 1990 on global soil erosion and its productivity impacts, Nelson (1988, p. 1) concluded that the evidence with respect to the rate, extent and severity of soil erosion around the world was "extraordinarily skimpy". Other specialists on land degradation issues had come to the same conclusion. Dregne (1988, p. 679) noted that estimates of land degradation, including his own, were based on "... little data and much informed opinion". Writing specifically of soil erosion and its productivity effects, Dregne (1988, p. 680) asserted that "... there is an abysmal lack of knowledge of where water and wind erosion have adversely affected crop yields". El-Swaify, Dangler and Armstrong (1982, p. 1), authors of what at the time was the most comprehensive published study of soil erosion in the developing countries, asserted that "... there is little or no documentation of the extent, impact or causes of soil erosion ... " in tropical environments.

Since these several authors wrote, two pioneering studies have been published that for the first time give promising first approximation estimates of the present extent and severity of human-induced degradation of agricultural land on a global scale [Dregne and Chou (1992); Oldeman, Hakkeling and Sombroeck (1990), conveniently summarized in Oldeman (1994)]. The Dregne and Chou study is an effort to extend the knowledge frontier with respect to the extent and productivity effects of human-induced land degradation beyond the narrow limits noted by Dregne (1988) and others.

Dregne and Chou (1992). Dregne and Chou estimated the spatial extent and productivity effects of land degradation in dry areas in most countries of the world, including all of the big ones. Dry areas are those in arid, semi-arid and dry sub-humid climatic zones. The estimates are for three kinds of land use: rainfed cropland, irrigated cropland, and range. Drawing on data prepared by the United Nations Food and Agriculture Organization [e.g., FAO (1990)], Dregne and Chou found 5.1 billion hectares of dryland in the three uses, 88% of it in range, 9% in rainfed crops, and 3% in irrigated crop production. Degradation of rainfed cropland is mainly by water and wind erosion. Irrigated land is degraded mainly by salts carried and deposited by irrigation return flows, and rangeland is degraded mainly by overgrazing, which results in increased erosion and a decline in the quality of vegetation for animal forage. To estimate the extent and severity of degradation on these three kinds of land use Dregne and Chou drew on the voluminous published and unpublished literature relevant to the subject (shaky though much of it was), and, importantly, on their own extensive experience in studying dryland agriculture.

They classified rainfed and irrigated cropland as slightly degraded (0–10% loss of productivity), moderately degraded (10–25% loss), severely degraded (25–50% loss) and very severely degraded (more than 50% loss). For rangeland the corresponding ranges of loss were 0–25%, 25–50%, 50–75% and more than 75%. The ranges for rangeland are wider than for rainfed and irrigated cropland because, according to Dregne and Chou, range analysts consider rangeland that has lost no more than 25% of its original productivity to be in good condition. The estimated productivity losses are independent of the technologies and management practices used on the land.

For each of the categories of severity of productivity loss Crosson (1995a) assumed that the loss was at the mid-point of the range given by Dregne and Chou. That is, for slightly degraded land the loss was assumed to be 5%, for moderately degraded land it was assumed to be 18%, and so on. Crosson then weighted these estimates of productivity loss by the amount of land in each degree-of-severity category in each of the three land uses to calculate the weighted average loss in each use. The averages were as follows:

Irrigated cropland	10.5%
Rainfed cropland	12.9%
Rangeland	43.0%

Finally, because in terms of lost productivity a loss of a hectare of irrigated land imposes a higher social cost than the loss of a hectare of rainfed cropland, which imposes a higher social cost than a loss of a hectare of rangeland, Crosson calculated the weighted average loss on the three kinds of land use taken together by weighting the percentage loss for each by its per hectare value of production. According to Dregne and Chou, these values (in prices around 1990) were $625 for irrigated land, $95 for rainfed cropland, and $17.50 for rangeland. This calculation showed that the average loss for the three land uses taken together was 12%. This is the cumulative loss over some period of time, which Dregne and Chou do not specify. But for most of the land the period must be several decades. Over three decades the average annual rate of loss would be 0.4%.

Oldeman (1994). This report by Oldeman is a summary of earlier work by him and associates at the Wageningen Agricultural University, The Netherlands [Oldeman, Hakkeling and Sombroeck (1990)]. The resulting estimates of global-scale soil degradation were based on responses to a survey of some 200 soil and environmental scientists around the world. The scientists were asked to estimate the extent of land degradation in their areas, using four categories of degradation: not degraded, lightly degraded, moderately degraded, and strongly + extremely degraded. The land included was in annual and permanent crops, permanent pasture, and forest/woodlands. Using FAO data, Oldeman, Hakkeling and Sombroeck found 8735 million hectares in these uses. The survey yielding the degradation estimates showed that 6770 hectares of this land was not degraded, while 1966 hectares (between 22 and 23% of the total) were degraded to some extent. Of the degraded land, 749 million hectares were lightly degraded, 911 million hectares were moderately degraded, and 305 million hectares were strongly/extremely degraded [Oldeman (1994, p. 112)]. Water and wind erosion were responsible for 84% of the degraded land; in what follows I assume that the distribution of the other 16% among the degree-of-degradation categories is the same as for wind and water erosion.

Oldeman, Hakkeling and Sombroeck (1990) did not estimate the amount of soil productivity loss in each of the degree-of-degradation categories. Rather they state the degradation consequences in terms of the difficulty of repairing the damage. Lightly degraded land can be restored by changes in on-farm management practices financed out of farmers' own resources; repair of damages to moderately degraded land is more difficult and would require some extent of public funding in addition to what farmers might contribute. Strongly degraded land is beyond repair without major publicly funded investments on a scale that likely would be beyond the financial capacity of some, if not many, developing countries. Extremely degraded land is beyond repair.

Crosson (1995a) assumed that the percentage productivity loss for each of the degree-of-degradation categories used by Oldeman, Hakkeling and Sombroeck (1990) was the same as in the work of Dregne and Chou (1992). That is, lightly degraded land has lost 5% of its productivity, moderately degraded land has lost 18% of its productivity and so on. Undergraded land – 6770 million hectares – is assumed to have lost no degradation-induced losses of productivity so is assigned a zero percentage of loss. Weighting these percentages by the total amount of land in each degradation category gives a weighted average loss of 4%. This is the *cumulative* productivity loss over the 45 years from the end of World War II and 1990 [Oldeman, Hakkeling and Sombroeck (1990)]. The average annual rate of loss is 0.1%.

The estimated productivity loss has to be low because, according to Oldeman (1994), only some 22–23% of the 8735 million hectares in crops, permanent pasture, and forest/woodland has suffered any degradation-induced losses. So in the calculation of the weighted average loss, 77–78% of the total amount of land gets a productivity loss of zero. Even if the productivity loss in each degradation category were double the amount assumed here, e.g., 10% for lightly degraded land, 35% for moderately degraded land and so on, the cumulative weighted average loss over the 45 years still would be only 8.3%. The average annual rate of loss would be 0.18%.

Comparison with other global estimates. There are no other estimates of global degradation-induced losses of soil productivity. But the global-scale estimate by Brown and Wolf (1984) of 26 billion tons per year of *excessive* soil erosion world-wide and another estimate by Pimentel et al. (1995) that global soil erosion is 75 billion tons per year probably imply much higher rates of degradation-induced productivity than estimated here. I gave above my reasons for thinking that the Brown and Wolf estimate is not to be taken seriously. The Pimentel et al. estimate is taken from Speth (1994), who is cited as the authority for their statement that "... about 80% of the world's agricultural land suffers moderate to severe erosion and 10% suffers slight to moderate erosion." Note that these estimates are far higher than the percentages for comparable degradation categories in Oldeman (1994). If the Speth-based percentages are correct, the degradation-induced losses of soil productivity would be substantially higher than I estimated above, based on Dregne and Chou (1992), Oldeman (1994) and Oldeman, Hakkeling and Sombroeck (1990).

Speth is widely recognized as a very able man, but he is not expert in matters related to land degradation. Dregne and Chou and Oldeman, Hakkeling and Sombroeck are such experts. I consider their work to be the most authoritative done to date on the issue of global-scale degradation-induced losses of soil productivity.

Do property rights account for the low global-scale degradation-induced losses of soil productivity? The U.S. and Europe are, of course, important parts of global agriculture, and I have already concluded that in those areas firm property rights in agricultural land probably do explain why degradation-induced losses of soil productivity in those areas are small. With respect to the rest of the world there is no comprehensive treatment of the property rights issue with respect to degradation of agricultural land. At least I know of no such treatment. There is, however, a substantial literature dealing with soil degradation and the relationships of that to property rights in some important countries in Asia and in parts of Sub-Saharan Africa. The following discussion is based on a sampling of that literature.

6. Soil degradation and property rights in parts of Asia

China and Indonesia. Lindert (2000) used time series on agricultural land quality in China and Indonesia to study changes in soil quality over time. In both countries the data ran more or less continuously from the 1930s through the 1980s, and in both of them the data were for the major agricultural regions. Lindert found that in both countries the soils lost some organic matter and nitrogen over the period covered but had increases in phosphorus and potassium. Changes in soil salinity and acidity were small. (Lindert followed traditional practice in defining soil quality in terms of the content per unit soil of these various materials.) Soil productivity, measured by crop output per hectare, rose in both countries. The relatively small losses of organic matter and nitrogen evidently were compensated by increased use of inorganic fertilizers. In the period covered population growth in both countries was high and the per capita amounts of agricultural land declined.

One of the productivity reducing characteristics of erosion over long periods is that it may carry away enough topsoil to reduce the depth of the crop-rooting zone. The more constricted space in which crop roots can take hold has adverse productivity effects. Moreover, the subsoil is generally less rich in nutrients and has smaller water-holding capacity than the topsoil, another avenue by which erosion reduces soil productivity.

Lindert's data permitted him to measure long-term trends in soil depth in both China and Indonesia. In this connection he writes: "The raw averages show that the topsoil layers did indeed become thinner in many parts of China and Indonesia.... A closer look at the original data, however, suggests that the topsoil losses are a false alarm. They are likely to reflect a change in the data gathering procedure rather than a true change in the soil." [This quote is not from Lindert's book but from a chapter in a later book, Lindert (2001, p. 141).] Then later in the same context he writes (p. 141): "Pending further tests the tentative conclusion is that the average net loss of topsoil in these two countries was close to zero."

Lindert's finding of no significant change in the quality of soils or in soil depth over about 50 years in China and Indonesia, combined with substantial increases in crop yields in both countries, suggests that in neither country has soil erosion been a significant problem. But with respect to China, Lindert's finding seems to be directly contradicted by other researchers. For example, Wen (1993, p. 63) writes: "About 6.7 million hectares, or 5% of total cultivated land, have become deserts. About 8 million hectares, or 6% are too saline for cultivation. About 42 million hectares of China's cultivated land, or one-third of the total cultivated land, are undergoing serious water and wind erosion. Only a quarter of the cultivated land is well managed and highly productive... ."

These numbers indicate that 40% of China's cultivated land is subject to some form of serious degradation. Wen then considers the erosion problem on China's loess plateau, an area of 53 million hectares, home to 70 million people and one of the countries important agricultural regions [Wen (1993, p. 65)]. The loess soils are very deep, some 100 meters on average. According to Wen, the region is subject to major erosion problems, and he presents a table (p. 65) showing that the erosion rate on 45% of the loess land exceeds 20 tons/hectare/yr. This is a high rate, certainly by U.S. standards. The soils are so deep that even at that rate it could take many years before the erosion would have an affect on crop yields. But over a period of some decades, one like that covered by Lindert's data, it seems likely that erosion in excess of 20 tons/hectare/yr would show up as a reduction in soil depth. As noted above, Lindert's data showed no such reduction.

There is an aspect to Wen's work that is somewhat confusing if one's interest, as in my case, is in the relationship between soil erosion and soil quality relevant to agriculture. On p. 67, Wen has a table that shows, among other things, total cropland and total eroded area in four of China's regions, including the loess plateau. The table indicates total cropland in the four regions to be 52 million hectares and the eroded area to be 137 million hectares, 2.6 times as much as the amount of cropland. One would like to know how much of the erosion is on cropland, but Wen does not provide that information.

Wen has no discussion of the property rights situation in China as it may bear on the soil erosion problem, although, as noted above, he does state in passing that only one-quarter of the country's cultivated land is well managed and highly productive. Lindert does not directly confront the land property rights issue either. He does have some discussion of how the process of agricultural development might promote a strengthening of property rights in the land, but he does not relate the discussion directly to China or Indonesia.

Li, Rozelle and Brandt (1998) studied the relationship between land tenure rights and incentives to invest in land improving measures by investigation of land management in 80 households in five north China villages. Each household held both privately owned land and "responsibility" land, i.e., collective land that farmers work under a contract with the village leaders binding the farmers to provide a certain quota of low-priced grain and/or cotton in exchange for the right to use the land. The authors note that tenure in the privately held land is generally secure and farmers can use it as they think best, including the right to make land swaps with neighboring farmers. Interviews by the authors "... revealed that many farmers in villages across China treat their land as if it were their own" (p. 64). The context of this statement suggests that it applies only to the privately held land.

The study of the 80 households showed that both inputs and crop yields were higher on the private land than on the "responsibility" land. Analysis showed that some of these differences were because the soils in the privately held land were of better quality than that in the "responsibility" land. Also, on average, the "responsibility" land was one-third more distant from the location of the household than the privately held land.

In conclusion, the authors assert (p. 69) that "by far the strongest, most robust finding is that the right to use land for long (or indefinite) periods of time encourages the use of land-saving investments". This finding is, of course, totally consistent with the conventional wisdom about the land tenancy/investment relationship. But the authors tell us nothing about the extent to which their finding can be generalized beyond the 80 households that they studied. Consequently, their work sheds little light on the question, to what extent might the property rights argument help to explain Lindert's finding that long-term soil erosion has done little, if any, damage to soil quality in China?

A brief statement in another source [Deininger and Feder (2001)] may throw some implicit light on the question: in China, how much agricultural land is operated under tenure systems that provide farmers incentive to invest in land conservation? Deininger and Feder note that around the world the transition from collective to private cultivation has historically been associated with big increases in productivity, and they cite China as an example. This is suggestive, but clearly not definitive.

Thailand. In their article on the relationship between farmland tenure security and investments in the land, Feder and Onchan (1987) write (p. 311): "Insecure ownership of land is a characteristic of many farmers in less developed countries in most regions of the world ... ." On its face, this statement could appear to provide a negative answer to the second question I have posed in this paper: are farmland tenure rights around the world secure enough to explain my finding [from Dregne and Chou (1992) and Oldeman, Hakkeling and Sombroeck (1990)] that over the past 4 or 5 decades land degradation has not seriously reduced global agricultural production capacity? But the challenge of the quoted Feder/Onchan statement to my question is attenuated, at least to some extent, by a subsequent statement (p. 312): "... ownership security enhances both the incentive to invest and the ability to implement investments [by increasing access to credit]. While this argument is intuitively plausible, *no research has rigorously estimated the effect of ownership security on farm investment*." (Emphasis added.) There are two comments that might be made about this statement. (a) The authors wrote it in 1987. They probably would not write it today because, as will be shown below in the discussion of literature bearing on Sub-Saharan Africa, many such studies have been done since 1987. (b) Given the italicized part of the statement, one must wonder how much authority should be given to the previously quoted statement about the weakness of the farmland tenure situation in the developing countries.

The Feder and Onchan study is focused on Thailand. Using data collected in villages in three Thai provinces they addressed the relationship of land tenure security to investments in land-conserving measures among two sets of Thai farmers. One set had secure property rights to the land they farmed; this set worked in close proximity to the other set, who were squatters on government-owned land, hence lacking in legal title to the land they farmed. That the two sets of farmers worked in the same area helped to control for soil and climate conditions.

Two kinds of land-conserving investments were considered: (a) bunding, by which fields are divided into sub-plots by raised earth walls. This gives better water (and presumably erosion) control, including increased retention of soil water. (b) Clearing the land of stumps. Feder and Onchan note that these investments require a commitment of financial resources.

The authors did a statistical analysis of the data they collected. In summing-up their findings they write (p. 317): "... For the pooled data [from the 3 provinces] the adoption of bunding and clearing are significantly higher on titled plots." Subsequent discussion suggests, however, that the reason is not that weak tenure rights weaken the *incentive* of squatters to invest but rather that it deprives them of *the financial wherewithal* to do so. In this connection Feder and Onchan note that although the squatters they studied "... face relatively small eviction risk ... their borrowing from cheap institutional sources is significantly lower than that of farmers with secure land ownership" (p. 318).

The Feder and Onchan study, like that of Li, Rozelle and Brandt (1998) on China, offers further confirmation of the importance of land property rights for incentives to invest in land conservation. But, also like the Li, Rozelle and Brandt study, that of Feder and Onchan throws little light on the question addressed here: on broad regional and global scales, how important are land property rights in explaining my finding that on those scales soil degradation has had little impact on agricultural productivity.

Sub-Saharan Africa. The literature review undertaken in preparation of this paper showed that the land tenure/investment incentive issue has been more explored for Sub-Saharan Africa (SSA) than for any other major region. Moreover, the earliest source cited here is 1993 and the latest is 2002, suggesting that this literature provides a reasonably up-to-date view of the tenure/investment incentive situation in SSA.

A common theme running through the cited literature is that the long-held conventional view that land degradation is a major obstacle to agricultural advance in SSA is wrong. Leach and Mearns (1996b) describe the conventional view as holding that in SSA land has been seriously damaged by overgrazing, desertification of drylands, deforestation, and soil erosion, mostly because of rapid population growth. Yet, they continue, the accumulated research assembled in their book suggests that the conventional wisdom "... may be deeply misleading" (p. 1). Some pages later they write (p. 5) "... received wisdom would have much of the blame for the vegetation change perceived by outsiders as environmental 'degradation' rest with local land-use practices, whether labeling them ignorant and indiscriminant or – more commonly – as ill-adapted to contemporary socio-economic and demographic pressures. In such accounts rural people's ecological knowledge is notable mostly by its absence, silenced before it is investigated." The Leach and Mearns (1996a) book marshals reports from scholars specialized in research on land-use issues in SSA. The cumulative impact of their work is that the conventional wisdom of these issues not only is "deeply misleading", it is mostly dead wrong.

One of the earlier studies in the literature cited here [Cleaver and Schrieber (1994)] found (not surprisingly) that farmers in SSA "... adopt soil conservation measures when they clearly perceive them to be in their own interest. In low resource and labor-constrained settings, and with risk-averse farmers, measures recommended for adoption must increase crop yields ... require little or no cash outlays, and conflict as little as possible with peak labor demands ... ."

"The shortage of labor has been one major reason for the poor record of many soil conservation programs. The other has been the perceived low rate of financial return to most of the methods that would be technically effective. Where, however, the labor/land ratio is high, as in parts of the East African highlands, various labor-intensive soil conservation techniques are financially attractive and, indeed, widely used. This suggests that farmers' willingness to undertake soil conservation measures will increase as population densities rise, as soil degradation and erosion problems intensify, and as policy reforms make intensive farming more profitable" (p. 129).

Place and Hazell (1993) studied results from a household survey done in Ghana, Kenya and Rwanda and found that "Rights which farmers hold over individual parcels of land vary widely, but in many cases are surprisingly privatized. Yet, with few exceptions, land rights are not found to be a significant factor in determining investments in land improvements, use of inputs, access to credit, or the productivity of land" (p. 10). Subsequently they speculate that one of the reasons for these findings was that "... there are more binding constraints [than tenure] on agricultural productivity (such as lack of improved technologies or adequate access to credit" (p. 19).

Unlike Place and Hazell in the countries they studied, Hayes, Roth and Zepeda (1997) in a study of the tenure/investment relationship in Gambia found that "Some of the relationships hypothesized between tenure security, investment and yields are corroborated. In particular, tenure security is found to enhance long-term investments, which in turn enhance yields" (p. 369).

Several studies in the cited literature make the interesting point that in some parts of SSA farmers will make investments in the land *because* their property right in the land is weak. In those areas, such investments tend to strengthen the right. Place and Hazell (1993) found this in a study of Malawi, Gray and Kevane (2001) found it in Burkina Faso, and Sjaastad and Bromley (1997) found it more generally in SSA. The latter argue that in some parts of that region "... a farmer with indigenous [i.e., informal] tenure will choose an investment schedule that maximizes the combined effects of productivity increase and [tenure] security increase; a farmer with freehold will be concerned only with productivity" (p. 556). Because of this rather curious situation of land rights, Sjaastad and Bromley argue that across SSA generally farmers with indigenous rights may actually invest more in land improvements than freehold farmers. Then in summing up their argument, Sjaastad and Bromley assert that in a dynamic development setting "Indigenous tenures may contain intrinsic mechanisms and incentives that, if allowed to unfold, trigger and sustain a transition to tenures that resemble the familiar Western institution of freehold" (p. 559).

Gichuki (2000a, 2000b) studied land management in the Makueni district of Kenya. In one of the reports Gichuki (2000a) notes that land tenure in the study area had changed from Crown Land in the 1940s to mainly individual ownership of small holdings, which varied in size from 0.8 to 40 hectares. Because of this shift in tenure, "...insecurity of rights to land (which are heritable) has not constrained investments" (2000a, p. 21). Then, on p. 22, Gichuki writes: "Improved land tenure has resulted in strengthening the security of the investments made in land improvement at the household level... ."

In the second article Gichuki (2000b) reported on a study of erosion on agricultural land in four regions of the Makueni district. In each region erosion observations were made on cropland, grazing land, footpaths and roads. In all four regions erosion on land in those uses varied from very low on 88% of the fields to low on the other 12%. Although cropland productivity differed among the regions only in one of them was erosion cited as a significant source of the difference. Gichuki observes that all of the farms visited in the regions had soil and water conservation measures in place, and that "Farmers were able to make long-term investments in soil conservation because of the granting of secure tenure" (p. 13).

Tiffen, Mortimore and Gichuki (1994) published the results of a long-term study of agricultural change and land use in the Machakos district of Kenya. They noted that in the 1930s the British authorities in Kenya deplored the condition of the land in Machakos, reporting that because of severe over-grazing the land had been seriously degraded and much of its productivity lost. With the passage of time, conditions in the district began to change. In the 1960s farmers there generally rejected a government project to build terraces to protect the land against erosion because they judged the particular technique proposed by the government to be inappropriate for local conditions. But the farmers then built different, in their judgment, more appropriate terraces using mostly their own resources. By the late 1980s population in the district had increased substantially over the previous 50 years, farmers had maintained the terraces and also widely planted trees, both for commercial and conservation purposes, and agricultural production and productivity had increased significantly, as had off-farm employment opportunities for farmers and their families. Tiffen, Mortimore and Gichuki (1994) attribute the agricultural transformation of the Machakos district to a process much like that described theoretically by Boserup (1965). Rising population pressure leads to increasing scarcity, hence value, of agricultural land, which, combined with increasing commercialization of agricultural output, strengthens farmer incentives to adopt more land-intensive practices, including soil and water conservation. The rising value of the land also creates pressure for better definition of land property rights and a shift away from traditional common property regimes toward greater emphasis on private rights.

Tiffen, Mortimore and Gichuki (1994) describe something like this process in Machakos, noting two critical aspects of it. One was the existence of a rising market for agricultural output in Nairobi, which was linked to Machakos by a good all-weather road. The other was the movement in the district toward strengthened private property rights in agricultural land. [Recall from above that Gichuki (2000a, 2000b) described the evolution of property rights, and consequent strengthening of incentives for investment in soil and water conservation in the Makunei district of Kenya in a way that sounds very similar to what they, Tiffen, Mortimore and Gichuki, found in Machakos district.]

Gray and Kevane (2001) tell a story about the evolution of land tenure rights and agricultural intensification in southwestern Burkina Faso that sounds very much like what happened in the Machakos and Makunei districts of Kenya. The authors begin by noting the conventional wisdom that soils in SSA are fragile "... and subject to catastrophic collapses in fertility" (p. 573). In the conventional wisdom account, they note, population growth is the main culprit in the land degradation scenario.

Gray and Kevane then cite 3 studies (2 for 1993 and 1 for 1994) that they say show "... that with growing populations and reductions in fallow time farmers [in southwestern Burkina Faso] are using techniques that substitute for the inability to fallow fields. Productivity is often augmented or restored with fertilizer, manure, agro-forestry or cover crops ..." (p. 574). Then, contrary to other evidence cited in this paper, Gray and Kevane refer to 3 studies that, they say, show that "... research has consistently failed to demonstrate impacts of titling and formal individualization on investment behavior .... This is particularly true for Burkina Faso, where studies have found no link between tenure status and agricultural practice" (p. 574).

Gray and Kevane were cited above as being among those writing on land use in SSA who have picked up on the fact that at least in some parts of that area farmers invest in the land precisely *because* their right in the land is weak and the investments are a way of strengthening it. With respect Burkina Faso they write that the process of agricultural intensification "... has implications for soil quality, but also for land rights .... By investing in quality farmers are simultaneously building land rights. A growing literature puts the individual actor in the dynamic position of creating land rights through cultivation and investment strategies" (pp. 574–575). Then "The longer one can stay on a field, whether one is a local or migrant farmer, the more difficult it is to take the land away and the less authority lineages and communities have over the field" (p. 575).

This account of the strong interest farmers in Burkina Faso (and, from other sources, elsewhere in SSA) have in acquiring secure property rights in the land raises a question of what to make of the previously quoted statement from Gray and Kevane that no research on land use in SSA had shown any relationship between tenure status and investments in the land.

Like Gray and Kevane, Niemeijer and Mazzucato (2002) studied land use and other agricultural practices in Burkina Faso. Also like Gray and Kevane, Niemeijer and Mazzucato begin their account with reference to the conventional wisdom's view that land in SSA is seriously degraded. They then assert that the studies on which this view is based are seriously limited because "... they have never been sufficiently validated with data on actual soil conditions and/or productivity to support their findings of widespread degradation and nutrient depletion" (p. 23). With respect to Burkina Faso (BF) they note that it is experiencing all the demographic and economic trends usually associated with soil degradation. In fact, BF has the highest rating of any Sahelian country on the Severity of Human-Induced Soil Degradation Index calculated by the World Economic Forum.

Population in BF more than doubled in the past 40 years and population density has reached more people per km^2 in some provinces, making it one of the most densely populated countries in the West African Sahel. There is little use of modern agricultural technology, e.g., in 1994 only 7 kg of fertilizer was used per hectare of cropland, and that was mostly on cotton, a largely commercially grown crop.

Despite little use generally of modern technology, yields of major crops do not appear to have fallen in BF over the last 40 years. In fact, between 1960 and 2000 rice yields about tripled and maize yields increased almost as much. These increases are attributed by Niemeijer and Mazzucato to increased irrigation, mechanization and fertilizer use in some parts of the country. Yields of sorghum, millet and groundnuts also increased, but not as much as for rice and maize.

In view of this yield experience, the authors assert that "It appears unlikely that the productivity of the land has significantly declined in the last few decades ..." (p. 25). In this connection they note that few soil degradation studies measure changes in soil properties over time, nor do they analyze these changes under long-term cultivation as practiced on small family-run farms. Such studies typically use models to study degradation/productivity relationships on experimental plots that do not "... match the diverse conditions and dynamic management practices found on small farms in the region" (p. 27).

In 1996 Niemeijer and Mazzucato did a survey of soil fertility in 2 villages in eastern BF and found that the results were "remarkably" similar to those found in a study done in the late 1960s for similar soil types and land uses. In 1996 they also did a study of 2 villages in different agro-ecological zones to examine differences among soils on cultivated land with respect to organic matter (OM), nitrogen (N), potassium (K), and phosphorus (P) on 4 kinds of land use. One kind was "compound fields", which are closest to the farm family home. These were the most intensively cultivated fields. Another kind was "village fields", which are further out, the third was "bush fields", which are the farthest out, and the fourth was land that had not been cultivated for at least 20 years, although it was cultivable. The authors found that all 3 of the nutrients were greater in the soils in the 3 kinds of cultivated land use than in the land that had not been cultivated. OM was slightly less on the compound and village fields than on the uncultivated land. All 3 nutrients and OM were higher on the most intensively cultivated compound fields than on the village fields, and they were higher on those fields than on the bush fields. On the basis of these findings Niemeijer and Mazzucato assert that cultivation on small family farms in BF (and by extension, elsewhere in SSA) does not necessarily lead to nutrient depletion, indeed, that it can actually increase fertility on intensively cultivated land.

In connection with the same study they collected "cultivation histories" from farmers in the region. They found that various soil and water conservation practices were more commonly employed at the time of the study than they had been in the past. This process was promoted by the emergence in the region of "social networks". These are "self-help" relationships among farmers that increase their access to the resources needed, for example, to undertake the investments in soil and water conservation. They also help farmers to diversify their livelihoods to take pressure off the land.

7. Conclusion

I am satisfied that the effects soil erosion and other forms of land degradation on soil quality and crop yields have been and are small in the U.S. and Europe. I am also satisfied that the strength of property rights in agricultural land in those areas is a sufficient explanation of that finding.

I am reasonably well satisfied that the work of Dregne and Chou (1992) and Oldeman, Hakkeling and Sombroeck (1990) demonstrates that the soil quality and productivity effects of soil erosion and other forms of land degradation are small on a global scale, although the empirical evidence for this conclusion is not as strong as it is for the U.S. and Europe [e.g., OECD (2001)].

And the evidence on land property rights around the world is too skimpy to conclude that the finding of low land degradation damage – if indeed that finding is correct – is attributable to the existence of strong rights for farmers most everywhere. The work of Lindert (2000) suggests that land degradation damage in China and Indonesia has been small over the past few decades, although other work in China seems to contradict this. In any case, Lindert does not treat the issue of property rights, and the work that has been done on that in China, although showing that where private property rights exist farmers invest in land conservation measures, I have found no information indicating how much Chinese agricultural land has that tenure status.

The work considered in Thailand suffers the same limitation for purposes of this chapter. The work showed that land tenure makes a difference in farmer incentives to invest in soil conservation, but it provided no information indicating how general across all of Thailand this conclusion could apply. Further research there and indeed elsewhere in the developing world could usefully inform policy making, as argued by Crosson and Anderson (1993) with regard to extending traditional agricultural research agendas further into natural resource management.

The substantial amount of research on the tenure/investment relationship in SSA suggests strongly, at least to me, that the conventional wisdom that land degradation is a serious problem in that region is mistaken. The research also suggested that although across SSA titled land rights are in short supply, indigenous rights seem to be sufficiently strong, at least in many parts of the region, to give farmers incentive to invest in soil conservation. The net outcome of the discussion of the tenure/investment relationship in SSA, in my judgment, is as follows: conditions there with respect to use of agricultural land are nowhere near as bad as many believe who not really familiar with the region. Moreover, there is evidence of an economic and social dynamic that could be moving the region toward a gradual strengthening of private property rights in agricultural land. If this happens, the soil conservation situation, already better than many believe, should improve even more.

References

Anderson, J.R., Thampapillai, J. (1990). Soil Conservation in Developing Countries: Project and Policy Intervention. Policy & Research Series, vol. 8. World Bank, Washington, DC.

Barrow, C.J. (1991). Land Degradation: Development and Breakdown of Terrestrial Environments. Cambridge University Press, Cambridge.

Bennett, H.H. (1939). Soil Conservation. McGraw-Hill, New York.

Bennett, H.H., Chapline, W.R. (1928). "Soil erosion: A national menace", US Department of Agriculture Circular No. 33. US Government Printing Office, Washington, DC.

Blaikie, P. (1985). The Political Economy of Soil Erosion in Developing Countries. Longman, London.

Boardman, J. (1998). "An average soil erosion rate for Europe: Myth or reality?". Journal of Soil and Water Conservation 53, 46–50.

Bollinne, A. (1982). "Etude et prévision de l'érosion des sols limoneux cultivés en moyenne Belgique", Ph.D. thesis. Université de Liège, Belgium.

Boserup, E. (1965). Population and Technological Change: A Study of Long-Term Trends. University of Chicago Press, Chicago.

Brown, L., Wolf, E. (1984). "Soil erosion: Quiet crisis in the world economy", Worldwatch Paper No. 60. Worldwatch Institute, Washington, DC.

Cleaver, K.M., Schrieber, G.A. (1994). Reversing the Spiral: The Population, Agriculture, and Environment Nexus in Sub-Saharan Africa, Directions in Development. World Bank, Washington, DC.

Crosson, P. (1995a). "Soil erosion and its on-farm productivity consequences: What do we know", Discussion Paper No. 95-29. Resources for the Future, Washington, DC.

Crosson, P. (1995b). "On-farm costs of soil erosion: What do we know?", Working Paper No. 95-27. Resources for the Future, Washington, DC.

Crosson, P. (1995c). "Soil erosion estimates and costs". Science 269, 461–463.

Crosson, P., Anderson, J.R. (1993). "Concerns for sustainability: Integration of natural resource and environmental issues for the research agendas of NARSs", Research Report No. 4. ISNAR, The Hague.

Crosson, P., Stout, A. (1983). The Productivity Effects of Cropland Erosion in the United States. Resources for the Future, Washington, DC.

Deininger, K., Feder, G. (2001). "Land institutions and land markets". In: Gardner, B., Rausser, G. (Eds.), Handbook of Agricultural Economics. Elsevier/North-Holland, Amsterdam, pp. 288–331.

Dregne, H. (1988). "Desertification of drylands". In: Unger, P., Sneed, T., Jordan, W., Jensen, R. (Eds.), Challenges in Dryland Agriculture: A Global Perspective, Proceedings of the International Conference on Dryland Farming. Texas Agricultural Experiment Station, Bushland/Amarillo, TX, pp. 678–680.

Dregne, H., Chou, N.-T. (1992). "Global desertification dimensions and costs". In: Dregne, H. (Ed.), Degradation and Restoration of Arid Lands. Texas Tech University, Lubbock, pp. 249–282.

El-Swaify, S., Dangler, E., Armstrong, C. (1982). Soil Erosion by Water in the Tropics. University of Hawaii Press, Honolulu.
FAO (1990). "Revised legend for the soil map of the world", World Soil Resources Report No. 60. FAO, Rome.
Feder, G., Onchan, T. (1987). "Land ownership security and farm productivity: Evidence from Thailand". Journal of Development Economics 24, 16–30.
GAO (1977). "To protect tomorrow's food supply, soil conservation needs priority attention", Department of Agriculture, Report to the Congress. U.S. General Accounting Office, Washington, DC.
Gichuki, F. (2000a). "Makueni district profile: Farm development 1946–1999", Working Paper No. 1. Drylands Research, Crewkerne, UK.
Gichuki, F. (2000b). "Makueni district profile: Soil management and conservation, 1989–1998", Working Paper No. 4. Drylands Research, Crewkerne, UK.
Gray, L., Kevane, M. (2001). "Evolving tenure rights and intensification in Southwestern Burkina Faso". World Development 29, 573–587.
Hayes, J., Roth, M., Zepeda, L. (1997). "Tenure security, investment and productivity in Gambian agriculture: A generalized probit analysis". American Journal of Agricultural Economics 79, 369–382.
Held, R.B., Clawson, M. (1965). Soil Conservation in Perspective. Johns Hopkins University Press, Baltimore, MD.
Hudson, N. (1981). Soil Conservation, New Edition. Batsford, London.
Lal, R., Hall, G.F., Miller, F.P. (1989). "Soil degradation. I. Basic processes". Land Degradation & Rehabilitation 1, 51–69.
Leach, M., Mearns, R. (Eds.) (1996a). The Lie of the Land: Challenging Received Wisdom on the African Environment. International African Institute with James Currey, Oxford.
Leach, M., Mearns, R. (1996b). "Challenging received wisdom in Africa". In: Leach, M., Mearns, R. (Eds.), The Lie of the Land: Challenging Received Wisdom on the African Environment. International African Institute with James Currey, Oxford, pp. 1–33.
Li, G., Rozelle, S., Brandt, L. (1998). "Tenure, land rights, and farmer investment incentives in China". Agricultural Economics 18, 1–20.
Lindert, P. (2000). Shifting Ground: The Changing Agricultural Soils of China and Indonesia. MIT Press, Cambridge, MA.
Lindert, P. (2001). "Soil change in agriculture in two developing countries". In: Alston, J.M., Pardey, P.G., Taylor, M.J. (Eds.), Agricultural Science Policy: Changing Global Agendas. Johns Hopkins University Press for IFPRI, Baltimore, MD, pp. 122–160.
Nelson, R. (1988). "Dryland management: The desertification problem", Environment Department Working Paper No. 8. World Bank, Washington, DC.
Niemeijer, D., Mazzucato, V. (2002). "Soil degradation in the West African Sahel: How serious is it?". Environment 44, 20–31.
OECD (2001). Environmental Indicators for Agriculture: Methods and Results, vol. 3. OECD, Paris.
Oldeman, R. (1994). "The global extent of soil degradation". In: Greenland, D.J., Szabolics, I. (Eds.), Soil Resilience and Sustainable Land Use. CAB International, Wallingford, pp. 99–118.
Oldeman, R., Hakkeling, R., Sombroeck, W. (1990). "World map of the status of human-induced soil degradation: An explanatory note". International Soil Information Reference Center, Wageningen, The Netherlands, and United Nations Environment Programme, Nairobi.
Pierce, F.J., Larson, W.E., Dowdy, R.H., Graham, W.A.P. (1984). "Productivity of soils: Assessing long-term changes due to erosion". Journal of Soil and Water Conservation 39, 131–136.
Pimentel, D. (1993). "Overview". In: Pimental, D. (Ed.), World Soil Erosion and Conservation. Cambridge University Press, Cambridge.
Pimentel, D., et al. (1995). "Environmental and economic costs of soil erosion and conservation benefits". Science 267, 1117–1123.
Place, F., Hazell, P. (1993). "Productivity effects of indigenous land tenure systems in Sub-Saharan Africa". American Journal of Agricultural Economics 75, 10–19.

Richter, G. (1983). "Aspects and problems of soil erosion hazard in the EEC countries". In: Prendergast, A.G. (Ed.), Soil Erosion. Abridged Proceedings of the Workshop on Soil Erosion and Conservation: Assessment of the Problems and the State of the Art in EEC Countries, Florence, Italy, 19–21 October 1982. EEC, Luxembourg, pp. 9–17.

Sjaastad, E., Bromley, D.W. (1997). "Indigenous land rights in Sub-Saharan Africa: Appropriation, security and investment demand". World Development 25, 549–562.

Soule, M.J., Tegene, A., Wiebe, K.D. (2000). "Land tenure and the adoption of conservation practices". American Journal of Agricultural Economics 82, 993–1005.

Speth, J.G. (1994). "Towards sustainable food security", 1993 Sir John Crawford Memorial Lecture. Consultative Group on International Agricultural Research, Washington, DC.

Tiffen, M., Mortimore, M., Gichuki, F. (1994). More People Less Erosion: Environmental Recovery in Kenya. Wiley, Chichester.

USDA (1989). U.S. Soil Conservation, 1989. The Second RCA Appraisal: Soil, Water, and Related Resources on Nonfederal Land in the United States. U.S. Department of Agriculture, Washington, DC.

Wen, D. (1993). "Soil erosion and conservation in China". In: Pimentel, D. (Ed.), World Soil Erosion and Conservation. Cambridge University Press, Cambridge, pp. 63–86.

WRI (1986). World Resources, 1986. World Resources Institute, Basic Books, New York.

Chapter 58

THE ECONOMICS OF WATER, IRRIGATION, AND DEVELOPMENT

KARINA SCHOENGOLD

Department of Agricultural Economics and School of Natural Resources, University of Nebraska, Lincoln, Lincoln, NE

DAVID ZILBERMAN

Department of Agricultural and Resource Economics and Giannini Foundation, University of California, Berkeley, Berkeley, CA

Contents

Handbook of Agricultural Economics, Volume 3
Edited by Robert Evenson and Prabhu Pingali

DOI: 10.1016/S1574-0072(06)03058-1

Abstract

The post-World War II era has witnessed a drastic increase in irrigation activities that have contributed substantially to the massive growth in agricultural production that enables humanity to feed its doubling population. However, a distinction has to be made between the overall positive contribution of irrigation to agricultural productivity and economic welfare and the significant amount of misallocation and mismanagement of resources that have accompanied the expansion of irrigation. In many cases, water resources have been overdeveloped; there has been overspending on capital; and significant costs in terms of loss of ecosystems, extinction of fish species, and contamination of water sources. This chapter provides an economic perspective on the contribution of irrigation and water resources to past agricultural development and future water resource management.

The efficiency of water use is affected by decisions made at many levels. The inefficiencies that can occur at different levels of water management are discussed in this chapter. The analysis first considers irrigation water use by an individual, and then moves to the importance of regional management. The importance of dynamic considerations about the future, and the role of interregional management are then covered.

Together, these sections present an economic framework for designing water institutions and policies to improve water resource allocation and prevent some of the current inefficiency in water resource systems. The second part of the chapter provides an overview of the benefits and costs that have been realized through agricultural water and irrigation projects in developing countries. There is a paucity of ex-post integrated assessments of these projects, so the chapter puts the pieces together, combining data with conceptual arguments.

Keywords

irrigation, water resources, developing countries, water project development

JEL classification: O13, Q1, Q25, Q5

1. Overview

The previous century has seen unprecedented growth in irrigation projects on a global level. The use of tube well irrigation has decreased the cost of using groundwater, and the subsidization of large reservoirs and canals has been used to achieve food security. Worldwide, irrigated land has increased from 50 mha (million hectares) in 1900 to 267 mha today, with much of this increase in developing countries [Gleick (2000)]. Currently 75% of all irrigated land is in developing countries. Irrigation has increased the amount of land under cultivation, and the yields on existing cropland. It has also allowed double cropping, and has decreased the uncertainty of relying on water supplied by rainfall.

Table 1 shows the growth in irrigated areas worldwide in recent decades. Certain regions such as Asia have benefited greatly from irrigation. The countries with the largest areas in irrigation are China, India, and the United States, which consistently contain

Table 1
Total irrigated land (in thousands of hectares) and percentage of arable land under irrigation

	Year			
	1965	1975	1985	1995
Regional totals				
Africa	7,795	9,010	10,331	12,388
	4.9%	5.2%	5.6%	6.1%
Asia	97,093	121,565	141,922	180,507
	21.8%	26.7%	28.9%	32.4%
Australia	1,274	1,469	1,700	2,400
	3.4%	3.5%	3.6%	5.2%
North & Central America	19,526	22,833	27,471	30,478
	7.6%	8.5%	10.0%	11.2%
South America	5,070	6,403	8,296	10,086
	5.9%	6.2%	7.6%	8.4%
Europe	9,401	12,704	16,018	26,150
	6.3%	9.0%	11.4%	8.4%
Individual countries				
China	33,587	47,782	44,584	49,859
	32.1%	47.5%	35.4%	37.0%
India	26,510	33,730	41,779	53,001
	16.3%	20.1%	24.7%	31.2%
United States	15,200	16,690	19,831	21,800
	8.5%	8.9%	10.4%	11.8%
World totals	150,155	188,637	225,686	262,304
	10.9%	13.3%	15.2%	17.3%

Source: FAOSTAT.

Table 2
Total potential irrigation area (in thousands of hectares)

	Potential area	Actual to potential percent
Africa	48,155	25.7%
Asia	282,826	63.8%
South America	59,575	16.9%

about half of the world's irrigated land. Other regions such as Africa have little land under irrigation. The world total shows a large increase in irrigated land, with close to a doubling in a 30-year time frame. In addition, Table 1 shows the percentage of arable cropped land that is irrigated. This percentage varies significantly between regions. For example, in 1995 Asia had 32.4% of total cropland under irrigation, while in Africa it was only 6.1%. Also, some of the countries, such as the United States and China, have had their share of arable land in irrigation remain relatively constant between 1965 and 1995, while in India this percentage has almost doubled.

While there is little land in irrigation in certain regions of the world, such as Africa, in some cases there is a significant amount of potential irrigated land. Table 2 shows the potential for irrigated land in Africa, Asia, and South America. One interesting thing to observe is that the ratio of actual to potential irrigated land is much greater in Asia than in Africa and South America. One conclusion that we can make from this table is that the future expansion of irrigated acreage is limited in Asia, but that there is significant potential in other developing regions of the world. However, the potential irrigated land is not evenly distributed across regions. This variation in Africa, and its implications for development and food security, is discussed in more detail in Rosegrant and Perez (1997).

An important concern for the future is the limited supply of fresh water. Recent years have seen a decline in the number of water projects built worldwide, because of environmental and cost concerns. Most of the areas that are good locations for water projects have already been developed, and more is known about the negative environmental effects of the construction of large dams and poorly managed irrigation systems. Evidence of this change can be seen in the projects funded by the World Bank. There has been a shift from the development of new irrigation projects to the improvement of existing irrigation facilities. An example of this type of project is the water-saving competition in the Aral Sea region sponsored by the World Bank and IWMI [Murray-Rust et al. (2003)].

Water resources are not distributed evenly around the globe, and arid regions will continue to have conflicts over water supplies. In addition, growing populations in developing countries are expected to increase total demand for food in the coming century. Those in developing countries are eating more meat products, and increasing demand for cereal crops as livestock feed as a result. Estimates by IFPRI show that to meet demand in 2020, world production of cereal crops will have to increase 40% over 1995

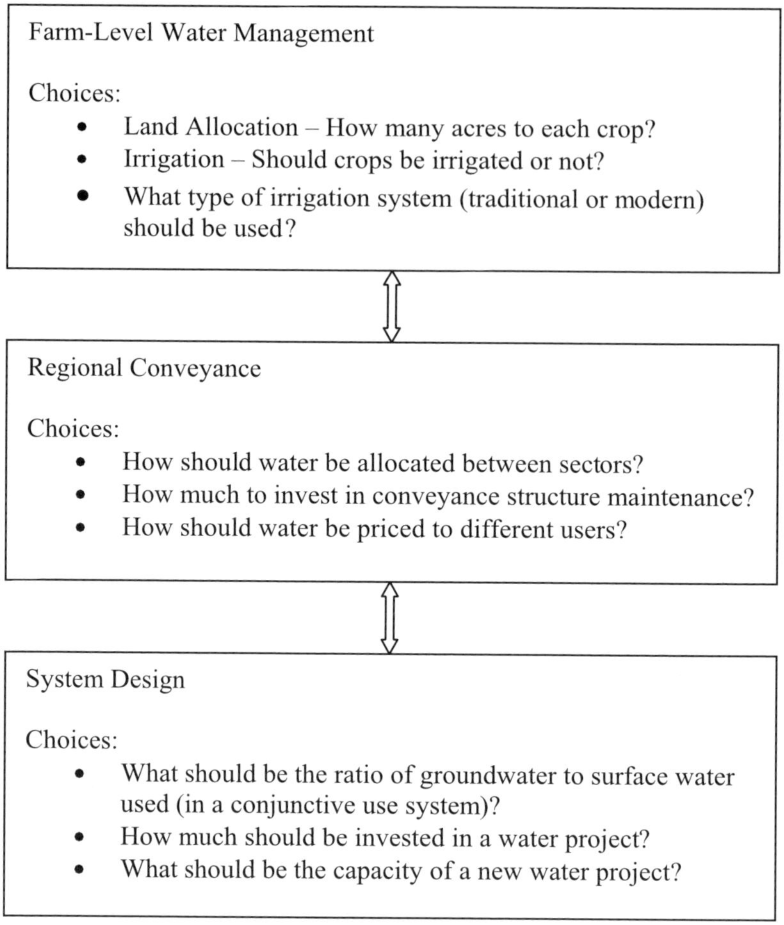

Diagram 1. The multiple levels of water system management.

levels. Better management of existing water systems, along with the use of more efficient irrigation technologies will be essential in upcoming decades. Thus, this chapter both assesses the performance of irrigation systems in the past and introduces a direction of water system reform for the future.

2. The multiple dimensions of water management

The efficiency of water use is affected by decisions at several levels of management. Diagram 1 illustrates some of the choices made at each level of management, and how these different levels are interrelated. In choosing the optimal system design, it is important to use a backward induction approach, and to base the system design on the expected responses at the levels of the region and farm.

2.1. Micro-level water management choices

Ultimately, the efficiency of irrigation systems is determined by farm level choices. These include choices of land allocation among crops, the extent to which these crops are irrigated, the use of non-water inputs, and the type of irrigation technologies. These choices are interdependent, and complete modeling of these choices is likely to be cumbersome. Therefore, it is here we discuss land allocation among activities; we first address the choice between rainfed and irrigated agriculture, and then move to the choice of a particular irrigation system.

2.1.1. Land allocation to irrigation at the farm level

There is an extensive literature on adoption of technology which is useful in analyzing the selection of acreage under irrigation [Feder, Just and Zilberman (1985); Feder and Umali (1993)]. This literature, to a large part, assumes that farmers are risk averse and constrained by credit availability. Driven by anecdotal evidence, most existing work assumes that adoption of irrigation reduces risk and increases yield but requires extra investment. The following model of a farmer's choice to use rainfed agriculture or put land in irrigation is adapted from Feder, Just and Zilberman (1985). The inclusion of a credit constraint in the model is of particular relevance to farmers in developing countries.

Suppose a farmer has $\bar{L}$ acres of land and can allocate it among two activities, irrigated and rainfed agriculture. Profit per acre under both is distributed normally where mean profit per acre under irrigation is μ_1 and the variance of profit is zero. The mean and variance of profit per acre under rainfed farming is μ_0 and σ_0^2, respectively. We denote L_0 as acreage under rainfed farming and $L_1 = \bar{L} - L_0$ as irrigated acreage. Irrigation has a fixed cost of K dollars and cost per acre of m dollars, and the farmer has a credit constraint of N dollars. Defining ϕ as a measure of risk aversion, we assume that the farmer has constant absolute risk aversion $\phi/2$ and thus his objective function is linear in the mean and variance of profit. If irrigation is selected but the credit constraints binds, acreage in irrigation is $L_1^* = (N - K)/m$. If credit is not constraining, and expected net profit per acre under irrigation is greater than rainfed farming, all the land will be irrigated ($L_1^* = \bar{L}$ if $\mu_1 - m - \mu_0 > 0$). Integrating this above condition, optimal acreage in irrigation is

$$L_1^* = \max\left\{0, \min\left(\bar{L}, \frac{N-K}{m}, \bar{L} + \frac{\mu_1 - m - \mu_0}{\phi\sigma_0^2}\right)\right\}.$$

Thus, irrigation will increase as the gain from irrigation is large, the risk reduction effect of irrigation is larger, costs of irrigation are smaller, and credit is less restrictive. From this result, we can conclude that the subsidization of financing irrigation investment is likely to increase acreage in irrigation, particularly as the yield gain and risk-reduction from irrigation increase.

2.1.2. Irrigation technology choice at the farm level

The previous section assumed that a farmer had the option to grow crops on rainfed land. In many places, rainfall is insufficient to grow any crop. In these cases a farmer must choose the type of irrigation technology to employ. Traditional irrigation methods, such as flood or furrow, use gravity to disperse water over a field. These methods have low costs of adoption, but are also relatively inefficient with water use. Modern technologies such as micro-sprinkler or drip irrigation have higher adoption costs, but deliver the water directly to the crop, applying water in a more precise fashion than traditional technologies.

To discuss the efficiency of different types of irrigation technology, we will use the notions of "effective water" and "applied water". Applied water is the total amount of water that is used by the farmer on the field, while effective water is the amount of water actually used by the crop. The difference between the two is due to evaporation and runoff, and irrigation efficiency is the ratio of effective water to applied water. We can use this to define the effective price of water as the price per unit of effective water. When irrigation efficiency is less than 100%, the effective price of water will be higher than the actual price. In addition to the irrigation technology, land quality characteristics such as the slope of the land and the water-holding capacity of the soil affect irrigation efficiency. Theoretical and empirical studies have shown that an increase in water price is positively correlated with adoption of precision irrigation technology [Caswell and Zilberman (1985, 1986); Dinar and Yaron (1992)].

According to Caswell and Zilberman (1986), under plausible conditions, modern irrigation technologies increase yields as well as save water in most cases, but the gains from this technology are reduced as land quality improves. This counterintuitive result is because differences in water holding capacity lead to differences in the effective price of water, where the effective price under traditional irrigation decreases as land quality improves. Therefore, the relative gains of a switch to precision irrigation are lower with high quality land. Adoption occurs when the yield and price saving effect of precision irrigation are greater than the fixed cost of the technology, thus we expect that modern technology will first be adopted in locations with low quality land such as steep hills and sandy soil. The details of this model are presented in Appendix A.

Another counterintuitive result of the analysis is that the availability of efficient irrigation technology can actually lead to a net increase in water use in a particular region. This is because there are two types of effects to consider; those at the intensive margin and those at the extensive margin. At the intensive margin, farmers that adopt efficient irrigation technology are likely to decrease total water use. However, there can also be a change at the extensive margin. Those with low quality land often find that it is not profitable to farm using traditional irrigation methods, since the effective price of water is high when irrigation efficiency is low. However, modern irrigation technology increases water use efficiency, decreasing the price of effective water. This can make it profitable to farm land that was left fallow under flood irrigation. Both the intensive and extensive changes in water use need to be evaluated with a change in water price or technology availability.

The increase in water use efficiency reduces unutilized water and thus with drip irrigation the problems of water buildup and waterlogging are diminished. Caswell, Lichtenberg and Zilberman (1990) show that when a penalty on drainage is introduced, adoption of sprinkler and drip irrigation are likely to accelerate. These technologies provide both an increase in productivity as well as a reduction in negative externalities, and their adoption will be enhanced by improved pricing of water and the introduction of drainage fees.

Providing the correct incentives for farmers to adopt efficient irrigation can have dramatic effects on water use. Switching from furrow or sprinkler irrigation to drip systems decreases water applications by up to 35% [Schoengold, Sunding and Moreno (2005)]. Global use of drip irrigation is twenty-eight times the level of the mid-1970s, but still accounts for less than 1% of world irrigated area, while sprinkler irrigation is used on 6% of irrigated land [Postel (1996)]. Improvement in water use efficiency is not limited to agriculture, and industrial and residential water users can also do a lot to improve the efficiency of their water use. With techniques available today, farmers could cut their water demands by 10–50%, industries by 40–90%, and cities by a third with no sacrifice of economic output or quality of life [Postel (1996)].[1]

2.1.3. Productivity of water

An important factor in determining the response of farmers to a change in water price is the shape of the function relating production output with water inputs. Following Caswell and Zilberman (1985) we define output per acre (Y) as a function of effective water (e), where effective water is the quantity used by the plant. This is equivalent to the product of the water-use efficiency parameter and applied water.

Some of the early work on water productivity was done by Hexem and Heady (1978), who use field experiments in the United States to estimate yield as a function of inputs including water and fertilizer. One commonly used production function in the economic literature is a Cobb–Douglas production function of the form $Y = Ae^{\delta}$, with a requirement that $\delta < 1$. While some work has shown that this representation is reasonably accurate at an aggregate level, econometric evidence has shown that this is a poor representation of the yield response of water at a more micro-level. There is evidence that a quadratic function, such as $Y = a+be-ce^2$ where $a, b, c > 0$, is a better representation of water productivity. This functional form has the property that above some level of input use, yields begin to decline. With an extreme weather shock, such as a flood, one can easily see how a field of crops is washed away, and the benefits of that additional

[1] While these values may be technically feasible, designing appropriate policies which provide the right incentives for individuals to change their behavior is difficult. As such, these levels of reductions are difficult to achieve in practice.

water are negative. Another commonly used function is the Von Liebig, which assumes water exhibits constant returns below some threshold level, and a zero return above that threshold. This takes a form such as $Y = Ae$ if $e \leqslant e^*$ and $Y = Ae^*$ if $e > e^*$. Berck and Helfand (1990) have shown that different choices of functional forms for production can be reconciled with certain assumptions about the heterogeneity of land quality. Existing work finds it is unclear which of these functional forms is the most accurate, and further work needs to be done on the subject.

In addition to the theoretical work done on the functional form of water productivity, empirical work has been done to estimate the returns to water in several locations. One study of the Syr-Darya River Basin finds the average return to water in the region is $\$0.11/m^3$. However, this value varies significantly throughout the area, and water use in non-saline areas is as much as five times higher than saline areas [Murray-Rust et al. (2003)]. There has also been work done on the relationship between high yielding varieties (HYV) and the productivity of water. Since HYV increase the marginal product of water, they have been found to also stimulate investment in irrigation [McKinsey and Evenson (2003)].

2.1.4. *Existence of low-capital efficient irrigation technologies*

Efficient irrigation technologies do not necessarily entail a high capital cost of adoption. There are examples from water-scarce areas that show the ingenuity of farmers in their ability to adapt to limited water supplies. One example is the leveling of farmland. Terracing of farmland has been used for thousands of years as a way of increasing the efficiency of applied water. A flat surface leads to less water runoff, and increased water use efficiency of the plant. Another method that has been used is the placement of clay pots below the ground level near the roots of tree crops. The porous clay permits the water to slowly drip from the pot, and provides a constant supply of water to the tree. One other example of a low-cost irrigation technology is the use of village tanks in India. Traditionally, villages in India have gathered rainwater in tanks, with each village having a system that designates how water is to be divided among users, and who is responsible for the upkeep of the system [Whitaker, Kerr and Shenoi (1997)]. There has also been a low-capital system of drip irrigation developed that is being used in parts of India. This system uses simple holes instead of emitters, and a cloth filter. Despite requiring a much lower investment in capital than most drip irrigation systems, it is remarkably efficient in water use [FAO (1999)]. The use of bucket drip irrigation, a method where water is delivered through drip tubes from an overhanging bucket, can reduce water use by as much as 50%.

2.2. *Regional allocation of water*

At a regional level, there are many aspects of water management that need to be addressed to improve the overall efficiency of a water system. In this section, we first discuss the initial choices made about a system, including the location and size of a water project, as well as the importance of financing the project. We then move to the discussion of important management choices of existing systems, such as conveyance, water trading, and water pricing.

2.2.1. The basic economics of oversized water projects

In the decision to construct a new water project, the benefits of the project must be compared with the costs. The large water projects in the Western United States were some of the first government-funded projects that required a benefit–cost analysis to be completed before the project was approved. Water projects funded by international agencies such as the World Bank also require such studies before approval. In addition to the decision on the location, choices about the size of a dam and conveyance system must be made. Economic theory has some insight into the choice of the optimal size of a dam. While dams provide many benefits through the supply of irrigation water, hydropower, and flood protection; the full costs of construction have often been ignored, both in the decision to build a dam and in the choice of the size of the water project. The externalities associated with construction are often ignored entirely, decreasing the perceived marginal cost of development. Also, it is often the case that development costs are subsidized, either by governments or international agencies. In these cases, the perceived costs of water development are below the true private costs.

A simple static model depicts the forces that lead to overinvestment in projects such as dams. Let W denote the capacity of a dam. The marginal market benefit to the surrounding region of building the dam and increasing the water supply are shown in the MB curve. The costs of building a dam can be broken down into two categories – direct capital and construction costs and externality costs. The marginal direct cost of building the dam is shown by the MPC curve, and the marginal social cost is shown by the MSC curve. The difference between these two curves accounts for the externalities associated with dam construction. These externalities include environmental costs such as the destruction of natural habitat and degradation of the soil, and other costs such as the welfare loss of displaced populations. Now suppose that construction is subsidized. Because of subsidies, the cost facing developers is often well below the full private costs, leaving the perceived cost of water development as shown by the subsidized MC curve.

The most important result of Graph 1 is that in cases where costs are subsidized and externalities are ignored, the dam capacity will be too large, and the marginal benefit of water supplied will be too low. If the full social cost of dam construction is taken into account, the optimal capacity of the dam will be W^*, and the marginal benefit will be at P^*.

It is also important to consider the relationship between storage capacity and other components of water delivery. The benefits of water development are a function of three activities – conveyance, management, and storage capacity. To some extent, these three activities can be considered substitutes for each other. When subsidies lead to a low relative cost of storage capacity, there is overinvestment in storage capacity and underinvestment in conveyance and management of irrigation systems. While it is clear

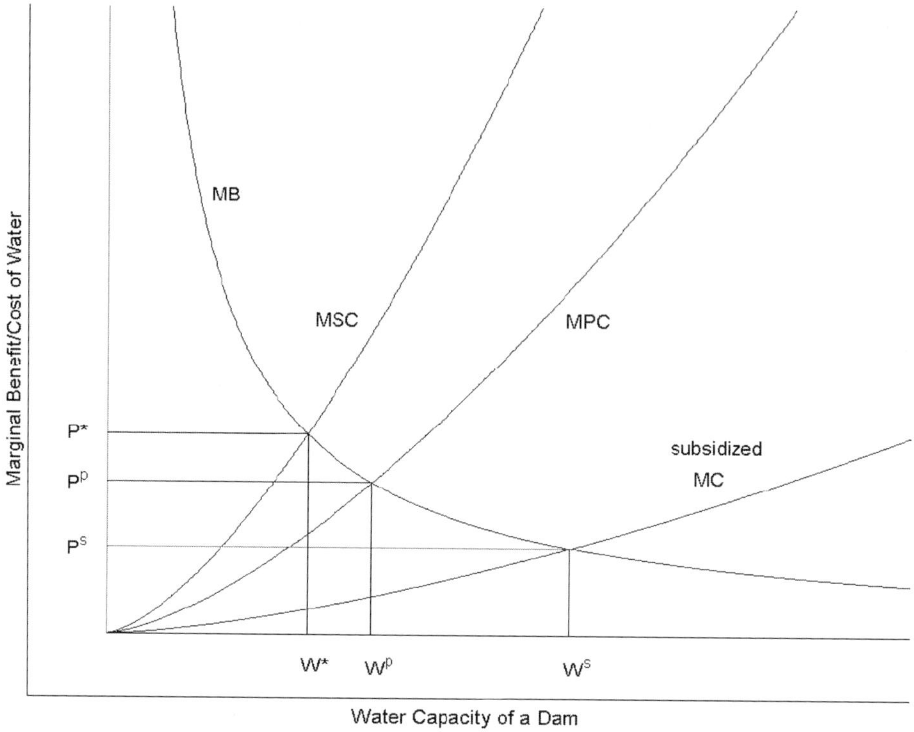

Graph 1. Effects of externalities and subsidies on water project capacity.

that irrigation and water development have provided tremendous benefits, the omission of the true costs has led to the construction of large dams, often in locations that are inappropriate for water project development because of fragile landscapes and ecosystems.

2.2.2. *Management of conveyance systems*

The construction of water conveyance systems is an important element of the overall efficiency of the system, as better management of conveyance systems reduces the need for new water projects. Many canal systems were built at a time when the costs of constructing an efficient distribution system were greater than the additional benefits. Various methods exist to improve the distribution of water. For example, lining the canals is one method that can limit the amount of water lost during conveyance. Another problem is poor maintenance of existing canal systems – over time there is deterioration, which leads to increased amounts of lost water. Poor management of irrigation systems leads to conveyance losses of up to 50% [Repetto (1986)].

Inefficiency also stems from the water lost to evaporation in canals and reservoirs. These problems have a disproportionate effect on the downstream users in a water system, creating equity problems among different water users. The maintenance of a canal system at one location has benefits to the local users; however it also has benefits to all of the downstream users of the water system. Because of this, canal maintenance provides a positive externality, as the social benefit of canal maintenance is greater than the private benefit to each water user. If these positive externalities are ignored, there will be too little investment in canal maintenance, leading to an inefficient water conveyance system. Chakravorty, Hochman and Zilberman (1995) show that without collective action (which leads to optimal investment and conveyance), canal systems will be shorter than optimal, with over-application of water close to the source and under-application far away. Transition to optimal conveyance will expand canals and production and will actually reduce the rental rate of lands that are upstream, even though the overall rent is likely to increase.

As discussed by Easter (1986), there has been a shift from the development of new water projects to better management of existing projects. This has led to an increased reliance on water user associations (WUAs). A WUA is a group of farmers who collectively manage and distribute their combined available water supply. A shift to management of water resources by the water users is being promoted as a means to improve conveyance systems, cost recovery, and the efficiency of water use.

In various places WUAs have existed side by side with publicly run irrigation systems for many years. Evidence suggests that higher yields, better conveyance structures, improved maintenance, greater efficiency, and a more reliable supply are associated with WUAs. One important question for economists concerns the effectiveness of different management strategies for a common resource; the irrigation system. In a study of Mexican farmer-managed irrigation systems, Dayton-Johnson (2000) investigates the incentives for an individual to provide collective maintenance labor under different WUA distributive rules. He finds that because of higher system wide costs, a system where labor requirements and water allocation are proportionally distributed may not be optimal. A better system is one of equal labor requirements and water allocation, with trading possible between members. He also finds that economic inequality among water users is positively correlated with a proportional distribution rule, evidence that wealthier landholders are able to push for a higher share of total water supply.

One country that now primarily uses WUAs to manage irrigation systems is Madagascar. An ordinance passed in 1990 requires water users to pay the costs of irrigation infrastructure, and the result has been an average cost recovery of 80–90%, well above most developing countries [Rabemanambola (1997)]. Another country with growing use of WUAs is India. Since seeing a decline in irrigation performance, the state of Andra Pradesh in India has created over 10,000 WUAs covering 3.7 mha of land. As Dayton-Johnson's results indicate, it does seem like some level of equality in land-holdings is necessary for the success of a WUA. Pakistan, where many areas have a few large landholders, has been less successful in the formation of WUAs. In Hubei, China, one goal of the shift to WUAs is financial autonomy. WUAs are required to purchase the water they use, giving them an incentive to conserve and use water efficiently [Easter (2000)].

2.2.3. *Political economy of water system management*

An understanding of the politics underlying water resource development and management is crucial for improvement in the future. Work by Rausser and Zusman (1991) shows that when those with political decision making authority place unequal weights (termed 'political power' by Rausser and Zusman) on different interest groups, the resulting water pricing and allocation methods are economically inefficient. Rausser (2000) extends this model into a multilateral bargaining model based on a Nash–Harsanyi bargaining framework. This model illustrates the tradeoffs between different interest groups who are concerned about water distribution and allocation.

One reason that has been offered to explain the poor management of conveyance structures in many public irrigation systems is termed the 'political economy of neglect'. This theory says that if agencies who fail to provide the necessary upkeep to their irrigation system are bailed out by a donor agency, there will be a lower incentive for them to provide efficient levels of maintenance. This describes the situation in many public irrigation systems. The funding for the initial costs of constructing the project often comes from non-governmental organizations (NGOs) such as the World Bank or the Asian Development Bank. This funding is often contingent on the recipient country managing the irrigation system so that revenues cover the operating costs of the system. However, the countries also know that if they fail to adequately maintain the irrigation systems, international agencies will provide additional funding. This provides an incentive for the public agency to neglect to provide adequate maintenance, creating a cycle of dependence on outside funding.

Another explanation for poor management and low quality service is discussed in Spiller and Savedoff (1999). Their paper looks at how government opportunism affects the efficient provision of water. Their paper focuses on countries in Latin America, but many of the conclusions have general implications. They discuss the emergence of low-level and high-level equilibria in water service provision. A low-level equilibrium refers to the case when government wants low water prices to keep their citizens happy. When water is provided either by public agencies, or private agencies that can be partially subsidized by the government, water prices are kept artificially low. This leads to limited service and poor infrastructure, and a public who is unwilling to pay higher prices for water service that they perceive as inefficient and low-quality. While it does not maximize social welfare, a low-level equilibrium is stable. A high-level equilibrium, one with higher water price, but also high-quality water service that is well-maintained improves social welfare. However, in cases where the government is short-sighted and had control over water service, it might not be stable. In their analysis of Latin American countries, Spiller and Savedoff identify several countries in each category. Honduras and Peru are examples of countries with low-level equilibria, while Mexico, Chile, and Argentina have high-level equilibria.

2.2.4. Transition from water rights to water markets

Water rights systems. In most parts of the world, the price paid by water users is well below the marginal value product of the water as an input. Existing estimates of the ratio of water charges to farmer benefits range from 26–33% in Korea to 5% in Nepal [Repetto (1986)]. Given the low price paid by users, demand would greatly exceed supply of water if it was allowed. Since water resources are scarce, and the price paid by users is below the input value, water must be allocated using a non-market mechanism. In many parts of the world, water is allocated using a "queuing" system [see Easter (1986) for an overview; Chambers (1988) for the Indian subpeninsula; and Lee (1990) for South America]. Queuing systems use either a historical or spatial basis to assign an order to the users of a water system. Two of the most common types of queuing systems are a prior appropriation system and a riparian rights system. The prior appropriation system is based on the principle of "first in time, first in right". Seniority in water rights is given to the first person to divert water for beneficial use. The riparian rights system gives any landowner with land adjacent to a water source the right to use that -water.

It is also common to have restrictions on trade within a watershed system (quite frequently of the form "use it or lose it"). In these systems, senior rights holders or upstream water users have little incentive to invest in water-saving irrigation technology, because they are assured of a stable water supply. These types of systems were established at a time when water was plentiful, and governments wanted to provide an incentive for private development and innovation. However, water in many systems is now over appropriated, and better management is essential to make the best use of a limited resource.

The transition to trading and markets. Both riparian and prior appropriation rights systems involve limitations on trade in water, leading to inefficiencies in water distribution. Neither type of system is economically efficient, as the water is not used in the activity where it earns the highest marginal value. Economic efficiency dictates that if transaction costs are low, either water markets or tradable permits are the best way to allocate water supplies [Burness and Quirk (1979); Coase (1960)]. These systems ensure that scarce water will flow to the user who earns the highest marginal value from the water. Graph 2 shows two farmers who earn a benefit from water of MB_1 and MB_2, respectively; however, farmer 1 has senior rights to water while the other (farmer 2) has junior rights. Total water available for a season is Z. A shift to a system of tradable water rights can increase the welfare of all parties involved, as shown in Graph 2.

With a prior appropriation system, senior rights holders have their demand fully satiated before junior rights holders receive any water. In Graph 2 the marginal benefit to farmer 1 of an additional unit of water is zero, while the marginal benefit to farmer 2 is P_2. If trading in water rights is allowed in the preceding model, there will be positive gains to society from trading. Farmer 1 will sell water to farmer 2 until the marginal benefit to both is P_1, and the increase in social welfare is the area of triangle ABC.

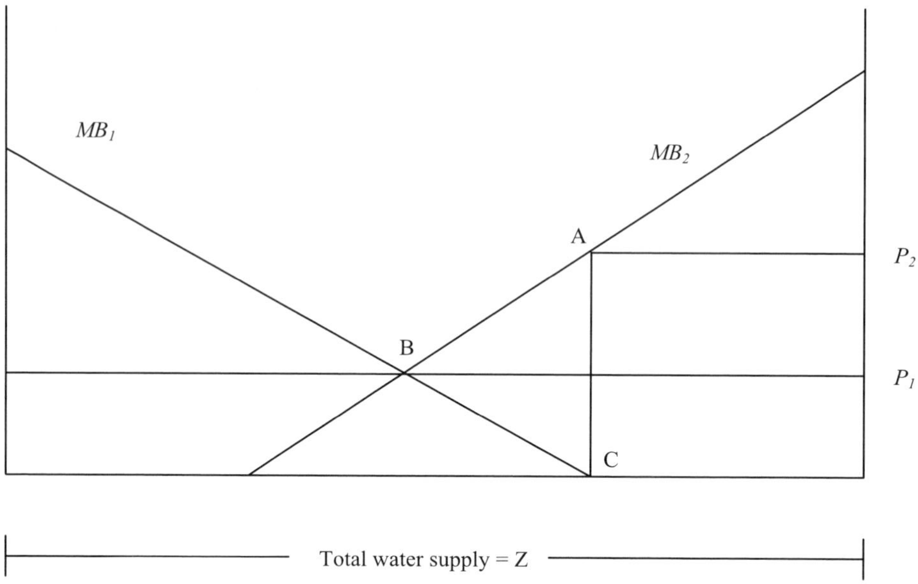

Graph 2. Gains from tradable water rights.

When transaction costs are introduced to the above model, the welfare gains of tradable permits will be reduced. If water is not very scarce, the transaction costs of trading water may be greater than the benefits. However, as demand for water expands over time and the shadow value of water increases, the benefits of trade will outweigh any transaction costs. Evidence for this is suggested by observations that in developed countries that allow water trading, trading activities increase significantly during drought years. Also, as discussed by Johansson (2000) [citing work by Renfro and Sparling (1986), Shah (1993), and Anderson and Snyder (1997)], informal water markets have repeatedly been developed under conditions of water scarcity.

There are alternative mechanisms of water trading that have to be considered when reforms are introduced. The first choice is whether to use a system of transferable permits or transfer ownership of water to the government agencies that will sell it in the market. Water users with senior rights will prefer transferable rights systems as they are able to earn the associated rents. A water agency might prefer water markets, as they earn the proceeds of water sales, and can use the revenue to improve service and management of water supplies. Brill, Hochman and Zilberman (1997) distinguish between passive and active water markets. In the case of passive water markets, water users buy and sell water to a regional water authority that controls water supply and conveyance. In the case of active markets, agents trade among themselves. Passive markets are more appropriate within regions and especially among water users that are served by the same utility, while active markets are appropriate between districts. Some form of passive trading within districts exists within many parts of the world.

Another choice is whether to only allow individuals to rent the right to use water on an annual basis or to allow complete transfer of ownership rights. In cases of infrequent droughts, renting the water rights to those with a high willingness to pay might be a better option than a permanent sale. In places with chronic water shortages, a rights holder might be better off with a sale of those rights. In addition, the permanent sale of water rights secures a future water supply for users. This can promote capital investment in the land that would not occur with an uncertain water supply.

A third decision is if out-of-basin trading among water users should be allowed. When water users in a single water basin are allowed to trade, the transaction costs, and especially the third party and environmental costs, will be lower. If water users are allowed to trade their rights outside of their water basin, concerns about third party effects must be addressed. These third parties may be individuals who use runoff or deep percolating water from the land, or the environmental benefits that accrue through the supply of residual fresh water. Addressing these issues may require limiting the quantity traded to the effective water, and not the applied water used by an individual. A discussion of the essential component in a water market is in Easter, Becker and Tsur (1997).

Examples of countries that have transitioned to water markets are Chile, South Africa, and Australia. Chile is probably the most well-known example of such a transition on a national scale. In 1981, Chile reformed its Water Code, and by doing so, changed the nature of water rights. After the change, water rights became completely separated from land ownership, and can be freely bought, sold, or rented. The government now has little control over water use, and most of the managerial decisions about conveyance systems and maintenance are made by private water users associations.

An interesting result of the shift to water markets in Chile is that few transactions have been observed in practice, while most of the transactions have been in combination with a sale of land (with water rights rarely sold separately from land). Part of the reason for this is the low value of land without water rights. There are also institutional reasons – at the time of the reform, there was a lot of uncertainty about the ownership of much of the water used. Much of the energy since the reform has gone into defining water rights, and some areas have seen 10 times as many water rights approvals as water sales [Bauer (1998)]. Clearly, well-defined water rights are a necessary condition for welfare-improving water sales. In some cases though, initial allocation of water is not far from optimal. However, even if only a small proportion of the total water used is being sold and these are final sales, the impact may be significant if the gain in productivity for this water is substantial. Eventually, as water rights are better defined, new actors enter the system, and conditions change, transactions will increase.

Another example of a country with a major change in its water law is South Africa. In 1998, the New South African Water Act changed the ownership of water from private to public; however, farmers still have private rights to use certain quantities of water. Transfers of water between users are allowed, although certain administrative policies must be followed. Nieuwoudt, Armitage and Backeberg (2001) discuss case studies of two agricultural regions to highlight the factors that can either lead to or impede an active water trading market. They find that despite costly administrative requirements, the

Lower Orange River area has an active water trading market. Reasons for this include water scarcity and a heterogeneous group of water users. Some of the farmers in the area grow table grapes, which are a high value crop, while others grow wine or raisin grapes, which earn a lower price. Since the marginal value of water is greater to growers of table grapes, the benefits of trading outweigh the transaction costs. In the second region (the Nkwaleni Valley), an active trading market has failed to emerge. Despite water scarcity, none of the water users have been willing to part with their water rights. The farmers in this region are fairly homogeneous, and all grow a combination of sugar cane and citrus. While some are willing to buy more water, if it was available, none are willing to sell. One clear implication of these results is that for water trading to be successful, there needs to be heterogeneity among potential water users.

Australia has also moved to a water-trading regime, and has decoupled ownership of land from the right to use water, in a similar manner as Chile. The shift from traditional water rights stemmed from a growing realization that greater flexibility was needed in water rights, and in particular, water resources are necessary in the natural habitat. A 1994 bill separated water rights from land ownership, and established a water allocation for environmental services and the development of water markets. The results of the change in Australia have been positive, and estimates are that the annual gains from the shift to tradable water rights are \$12 million in Victoria, and \$60–\$100 million in New South Wales [ACIL (2003)]. Despite these gains, there are still some barriers that have been identified as an impediment to the highest possible returns to tradable water rights. One of these impediments is a limitation on the lease of water-use rights. Water rights can be permanently sold in all states of the country, but some states still have a restriction on short-term (i.e., one year) leases of those rights. Another aspect that has been identified as a limitation on the benefits of trading is the lack of an options market in water resources. The elimination of these barriers will only increase the benefits already realized in Australia.

2.2.5. *Water pricing systems*

The costs of providing irrigation water include a fixed cost of operation and maintenance (O&M) and a variable cost, which depends on the quantity of water supplied. In addition, there is a capital cost of constructing a water project. There are many pricing systems used for recovering some or all of these costs. In most countries, the revenues received fall far short of the costs of supplying irrigation water to users, and often do not even attempt to recover the initial capital costs. Recovery of operation and maintenance costs ranges from a low of 20–30% in India and Pakistan to a high of close to 75% in Madagascar [Dinar and Subramanian (1997)]. In some areas of India, receipts even fail to cover the administrative costs of collection [Saleth (1996)].

Water pricing systems can be designed to provide an incentive for water users to adopt water-conserving technologies, or to alter the amount of land under cultivation.[2] A vol-

[2] For a more detailed discussion of irrigation water pricing systems, see Johansson (2000).

umetric fee provides an incentive to limit water use, while a per-hectare fee provides an incentive to cultivate agricultural land more intensively. Some of the most common pricing systems are per-hectare fees, increasing or decreasing block rates, and volumetric fees. These rates can either be fixed or depend on the area and time of year. Many systems combine these; for example, charging a per-hectare fee for access to water, and then a reduced volumetric fee for water delivered. This is the type of pricing system used in Brazil for irrigation water. Irrigation water is mostly metered in Brazil, and the irrigation law requires that the price of irrigation water be the sum of two charges. The per-hectare charge is designed to repay the capital costs of the project, which are calculated using a 50-year repayment period and a subsidized interest rate. The volumetric fee is designed to repay the operation and maintenance costs of the water project. However, the revenues from this are unpredictable, and in practice have failed to cover the costs of water projects [Todt de Azevado (1997)].

Inaccurate volumetric measurement. One source of inefficiency in water pricing stems from the inability to measure the quantity of water an individual uses. In many areas of both the developed and the developing world, the cost of installing metering devices to accurately measure water use by individuals is prohibitive. Various pricing systems have been developed as an alternative to volumetric pricing. Overwhelmingly, developing countries use a per-hectare water fee, if they charge at all. One country that used per area pricing is Pakistan. In Pakistan, water charges are levied on a per unit area basis, and vary across region, crop, and season. However, the price variation across crops is not related to either the water requirements or the profitability of the crop. Other countries, such as Egypt and Indonesia, do not charge farmers anything for the water they use but require farmers to maintain and operate the irrigation canal system. One commonly used pricing scheme is based on the duration of water delivery. This system can approximate a volumetric measure using an expected quantity per hour.

Subsidization of water delivery costs. While precision irrigation technology can dramatically reduce water use, its adoption is minimal. One reason for this is that the price of irrigation water generally does not reflect the scarcity value of the water. Irrigation water is subsidized in many regions, and the price often does not even reflect the cost of delivery, let alone the shadow value of a scarce resource. An example of inefficient pricing can be seen in India, where from 1983 to 1986, the estimated working expenses of major water projects was 2.2 times the gross revenue collected from the water users [Saleth (1996)]. Using 1987 data, a study of six Asian countries showed that the irrigation charge as a percentage of total cost ranged from 1.0% to 22.5% [Repetto (1986)]. The elimination of subsidies on water delivery will promote the adoption of precision irrigation, which will decrease water use, increase yields, and reduce environmental externalities such as water logging and salinization.

Improved pricing and water theft. Another benefit of improved water pricing policies is discussed by Ray and Williams (1999), who explain the prevalence of water theft on shared canals in India. Upstream water users are able to steal water meant for downstream users, and the penalties, if they exist, are usually some type of bribe to the inspector. Using a linear programming model, they show the effects of various pricing policies on farms along the canal. Eliminating price supports and water subsidies increases social welfare, but the gains are not uniform along the canal. Without water theft, farmers at all points along the canal have higher revenues with subsidized prices. However, when water theft is taken into account, farmers at the head of the canal lose, while those in the middle gain from a shift to non-subsidized water and output prices. Those at the tail end of the canal are slightly better off with subsidies, but the loss to them of improved pricing is minimal.

2.2.6. Groundwater management

Groundwater as an open-access resource. When property rights to a natural resource are poorly defined, there is often a problem of open access. In cases where the resource is limited in supply, users of the resource will not take into account the effects of their use on the future availability and cost of the resource to other users. One of the biggest obstacles to the optimal management of groundwater systems is the open access problem. Since groundwater is rarely regulated, anyone has the ability to dig a well and pump water for personal use. However, since the same groundwater table is available to many users, each user inflicts an externality on others, as a greater level of water extracted reduces availability to other users in the future.

Subsidization of groundwater pumping costs. One obstacle to the efficient management of groundwater is the subsidization of pumping costs. The main cost of pumping groundwater is the power required to lift the water to the surface. In many countries electricity is subsidized, which decreases the marginal cost of pumping, and leads to increased extraction of groundwater. Two countries with subsidization of electricity costs are India and Pakistan, and this subsidization is part of the reason for the overdraft of groundwater that is occurring in these countries. From 1951 to 1986, the use of tank irrigation in India fell slightly, while the use of canal irrigation and well irrigation increased dramatically. Tank and canal irrigation depend on surface water while well irrigation relies on groundwater supplies. The amount of land under canal irrigation has increased from approximately eight thousand to fifteen thousand hectares, while the land under well irrigation has increased from 6500 to 20,000 hectares, an increase of over 300%. This is partly due to technological improvements that make digging wells and pumping water easier, but it is also due to the low costs paid for pumping of water. Electricity users pay a low flat rate, almost eliminating the marginal cost of groundwater pumping [Whitaker, Kerr and Shenoi (1997)].

Introduction of efficient groundwater pricing. Because of the externality imposed on other water users, the elimination of electricity subsidies still leads to a sub-optimal groundwater price. The theory of exhaustible resources dictates that the price of groundwater should equal the sum of the cost of extraction and the user cost, with the user cost equal to the opportunity cost [Hotelling (1931); Devarajan and Fisher (1981)]. Appendix B presents a formal model that shows that optimal use of groundwater requires equating the marginal benefit of water with the sum of the pumping cost and the user cost. The user cost measures the loss of future benefits due to depletion and the increase in future pumping costs associated with depleted stock. A first-best solution would be to impose a tax equal to the user cost on every acre foot of groundwater extracted [Shah, Zilberman and Chakravorty (1993); Howe (2002)]. However, the monitoring and enforcement of a tax like this would be impossible with the cost and availability of currently available technology. As discussed in Shah, Zilberman and Chakravorty (1993), a second-best solution would be to base the tax on the irrigation technology and crop choice.

2.2.7. *Between sector allocations of water*

One area we have neglected to mention earlier is the interaction between agricultural water users and other sectors, such as urban and industrial groups. Many times there not only are misallocations of water among farmers, but also between sectors. With limited water supplies, competing interests between user groups become important. Among these three sectors, agriculture uses the lions share of the water supply, despite the fact that it often earns the lowest marginal value of water. As populations increase, pressures to supply an adequate amount of water for domestic and industrial purposes also increase, causing conflicts between sectors. This has been true for over 100 years in places such as California. In Chile, growing cities such as Santiago have bought water rights from agricultural users to supply urban residents. However, an adequate solution to the question of between-sector allocations is more complicated than a simple transfer of water from agriculture to the urban sector. For example, a study of Hyderabad City, the capital of Andhra Pradesh in India finds that improvements in the pricing structure of urban water could lead to more efficient urban water allocations, removing the need for costly transfers from the agricultural sector [Saleth and Dinar (1997)]. Also, differences in water quality requirements exist between sectors. Much of the water used in agriculture would require further treatment for use in other sectors.

2.2.8. *Use of non-traditional water sources*

As traditional water supply sources have become scarcer, there is growing use of non-traditional sources of water. These include the reuse and recycling of wastewater, and desalination of ocean water. In the Western United States and parts of Africa and the Middle East, there has been a growth in the use of reclaimed wastewater for industrial, agricultural, and commercial uses [Gleick (2000)]. Reclaimed water may be produced at a cost of 30 to 40 cents per cubic meter and will be competitive with other sources of water in Israel and Jordan. In Israel, partially reclaimed water is used extensively in production of industrial crops such as cotton. Crops that can tolerate saline water are able to reuse the water that was initially applied on other crops. Another option is desalination of ocean water. While still expensive, desalination has begun to be used in water-scarce regions such as North Africa and the Middle East, and the world's 7500 desalting plants can produce 0.1% of the world's water use [Weber (1991)].

Rhodes and Dinar (1991) present results that suggest that for crops such as cotton and certain vegetables, yield levels can be maintained if high quality water is used early in the life of a plant and more saline water is applied toward the end of the season. Their approach will enable water planners to take advantage of drainage water and other low-quality water, but will still require maintaining inventories of water of various qualities. Amir and Fisher (2000) explain that farmers in the Jezreel Valley of Israel use both high quality freshwater and brackish reclaimed water in crop production. An arbitrary policy to limit production of low-value crops such as cotton does increase the average return of water, but it also limits the ability of producers to optimize their use of both types of water sources. This evidence shows there is a benefit to having multiple qualities of water available for different end uses. However, this option requires evaluating the economic tradeoff between the cost of separate storage and the cost of bringing all water quality to the highest standard.

2.3. *Intertemporal aspects of water*

2.3.1. *Dynamic consideration and uncertainty*

The previous section presented a model of the optimal size of a water project using a static framework. This is useful, but neglects some of the dynamic considerations that are important. A water project is planned not just for a single period, but for many years. Dynamic considerations include calculations of future benefits and costs, the choice of an appropriate discount rate, and population growth. Because of the high rate of population growth in many developing countries, it might be optimal to choose a larger water capacity than current demand indicates.

One source of uncertainty comes from expectations about future demand for water, which is often difficult to accurately predict. If developers assume that demand for water inputs will stay constant after the construction of a water project, the chosen supply level could be either too high or too low. Water demand could decrease for several reasons after the construction of a water project. One reason is that crop yields in irrigated areas are higher than in rainfed areas, and higher benefits per unit of water might reduce total demand for water. Another factor is the choice of irrigation technology. If farmers adopt precision irrigation technology that is more water efficient, this could also decrease the total demand for water after a water system is built. There are also several reasons for a potential increase in water demand. Many water projects are built in countries with high rates of population growth, which can increase demand for water. Water projects and the resulting employment opportunities can also increase migration into the developed area. In addition, arid areas that otherwise are unproductive are able to grow crops after water development, leading to an increase in water demand for agricultural uses. While the direction of the shift in water demand is site-specific, if constant future water demand is presumed, the resulting dam size is usually suboptimal.

In a simplest form, the decision in designing a water project is related to construction of capacity to convey a certain amount of water, from a source to a destination [see Chakravorty, Hochman and Zilberman (1995)]. Let $\overline{W}$ be the upper bound of water that can be diverted during a period and the fixed cost of the project is $f(\overline{W})$. At period t, the amount of water utilized is $W_t \leqslant \overline{W}$. The water provides benefits of $B(W_t, \varepsilon_t)$ where ε_t is a random variable.

The annual cost of the water is $c(W_t)$ (it includes both direct and externality costs). Assuming a project design for T years and discount rate of r, the optimal size of the project is determined by maximizing discounted expected net benefits, i.e.,

$$\max_{\overline{W}, W_t} \int_0^T e^{-rt} E\big\{B(W_t, \varepsilon_t) - c(W_t)\big\}\,\mathrm{d}t - f(\overline{W}) \tag{1}$$

s.t. $W_t \leqslant \overline{W}$.

For an infinite planning horizon and identical random element, $\varepsilon_t = \varepsilon$, the water use at each period is $W_t = \overline{W}$ and the optimal design problem is reduced to

$$\max_{\overline{W}} \frac{E[B(\overline{W}, \varepsilon)] - C(\overline{W})}{r} - f(\overline{W}),$$

where $E[B(\overline{W}, \varepsilon)]$ is the expected benefit per period and $N(\overline{W}) = E[B(\overline{W}, \varepsilon)] - C(\overline{W})$ is the net expected benefit per period. Optimal capacity is at the level when the marginal net expected benefit $MB(\overline{W}) = \partial N / \partial \overline{W}$ is equal to the marginal cost of capacity $MC(\overline{W}) = \partial f / \partial \overline{W}$ times the interest rate, i.e., when

$$MB(\overline{W}) = MC(\overline{W}). \tag{2}$$

There is a vast literature on the appropriate discount rate for project development, and we will not address this point here [see Arrow (1997) for an overview]. Low discount rates place a greater weight on future costs and benefits (compared to current costs and benefits) than a high discount rate. In cases where the interest rate is subsidized, such as when a donor agency expects repayment of the principle with no interest, using Equation (2) will lead to overinvestment in projects and diversion capacity. Failure to account for all costs, including externalities, leads to the same result. It is not necessarily optimal for the project to operate at full capacity in every period. Suppose that the random factor ε_t does not have identical independent distribution at all periods and instead has the same mean but its variability increases over time. This could occur if uncertainty about benefits is greater for periods further in the future. For simplicity, assume that ε_t is normal and is distributed with mean μ and variance σ_t^2 and expected benefit is of the form $B(\overline{W}, \varepsilon_t) = a\mu\overline{W}_t + b\overline{W}^2\sigma_t^2$.

The marginal benefit of additional capacity increases with the random effect in cases when it represents temperature and the gain from greater capacity is higher with a positive probability of increased climate variability.

If the variance increases substantially over time, optimal water use will be below capacity at an early period and will reach full capacity at time t^*. Thus for $t \leqslant t^*$, $W_t < \overline{W}$, and $W_t = \overline{W}$ for $t \geqslant t^*$.

The stochastic element ε_t may represent random natural phenomena, but in some cases it represents uncertainty about the key parameters of the system at the time when the design of the dam or other projects is made. Suppose that $\varepsilon_t = \bar{\varepsilon} + \eta_t$ where $\bar{\varepsilon}$ represents true randomness and η_t represents a random effect of lack of knowledge. Extra time that allows for learning can reduce both mean and variance of η_t.

Traditional cost–benefit analysis asks if a project should be built or not. It says that if the net present value of the project is positive, then it should be built, and if it is negative, it should not. This ignores a third possibility – the option of waiting. If the value people place on the benefits of this ecosystem is uncertain, then waiting to build the project can allow further information to be learned about these benefits as increased knowledge becomes available.

Arrow and Fisher (1974) and more recently Dixit and Pindyck (1994) develop models that suggest that in these cases the decision maker may consider delaying the decision about optimal project design so that more information can be learned. They not only look at the question 'to build or not to build', but they also consider the importance of when to build. Delaying building a project by one or two periods may lead to a loss of benefits in these periods but will lead to a future gain as more information is taken into account. This work shows that if the gains from acquiring new information are greater than the foregone benefits of current construction, it is better to delay construction of a new project. The gain from the ability not to make an immediate decision is referred to as "option value". In particular, in cases when there is uncertainty about the productivity of water due to a new technology or uncertainty about environmental impacts, the option value of waiting may be quite high and there may be significant gain from delay. Because of this, a positive net present value of a benefit–cost analysis is a necessary, but not a sufficient condition for construction.

Zhao and Zilberman (1999) extend this analysis to consider projects where restoration is costly but feasible. This is more realistic for water development. Dams are being removed from many sites worldwide, and natural habitats are being restored. They find that in some cases, it might be better to construct a new project even if there is a chance it will lead to costly restoration in the future. This could happen if the expected benefits of a project are larger than the expected future restoration costs.

2.3.2. *Waterlogging and drainage*

A solution to the problem of waterlogging should combine two elements – a functioning drainage system and the use of more efficient irrigation technology. Various details regarding the development of a plan to manage drainage are discussed in Dinar and Zilberman (1991). The construction of a drainage system can decrease levels of waterlogging in the soil. A well-functioning drainage system can allow an otherwise exhaustible soil resource to become sustainable. While effective, this has problems of its own. The construction of a drainage system can be very expensive, and the drained water has to be deposited in an area where the saline water will not have negative environmental effects. It may be best to combine a limited drainage system with the use of efficient irrigation technology, limiting the need for drainage and deposit of stored water [see Chakravorty, Hochman and Zilberman (1995)]. While drainage and waterlogging are problems in many areas of the world, quantitative data on the prevalence of these problems are not widely available for all regions. However, areas such as Asia and South America have very good data available. In China, 24.6 million hectares are susceptible to waterlogging, with drainage equipment on 20.3 million hectares. In the former USSR, 12% of the cropped land has been drained, although this varies from 6% in the Russian Federation to over 100% in the Baltic States.[3] In Mexico, over 5.2 million hectares have been drained for agriculture, along with 1.3 million hectares in Brazil, figures which represent 19.1 and 2.0% of the arable land, respectively.[4]

The following model illustrates the impact of drainage consideration on project evaluation. Suppose the per period net benefit of water is given by $B(W_t, S_t)$, where S_t is the stock of water trapped underground at time t, while $f(\overline{W})$ is the cost of constructing a water project of capacity $\overline{W}$. Let a fraction of the water be percolating and generate a stock of rising water that eventually hampers production. The initial stock is S_0, and the equation of motion is $\dot{S} = \alpha W_t$. The productivity of water declines as S_t, the stock of water trapped underground, rises. In this case the optimal water project design problem is

$$\max \int_0^\infty e^{-rt} B(W_t, S_t)\,\mathrm{d}t - f(\overline{W})$$

subject to

$$\dot{S} = \alpha W_t$$

and

$$W_t \leqslant \overline{W}.$$

Using the technique in Hochman and Zilberman (1985), the optimal solution to this problem is such that an optimal capacity $\overline{W}^*$ is established. For an initial period water diversion is constrained by the capacity, but beyond a critical point water deliveries decline over time as the user cost (associated with the extra waterlogging cost) reduces the net benefit of water use. A lower capacity to accumulate waterlogging and higher α (the fraction of water that contributes to waterlogging) will reduce the water project capacity and water deliveries. Further details on the dynamics of drainage management are presented in Tsur (1991).

[3] In this area the drained area is greater than the total cropped area due to a need to use drainage for construction sites.

[4] All of this data is available from AQUASTAT, 2003, from the Land and Water Division of the FAO.

As suggested by Van Schilfgaarde (1991), water project designers have ignored the drainage consideration and, as a result, the benefits of water projects have been overstated, and their capacity exceeded the socially optimal level. If the cost of waterlogging is low at an early period of a water project, the construction of a drainage canal can be delayed to year t_D and, once drainage facilities are introduced, the dynamics of water use may change. Specifically, both t_D and D, the drainage capacity, may be policy variables. Let the cost of the drainage capacity be $C_D(D)$. When drainage is introduced, equation of motion becomes

$$\dot{S} = \alpha W_t - D_t$$

and the optimization problem is

$$\max_{\overline{W}, W_t, t_D, D_t} \int_0^\infty e^{-rt} B(W_t, S_t)\,\mathrm{d}t - f(\overline{W}) - e^{-rt_D} C_D(D)$$

subject to

$$\dot{S} = \alpha W_t \quad \text{for } t < t_D,$$
$$\dot{S} = \alpha W_t - D_t \quad \text{for } t > t_D,$$
$$W_t \leqslant \overline{W}.$$

A lower cost of drainage will tend to increase $\overline{W}$ and water use at every period. When the cost of drainage is sufficiently low, the system may reach a steady state when $W_t = \overline{W}$ with all the infiltrating water being drained to prevent any buildup of underground water stock.

2.4. *Interregional choices*

2.4.1. *Conflicts and cooperation over water*

In many places, water sources cross political boundaries, and agreements are necessary to determine not only the division of water between user groups, but also the allowable activities and levels of pollutants in that water. International dialog and agreements are necessary in many areas to protect both the allocation and quality of freshwater resources. While it has often been argued that conflicts over water supplies are increasingly likely as populations increase, and existing freshwater supplies are over appropriated, work by Wolf (1998) suggests otherwise. Wolf finds that the number of agreements to cooperate on water management is many times greater than the number of conflicts. In addition, Wolf outlines the necessary conditions for an armed conflict over water to emerge, and finds that there are few possible sites that meet the criteria. Work by Franklin Fisher and the Middle East Water Project has developed the WAS (Water Allocation System), a model of the potential gains from the trade of water between Israel, Jordan, and the Palestinians [Fisher (2001)]. Their model finds that there are potentially significant gains from the trade of water between the Israeli and the Palestinian governments, regardless of the initial allocation, however, the paper also discusses some of the political and security reasons that such trade might not occur.

Joint cooperation is necessary to maintain or improve quality of water, in addition to agreements over quantity allocation. Several examples exist of joint cooperation between regions to improve water quality. For example, in 1972, Canada and the United States signed the Great Lakes Water Quality Agreement, making both countries responsible for activities that affect the water quality in the Great Lakes. This agreement, and the ongoing dialog it began between countries, has been at least partially responsible for the dramatic increase in water quality of the Great Lakes [Botts and Muldoon (1996)]. Another example of such an agreement is the Chesapeake Bay Agreement, designed to improve water quality in the Chesapeake Bay. This agreement was signed by Maryland, Virginia, Pennsylvania, and the District of Colombia; and was designed to reduce nutrient levels in the water 40% below a 1985 benchmark [Bockstael and Bell (1998); McConnell and Strand (1998)].

2.4.2. Trade and the concept of "virtual water"

Water scientists have traditionally assumed that annual per-capita requirements for water are 1000 m^3 [Gleick (2000)], a target which leaves many developing countries with a severe water shortage. For example, the annual per-capita water supply in Jordan is only 100 m^3. However, the 1000 m^3 requirement is an average amount, and assumes self-sufficiency in food production and, in particular, in grains needed to feed humans and livestock. There is significant heterogeneity and availability of water ranges from 5000 m^3 in Canada and Northern Europe to 100 m^3 in Jordan.

Trade can alleviate some of the water constraints. Countries with limited water resources may produce high value goods for export that enable them to purchase grains that are water intensive but cheap. Thus, water scientists introduce the notion of virtual water. For example, if every acre-foot of water put into tomatoes earns $500, while every acre-foot of water put into wheat earns $20, then an acre-foot used to grow tomatoes is worth 25 acre-feet in wheat. The idea of "virtual water" is that if a society can generate enough value (through the use of their available water) to get 1000 m^3 worth of food per capita, then that society has enough virtual water. This could be accomplished if water-scarce countries concentrate on exporting non-agricultural commercial products or growing high value crops for export (like flowers or produce) and then use the revenues to import staple crops like grains. Even though water itself is not tradable across nations, this allows countries to substitute trade in goods produced for direct trade in water. An example of a water scarce country with a shift toward high value crops is Yemen. Yemen has actively pursued a policy of subsidizing imported cereal products instead of supporting its own production, and consequently imports three-quarters of its cereal crops. Between 1970 and 1996, agricultural land used for cereal crops decreased from 85% to 61% of cultivated land, while the share of cash crops increased from 3% to 14% [Ward (2000)].

3. The benefits and costs of irrigation

3.1. Benefits of irrigation

3.1.1. Contribution of irrigation to agricultural productivity

Increased supplies of irrigation water have been instrumental in feeding the populations of developing countries in the last 50 years. Irrigation water has increased food security and improved living standards in many parts of the world. Fifty years ago it was common to hear concerns of food shortages and mass starvation, and while malnutrition is still a concern in many countries, the reason is not an insufficient global food supply. In fact, in the early 1990s, nearly 80% of malnourished children lived in countries that produced food surpluses, evidence that the cause of malnutrition is a lack of sufficient income by households to purchase food, not a lack of supply [FAO (1999)]. A report by IFPRI shows that between 1967 and 1997, global cereal production increased 84% at a time when population increased by 67% and that malnutrition among children under the age of five in developing countries declined from an aggregate rate of over 45% to 31% during this period. India, a historically impoverished country, has not had a major famine since the 1960s.

There are a number of reasons for this increase in food production, including high yield varieties of seed and increased use of fertilizers. However, the role of water development in providing irrigation water to cropland has also been significant. Benefits include the expansion of food supply, stabilization of water supply, flood protection, and the improved welfare of some indigenous populations.

3.1.2. Food supply expansion

Irrigation and agricultural land expansion. One benefit of water projects is an expansion in the feasible land base for agricultural production. Many regions with high quality soils have a Mediterranean climate and receive rainfall during the winter months when it cannot be used for crop production. For these areas, the development of reservoirs allows water to be stored during the rainy time of the year, and then used for farming when precipitation is low. Canals allow water to be transported from water-rich to arid areas, where it can be used for crop production.

Irrigation and increased crop yields. There is indisputable evidence that irrigating land leads to increased productivity. One acre of irrigated cropland is worth multiple acres of rainfed cropland. Globally, 40% of food is produced on irrigated land, which makes up only 17% of the land being cultivated. Dregne and Chou (1992) estimate the value of production of irrigated cropland at \$625/ha/year, compared to \$95/ha/year for rainfed cropland and \$17.50/ha/year for rangelands. In Asia, yields from most crops have increased 100–400% after irrigation [FAO (1996)]. Irrigation allows farmers to apply water at the most beneficial times for the crop, instead of being subject to the erratic timing of rainfall. One recent study using Indian production data from 1956 through 1987 shows that irrigation affects total factor productivity (TFP) beyond the input value of the water [Evenson, Pray and Rosegrant (1999)].

Irrigation and double cropping of land. Another benefit of reservoirs is that stored water can be used for double cropping of fields. There are many tropic and sub-tropic areas that are consistently warm and have seasonal rains for a portion of the year, but remain dry for the other portion of the year. The ability to store water during the rainy season for use in the dry season could allow a farmer to move from a single harvest per year to two or three. An example of this occurs is in the central plain of the main island of the Philippines. This area has a rainy season from mid-June into November, and more than 70% of the total rainfall falls in a four-month period. Water storage systems have allowed the region to have two cropping seasons in a year – the first is mainly dependent on rainwater, with supplemental irrigation, while the second, from December to May, is almost entirely dependent on irrigation water [Ferguson (1992)]. Although statistics are generally not available, there is anecdotal evidence that the expansion of double cropping has allowed land to be saved for nature, instead of developed for agricultural production.

3.1.3. Welfare improvements

Irrigation, employment opportunities, and income. Employment opportunities in many regions have increased after the development of irrigation systems. This can occur because additional labor in planting and harvesting is needed for new land brought into production, for land that is being double cropped, or for industries that support agricultural production. One example of this occurred in Borletar, Nepal. The construction of a large public works project during the 1980s has doubled total labor demand in the region, improving productivity and welfare. Production potential has increased by 300% and income by 600%, leading to increased food security for the indigenous population [FAO (1999)]. A 1997 study in Kenya and Zimbabwe showed that the average net increase in income from irrigation was $150–$1000 per family farm [FAO (1999)]. Growth in agricultural productivity also has a multiplier effect, providing benefits for non-agricultural sectors as well. In India, the value of non-agricultural output increases by 2.19 times the value of increases in irrigated production output [Hazell and Haggblade (1990)].

Irrigation and land values. Land values in a region are a function of the productive potential of the land. The development of irrigation systems allows farmers to grow higher yields of existing crops, or more profitable cash crops. Because of this, the benefits to landholders of irrigation development can be large. An example of this can be observed in the land supported by the Loskop Irrigation Scheme in South Africa. Non-irrigated grazing land in the area is worth between R1000/ha and R1500/ha while land with irrigation pivots is worth R10,000/ha [Tsur et al. (2004)].

3.1.4. Irrigation supply stabilization

The construction of a water storage and conveyance system decreases the risk associated with stochastic rainfall. Farmers are better able to plan their cropping patterns with a reliable water supply. The planting of certain crops, such as tree crops, requires the assurance of a sufficient water supply and may not be an economically rational choice for farmers before water development. Irrigation also allows farmers to apply water at the times that are most beneficial for the crop, instead of being subject to the variation in rainfall. The following example illustrates this point.

Due to weather shocks, the water supply is stochastic. During dry years, which occur with probability α, the available water supply is W_L, while during wet years, which occur with probability $(1 - \alpha)$, water supply is W_H. Since the choice of crop and irrigation technology must be made before the weather is observed, farmers must make these choices under uncertainty. If farmers are only assured of receiving a water supply of W_L ex-ante, then they might be unwilling to invest in high-value crops such as fruit and nut trees, or vine crops; as these crops require a minimum level of water each year. If an irrigation system and reservoir is developed, then farmers can rely on receiving a water supply of $\overline{W}$ in every year, where $\overline{W} = \alpha W_L + (1 - \alpha) W_H$. The removal of uncertainty from the water supply allows the farmers to improve their welfare through their decisions on both crop choice and irrigation technology.

3.1.5. Environmental benefits

Irrigation and deforestation. The expansion of agriculture is a primary cause of deforestation in developing countries. For example, between 1975 and 1988 the forested area in Northeast Thailand decreased by almost 50% because of growth in cassava production [Siamwalla (1997)]. Increasing food production in a region requires either more intensive use of existing cropland or an expansion of agriculture onto new cropland. Over time, production increases are essential because of larger populations, higher standards of living, and increased meat consumption. Using high-yield varieties of crops increases output on existing cropland, and irrigation is a necessary input into many high-yield varieties of crops in production. While deforestation is still an important concern worldwide, one would expect that without the benefit of irrigation, the remaining forest cover today would be less than we observe.

3.1.6. Benefits of the conjunctive use of groundwater and surface water

There is a large amount of literature on the benefits of conjunctive use of surface water and groundwater [Burt (1964); O'Mara (1988); Fisher et al. (1995)]. These benefits accrue because of the different nature of the resources. Surface water usually has lower delivery and extraction costs, but is subject to variability in supply. Groundwater can be expensive to pump, but has a reliable supply. In aquifers with recharge, the use of surface water during years of high precipitation can recharge an existing aquifer and decrease future overdraft of groundwater supplies. In aquifers without recharge, the availability of surface water for irrigation can be a substitute for nonrenewable groundwater supplies. In either case, the conjunctive use of the two sources can decrease the risk associated with a stochastic surface water supply. Arvin Edison Water and Storage District (AEWSD), located in California's Central Valley, provides a model of beneficial conjunctive use. AEWSD utilizes underground water banking in their water management plan. In wet years when they receive large quantities of surface water, they store some of it underground, and then pump this stored water during dry years, when the surface water supply is insufficient to meet district demand. Tsur (1997) estimates the value of this supply stabilization by the district to be $488,523 per year, a value equal to 47% of the total value of groundwater.

3.1.7. *Benefits of flood control*

A major purpose cited for the construction of many dams is flood control. While floods are rare occurrences in many areas, they have high costs when they do happen. Floods can cause tremendous damage – destroying property, killing people, and ruining environmental habitats. Dams have been instrumental in reducing these costs. The World Register on Dams shows that 17.3% of large dams report flood control as a main purpose. The majority of these dams are in developed countries (United States, Europe, and Japan make up a large proportion of the total); however developing countries have shared in some of these benefits as well.

One of the difficulties in measuring the value of the flood control benefits of a dam is that the benefits are probabilistic. When a dam is constructed, it is impossible to predict in which years there will be floods, and how damaging those floods will be. Because of this, a cost–benefit analysis of a proposed dam must use an expected value for the benefits of flood control. As discussed by Krutilla (1966), a dam that reduces the probability of flood damage to zero will not be feasible in a traditional cost–benefit analysis or economically optimal, due to the necessarily high costs.

3.2. *Costs of irrigation*

Despite the benefits discussed in the preceding section, there have also been many negative impacts of water projects. There have been financial, environmental, and social costs of developing water systems. Environmental problems include habitat destruction and a decrease in water quality while social costs include the displacement of indigenous populations, and increased occurrences of waterborne diseases that affect those populations.

3.2.1. *Capital costs*

In deciding whether a project is worth undertaking, it is important to weigh the anticipated benefits against the expected costs. Historically, the capital costs of constructing water projects have been consistently underestimated. A recent study of 81 large dams by the World Commission on Dams found that the average cost overrun was 56%. In addition, ex-ante predictions of the benefits of water projects have often been overly optimistic. This combination of factors has resulted in observations that the internal rate of return to most water projects is well below the expected rate of return, although most of the return rates are still positive. This result varies by region; investment costs for irrigation projects in West Africa have averaged over three times more per hectare irrigated than projects in Asia. The West African region has not used double cropping methods and has had poor management of water supplies. Because of this, returns to most of the West African projects have been negative [Matlon and Adesina (1997)].

In addition, the rates of return have been declining over time. Postel (1999) reviews the result of a World Bank study that shows the cost of irrigation has increased substantially since the 1970s. The study of more than 190 bank-funded projects found that irrigation development now averages $480,000 per square km. This cost varies by location – the capital cost for new irrigation capacity in China is $150,000 per square km, while the capital costs in Africa are $1,000,000–2,000,000 per square km. There are a few reasons for this increase in the cost of irrigation development. The best sites for water projects have already been developed, and those that remain are increasingly expensive. Also, improved knowledge about the environmental impacts of dam construction has led to requirement of detailed environmental impact reports and damage mitigation before the approval of many projects.

3.2.2. *Environmental costs*

Habitat destruction. The construction of a large dam causes changes in a river ecosystem. There are changes in stream flow, water temperature, and water quality. These changes affect the flora and fauna living in a river basin area. Fish species that live in warmer waters might not survive the cold waters below a dam site, or species that thrive in flowing waters may not survive in the still water of a reservoir.

Blocking migration of native species. Many river systems are used by species of migratory fish, such as salmon. In the course of their lifetime, salmon species are born upstream, swim down a river, and eventually return upstream to mate and reproduce. The construction of large dams can block the routes used by these fish, and affect their reproductive behavior. One example of this occurred on the Porto Primavera Dam in Brazil. Construction of this dam obstructed the migration of native fish species, and led to an 80% decrease in upstream fish catch [WCD (2000)]. Decreases like this not only affect the health of the species but also the welfare of people who depend on the fish species for their consumption or livelihood.

3.2.3. *Dynamic costs of water resources*

The development of irrigation projects had allowed crop production on otherwise arid lands. This has had many benefits, including expanding output and increasing land values. However, there are environmental problems that have occurred over time as the amount of land being irrigated has expanded. These costs include increased salinity levels in fresh water sources, and waterlogging and salinization of soil.

Increased salinity levels in freshwater supplies. The development of irrigation can increase the salinity levels of existing lakes and rivers. This happens when water that formerly ran into a freshwater lake is diverted, or when water withdrawals from a river are too great. With a river basin that flows into a sea, if water withdrawals are too great, the salt water from the sea can recede into the river basin. Over time this can lead to changes in the ecological balance of a river or lake and the species that it supports.

One area where irrigation has led to environmental disaster is in the Aral Sea, located between Uzbekistan and Kazakhstan. The ecological balance of the habitat has been destroyed and an industry that employed many people has been wiped out. The two rivers that feed into the Aral Sea are the Amu Darya and the Syr Darya. The area has been a site of irrigated agriculture for centuries, but until the last century this has been at a sustainable level. In the last century, the region became a large producer of cotton, an export crop for the USSR. In 1956, construction of the Kara Kum Canal was completed, a project that diverted water to be used to increase cotton supplies. Between 1962 and 1994, the volume of water in the sea was reduced by 75% and the salinity level of the sea has increased from 10 to over 100 grams per liter. This has taken a toll on the wildlife that lives in the area. The Aral Sea used to be a thriving site for the fishing industry, employing 60,000 individuals. This industry has been entirely wiped out, with many of the fish species disappearing [Murray-Rust et al. (2003); Calder and Lee (1995)]. Another example occurs in the Periyar River Basin in Kerala, India. On this river basin, a system of dams has increased freshwater withdrawals from the river. Because of this, seawater intrudes nearly 20 miles up the river system during the dry season, which has forced seasonal closures of factories that are dependent on river water [Repetto (1986)].

Waterlogging and salinization of land. Waterlogging and salinization are two problems related to the productivity of land that often occur together. Salinization occurs when the salt content of the soil increases, affecting the productivity of the land and limiting the crop choice of a grower. This is particularly a problem in lands that are arid or semi-arid. In arid regions, there is little rainfall to dissolve the salts in the soil. When water is applied without proper drainage, the evaporation in arid climates can quickly lead to high levels of salt in the soil, reducing the yield potential of the land. Another type of problem that can occur on irrigated lands is known as "waterlogging". This can happen if there is a layer of rock that forms a barrier, through which the water cannot escape. Over time, the water can accumulate and reach the root zone of the plants, making agricultural production impossible. Waterlogging eventually leads to the salinization of the soil, as water evaporates and the salt content of the soil increases. Estimates are that 20% of the irrigated land worldwide is affected by salinity levels in the soil, and that 1.5 million hectares are taken out of production each year as a result of high salinity levels in the soil. The costs of this are significant. One estimate is that salinization costs the world's farmers $11 billion per year in lost income [Postel (1999)]. However, this estimate does not include the general equilibrium effects of an increase in output price due to lower output, so it should be considered an upper bound.

One location in which waterlogging and soil salinization is a serious problem is the Indus Basin in Pakistan. In Pakistan, about 38% of the irrigated area is waterlogged. The problems are worst in the Sindh Province of the Indus Basin, which contains more than half of the area affected by waterlogging and soil salinization. This area has seen a decline of 40–60% in crop production as a result of these problems [Wambia (2000)].

Decreased levels of sediment and nutrients in water. One benefit of river systems is the movement of sediment and nutrients. Sediment that is moved downstream by the river can replace eroding soil, and provide beneficial nutrients to downstream cropland. The construction of a dam in a river system can trap sediment and nutrients behind the dam, degrading the quality of the downstream river system.

An example of this is on the Nile River in Egypt. Traditionally, the Nile River would flood each year, irrigating the banks of the river, and replacing eroding soil with new sediment. The new sediment not only kept the land from eroding, it also added nutrients to the soil. Since the construction of the Aswan Dam in southern Egypt, most of the sediment in the river is caught behind the dam and is not released downstream. There have been a few problems because of this. The lack of sufficient sediment is causing erosion in the coastline of the Nile Delta by 5–8 meters per year, and the removal of a natural source of nutrients has required farmers to increase their use of fertilizers.

Contamination of water supplies. Water supply contamination from agriculture can occur from several sources, including animal waste, or fertilizer and pesticide runoff. Using water that has been contaminated with animal waste for domestic uses can cause diseases such as diarrhea, hepatitis, or typhoid fever. More than one-third of the world's population lacks access to basic sanitation, and most of these people live in developing countries. Over half of China's population consumes water that exceeds the maximum permissible limits on human and animal waste, and an estimated 80% of the diseases and one-third of deaths in developing countries are caused by consumption of contaminated water.

As agricultural runoff is a nonpoint source of water pollution, its regulation poses challenges. In comparison to point source pollutants, the control of nonpoint source pollutants is more difficult, as individual emission levels cannot be directly measured, limiting the choice of policy instruments [Shortle and Horan (2001)]. Nonpoint source pollution control must be achieved through an indirect measure, necessitating a second-best outcome in efficiency. A possible policy is to subsidize modern irrigation technologies which reduce agricultural drainage flows, or to tax fertilizer inputs that lead to water quality degradation.

3.2.4. Social concerns

Waterborne diseases. In many places, large dams and irrigation projects have been blamed for public heath problems, including increased incidences of diseases such as malaria, diarrhea, cholera, typhoid, schistosomiasis, and river blindness. For example, higher levels of the snail host in irrigation canals have led to the increased occurrences of schistosomiasis in the Senegal River Valley and the Niger River Basin [Matlon and Adesina (1997)]. However, there is evidence that many of these cases have been the result of poor planning, and not a necessary effect of dam construction. Often, increased vector breeding occurs in fields and not in the dams and canals [Von Braun (1997)]. Incorporating public health concerns into the planning of a new water project can reduce the impact of the project. For example, a new reservoir can be an attractive breeding ground for mosquitoes, which can lead to the spread of malaria. Using sprays for pest control can decrease this risk. In areas where this risk has been ignored, such as the Senegal River Valley and the Kou Valley in Burkina Faso, there have been increased outbreaks of malaria in the regions. In addition, there have been areas where the incidence of malaria and other waterborne diseases actually decreases after the development of irrigation projects.

Further evidence that the effect of irrigation on public health is ambiguous comes from the work of public health researchers, who have found a range of outcomes when studying the impact of irrigation development on disease incidence. One study from the Tigray region of Ethiopia compared the incidence of malaria in villages located near dam sites (less than 3 km) to villages at similar altitudes located far from dam sites (more than 8 km) [Ghebreyesus et al. (1999)]. This study compared the incidence of disease at various times of the year in children under the age of 10. In all cases, the incidence of malaria was greater in the at-risk villages than in the control villages, and this difference was statistically significant. However, Ijumba and Lindsay (2001) review many studies from Africa and find that irrigation development does not always lead to a higher incidence of malaria, and can actually decrease incidence under certain situations. They find that this result varies by location, and while irrigation development increases the incidence of malaria in highland regions where populations lack any immunity, in many parts of Sub-Saharan Africa irrigation development can actually decrease malaria incidence. Ijumba and Lindsay (2001) also discuss other factors that affect the incidence of malaria and are also closely related to the development of irrigation systems. One factor is population migration. The development of irrigation systems and the resulting employment opportunities can lead to an inflow of people, many of whom may lack any resistance to malaria. Another factor is increased wealth, which can be a result of irrigation development. Increased wealth allows access to antimalarial drugs and prevention techniques such as bed nets. This factor is one of the explanations for the decreases in the incidence of malaria observed in some locations after irrigation development.

Displacement of human populations. The development of water projects in the last century has led to the displacement of 40–80 million people. In addition to their physical displacement, it has also often resulted in forced lifestyle changes. Between 1950 and 1990, 26 to 58 million people were displaced in China and India (two of the major dam building nations). Compensation for these forced resettlements has been minimal, if it occurs at all. Resettlement plans regularly fail to take into account the loss of a viable livelihood in addition to the loss of physical land, often leaving resettled populations worse off than before dam construction. For example, one study found that 72% of the 32,000 people displaced by the Kedung Ombo Dam in Indonesia were worse off after resettlement [WCD (2000)]. The construction of the Liu-Yan-Ba Dam on the Yellow River in China forced the resettlement of 40,000 people from fertile valleys to unproductive wind-blown highlands, resulting in extreme poverty for many of the resettled people [WCD (2000)].

3.2.5. Overuse of groundwater resources

Irrigated agriculture relies both on ground and surface water. Most of the large-scale irrigation projects divert surface water, but a significant proportion of the new land under irrigation in the last century is from the pumping of groundwater. In many situations groundwater resources are renewable and are replenished by rainstorms. Sometimes, as in the case of Libyan Desert, aquifers where fossil water is being mined are not replenished. Libya's plan to extract 2.2 km^3 per year from a desert aquifer is estimated to deplete the aquifer in 40–60 years [Postel (1999)]. Worldwide, as much as 8% of food crops grow on farms that use groundwater faster than the aquifers are replenished [Postel (1999)]. For example, the Punjab region of India is rapidly depleting its groundwater reserves. Punjab is a major production region of India, and most of the crops produced are cereal grains, such as rice and wheat. The past several decades have seen groundwater levels dropping at 25–30 cm per year. At groundwater depths below 15 meters, the commonly used tubewells will not function, and a well must be abandoned. The percentage of land where the water table is below 10 meters has increased from 3% to 46% between 1973 and 1994.[5] This overuse of groundwater threatens the future of the area and the national goal of food security.

In some areas such as Jakarta and Bangkok, the overdraft of aquifers is leading to a sinking of the ground level above the aquifer. In Bangkok, one-third of the city is below sea level. The fall in the ground level has led to increased damage from floods and higher costs of flood protection [Barker and Molle (2002)].

Another problem that can occur with overdraft of coastal aquifers is seawater intrusion into the aquifer. If the water table of the aquifer is drawn down too low, seawater from the adjacent ocean can enter the system; increasing the salinity level of the fresh water remaining in the aquifer. For irrigators relying on the available groundwater, this

[5] *Source*: Water Resources Directorate, Chandigarh, Punjab.

can limit the crop choice to those that can withstand high salinity levels of applied water. One area where this is a problem is in the Gaza Strip, which lies between Israel and the Mediterranean Sea. Gaza relies entirely on groundwater for its freshwater supply. Increased pumping has lowered the levels of the aquifers located in Gaza, and has allowed the intrusion of seawater. Citrus crops, which have traditionally been a source of revenue for the area, are intolerant of high salt levels in water, and there has been a decrease in both the yields and the quality of the crop. In some parts, high salinity levels have forced a change from citrus crops to other more salt-tolerant fruits and vegetables.

4. Conclusion

Irrigation was the source of more than 50% of the increase in global food production during 1965–1985 [Gardner (1996)] and more than 60% of the value of Asian food crops comes from irrigated land [Hinrichsen (2000)]. Irrigation in the last half of the twentieth century took advantage of most opportunities for diversion of water and in some situations, exploited non-renewable water resources. The environmental benefits of a sufficient fresh water supply for ecosystems are much better understood now than 50 years ago. Despite a growing concern about the third-party effects of water projects, there is a challenge to increase food supplies by at least 40% in the next 50 years, due to growing populations and changing preferences. Increased productivity should not come by the expansion of water supplies but by increased productivity of existing sources. That can be achieved through reform of water design and management systems. In particular, reform should include increased reliance on cost–benefit analysis for water projects, emphasis on appropriate design and management of conveyance facilities, and use of mechanisms that establish the price of water to represent the marginal cost of extraction, user costs, and environmental costs. Correcting these institutional problems is a necessary step to improve water quality and increase the supply of effective water.

The growing use of water user associations (WUAs) is a positive step toward the improvement of water management systems. Experience with trading in water suggests that it can improve efficiency as long as attention is paid to issues of third-party effects. Water quality issues should be addressed more by incentives to limit pollution. Current technologies allow the maintenance of yields with significant reduction of water use, but technology may be costly and many are in their infancy. New wireless technologies and improved power of computers that can reach even the most remote areas may suggest that the challenge of research is to develop water use management technology that is affordable by the poor, as well as mechanisms to enhance adoption of these technologies. Effective policies, pricing and management of water is one of the major challenges that society is facing as we enter the new millennium.

Appendix A

Below we present a simple model of irrigation technology choice, as developed by Caswell and Zilberman (1986). Consider an area with a fixed amount of heterogeneous quality land that grows a single crop. Let y denote the yield per acre, and e the effective water per acre. Output is given by a constant-returns-to-scale production function, $y = f(e)$. The applied water per acre under technology i is a_i and α is the land quality index, which assumes values from 0 to 1. Assume that there are two technologies: a traditional technology $(i = 0)$ and a modern technology $(i = 1)$. Irrigation effectiveness is defined as $h_i(\alpha) = e_i(\alpha)/a_i(\alpha)$ and for each α, $1 > h_1 > h_0 > 0$. The cost per acre associated with each technology is k_i. This cost includes annualized repayment of investment costs and annual operating costs. The modern technology is assumed to be more capital-intensive, so that $k_1 > k_0$.

The profit-maximizing choice of water application rate and irrigation technology is solved via a two-stage procedure. First the optimal amount of water for each technology is chosen and then the more profitable irrigation technology. Let $\Pi_i(a)$ denote the quasi-rent (exclusive of land rent) per acre of technology i, determined according to the following choice problem:

$$\Pi_i = \max_i \{ Pf(h_i(\alpha) \cdot a_i) - wa_i - k_i \},$$

where P is the output price and w the price of applied water. The first-order condition is

$$Pf'h_i - w = 0.$$

The price of effective water is the price of applied water divided by the irrigation efficiency (w/h_i), so optimal production occurs where the marginal product of applied water is equal to the price of effective water: $Pf' = w/h_i$. The price of effective water is lower under the modern technology due to the higher irrigation efficiency; therefore higher levels of effective water will be used and higher yields may be obtained.

The optimal water application under each technology determines the quasi-rent associated with the technology (Π_i), and the technology with the highest quasi-rent is selected, assuming it is non-negative. The quasi-rent difference between the two technologies can be written as

$$\Delta\Pi = P\Delta y - w\Delta a - \Delta k.$$

As shown by Graph 3, the quasi-rent difference can either be positive or negative. In the graph, $f^0(\alpha)$ represents the profit earned by the traditional irrigation technology, as a function of land quality, while $f^1(\alpha)$ represents the profit earned by the modern irrigation technology. The parameter indicates the quality of the land. There is a single value of the parameter that separates optimal irrigation technology by quality of land. For $\alpha < \alpha^0$, it is more profitable to use the modern, efficient irrigation technology. For $\alpha > \alpha^0$, a high land quality already results in a high level of water efficiency, resulting in higher profits from the traditional technology.

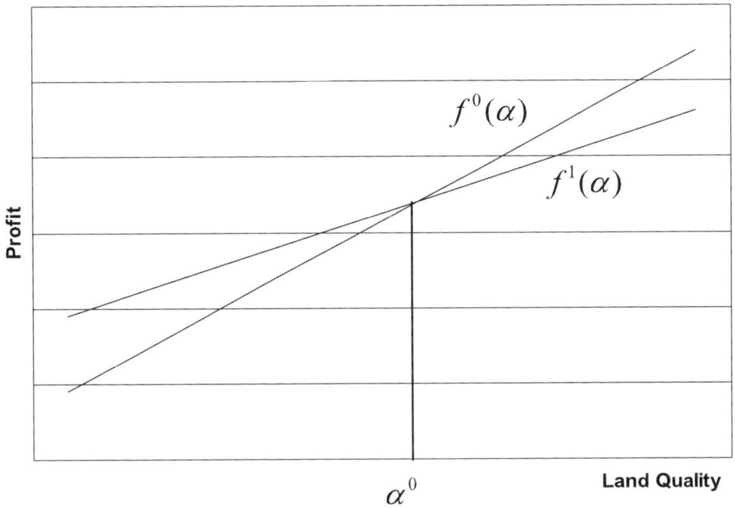

Graph 3. Comparison of modern and traditional technologies.

Inclusion of environmental costs of water runoff

This model can be extended [Caswell, Lichtenberg and Zilberman (1990)] to illustrate how irrigation technology choice affects the generation of negative environmental externalities in the form of agricultural drainage water. Irrigation water that is not used by crops is a major source of pollution, as it may result in waterlogging, salinization of soil, and pesticide runoff. By extending our simple model of technology choice and water use, we gain insight into the incentives for farmers to reduce agricultural drainage flows.

Let the pollution coefficient associated with water residuals be $g_i(\alpha)$, which is the fraction of water applied by technology i, on land of quality α, that is not utilized by the crop and which is environmentally damaging. The pollution coefficient is defined as

$$g_i(\alpha) \leqslant 1 - h_i(\alpha).$$

Since the modern technology is more water efficient, it is reasonable to assume that it has a lower pollution coefficient, i.e., $g_1(\alpha) < g_0(\alpha)$.

If the producer bears the costs associated with the pollution arising from water residual accumulation, the individual's profit maximization problem becomes

$$\Pi_i(\alpha) = \max_{a_i}\bigl\{Pf\bigl(h_i(\alpha)\cdot\alpha\bigr) - wa_i - k_i - \bigl(x \cdot g_i(\alpha)\bigr)\bigr\},$$

where x denotes the cost per unit of pollution. Usually this cost is a production externality that is not incorporated by farmers in their water use decisions. However, one could imagine the imposition of a pollution tax associated with water residuals.

The imposition of a pollution tax increases the profitability of adopting the water conserving technology, especially in situations where the initial costs of pollution per unit of water are large relative to water price. As shown in Graph 3, as land quality increases, the benefit of modern technology adoption decreases and the quasi-rent differential between the two technologies declines.

The modern technology will be selected in cases where the increased profits from higher yields or lower water costs offset the higher costs associated with adoption of the technology. These results indicate that modern technology adoption will increase with increasing water or output prices. In addition, modern technology adoption is more likely to occur with poor land quality, due to the high price of effective water under the traditional technology, and the land-augmenting qualities of the modern technology. The impact of modern technology adoption on aggregate applied water use levels depends on the elasticity of the marginal productivity of water (EMP), which measures how responsive the crop is to further irrigation.[6] Under most conditions, adoption results in both a decrease in overall water use and an increase in crop yields.

Appendix B

The following model is adapted from Provencher and Burt (1993). It shows the difference between the decisions made by a social planner and the decisions made by individuals in their use of a nonrenewable common property resource.

A region overlying a nonrenewable aquifer has N identical water users. In each period, each user withdraws u_t units of groundwater for use. The total available stock of water at time t is S_t, and the per-unit cost of pumping groundwater is $C(S_t)$, with $C' < 0$. The benefit that each user receives from the use of u_t units of groundwater is $B(u_t)$. We assume that $B' > 0$, and that $B'' < 0$. Since the aquifer has no recharge, the equation of motion for the available stock of groundwater is $S_{t+1} = S_t - N \cdot u_t$. The current value of the net benefit to each user in period t of using u_t units of water is $B(u_t) - u_t \cdot C(S_t)$.

B.1. Social planner's decision

Let $V(S_t)$ be the value at time t of the future net benefits to a single water user. Using the dynamic programming methodology, a social planner will want to solve the following:

$$N \cdot V(S_t) = \max_{u_t} N\big[B(u_t) - u_t \cdot C(S_t) + \beta \cdot V(S_{t+1})\big]$$

$$\text{s.t.} \quad S_{t+1} = S_t - N \cdot u_t.$$

[6] EMP is defined as $\varepsilon_i(e) = -f''(e_i) \cdot e_i / f'(e_i)$.

Solving this yields the following condition:

$$\frac{\partial B}{\partial u_t} - C(S_t) = \beta\left\{\frac{\partial B}{\partial u_{t+1}} - C(S_{t+1}) - N \cdot u_{t+1} \cdot \frac{\partial C}{\partial S_{t+1}}\right\}.$$

The left side of this equation is the net benefit of extraction of one more unit of groundwater in period t, while the right side is the discounted future benefit, taking into account the increased costs in the future that result from pumping groundwater today. This condition takes into account the additional future costs faced by all users of the aquifer, not just one individual.

B.2. Individual user's decision

For an individual decision maker, $\widetilde{V}(S_t)$ is the value at time t of future net benefits to a single water user. However, when an individual makes their decision about water use, they consider the decisions of other users as given. From an individual's perspective, the equation of motion governing available stock is $S_{t+1} = S_t - (N-1) \cdot u_t^* - u_t$, where u_t^* is the quantity of water used by each of the other growers. Using the dynamic programming framework, an individual will want to solve the following:

$$\widetilde{V}(S_t) = \max_{u_t}\left[B(u_t) - u_t \cdot C(S_t) + \beta \cdot \widetilde{V}(S_{t+1})\right]$$

$$\text{s.t.} \quad S_{t+1} = S_t - (N-1) \cdot u_t^* - u_t.$$

Solving this yields the following condition:

$$\frac{\partial B}{\partial u_t} - C(S_t) = \beta\left\{\frac{\partial B}{\partial u_{t+1}} - C(S_{t+1}) - u_{t+1} \cdot \frac{\partial C}{\partial S_{t+1}}\right\}.$$

Comparing the result from the social planner and the individual, we see that the social planner takes full account of the impact of withdrawing water today on future costs. The individual assumes that the actions of other are given both in the present and in the future. Therefore the individual ignores the impact of others, and only considers the impact of his/her own water use on his/her own future water costs. This results in each individual extracting too much groundwater per period.

References

ACIL Tasman (2003). "Water trading in Australia current and prospective products: Current trends and prospective instruments to improve water markets", Prepared for the Water Reform Working Group.

Amir, I., Fisher, F.M. (2000). "Response of near-optimal agricultural production to water policies". Agricultural Systems 64, 115–130.

Anderson, T., Snyder, P.S. (1997). Water Markets: Priming the Invisible Pump. Cato Institute, Washington, DC.

Arrow, K. (1997). "Intergenerational equity and the role of discounting in long term social investment", Working Paper No. 97-005. Department of Economics, Stanford University.

Arrow, K., Fisher, A. (1974). "Environmental preservation, uncertainty, and irreversibility". Quarterly Journal of Economics 88, 312–319.
Barker, R., Molle, F. (2002). "Perspectives on Asian irrigation", Presented at the Conference on Asian Irrigation in Transition – Responding to the Challenges Ahead. Asian Institute of Technology, Bangkok, Thailand.
Bauer, C.J. (1998). Against the Current: Privatization, Water Markets, and the State in Chile. Kluwer Academic Publishers, Norwell, MA.
Berck, P., Helfand, G. (1990). "Reconciling the von Liebig and differentiable crop production functions". American Journal of Agricultural Economics 72, 985–996.
Bockstael, N., Bell, K. (1998). "Land use patterns and water quality: The effect of differential land management controls". In: Just, R., Netanyahu, S. (Eds.), Conflict and Cooperation on Trans-Boundary Water Resources. Kluwer Academic Publishers, Norwell, MA.
Botts, L., Muldoon, P. (1996). "The Great Lakes water quality agreement: Its past successes and uncertain future". Dartmouth College, Hanover, NH.
Brill, E., Hochman, E., Zilberman, D. (1997). "Allocation and pricing at the water district level". American Journal of Agricultural Economics 79, 952–963.
Burness, H.S., Quirk, J.P. (1979). "Appropriative water rights and the efficient allocation of resources". American Economic Review 69, 25–37.
Burt, O.R. (1964). "The economics of conjunctive use of ground and surface water". Hilgardia 36, 25–41.
Calder, J., Lee, J. (1995). "Aral Sea and defense issues", Inventory of Conflict and Environment (ICE) Case Study No. 69. American University, Washington, DC.
Caswell, M., Lichtenberg, E., Zilberman, D. (1990). "The effects of pricing policies on water conservation and drainage". American Journal of Agricultural Economics 72, 883–890.
Caswell, M., Zilberman, D. (1985). "The choice of irrigation technologies in California". American Journal of Agricultural Economics 67, 224–234.
Caswell, M., Zilberman, D. (1986). "The effects of well depth and land quality on the choice of irrigation technology". American Journal of Agricultural Economics 68, 798–811.
Chakravorty, U., Hochman, E., Zilberman, D. (1995). "A spatial model of optimal water conveyance". Journal of Environmental Economics and Management 28, 25–41.
Chambers, R. (1988). Managing Canal Irrigation: Practical Analysis from South Asia. Cambridge University Press, Cambridge, MA.
Coase, R.H. (1960). "The problem of social cost". Journal of Law and Economics 3, 1–44.
Dayton-Johnson, J. (2000). "Choosing rules to govern the commons: A model with evidence from Mexico". Journal of Economic Behavior and Organization 42, 19–41.
Devarajan, S., Fisher, A.C. (1981). "Hotelling's 'economics of exhaustible resources': Fifty years later". Journal of Economic Literature 19, 65–73.
Dinar, A., Subramanian, A. (1997). "Water pricing experiences: An international perspective", World Bank Technical Paper No. 386. The World Bank, Washington, DC.
Dinar, A., Yaron, D. (1992). "Adoption and abandonment of irrigation technologies". Agricultural Economics 6, 315–332.
Dinar, A., Zilberman, D. (Eds.) (1991). The Economics and Management of Water and Drainage in Agriculture. Kluwer Academic Publishers, Norwell, MA.
Dixit, A., Pindyck, R. (1994). Investment under Uncertainty. Princeton University Press, Princeton, NJ.
Dregne, H.E., Chou, N.T. (1992). "Global desertification dimensions and costs". In: Degradation and Restoration of Arid Lands. Texas Tech. University, Lubbock, TX.
Easter, K.W. (1986). Irrigation, Investment, Technology, and Management Strategies for Development. Westview Press, Boulder, CO.
Easter, K.W. (2000). "Asia's irrigation management in transition: A paradigm shift faces high transaction costs". Review of Agricultural Economics 22, 370–388.
Easter, K.W., Becker, N., Tsur, Y. (1997). "Economic mechanisms for managing water resources: Pricing, permits, and markets". In: Biswas, A.K. (Ed.), Water Resources: Environmental Planning, Management, and Development. McGraw-Hill, New York.

Evenson, R.E., Pray, C.E., Rosegrant, M.W. (1999). "Agricultural research and productivity growth in India", IFPRI Research Report No. 109. International Food Policy Research Institute, Washington, DC.
FAO (1996). Agriculture and Food Security. World Food Summit. Food and Agricultural Organization of the United Nations, Rome.
FAO (1999). Poverty Reduction and Irrigated Agriculture. International Programme for Technology and Research in Irrigation and Drainage. Food and Agricultural Organization of the United Nations, Rome.
Feder, G., Just, R.E., Zilberman, D. (1985). "Adoption of agricultural innovations in developing countries: A survey". Economic Development and Cultural Change 33, 255–298.
Feder, G., Umali, D.L. (1993). "The adoption of agricultural innovations: A review". Technological Forecasting and Social Change 43, 215–239.
Ferguson, C.A. (1992). "Water allocation, inefficiency and inequity in a government irrigation system". Journal of Development Economics 38, 165–182.
Fisher, A., Fullerton, D., Hatch, N., Reinelt, P. (1995). "Alternatives for managing drought: A comparative cost analysis". Journal of Environmental Economics and Management 29, 304–320.
Fisher, F.M. (2001). "Water: Casus Belli or source of cooperation?", Working paper.
Gardner, G. (1996). "Shrinking fields: Cropland loss in a world of eight billion", World Watch Paper No. 131. World Watch Institute, Washington, DC.
Ghebreyesus, T.A., Haile, M., Witten, K.H., Getachew, A., Yohannes, A.M., Yohannes, M., Teklehaimanot, H.D., Lindsay, S.W., Byass, P. (1999). "Incidence of malaria among children living near dams in northern Ethiopia: Community based incidence survey". BMJ 319, 663–666.
Gleick, P.H. (2000). "The changing water paradigm: A look at twenty-first century water resources development". Water International 25, 127–138.
Hazell, P.B.R., Haggblade, S. (1990). "Rural–urban growth linkages in India", Background paper prepared for the 1991 Country Memorandum for India. World Bank, Washington, DC.
Hexem, R.W., Heady, E.O. (1978). Water Production Functions for Irrigated Agriculture. The Iowa State University Press, Ames, IA.
Hinrichsen, D. (Contributing Ed.) (2000). People and the Planet (online), http://www.peopleandplanet.net/doc.php?id=386, September 15.
Hochman, E., Zilberman, D. (1985). "Optimal exploitation of energy resources: Solar power and electricity generation in below sea level basins". Journal of Environmental Economics and Management 12, 322–337.
Hotelling, H. (1931). "The economics of exhaustible resources". Journal of Political Economy 39, 137–175.
Howe, C. (2002). "Policy issues and institutional impediments in the management of groundwater: Lessons from case studies". Environment and Development Economics 7, 625–641.
Ijumba, J.N., Lindsay, S.W. (2001). "Impact of irrigation on malaria in Africa: Paddies paradox". Medical and Veterinary Entomology 15, 1–11.
Johansson, R.C. (2000). "Pricing irrigation water: A survey", Policy Research Working Paper No. 2449. The World Bank, Washington, DC.
Krutilla, J.V. (1966). "An economic approach to coping with flood damage". Water Resources Research 2, 183–190.
Lee, T.R. (1990). Water Resources Management in Latin America and the Caribbean. Westview Press, Boulder, CO.
Matlon, P.J., Adesina, A.A. (1997). "Agricultural growth and sustainability: Prospects for semi-arid West Africa". In: Vosti, S.A., Reardon, T. (Eds.), Sustainability, Growth, and Poverty Alleviation. Johns Hopkins University Press, Baltimore, MD.
McConnell, K.E., Strand, I. (1998). "Inter-jurisdictional water quality management: Experiences from Chesapeake Bay nutrient management". In: Just, R., Netanyahu, S. (Eds.), Conflict and Cooperation on Trans-Boundary Water Resources. Kluwer Academic Publishers, Norwell, MA.
McKinsey, J.W., Evenson, R.E. (2003). "Crop genetic improvement impacts on Indian agriculture". In: Evenson, R.E., Gollin, D. (Eds.), Crop Variety Improvement and Its Effect on Productivity: The Impact of International Agricultural Research. CABI Publishing, Wallingford, UK.

Murray-Rust, H., Abdullaev, I., ul Hassan, M., Horinkova, V. (2003). "Water productivity in the Syr-Darya River basin", Research Report No. 67. International Water Management Institute, Colombo, Sri Lanka.

Nieuwoudt, W.L., Armitage, R.M., Backeberg, G.R. (2001). "Incentives and institutions of water market transfers in South Africa: Two case studies", In: Proceedings of the 7th Conference of the International Water and Resource Economics Consortium, Girona, Spain.

O'Mara, G.T. (1988). "Efficiency in irrigation: The conjunctive use of surface and groundwater resources". The World Bank, Washington, DC.

Postel, S. (1996). "Forging a sustainable water strategy". In: Starke, L. (Ed.), State of the World 1996. W.W. Norton and Co., New York.

Postel, S. (1999). Pillar of Sand: Can the Irrigation Miracle Last? W.W. Norton and Co., New York.

Provencher, B., Burt, O. (1993). "The externalities associated with the common property exploitation of groundwater". Journal of Environmental Economics and Management 24, 139–158.

Rabemanambola, F. (1997). "Madagascar". In: Dinar, A., Subramanian, A. (Eds.), Water Pricing Experiences: An International Perspective, World Bank Technical Paper No. 386. The World Bank, Washington, DC.

Rausser, G.C. (2000). "Collective choice in water resource systems". In: Dinar, A. (Ed.), The Political Economy of Water Pricing Reforms. Oxford University Press, New York.

Rausser, G.C., Zusman, P. (1991). "Organizational failure and the political economy of water resources management". In: Dinar, A., Zilberman, D. (Eds.), The Economics and Management of Water and Drainage in Agriculture. Kluwer Academic Publishers, Norwell, MA.

Ray, I., Williams, J. (1999). "Evaluation of price policy in the presence of water theft". American Journal of Agricultural Economics 81, 928–941.

Renfro, R.Z.H., Sparling, E.W. (1986). "Private tubewell and canal water trade on Pakistan Punjab watercourses". In: Easter, K.W. (Ed.), Irrigation Investment, Technology, and Management Strategies for Development. Westview Press, Boulder, CO.

Repetto, R. (1986). "Skimming the water: Rent-seeking and the performance of public irrigation systems", World Resources Institute Research Report No. 4. World Resources Institute, Washington, DC.

Rhodes, J.D., Dinar, A. (1991). "Reuse of agricultural drainage water to maximize the beneficial use of multiple water supplies for irrigation". In: Dinar, A., Zilberman, D. (Eds.), The Economics and Management of Water and Drainage in Agriculture. Kluwer Academic Publishers, Norwell, MA.

Rosegrant, M.W., Perez, N.D. (1997). "Water resources development in Africa: A review and synthesis of issues, potentials, and strategies for the future", EPTD Discussion Paper No. 28. Environment and Production Technology Division, International Food Policy Research Institute, Washington, DC.

Saleth, R.M. (1996). Water Institutions in India: Economics, Law and Policy. Commonwealth Publishers, New Delhi.

Saleth, R.M., Dinar, A. (1997). "Satisfying urban thirst: Water supply augmentation and pricing policy in Hyderabad City, India", World Bank Technical Paper No. 395. The World Bank, Washington, DC.

Schoengold, K., Sunding, D.L., Moreno, G. (2005). "Panel estimation of a water demand function based on an episode of rate reform", Working paper. University of California, Berkeley.

Shah, F.A., Zilberman, D., Chakravorty, U. (1993). "Water rights doctrines and technology adoption". In: Hoff, K., Braverman, A., Stiglitz, J.E. (Eds.), The Economics of Rural Organization: Theory, Practice, and Policy. Oxford University Press, New York.

Shah, T. (1993). Groundwater Markets and Irrigation Development: Political Economy and Practical Policy. Oxford University Press, Bombay.

Shortle, J.S., Horan, R.D. (2001). "The economics of nonpoint pollution control". Journal of Economic Surveys 15, 255–289.

Siamwalla, A. (1997). "The relationship between trade and environment, with special reference to agriculture". In: Vosti, S.A., Reardon, T. (Eds.), Sustainability, Growth, and Poverty Alleviation. Johns Hopkins University Press, Baltimore, MD.

Spiller, P., Savedoff, W. (1999). "Government opportunism and the provision of water". In: Spiller, P., Savedoff, W. (Eds.), Spilled Water: Institutional Commitment in the Provision of Water Services. Inter-American Development Bank, Washington, DC.

Todt de Azevado, L.G. (1997). "Brazil". In: Dinar, A., Subramanian, A. (Eds.), Water Pricing Experiences: An International Perspective, World Bank Technical Paper No. 386. The World Bank, Washington, DC.
Tsur, Y. (1991). "Managing drainage problems in a conjunctive ground and surface water system". In: Dinar, A., Zilberman, D. (Eds.), The Economics and Management of Water and Drainage in Agriculture. Kluwer Academic Publishers, Norwell, MA.
Tsur, Y. (1997). "The economics of conjunctive ground and surface water irrigation systems: Basic principles and empirical evidence from Southern California". In: Parker, D.D., Tsur, Y. (Eds.), Decentralization and Coordination of Water Resource Management. Kluwer Academic Publishers, Norwell, MA.
Tsur, Y., Roe, T., Doukkali, R., Dinar, A. (2004). Pricing Irrigation Water: Efficiency, Implementation Cost, and Equity. Resources for the Future Press, Washington, DC.
Van Schilfgaarde, J. (1991). "Preface". In: Dinar, A., Zilberman, D. (Eds.), The Economics and Management of Water and Drainage in Agriculture. Kluwer Academic Publishers, Norwell, MA.
Von Braun, J. (1997). "The links between agricultural growth, environmental degradation, and nutrition and health: Implications for policy and research". In: Vosti, S.A., Reardon, T. (Eds.), Sustainability, Growth, and Poverty Alleviation. Johns Hopkins University Press, Baltimore, MD.
Wambia, J.M. (2000). "The political economy of water resources institutional reform in Pakistan". In: Dinar, A. (Ed.), The Political Economy of Water Pricing Reforms. Oxford University Press, New York.
Ward, C. (2000). "The political economy of irrigation water pricing in Yemen". In: Dinar, A. (Ed.), The Political Economy of Water Pricing Reforms. Oxford University Press, New York.
WCD (2000). "Dams and development: A new framework for decision making", The Report of the World Commission on Dams.
Weber, P. (1991). "Desalination's appeal evaporates", World Watch Institute Technical Paper 4:6. World Watch Institute, Washington, DC.
Whitaker, M.L., Kerr, J.M., Shenoi, P.V. (1997). "Agricultural sustainability, growth, and poverty alleviation in the Indian semi-arid tropics". In: Vosti, S.A., Reardon, T. (Eds.), Sustainability, Growth, and Poverty Alleviation. Johns Hopkins University Press, Baltimore, MD.
Wolf, A.T. (1998). "Conflict and cooperation along international waterways". Water Policy 1, 251–265.
Zhao, J., Zilberman, D. (1999). "Irreversibility and restoration in natural resource development". Oxford Economic Papers 51, 559–573.

Chapter 59

LAND USE: FOREST, AGRICULTURE, AND BIODIVERSITY COMPETITION*

ROGER A. SEDJO

Resources for the Future, Washington, DC

R. DAVID SIMPSON

National Center for Environmental Economics, EPA, Washington, DC

Contents

* Paper presented to the Bretton Woods Conference on Agricultural Producers and Productivity, Bretton Woods, NH, September 29–October 1, 2002.

Handbook of Agricultural Economics, Volume 3
Edited by Robert Evenson and Prabhu Pingali

DOI: 10.1016/S1574-0072(06)03059-3

Abstract

Since at least the time of von Thünen's contributions to the subject, economists have been interested in explaining land use in the hinterlands. This interest has grown with increasing societal demands on remaining forested areas and concern for the nonmarket resources generated by such habitats. However, the situation is not nearly as dire as one might expect based on the more alarming pronouncements. Despite both economic and population growth, forest areas in much of the developed world have been increasing, not declining. Improvements in growing and processing technology, in combination with increased concern for forest resources, have enabled increases in productivity that have largely offset growth in demand. While the record has not been quite as good in the developing countries, productivity growth and a slowing in the rate of population increase are also reducing pressures on forests in poorer nations.

While these trends are promising, concern remains with the decline in natural habitats, particularly in moist tropical forests, where biological diversity is most concentrated. Biologists and conservation advocates have advanced a number of arguments both that "biodiversity" is imperiled as forests are felled and that the values of the services generated by such systems justify their preservation. While often plausible, these arguments remain largely unproved. It seems unlikely that prospects for commercializing the products and services of forest ecosystems will prove sufficient to motivate their conservation on a large scale. Other instruments for conservation must, then, be employed if areas under intense pressure for conversion are to be preserved.

Keywords

biological diversity, forestry, land use, productivity

JEL classification: Q15, Q23, Q57

Well over 100 years ago, Johann Heinrich von Thünen laid out what is probably still the most useful economic theory of land use with his concentric model. The 19th century German landowner published a book in three parts (1826, 1850, and 1863) that described land use patterns as concentric circles. Applying the concept of the diminishing productivity of land on which David Ricardo based his theory of rent, von Thünen explained agricultural production in the neighborhood of a city.

Von Thünen views a city's population as centered in a circle, surrounded by agricultural lands like a donut. Forestland comprises a more distant donut surrounding the agricultural lands. Since he abstracts from soil productivity and contours of the land, von Thünen focuses our attention on location, with location *vis à vis* markets (the population) being the critical component. High valued (value per unit weight) labor intensive products, e.g., crops, tend to be located near the markets, while low valued land intensive products, pasture and timber, are located farther away from the market.

Commodities required to be consumed fresh, and those costly to transport, would be produced nearest the town. In broad rings, other goods were produced at greater and greater distances from the town as their nature and value made them more and more able to bear the time and cost involved in transportation.

Subsequently, it has been recognized that other factors count, including land productivity, the nature of the terrain, natural barriers and man-made transport connections.[1] In the simple world of this von Thünen model the agricultural area would tend to expand should the population rise; conversely, it would tend to contract should agricultural innovation increase productivity (at least under the conventional wisdom that the demand for agricultural products is inelastic). Shifts would occur among land uses reflecting changing incomes, tastes and relative cross-sector productivity. This model, adjusted for land quality differences, continues to usefully explain land use patterns even today.

However, in the von Thünen model there is no explicit appreciation for environmental and ecological values. Concern is now growing for values generated from the forest such as flood protection, erosion control, and wildlife habitat [see, e.g., Daily (1997)]. These considerations, which we will cluster under the rubric of "biodiversity" and "ecosystem service" values motivate revisiting von Thünen's model.

In von Thünen's day, one might have reasonably regarded these commodities as having been either in excess supply, or at least provided by areas of natural habitat large enough as to have obviated concerns with marginal degradation. This may no longer be true. We are now, as a society, more concerned over the loss of certain types of habitat, e.g., old growth forests and wetlands. Such areas are increasingly being designated for "protection" and set aside from conversion to other uses as parks and wilderness areas.

In the future, the demand for the environmental services provided by land now used, or that could be used, in agriculture likely will grow much faster than the demand for

[1] In recent years economists have rediscovered geography, and a "new economic geography" seeks to explain why people are clustered in the landscape, rather than spread over it evenly [see Paul Krugman's survey (1998)]. The insights of this literature, while important, are not necessarily relevant to our inquiry here, however.

Table 1
Global land use: selected regions and years (million ha)

Year	Global Arable land (permanent crops)	Africa Arable land (permanent crops)	North America	South America	Asia	Europe
1971	1457 (103.3)	214 (n.a.)	271 (n.a)	84 (n.a.)	463 (n.a.)	372 (n.a.)
2001	1532 (130.4)	209 (26)	266 (85)	126 (14)	569 (61)	305 (17)

Source: FAO Annual Production Yearbook, selected issues.

food and fiber. An implication drawn by Crosson (1990) is that there should be a shift in emphasis in agricultural research from development of new technologies to increase on-farm productivity and knowledge to a focus on institutional forms most likely to result in improved management of environmental resources. In this paper, we document the existence of lands for habitat and note the tendency to move toward the protection and codification of these lands.

1. Global land resources: An overview

Table 1 presents FAO estimates of land uses at recent selected dates. The data show a gradual rise in recent years in lands under permanent crops and an associated increase in arable lands. Also, permanent meadows and pastures are increasing. Forestlands have remained roughly constant, while irrigated land has increased substantially.

1.1. Croplands and pasture

Table 1 presents FAO data on land use covering a 30-year period. While the data is aggregated and some data is not collected in recent years, perhaps what is most remarkable is the relative stability of the figures. Globally, there is a gradual increase in the arable and land under permanent crops category, due in part to increases in the area of permanent crops. However, arable land has shown a decline in certain regions.

1.2. Forest lands

The FAO Forest Resources Assessment (2000) indicates the world's forest cover total 3.86 billion ha in 2000, or 29% of the world's land area. Table 2 shows forest as distributed among tropical forests (47%); subtropical forests (9%), temperate forests (11%) and boreal forests (33%). The area of temperate forest worldwide covers a land area

Table 2
Forest cover in 2000 (million ha)

Total	Tropical	Subtropical	Temperate	Boreal
3869	47%	9%	11%	33%

Source: FAO (2000).

Table 3
Forest cover in 2000 by selected region in 2000 (million ha)

Total	Asia	Africa	Latin America[a]
3869	547.8	649.9	1064

[a]Includes South America, Central America and the Caribbean.

roughly the size of North America, while the area of tropical forest covers and area roughly the size of South America, which is roughly 10% smaller than North America. Table 3 indicates the area of forest in 2000 for selected regions.

Although the total areas of forest in the various land uses change only slowly, the use to which a particular parcel of land is put can change quickly. Additionally, socio-economic forces can conspire to dramatically change the land use in entire regions. New England and the southeastern United States are examples. Barnett (1980) reports that since about 1850, New England has experienced the continuing replacement of its agricultural lands by forests. The opening of the agricultural lands of the U.S. Midwest and the accompanying transportation networks made much of New England agricultural uncompetitive, leading to the abandonment of agricultural lands and their gradual natural regeneration into forests. Although the native peoples are now recognized as having a substantial impact on the land, including significant forest clearing, the arrival of European colonists was followed by gradual further land conversion to farming in the 17th, 18th and especially the 19th century.

This story is repeated in the U.S. South where agricultural, particularly tobacco and cotton, led to the clearing of the forest only to be reversed as these crops declined and forests were reestablished on the land. As agriculture shifted from the eastern part of the United States inland to the highly fertile grasslands and prairies of the Midwest, agriculture was abandoned on marginal farmlands of the eastern seaboard.

Much of Europe has had a similar experience. For example, the French forest is estimated to have been reduced by one-half between 1000 and 1300 [UNECE (1996)]. However, this trend was subsequently reversed. Over the 19th century the area of French forests increased by 50% and this increase was repeated over the 20th century with an additional 50% increase in net forest area in France. The ebb and tide of forests in various developing countries throughout the world is documents in Tucker and Richards

(1983) and Richards and Tucker (1988). Thus, the notion of land use stability is not reflective of much of the world's experience, particularly over periods of many decades or centuries.

2. Pressure on the resources

2.1. Agriculture and land conversion

Much of human activity in the past several hundred years has been oriented to increasing the area of tillable land and increasing human ability to till large areas [Hayami and Ruttan (1985)]. This has been particularly true in land abundant regions. However, Richards and Tucker (1988) report that in North America and Europe, essentially all of the land expansion ended before 1920. For East Asia, he reports most of the expansion ended by 1920. In other regions however, including the Soviet Union, South Asia, South American and Africa, substantial expansion of areas in regular cropping continued well into the 20th century. This finding is consistent with the data reported above by the FAO.

The effect of the continuation of substantial expansion of regular cropping into undeveloped areas has been to reduce the area of other land uses including wild forest and other natural vegetation as these lands were converted. Land conversion to agriculture is still regarded as the major force driving tropical deforestation.

Numerous studies have suggested that the rate of land conversion has been socially excessive due to either governmentally provided incentives [e.g., Binswanger (1987)] and/or the absence of well defined property rights [Mendelsohn (1994); Forest Trends (2002)], which results in a commons that is exploited at an excessive rate. Examples of this include the Homestead Act in the United States, which in effect subsidized land conversion, Brazilian governmental policies to promote "development" in the Amazon, and World Bank projects in Rondonia (in Brazil) that subsidized massive land conversion form forests to agriculture. However, Berck (1979) argues that the rate of harvest was not excessive in the case of the US. Also, Johnson and Libecape (1980) argue that the rapid drawdown of the Lake States forests was not socially excessive in the latter part of the 19th and beginning of the 20th centuries.

2.2. Timber supply and demand

Pressure on forestlands comes primarily from land use conversions away from forestry and to other uses. Forests have traditionally been viewed as the lowest-use value land, at least in the European context. They were viewed as available for conversion to pasture and cropping as well as providing lands for development of various types.[2] In addition

[2] It is interesting to note, in fact, that title to land is, in most nations, established by "improving" it. Such improvements might include the cutting trees, fencing, plowing, planting, and erecting structures [Alston,

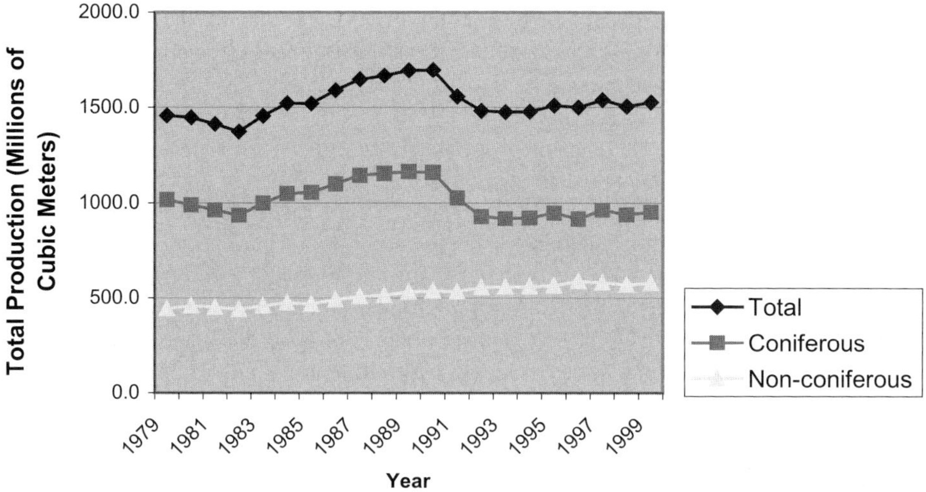

Figure 1. World industrial roundwood production. *Source*: FAO, Rome.

to land use conversion, concerns have been raised in recent decades about the effects of intensive and excessive logging. Traditionally, the concern has been that the reduction in the area of forest will impact negatively on the supply of wood. Projections of future timber famine have given rise to efforts to preserve and maintain lands in forests. Interestingly, timber was the only resource[3] found to be experiencing increasing scarcity as reflected in rising real prices in the RFF research of the 1960s and 70s. However, the multitude of predicted timber famines have never materialized [Clawson (1979)]. Furthermore, despite more recent predictions of rising prices and growing scarcity [Barney (1982)], in recent decades overall production of industrial wood has stagnated (Figure 1)[4] while real prices generally have remained flat. Note that this stagnation has occurred during a period of remarkable economic expansion, including the expansion of China.

The usual explanation for the stagnation has been not the absence of supply potential, since global prices have also tended to stagnate, but rather a lack of effective demand.

Libecap and Schneider (1996)]. Such standards codified John Locke's (1980 [1690]) theory of property rights as tied to the labor of the individual claiming them. More generally, the modern economic theory of property rights [Barzel (1997)] holds that they come to be established when the value of the things to which they adhere justifies the expense of enforcing rights of ownership. Things in excess supply – such as forests several hundred years ago – are cheap and do not justify the expense of recording their ownership. The exceptions prove the rule. In parts of Tanzania where forests are scarce property rights are established by planting trees rather than felling them.

[3] Fish may be another resource found to experience increasing scarcity.

[4] Data on fuelwood production is much less reliable than that of industrial wood. However, FAO data indicated that fuelwood production/consumption has shown no change between 1988 and 2000.

Harvests from most of the world's forests are far less than forest growth (OECD selected years), leaving a substantial wood surplus. Finally, the advent of intensively managed planted forests, with growth capacities 5 to 10 times that of nature forests give promise of more wood producing capacity even as the harvesting of natural forests is declining [Sedjo and Botkin (1997)]. In fact, one may argue that intensive management is a response to decreased natural forest availability [Hyde, Newman and Seldon (1992)]. Furthermore, the recent FAO report has indicated that the rate of forest plantation establishment has been very rapid, with close to 50 million ha of plantation, or nearly 30%, being 10 years old or less. Many projections future timber supply (Figure 3) forecast a growing role for plantation wood from nontraditional producers of timber.

2.3. *Deforestation*

There has still been much concern over the extent of deforestation in recent years and numerous studies have tried to develop accurate estimates of changes. While the forest and woodlands category of Figure 1 shows little change over the period of the 1970s and 1980s, other data suggest forest decline during that period. More recently, the Global Forest Resources Assessment [FAO (2000)] estimated the net annual loss of forest area worldwide in the 1990s at 9.4 million ha, with the tropics experiencing a decrease and the temperate regions experiencing a modest increase in forest area. However, this was 5.2 million ha less deforestation than had earlier been experienced in the 1980s [FAO (1993)]. For the humid tropics the data found a net annual loss of 6.4 million ha. A recent study [Achard et al. (2002)], suggests that the official reports on tropical deforestation are overstated. Using satellite imagery the study found that the world lost an average of 5.8 million ha of humid tropical forest. After adjusting for regrowth, the study found a net loss of 4.9 million ha per year, or a reduction of 23% lower than the generally accepted rate.

Additionally, some of the longer term estimates, over many decades or centuries, of the extent of deforestation may be excessive. For example, Fairhead and Leach (1998) present evidence that in the case of Africa, the initial forest cover was probably assumed to be more extensive than it actually was. They argue that many areas in Africa, believed to have been forested until relatively recently, actually were deforested long ago as part of the traditional agricultural systems that had been practiced. A point here is that the measurement of rates of forest decline (expansion), both currently and historically, are still subject to discussion and uncertainty.

From the data two largely accepted conclusions emerge: First, that deforestation is being experienced in the tropics. Second, the temperate and boreal regions have experienced net reforestation, often due to the abandonment of agricultural activities in those areas. The total net result is that global forests are experiencing some degree of net deforestation, although the precise amount is still the subject of debate and the impacts do not appear to be captured in many of the overall statistics. Additionally, some portion of the net reforestation is in the form of planted forests, although this is still a small fraction of total forested areas, perhaps 5%.

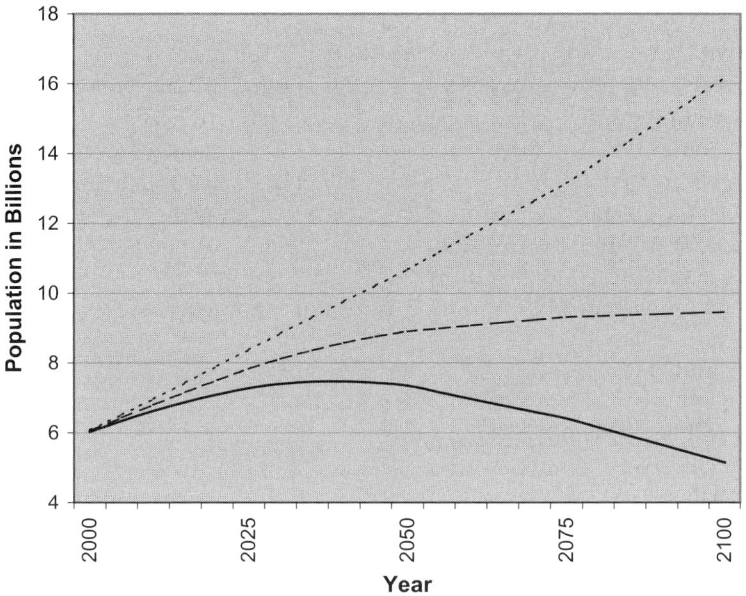

Figure 2. UN world population projections (high, medium and low). *Source*: World Population Prospects, 1998 Revision, United Nations.

2.4. Population

Finally, there is the question of the future world population. Over the past several decades the demographic focus has tended to be on the rapid growth of global population. It is note worthy that many of the projections (Figure 2) are now anticipating not only a reduction in growth, but an absolute decline in world population by or before the middle part of the 21st century. Hence, based on existing trends, it can be argued that the world's forests appear more than adequate to the task of providing for the world's wood needs into the indefinite future. It is likely to be also true that existing agricultural lands, with the application of appropriate technology, are adequate to meet growing world population needs for food product through the middle of the 21st century and beyond. For example, in the U.S. agricultural productivity has grown at close to 2% per year over last several decades.

Although the world has experienced dramatic population increases in the past few centuries, there are reasons to believe that this growth could be coming to an end. Recent UN population projections, reflecting dramatic decreases in fertility rates throughout the world, indicated global populations peaking and declining thereafter (Figure 2). Not only fertility rates dropped sharply in the developed world of North America, Europe and the more industrial countries of Asia including China (Japan's population is expected to decline in absolute terms beginning 2007), but populations in parts of Africa,

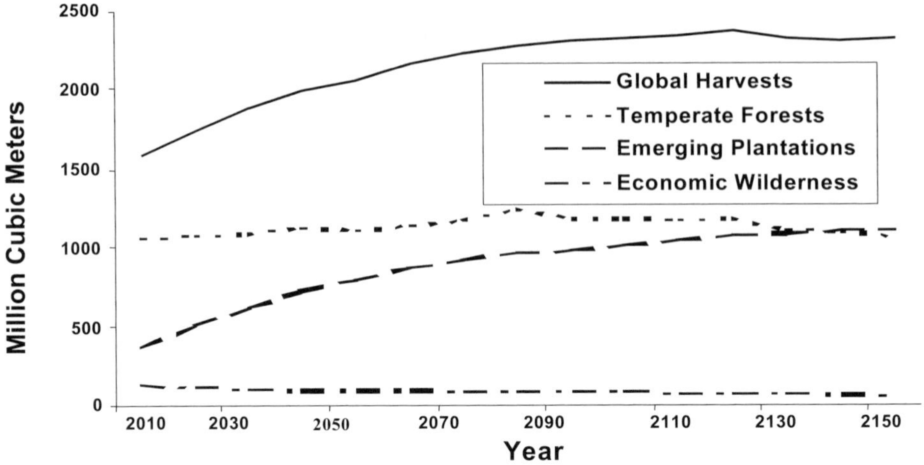

Figure 3. Projection of annual timber. *Source*: Sohngen, Mendelsohn and Sedjo (1999).

e.g., South Africa, Botswana and Zimbabwe are declining also, albeit for other reasons, i.e. AIDS and dramatically increased mortality rates. However, while fertility rates are declining everywhere, some areas of the world, e.g., South Asia, Latin American and much of Africa are still experiencing fertility rates well above those associated with stable populations. Thus, even if the future world experiences stable populations, the composition of this will certainly undergo dramatic change.

However, a world with stable or declining populations and a changing population mix, in an environment of continuing technological improvement, is likely to look quite different from the world with growing populations, with which we have grown so accustomed over the past two centuries. These changes, no doubt, will have substantial implications on land use.

3. Protected areas

The WRI recently prepared data on land conversion and land protection (Table 4). Not surprisingly, land conversion has been substantial in much of the global. However, in every region, often the vast majority of the land area has not been converted.

Table 5 takes the material of Table 4 and adds the McCloskey and Spalding estimates of wild areas. This perspective suggests that there are large areas that have not been designated as protected areas but that nevertheless retain the features of wild areas in that they have not been converted from natural habitat.

It is notable that the combination of Europe and Russia into a single category, as in Table 4, gives the impression of minimal protection area across the two entities. However, reporting the wild areas separately, as done by McCloskey and Spalding, reveal the

Table 4
Converting and protecting land

Region	Land conversion	Land protection
NA	27%	11.1%
Europe and Russia	35%	4.7%
Asia	44%	6.0%
CA Caribbean	28%	6.1%
South America	33%	7.4%
Middle East/North Africa	12%	2.1%
Sub-Saharan Africa	25%	6.0%
Oceania	9%	7.1%

Source: WRI as reported in the NYT, August 20, 2002, p. D4.

Table 5
The extent of wild areas in the world by major regions

Region	Land conversion	Land protection	Wild area
NA	27%	11.1%	37%
Asia	44%	6.0%	14%
South America	33%	7.4%	21%
Oceania	9%	7.1%	28%
Europe			3%
Soviet Union			34%[a]

Source: See Table 3 and McCloskey and Spalding (1989).
[a]Note that Tables 3 and 4 are not entirely consistent since the former Soviet Union includes countries not now part of Russia.

limited wild area in Europe (3%) and large amounts of wild areas in the former Soviet Union (34%).

The above estimates or wild areas are consistent with the notion that habitat protection is provided through inaccessibility. The FAO estimates that about one-half of the world's forest area is economically inaccessible and therefore unavailable for timber harvests under normal circumstances (see Table 6). This includes huge forest areas in all of the regions of the world including the OECD countries. Europe, however, has the smallest inaccessible area. Such a situation affords a large measure of protection from commercial harvests or development in these forests unless the areas are opened. The extent to which these areas become more accessible via road-building will, no doubt, affect future exploitation.

More generally, the number of protected areas worldwide is growing rapidly. Additional data indicates that the area in protected status has increased dramatically in recent years. Reid and Miller (1993) report that the cumulative world area under protected sta-

Table 6
Inaccessible forest

Region	Area of forest unavailable for timber supply (million ha)
Africa	233
Asia	177
Oceania	61
Europe	20
Russia	166
North America	238
Central America	49
South America	709
Total	1653

Source: FAO Global Fiber Supply Model, 1996.

tus has risen from a negligible area in 1900 to about 5 million square kilometers by the late 1980s, almost 4 million square kilometers of which was added after 1970. By 2003 the area in protected area status was 18 million km^2. Of this, about 17 million km^2 were terrestrial reserves or 11.5% of the Earth's land surface. For 9 of 14 major terrestrial biomes, this is substantially greater that the 10% target proposed a decade earlier [Chape et al. (2003)].

While protected areas are generally viewed positively from the perspective of the environment, protected areas can displace human populations and reduce the potential for producing livelihood from the land.

4. Activities to relieve pressures

4.1. Biotechnology

Biotechnology in agriculture and forestry could have a substantial effect on land use over the long term. The potential of biotechnology and transgenic plants in agriculture to reduce the requirements for other inputs (such as pesticides), lower inputs costs, and increase yields is being well established. Similarly, substantial potential to increase productivity and output quality also exists in forestry [Sedjo (2004)]. However, biotechnology must overcome public concerns about its safety and it effects upon the environment.

4.2. Forest certification

Forest certification is a very new phenomenon but one that could have a substantial effect on land management. One of the most notable outcomes of the Rio Earth Summit of 1992 has been the strong expansion of forest certification [Cashore, Auld and Newson (2004)]. Although the concept was designed originally due to concerns about tropical forests, the concept is now being most widely applied to forests applied to forests in the temperate world.

Forest certification is largely a private-sector activity, involving both nongovernmental and private profit-making entities. The objective is to have forest management practices "certified" as meeting some minimum standard consistent with "best management practices", sustainability" or "well managed". This approach allows for timber harvests but also provides that management generate desired ecological outcomes including sustainability of not only timber production, but also the nontimber outputs of the forest, including maintenance of ecological systems and the preservation of habitat for threatened and endangered species.

4.3. A vision of future land use

The world's land is allocated to agriculture, forest, urban, and natural habitats, and so forth. Victor and Aussebel (2000) present a recent vision of future land use. Focusing on the issue of conservation and forested areas, they provide a scenario for 2050 in which the forested area remains much as it is today, due to the substitution of wood from planted forests and forest cropping for wood from natural forests, such has been the mode for millennium. This result is an outgrowth of their assumptions regarding productivity increases in both agriculture and forestry. Their implications of increased productivity in agriculture and forestry on land use are not new. Others in agriculture [Crosson (1990)] and forestry [Sedjo and Lyon (1991); Sedjo and Botkin (1997)] have tracked these changes and noted their implications for land use.[5]

4.4. Developed and developing world experience

It is clear that in much of the developed world the problem in agriculture is one of excess production. As is well known the issue here typically relates to the well being of the agricultural community and systems of subsidies are commonly enacted that transfer wealth from the nonagricultural community to the agricultural community, albeit at a cost in efficiency. Such a system of incentives typically results in excess land in agriculture compared to other uses. This effect is offset to some extent by subsidy programs designed to promote tree growing (and various forms of land set-aside). However, the

[5] This perception is not without its detractors. Angelsen and Kaimowitz (2001) have argued that if agriculture technology results in lowering the supply curve (costs) and if demand for agriculture is elastic, e.g., greater than one, cost-reducing technology could result in an increase in the quantity demanded, thereby potentially increasing agricultural pressures on the land. They cite the demand for grains as an input to beef production. However, a more complete model would recognize that grain feed substitutes for grazing, thereby having land use implications beyond grain production.

incentives probably do relatively little to provide market protection to prevent the conversion of lands providing conservation and environmental services to farm or tree farm uses. The argument can certainly be made that farm programs have provided incentives for converting lands to agriculture that are marginal as such but may be socially valuable for their conservation values. Thus, agriculture programs have probably been detrimental to conservation and environmental values.

In many parts of the developing world, however, the reverse may be true, with price controls and subsidized agricultural imports providing disincentives to domestic production and investment in the agricultural sector. There is now some reason to believe that this approach is being eliminated in some countries in the developing world, e.g., India.

5. Land use and its impacts on natural habitats

In recent years, one of the most common arguments heard for the preservation of natural forests is that they are home to most of the world's biological diversity. "Biodiversity", as it is commonly abbreviated, refers to the variety of life on Earth. The term is difficult to define operationally. While we might prefer to maintain "more diversity", there is little agreement as to when one collection of organisms is "more diverse" than another. While both biologists [Vane-Wright, Humphries and Williams (1991)] and economists [Weitzman (1992)] agree that species least closely related to other living species should receive special attention in conservation efforts, it can be difficult to put such a principle into practice. Biologists estimate that as many as 90% of living species are as yet not identified. We may, then, be losing things we did not even know were there.

In practice, biodiversity is typically measured by number of species – "[t]he best ...the only unit we have" [Bisby et al. (1995)]. A number of factors are believed to be responsible for the extinction of species. Humans have hunted some to extinction, for instance, the passenger pigeon. Others are imperiled by chemical pollution. Climate change could doom many sessile species. Competition from exotic organisms, infection by exotic diseases, and infestation by exotic pests have driven many species, especially those in island habitats to extinction. Of all the threats to biodiversity, though, the most serious is thought to come from habitat conversion [Wilson (1992); Primack (2000)].

Biodiversity tends to be greatest in the tropics. It has been estimated that as many as 70% of the world's living land species can be found on as little as 2.4% of the Earth's terrestrial area. Species diversity declines toward the poles. Economic progress tends to follow the opposite pattern. Hence, the challenge of preserving biodiversity (again, with the important proviso that that term is defined operationally as "numbers of species") is largely to prevent habitat conversion in the countries of the developing tropics in which institutions tend to be weak or inchoate and growing populations exert powerful pressure for change.

The economics of such pressures are straightforward in some senses but complicated in others. They are straightforward in that a larger (and in many cases, wealthier in aggregate) population located in an expanding urban center expands the size of the concentric rings of the von Thünen city. The densely populated area devoted to urban residences increases. The belt of agricultural land around broadens. Production from forests takes place farther and farther from urban centers. The frontier beyond which wilderness lies grows more distant.

Existing cities grow, fueled by waves of migration from the hinterlands to the center that are driven by as-yet poorly understood forces. New centers come to be established as critical masses of population coalesce. Government policy may drive some of the dispersion of population into new areas. The government of Brazil, for example, encouraged the settlement of the Amazon for political reasons [Binswanger (1987)]. The expansion of economic activity, and of forest clearing in particular, is often felt to be associated with the provision of public infrastructure, especially roads [Pfaff (1999); Chomitz and Gray (1996)]. Any number of "perverse subsidies" are cited and decried for their effects on both the economies of and ecologies of nations implementing them [Myers and Kent (2001)].

How serious is species loss? The conventional wisdom holds that species losses have been substantial. The basis for this is usually found in the concept of island biogeography [MacArthur and Wilson (1967)], which hypothesizes a species–area relationship where the number of species is a positive but concave function of the area of habitat. A reduction in the area of habitat would be associated with a loss of species. Also, fragmentation, which results in isolation, should be associated with species loss. Both habitat loss and fragmentation have occurred.

However, the application of the island biogeography hypothesis has been criticized on two accounts: statistical and ecological grounds, as well as empirical grounds. In the first case, even if the hypothesized relationship is correct, the precise natural of the empirical relationship between habitat size and the rate of species is unknown and likely will vary with reserve-specific conditions. Second, there is substantial empirical evidence that conflicts with the simple species–area relationship hypothesis. For example, although the hypothesis predicts that fragmented habitat would support fewer species, many exceptions to this hypothesis have been found [e.g., Debinski and Holt (2000)]. Additionally, although it is generally accepted that, other things equal, the number of species in a reserve is an increasing function of area, more specific predictions have been opened to considerable debate [Doak and Mills (1994)]. An example is the assertion [Myers (1979)] that the potential losses could be massive. Myers (1979 p. 43) estimates that a 90% reduction in habitat would reduce the number of species by 50%. However, the empirical evidence for massive species losses is much weaker. For example, Lugo (1988) notes that in Puerto Rico, the massive forest conversion did not lead to a correspondingly massive extinction, certainly nowhere near the 50% alluded to by Myers.

More generally, reliance on projections from biogeographic models is problematic. As Reid states, relatively few attempts have been made to rigorously document the magnitude of species losses and attempts to enumerate species losses generally document relatively few extinctions. Reid (1992, p. 55) notes that only 60 bird and mammals are know to have become extinct between 1900 and 1990. In the continental United States, despite dramatic forest disturbances in the eastern forests over the last two centuries, only 3 forest birds went extinct [Simberloff (1992, p. 85)]. Thus, although habitat loss is likely to be related to an increased rate of species losses, in the absence of disastrous habitat losses, which is not now happening on a global scale, the evidence available does not suggest catastrophic species losses.

6. The economics of natural habitats

What is the optimal clearing of natural habitats that are believed to shelter the majority of the world's biodiversity? The question is complicated by a number of factors. The first is that the values of biodiversity are not well understood or documented. Evidence from the natural sciences shows that more diverse systems tend to be more "productive" in certain biological senses[6] and more resilient [Tilman and Downing (1994)], but it is not clear whether that "productivity" is important in any economically relevant sense. Similarly, although it is almost certainly the case that there are substantial redundancies in ecosystems, we generally do not know the degree of resilience provided diverse natural ecosystems by enhanced habitat protection provides important insurances services.

Moreover, many of the services that diverse natural ecosystems provide are likely to be public goods. Hence, they are not traded in existing markets, and therefore do not have observable prices. Economists and others who have investigated the range of values afforded by biodiversity and natural ecosystems often identify a broad spectrum of contributions [see, e.g., Pearce and Moran (1994); Daily (1997)]. Some, such as the services provided by insects and other small animals in pest control and pollination, may be relatively local in scale.[7] Others, such as the erosion protection and flood control functions of wetlands, can affect far distant areas [Heal et al. (2001)]. Finally, some global public goods may arise from diversity and diverse natural ecosystems. These could include the carbon sequestration provided by intact forests and other ecosystems, as well as the existence value associated with the continued survival of imperiled species [see, e.g., Freeman (1993) for a discussion of existence values and their skeptics]. Not surprisingly, one of the conclusions emerging from the limited literature available on the economic valuation of biodiversity and ecosystem services is that the "values" perceived depend crucially on the scale at which the analysis is conducted [Kremen et al. (2000)].

It is surprising that there is as little careful empirical work on the economic values of biodiversity and ecosystem services as there is. The issues have drawn the attention

[6] Even this finding is in some dispute. A group of critics [Wardle et al. (2000)] assert that experimental ecosystems with planted with more species are found to be productive not because a *collection* of species is more productive so much as because the more species are planted, the more likely it is that the single species best suited for the conditions will be included among them, and grow dominant.

[7] Even here, though, current biology points to complex considerations. A population of bees may fertilize a single farmer's field, but that population comprises part of a network of metapopulations that stand ready to recolonize areas suffering local extinctions.

of economists as distinguished as Kenneth Arrow [Arrow et al. (1995); Daily et al. (2000)], Partha Dasgupta [Dasgupta, Levin and Lubchenko (2000); Daily et al. (2000)], Geoffrey Heal (2000, 2002), and Martin Weitzman (1992, 1998, 2000). It is revealing, though, that none of these individuals has attempted to reduce theory to practice by hazarding monetary estimates of value; Heal (2000), in fact, goes so far as to claim that the valuation of biodiversity is neither a necessary nor a sufficient condition for its maintenance.[8] Such work as has been published in leading economics journals on the topic is revealing in its lack of consensus. Two papers on essentially the same topic published in the same journal and using the same data derived estimates that differ by three orders of magnitude for the value of the "marginal hectare" of biologically diverse habitat [Simpson, Sedjo and Reid (1996); Rausser and Small (2000); see also Craft and Simpson (2001), who show that social and private values may diverge radically].[9]

The boldest attempts to estimate biodiversity and ecosystem service values have been launched by interdisciplinary groups. The most celebrated (although considerably less "celebratory" adjectives have also been applied) attempt was made by Costanza et al. (1997). In this piece, the authors derived a central estimate of some US$33 trillion per year as the value of the services provided by the world's ecosystems and natural capital. A number of authors have criticized the effort [including Ayres (1997); Smith (1997); Toman (1998); Freeman (1998); and Pearce (1998)]. Objections have included charges that Costanza and his colleagues confused total and marginal values, double-counted, and made questionable interpretations of earlier studies. Perhaps the most trenchant criticism made is simply that the numbers Costanza and his coauthors derived have no meaning. It is a logical impossibility to assign a willingness to pay for ecosystem services greater than the income of the world. At the same time, asking "What is the value of everything?" is not a well-defined question. To borrow Michael Toman's memorable phrase, Costanza and his colleagues arrived at a "serious underestimate of infinity".

Work directed toward the valuation of biodiversity and diverse natural ecosystems has taken a number of approaches. A number of papers have taken an approach that treats biodiversity as potential redundancy. Economic value is, of course, value on the margin. Hence the value of a "marginal species" is determined both by the likelihood that it provides a certain service (the "recipe" for a new product, the provision of an essential ecosystem service, resistance to a certain agricultural pest, or the key to enhanced agricultural yield) and the likelihood that another source of an equivalent service will

[8] The argument that valuation is not sufficient for action is straightforward. Conventional wisdom has it that biodiversity is, in many aspects, a public good. Private actors will underprovide it. The collective action problem must also be overcome. It is less clear that valuation is not necessary. Someone must decide it is worthwhile to maintain biodiversity if it is to be maintained. This may not require a formal valuation exercise, but it must proceed from an informal one.

[9] The papers do, however, consider subtly different questions. Both agree that most of the world's remaining biologically diverse habitats are worth essentially zero in the application they consider, the search for new pharmaceutical products. (However, the much higher estimate of the value of the marginal hectare uses additional a priori information to stratify the data.)

not be available at a lower cost. This concept is developed in the ecological literature by Naeem (1998) and Tilman, Lehman and Bristow (1997), and among economists by Weitzman (1992, 1998), Simpson, Sedjo and Reid (1996), Evenson and Gollin (1997), Gollin, Smale and Skovmand (2000), and Rausser and Small (2000). As was indicated above in reference to the Simpson, Sedjo, and Reid paper, on one hand, and the Rausser and Small work, on the other, the methodological similarities the papers illustrate do not necessarily indicate quantitative agreement among their estimates of value.

Some studies have considered the impact of ecological conditions on the harvest of biological resources. Studies in this vein include Lynne, Conroy and Prochaska (1981), Ellis and Fisher (1987), Bell (1998), and Barbier and Strand (1998). There is, in fact a curiously voluminous literature on the services provided by mangrove systems to fisheries – large enough, in fact, to have motivated its own survey articles [Barbier (2000)]. Barbier and Strand offer an interesting observation: the value of mangrove ecosystems on fisheries depends in large part on the management of the fishery [this general point was first established by Freeman (1991)]. Barbier and Strand conclude that the open access problem is of far greater practical significance than is the ecosystem degradation issue.

Another area in which there has been a perhaps surprising volume of work concerns the interactions among deforestation, erosion, and the deposition of silt in reservoirs. Several authors have estimated the values of standing forests in preventing siltation [Magrath and Grosh (1985); Crowder (1987); Mahmood (1987); Southgate and Macke (1989)].

Still, detailed studies of the values provided by diverse natural ecosystems remain largely conspicuous by their absence. This is probably due largely to the econometric problems researchers encounter. The most straightforward approach to estimating the value of ecosystem services provided by one element of the landscape would be to compute their contribution to the values generated at another. This might be done through production, cost or profit function estimation. Conceptually, this is a straightforward proposition. Practically, it is not. Such approaches have been attempted [see, e.g., Acharya (2000) and Pattanayak and Kramer (2002)], but have often foundered for want of adequate data.

An equivalent approach would involve estimating anticipated profits as reflected in the market value of land. Hedonic regressions relating the price of land to the condition and configuration of surrounding land areas have also been conducted. Examples include Anderson and Cordell (1988), Lupi, Graham-Tomasi and Taff (1991), Garrod and Willis (1992), Cheshire and Sheppard (1995), Geoghegan, Wainger and Bockstael (1997), Powe et al. (1997), Tyrväinen and Miettinen (2000), Mahan, Polasky and Adams (2000), and Irwin and Bockstael (2001). While some of these studies have found positive effects, the results are somewhat mixed. Moreover, econometric methods have evolved considerably over time, a fact noted by some of the later authors in distinguishing their results from those of their predecessors.

The development of a better statistical understanding of the problem has underscored the difficulties inherent in doing work that is both credible and relevant to policy, however. Briefly, the econometrician is typically presented with data in which two phenomena are reflected. The first is the price, or equivalently, the discounted present value of the stream of earnings to which a particular parcel can be expected to give rise. The second is the choice of whether or not certain parcels have been converted from their "natural" state so as to themselves generate higher earnings as farms or homes. Both are determined simultaneously, a fact that makes their estimation difficult. The way out of this statistical bind is to develop instrumental variables that explain the land conversion decision [see Irwin and Bockstael (2001) for a more complete discussion]. Less formally, this means that one should identify the reasons for which natural land has stayed that way. Suppose, however, that there are clear indications as to why land has not been converted – it may, for example, be too steep or too swampy. Then it may be that there is little danger that such land *will* be converted, and hence no real policy issue. This begs the question, can we only arrive at estimates of the value of protected habitats when there is little danger of their conversion? We might also note in passing that some of the advantages claimed for natural ecosystems involve their ability to prevent the occurrence of catastrophic events, such as floods and landslides. It may be somewhat optimistic to suppose that land markets automatically and immediately reflect changes in the risks of such events, especially in areas in which they may not presently be common.

As remaining natural habitats shelter biodiversity, one of the things an economist setting out to value their services might be most interested in concerns the value of that biodiversity. The value attached to the existence of a species is, by and large – and perhaps redundantly – an "existence value".[10] Existence values, are, in turn, things for which no "behavioral trail" [the term is Douglas Larson's (1993)] can be found linking them to the purchase of market goods. To borrow another phrase, they are A. Myrick Freeman's (1993) "impossible case" in which there is no shift in the supply or demand or purchased goods from which to infer values.

This leaves only the controversial expedient of asking questions – the "contingent valuation method". Even one of that approach's most eloquent advocates writes: "The economics profession as a whole seems to regard the method as seriously flawed" [Smith (1997) succinct statements of those perceived flaws, see Diamond and Hausman (1994) and Diamond (1996)]. In the context of biodiversity specifically, the problem concerns the credibility of answers posed to hypothetical questions. For instance, Gardner Brown and Jason Shogren report the results of 18 studies of willingness to pay to preserve different species. The results, as the authors report, indicate that "the average person was willing to pay about $1000 to protect 18 different species" [Brown and Shogren (1998, p. 12)]. To quote the authors again, "Many will find these figures suspiciously high". This is especially true, as other researchers have found significant

[10] This is not entirely true. To the extent that individual species do have potential instrumental values, a la Tilman, Lehman and Bristow (1997), Simpson, Sedjo and Reid (1996), Rausser and Small (2000), or Craft and Simpson (2001), they may not be wholly intangible. They are, however, for reasons we have already discussed, difficult to measure.

discrepancies between respondents' *stated* values and their *actual* behavior when required to "put their money where their mouth is" [Seip and Strand (1992); Polasky, Gainutdinova and Kerkvliet (1996)].

7. Institutional considerations in conservation policy

There is little agreement among authors as to the values to which natural habitats and the biological diversity they shelter give rise. What, then, are appropriate incentives for conserving habitat, especially in the developing countries in which it is most imperiled?

There is great controversy surrounding this question as well. The controversy arises from several sources. The 3000-year history of "protected areas" is replete with accounts of hunting reserves and pleasure gardens established by royal fiat and enforced with often draconian measures [see, e.g., Davenport and Rao (2002)]. While new nations claimed a more democratic basis for the choices they made in setting lands aside, there were still concerns with the treatment their immigrant majorities afforded to native peoples [Colchester (1996)]. A major change in direction occurred in the 1980s when several international conservation organizations decided to emphasize "sustainable use" strategies for areas of natural habitat designated for protection [IUCN, WWF, UNEP (1981, 1991)]. Their focus turned from a "fences and fines" approach in which local peoples were excluded (after, in some cases, previously having been evicted) from protected areas to a philosophy of "use it or lose it". Protected areas could be maintained, the new view held, if local people could make enough money from conducting ecologically benign operations within them to compensate themselves for the opportunity costs of not converting habitat to other uses.

This new view also set the stage for conflict between the adherents of different views of human nature. Are indigenous peoples "noble savages" and natural conservationists who will, if left to their own devices, preserve the biological resources with which they have coexisted for centuries? Or will they, if freed by medicine and technology from the Malthusian cycle, quickly devastate their natural environments? Some societies have maintained traditional use patterns over centuries of development [see, for example, Ostrom (1990)]. However, many doubt that this phenomenon generalizes widely [see van Schaik and Rijksen (2002)].

Finally, there is a controversy regarding what can be and should be doable in biodiversity conservation. Conserving the world's richest remaining areas of endemic biodiversity will not be cheap. It seems reasonable to suppose that at least $10 billion has already been spent to conserve biodiversity around the globe.[11] Estimates of what more is required depend on the ambition of the objective. A group at Conservation International

[11] This estimate might be derived as follows. Castro and Locker (2000) estimate that some three-and-a-quarter-billion dollars were spent by all public and private funders to preserve biodiversity in Latin America between 1990 and 1997. It seems reasonable to suppose that spending would have continued at the same yearly rate since then. So suppose that $5 billion were spent in Latin America. The World Bank, on its own

"cautiously estimate the cost of protecting 70% of global biodiversity at $19 billion above current expenditures" [Bruner, Rice and da Fonseca (2002)]. Other estimates are higher. Bruner and his colleagues were computing the costs of acquiring only about 2% of the Earth's terrestrial surface. If that objective were expanded to encompass between 10 and 15%, the price tag might be as high as half a trillion dollars [James, Gaston and Balmford (2001)]. An economist would ask whether such an area and expenditure would be economically efficient.

Also, can such sums be raised? Some believe they cannot [see, e.g., Forest Trends (2002)]. If one adopts this view, there are only two options for conserving the majority of the world's imperiled biodiversity. The first is to ride roughshod over the rights of the local peoples who now control habitats at risk. Whether this could even be done at a cost lower than those we have cited above is open to question. Despite the assertions – we would say caricatures – of some in the debate on conservation policy, nobody seems seriously to be suggesting this strategy.[12]

The only hope for preserving substantially all of the Earth's biodiversity is, in the view of such pessimists, to acquaint the people of the world whose actions decide the fate of imperiled habitats with the values they will realize by preserving them. It is somewhat ironic, or perhaps misleading, to assign labels such as "optimists" and "pessimists" in this debate. Those with a skeptical view of human nature (for instance, many economists) will find it wildly optimistic to suppose that the hearts and minds of local peoples will be changed so as to instill in them a new appreciation of the natural ecosystems around them. Of course, there is also the "free rider" problem associated with individual behavior and public goods. Closely related, some will find it equally optimistic to suppose that conservation planners will be able to identify and overcome the market failures that purportedly prevent the establishment of more eco-friendly enterprises in the developing tropics.

Economists have been, by and large, on the sidelines of what has become a very acrimonious debate among biologists and anthropologists [the extremes are well represented by Terborgh (1999) and Wilshusen et al. (2002)] over conservation strategy. Some economists have, however, extolled the straightforward logic that, since conservation donors can only "get what they pay for", they ought to "pay for what they want to get". That is, payments for the conservation of habitat per se are likely to be more

account and in its administration of the Global Environment Facility (GEF), is one of the leading funders of biodiversity preservation. It committed about $1.7 billion in biodiversity related funding worldwide between 1988 and 1997. About half of the World Bank's biodiversity-related expenditures have been allocated to Latin America and the Caribbean. This ratio might be common to most funders. So, if $5 billion represents roughly half the world total, we would have our answer.

[12] The most problematic statements in this regard probably come from ecologist John Terborgh (1999) who has proposed as his "most radical option" that armed "naturekeeping" troops akin to "peacekeeping" forces be stationed in tropical countries. Other authors have reacted predictably to the suggestion that conservation advocates resort to "authoritarian protectionism that would ... precipitate resistance and conflict" [Wilshusen et al. (2002)].

effective in achieving conservation goals than are "integrated conservation and development projects" (ICDPs) that attempt to achieve conservation by subsidizing "sustainable use" projects such as ecotourism, bioprospecting, and the collection of nontimber forest products. Simpson and Sedjo (1996), Southgate (1998), Barrett and Arcese (1995), and Ferraro (2001) make informal arguments to this effect; they often draw on a policy and project evaluation literature citing the inadequacies of ICDPs [Wells, Brandon and Hannah (1992); World Bank (1999)]. A preliminary attempt at formalization is reported in Ferraro and Simpson (2002), in which it is shown that the concavity of the profit function in input prices is a sufficient condition for conservation donors to find "direct" incentives more effective than "indirect".

Economists have contributed to two other branches of literature relevant to biodiversity conservation policy. First, there is work by economists, but largely published outside of the formal economics genre on the benefits of biodiversity and natural systems. Peters, Gentry and Mendelsohn (1989) argued that natural areas can be far more profitably exploited for the sustainable harvest of nontimber forest products than for timber and agriculture. This paper has been both highly influential and highly controversial, however, and other work has failed to replicate its claims [Godoy, Lubowski and Markandya (1993)]. Questions arise both as to the "thinness" of markets for these products and the ecological effects of nontimber harvests – for example, there affect on animal populations. Other studies claim that natural environments can be more profitably exploited for ecotourism and large game viewing and hunting than agriculture.

The observation that *some* natural systems can "pay for themselves" has been touted by a curious combination of writers. Gretchen Daily and Katherine Ellison (2002) write: "Private enterprise cannot substitute for governments ... government regulation is called for to kick-start and supervise the profound economic transformation needed... ." Terry Anderson and Donald Leal (2001) write: "Expanding [market] processes to include natural resources and environmental amenities offers the only possibility for improving environmental quality, raising living standards, and ... expanding individual liberty." Yet these authors, though they occupy opposite ends of the ideological spectrum, are unanimous in their praise of what the former term "visionaries" and the latter "enviro-capitalists" [see also Heal (2000, 2002)], who bundle public and private goods. The unresolved question is whether such bundling will solve problems associated with biodiversity and ecosystem services [see Simpson (2002)]. Theory would suggest that it will not: public goods may remain undersupplied when private ones are "optimally" allocated. Of course, there are a number of unresolved issues of political economy that might temper such conclusions.

Given the uncertainty that still surrounds the valuation of biodiversity and ecosystem services, it is also surprising that there is as much emphasis as there is on the role of property rights *per se* as a conservation strategy. However, there is agreement among parties that disagree strongly on other matters of strategy that strengthening tenure should be part of the conservation agenda [see, e.g., van Schaik et al. (2002); Forest Trends (2002)]. This is curious. Enhanced security of tenure assures only that the owner of a property will put it to its highest and best long-term *private* use. Absent better information about what those values might be, it seems wishful thinking in many cases to suppose that sustainable uses will win out. As a well-established literature clearly demonstrates, secure private ownership is no guarantee of long-run sustainable use [see, e.g., Clark (1991) for the conditions under which extinguishing a biological resource is optimal].

The economic theory of property rights, as explicated by Yoram Barzel (1997), for example, holds that property rights come to be defined when the benefits of their existence justify the costs of their enforcement. It has been persuasively argued that informal systems of property rights do, in fact, exist over much of the developing world [Alcorn (1996); de Soto (2001)]. It remains an open question whether informal claims come to be strengthened as a consequence of the increased advantages that can be realized from doing so in a wealthier society [e.g., Alston, Libecap and Schneider (1996)] or remain unregistered due to bureaucratic inertia and political scheming [de Soto (2001); Barrett et al. (2001)].

A rhetorical question poses the practical issue clearly, though. Property rights in much of the developing world remain informal and inchoate. Ecological degradation continues as more natural habitats are destroyed. The question is, does the degradation proceed from a want of property rights, or is the absence of property rights itself evidence that alternative, "sustainable" uses are not economically attractive?

8. Summary and conclusions

Aggregate land uses globally have remained relatively stable over the past two to three decades. Land use in agriculture has increased very modestly, although irrigated land has increased substantially. Recent productivity trends in agricultural production suggest adequate future food availability to meet global requirements. Despite concerns about deforestation, rates of tropical deforestation have declined, while temperate and boreal forests are expanding modestly. Global timber supply and demand appear well in balance with little indication of future scarcities or substantial price increases. Nevertheless, the area in forest globally continues to decrease modestly although programs such as certified forest management and biotechnology are offered as vehicles to both reduce environmental damages to forest and increase the availability to industrial wood. Globally, substantial areas remain relatively pristine by virtue of their inaccessibility. Furthermore, although still modest on a global level, the land area in protected status has increased at a rapid rate in recent decades, with significant additions occurring on a regular basis. Recent trends in fertility rates have exhibited a significant decline and some projects see the world's population stabilizing about 2040–2050, and perhaps declining thereafter.

Despite these generally positive trends, great concern remains with the preservation of the natural habitats on which the bulk of the world's biodiversity depends. Efforts to save these habitats and their complements of natural species are hampered by a number of considerations, however. There is still less-than-complete agreement as to the objective to be pursued. Moreover, conservation practitioners and donors disagree as to which tactics will be most effective in conserving imperiled habitat even when they agree as to the necessity of habitat preservation. By and large, economists have not provided much useful guidance to conservation practitioners. While it would be straightforward to apply well-understood principles of economic valuation to the services of ecosystems and biodiversity, the data with which to conduct such exercises simply does not exist. Such efforts as have been undertaken to value ecological assets are often controversial, have not converged to consensus estimates, and generally have not penetrated the more prestigious journals. Economists may be better prepared to advise the conservation community on cost-effective conservation approaches than they are to provide a full-fledged benefit–cost analysis.

References

Achard, F., Eva, H.D., Stibig, H.-J., Mayaux, P., Gallego, J., Richards, T., Malingreau, J.-P. (2002). "Determination of deforestation rates of the world's humid tropical forests". Science 297 (5583), 999–1002.

Acharya, G. (2000). "Approaches to valuing the hidden ecological services of wetland ecosystems". Ecological Economics 35, 63–74.

Alcorn, J.B. (1996). "Indigenous resource management systems". In: Borrini-Feyerabend, G. (Ed.), Beyond Fences: Seeking Social Sustainability in Conservation. International Union for the Conservation of Nature, Gland, Switzerland.

Alston, L.J., Libecap, G.D., Schneider, R. (1996). "The demand and supply of property rights on the frontier: The cases of North America and Brazil". In: Anderson, T.L., Hill, P.J. (Eds.), The Privatization Process; A Worldwide Perspective. Rowman & Littlefield, Lanham, MD, pp. 203–236.

Anderson, L., Cordell, H. (1988). "Influence of trees on residential property values in Athens, Georgia". Landscape and Urban Planning 15, 153–164.

Anderson, T., Leal, D. (2001). Free Market Environmentalism, revised ed. Cambridge University Press, Cambridge.

Angelsen, A., Kaimowitz, D. (Eds.) (2001). Agricultural Technologies and Tropical Deforestation. CABI, Wallingford.

Arrow, K., Bolin, B., Costanza, R., Dasgupta, P., Folke, C., Holling, C.S., Jansson, B.-O., Levin, S., Mäler, K.-G., Perrings, C., Pimental, D. (1995). "Economic growth, carrying capacity, and the environment". Science 268, 52–521.

Ayres, R.U. (1997). "On the economic valuation of ecosystem services". INSEAD, Fontainebleau.

Barbier, E.B. (2000). "Valuing the environment as input: Review of applications to mangrove–fisheries linkages". Ecological Economics 35, 47–61.

Barbier, E.B., Strand, I. (1998). "Valuing mangrove–fishery linkages: A case study of Campeche, Mexico". Environmental and Resource Economics 12 (2), 151–166.

Barney, G.O. (1982). The Global 2000 Report to the President. Penguin Books, New York.

Barrett, C.B., Arcese, P. (1995). "Are ICDPs sustainable? On the conservation of large mammals in Sub-Saharan Africa". World Development 23 (7), 1073–1085.

Barrett, C.B., Brandon, K., Gibson, C., Gjertsen, H. (2001). "Conserving tropical biodiversity amid weak institutions". BioScience 51 (6), 497–502.

Barrett, J.W. (Ed.) (1980). Regional Silviculture of the United States. John Wiley and Sons, New York.

Barzel, Y. (1997). The Economic Theory of Property Rights, second ed. Cambridge University Press, Cambridge.

Bell, F.W. (1998). "The state of nature as an input into the production function". Journal of Economic Research 3, 1–20.
Berck, P. (1979). "The economics of timber: A renewable resource in the long-run". Bell Journal of Economics 10, 447–462.
Binswanger, H. (1987). "Fiscal and legal incentives with environmental effects on the Brazilian Amazon". ARDD, World Bank, Washington, DC.
Bisby, F.A., Coddington, J., Thorpe, J.P., Smartt, J., Hengeveld, R., Edwards, P.J., Duffield, S.J. (1995). "Characterization of biodiversity". In: Heywood, V.H. (Ed.), Global Biodiversity Assessment. Cambridge University Press, Cambridge, Ch. 2, pp. 21–106.
Brown Jr., G.M., Shogren, J. (1998). "Economics of the endangered species act". Journal of Economic Perspectives 12 (3), 3–20.
Bruner, A., Rice, R.E., da Fonseca, G.A.B. (2002). "An estimate of the cost of conservation". Supplementary material to Pimm, S., et al., "Can we defy nature's end?". Science 293, 2207–2208. Available at: http://www.sciencemagazine.org/cgi/content/full/293/5538/2207/DC1 (accessed 18 August 2002).
Cashore, B., Auld, G., Newson, D. (2004). Governing Through Markets. Yale University Press, New Haven, CT.
Castro, G., Locker, I. (2000). "Mapping conservation investments: An assessment of biodiversity funding in Latin America the Caribbean". Biodiversity Support Program, Washington, DC.
Chape, S., Fish, L., Fox, P., Spalding, M. (2003). United Nations List of Protected Areas. IUCN/UNEP, Gland, Switzerland/Cambridge, UK.
Cheshire, P., Sheppard, S. (1995). "On the price of land and the value of amenities". Economica 62 (246), 247–267.
Chomitz, K.M., Gray, D.A. (1996). "Roads, land use, and deforestation: A spatial model applied to Belize". World Bank Economic Review 10 (3), 487–512.
Clark, C. (1991). Mathematical Bioeconomics, second ed. John Wiley, New York.
Clawson, M. (1979). "Forest in the long-sweep of American history". Science 204 (15), 1168–1174.
Colchester, M. (1996). "Beyond 'participation': Indigenous peoples, biological diversity conservation and protected area management". Unasylva 186, 64.
Costanza, R., d'Arge, R., de Groot, R., Farber, S., Grasso, M., Hannon, B., Limburg, K., Naeem, S., O'Neill, R., Paruelo, J., Raskin, R., Sutton, P., van den Belt, M. (1997). "The value of the world's ecosystem services and natural capital". Nature 387, 253–260.
Craft, A.B., Simpson, R.D. (2001). "The social value of biodiversity in new pharmaceutical product research". Environmental and Resource Economics 18 (1), 1–17.
Crosson, P.R. (1990). "Cropland and soils: Past performance and policy challenges". In: Frederick, K.D., Sedjo, R.A. (Eds.), America's Renewable Resources. RFF, Washington, DC. Ch. 5.
Crowder, B.M. (1987). "Economic costs of reservoir sedimentation: A regional approach to estimating cropland erosion damages". Journal of Soil and Water Conservation (May–June), 194–197.
Daily, G.C. (Ed.) (1997). Nature's Services: Society Dependence on Natural Ecosystems. Island Press, Washington, DC.
Daily, G., Ellison, K. (2002). The New Economy of Nature: The Quest to Make Conservation Profitable. Island Press, Washington, DC.
Daily, G., Söderqvist, T., Aniyar, S., Arrow, K., Dasgupta, P., Ehrlich, P., Folke, C., Jansson, A.-M., Jansson, B.-O., Kautsky, N., Levin, S., Lubchenco, J., Mäler, K.-G., Simpson, D., Starrett, D., Tilman, D., Walker, B. (2000). "The value of nature and the nature of value". Science 289, 395–396.
Dasgupta, P., Levin, S., Lubchenko, J. (2000). "Economic pathways to ecological sustainability". BioScience 50, 409–435.
Davenport, L., Rao, M. (2002). "The history of protection: Paradoxes of the past and challenges for the future". In: Making Parks Work: Past, Present and Future. Island Press, Washington, DC, Ch. 3, pp. 30–50.
Debinski, D.M., Holt, R.D. (2000). "A survey and overview of habitat fragmentation experiments". Conservation Biology 14 (2), 342–355.

de Soto, H. (2001). The Mystery of Capital: Why Capitalism Succeeds in the West and Fails Everywhere Else. Basic Books, New York.
Diamond, P. (1996). "Testing the internal consistency of contingent valuation surveys". Journal of Environmental Economics and Management 30 (3), 337–347.
Diamond, P.A., Hausman, J.A. (1994). "Contingent valuation: Is some number better than no number?". Journal of Economic Perspectives 8 (4), 45–64.
Doak, D.F., Mills, L.S. (1994). "A useful role for theory in conservation". Ecology 75 (3), 615–626.
Ellis, G.M., Fisher, A.C. (1987). "Valuing the environment as an input". Journal of Environmental Management 25, 149–156.
Evenson, R.E., Gollin, D. (1997). "Genetic resources, international organizations, and improvement in rice varieties". Economic Development and Cultural Change 45 (3), 471–500.
Fairhead, J., Leach, M. (1998). Reframing Deforestation: Global Analysis and Local Realities: Studies in West Africa. Routledge, New York.
FAO (1993). Forest Resource Assessment 1990: Tropical Countries. FAO, Rome.
FAO (2000). Forest Resources Assessment (2000) Summary Report. FAO, Rome.
Ferraro, P.J. (2001). "Global habitat protection: Limitations of development interventions and a role for conservation performance payments". Conservation Biology 15 (4), 990–1000.
Ferraro, P.J., Simpson, R.D. (2002). "The cost–effectiveness of conservation payments". Land Economics 78 (3), 339–353.
Forest Trends (2002). "To Johannesburg and beyond: Strategic options to advance the conservation of natural forests", Discussion Paper for the GEF Forest Roundtable, 11 March 2002, New York. Available at: http://www.foresttrends.org/whoweare/pdf/grf_revisedfinal1.pdf.
Freeman III, A.M. (1991). "Valuing environmental resources under alternative regimes". Ecological Economics 3, 247–256.
Freeman III, A.M. (1993). The Measurement of Environmental and Resource Values: Theory and Methods. Resources for the Future, Washington, DC.
Freeman III, A.M. (1998). "Notes for valuing nature workshop: Commentary on methods and data used by Costanza, et al.", Paper prepared for the Bowdoin College Environmental Studies Program Workshop on Valuing Nature, August 1998.
Garrod, G., Willis, K. (1992). "Valuing goods' characteristics: An application of the hedonic price method to environmental attributes". Journal of Environmental Management 34, 59–76.
Geoghegan, J., Wainger, L.A., Bockstael, N.E. (1997). "Spatial landscape indices in a hedonic framework: An ecological analysis using GIS". Ecological Economics 23, 251–264.
Godoy, R., Lubowski, R., Markandya, A. (1993). "A method for the economic valuation of non-timber tropical forest products". Economic Botany 47 (3), 220–233.
Gollin, D., Smale, M., Skovmand, B. (2000). "Searching an *ex situ* collection of wheat genetic resources". American Journal of Agricultural Economics 82 (4), 812–827.
Hayami, Y., Ruttan, V.W. (1985). Agricultural Development: An International Perspective. Johns Hopkins University Press, Baltimore, MD. See p. 115.
Heal, G.M. (2000). Nature and the Marketplace. Island Press, Washington, DC.
Heal, G. (2002). "Biodiversity and globalization", Paper presented at Kiel Week Conference, Institute for World Economics, Kiel, Germany, June 2002.
Heal, G., Daily, G.C., Ehrlich, P.R., Salzman, J., Boggs, C., Hellman, J., Hughes, J., Kremens, C., Ricketts, T. (2001). "Protecting natural capital through ecosystem service districts". Stanford Environmental Law Journal 20 (2), 333–364.
Hyde, W.F., Newman, D.H., Seldon, B.J. (1992). The Economic Benefits of Forestry Research. Iowa State University Press, Ames, IA.
International Union for the Conservation of Nature (IUCN), United Nations Environment Program (UNEP) and World Wildlife Foundation (WWF) (1981). World Conservation Strategy: Living Resource Conservation for Sustainable Development. IUCN, UNEP, WWF, Gland.

International Union for the Conservation of Nature (IUCN), United Nations Environment Program (UNEP) and World Wildlife Foundation (WWF) (1991). Caring for the Earth: A Strategy for Sustainable Living. IUCN, UNEP, WWF, Gland.

Irwin, E.G., Bockstael, N.E. (2001). "The problem of identifying land use spillovers: Measuring the effects of open space on residential property values". American Journal of Agricultural Economics 83 (3), 698–704.

James, A., Gaston, K.J., Balmford, A. (2001). "Can we afford to conserve biodiversity". BioScience 51 (1), 43–52.

Johnson, R., Libecape, G. (1980). "Efficient markets and Great Lakes timber: A conservation issue re-examined". Explorations in Economic History 17, 372–385.

Kremen, C., Niles, J., Dalton, M., Daily, G., Ehrlich, P., Fay, P., Grewal, D., Guillery, R.P. (2000). "Economic incentives for rain forest conservation across scales". Science 288, 1828–1832.

Krugman, P. (1998). "Space: The final frontier". Journal of Economic Perspectives 12 (2), 161–174.

Larson, D.M. (1993). "On measuring existence value". Land Economics 69 (4), 377–388.

Lugo, A.E. (1988). "Estimating reductions in the diversity of tropical forest species". In: Biodiversity and Ecosystem Function. National Academy Press, Washington, DC.

Lupi Jr., F., Graham-Tomasi, T., Taff, S.J. (1991). "A hedonic approach to urban wetland valuation", Staff Paper No. P91-8. Department of Agricultural and Applied Economics, University of Minnesota.

Lynne, G.D., Conroy, P., Prochaska, F.J. (1981). "Economic valuation of marsh areas for marine production processes". Journal of Environmental Economics and Management 8, 175–186.

MacArthur, R.H., Wilson, E.O. (1967). The Theory of Island Biogeography. Monographs in Population Biology. Princeton University Press, Princeton, NJ.

Magrath, W.B., Grosh, M.E. (1985). "A dynamic approach to watershed management with emphasis on soil erosion", Working Paper No. 336. Department of Agricultural Economics, Cornell University.

Mahan, B., Polasky, S., Adams, R.M. (2000). "Valuing urban wetlands: A property price approach". Land Economics 76, 100–113.

Mahmood, K. (1987). "Reservoir sedimentation: Impact, extent, and mitigation", World Bank Technical Paper No. 71. World Bank, Washington, DC.

McCloskey, J.M., Spalding, H. (1989). "A reconnaissance-level inventory of the amount of wilderness remaining in the world". Ambio 18 (4), 222–226.

Mendelsohn, R. (1994). "Property rights and tropical deforestation". Oxford Economic Papers 46 (Suppl.), 750–756.

Myers, N. (1979). The Sinking Ark. Pergamon, New York.

Myers, N., Kent, J. (2001). Perverse Subsidies: How Tax Dollars Can Undercut the Environment and the Economy. International Institute for Sustainable Development, London.

Naeem, S. (1998). "Species redundancy and ecosystem reliability". Conservation Biology 12, 39–45.

Ostrom, E. (1990). Governing the Commons: The Evolution of Institutions for Collective Action. Cambridge University Press, New York.

Pattanayak, S., Kramer, R. (2002). "Worth of watersheds: A producer surplus approach for valuing drought control in Eastern Indonesia". Environment and Development Economics 6 (1), 123–145.

Pearce, D. (1998). "Auditing the Earth". Environment 40 (2), 23–28.

Pearce, D., Moran, D. (1994). The Economic Value of Biodiversity. Earthscan, London.

Peters, C.M., Gentry, A.H., Mendelsohn, R.O. (1989). "Valuation of an Amazonian rainforest". Nature 339 (6227), 655–656.

Pfaff, A.S.P. (1999). "What drives deforestation in the Brazilian Amazon? Evidence from satellite and socioeconomic data". Journal of Environmental Economics and Management 37 (1), 26–43.

Polasky, S., Gainutdinova, O., Kerkvliet, J. (1996). "Comparing CV responses with voting behavior: Open-space survey and referendum in Corvallis, Oregon", Western Regional Research Publication, W-33, Benefits and Costs Transfer in Natural Resource Planning, Ninth Interim Report, pp. 105–129.

Powe, N., Garrod, G., Brunsdon, C., Willis, K. (1997). "Using a geographical information system to estimate an hedonic price model of the benefits of woodland access". Forestry 70, 139–149.

Primack, R. (2000). "Extinction, causes of". In: Levin, S. (Ed.), Encyclopedia of Biodiversity, vol. 2. Academic Press, San Diego, pp. 697–713.

Rausser, G., Small, A.A. (2000). "Valuing research leads: Bioprospecting and the value of genetic resources". Journal of Political Economy 108 (1), 173–206.
Reid, W.V. (1992). "How many species will there be?". In: Whitmore, R.H., Sayer, J.A. (Eds.), Tropical Deforestation and Species Extinction. Chapman and Hall, New York, pp. 55–74.
Reid, W.V., Miller, K.R. (1993). Keeping Options Alive: The Scientific Basis for Conserving Biodiversity. WRI, Washington, p. 71.
Richards, J.F., Tucker, R.P. (Eds.) (1988). World Deforestation in the Twentieth Century. Duke Press Policy Studies. Duke University Press, Durham.
Sedjo, R.A. (2004). "The economic contribution of biotechnology and forest plantations in global wood supply and forest conservation". In: Strauss, S.H., Bradshaw, H.D. (Eds.), Forest Biotechnologies: Technical Capabilities, Ecological Questions, and Social Issues in Genetic Engineering of Plantation Trees. Resources for the Future, Washington, DC. Ch. 2.
Sedjo, R.A., Botkin, D. (1997). "Using forest plantations to spare natural forests". Environment 30 (10).
Sedjo, R.A., Lyon, K.S. (1991). The Long-Term Adequacy of World Timber Supply. RFF, Washington, DC.
Seip, K., Strand, J. (1992). "Willingness to pay for environmental goods in Norway: A contingent valuation study with real payments". Environmental and Resource Economics 2, 91–106.
Simberloff, D. (1992). "Do species – are curves predict extinction in fragmented forests?". In: Whitmore, R.H., Sayer, J.A. (Eds.), Tropical Deforestation and Species Extinction. Chapman and Hall, New York, pp. 55–74.
Simpson, R.D. (2002). "Review of *The New Economy of Nature*: *The Quest to Make Conservation Profitable* by Gretchen Daily and Katherine Ellison (Washington: Island Press)". Issues in Science and Technology 18 (4), 84–87.
Simpson, R.D., Sedjo, R.A. (1996). "Paying for the conservation of endangered ecosystems: A comparison of direct and indirect approaches". Environment and Development Economics 1, 241–257.
Simpson, R.D., Sedjo, R.A., Reid, J.W. (1996). "Valuing biodiversity for use in pharmaceutical research". Journal of Political Economy 104, 163–185.
Smith, V.K. (1997). "Mispriced planet". Regulation 20 (3), 16–17.
Sohngen, B., Mendelsohn, R., Sedjo, R. (1999). "Forest management, conservation, and global timber markets". American Journal of Agricultural Economics 81 (1), 1–18.
Southgate, D. (1998). Tropical Forest Conservation: An Economic Assessment of the Alternatives in Latin America. Oxford University Press, New York.
Southgate, D., Macke, R. (1989). "The downstream benefits of soil conservation in third world hydroelectric watersheds". Land Economics 65 (1), 38–48.
Terborgh, J. (1999). Requiem for Nature. Island Press, Washington, DC.
Tilman, D., Downing, J.A. (1994). "Biodiversity and stability in grasslands". Nature 367, 363–365.
Tilman, D., Lehman, C.L., Bristow, C.E. (1997). "Plant diversity and ecosystem productivity: Theoretical considerations". Proceedings of the National Academy of Sciences 94, 1857–1861.
Toman, M.A. (1998). "Why not to calculate the value of the world's ecosystem services and natural capital". Ecological Economics 25 (1), 57–60.
Tucker, R.P., Richards, J.F. (1983). Global Deforestation and the Nineteenth-Century World Economy. Duke Press Policy Studies. Duke University Press, Durham.
Tyrväinen, L., Miettinen, A. (2000). "Property prices and urban forest amenities". Journal of Environmental Economics and Management 39, 205–223.
UNECE (1996). "Long-term historical changes in the forest resource", Geneva Timber and Forest Study Paper 10 ECE/TIM/SP/10. UN Economic Commission for Europe and the FAO, Timber Section, Geneva, Switzerland.
Vane-Wright, R.I., Humphries, D.J., Williams, P.H. (1991). "What to protect? Systematics and the agony of choice". Biological Conservation 55, 235–254.
van Schaik, C., Rijksen, H.D. (2002). "Integrated conservation and development projects: Problems and potential". In: Making Parks Work: Past, Present and Future. Island Press, Washington, DC, Ch. 2, pp. 15–29.

van Schaik, C., Terborgh, J., Davenport, L., Rao, M. (2002). "Making parks work: Past, present, and future". In: Making Parks Work: Past, Present and Future. Island Press, Washington, DC, Ch. 32, pp. 468–481.
Victor, D., Ausubel, J. (2000). "Restoring the forests". Foreign Affairs (November–December).
von Thünen, J. (1826, 1850, and 1863). Derlsolierte Staat in Beziehung auf Landwirtschaft und Nationalokoenomie.
Wardle, D.A., Huston, M.A., Grime, J.P., Berendse, F., Garnier, E., Lauenroth, W.K., Setälä, H., Wilson, S.D. (2000). "Biodiversity and ecosystem function: An issue in ecology". Bulletin of the Ecological Society of America 81, 235–239.
Weitzman, M.L. (1992). "On diversity". Quarterly Journal of Economics 107, 363–406.
Weitzman, M.L. (1998). "The Noah's Ark problem". Econometrica 66 (6), 1279–1298.
Weitzman, M.L. (2000). "Economic profitability vs ecological entropy". Quarterly Journal of Economics 115 (1), 237–263.
Wells, M., Brandon, K., Hannah, L. (1992). "People and parks: Linking protected area management with local communities". World Bank, World Wildlife Fund, and U.S. Agency for International Development, Washington, DC.
Wilshusen, P.R., Brechin, S.R., Fortwangler, C.L., West, P.C. (2002). "Reinventing the square wheel: Critique of a resurgent 'protection paradigm' in international biodiversity conservation". Society and Natural Resources 15 (1), 17–40.
Wilson, E.O. (1992). The Diversity of Life. Belknap, Cambridge, MA.
World Bank (1999). "Investing in biodiversity: A review of Indonesia's integrated conservation and development projects". Indonesia and Pacific Islands Country Department, World Bank, Washington, DC.

Chapter 60

PAST CLIMATE CHANGE IMPACTS ON AGRICULTURE*

ROBERT MENDELSOHN

Yale University, New Haven, CT

Contents

* I wish to thank Professor Robert Evenson for proposing the topic of this paper and his advice.

Handbook of Agricultural Economics, Volume 3
Edited by Robert Evenson and Prabhu Pingali

DOI: 10.1016/S1574-0072(06)03060-X

Abstract

This paper examines the likely impact on agriculture of the climate change which has already taken place between 1960 and 2000. Accumulating greenhouse gases have caused global temperatures to rise approximately 0.25 °C during this period and for precipitation patterns to shift. Using cross-sectional and crop simulation evidence, temperature, precipitation, and carbon dioxide response functions are used to calculate the impacts on agriculture.

Temperature and precipitation changes together have caused estimated global impacts ranging from a loss of 0.05% to a gain of 0.9% of agricultural GDP. Including carbon fertilization effects, historic climate change is estimated to have caused a 2–4% increase in global production. Given the rapid increase in agricultural production over the last 40 years, the contribution of climate change to the overall growth of agriculture has been small, contributing between 2.6% and 5.4% of overall growth. This effect has been larger in mid to high latitude countries where climate change is estimated to have caused 4–7% of historic agricultural growth and smaller in low latitude countries where the climate change has contributed between 0.6% and 3% of the growth.

Future climate changes which are expected to be much larger, may well have very different effects than past climate changes. Future changes may continue to be slightly beneficial to global agriculture up to about 2.5 °C, but they will eventually become harmful and cause reductions in global production.

Keywords

climate change, agriculture, environmental valuation, impact analysis

JEL classification: Q10, Q51, Q53, Q54

1. Introduction

There has been extensive debate concerning what impacts future climate change will cause to a host of sectors including agriculture, forestry, water, energy, coasts, ecosystems, and health [Pearce et al. (1996); Watson et al. (1996); McCarthy et al. (2001)]. With forecasts of potentially significant increases in temperature ranging from 1.4 to 5.8 °C by 2100 [Houghton et al. (2001)], the general consensus is that future warming could have very large effects on the above sectors [Pearce et al. (1996); McCarthy et al. (2001)]. However, despite the substantial interest in historic changes in climate, there is a surprising absence of impact studies on historic effects. Partially this is because historic warming has been very slight. Over the last hundred years, scientists believe that the globe has warmed only 0.5 °C [Houghton et al. (2001)]. Given the extensive changes in population, economic activity, capital investment, and technology, it has not been easy to isolate the impact of this small climate change. For example, world population has increased from 1.6 billion to 6.3 billion over the twentieth century and global GDP has increased from $7 to $33 trillion [Nordhaus (1997)]. The change in the global temperature from about 17.9 to 18.4 °C is a comparatively small change.[1]

This paper attempts to address the absence of historical impact analysis by evaluating the impact of climate change on global agriculture over the last 40 years. According to the Third Assessment Report (TAR) of the Intergovernmental Panel on Climate Change (IPCC), global temperatures have increased approximately 0.25 °C in the last forty years [Houghton et al. (2001)]. These changes have not been uniform with slightly more warming in the high latitudes and winter months. There have also been slight changes in precipitation with increases in the mid and high latitudes and decreases in some tropical regions [Houghton et al. (2001)]. Finally, the level of carbon dioxide has increased from 317 ppm in 1960 to 367 ppm in 2000 [Houghton et al. (2001)]. All three of these changes have already had effects on agriculture between 1960 and 2000. However, the climate change has been very slow and there have been other changes from land conversion, capital investment, and technological change. The climate impacts are consequently difficult to isolate over time since they are not the only things changing.

In this paper, we wish to determine what fraction of the change in production over this time period to attribute to climate and CO_2 fertilization. Looking at world-wide production, agricultural production has increased from $364 billion in 1960 to $1380.0 billion in 2000 or 380% [World Bank (2003)]. Most of this global increase can be explained by the technologically driven increase in agricultural productivity per hectare of 2% per year. This alone would predict that production would increase by 222% over this period. However, the increase in agricultural production is not uniform across every region. Agriculture in the countries of the OECD in 2100 is 7 times larger, Oceanic

[1] The current global average temperature of 18.4 °C was computed as a population weighted average of annual surface temperatures from Leeman and Cramer (1991). The value does not reflect the temperature at sea.

agriculture has increased by 5.5 times, Latin American and Asian agriculture have increased by 3.9 times, USSR and Eastern Europe have increased 2.5 times, but African agriculture is only 1.5 times its 1960 production levels. Have the changes in climate and CO_2 contributed to these observed changes in agricultural performance? Because the USSR and the OECD countries lie in the mid to high latitudes, they are currently cooler and would therefore benefit from warming. In contrast, the low latitude countries of Latin America, Asia, Oceania, and Africa are much warmer so that warming should be more harmful. Warming could partially explain these regional differences. Further, the overall increase in CO_2 should have caused carbon fertilization. However, if carbon fertilization increases crop productivity proportionally, it will have a bigger effect on modern farms than on labor-intensive farms.

In order to quantify what effect climate change and CO_2 increases have had from 1960 to 2000, we rely on climate sensitivity functions calibrated from agronomic–economic models and cross-sectional evidence. These are the two primary methods the impact literature has used to measure climate sensitivity. The agronomic–economic approach begins with experiments on crops under different controlled conditions. By changing temperature and CO_2 levels, agronomists have been able to learn how these two factors change crop yields [see, for example, Rosenzweig and Parry (1994)]. These changes on crop yields are then fed into economic models that explain crop choice, productivity, and market prices [see Adams et al. (1990, 1993, 1999)]. The economic models then predict welfare responses. The cross-sectional approach looks at farm performance across a range of climate zones [Mendelsohn, Nordhaus and Shaw (1994, 1999); Mendelsohn, Dinar and Sanghi (2001)]. By examining how farm value or net revenue changes across climates, one can learn about the long term climate sensitivity of farms. Both approaches have strengths and weaknesses. The strength of the agronomic–economic model is that it is based on carefully controlled experiments. This is a weakness of the cross-sectional approach because it is vulnerable to unwanted variation. The agronomic–economic approach also is capable of exploring phenomenon not yet seen on the planet such as higher CO_2 levels. Again this is a weakness of the cross-sectional approach which must rely on the experiments to predict carbon fertilization effects. A third strength of the agronomic–economic model is that mechanisms are very clear. All micro changes can be observed and explained by the model. Again this is a weakness of at least the Ricardian model which is more of a black box approach, linking climate directly to net income. The cross-sectional approach, however, has its merits as well. The technique relies upon comparisons over vast landscapes so that it is representative of actual farm conditions. This is a weakness of the agronomic–economic model, which often has to extrapolate from limited field tests to the landscape as a whole. Perhaps the most important strength of the cross-sectional approach is that it takes into account adaptation because every farmer has carefully adapted to local conditions. The agronomic–economic model only includes adaptation to the extent that the modeler thinks of it. The cross-sectional model therefore does a much better job of including efficient adaptation.

Using these two independent approaches to measure climate sensitivity, we predict the magnitude of agricultural impacts from climate and CO_2 changes over the last 40 years. Because climate effects vary by location, we calculate these effects independently for each country. Although the carbon fertilization effects are universal, we assume that they are proportional to production so that more productive sites get larger benefits. We calculate agricultural impacts using both the cross-sectional and experimental evidence. We also examine the effect of climate alone and then with fertilization. We then compare the results of these climate impacts with the observed overall changes in agriculture. Although we calibrate these effects for each individual country, we emphasize the regional outcomes in this paper.

2. Historic climate change

The historic changes in climate over the last 100 years have been the subject of intense scrutiny as scientists have sought to demonstrate that increasing greenhouse gas concentrations have already led to warming [Houghton et al. (2001)]. The consensus of analysis of surface temperatures across the globe is that temperatures have increased by 0.5 °C over the last 100 years.

In this paper, we focus on the climate change that has occurred between 1960 and 2000. According to the Third Assessment Report of the IPCC, temperatures have increased 0.25 °C since 1960 [Houghton et al. (2001)]. The warming has not been uniform as can be seen in Table 1. Warming has been higher in the high latitudes, 0.6 °C in the last 40 years. Warming has been slightly lower in the mid latitudes, averaging about 0.4 °C. Warming has been lower in the low latitudes, averaging about 0.2 °C in Asia and Latin America and about 0.1 °C in Africa and Oceania. Further warming, has been higher in the winter months than the rest of the year. Because these observed changes

Table 1
Estimated regional climate change from 1960 to 2000

Region	Temperature (Δ °C)	Precipitation (Δ%)
Africa	0.10	−0.008
Asia	0.20	0.011
Oceania	0.10	0.013
Latin America	0.25	−0.001
N. America	0.40	0.015
USSR + E. Europe	0.60	0.015
W. Europe	0.56	0.015
Globe	0.25	0.008

Source: Houghton et al. (2001), Figure 2.9 (p. 116), Figure 2.10 (p. 117), and Figure 2.25 (p. 144).

in temperature are consistent with the predicted effects of greenhouse gases, we have assumed that greenhouse gases have caused these regionally specific changes in temperature.

In addition, there have been some changes in precipitation over the last 40 years [Houghton et al. (2001)]. The mid and high latitudes have been getting wetter. In contrast, several tropical regions have been getting drier. Scientists are less confident that greenhouse gases have caused these regional precipitation patterns over the last 40 years although the models do predict an overall increase in precipitation. In this analysis, we assume that the 1.5% increase in precipitation observed for the mid and high latitudes is caused by greenhouse gases (see Table 1). We also assume that greenhouse gases have caused a 0.1% decline in precipitation in South America and a 0.8% decline in precipitation in Africa.

Finally, there is strong evidence that global levels of CO_2 have increased over the last 40 years. Measurements of CO_2 in Hawaii suggest that 1960 atmospheric concentrations were equal to 315 ppm and 2000 measurements are approximately 367 ppm [Houghton et al. (2001)]. Atmospheric scientists argue that these CO_2 concentrations are mixed evenly across the planet [Houghton et al. (2001)].

The changes in climate assumed in this analysis are thus regionally accurate. However, the analysis does not calculate historic rates of change for each country. A careful analysis of the climate records of each country over time would probably produce some fine distinctions in the pattern of change for each country. This analysis has merely tried to obtain regional measurements of average change.

3. Climate and carbon sensitivity

Climate sensitivity was captured in the Global Impact Model (GIM) [Mendelsohn et al. (2000)]. GIM contains two climate impact sensitivity models: the cross-sectional model and the experimental model [see Mendelsohn and Schlesinger (1999)]. These sensitivities come from two different sets of empirical studies each with their own strengths and weaknesses [see Mendelsohn and Neumann (1999); Mendelsohn (2001)]. Because each experimental site is expensive, the experimental studies are often limited to just a few sites. The experimental approach has to work hard to make the results representative. The cross-sectional approach, in contrast, is generally performed across the entire sector and so is automatically representative. The experimental approach includes other factors only to the extent that the modeler knows to include them. The experimental approach has been criticized for not including human adaptation and possibly ecosystem adaptations (such as insects and disease) as well. The cross-sectional approach includes these factors because they are built in to what is happening at each place today. However, the cross-sectional approach has its own weaknesses. It is difficult to control for unwanted factors that may be correlated with climate. It is easy for unwanted factors to influence the results raising questions of cause and effect. The experimental approach with its carefully controlled experiments does not have this problem. The cross-sectional approach also cannot predict the effect of factors that have not yet appeared. For example, the cross-sectional approach cannot predict the consequences of higher CO_2 levels because every site in the cross section has the same CO_2 level. The experimental approach can create these new conditions. Because the strengths and weaknesses of the two approaches are so very different, both approaches should be applied whenever practical. The two methods check each other so that if results differ, there may still be a problem that one or both methods have not yet adjusted for. However, if the results are the same, scientists can have confidence that they have gotten reliable results since the two methods rely on such different assumptions.

Both the cross-sectional results and the experimental results used in this study include adaptation. The results are more optimistic than earlier literature, which limited farmer responses to climate change. Although one could include these more pessimistic results simply because they were published, we choose not to in this study because they are essentially flawed. By allowing farmers not to make any changes, the earlier studies exaggerate how harmful warming will be. Perhaps more important, the earlier studies often did not recognize that warming is likely to be beneficial to cool environments. That is, farmers in places that are currently cool could actually make changes, for example choose new crops once it warms, that would make them better off than they are now. Because this is private adaptation, there is every reason to expect that it will be done [Mendelsohn et al. (2000)]. Of course, some adaptations may or may not be done. For example, governments will need to reallocate water efficiently to the highest valued user. This may or may not happen. One point is very clear, however, adaptation is very important. Studies that did not include adaptation seriously overestimated.climate damages in the past.

Developing country agricultural estimates are based on research from India and Brazil [Mendelsohn, Dinar and Sanghi (2001)]. This research indicates that developing countries have steeper climate response functions than OECD countries. That is, agriculture in developing countries has lower productivity but also a greater sensitivity to warming. As has been noted in the adaptation literature, many developing countries may not have the institutions and capacity to adapt. It is also clear that developing country agriculture, with its labor-intensive technology, is simply more climate sensitive than more capital-intensive methods of farming [Mendelsohn, Dinar and Sanghi (2001)].

The cross-sectional response function comes from an empirical analysis that compares how farms perform in different climates today. Generally, the empirical analysis indicates that economic performance is hill-shaped but relatively flat with respect to temperature. Agriculture in countries that are currently cool will improve whereas farming in countries that are hot will decline with warming. The high point of the hill is a temperate climate found often in the mid latitudes. Precipitation has a beneficial impact on agriculture in general. This impact declines as precipitation increases. Both temperature and precipitation are captured in seasonal models focused on January, April, July, and October monthly measurements.

The experimental response function is generated from the outcome of the predictions of agronomic–economic models [Adams et al. (1999)]. These complex impact models generate results for specific climate scenarios. Looking across these outcomes, one can generate a reduced form model that reflects the underlying experimental model results. The resulting response functions are more steeply hill-shaped than the cross-sectional model with respect to temperature and precipitation. We believe that the difference in the response functions between the experimental results and the cross-sectional results is partially explained by how the two approaches handle adaptation. However, the two models make a number of different assumptions about farming so that there could be other explanations as well. In this analysis, we assume that both response functions are equally plausible.

Because both response functions are hill shaped with respect to temperature, countries that are cool, on the left hand side of the hill, will gain from warming and countries that are hot, on the right hand side, will lose. Exactly where each country stands will vary. Countries that are temperate will see small positive or negative impacts depending upon the climate forecast and sector. Tropical countries will largely be damaged. High latitude countries will largely benefit from warming. Precipitation also has a quadratic relationship. More precipitation is generally beneficial to agriculture. Both models also respond to carbon dioxide levels. Both sensitivity models assume that carbon fertilization will increase productivity in agriculture. This effect is introduced in log form so that higher CO_2 increases benefits but at a diminishing rate. A doubling of CO_2 is assumed to increase crop productivity by 25% [Reilly et al. (1996)].

The importance of carbon fertilization has long been recognized in the literature [Adams et al. (1990); Reilly et al. (1996)]. In order to demonstrate this effect, many impact studies have shown what happens to agriculture with and without fertilization. This has sometimes confused readers into thinking that it is uncertain whether or not there will be carbon fertilization. There is sufficient empirical evidence at this point to conclude that carbon fertilization will occur. Both experiments done in laboratories and experiments done in fields concur that crop yields will increase with heightened CO_2. This effect, however, is not uniform. C3 crops are expected to increase yields by 30% with a doubling of CO_2 [Reilly et al. (1996)]. C4 crops are less much responsive to CO_2 [Reilly et al. (1996)]. Even within C3 crops, however, responses can vary across crops from +20% to almost 100% with a doubling of CO_2.

GIM makes use of economic data that is only readily available at the national level. Each country has numerous characteristics such as arable land, cropland, population, and GDP that also play a role in determining country-specific impacts [see Mendelsohn et al. (2000)]. This national data permits impacts to be disaggregated to each country. This is a large improvement over the integrated assessment literature which focuses on much more aggregated sets of countries [see, for example, Nordhaus and Boyer (2000)]. These detailed forecasts allow readers to see distributional impacts across countries in the world. However, impacts within countries remain opaque. The model cannot predict how different regions within a country may fare. For example, it is expected that impacts in the northern and southern regions of the United States will be quite different with more severe warming scenarios [Mendelsohn (2001)]. As research in this area continues, hopefully, subnational impacts will be available in the future.

The agronomic–economic model at the core of the experimental research comes from a series of studies by Adams et al. (1990, 1993, 1999). Without carbon fertilization, the agronomy results underlying the model suggest that climate change will be harmful to the United States. When carbon fertilization is included in the results, the model, however, clearly indicates that warming is beneficial to the United States. The cross-sectional results come from the Ricardian model [Mendelsohn, Nordhaus and Shaw (1994)]. This model also suggests that climate change alone will be harmful although much less harmful than the agronomic predictions. With carbon fertilization, the Ricardian results suggest that warming will be beneficial to the United States. Through a series of comments and replies, several potential problems were raised and addressed concerning the Ricardian method. One criticism is that irrigation was not properly taken into account [Cline (1996)]. However, even when percent irrigated land is included in the model, there is no change in results [Mendelsohn and Nordhaus (1996)]. Further, it has more recently been shown that adding surface water, while important, also has little effect on the results [Mendelsohn and Dinar (2003)]. The Ricardian model also does not capture changes in prices. By assuming constant prices, the model underestimates the damages of reductions and overestimates the benefits of increases in productivity [Cline (1996)]. However, this bias is generally going to be small in most cases largely because world prices are not expected to change much [Mendelsohn and Nordhaus (1996)]. Another concern was that the cost of adaptation was not measured [Quiggin and Horowitz (1999)]. The Ricardian method is a long run model and does not capture short run adjustments to such phenomenon as yearly weather. However, the model measures the impact of climate on land value, so that it will capture not only effects on yields but also effects on costs [Mendelsohn and Nordhaus (1999)]. A final criticism that has been raised concerns the limitations of the model to measure effects accurately only within the range of the sample. This limitation, for example, explains why the Ricardian model does not capture carbon fertilization effects, which will be the same at any one time across the whole sample. However, by carefully building large data sets across many climate zones, the approach is able to examine a wide range of climates. Any foreseeable climate change in the next century is well within the range of the data.

4. Results

We begin by using the climate response functions in GIM to evaluate the impact of the observed changes in regional temperature, precipitation, and carbon dioxide. We provide two estimates of impacts, one from the cross-sectional evidence and another from the experimental-simulation evidence. Estimates of impacts computed for each country are shown in Appendix A. The paper concentrates on the results at the regional level.

Table 2
Impact of climate change from 1960 to 2000 on global agriculture, no carbon fertilization (millions USD)

Regions	Cross-sectional	Experimental
Latin America	−371 (−0.23%)	−2018 (−1.26%)
Africa	−252 (−0.26%)	−1648 (−1.69%)
Asia	−1647 (−0.27%)	−1637 (−0.27%)
Oceania	−31 (−0.15%)	150 (0.71%)
N. America	−561 (−0.29%)	3603 (1.86%)
W. Europe	143 (0.07%)	9216 (4.20%)
USSR + East	2033 (2.87%)	4876 (6.88%)
Globe	−686 (−0.05%)	12542 (0.91%)
OECD	−371 (−0.06%)	15721 (2.75%)
ROW	−2349 (−0.32%)	−8055 (−1.09%)

Percentages in parenthesis reflect impacts as a fraction of agricultural GDP. The USSR and Eastern Europe, OECD, and ROW add up to global estimates.

Table 2 isolates the effects of climate change alone. That is, the impacts are calculated just from changes in temperature and precipitation. The results show that the impact of climate change over the last 40 years has potentially been substantial. The cross-sectional model predicts it resulted in a loss of almost 700 million USD world-wide whereas the experimental model predicts a net gain of 12,500 million USD. However, the global effect is a small fraction of agricultural GDP, just −0.05% to +0.91%. Global impacts, however, reflect losses in certain regions and gains in other regions. Regional impacts are more dramatic than these net global impacts. With the cross-sectional evidence, damages range from −0.15% to −0.27% across tropical regions. North America is predicted to have losses of −0.29%. In contrast, historic climate change has caused agricultural gains in Western Europe, the former Soviet Union, and Eastern Europe. The size of these gains almost offsets the losses in the rest of the world. The experimental evidence provides even more dramatic regional effects across tropical countries with losses ranging from −0.27% to −1.7%. Oceania and North America are predicted to gain in the experimental results because of increases in precipitation. Western and Eastern Europe and the former Soviet Union are also predicted to enjoy large gains of 4.2% to 6.9% under the experimental model.

Table 3
Impact of climate change from 1960 to 2000 on global agriculture, with carbon fertilization (millions USD)

Regions	Cross-sectional	Experimental
Latin America	3527 (2.19%)	2901 (1.81%)
Africa	857 (0.88%)	1339 (1.37%)
Asia	1362 (0.22%)	17213 (2.79%)
Oceania	1715 (8.06%)	801 (3.77%)
N. America	7728 (3.99%)	9536 (4.92%)
W. Europe	3286 (1.50%)	15938 (7.26%)
USSR + East	8237 (11.62%)	7045 (9.94%)
Globe	26711 (1.94%)	54773 (3.97%)
OECD	15672 (2.75%)	33187 (5.81%)
ROW	2802 (0.38%)	14541 (1.97%)

Percentages in parenthesis reflect impacts as a fraction of agricultural GDP. The USSR and Eastern Europe, OECD, and ROW add up to global estimates.

Although these small percentage changes in production pale against the observed changes in agriculture, they do help explain some of the regional variation in observed growth rates. Tropical nations have been hurt by warming, which corresponds to their lower overall growth rate. Developed countries, in contrast, have either suffered smaller damages or enjoyed substantial gains from warming. Although these climate impacts have been small, they are at least in the same direction as observed effects across regions.

Table 3 adds the effects of carbon fertilization and climate change. The table reports the combined effects of temperature, precipitation, and carbon dioxide changes. Carbon fertilization has a large predicted consequence. For the globe, greenhouse gases have increased agricultural production between 2% and 4%. Carbon fertilization clearly swamps the climate changes alone. At the regional level, harmful impacts disappear and become at least somewhat beneficial. In the cross-sectional model, benefits in tropical regions range from 0.2% to 2%. Oceania enjoys even higher rates of change as temperatures change little and precipitation and CO_2 increases. The gains to the mid and high latitudes increase substantially. The former Soviet Union and Eastern Europe gain almost 12% in this scenario. Although both models predict beneficial effects, the exper-

Table 4
Percent of agricultural change from 1960 to 2000 due to climate and carbon fertilization

Regions	Cross-sectional	Experimental
Latin America	(2.9%)	(2.4%)
Africa	(2.8%)	(4.4%)
Asia	(0.3%)	(3.8%)
Oceania	(9.9%)	(4.6%)
N. America	(4.8%)	(5.9%)
W. Europe	(1.8%)	(8.5%)
USSR + East	(19.0%)	(16.3%)
Globe	(2.6%)	(5.4%)
OECD	(3.2%)	(6.8%)
ROW	(0.6%)	(3.0%)

imental model predicts larger carbon fertilization effects than the cross-sectional model. The experimental results assume that the increased productivity at the farm level results in a proportional increase in overall agricultural GDP. The cross-sectional model, in contrast, assumes that only farm values increase proportionally. The results in Table 2 are consistent with global warming predictions reported in the literature. If warming turns out to be small, the literature generally finds that carbon fertilization will dominate climate effects and lead to overall agricultural benefits [Reilly et al. (1996); Mendelsohn et al. (2000); Mendelsohn (2003)].

Appendix A shows what happens to each country with warming and carbon fertilization. Obviously, the results described in Table 3 capture the regional patterns of these changes. It is particularly striking how polar countries benefit from the warming that has taken place to date. Another result that is not apparent in the regional results concerns what happens to island nations. Although in absolute magnitude, the effects on agriculture are small, the impacts on small island countries are consistently more harmful.

Table 4 evaluates how large the climate signal has been relative to the overall observed changes in agriculture. Carbon fertilization in particular and climate as well have contributed to the observed changes in agriculture. Climate and especially CO_2 are estimated to be responsible for between 2.6% to 5.4% of the global increase in production over the last 40 years. The climate impact in selected regions has been even more important. Between 16% and 19% of the increase in production in the former USSR and Eastern European region has been due to warming and carbon fertilization. Warming and carbon fertilization are responsible for between 0.6% and 3% of the increased production in developing countries. Climate change and carbon fertilization has also made a contribution to the growth of agriculture in developed countries accounting for between 3% and 7% of their growth between 1960 and 2000. There is little doubt but that recent climate change has affected world agriculture, primarily through carbon fertilization. However, it has been difficult to see because of the many other dramatic changes that are also occurring in this sector.

5. Conclusion

This study examined the effect that climate change and carbon fertilization between 1960 and 2000 had on agricultural GDP across the world. Although the change in climate has been small, it has been enough to have effects in each region. The small increase in carbon dioxide has also had effects fertilizing plants around the world. These effects have not been visible because there have been many other dramatic changes in agriculture, especially in rapid technical change. Nonetheless, both cross-sectional and experimental evidence suggest that there has been a substantial effect in the last 40 years. Whereas the climate alone has either reduced global productivity very slightly (0.05%) or increased it slightly (0.9%), higher carbon dioxide levels have produced an overall beneficial effect. On net, climate change and carbon fertilization has increased world production from 2% to 4%.

The study reveals that the effects of climate and carbon dioxide have varied sharply across regions. Warming alone has generally reduced the productivity of tropical farms while it has increased the productivity of most mid to high latitude farms. Carbon fertilization has increased productivity worldwide but it is likely that more productive sites have enjoyed larger benefits. On net, greenhouse gases increased productivity in every region. The percentage gains from warming, however, have been larger in developed countries (3–6%) compared to developing countries (0.4–2%).

The study shows that the historic changes in climate over the last 40 years have already affected agriculture across the world. The greenhouse effect is responsible for between 2.6% and 5.4% of the increase in agricultural production between 1960 and 2000. Most of this impact is due to the beneficial impacts of carbon fertilization. However, climate has also made some small contributions, generally helping mid and high latitude countries and slightly damaging low latitude countries. The study shows that climate change can be important even without dramatic events such as hurricanes and tropical storms. In fact, it is the gradual but persistent effects of greenhouse gases that may well turn out to be the most important.

Even if the small climate changes of the last 40 years have had little effect on agriculture, this does not imply that future climate changes are safe to ignore. First, future climate changes are expected to be much larger, from 1.4 to 5 °C, in comparison to the 0.5 °C of the last 40 years. Second, the impacts of larger changes may be much more severe. There is evidence that damages in agriculture (especially where it is already too warm) will increase more than proportionally with temperature [Mendelsohn, Nordhaus and Shaw (1994); Mendelsohn and Schlesinger (1999)]. Third, water may also become scarce so that there may be limits to the extent that agriculture can depend upon irrigation [Mendelsohn and Neumann (1999)].

Appendix A

Table A.1
Historic country impacts of climate change plus carbon fertilization, 1960–2000 (millions USD)

	Experimental	Cross-sectional
Afghanistan	186	−8
	(3.82)	(−0.17)
Albania	107	−3
	(6.61)	(−0.23)
Algeria	203	202
	(2.29)	(2.43)
Angola	6	4
	(1.28)	(0.87)
Antigua/Barbuda	1	0
	(3.43)	(0.07)
Argentina	355	886
	(2.27)	(6.05)
Armenia	106	−3
	(9.24)	(−0.24)
Australia	469	1579
	(3.32)	(11.66)
Austria	547	69
	(9.61)	(1.29)
Azerbaijan	47	−10
	(6.53)	(−1.44)
Bahamas	4	0
	(2.65)	(0.01)
Bahrain	1	−0
	(1.18)	(−0.02)
Bangladesh	224	−19
	(1.49)	(−0.14)
Barbados	5	0
	(3.06)	(0.01)
Belarus	193	289
	(9.76)	(15.53)
Belgium	249	32
	(7.33)	(1.00)
Belize	0	0
	(0.22)	(0.01)
Benin	5	−1
	(0.52)	(−0.12)
Bhutan	6	−0
	(3.21)	(0.00)
Bolivia	22	4
	(1.86)	(0.40)
Bosnia/Herzegovina	81	32
	(7.77)	(3.32)

(*continued on next page*)

Table A.1
(*continued*)

	Experimental	Cross-sectional
Botswana	22	9
	(1.70)	(3.43)
Brazil	722	1359
	(0.96)	(1.92)
Brunei Darussalam	3	0
	(2.38)	(0.01)
Bulgaria	35	148
	(6.46)	(29.29)
Burkina Faso	5	−2
	(0.55)	(−0.23)
Burundi	7	2
	(1.47)	(0.47)
Cambodia	11	−6
	(0.58)	(−0.33)
Cameroon	26	−1
	(0.55)	(−0.02)
Canada	1277	1995
	(8.46)	(14.11)
Cape Verde	1	0
	(1.52)	(0.03)
Central African Rep.	5	−0
	(0.69)	(−0.05)
Chad	5	−1
	(0.70)	(−0.21)
Chile	230	149
	(3.34)	(2.31)
China	7602	−128
	(3.80)	(−0.07)
Columbia	155	132
	(0.79)	(0.72)
Comoros	0	0
	(0.53)	(0.05)
Congo	9	0
	(0.69)	(0.01)
Costa Rica	79	12
	(4.51)	(0.73)
Côte d'Ivoire	18	−1
	(0.50)	(−0.04)
Croatia	150	45
	(6.93)	(2.24)
Cuba	20	−0
	(2.56)	(−0.02)
Cyprus	60	2
	(2.61)	(0.11)

(*continued on next page*)

Table A.1
(*continued*)

	Experimental	Cross-sectional
Czech Rep.	208	150
	(8.83)	(6.80)
Dem. People Rep. Korea	149	3
	(5.91)	(0.14)
Denmark	569	109
	(8.65)	(1.76)
Djibouti	0	0
	(0.86)	(0.00)
Dominica	2	0
	(4.84)	(0.07)
Dominican Rep.	58	32
	(2.87)	(1.67)
Ecuador	26	5
	(1.22)	(0.27)
Egypt	218	4
	(1.87)	(0.03)
El Salvador	36	14
	(3.07)	(1.24)
Equatorial Guinea	1	0
	(0.42)	(0.06)
Estonia	42	52
	(10.84)	(14.26)
Ethiopia	66	20
	(1.61)	(0.51)
Fiji	1	0
	(0.14)	(0.01)
Finland	858	119
	(12.15)	(1.80)
France	3767	707
	(6.64)	(1.33)
Gabon	3	10
	(0.46)	(1.81)
Gambia	1	−0
	(0.63)	(−0.12)
Georgia	32	2
	(6.06)	(0.37)
Germany	2278	516
	(7.95)	(1.92)
Ghana	16	−2
	(0.52)	(−0.07)
Greece	510	99
	(3.82)	(0.79)
Guatemala	182	1
	(4.16)	(0.02)
Guinea	4	−0
	(0.38)	(−0.02)

(*continued on next page*)

Table A.1
(*continued*)

	Experimental	Cross-sectional
Guinea-Bissau	1	−0
	(0.44)	(−0.10)
Guyana	0	9
	(0.05)	(3.54)
Haiti	26	0
	(2.77)	(0.01)
Honduras	29	0
	(3.26)	(0.04)
Hungary	162	206
	(6.98)	(9.49)
Iceland	26	0
	(13.99)	(0.03)
India	−657	−353
	(−0.53)	(−0.05)
Indonesia	1311	−1
	(3.46)	(0.00)
Iran	634	482
	(2.97)	(2.30)
Iraq	28	8
	(2.05)	(0.63)
Ireland	367	50
	(8.07)	(1.17)
Israel	103	4
	(2.64)	(0.12)
Italy	1408	383
	(5.48)	(1.59)
Jamaica	15	0
	(3.70)	(0.02)
Japan	2904	154
	(4.79)	(0.27)
Jordon	9	1
	(2.58)	(0.36)
Kazakhstan	234	−187
	(9.76)	(8.33)
Kenya	36	8
	(1.41)	(0.34)
Kuwait	258	0
	(1.48)	(0.00)
Kyrgyzstan	82	−8
	(10.12)	(−1.00)
Lao People Dem. Rep.	−50	−4
	(−3.71)	(−0.30)
Latvia	34	80
	(10.37)	(25.75)
Lebanon	53	1
	(3.33)	(0.09)

(*continued on next page*)

Table A.1
(*continued*)

	Experimental	Cross-sectional
Lesotho	3	1
	(2.70)	(0.77)
Liberia	1	−0
	(0.16)	(−0.01)
Libya	20	40
	(1.88)	(4.22)
Lithuania	64	139
	(9.87)	(22.81)
Luxembourg	22	4
	(8.26)	(1.62)
Macedonia	44	27
	(7.34)	(4.90)
Madagascar	16	4
	(1.18)	(0.34)
Malawi	9	0
	(1.27)	(0.36)
Malaysia	517	127
	(3.62)	(0.95)
Maldives	−0	−0
	(−0.08)	(0.00)
Mali	9	−1
	(0.60)	(−0.05)
Malta	−0	−0
	(−0.33)	(−0.12)
Mauritania	3	−0
	(0.79)	(−0.02)
Mauritius	2	2
	(0.67)	(0.82)
Mexico	610	722
	(3.07)	(3.87)
Moldova	58	−11
	(7.18)	(−1.48)
Mongolia	54	−4
	(15.70)	(−1.27)
Morocco	143	31
	(2.28)	(0.53)
Mozambique	13	−0
	(1.11)	(0.00)
Myanmar	20	−18
	(1.96)	(−1.88)
Namibia	8	16
	(1.69)	(3.48)
Nepal	48	−5
	(1.97)	(−0.22)
Netherlands	1179	36
	(7.42)	(0.24)

(*continued on next page*)

Table A.1
(*continued*)

	Experimental	Cross-sectional
New Zealand	255	136
	(4.36)	(2.47)
Nicaragua	48	−0
	(3.00)	(−0.01)
Niger	7	−1
	(0.69)	(−0.11)
Nigeria	67	−16
	(0.50)	(−0.12)
Norway	441	45
	(12.12)	(1.33)
Oman	7	0
	(1.43)	(0.00)
Pakistan	123	−48
	(0.63)	(−0.26)
Panama	26	12
	(3.66)	(1.76)
Papua New Guinea	67	0
	(4.26)	(0.01)
Paraguay	30	53
	(1.03)	(1.96)
Peru	158	19
	(2.44)	(0.31)
Philippines	546	−1
	(3.41)	(−0.01)
Poland	524	647
	(8.63)	(11.37)
Portugal	175	79
	(3.18)	(1.53)
Qatar	1	−0
	(1.17)	(−0.02)
Rep. of Korea	1315	73
	(4.97)	(0.29)
Romania	355	407
	(7.50)	(9.16)
Russia	3315	6387
	(12.35)	(25.37)
Rwanda	17	3
	(1.64)	(0.30)
St. Lucia	2	0
	(3.31)	(0.04)
St. Vincent/Grenada	1	0
	(2.57)	(0.04)
Sao Tome/Principe	0	0
	(3.64)	(0.00)
Saudi Arabia	172	49
	(1.64)	(0.50)

(*continued on next page*)

Table A.1
(*continued*)

	Experimental	Cross-sectional
Senegal	9 (0.77)	−1 (−0.11)
Serbia/Montenegro	181 (7.17)	−21 (−0.88)
Seychelles	0 (0.13)	0 (0.11)
Sierra Leone	0 (0.01)	−0 (−0.03)
Singapore	0 (0.00)	0 (0.00)
Slovakia	109 (8.91)	72 (6.23)
Slovenia	61 (8.17)	12 (1.77)
Solomon Islands	−1 (−0.23)	0 (0.00)
Somalia	10 (0.97)	−1 (−0.07)
South Africa	112 (2.04)	399 (7.73)
Spain	1104 (3.68)	607 (2.16)
Sri Lanka	43 (1.14)	−3 (−0.10)
Sudan	35 (0.79)	−8 (−0.20)
Suriname	0 (0.11)	0 (0.00)
Swaziland	3 (1.58)	0 (0.10)
Sweden	624 (10.53)	137 (2.46)
Switzerland	712 (9.94)	18 (0.27)
Syria	105 (2.85)	19 (0.54)
Taiwan	84 (2.82)	3 (0.12)
Tajikistan	39 (7.93)	−5 (−0.11)
Thailand	88 (0.44)	148 (0.79)
Togo	4 (0.54)	−1 (−0.20)

(*continued on next page*)

Table A.1
(*continued*)

	Experimental	Cross-sectional
Trinidad/Tobago	4	0
	(2.90)	(0.01)
Tunisia	69	118
	(2.11)	(3.85)
Turkey	1145	919
	(4.02)	(3.44)
Turkmenistan	52	−8
	(2.49)	(−0.39)
Uganda	43	8
	(1.19)	(0.23)
Ukraine	505	−178
	(8.21)	(−3.08)
United Arab Emirates	7	0
	(1.24)	(0.00)
United Kingdom	1103	276
	(7.92)	(2.11)
United Rep. Tanzania	41	3
	(1.24)	(0.09)
United States	8259	5733
	(4.31)	(3.19)
Uruguay	24	40
	(1.81)	(3.28)
Uzbekistan	223	−26
	(5.16)	(−0.64)
Vanuatu	0	0
	(0.22)	(0.00)
Venezuela	33	77
	(0.77)	(1.92)
Vietnam	97	−13
	(1.25)	(−0.19)
Western Samoa	−0	0
	(−0.35)	(0.05)
Yemen	22	1
	(2.20)	(0.07)
Zaire	25	6
	(0.93)	(0.24)
Zambia	14	2
	(1.37)	(0.21)
Zimbabwe	15	1
	(1.64)	(0.10)

Percentage change in parenthesis is fraction of agricultural GDP.

References

Adams, R., Rosenzweig, C., Pearl, R., Ritchie, J., McCarl, B., Glyer, J., Curry, R., Jones, J., Boote, K., Allen, L. (1990). "Global climate change and US agriculture". Nature 345, 219–224.

Adams, R., Fleming, R., Chang, C., McCarl, B., Rosenzweig, C. (1993). A Reassessment of the Economic Effects of Global Climate Change in US Agriculture. US Environmental Protection Agency, Washington, DC.

Adams, R., McCarl, B., Segerson, K., Rosenzweig, C., Bryant, K., Dixon, B., Connor, R., Evenson, R., Ojima, D. (1999). "The economic effects of climate change on US agriculture". In: Mendelsohn, R., Neumann, J. (Eds.), The Impact of Climate Change on the United States Economy. Cambridge University Press, Cambridge, UK.

Cline, W. (1996). "The impact of global warming on agriculture: Comment". American Economic Review 86, 1309–1312.

Houghton, J., Ding, Y., Griggs, D., Noguer, M., van der Linden, P., Dai, X., Maskell, K., Johnson, C. (Eds.) (2001). Climate Change 2001: The Scientific Basis. Third Assessment Report of the Intergovernmental Panel on Climate Change (IPCC). Cambridge University Press, Cambridge, UK.

Leeman, R., Cramer, W.P. (1991). "The IIASA database for mean monthly values of temperature, precipitation and cloudiness of a global terrestrial grid". International Institute for Applied Systems Analysis (IIASA), RR-91-18.

McCarthy, J., Canzani, O., Leary, N., Dokken, D., White, K. (Eds.) (2001). Climate Change 2001: Impacts, Adaptation, and Vulnerability. Third Assessment Report (TAR) of the Intergovernmental Panel on Climate Change (IPCC). Cambridge University Press, Cambridge, UK.

Mendelsohn, R. (Ed.) (2001). Global Warming and the American Economy: A Regional Assessment of Climate Change. Edward Elgar Publishing, UK.

Mendelsohn, R. (2003). "Assessing the market damages from climate change". In: Griffin, J. (Ed.), Global Climate Change: The Science, Economics, and Politics. Edward Elgar Publishing, UK, pp. 92–113.

Mendelsohn, R., Dinar, A. (2003). "Climate, water, and agriculture". Land Economics 79, 328–341.

Mendelsohn, R., Dinar, A., Sanghi, A. (2001). "The effect of development on the climate sensitivity of agriculture". Environment and Development Economics 6, 85–101.

Mendelsohn, R., Neumann, J. (Eds.) (1999). The Economic Impact of Climate Change on the Economy of the United States. Cambridge University Press, Cambridge, UK.

Mendelsohn, R., Nordhaus, W. (1996). "The impact of global warming on agriculture: Reply". American Economic Review 86, 1312–1315.

Mendelsohn, R., Nordhaus, W. (1999). "Reply to Quiggin and Horowitz". American Economic Review 89, 1046–1048.

Mendelsohn, R., Nordhaus, W., Shaw, D. (1994). "Measuring the impact of global warming on agriculture". American Economic Review 84, 753–771.

Mendelsohn, R., Nordhaus, W., Shaw, D. (1999). "Impact of climate variation on US agriculture". In: Mendelsohn, R., Neumann, J. (Eds.), The Economic Impact of Climate Change on the Economy of the United States. Cambridge University Press, Cambridge, UK.

Mendelsohn, R., Schlesinger, M. (1999). "Climate response functions". Ambio 28, 362–366.

Mendelsohn, R., Morrison, W., Schlesinger, M., Andronova, N. (2000). "Country-specific market impacts from climate change". Climatic Change 45, 553–569.

Nordhaus, W. (1997). "Do real wages and output series capture reality? The history of lighting suggests not". In: Bresnahan, T., Gordon, R. (Eds.), The Economics of New Goods. University of Chicago Press, Chicago.

Nordhaus, W., Boyer, J. (2000). Warming the World: Economic Models of Climate Change. MIT Press, Cambridge, MA.

Pearce, D., Cline, W., Achanta, A., Fankhauser, S., Pachauri, R., Tol, R., Vellinga, P. (1996). "The social cost of climate change: Greenhouse damage the benefits of control". In: Bruce, J., Lee, H., Haites, E. (Eds.), Climate Change 1995: Economic and Social Dimensions of Climate Change. Cambridge University Press, Cambridge, UK.

Quiggin, J., Horowitz, J. (1999). "The impact of global warming on agriculture: A Ricardian analysis: A comment". American Economic Review 89, 1044–1045.

Reilly, J., Baethgen, W., Chege, F., van de Geijn, S., Enda, L., Iglesias, A., Kenny, G., Patterson, D., Rogasik, J., Rotter, R., Rosenzweig, C., Sombroek, W., Westbrook, J. (1996). "Agriculture in a changing climate: Impacts and adaptations". In: Watson, R., Zinyowera, M., Moss, R., Dokken, D. (Eds.), Climate Change 1995: Impacts, Adaptations, and Mitigation of Climate Change: Scientific-Technical Analyses. Cambridge University Press, Cambridge, UK.

Rosenzweig, C., Parry, M. (1994). "Potential impact of climate change on world food supply". Nature 367, 133–138.

Watson, R., Zinyowera, M., Moss, R., Dokken, D. (Eds.) (1996). Climate Change 1995: Impacts, Adaptations, and Mitigation of Climate Change: Scientific-Technical Analyses. Cambridge University Press, Cambridge, UK.

World Bank (2003). World Development Indicators 2003. http://www.worldbank.org/data/.

AUTHOR INDEX

n indicates citation in a footnote.

SUBJECT INDEX